We would like to dedicate this book to our families:
David and Sherah
Blaine, Blaire, and Trey

This book would not have been a reality had it not been for their patience, understanding, support, and love.

Tell me, and I'll forget,
Show me, and I may not remember.
Involve me, and I'll understand.

Native American saying

Courtesy Texas Instruments Incorporated

Beginning and Intermediate Algebra with Graphing Calculators

An Integrated Approach

Beginning and Intermediate Algebra with Graphing Calculators

An Integrated Approach

Deborah Jolly Cochener
Austin Peay State University

Bonnie MacLean Hodge
Austin Peay State University

R. David Gustafson
Rock Valley College

Brooks/Cole
Thomson Learning™

Australia • Canada • Mexico • Singapore • Spain • United Kingdom • United States

Publisher: *Robert W. Pirtle*
Marketing Manager: *Leah Thomson*
Advertising: *Samantha Cabaluna*
Editorial Assistant: *Erin Wickersham*
Production Editor: *Ellen Brownstein*
Production Service: *Clarinda Publication Services*
Permissions Editor: *Mary Kay Hancharick*

Interior Design: *John Edeen*
Cover Design: *Vernon T. Boes, Laurie Albrecht*
Cover Photo: *Ed Young*
Interior Illustration: *Lori Heckelman*
Print Buyer: *Vena Dyer*
Typesetting: *The Clarinda Company*
Printing and Binding: *Von Hoffmann Press*

For more information, contact:
BROOKS/COLE
511 Forest Lodge Road
Pacific Grove, CA 93950 USA
www.brookscole.com

For permission to use material from this work, contact us by
Web: www.thomsonrights.com
fax: 1-800-730-2215
phone: 1-800-730-2214

Printed in United States of America

10 9 8 7 6 5 4 3 2 1

Library of Congress Cataloging-in-Publication Data

Beginning and intermediate algebra with graphing calculators : an
    integrated approach / Deborah J. Cochener, Bonnie Hodge, R. David
    Gustafson.
        p.   cm.
    Includes index.
    ISBN 0-534-35942-6
    1. Algebra.   I. Cochener, Deborah J.   II. Hodge, Bonnie M.
III. Gustafson, R. David (Roy David).
QA152.2.B4174   2000
512.9—dc21   99-31325
                                                    CIP

Photo credits: p. 2, The British Museum; **p. 14,** Texas Instruments Incorporated; **p. 17,** Texas Instruments Incorporated; **p. 61,** ©Corbis; **p. 105,** ©Corbis; **p. 183,** ©Corbis; **p. 217,** ©Corbis; **p. 261,** ©Corbis; **p. 314,** ©Corbis; **p. 392,** ©Corbis; **p. 448,** ©Corbis; **p. 501,** ©Corbis; **p. 559,** ©Corbis; **p. 569,** ©1999, The Leaf-Chronicle; **p. 612,** Archaeological Consulting/Gary Breschini and Trudy Haversat; **p. 619,** ©Corbis; **p. 673,** ©Corbis.

CONTENTS

4 An Extended Look at Applications and Absolute Value Equations   185

5 Factoring   219

6 Rational Expressions and Equations   263

7 Making the Transition from Elementary to Intermediate Algebra   316

To the Instructor and Student

Beginning and Intermediate Algebra with Graphing Calculators: An Integrated Approach combines the topics of beginning and intermediate algebra. This type of book has several advantages:

- By combining topics, much overlap and redundancy can be eliminated. The instructor will have time to teach for mastery of the material.
- For many students, the purchase of a single book will save money.
- A combination approach in one book will enable some colleges to cut back on the number of hours needed for mathematics remediation.

However, there are three concerns inherent in a combined approach:

- The first half of the book must include enough beginning algebra to ensure that students who complete the first half of the book and then transfer to another college will have the necessary prerequisite to enroll in an intermediate algebra course.
- The beginning algebra material should not get too difficult too fast.
- Intermediate algebra students beginning in the second half of the book must get some review of basic topics so that they can compete with students who are continuing from the first course.

Unlike many other texts, this book uses an *integrated approach*, which addresses each of the previous three concerns by

- including a full course in beginning algebra in the first six chapters,
- delaying the presentation of intermediate algebra topics until Chapter 7 or later, and
- providing a quick review of basic topics for those who begin in the second half of the book.

The approach throughout the text is fourfold and cyclic, using

- numerical approaches, specifically through the use of tables;
- graphical approaches, done both on graph paper and with the graphing calculator, providing students with opportunities to model data and to interpret models;
- symbolic approaches, which help students link the numerical and graphical approaches;
- verbal approaches in which students are required to express mathematical concepts in their own words; and

- continuous requests for students to demonstrate proficiency in the evaluation of expressions, the simplification of expressions, the graphing of functions, and the solving of equations.

Every effort has been made to provide the student with opportunities to become active participants in the formulation of their algebraic foundation, so that they may apply the principles of mathematics to other disciplines. The calculator is presented as a tool to be used for the confirmation of analytical processes and for the discovery of numerical and graphical patterns that lead to the formulation of fundamental algorithms.

■ GOALS OF THE BOOK

In addition to using a truly integrated approach, our goal has been to write a book that

- is enjoyable to read.
- is easy to understand.
- is relevant.
- will develop the necessary skills for success in future academic courses or on the job.

The NCTM standards, the AMATYC Crossroads, and the current trends in mathematics reform have been reinforced by emphasizing graphing and problem solving.

■ CALCULATORS

The use of a Texas Instruments TI-83 or TI-86 graphing calculator is assumed throughout the text. Students who use the TI-82, TI-83 Plus, or TI-85 will have no problem making the transition. We believe that students should learn calculator skills at this level of mathematics so they will be better able to use the technology in science and business classes as they progress through their individual curriculums.

■ FEATURES

Technology sections provide extensive ▶ keystroking information for Texas Instruments TI-83 and TI-86 graphing calculators. The information is also applicable, with slight modifications, to the TI-82 and TI-85 graphing calculators. Throughout the text the graphing calculator is used to verify results. Graphing calculator information is provided early and continues throughout the text. These sections free the instructor from having to provide keystroking information for the different calculators.

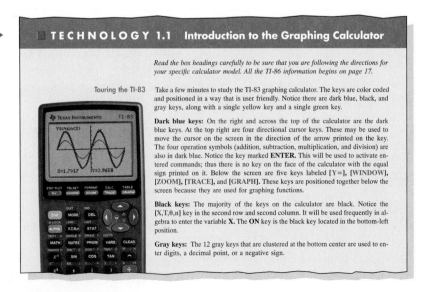

■ TECHNOLOGY 1.1   Introduction to the Graphing Calculator

Read the box headings carefully to be sure that you are following the directions for your specific calculator model. All the TI-86 information begins on page 17.

Touring the TI-83   Take a few minutes to study the TI-83 graphing calculator. The keys are color coded and positioned in a way that is user friendly. Notice there are dark blue, black, and gray keys, along with a single yellow key and a single green key.

Dark blue keys: On the right and across the top of the calculator are the dark blue keys. At the top right are four directional cursor keys. These may be used to move the cursor on the screen in the direction of the arrow printed on the key. The four operation symbols (addition, subtraction, multiplication, and division) are also in dark blue. Notice the key marked **ENTER.** This will be used to activate entered commands; thus there is no key on the face of the calculator with the equal sign printed on it. Below the screen are five keys labeled [Y=], [WINDOW], [ZOOM], [TRACE], and [GRAPH]. These keys are positioned together below the screen because they are used for graphing functions.

Black keys: The majority of the keys on the calculator are black. Notice the [X,T,θ,n] key in the second row and second column. It will be used frequently in algebra to enter the variable **X.** The **ON** key is the black key located in the bottom-left position.

Gray keys: The 12 gray keys that are clustered at the bottom center are used to enter digits, a decimal point, or a negative sign.

Examples provide both detailed ▶
algebraic solutions as well as calculator
confirmations of analytical work.

EXAMPLE 5   Solve $2(x + 2) - x = 10 - x$.

Solution   *Algebraic solution:* $2x + 4 - x = 10 - x$
$$x + 4 = 10 - x$$
$$2x + 4 = 10$$
$$2x = 6$$
$$x = 3$$

Graphic solution: $y1 = 2(x + 2) - x$
$$y2 = 10 - x$$

Notice the display at the bottom of the screen. This means that when the variable x has a value of 3, the expressions entered at the y = prompts evaluate to 7.

Check: $2(3) + 4 - 3 ? 10 - 3$
$$6 + 4 - 3 ? 7$$
$$7 = 7$$

Calculator screens (using the TI-83) ▶
are provided throughout the text so
that students can confirm their
calculator screens.

Properties and procedures are boxed ▶
for easy reference.

Power to a Power Rule
If m and n are natural numbers, then
$$(x^m)^n = x^{mn}$$

Exploring the Concept features allow ▶
students to explore the development
of algorithms before they are
formally presented.

EXPLORING THE CONCEPT   Use your graphing calculator to explore the relationship between each pair of numbers.

a. $(2^4)^5$; $(2^5)^4$

b. $(4^2)^4$; $(4^4)^2$

c. Which property justifies the results of each pair of exponential expressions?

```
(2^4)^5
         1048576
(2^5)^4
         1048576
```

```
(4²)^4
         65536
(4^4)²
         65536
```

Since $(2^4)^5 = 2^{4 \cdot 5}$ and $(2^5)^4 = 2^{5 \cdot 4}$, it is the commutative property of multiplication that justifies the results. Thus, $(x^m)^n = (x^n)^m$.

When extensive keystroking information
is not necessary, Technology Tips are
provided in the margin of the text to
clarify the technology for the student. ▶
When more information is needed,
these tips refer the student to the
appropriate technology section.

Teaching Tips for the instructor are ▶
provided throughout the text.

■ TECHNOLOGY TIP
Lines and curves that are graphed with the graphing calculator are actually straight lines and smooth curves. It is the resolution of the calculator screen that makes them appear "jagged."

■ TEACHING TIP
Display the calculator-generated graph on the chalkboard and then plot the ordered pairs from $x = 2$ through $x = 4$ to emphasize that a line is an infinite set of collinear points.

You may now access the TABLE feature of the calculator to compare your table of values, as shown in Figure 3-9, to those of the calculator.

If you access the TRACE feature you can use the left and right cursor keys to trace along the line. You will notice that the coordinate values are cumbersome decimal values.

For this reason, we will *not* use the standard viewing window when we want to trace (see "Technology 3.4: Graphing Basics"). The calculator has a preset window called ZInteger that yields integer values when tracing. To set these values manually, enter $[-47, 47]$ by $[-31, 31]$ on your WINDOW screen now, with x and y scales equal to 10. Trace and observe the x and y coordinates.

24   CHAPTER 1   BASIC ARITHMETIC AND GRAPHING CALCULATOR CONCEPTS

Video icons show which examples ▶
are taught on videotape. Examples
provide both analytical solutions and
calculator verifications.

EXAMPLE 6   If $x = 4$ and $y = 3$, evaluate $\dfrac{3x^2 - 2y}{2(x + y)}$.

Solution

■ STUDY TIP
Although you are only on Section 1.2, you should begin to prepare NOW for your first major exam. When you finish tonight's homework (and as you complete each subsequent homework assignment) go back and work a few problems from each of the Exercise Sets that have been previously assigned. If you do this, you will find that much of your preparation for the test will have been accomplished by the time your instructor announces the test date.

Study Tips are provided to guide the ▶
student through a comprehensive
study plan.

$$\frac{3x^2 - 2y}{2(x + y)} = \frac{3(4^2) - 2(3)}{2(4 + 3)}$$ Substitute 4 for x and 3 for y.

$$= \frac{3(16) - 2(3)}{2(7)}$$ Find the value of 4^2 in the numerator and do the addition in the denominator.

$$= \frac{48 - 6}{14}$$ Do the multiplications.

$$= \frac{42}{14}$$ Do the subtraction.

$$= 3$$

■ **NUMBER LINE**

The basic operations of addition, subtraction, multiplication, and division can be applied to all real numbers. Up to this point, it is possible that your experience has led
you to deal only with the set of positive natural numbers when performing these operations. As we address these basic operations in the next two sections, the rules of
operations with signed numbers will focus on the set of integers. However, these results are applicable to *all* real numbers.

We use the number line shown in Figure 1-3 to represent sets of numbers.

FIGURE 1-3

The number line continues forever to the left and to the right. Numbers to the left
of 0 are negative, and we say that the line continues to negative infinity. Numbers
to the right of 0 are positive, and we say that the line continues to positive infinity.

 WARNING!   The number 0 is neither positive nor negative.

WARNING! boxes alert students to ▶
common errors and misconceptions.

EXERCISE 2.3

VOCABULARY   In Exercises 1–2, fill in the blank to make a true statement.

1. The _____ property allows the expression
$3(x + y)$ to be calculated as $3x + 3y$.

2. _____ terms have identical variable factors.

CONCEPTS   In Exercises 3–10 tell whether the terms are like or unlike terms. If they are like terms, find their sum.

3. $3y, 4y$
4. $3x^2, 5x^2$
5. $3x, 3y$
6. $3x^2, 6x$
7. $3x^3, 4x^3, 6x^3$
8. $-2y^4, -6y^4, 10y^4$
9. $32x^5y^3, -21x^5y^3, -11x^5y^3$
10. $-x^2y, xy, 3xy^2$

In Exercises 11–18, simplify each expression, if possible. To simplify means to combine like terms.

11. $4y + 5y$
12. $-2x + 3x$
13. $-8t^2 - 4t^2$
14. $15x^2 + 10x^2$
15. $32u^3 - 16u^3$
16. $25xy^2 - 7xy^2$
17. $18x^5y^2 - 11x^5y^2$
18. $17x^6y - 22x^6y$

PRACTICE   In Exercises 19–24, simplify each expression, if possible.

19. $3rst + 4rst + 7rst$
20. $-2ab + 7ab - 3ab$
21. $-4a^2bc + 5a^2bc - 7a^2bc$
22. $3x^2y - 5xy^2 + 7x^2y$
23. $3mn^3 - 5mn^3 + 3m^3n$
24. $6abc + 6bca + 7cba$

Vocabulary exercises are provided at the ▶
beginning of each exercise set.

Concepts exercises are provided in each ▶
exercise set.

Practice problems are provided in each ▶
exercise set.

APPLICATIONS

✎**68. Course load**   A man enrolls in college for c hours of
credit, and his sister enrolls for 4 more hours than her
brother. Write an expression that represents the number
of hours the sister is taking. Could the sister be taking
3 hours of courses? Why or why not?

69. Mileage   An antique Ford has 25,000 more miles on
its odometer than a new car. If the new car has
traveled m miles, find an expression that represents the
mileage on the Ford.

70. T-bills   Write an expression that represents the value
of t T-bills, each worth $9987.

Applications that use real-life scenarios
appear in every exercise set. ▶
Writing icons denote writing exercises. ▶

REVIEW

27. Determine the dimensions of the viewing window that
could simultaneously display the following groups of
points.
 a. $(2, 3), (-11, 3), (-15, -4),$ and $(9, -20)$

 b. $(-18, 13.4), (16.5, -4)$ and $(25, 30)$

 c. $(50, -50), (-75, 15),$ and $(25, -80)$

✎**28.** Graph the linear equation $y = x - 15$ in both the
standard (ZStandard) viewing window and in the
integer (ZInteger) viewing window and explain the
difference in what you see.

29. Graph each of the following functions in the integer
viewing window and TRACE to determine the
coordinates of the x-intercepts.
 a. $y = x^2 - 14x + 45$ **b.** $y = x^2 - x - 56$

 c. $y = x^2 + 11x + 24$

✎**30.** Trace to find the y-intercept of each of the functions
defined in Exercise 29. Compare each of the y-values
of the intercepts with the constant value of the
function. What do you observe?

Review exercises appear at the end of ▶
exercise sets.

Chapter Summaries are provided in ▶ a format that presents the student with both the text information and representative problem exercises to maximize successful review strategies. Answers to all of these problems appear in the back of the text.

---

178   CHAPTER 3   MODELING LINEAR EQUATIONS IN ONE AND TWO VARIABLES

CHAPTER SUMMARY

CONCEPTS	REVIEW EXERCISES

Section 3.1 | **Introduction to Linear Equations in One Variable**

Any real number can be added to (or subtracted from) both sides of an equation to form another equation with the same solution(s) as the original equation.

1. Tell whether the indicated number is a solution of the given equation.
   a. $3x + 7 = 1; -2$
   b. $5 - 2x = 3; -1$
   c. $2(x + 3) = x; -3$
   d. $5(3 - x) = 2 - 4x; 13$

2. Solve each equation algebraically, showing all steps of your work. Check all results in the original equation.
   a. $3x + 7 = 1$
   b. $7 - 9x = 8$

Both sides of an equation can be multiplied by (or divided by) any *nonzero* real number to form another equation with the same solution(s) as the original equation.

   c. $\dfrac{x + 3}{4} = 2$
   d. $\dfrac{x - 7}{2} = -2$
   e. $\dfrac{x}{2} + \dfrac{7}{2} = 3$
   f. $\dfrac{x}{3} - 3 = 2$

■ Chapter Test

1. Use exponents to rewrite $2\,x\,x\,x\,x\,y\,y\,y\,y$

In Problems 2–4, write each expression as an expression containing only one exponent.

2. $y^2(y\,y^3)$
3. $(2x^3)^5(x^2)^3$
4. $(2r\,r^2\,r^3)^3$

5. Simplify: $3x^0$.
6. Identify $3x^2 + 2$ as a monomial, a binomial, or a trinomial.

7. Find the degree of the polynomial $3x^2y^3z^4 + 2x^3y^2z - 5x^2y^3z^5$.
8. If $P(x) = x^2 + x - 2$, find $P(-2)$.

9. Simplify: $(xy)^2 + 5x^2y^2 - (3x)^2y^2$.
10. Add: $(3x^3 + 4x^2 - x - 7) + (2x^3 - 2x^2 + 3x + 2)$.

11. Subtract: $(2x^2 - 7x + 3) - (3x^2 - 2x - 1)$.
12. Simplify: $-6(x - y) + 2(x + y) - 3(x + 2y)$.

13. Simplify: $-2(x^2 + 3x - 1) - 3(x^2 - x + 2) + 5(x^2 + 2)$.

In Problems 14–17, find each product.

14. $(-2x^3)(2x^2y)$
15. $3y^2(y^2 - 2y + 3)$
16. $(2x - 5)(3x + 4)$
17. $(2x - 3)(x^2 - 2x + 4)$

18. Simplify the fraction: $\dfrac{8x^2y^3z^4}{16x^3y^2z^4}$.
19. Do the division: $\dfrac{6a^2 - 12b^2}{24ab}$.
20. Divide: $2x + 3\overline{)2x^2 - x - 6}$.

■ Cumulative Review: Chapters 1 and 2

VOCABULARY / CONCEPTS

1. A number greater than one, whose only factors are one and itself is called a _____ number.

2. Explain, in a step-by-step format, how to find the least common denominator for two fractions.

   Following your step-by-step directions, find the least common denominator for $\frac{5}{24}$ and $\frac{7}{18}$.

3. Explain the difference between the expressions $4x$ and x^4.
4. Explain the difference between *evaluate* and *simplify* when applied to the expression $3x + 5x$.

5. Determine the absolute value of each expression and verify with your calculator.
   a. $|-3|$
   b. $-|-3|$
   c. $-|3|$

6. What rule supports the statement $-(-5) = 5$?

7. Explain why $\frac{8}{0}$ is undefined and $\frac{0}{x} = 0$ provided $x \neq 0$.

8. Which of
   a. (-4)
   c. $-4(-3$

◀ Chapter Tests (with answers to all exercises) are provided at the end of each chapter.

◀ Cumulative Reviews (with answers to all exercises) appear at the end of every even-numbered chapter.

■ ■ ■ ■ ■ ■   **CAREERS & MATHEMATICS**

Pilot   Pilots are highly trained professionals who fly airplanes and helicopters to carry out a wide variety of tasks. Most pilots transport passengers and cargo, but some are involved in crop dusting, spreading seed, testing aircraft, fire-fighting, tracking criminals, traffic reporting, and rescue and evacuation.

Each chapter contains a Careers & ▶ Mathematics feature that relates mathematics to career applications.

Qualifications   All pilots paid to transport passengers or cargo must have a commercial pilot's license with an instrument rating. They must also have good health, vision correctable to 20/20, good hearing, and no physical handicaps that would interfere with safety. Most employers also require at least 2 years of college, and many prefer college graduates. Requirements for a private pilot's license are less restrictive.

Job Outlook   Because pilots like their jobs and few change careers, competition for employment is stiff. Over the next decade, however, retirements will provide job openings for qualified applicants.

Example Application   The air passing over the curved top of the wing in Illustration 1 moves faster than the air beneath, because it has a greater distance to travel. This faster-moving air exerts less pressure on the wing than the air beneath, causing *lift*.

Every chapter includes a Perspective ▶
feature that provides interesting
mathematical and historical information.

■ PERSPECTIVE

All Dark Things

The ancient Egyptians developed two systems of writing. Each symbol in one system, hieroglyphics, was a picture of an object. Because hieroglyphic writing was usually inscribed on stone, many examples still survive. For daily purposes, Egyptians used hieratic writing. Simpler than hieroglyphics, hieratic writings were done with ink on papyrus sheets. Papyrus, made from plants, is very delicate and quickly dries and crumbles. Few Egyptian papyri survive.

Those that do survive provide important clues to the content of ancient mathematics. One, the Rhind Papyrus, was discovered in 1858 by a British archeologist, Henry Rhind. Also known as the Ahmes Papyrus after its ancient author, it begins with a description of its contents: *Directions for Obtaining the Knowledge of All Dark Things.*

The Ahmes Papyrus and another, the Moscow Papyrus, together contain 110 mathematical problems and their solutions. Many of these were probably for education, beause they represented situations that the scribes, priests, and other workers in government and temple administration were expected to be able to solve.

The Ahmes Papyrus (the British Museum)

■ ANCILLARIES FOR THE INSTRUCTOR

Annotated Instructor's Edition

Free to professors when the text is adopted, the Annotated Instructor's Edition includes the complete text of the Student Edition, along with answers to all problems printed in blue next to the problems. Teaching Tips are included in the margins.

Instructor's Resource Manual

This manual contains complete, worked-out solutions for all exercises in the text, as well as printed test forms with answers and multiple-choice and free-response questions keyed to the text. Other features include:

- chapter tests for each chapter.
- final exams for both beginning and intermediate algebra courses.
- pages that can be made into transparencies for classroom use.
- suggested coverage time for each section.
- Ideas to Emphasize—main ideas to be covered in class.
- Discussion Ideas—topics for class discussion.
- Teaching Tips—alternate approaches or examples.
- Technology Notes—help with graphing calculator use.
- Points to Ponder—various types of helpful information for the instructor.

Thomson Learning Testing Tools CD

This fully integrated suite of programs allows a professor to easily post course information, office hours, lesson information, assignments, sample tests, and link to rich Web content, including review and enrichment material from Brooks/Cole. Updates are quick and easy, and customer support is available 24 hours a day, seven days a week.

Text-Specific Video Series

Free to schools when the text is adopted, this video series works through one example from each section of the text (designated by a video icon in both the Student Edition and the Annotated Instructor's Edition). Two additional examples are provided for every section.

Demo Video

The demonstration video is approximately 30 minutes long and features outtakes from the text-specific video series.

Interactive Algebra 3.0

This extremely intuitive, text-specific student tutorial is offered in student and instructor versions. It offers a mathematical guide to students of any technological proficiency and a tracking program for professors to follow their students' progress.

ANCILLARIES FOR THE STUDENT

Student Solutions Manual

This manual contains complete, worked-out solutions to all odd-numbered exercises in the text.

Greatest Hits Tutorial Video

This student video features the concepts and skills students traditionally have the most difficulty comprehending. The tape includes examples, chosen by authors, from each chapter.

Acknowledgments

We are grateful to the following people, who reviewed the manuscript at various stages of development. They all had valuable suggestions that have been incorporated into the text.

Gladys H. Crates
Chattanooga State Community College

Susan Jones
Nashville State Technical College

Ruth Ann Edwards
Craven Community College

Kathryn C. Wetzel
Amarillo College

Laurette Blakey Foster
Prairie View A & M University

We are grateful to our editor, Robert Pirtle, for his encouragement and support throughout this project. Thanks to Jennifer Huber for her encouragement early on in this project. We are also grateful to Ellen Brownstein who managed the production process. We want to express our many thanks to Trish Finley and The Clarinda Company for their skillful project management and outstanding composition, respectively, and to Trish Cabral for producing the videos that accompany the text.

We are particularly grateful to David Cochener for his meticulous attention to detail in reading the manuscript and working problems.

Deborah J. Cochener
Bonnie M. Hodge
R. David Gustafson

Basic Arithmetic and Graphing Calculator Concepts

1

■ PERSPECTIVE

All Dark Things

The ancient Egyptians developed two systems of writing. Each symbol in one system, hieroglyphics, was a picture of an object. Because hieroglyphic writing was usually inscribed on stone, many examples still survive. For daily purposes, Egyptians used hieratic writing. Simpler than hieroglyphics, hieratic writings were done with ink on papyrus sheets. Papyrus, made from plants, is very delicate and quickly dries and crumbles. Few Egyptian papyri survive.

Those that do survive provide important clues to the content of ancient mathematics. One, the Rhind Papyrus, was discovered in 1858 by a British archeologist, Henry Rhind. Also known as the Ahmes Papyrus after its ancient author, it begins with a description of its contents: *Directions for Obtaining the Knowledge of All Dark Things.*

The Ahmes Papyrus and another, the Moscow Papyrus, together contain 110 mathematical problems and their solutions. Many of these were probably for education, beause they represented situations that the scribes, priests, and other workers in government and temple administration were expected to be able to solve.

The Ahmes Papyrus (the British Museum)

Algebra is an extension of arithmetic and allows us to reason from the specific to the general. Learning algebra provides us with powerful problem-solving strategies and enhances our logical thinking skills. In algebra, the operations of addition, subtraction, multiplication, and division are performed on both constants and variables, with the understanding that the variables may be replaced by constants. This leads to solutions of problems that would be difficult or impossible for us to solve by using arithmetic alone. Because humans are visual thinkers, our goal will be to draw pictures wherever possible to help our understanding. That is why we will use a graphing calculator as we develop our problem-solving skills and tools.

1.1 Review of Fractions

■ SETS OF NUMBERS ■ SIMPLIFYING FRACTIONS ■ MULTIPLYING FRACTIONS ■ DIVIDING FRACTIONS ■ ADDING FRACTIONS ■ SUBTRACTING FRACTIONS ■ MIXED NUMBERS ■ DECIMAL FRACTIONS

■ SETS OF NUMBERS

■ TECHNOLOGY TIP
Technology Tips are provided throughout this text to help you support your algebraic work with the graphing calculator. If we feel that you might need extensive help with your calculator, we will refer you to an extended explanation at the end of the section. (See "Technology 1.1: Introduction to the Graphing Calculator," page 14.)

We begin our discussion with the set of real numbers. A **set** is a well-defined collection of objects. For example, the set {1, 2, 3, 4, 5} contains the numbers 1, 2, 3, 4, and 5. The members, or elements, in a set are listed within braces { }.

Two basic sets of numbers are the **natural numbers** and the **whole numbers.**

Natural Numbers
The **natural numbers** (or the **positive integers**) are the numbers 1, 2, 3, 4, 5, ... This would be written as {1, 2, 3, 4, 5, . . .} when using set notation.

Whole Numbers
The **whole numbers** are the numbers 0, 1, 2, 3, 4, 5, ... This would be written as {0, 1, 2, 3, 4, 5, . . .} when using set notation.

The three dots in the definitions indicate that the lists continue forever.

We can use whole numbers to describe many real-life situations. For example, some cars get 30 miles per gallon of gas, some students might have paid $1750 in tuition, and the temperature can be 0°.

Integers
The **integers** are the numbers . . . , −5, −4, −3, −2, −1, 0, 1, 2, 3, 4, 5, ... This would be written as {. . . , −5, −4, −3, −2, −1, 0, 1, 2, 3, 4, 5, . . .} when using set notation.

Numbers that show a loss or a downward direction are called **negative integers,** denoted as −1, −2, −3, and so on. For example, a debt of $1500 can be denoted as −$1500, and a temperature of 20° below zero can be denoted as −20°. The set of negative integers together with the set of whole numbers form the set of **integers.**

Because the set of natural numbers and the set of whole numbers are included within the set of integers, we say that these sets are **subsets** of the set of integers.

Integers cannot describe every real-life situation. For example, a student might study $3\frac{1}{2}$ hours, or a television might cost $217.36. To describe these situations, we need fractions. A **fraction** that can be expressed as a quotient of integers, $\frac{p}{q}$, where $q \neq 0$, is called a **rational number.**

Rational Numbers
A **rational number** is any number that can be written as a fraction with an integer in its numerator and a nonzero integer in its denominator.

Fractions can also be used to indicate division. For example, the division of $8 \div 2$ can be written as a fraction, $\frac{8}{2}$. Fractions are also used to indicate parts of a whole. In Figure 1-1, a rectangle has been divided into 5 equal parts, and 3 of the parts are shaded. The fraction $\frac{3}{5}$ indicates how much of the figure is shaded. The denominator of the fraction shows the total number of equal parts into which the whole is divided, and the numerator shows the number of these equal parts that are being considered.

Some examples of rational numbers are

FIGURE 1-1

$$\frac{3}{2}, \frac{17}{12}, -\frac{43}{8}, 0.25, \quad \text{and} \quad -0.66666\ldots.$$

The decimals 0.25 and −0.66666 . . . are rational numbers because they can be written as fractions: 0.25 is the fraction $\frac{1}{4}$ and −0.66666 . . . is the fraction $-\frac{2}{3}$.

Because every integer can be written as a fraction with a denominator of 1, every integer is also a rational number. Because every integer is a rational number, the set of integers is a subset of the rational numbers.

■ **TECHNOLOGY TIP**
Your calculator will display an error message if you attempt to divide by 0.

WARNING! Because division by 0 is undefined, expressions such as $\frac{6}{0}$ and $\frac{0}{0}$ do not represent any number.

■ SIMPLIFYING FRACTIONS

A **prime number** is a whole number greater than 1 whose only factors are 1 and the number itself. The set of prime numbers can be represented as {2, 3, 5, 7, 11, 13, 17, 19, . . .}. A fraction is in **lowest terms** when the numerator and the denominator have no common factors (other than 1). To reduce the fraction to lowest terms (often called *simplifying*) we can factor the numerator and the denominator into **primes** and divide out common factors. The following example illustrates the **fundamental property of fractions.**

The Fundamental Property of Fractions

If a represents a real number, and b and x represent nonzero real numbers, then

$$\frac{a \cdot x}{b \cdot x} = \frac{a}{b}.$$

The fraction $\frac{6}{15}$ IS NOT in lowest terms because 6 and 15 have a common factor of 3.

$$\frac{6}{15} = \frac{2 \cdot 3}{5 \cdot 3}$$ Factor the numerator and denominator into a product of prime factors.

$$= \frac{2 \cdot \overset{1}{\cancel{3}}}{5 \cdot \underset{1}{\cancel{3}}}$$ Divide out the common factor of 3.

$$= \frac{2}{5}$$

Therefore, $\frac{6}{15}$ simplifies to $\frac{2}{5}$. From Figure 1-2 we can see that $\frac{6}{15}$ and $\frac{2}{5}$ are equal fractions because they represent the same part of the rectangle.

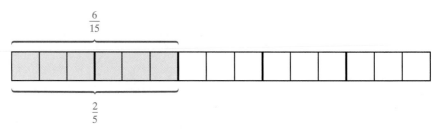

FIGURE 1-2

EXAMPLE 1

a. To simplify (reduce) $\frac{6}{30}$, we factor the numerator and denominator into a product of primes and divide out (cancel) the common factors of 2 and 3.

$$\frac{6}{30} = \frac{\overset{1}{\cancel{2}} \cdot \overset{1}{\cancel{3}}}{\underset{1}{\cancel{2}} \cdot \underset{1}{\cancel{3}} \cdot 5} = \frac{1}{5}$$

b. To simplify (reduce) $\frac{33}{40}$, we factor the numerator and denominator into a product of primes. Because the numerator and denominator have no common factors, the fraction $\frac{33}{40}$ is simplified.

$$\frac{33}{40} = \frac{3 \cdot 11}{2 \cdot 2 \cdot 2 \cdot 5} = \frac{33}{40}$$

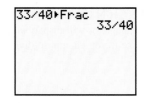

The number 1 is the **identity element for multiplication.** This means that *any* number or expression multiplied by 1 gives a result equal to the original number or expression. Symbolically, this may be stated as $a \cdot 1 = a$.

Basic Properties of Fractions

1. $\dfrac{a}{0}$ is undefined. **2.** If $a \neq 0$, then $\dfrac{0}{a} = 0$.

3. $\dfrac{a}{1} = a$. **4.** If $a \neq 0$, then $\dfrac{a}{a} = 1$.

WARNING! Remember that a fraction is simplified only when its numerator and denominator have no common factors.

■ Multiplying Fractions

Multiplying Fractions

If a, b, c, and d represent real numbers, then $\dfrac{a}{b} \cdot \dfrac{c}{d} = \dfrac{a \cdot c}{b \cdot d}$ ($b \neq 0$ and $d \neq 0$).

Simply stated, when multiplying two fractions, write the product of the numerators over the product of the denominators. Remember, you must then reduce the fraction to lowest terms. Fractions such as $\frac{8}{21}$ with a numerator that is smaller than the denominator are called **proper fractions.** Fractions such as $\frac{8}{3}$ with a numerator that is larger than the denominator are called **improper fractions.**

EXAMPLE 2   Multiply: $\dfrac{3}{7} \cdot \dfrac{13}{5}$

Solution
$$\frac{3}{7} \cdot \frac{13}{5} = \frac{3 \cdot 13}{7 \cdot 5}$$
$$= \frac{39}{35}$$

■ **TECHNOLOGY TIP**
To ensure that the appropriate order of operations is followed, each fraction of a multiplication or division problem should be enclosed within parentheses. (When using the graphing calculator, it is suggested that students initially enclose all fractions within parentheses.)

Notice, in this particular problem, the numerators and denominators of the original fractions are prime numbers. It is therefore very easy to see that the fraction, $\frac{39}{35}$, is simplified.

EXAMPLE 3   **European travel**   Out of 36 students in a history class, three-fourths have signed up for a trip to Europe. There are 30 places available on the chartered flight. Will there be room for one more student?

Solution   We first find three-fourths of 36 by multiplying 36 by $\frac{3}{4}$.

$$\frac{3}{4} \cdot 36 = \frac{3}{4} \cdot \frac{36}{1}$$ Write 36 as $\frac{36}{1}$.

$$= \frac{3 \cdot 36}{4 \cdot 1}$$ Multiply the numerators and multiply the denominators.

$$= \frac{3 \cdot \overbrace{3 \cdot 3 \cdot 2 \cdot 2}^{36}}{\underbrace{2 \cdot 2 \cdot 1}_{4}}$$ To simplify fractions, factor the numerator and denominator.

$$= \frac{3 \cdot 3 \cdot 3 \cdot \overset{1}{\cancel{2}} \cdot \overset{1}{\cancel{2}}}{\underset{1}{\cancel{2}} \cdot \underset{1}{\cancel{2}} \cdot 1}$$ Divide out the common factors of 2.

(3/4)*36▸Frac
                    27

$$= \frac{27}{1}$$
$$= 27$$

Twenty-seven students plan to go on the trip. Because there is room for 30 passengers, there is plenty of room for one more.   ■

■ DIVIDING FRACTIONS

One number is called the **reciprocal** (or **multiplicative inverse**) of another if their product is 1. For example, $\frac{3}{5}$ is the reciprocal of $\frac{5}{3}$ because

$$\frac{3}{5} \cdot \frac{5}{3} = \frac{15}{15} = 1$$

Dividing Fractions
To divide two fractions, we multiply the first fraction by the reciprocal of the second fraction. In symbols, if a, b, c, and d are real numbers, then

$$\frac{a}{b} \div \frac{c}{d} = \frac{a}{b} \cdot \frac{d}{c} = \frac{a \cdot d}{b \cdot c} \qquad (b \neq 0,\ c \neq 0,\ \text{and}\ d \neq 0)$$

EXAMPLE 4 **a.** $\dfrac{3}{5} \div \dfrac{6}{5} = \dfrac{3}{5} \cdot \dfrac{5}{6}$ Multiply $\frac{3}{5}$ by the reciprocal of $\frac{6}{5}$.

$= \dfrac{3 \cdot 5}{5 \cdot 6}$ Multiply the numerators and multiply the denominators.

$= \dfrac{3 \cdot 5}{5 \cdot 2 \cdot 3}$ Factor the denominator.

$= \dfrac{\overset{1}{\cancel{3}} \cdot \overset{1}{\cancel{5}}}{\underset{1}{\cancel{5}} \cdot 2 \cdot \underset{1}{\cancel{3}}}$ Divide out the common factors of 3 and 5.

$= \dfrac{1}{2}$

b. $\dfrac{15}{7} \div 10 = \dfrac{15}{7} \div \dfrac{10}{1}$ Write 10 as the improper fraction $\frac{10}{1}$.

$= \dfrac{15}{7} \cdot \dfrac{1}{10}$ Multiply $\frac{15}{7}$ by the reciprocal of $\frac{10}{1}$.

$= \dfrac{15 \cdot 1}{7 \cdot 10}$ Multiply the numerators and multiply the denominators.

$= \dfrac{3 \cdot \overset{1}{\cancel{5}}}{7 \cdot 2 \cdot \underset{1}{\cancel{5}}}$ Factor the numerator and denominator.

$= \dfrac{3}{14}$ Divide out the common factor of 5.

■

■ STUDY TIP
Mathematics cannot be learned by passive observation; it requires active participation on your part. Take the time now to make a 3×5 note card for each boldfaced definition that has been presented thus far. Place the word(s) on the front of the card and the definition on the back. As you work on learning these definitions, tackle only a few at each study session and take advantage of all those "hidden" study times: waiting for a professor to begin class, waiting for a traffic light to change, waiting for your child to come out of school, etc. These brief study moments are excellent opportunities for painlessly learning the properties and definitions that are fundamental to a mathematics course.

■ TECHNOLOGY TIP
When entering addition or subtraction expressions involving fractions, it is not necessary to enclose each fraction in parentheses. When $\frac{3}{7} + \frac{2}{7}$ is entered as 3/7 + 2/7 ►Frac, the order of operations guarantees that the division operations of 3 divided by 7 and 2 divided by 7 will be performed before the operation of addition.

■ ADDING FRACTIONS

Adding Fractions with the Same Denominator
To add two fractions with the same denominator, we add the numerators and keep the common denominator. In symbols, if a, b, and d are real numbers, then

$$\frac{a}{d} + \frac{b}{d} = \frac{a + b}{d} \qquad (d \neq 0)$$

For example,

$$\frac{3}{7} + \frac{2}{7} = \frac{3+2}{7}$$ Add the numerators and keep the common denominator.

$$= \frac{5}{7}$$

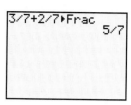

To add fractions that do not have common denominators, we can use the rule above, after rewriting the fractions with least common denominators. The **least common denominator** is the smallest number that can be divided evenly by two or more numbers.

$$\frac{1}{3} + \frac{1}{5}$$ The common denominator is 15.

Thus each fraction must be rewritten with the denominator of 15.

$$\frac{1}{3} = \frac{1 \cdot 5}{3 \cdot 5} = \frac{5}{15} \quad \text{and} \quad \frac{1}{5} = \frac{1 \cdot 3}{5 \cdot 3} = \frac{3}{15}$$

Now add the two new fractions that are equivalent to the original fractions:

$$\frac{1}{3} + \frac{1}{5} = \frac{5}{15} + \frac{3}{15} = \frac{5+3}{15} = \frac{8}{15}$$

EXAMPLE 5   Add $\dfrac{3}{10}$ and $\dfrac{5}{28}$.

Solution   To find the least common denominator (LCD), we find the prime factorization of both denominators and use each prime factor the greatest number of times it appears in either factorization:

$$\left.\begin{array}{l} 10 = 2 \cdot 5 \\ 28 = 2 \cdot 2 \cdot 7 \end{array}\right\} \text{LCD} = 2 \cdot 2 \cdot 5 \cdot 7 = 140$$

Because 140 is the smallest number that 10 and 28 divide exactly, we write both fractions as fractions with the least common denominator of 140.

$$\frac{3}{10} + \frac{5}{28} = \frac{3 \cdot 14}{10 \cdot 14} + \frac{5 \cdot 5}{28 \cdot 5}$$ Write each fraction as a fraction with a denominator of 140.

$$= \frac{42}{140} + \frac{25}{140}$$

$$= \frac{42 + 25}{140}$$ Add numerators and keep the denominator.

$$= \frac{67}{140}$$

Since 67 is a prime number that does not divide 140, $\frac{67}{140}$ is in lowest terms and cannot be simplified.

Verify with your calculator.

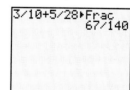

SUBTRACTING FRACTIONS

Subtracting Fractions with the Same Denominator

To subtract two fractions with the same denominator, we subtract their numerators and keep their common denominator. In symbols, if a, b, and d are real numbers, then

$$\frac{a}{d} - \frac{b}{d} = \frac{a-b}{d} \qquad (d \neq 0)$$

For example,

$$\frac{7}{9} - \frac{2}{9} = \frac{7 - 2}{9}$$

$$= \frac{5}{9}$$

■ **TECHNOLOGY TIP**

Once again, it is not necessary to enclose the fractions of a subtraction problem in parentheses. The order of operations guarantees that the fraction division operations are performed before the operation of subtraction.

To subtract fractions with unlike denominators, we write them as equivalent fractions with a common denominator. For example, to subtract $\frac{2}{5}$ from $\frac{3}{4}$, find the LCD of 20 and proceed as follows:

$$\frac{3}{4} - \frac{2}{5} = \frac{3 \cdot 5}{4 \cdot 5} - \frac{2 \cdot 4}{5 \cdot 4}$$

$$= \frac{15}{20} - \frac{8}{20}$$

$$= \frac{15 - 8}{20} \qquad \text{Subtract the numerators and keep the denominator.}$$

$$= \frac{7}{20}$$

EXAMPLE 6   Subtract 5 from $\frac{23}{3}$.

Solution

$$\frac{23}{3} - 5 = \frac{23}{3} - \frac{5}{1} \qquad \text{Write 5 as the improper fraction } \frac{5}{1}.$$

$$= \frac{23}{3} - \frac{5 \cdot 3}{1 \cdot 3} \qquad \text{Write } \frac{5}{1} \text{ as a fraction with a denominator of 3.}$$

$$= \frac{23}{3} - \frac{15}{3}$$

$$= \frac{23 - 15}{3} \qquad \text{Subtract the numerators and keep the denominator.}$$

$$= \frac{8}{3}$$

■ **MIXED NUMBERS**

The **mixed number** $3\frac{1}{2}$ represents the sum of 3 and $\frac{1}{2}$. We can write $3\frac{1}{2}$ as an improper fraction as follows:

$$3\frac{1}{2} = 3 + \frac{1}{2}$$

$$= \frac{6}{2} + \frac{1}{2} \qquad 3 = \frac{6}{2}.$$

$$= \frac{6 + 1}{2} \qquad \text{Add the numerators and keep the denominator.}$$

$$= \frac{7}{2}$$

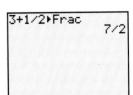

EXAMPLE 7   Add $2\frac{1}{4}$ and $1\frac{1}{3}$.

Solution   We can change each mixed number to an improper fraction

$$2\frac{1}{4} = 2 + \frac{1}{4} \qquad\qquad 1\frac{1}{3} = 1 + \frac{1}{3}$$

$$= \frac{8}{4} + \frac{1}{4} \qquad\qquad = \frac{3}{3} + \frac{1}{3}$$

$$= \frac{9}{4} \qquad\qquad\qquad = \frac{4}{3}$$

■ TECHNOLOGY TIP

Before performing operations with fractions that are mixed numbers, you may prefer to convert the mixed numbers to improper fractions first.

and then add the fractions.

$$2\frac{1}{4} + 1\frac{1}{3} = \frac{9}{4} + \frac{4}{3}$$

$$= \frac{9 \cdot 3}{4 \cdot 3} + \frac{4 \cdot 4}{3 \cdot 4}$$

$$= \frac{27}{12} + \frac{16}{12}$$

$$= \frac{43}{12}$$

Finally, we change $\frac{43}{12}$ to a mixed number.

$$\frac{43}{12} = 3 + \frac{7}{12} = 3\frac{7}{12}$$

 WARNING!   Many students in beginning algebra classes believe a fraction is not simplified when left as an improper fraction. A fraction is considered simplified when the numerator and denominator have NO common factors. Improper fractions are acceptable.

EXAMPLE 8   **Fencing land**   The three sides of a triangular piece of land measure $33\frac{1}{4}$, $57\frac{3}{4}$, and $72\frac{1}{2}$ meters. How much fencing will be needed to enclose the area?

Solution   We can find the sum of the lengths by adding the whole number parts and the fractional parts of the dimensions separately:

$$33\frac{1}{4} + 57\frac{3}{4} + 72\frac{1}{2} = 33 + 57 + 72 + \frac{1}{4} + \frac{3}{4} + \frac{1}{2}$$

$$= 162 + \frac{1}{4} + \frac{3}{4} + \frac{2}{4} \qquad \text{Change } \tfrac{1}{2} \text{ to } \tfrac{2}{4} \text{ to obtain a common denominator.}$$

$$= 162 + \frac{6}{4} \qquad\qquad \text{Add the fractions by adding the numerators and keeping the denominator.}$$

$$= 162 + \frac{3}{2} \qquad\qquad \frac{6}{4} = \frac{2 \cdot 3}{2 \cdot 2} = \frac{\overset{1}{\cancel{2}} \cdot 3}{\underset{1}{\cancel{2}} \cdot 2} = \frac{3}{2}$$

$$= 162 + 1\frac{1}{2} \qquad\qquad \text{Change } \tfrac{3}{2} \text{ to a mixed number.}$$

$$= 163\frac{1}{2}$$

```
33+57+72+1/4+3/4
+1/2▶Frac
            327/2
```

It will require $163\frac{1}{2}$ meters of fencing to enclose the triangular area.

Due to the size of the mixed numbers, it was easier to enter the numbers as displayed instead of first converting to improper fractions.

Commutative Properties

If a and b are real numbers, then

$a + b = b + a$ Commutative property of addition.

$ab = ba$ Commutative property of multiplication.

In the previous example, $33\frac{1}{4} + 57\frac{3}{4} + 72\frac{1}{2}$, is rewritten as $33 + 57 + 72 + \frac{1}{4} + \frac{3}{4} + \frac{1}{2}$ using the Commutative Property of Addition.

EXAMPLE 9 When multiplication or division involves mixed numbers, it is best to convert the mixed number to an improper fraction.

a. $\left(4\frac{1}{3}\right)\left(2\frac{3}{5}\right)$

b. $\left(5\frac{1}{2}\right) \div \left(1\frac{3}{5}\right)$

Solution **a.** $\left(4\frac{1}{3}\right)\left(2\frac{3}{5}\right) = \frac{13}{3} \cdot \frac{13}{5} = \frac{13 \cdot 13}{3 \cdot 5} = \frac{169}{15}$

b. $\left(5\frac{1}{2}\right)\left(1\frac{3}{5}\right) = \frac{11}{2} \div \frac{8}{5} = \frac{11}{2} \cdot \frac{5}{8} = \frac{11 \cdot 5}{2 \cdot 8} = \frac{55}{16}$

■ **TECHNOLOGY TIP**
Mixed numbers such as $\left(4\frac{1}{3}\right)\left(2\frac{3}{5}\right)$ can also be entered on the calculator as $(4 + 1/3)(2 + 3/5)$.

```
(4+1/3)(2+3/5)▶F
rac
           169/15
```

```
(13/3)(13/5)▶Fra
c
           169/15
(11/2)/(8/5)▶Fra
c
            55/16
```

■ DECIMAL FRACTIONS

Rational numbers can always be converted to decimal fractions. For example, to change $\frac{1}{4}$ and $\frac{5}{22}$ to decimal fractions, we use long division:

$$\begin{array}{r} 0.25 \\ 4\overline{)1.00} \\ \underline{8} \\ 20 \\ \underline{20} \\ 0 \end{array} \qquad \begin{array}{r} 0.22727\ldots \\ 22\overline{)5.00000} \\ \underline{4\,4} \\ 60 \\ \underline{44} \\ 160 \\ \underline{154} \\ 60 \\ \underline{44} \\ 16 \end{array}$$

The decimal fraction 0.25 is called a **terminating decimal**, and the decimal fraction 0.2272727 . . . (often written as $0.2\overline{27}$ is called a **repeating decimal** because it repeats a block of digits, 27. Every rational number can be converted to either a terminating or a repeating decimal.

■ **TECHNOLOGY TIP**
Calculator displays are limited to a finite number of digits. The display will be rounded, not displayed with the bar.

Terminating Decimals	*Repeating Decimals*	
$\frac{1}{2} = 0.5$	$\frac{1}{3} = 0.33333\ldots$	*or* $0.\overline{3}$
$\frac{3}{4} = 0.75$	$\frac{1}{6} = 0.16666\ldots$	*or* $0.1\overline{6}$
$\frac{5}{8} = 0.625$	$\frac{5}{22} = 0.2272727\ldots$	*or* $0.2\overline{27}$

```
1/3     .3333333333
1/6     .1666666667
5/22    .2272727273
```

■ **STUDY TIP**
A fraction that is reduced to lowest terms will convert to a terminating decimal provided the denominator has as its only factors powers of 2 and/or 5.

When performing computations with fractions it is acceptable to use the *equivalent* terminating decimal. For example $\frac{1}{2}$ or 0.5. DO NOT round repeating decimals to replace fractions when computing: $\frac{1}{3} \neq .3$!

Some numbers cannot be expressed as a quotient of integers or as a terminating or repeating decimal. Such numbers are called **irrational** numbers. Examples of these types of numbers would be $\sqrt{2}$ and π. We will discuss irrational numbers in more detail in Section 1.5. It is important to mention irrational numbers at this point because the set of irrational numbers combined with the set of rational numbers determine the set of **real** numbers. All of the properties defined thus far are applicable to the set of real numbers.

Real Numbers
A real number is any number that is either rational or irrational.

EXERCISE 1.1

VOCABULARY *In Exercises 1–6, fill in the blank to make the statement true.*

1. The _____ numbers or positive integers are the set of numbers {1, 2, 3, . . .}.

2. The _____ numbers are the set of numbers {0, 1, 2, 3, . . .}.

3. A number that can be written as the quotient of two integers is called a _____ number.

4. A number greater than 1 whose only factors are 1 and itself is called a _____ number.

5. The number 1 is the _____ element for multiplication, and 0 is the _____ element for addition.

6. Another name for reciprocal is _____.

CONCEPTS

7. Identify the rational numbers that are in lowest terms.

   a. $\dfrac{4}{5}$ **b.** $\dfrac{6}{8}$ **c.** $\dfrac{5}{2}$

8. Simplify.

   a. $\dfrac{5}{0}$ **b.** $\dfrac{9}{5}$

9. Which of the following are correct?

   a. $\dfrac{4}{5} \div \dfrac{2}{3} = \dfrac{5}{4} \cdot \dfrac{2}{3}$ **b.** $\dfrac{3}{7} \div 10 = \dfrac{3}{7} \cdot \dfrac{1}{10}$

   c. $1\dfrac{1}{2} \div \dfrac{2}{5} = \dfrac{3}{2} \cdot \dfrac{5}{2}$ **d.** $\dfrac{4}{5} \div 4\dfrac{2}{3} = \dfrac{4}{5} \cdot 4\dfrac{3}{2}$

10. Determine the least common denominator of each pair.

    a. $\dfrac{4}{5} + \dfrac{3}{2}$ **b.** $\dfrac{7}{8} + \dfrac{5}{16}$ **c.** $\dfrac{5}{9} + \dfrac{7}{15}$

✍**11.** Describe how you would find the least common denominator of two fractions.

12. Which of the following are displayed correctly on the calculator?

    a. $\dfrac{4}{5} + \dfrac{3}{7}$ **b.** $\dfrac{4}{5} \cdot \dfrac{3}{7}$

   c. $\dfrac{4}{5} \div \dfrac{3}{7}$

✍**13.** Explain why the commutative property is not valid for division.

✍**14.** Enter the fraction 2/3 on the calculator. Is the decimal number displayed equal to 2/3? Explain.

✍**15.** Explain how to convert a mixed number such as $4\frac{17}{19}$ to an improper fraction using only the calculator.

✍**16.** Explain how to convert an improper fraction into a mixed number.

PRACTICE *In Exercises 17–20, reduce each fraction to lowest terms by factoring numerators and denominators into primes and dividing out common factors. Use the* ➤*Frac option of the graphing calculator to verify your results.*

17. $\dfrac{24}{18}$ **18.** $\dfrac{35}{14}$ **19.** $\dfrac{72}{64}$ **20.** $\dfrac{26}{21}$

In Exercises 21–44, perform the indicated operation. Make sure your result is reduced to lowest terms. Use the ➤*Frac option of the graphing calculator to verify your results.*

21. $\dfrac{5}{12} \cdot \dfrac{18}{5}$ **22.** $\dfrac{5}{4} \cdot \dfrac{12}{10}$ **23.** $12 \cdot \dfrac{5}{6}$ **24.** $9 \cdot \dfrac{7}{12}$

25. $2\dfrac{2}{3} \cdot 1\dfrac{5}{8}$ **26.** $8\dfrac{2}{9} \cdot 7\dfrac{3}{4}$ **27.** $\dfrac{2}{12} \div \dfrac{8}{13}$ **28.** $\dfrac{4}{7} \div \dfrac{20}{21}$

29. $6 \div \dfrac{3}{14}$ **30.** $23 \div \dfrac{46}{5}$ **31.** $1\dfrac{12}{30} \div 7$ **32.** $4\dfrac{2}{8} \div 17$

33. $\dfrac{3}{5} + \dfrac{3}{5}$ **34.** $\dfrac{4}{7} - \dfrac{2}{7}$ **35.** $\dfrac{1}{6} + \dfrac{1}{24}$ **36.** $\dfrac{17}{25} - \dfrac{2}{5}$

37. $\dfrac{5}{14} - \dfrac{4}{21}$ **38.** $\dfrac{2}{33} + \dfrac{3}{22}$ **39.** $\dfrac{17}{3} + 4$ **40.** $\dfrac{13}{9} - 1$

41. $4\dfrac{3}{5} + \dfrac{3}{5}$ **42.** $2\dfrac{1}{8} + \dfrac{3}{8}$ **43.** $3\dfrac{3}{4} - 2\dfrac{1}{2}$ **44.** $15\dfrac{5}{6} + 11\dfrac{5}{8}$

✑**45.** Explain why $\frac{2}{3} \div \frac{4}{5}$ must be entered as (2/3)(4/5) on the calculator.

APPLICATIONS *In Exercises 46–50, use the admissions profile in Illustration 1. The admissions profile for the class of 2000 for Vanderbilt University is given in Illustration 1. Write an arithmetic expression to model the given question/ direction and determine the result.*

46. What fractional part of the accepted applications were enrolled in Vanderbilt? Use your graphing calculator to express this fraction in lowest terms.

47. What fractional part of the number of applicants were actually enrolled in Vanderbilt? Use your graphing calculator to express this fraction in lowest terms.

48. How many of the applicants attended public high school?

49. How many of the applicants attended an independent high school?

50. How many of the applicants attended religious high school?

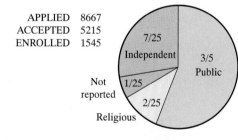

ADMISSIONS PROFILE FOR THE CLASS OF 2000 APPLICATION ACTIVITY

APPLIED   8667
ACCEPTED   5215
ENROLLED   1545

ILLUSTRATION 1

In Exercises 51–56, write an arithmetic expression to model the direction/question and determine the result.

51. Data analysis Seven hundred fifty students at the local community college were surveyed when they enrolled for classes to determine how they learn about Web sites. The survey determined that one-fifth of the students learn about Web sites from their friends.

a. Write an arithmetic expression to determine how many students learned about Web sites from their friends. Simplify the expression.

b. Write an arithmetic expression to determine how many students learned about Web sites from other sources. Simplify the expression.

52. Data analysis Three hundred fifteen students graduated from Northeast High School. One-fifteenth of these students all attended the same state university.

a. Write an arithmetic expression to determine how many students went to this university. Simplify the expression.

b. Write an arithmetic expression to determine how many students did NOT attend this university. Simplify the expression.

c. One-third of the class did not attend any school of higher education at all. Write an arithmetic expression to determine how many students did not pursue higher education. Simplify the expression.

53. **Energy consumption**   Illustration 2 indicates the fractional part of energy consumption in U.S. households. What fractional part of the energy consumption is represented by the combination of home heating and water heating?

54. **Energy consumption**   Illustration 2 in Exercise 53 indicates the fractional part of energy consumption in U.S. households. What fractional part of the energy consumption is represented by the combination of air conditioning and cooking/refrigeration?

55. **Revenue**   Based on the federal government revenues for the fiscal year 1998 in Illustration 3, what fractional part of the revenues comes from sources other than borrowing and corporate income taxes?

56. **Revenue**   Based on the federal government outlays for the fiscal year 1998 as displayed in Illustration 3, what fractional part of the outlay is determined from sources other than federal operations and national defense?

HOUSEHOLD ENERGY USE
(How energy is used in U.S. households, in percentage)

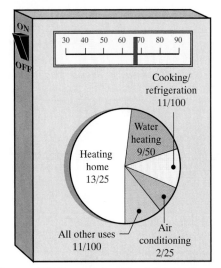

SOURCE: U.S. Energy Information Administration

ILLUSTRATION 2

Federal Government Dollar, FY 1998

Revenues: $1,415.5 billion

Excise taxes
3/100
Other
1/25
Corporate
income taxes
1/10
Social
insurance
receipts
8/25
Borrowing
3/25
Individual
income taxes
39/100

Outlays $1,612.1 billion

Other federal
operations
1/20
Grants to states
and localities
3/20
Direct
benefit
payments
for
individuals
12/25
National
defense
4/25
Net interest
4/25

SOURCE: Office of Management and Budget, *Budget of the United States Government, FY 1998 (1997)*

ILLUSTRATION 3

Read the box headings carefully to be sure that you are following the directions for your specific calculator model. All the TI-86 information begins on page 17.

Touring the TI-83

Courtesy Texas Instruments Incorporated

Take a few minutes to study the TI-83 graphing calculator. The keys are color coded and positioned in a way that is user friendly. Notice there are dark blue, black, and gray keys, along with a single yellow key and a single green key.

Dark blue keys: On the right and across the top of the calculator are the dark blue keys. At the top right are four directional cursor keys. These may be used to move the cursor on the screen in the direction of the arrow printed on the key. The four operation symbols (addition, subtraction, multiplication, and division) are also in dark blue. Notice the key marked **ENTER.** This will be used to activate entered commands; thus there is no key on the face of the calculator with the equal sign printed on it. Below the screen are five keys labeled [Y=], [WINDOW], [ZOOM], [TRACE], and [GRAPH]. These keys are positioned together below the screen because they are used for graphing functions.

Black keys: The majority of the keys on the calculator are black. Notice the [X,T,θ,n] key in the second row and second column. It will be used frequently in algebra to enter the variable **X.** The **ON** key is the black key located in the bottom-left position.

Gray keys: The 12 gray keys that are clustered at the bottom center are used to enter digits, a decimal point, or a negative sign.

Yellow key: The yellow key is the **2nd** key located in the upper-left position.

To access a symbol in yellow (printed above any of the keys) first press the yellow [2nd] key, and then the key BELOW the symbol (function) to be accessed. For example, to turn the calculator **OFF,** notice that the word **OFF** is printed in yellow above the **ON** key. Therefore, press the **2nd** key and the **ON** key to turn the calculator off. This would be denoted as [2nd] [OFF]. The word "2nd" is the cue to look above the keypad for the command. These keystrokes are done *sequentially,* not simultaneously.

Green key: Alphabet letters (printed in green above some of the keys), or any other symbol and/or word printed in green above a key, are accessed by first pressing the green **ALPHA** key and then the key below the desired letter/symbol/word. Again, the keystrokes are sequential.

Catalog feature: Press [2nd] [CATALOG] to display an alphabetical list of available calculator operations. Use the [▲] and [▼] cursor keys to scroll through this list. Operations may be accessed by placing the pointer adjacent to the operation and pressing [ENTER]. To exit the catalog, press [CLEAR].

Note: The TI-83 has an Automatic Power Down (APD) feature that turns the calculator off when no keys have been pressed for several minutes. When this happens, press [ON] to access the last screen used.

Let's Get Started!: TI-83

Turn the calculator on by pressing [ON]. If the display is not clear, press [2nd] [▲] to darken the screen or [2nd] [▼] to lighten the screen. Notice that when the [2nd] key is pressed, an arrow pointing up appears on the blinking cursor.

To ensure the calculator is in the desired mode, press [MODE]. All of the options on the far left should be highlighted. If not, use the [▼] to place the blinking cursor on the appropriate entry and press [ENTER]. Exit MODE by pressing [CLEAR]. This accesses the home screen, which is where expressions are entered. Press [CLEAR] until the screen is cleared except for the blinking cursor in the top left corner.

Integer Operations: TI-83   When entering integers on the calculator, you must differentiate between a subtraction sign and a negative sign. Notice that the subtraction sign appears as a blue key on the right side of the calculator, whereas the negative sign appears as a gray key next to the [ENTER] key and is labeled (−).

EXAMPLE   **Simplify:** $-8 - 2$ **Keystrokes:** [(−)] [8] [−] [2] [ENTER]
Screen display:

Observe the difference in the *size* and *position* of the negative sign as compared to the subtraction sign.

Absolute Value: TI-83   Absolute value is accessed by pressing [MATH] [▶] (NUM) [1:abs(] or through the [CATALOG]; press [2nd] [CATALOG] [ENTER]. The absolute value operation is displayed as **abs(**. Be sure to enter the right parenthesis.

EXAMPLE   **Simplify:** $|-3 - 2|$ **Keystrokes:** [2nd] [CATALOG] [ENTER] [(−)] [3]
[−] [2] [)] [ENTER]
Screen display:

Note: This would be read as "the absolute value of the quantity negative three minus two." Keep this in mind as you enter the expression.

Square Roots: TI-83   The square root symbol is located above the [x^2] key. To access it, press [2nd] $\left[\sqrt{}\right]$.

EXAMPLE   **Simplify:** $\sqrt{25}$ **Keystrokes:** [2nd] $\left[\sqrt{}\right]$ [2] [5] [)] [ENTER]
Screen display:

Powers of Numbers: TI-83   A number may be squared (raised to the second power) either by pressing the [x^2] key after entering the number or by using the caret [^] and then entering the desired exponent.

EXAMPLE   **Simplify:** 4^2
Keystrokes: [4] [x^2] [ENTER]   or   **Keystrokes:** [4] [^] [2] [ENTER]
Screen display: **Screen display:**

To raise to the third power (or higher), use the caret key.

EXAMPLE   **Simplify:** $4(3)^5$ **Keystrokes:** [4] [(] [3] [)] [^] [5]

Screen display:

Fraction Basics: TI-83

The calculator can be used to perform arithmetic operations with fractions. Pressing the [MATH] key reveals a math menu. Take a minute to use the [▼] cursor to scroll down the menu. Notice that there are ten options available. Use of the first option, **1:▶Frac,** will now be illustrated.

Enter the fraction 1/2 by pressing [1] [÷] [2] [MATH]. Under the MATH menu, **1:▶Frac** is highlighted. Because it is highlighted, press [ENTER] to select option 1. Press [ENTER] again to activate the "convert to a fraction" command. Notice that the calculator simply displays the fraction.

Reenter 1/2 by using the following keystrokes: [1] [÷] [2] [ENTER]. This time the MATH menu and the ▶**Frac** command were not accessed. The graphing calculator converts the expression to a decimal (as will any scientific calculator). However, if a fraction is desired rather than a decimal, press [MATH] [ENTER] (again, the ▶**Frac** option is chosen because it is highlighted). The calculator now displays **Ans ▶Frac,** and the decimal is converted back to the fractional form once [ENTER] is pressed.

Note: From this point on, the ▶**Frac** option will be denoted by the keystrokes [MATH] [1:▶Frac].

Enter the fraction 12/24 and access the ▶**FRAC** option. How does the calculator display this fraction? The graphing calculator will *always* display fractions in reduced form.

Read the mixed number "$-3\frac{1}{2}$" aloud. Did you say the word "and"? This is the clue as to the way mixed numbers are entered on the calculator. Enter $-3\frac{1}{2}$ by pressing [(−)] [3] [+] [(−)] [1] [÷] [2] [MATH] [1:▶Frac] [ENTER]. The display should correspond to the one at the left. Because the entire fraction is negative, both the integer part and the rational part must be entered as negative numbers. This can become cumbersome; therefore, we suggest that mixed numbers be entered as improper fractions.

Order of Operations: TI-83

The calculator performs the operations of multiplication and division before the operations of addition and subtraction. Moreover, it does the multiplication and division in the order in which they appear in the expressions (from left to right). It then performs the additions and subtractions in the order in which they appear from left to right. Which of the screens below is the correct display for the problem $\frac{8 + 16}{2}$?

The correct screen is "b." The problem should be read as "the quantity 8 plus 16 divided by 2."

Addition and Subtraction of Fractions: TI-83

Examining the problem as displayed by the calculator, 1/2 + 1/3, we see that the order of operations clearly dictates that the two division operations be completed before the addition. Therefore, no parentheses are necessary when entering this problem on the calculator.

EXAMPLE $\dfrac{1}{2} + \dfrac{1}{3}$

Keystrokes: [1] [÷] [2] [+] [1] [÷] [3] [MATH]
[1:▶Frac] [ENTER]

Screen display:

Multiplication and Division of Fractions: TI-83

Parentheses are critical on this problem to ensure that the calculator addresses the division of two fractions. If the problem were entered with no parentheses, 2/3/1/5, the calculator would perform the operations of division from left to right: 2 divided by 3 = 2/3, 2/3 divided by 1 = 2/3, 2/3 divided by 5 = 2/15.

EXAMPLE $\dfrac{2}{3} \div \dfrac{1}{5}$

Keystrokes: [(] [2] [÷] [3] [)] [÷] [(] [1] [÷] [5] [)]
[MATH] [1:▶Frac] [ENTER]

Screen display:

Touring the TI-86

Courtesy Texas Instruments Incorporated

Take a few minutes to study the TI-86 graphing calculator. The keys are color coded and positioned in a way that is user friendly.

Gray keys: The 12 gray keys clustered at the bottom center are used to enter digits, a decimal point, and a negative sign. At the top right are 4 directional cursor keys. These may be used to move the cursor on the screen in the direction of the arrow.

Black keys: The majority of the keys on the calculator are black. The four operation symbols (division, multiplication, subtraction, and addition) are located in the column on the far right. Notice the key marked **ENTER** at the bottom-right position. This will be used to activate commands that have been entered. The equal sign on the face of the calculator is NOT used for computation. The key marked **x-VAR** in the second row, second column will be used frequently to enter the variable **x.** The **ON** key is in the bottom-left position.

Below the screen are 5 black keys labeled **F1, F2, F3, F4,** and **F5.** These are menu keys and will be addressed as they are needed.

Yellow-orange key: The only key of this color is located in the top-left position and is labeled **2nd.**

Blue key: This key is labeled **ALPHA** and is located at the top left of the keypad.

Above most of the keys are words and/or symbols printed in either yellow-orange or blue. To access a symbol in yellow-orange (printed above any of the keys), first press the yellow-orange [2nd] key, and then the key BELOW the symbol (function) you wish to access. For example, to turn the calculator **OFF,** notice that the word **OFF** is printed in yellow-orange above the **ON** key. Therefore, press the **2nd** key and the **ON** key to turn the calculator off. These keystrokes are done *sequentially,* not simultaneously. Thus, the previous command for turning the calculator off would appear as **[2nd] [OFF].** The word "2nd" is the cue to look above the keypad for the command. A function that is accessed from a menu will be written in parentheses, ( ). When a menu key is indicated, the current function of the key will follow in parentheses to correspond to the display at the bottom of the screen. For example, press **[GRAPH]** to display the graph menu. The notation **[F2] (WIND)** denotes access to the WINDOW submenu. Users should be aware that the menu denoted as WIND on the TI-86 corresponds to the WINDOW menu referred to for the TI-83. To remove the menu at the bottom of the screen, press [EXIT].

To access a symbol printed in blue above some of the keys, first press the blue **ALPHA** key and then the key *below* the desired letter or symbol. Again, the keystrokes are sequential.

*Note: The TI-86 has an **A**utomatic **P**ower **D**own (APD) feature that turns the calculator off when no keys have been pressed for several minutes. When this happens, press [ON] to access the last screen used.*

Let's Get Started!: TI-86

Turn the calculator on by pressing [ON]. If the display is not clear, press [2nd] and hold down [▲] to darken the screen or [2nd] [▼] to lighten it. Notice that when the [2nd] key is pressed, an arrow pointing up appears on the blinking cursor.

To ensure the calculator is in the desired mode, press [2nd] [MODE]. All of the options on the far left should be highlighted. If not, use the [▼] to place the blinking cursor on the appropriate entry and press [ENTER]. Exit **MODE** by pressing [EXIT], [CLEAR] or [2nd] [QUIT]. This accesses the home screen, which is where expressions are entered.

Integer Operations: TI-86

When entering integers on the calculator, you must differentiate between a subtraction sign and a negative sign. Notice the subtraction sign is in the right column, whereas the negative sign appears as a gray key next to the [ENTER] key and is labeled (−).

EXAMPLE **Simplify: −8 − 2**   **Keystrokes:** [(−)] [8] [−] [2] [ENTER]
Screen display:

```
-8-2
                    -10
```

Observe the difference in *size* and *position* of the negative sign as compared to the subtraction sign.

Absolute Value: TI-86

The TI-86 does not have absolute value on its keypad. However, a key can be created using the [CUSTOM] key. Up to 15 frequently used functions can be customized (and accessed with only two keystrokes) provided the function desired is listed under **CATALOG**. To customize a function, press [2nd] [CATALOG] [F1](CATLG) followed by [F3](CUSTM). Place the arrow next to **abs** in the list, then press [F1]. The function **abs** is now listed under **PAGE▼**. The same procedure can be used to add other functions to the **CUSTOM** menu as necessary. Each time, select an open slot in the menu by choosing the menu key below the open slot. Pressing [MORE] accesses additional slots for customizing. When finished, press [EXIT] until the menus at the bottom of the screen clear and the home screen is displayed.

Pressing [CUSTOM] reveals the customized menu; to access a customized function, simply press the **F** key below the desired function. Pressing [EXIT] removes this menu.

EXAMPLE **Simplify |−3 − 2|** (Make sure you are at the home screen; press [EXIT] if necessary.)

Keystrokes:
[CUSTOM] [F1](abs) [(] [(−)] [3] [−] [2] [)] [ENTER]

Note: Absolute value can also be accessed by pressing [2nd] [MATH] [F1](NUM) [F5](abs).

Screen display:

Square Roots: TI-86

The square root symbol is located above the [x²] key. To access it, press [2nd] [√].

EXAMPLE **Simplify: √25**   **Keystrokes: [2nd] [√] [2] [5] [)] [ENTER]**

Screen display:

Powers of Numbers: TI-86

A number may be squared (raised to the second power) either by pressing the [x²] key after entering the number or by using the caret [^] and then entering the desired exponent.

EXAMPLE **Simplify: 4^2**

Keystrokes: [4] [x²] [ENTER]   or   **Keystrokes: [4] [^] [2] [ENTER]**

Screen display: **Screen display:**

To raise to the third power (or higher), use the caret key.

EXAMPLE **Simplify: $4(3)^5$**   **Keystrokes: [4] [(] [3] [)] [^] [5]**

Screen Display:

Fraction Basics: TI-86

Since the ►**Frac** option is used frequently, it will be added to the CUSTOM menu. To do this, press [2nd] [CATLG-VARS] [F1](CATLG) [F3](CUSTM). Cursor down the list to place the arrow next to ►**Frac,** found near the bottom of the list. Press [F2] or another open **F** key to create the custom key. (*Hint:* Since ►**Frac** was found at the bottom of the list, pressing [▲] [2nd] [M2](PAGE▲) until the arrow is positioned next to the desired selection would be quicker than scrolling through the entire list.)

Note: To delete a customized entry, press [2nd] [CUSTOM] [F4](BLANK) and then the **F** key that contains the command to be deleted.

Press [EXIT] until the blinking cursor is displayed at the HOME screen.

Enter the decimal number 0.42 by pressing [.] [4] [2] [CUSTOM] [F2] (►**Frac**) [ENTER]. Notice that the calculator displays the fraction reduced to lowest terms.

Recall that a mixed number is actually the sum of an integer and a fraction. Therefore, to enter a mixed number, simply enter the expression as an indicated sum. To enter $-3\frac{1}{2}$ and express it as a fraction, press [(−)] [3] [+] [(−)] [1] [÷] [2] [CUSTOM] [F2](►**Frac**). Because the entire fraction is negative, both the integer part and the rational part must be entered as negative numbers. Your display should correspond to the one at the left.

Order of Operations: TI-86

The calculator performs the operations of multiplication and division before the operations of addition and subtraction. Moreover, it does the multiplication and division in the order in which they appear in the expressions (from left to right). It then performs the additions and subtractions in the order in which they appear from left to right. Which of the screens below is the correct display for the problem $\dfrac{8+16}{2}$?

a.
```
8+16/2█
```

b.
```
(8+16)/2
```

The correct screen is "b." The problem should be read as "the quantity 8 plus 16 divided by 2."

Addition and Subtraction of Fractions: TI-86

Examining the problem as displayed by the calculator, 1/2 + 1/3, we see that the order of operations clearly dictates that the two division operations be completed before the addition. Therefore, no parentheses are necessary when entering this problem on the calculator.

EXAMPLE $\dfrac{1}{2} + \dfrac{1}{3}$

Keystrokes: [1] [÷] [2] [+] [1] [÷] [3] [CUSTOM] [F2](:►Frac) [ENTER]

Screen display:
```
1/2+1/3►Frac
                5/6
█

 abs │►Frac│    │    │    ►
```

Multiplication and Division of Fractions: TI-86

Parentheses are critical on this problem to ensure that the calculator addresses the division of two fractions. If the problem were entered with no parentheses, 2/3/1/5, the calculator would perform the operations of division from left to right: 2 divided by 3 = 2/3, 2/3 divided by 1 = 2/3, 2/3 divided by 5 = 2/15.

EXAMPLE $\dfrac{2}{3} \div \dfrac{1}{5}$

Keystrokes: [(] [2] [÷] [3] [)] [÷] [(] [1] [÷] [5] [)] [CUSTOM] [F1](:►Frac) [ENTER]

Screen display:
```
(2/3)/(1/5)►Frac
                10/3

 abs │►Frac│    │    │    ►
```

1.2   Exponents and Order of Operations

■ NATURAL-NUMBER EXPONENTS   ■ ORDER OF OPERATIONS

■ **TECHNOLOGY TIP**

Your graphing calculator has an x^2 key for raising numbers to the second power. For exponents larger than 2, you will need to use the caret, $\wedge$, key. Thus, 3 squared may be entered as 3^2 using the x^2 key or as $3\wedge2$ using the caret key. (See "Technology 1.1: Introduction to the Graphing Calculator, page 14.)

■ NATURAL-NUMBER EXPONENTS

To show how many times a number is to be used as a factor in a product, we use an **exponent.** For example, 3 is the exponent in the expression 2^3. The exponent indicates that 2 is to be used as a factor three times:

$$2^3 = \overset{\text{3 factors of 2}}{\overbrace{2 \cdot 2 \cdot 2}}$$
$$= 8$$

> **WARNING!** Note that $2^3 = 8$. This is not the same as $2 \cdot 3 = 6$.

In algebra, we use letters, called **variables,** to represent real numbers. For example:

If x represents the number 4, then $x = 4$.

> **WARNING!** We usually do not use a times sign, $\times$, to indicate multiplication. It might be mistaken for the variable x.

The exponent 5 in the expression x^5 indicates that x is to be used as a factor five times.

$$x^5 = \overbrace{x \cdot x \cdot x \cdot x \cdot x}^{5 \text{ factors of } x}$$

In the expression x^5 (called an **exponential expression** or a **power of x**), 5 is the **exponent** and x is the **base.** In expressions such as x or y, the exponent is understood to be 1:

$$x^1 = x \qquad \text{and} \qquad y^1 = y$$

In general, we have the following definition.

> **Natural-Number Exponent**
> If n is a natural number, then
> $$x^n = \overbrace{x \cdot x \cdot x \cdot \ldots \cdot x}^{n \text{ factors of } x}$$

EXAMPLE 1   **a.** $4^2 = 4 \cdot 4 = 16$ Read 4^2 as "4 squared" or as "4 to the second power."

b. $5^3 = 5 \cdot 5 \cdot 5 = 125$ Read 5^3 as "5 cubed" or as "5 to the third power."

c. $6^4 = 6 \cdot 6 \cdot 6 \cdot 6 = 1296$ Read 6^4 as "6 to the fourth power."

The base of an exponential expression can be a variable.

EXAMPLE 2   **a.** $y^6 = y \cdot y \cdot y \cdot y \cdot y \cdot y$ Read y^6 as "y to the sixth power."

b. $x^3 = x \cdot x \cdot x$ Read x^3 as "x cubed" or "x to the third power."

c. $z^2 = z \cdot z$ Read z^2 as "z squared" or as "z to the second power."

d. $a^1 = a$ Read a^1 as "a to the first power."

■ ORDER OF OPERATIONS

The order in which we do arithmetic is important. For example, if we evaluate $2 + 3 \cdot 4$ by multiplying first, we get a result of 14. However, if we add first, we get a result of 20.

$$2 + 3 \cdot 4 = 2 + 12 \qquad 2 + 3 \cdot 4 = 5 \cdot 4$$
$$= 14 \qquad\qquad\qquad = 20$$

To eliminate the possibility of getting different answers, we will agree to do multiplications before additions. The correct calculation of $2 + 3 \cdot 4$ is

$$2 + 3 \cdot 4 = 2 + 12$$
$$= 14$$

■ **TECHNOLOGY TIP**
Regardless of the displayed grouping symbols, you must enter **only** the parentheses on your graphing calculator. The calculator **reserves** brackets [ ] for matrix operations and braces { } for list operations. This means that a problem may have nested layers of parentheses such as ((( ))).

To indicate additions should be done before multiplications, we use **grouping symbols** such as parentheses ( ), brackets [ ], or braces { }. In the expression $(2 + 3)4$, the parentheses indicate that the addition should be done first:

$$(2 + 3)4 = 5 \cdot 4$$
$$= 20$$

Unless grouping symbols indicate otherwise, exponential expressions are always evaluated before multiplications. The expression $5 + 4 \cdot 3^2$ should be evaluated in the following way:

$$5 + 4 \cdot 3^2 = 5 + 4 \cdot 9$$
$$= 5 + 36$$
$$= 41$$

To guarantee that calculations will have a single correct result, we use the following set of **priority rules.**

■ **TECHNOLOGY TIP**
These "priority rules" are designated as **E**quation **O**perating **S**ystems in your calculator manual. Read the page of your manual carefully to determine if your calculator gives "implied" multiplication a higher priority than the order of operations that are listed at the right.

Order of Mathematical Operations
Use the following steps to do all calculations within each pair of grouping symbols, working from the innermost pair to the outermost pair.
1. Find the values of any exponential expressions.
2. Do all multiplications and divisions as they are encountered while working from left to right.
3. Do all additions and subtractions as they are encountered while working from left to right.

   When all grouping symbols have been removed, repeat the rules above to finish the calculation.

   In a fraction, simplify the numerator and the denominator separately. Then perform the division indicated by the fraction bar, if possible.

WARNING!   Note that $4(2)^3 \neq (4 \cdot 2)^3$:

$$4(2)^3 = 4 \cdot 2 \cdot 2 \cdot 2 = 4(8) = 32 \qquad \text{and} \qquad (4 \cdot 2)^3 = 8^3 = 8 \cdot 8 \cdot 8 = 512$$

```
4(2)^3
               32
(4*2)^3
               512
```

Likewise, $4x^3 \neq (4x)^3$ because

$$4x^3 = 4 \cdot x \cdot x \cdot x \qquad \text{and} \qquad (4x)^3 = (4x)(4x)(4x) = 64x^3$$

EXAMPLE 3   Evaluate $5^3 + 2(8 - 3 \cdot 2)$.

Solution   We do the work within the parentheses first and then simplify.

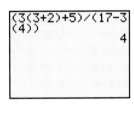

$$5^3 + 2(8 - 3 \cdot 2) = 5^3 + 2(8 - 6)$$ Do the multiplication within the parentheses.

$$= 5^3 + 2(2)$$ Do the subtraction within the parentheses.

$$= 125 + 2(2)$$ Find the value of the exponential expression.

$$= 125 + 4$$ Do the multiplication.

$$= 129$$ Do the addition.   ■

EXAMPLE 4   Evaluate $\dfrac{3(3 + 2) + 5}{17 - 3(4)}$.

Solution   We simplify the numerator and the denominator of the fraction separately and then simplify the fraction.

■ **TECHNOLOGY TIP**
There is no way to "draw" the long fraction bar that acts as a grouping symbol in the expression $\dfrac{3(3 + 2) + 5}{17 - 3(4)}$.
Parentheses must be used to designate the numerator and denominator.

$$\frac{3(3 + 2) + 5}{17 - 3(4)} = \frac{3(5) + 5}{17 - 3(4)}$$

$$= \frac{15 + 5}{17 - 12}$$

$$= \frac{20}{5}$$

$$= 4$$

The series of equality statements that show $\dfrac{3(3 + 2) + 5}{17 - 3(4)}$ is ultimately equal to 4 demonstrates the **transitive property of equality.**

> **Transitive Property of Equality**
> If $a = b$ and $b = c$, then $a = c$.

Example 4 illustrates the **simplification** of an arithmetic expression using the order of operations. The example below illustrates the **evaluation** of an algebraic expression. The word *evaluation* means to find the value of an expression for given values of the variable(s). This is done by substituting those values in for the variable and then simplifying the resulting arithmetic expression.

EXAMPLE 5   If $x = 3$ and $y = 4$, evaluate

a. $3y + x^2$ and   **b.** $3(y + x^2)$.

Solution   **a.** $3y + x^2 = 3(4) + 3^2$ Substitute 3 for x and 4 for y.

$$= 3(4) + 9$$ Evaluate the exponential expression.

$$= 12 + 9$$ Do the multiplication.

$$= 21$$ Do the addition.

■ **TECHNOLOGY TIP**
Use the STO▶ function to store variable values and then compute the value of the expression. The colon key can be used to separate commands so that the complete problem may be keyed in before pressing ENTER. (See "Technology 1.2: Using the STOre Key," page 27.)

b. $3(y + x^2) = 3(4 + 3^2)$ Substitute 3 for x and 4 for y.

$$= 3(4 + 9)$$ Evaluate the exponential expression.

$$= 3(13)$$ Parentheses indicate that addition is performed next.

$$= 39$$ Do the multiplication.   ■

EXAMPLE 6   If $x = 4$ and $y = 3$, evaluate $\dfrac{3x^2 - 2y}{2(x + y)}$.

Solution

$$\frac{3x^2 - 2y}{2(x + y)} = \frac{3(4^2) - 2(3)}{2(4 + 3)}$$ Substitute 4 for x and 3 for y.

$$= \frac{3(16) - 2(3)}{2(7)}$$ Find the value of 4^2 in the numerator and do the addition in the denominator.

$$= \frac{48 - 6}{14}$$ Do the multiplications.

$$= \frac{42}{14}$$ Do the subtraction.

$$= 3$$

■ STUDY TIP

Although you are only on Section 1.2, you should begin to prepare NOW for your first major exam. When you finish tonight's homework (and as you complete each subsequent homework assignment) go back and work a few problems from each of the Exercise Sets that have been previously assigned. If you do this, you will find that much of your preparation for the test will have been accomplished by the time your instructor announces the test date.

EXAMPLE 7   **Test scores**   A student made the following scores on the four tests given during the fall semester in his elementary algebra class:

80, 72, 88, 60

What is his average going into the final?

Solution   To find an average, simply add the four scores and divide by 4 (the number of scores added).

$$\frac{80 + 72 + 88 + 60}{4} = \frac{300}{4}$$ Simplify the numerator.

$$= 75$$ Divide.

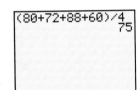

To confirm this result with your calculator, make sure you enclose the entire numerator in parentheses so that the entire numerator is simplified before the division is performed.

EXERCISE 1.2

VOCABULARY   *In Exercises 1–4, fill in the blank to make a true statement.*

1. An _____ is used to show how many times a number is used as a factor.

2. _____ are used in algebra to represent real numbers.

3. In the exponential expression x^3, x is the _____ and 3 is the _____.

4. The _____ property of equality states *if $a = b$ and $b = c$, then $a = c$.*

CONCEPTS

5. Write each expression as a product of several factors using the definition of natural-number exponent.
   a. x^2 **b.** $3z^4$
   c. $(5x)^2$ **d.** $5(2x)^3$

✎**6.** The symbols $3x$ and x^3 have different meanings. Explain.

✎**7.** Students often say that x^n means, "x multiplied by itself n times." Explain why this is not correct.

✍ **8.** If x were greater than 1, would raising x to higher and higher powers produce bigger numbers or smaller?

✍ **9.** What would happen in Exercise 8 if x was a number between 0 and 1?

10. The following grouping symbols are used in problems involving the order of operations rules: ( ), { }, [ ].

Which of these symbols cannot be used on the calculator when performing orders of operations?

✍**11.** Explain the difference between the directions *evaluate* and the directions *simplify.*

PRACTICE   *In Exercises 12–16, find the value of each expression.*

12. 4^2 **13.** 5^2 **14.** 6^3 **15.** 7^3 **16.** 2^4

In Exercises 17–23, find the value of each expression if $x = 3$ and $y = 2$. Verify your computations using the STOre feature on the calculator. Copy the screen display to justify your work.

17. $4x^2$ **18.** $4y^3$ **19.** $(5y)^3$ **20.** $(2y)^4$

21. $2x^y$ **22.** $3y^x$ **23.** $(3y)x$

In Exercises 24–51, simplify each expression by doing the indicated operations. Verify your computations with the calculator. Copy the screen display to justify your work.

24. $3 \cdot 5 - 4$ **25.** $4 \cdot 6 + 5$ **26.** $3(5 - 4)$ **27.** $4(6 + 5)$

28. $3 + 5^2$ **29.** $4^2 - 2^2$ **30.** $(3 + 5)^2$ **31.** $(5 - 2)^3$

32. $2 + 3 \cdot 5 - 4$ **33.** $12 + 2 \cdot 3 + 2$ **34.** $64 \div (3 + 1)$ **35.** $16 \div (5 + 3)$

36. $(7 + 9) \div (2 \cdot 4)$ **37.** $(7 + 9) \div 2 \cdot 4$ **38.** $24 \div 4 \cdot 3 + 3$ **39.** $36 \div 9 \cdot 4 - 2$

40. $(100 \div 10) \cdot (10 \div 100)$ **41.** $100 \div [10 \cdot (10 \div 100)]$ **42.** $3^2 + 2(1 + 4) - 2$

43. $4 \cdot 3 + 2(5 - 2) - 2^3$ **44.** $5^2 - (7 - 3)^2$ **45.** $3^3 - (3 - 1)^3$

46. $(2 \cdot 3 - 4)^3$ **47.** $(3 \cdot 5 - 2 \cdot 6)^2$

48. $\dfrac{(3 + 5)^2 + 2}{2(8 - 5)}$ **49.** $\dfrac{25 - (2 \cdot 3 - 1)}{2 \cdot 9 - 8}$ **50.** $\dfrac{(5 - 3)^2 + 2}{4^2 - (8 + 2)}$ **51.** $\dfrac{(4^2 - 2) \div 7}{5(2 + 4) - 3^2}$

In Exercises 52–61, evaluate each expression given that $x = 3$, $y = 2$, and $z = 4$. Verify your computation by using the STOre feature on your calculator.

52. $2x - y$ **53.** $2z + y$ **54.** $(3 + x)y$ **55.** $(4 + z)y$

56. $xyz + z^2 - 4x$ **57.** $zx + y^2 - 2z$ **58.** $3x^2 + 2y^2$ **59.** $3x^2 + (2y)^2$

60. $\dfrac{2x + y^2}{y + 2z}$ **61.** $\dfrac{2z^2 - y}{2x - y^2}$

APPLICATIONS   *In Exercises 62–63, find the area of a square whose side has the given measure. The formula for the area of a square is $A = s^2$.*

62. $s = 12$ **63.** $s = 13$

In Exercises 64–65, find the volume of a cube whose sides have the given measure. The formula for the volume of a cube is $V = s^3$.

64. $s = 8$ **65.** $s = 6$

In Exercises 66–70, write a numerical expression to model the situation and evaluate using the order of operations.

66. Averaging grades   A student earns grades of 77, 85, 83, and 87 on four major exams. What is his average in the course?

67. College enrollment   Full-time enrollments for a regional group of community colleges are 8791, 2385, 5468, 6101, 1282, and 7181 students for the fall term. What is the total full-time enrollment for the community colleges for the fall term?

68. Entertainment   A family of four spends Saturday at a local amusement park. Entrance fees are $28.75, cokes are $2.50, burgers are $4.25, and chips are $1.50. If the family eats lunch consisting of drinks, burgers, and chips and then takes a break in the afternoon for everyone to have a coke, what is the total cost of the outing?

69. Entertainment   An afternoon matinee birthday party for 12 middle school children costs $45 for tickets, $30 for popcorn, and $21 for drinks. What is the average cost per child?

70. Advertising   An advertisement in the classified section of the newspaper reads:

Wanted: person at least 18 years old to deliver phone books. Earn $50 plus $.15 per book delivered.

If 1000 phone books are delivered, how much is earned?

REVIEW   In Exercises 71–74, perform the indicated operation. Reduce answers and verify with your calculator.

71. $\dfrac{7}{8} + \dfrac{7}{12}$

72. $\dfrac{6}{11} \div \dfrac{9}{8}$

73. $5\dfrac{3}{8} - 2\dfrac{5}{6}$

74. Enrollment   If there are 21 girls in Sherah's sorority's pledge class and $\frac{5}{7}$ of them went to private school when in high school, how many of the pledges went to private school?

The graphing calculator can be used to evaluate expressions. To "evaluate an expression" means to find the value of the expression for assigned values of the variable. Evaluating with a calculator can be accomplished in two ways. First, by substituting the values in by hand and then entering the resulting expression in the calculator. The second approach is to store the values under different variable names by using the STO► key. Remember, read the box headings carefully to be certain you are following the directions for your specific calculator model. All the TI-86 information begins at the bottom of this page.

STOre: TI-83

To store a value, simply enter the value, press [STO►], and then the desired variable. For example, to store $x = 5$, press [5] [STO►] [X,T,θ,n] [ENTER]. The value 5 is now stored under the variable x. This value will remain stored in x until another value replaces it. To check and see what value is actually stored under a specific variable, enter the variable at the prompt (blinking cursor) and then press [ENTER]. The value is then displayed on the screen.

Note: Because **x** is used as a variable so often in algebra, it has its own key on the TI-83. The calculator is in function MODE; therefore, when [X,T,θ,n] is pressed the screen displays the variable **x.**

EXAMPLE

Evaluate $3x^2 + 6x + 2$ when $x = -11$.

Solution: Store $x = -11$, by pressing [(−)] [1] [1] [STO►] [X,T,θ,n]. The colon ":" key (located above the gray decimal key) is used to separate commands that are to be entered on the same line, so now press [ALPHA] [:] before entering the expression to be evaluated.

```
-11→X:3X²+6X+2
              299
```

Press [3] [X,T,n] [x²] [+] [6] [X,T,n] [+] [2] to enter the expression. At this point, the calculator has been instructed to store -11 in x and then evaluate $3x^2 + 6x + 2$ for this value of x. The calculator will not perform the instructions until [ENTER] is pressed. The polynomial evaluates to 299. Your screen should look like the one at the left.

To store a value under a variable other than **x,** the [ALPHA] key must be used to access the letters of the alphabet. Pressing the [ALPHA] key, followed by another key, allows access to the uppercase letters and symbols written in green above the keypads.

EXAMPLE

Evaluate $a^2 - 3a + 5$ when $a = -7$.

Keystrokes: [(−)] [7] [STO►] [ALPHA] [A] [ALPHA] [:] [ALPHA] [A] [x²] [−] [3] [ALPHA] [A] [+] [5] [ENTER]

Screen Display:

```
-7→A:A²-3A+5
            75
■
```

Note: Be sure to use the gray [(−)] key for the negative sign on a number and use the blue [−] key to indicate subtraction.

STOre: TI-86

The [x-VAR] key is used to display the variable **x.** Note that the variable **x** is displayed as a lowercase letter on the TI-86, instead of an uppercase **x** as on the TI-83 screens that appear in this text.

To store a value, simply enter the value, press [STO►], and then press the desired variable. For example, to store $x = 5$, press [5] [STO►] [x-VAR] [ENTER]. The value 5 is now stored under the variable **x.** This value will remain stored in **x** until another value replaces it. To determine the value that is actually stored under a specific variable, enter the variable at the prompt (blinking cursor) and then press [ENTER]. The value is then displayed on the screen.

Note: Pressing the **STO➤** key automatically initiates the **[ALPHA]** key, allowing access to the remaining letters of the alphabet. Press **STO➤** and observe that the blinking cursor has an **A** in it for **ALPHA**. Letters will automatically be recorded in uppercase format. To disengage the ALPHA feature, press the **[ALPHA]** key.

EXAMPLE   **Evaluate $3x^2 + 6x + 2$ when $x = -11$.**

Solution: Store $x = -11$, by pressing **[(−)] [1] [1] [STO➤] [x-VAR]**. The colon ":" key (located above the gray decimal key) is used to separate commands that are to be entered on the same line. Press **[2nd] [:]** before entering the expression to be evaluated. At this point, you should note that the blinking cursor has an **A** in it for ALPHA. Disengage the ALPHA feature by pressing **[ALPHA]** now. Press **[3] [x-VAR] [x²] [+] [6] [x-VAR] [+] [2]** to enter the expression that is to be evaluated. At this point, the calculator has been instructed to store -11 in **x** and to evaluate $3x^2 + 6x + 2$ for this value of **x.** The calculator will not perform the instructions until the **[ENTER]** key is pressed. The polynomial evaluates to 299. Your screen should look like the one above.

The TI-86 has both uppercase and lowercase alphabet letters that can be used as variables. To access a lowercase letter when using the **STO➤** feature, press **[2nd] [ALPHA]** after pressing the **[STO➤]** key.

When not using the **STO➤** feature, uppercase letters are accessed by **[ALPHA]** and lowercase letters are accessed by **[2nd] [ALPHA]**.

EXAMPLE   **Evaluate $a^2 - 3a + 5$ when $a = -7$.**

The keystrokes would need to be **[(−)] [7] [STO➤] [A] [2nd] [:] [A]** *(observe that the cursor is still blinking with an uppercase A inside of it)* **[ALPHA]** *(the ALPHA feature is now disengaged)* **[x²] [−] [3] [ALPHA] [A] [+] [5] [ENTER]**.

The TI-86 has the capability of recognizing combinations of letters as independent variables. For example, the expression $4AC$ means "4 times A times C." However, the TI-86 would recognize AC as representing only *one* unknown, not a product of two unknowns. Thus, the expression "4 times A times C" must be displayed on the TI-86 as 4A*C or 4A C. Variables can be separated with a space to denote the multiplication of two different quantities. By being able to combine letters to form new variables, the TI-86 has a limitless list of variables available for use. Uppercase letters are reserved for user variable names. Note that the TI-86 would recognize 4a*c as an entirely different product because the variables are lowercase.

Lowercase letters should be used with care as some are reserved for system variables. Examples of these are:

a	coefficient of regression
b	coefficient of regression
c	speed of light
e	natural log base
g	force of gravity
h	Planck's constant
k	Boltzman's constant
n	number of items in a sample
u	atomic mass unit

The variables r, t, x, y, and θ are updated by graph coordinates based on the graphing mode.

1.3   Adding and Subtracting Real Numbers

■ NUMBER LINE ■ ABSOLUTE VALUE OF A REAL NUMBER ■ ADDING REAL NUMBERS
■ SUBTRACTING REAL NUMBERS

■ NUMBER LINE

■ **TECHNOLOGY TIP**
Your graphing calculator has both a subtraction key and a negative key. The negative key is displayed as $(-)$. Be sure you use this key to designate negative numbers.

The basic operations of addition, subtraction, multiplication, and division can be applied to all real numbers. Up to this point, it is possible that your experience has led you to deal only with the set of positive natural numbers when performing these operations. As we address these basic operations in the next two sections, the rules of operations with signed numbers will focus on the set of integers. However, these results are applicable to *all* real numbers.

We use the number line shown in Figure 1-3 to represent sets of numbers.

FIGURE 1-3

The number line continues forever to the left and to the right. Numbers to the left of 0 are negative, and we say that the line continues to negative infinity. Numbers to the right of 0 are positive, and we say that the line continues to positive infinity.

WARNING!   The number 0 is neither positive nor negative.

■ ABSOLUTE VALUE OF A REAL NUMBER

■ **STUDY TIP**
When taking notes, leave a 2-inch margin on the left. This space allows you to add any comments you may need later. Review your notes as soon as possible when the class is over.

On a number line, the distance between a number x and 0 is called the **absolute value** of x. For example, the distance between 5 and 0 is 5 units (see Figure 1-4). Thus, the absolute value of 5, denoted as $|5|$, is 5:

$|5| = 5$ Read as "the absolute value of 5 is 5."

The distance between -6 and 0 is 6. Thus,

$|-6| = 6$ Read as "the absolute value of -6 is 6."

FIGURE 1-4

Because the absolute value of a real number represents the distance on the number line that the number is from 0, without regard to direction, the absolute value of every real number is either positive or 0. That is, if x is a real number, then

$|x| \geq 0$

EXAMPLE 1   Evaluate   **a.** $|6|$,   **b.** $|-3|$,   **c.** $|0|$ **d.** $|2 + 5|$

Solution   **a.** $|6| = 6$, because 6 is 6 units from 0.

b. $|-3| = 3$, because -3 is 3 units from 0.

c. $|0| = 0$, because 0 is 0 units from 0.

d. $|2 + 5| = |7| = 7$

Each of the previous problems can be verified with your graphing calculator. Locate the **ABS** key on your calculator or look for the **ABS** function under the MATH menu. Each example is verified below:

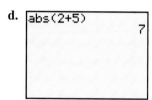

Example **d,** $|2 + 5|$, should be read as "the absolute value of the quantity 2 plus 5." For this reason, the parentheses are critical. Absolute value symbols serve as a grouping symbol.   ■

■ ADDING REAL NUMBERS

Because the positive direction on the number line is to the right, positive numbers can be represented by arrows pointing to the right. Negative numbers can be represented by arrows pointing to the left.

Using this idea of direction, the following examples can be represented with a number line interpretation.

$2 + 3 = 5$

$(-2) + (-3) = -5$

$(-6) + 2 = -4$

$7 + (-4) = 3$

EXPLORING THE CONCEPT   Use your calculator to determine each pair of sums and respond to the questions that follow.

a. $4.5 + 7.2$

$$\left(-\frac{1}{2}\right) + \left(-\frac{2}{3}\right)$$

How do the signs of each of the addends relate to the sign of the sum?

b. $-8.2 + 5.9$

$$\frac{1}{5} + \left(-\frac{8}{7}\right)$$

Compare the absolute values of the terms. Based on the absolute value, which term is used to designate the sign of the sum?

c. $3.7 + (-1.9)$

$$\left(-\frac{1}{8}\right) + \frac{7}{8}$$

Again, compare the absolute values of the terms. Based on the absolute value, which term is being used to designate the sign of the sum?

d. $-6 + 6$

$$\left(\frac{1}{4}\right) + \left(-\frac{1}{4}\right)$$

What happens when you add a number and its opposite?

Conclusions   **a.**

```
4.5+7.2
            11.7
-1/2+-2/3▶Frac
            -7/6
```

The signs of the terms are the same and the sum retains this same sign.

b.

```
-8.2+5.9
            -2.3
1/5+-8/7▶Frac
            -33/35
```

The term with the larger absolute value is used to determine the sign of the sum.

c.

```
3.7+-1.9
            1.8
-1/8+7/8▶Frac
            3/4
```

The term with the larger absolute value is used to determine the sign of the sum.

d.

```
-6+6
            0
(1/4)+(-1/4)
            0
```

Adding a number and its opposite produces a sum of 0.

Additive Inverse
Because $a + (-a) = 0$, the numbers a and $-a$ are called **negatives** or **additive inverses**.

Adding Real Numbers
1. To find the sum of two real numbers with the same sign, we add their absolute values and keep their common sign.
2. To find the sum of two real numbers with unlike signs, we subtract their absolute values (the smaller from the larger) and use the sign of the number with the larger absolute value.

EXAMPLE 2   **a.** $(+4) + (+6) = +(4 + 6)$

$= 10$

b. $(-4) + (-6) = -(4 + 6)$

$= -10$

c. $-\dfrac{1}{2} + \left(-\dfrac{3}{2}\right) = -\left(\dfrac{1}{2} + \dfrac{3}{2}\right)$

$= -\dfrac{4}{2}$

$= -2$

a.
```
4+6
            10
```
b.
```
-4+-6
           -10
```
c.
```
-1/2+(-3/2)▶Frac
            -2
```

■

EXAMPLE 3   **a.** $6 + (-9) = -(9 - 6)$

$= -3$

b. $-\dfrac{2}{3} + \left(\dfrac{1}{2}\right) = -\left(\dfrac{2}{3} - \dfrac{1}{2}\right)$

$= -\left(\dfrac{4}{6} - \dfrac{3}{6}\right)$

$= -\dfrac{1}{6}$

a.
```
6+-9
           -3
```
b.
```
-2/3+(1/2)▶Frac
          -1/6
```

■

EXAMPLE 4   **a.** $[(+3) + (-7)] + (-4) = [-4] + (-4)$ Do the work within the brackets first.

$= -8$

b. $3 + [(-7) + (-4)] = 3 + [-11]$ Do the work within the brackets first.

$= -8$

a.
```
(3+-7)+-4
           -8
```
b.
```
3+(-7+-4)
           -8
```

■

■ **TECHNOLOGY TIP**
1. Remember to use ONLY parentheses as grouping symbols on the calculator.
2. Textbooks will frequently place parentheses around a negative number: $5 + (-4)$. This is done for clarity to distinguish between the negative sign on the 4 and the operation symbol that precedes it. The calculator needs no such distinction since the negative sign is smaller and elevated when compared to the operation symbol. The negative sign on the calculator is also reserved for the operation of negation only.

The **associative property of addition** guarantees that three real numbers can be regrouped in an indicated addition.

Associative Property of Addition
If a, b, and c are real numbers, then $(a + b) + c = a + (b + c)$.

EXAMPLE 5   If $x = -4$, $y = 5$, and $z = -13$, evaluate   **a.** $x + y$ and   **b.** $2y + z$.

Solution   We substitute -4 for x, 5 for y, and -13 for z and simplify.

a. $x + y = (-4) + (5)$

$= 1$

b. $2y + z = 2 \cdot 5 + (-13)$

$= 10 + (-13)$

$= -3$

a.
```
-4→X:5→Y:-13→Z:X
+Y
            1
```
b.
```
2Y+Z
           -3
```

■

■ **TECHNOLOGY TIP**
Because all the variable values for both examples were initially stored for the first example, we did not have to reenter the values for the second example.

Words and phrases such as *found, gain, credit, up, increase, forward, rises, in the future,* and *to the right* indicate a positive direction. Words and phrases such as *lost, loss, debit, down, decrease, backward, falls, in the past,* and *to the left* indicate a negative direction.

EXAMPLE 6   **Account balance**   The treasurer of a math club opens a checking account by depositing $350 in the bank. The bank debits the account $9 for check printing, and the treasurer writes a check for $22. Find the account balance after these transactions.

Solution   The deposit can be represented by $+350$. The debit of $9 can be represented by -9, and the check written for $22 can be represented by -22. The balance in the account after these transactions is the sum of 350, -9, and -22.

$$350 + (-9) + (-22) = 341 + (-22) \qquad \text{Work from left to right.}$$
$$= 319$$

The account balance is $319.

■ SUBTRACTING REAL NUMBERS

The numbers 4 and -4 are opposites or additive inverses of one another. In fact, -4 can be read as either "negative 4" or "the opposite of 4." Thus, $-(-4)$ could be read as "the opposite of negative 4." This idea of a double negative will be important as we discuss subtracting real numbers.

> **Double Negative Rule**
> If a is a real number, then $-(-a) = a$.

EXPLORING THE CONCEPT   Perform each of the addition problems and the corresponding subtraction problem on your graphing calculator and then respond to the questions.

	Addition	*Subtraction*
a.	$(-6) + 5$	$(-6) - (-5)$
b.	$6 + 5$	$6 - (-5)$
c.	$(-6) + (-5)$	$(-6) - 5$
d.	$6 + (-5)$	$6 - 5$

e. How is each pair of addition and subtraction problems the same? different?

f. How does the double negative rule apply in **a** and **b**?

Conclusions   **a.**
```
-6+5
          -1
-6--5
          -1
```
b.
```
6+5
          11
6--5
          11
```

c.
```
-6+-5
           -11
-6-5
           -11
```
d.
```
6+-5
           1
6-5
           1
```

e. The results are the same in each pair of problems. They are different in the following ways:

   a and **b:** you are adding a positive versus subtracting a negative.
   c and **d:** you are adding a negative versus subtracting a positive.

f. Subtracting a negative 5 is the same as adding a positive 5: $-(-5) = +5$ ■

Subtracting a negative number is the same as adding the opposite number.

Subtracting Real Numbers
If a and b are two real numbers, then $a - b = a + (-b)$.

EXAMPLE 7   Evaluate   **a.** $12 - 4$,   **b.** $-13 - 5$,   and   **c.** $-14 - (-6)$.

Solution   **a.** $12 - 4 = 12 + (-4)$ **b.** $-13 - 5 = -13 + (-5)$
                   $= 8$ $= -18$

c. $-14 - (-6) = -14 + 6$ Use the double negative rule.
                  $= -8$

a.
```
12-4
           8
```
b.
```
-13-5
          -18
```
c.
```
-14--6
          -8
```
■

EXAMPLE 8   If $x = -5$ and $y = -3$, evaluate   **a.** $\dfrac{y - x}{7 + x}$ and   **b.** $\dfrac{6 + x}{y - x} - \dfrac{y - 4}{7 + x}$.

Solution   We can substitute -5 for x and -3 for y into each expression and simplify.

a. $\dfrac{y - x}{7 + x} = \dfrac{-3 - (-5)}{7 + (-5)}$

$= \dfrac{-3 + [-(-5)]}{2}$ To subtract -5, add
                                the opposite of -5.

$= \dfrac{-3 + 5}{2}$ $-(-5) = +5$

$= \dfrac{2}{2}$

$= 1$

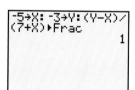
```
-5→X: -3→Y: (Y-X)/
(7+X)▸Frac
           1
```

b. $\dfrac{6+x}{y-x} - \dfrac{y-4}{7+x} = \dfrac{6+(-5)}{-3-(-5)} - \dfrac{-3-4}{7+(-5)}$

$= \dfrac{1}{-3+5} - \dfrac{-3+(-4)}{2}$ $-(-5) = +5$

$= \dfrac{1}{2} - \dfrac{-7}{2}$

$= \dfrac{1-(-7)}{2}$

$= \dfrac{1+[-(-7)]}{2}$ To subtract -7, add the opposite of -7.

$= \dfrac{1+7}{2}$ $-(-7) = +7$

$= \dfrac{8}{2}$

$= 4$ ■

EXAMPLE 9   Simplify   **a.** $3 - [4 + (-6)]$ and   **b.** $[-5 + (-3)] - [-2 - (+5)]$.

Solution   **a.** $3 - [4 + (-6)] = 3 - (-2)$ Do the addition within the brackets first.

$= 3 + [-(-2)]$ To subtract -2, add the opposite of -2.

$= 3 + 2$ $-(-2) = 2$

$= 5$

```
3-(4+-6)
                5
```

b. $[-5 + (-3)] - [-2 - (+5)]$

$= [-5 + (-3)] - [-2 + (-5)]$ To subtract $+5$, add the opposite of 5.

$= -8 - (-7)$ Do the work within the brackets.

$= -8 + [-(-7)]$ To subtract -7, add the opposite of -7.

$= -8 + 7$ $-(-7) = 7$

$= -1$

```
(-5+-3)-(-2-5)
                -1
```

    ■

EXAMPLE 10   Simplify $[-6 - (-4)] - [-7 + 10]$.

Solution $[-6 - (-4)] - [-7 + 10] = [-6 + 4] - [-7 + 10]$

$= -2 - 3$

$= -2 + (-3)$

$= -5$

    ■

EXAMPLE 11   **Temperature change** At noon the temperature was 7° above zero. At midnight the temperature was 4° below zero. Find the difference between these two temperatures.

Solution   A temperature of 7° above zero can be represented as $+7$. A temperature of 4° below zero can be represented as -4. To find the difference between these two temperatures, we set up a subtraction problem and simplify.

$$7 - (-4) = 7 + [-(-4)] \qquad \text{To subtract } -4, \text{ add the opposite of } -4.$$
$$= 7 + 4 \qquad\qquad -(-4) = 4$$
$$= 11$$

The difference between the temperatures is 11°. Figure 1-5 shows this difference.

```
7--4
              11
```

FIGURE 1-5

EXERCISE 1.3

VOCABULARY *In Exercises 1–4, fill in the blanks to make a true statement.*

1. The distance between 0 and a number x on the number line is called the _____ of x.

2. If $a + (-a) = 0$, then a and $-a$ are called additive _____.

3. The _____ property of addition guarantees that $(a + b) + c = a + (b + c)$.

4. The _____ rule, $-(-a) = a$, is used in the subtraction of real numbers.

CONCEPTS

5. Determine each absolute value. Verify with your calculator, copying the screen display to justify your work.
   a. $|-4|$ b. $-|-6|$ c. $-|5|$

6. Find each sum. Verify with your calculator, copying the screen display to justify your work.
   a. $4 + 11$ b. $(-4) + (-11)$
   c. $(-4) + 11$ d. $4 + (-11)$

7. Find each difference. Verify with your calculator, copying the screen display to justify your work.
   a. $8 - 4$ b. $-8 - 4$
   c. $8 - (-4)$ d. $-8 - (-4)$

✍ 8. Explain the process for adding integers.

✍ 9. Explain the process for subtracting integers.

✍10. Many students believe that the operation of absolute value always means the result will be positive. Describe a situation when this will not be the case.

PRACTICE *In Exercises 11–20, find each sum and verify with your calculator. Copy your screen display to justify your work.*

11. $|9 + (-11)|$ 12. $|10 + (-13)|$ 13. $(-5) + (-7)$ 14. $(-6) + (-4)$ 15. $(-0.4) + 0.9$

16. $(-1.2) + -5.3$ 17. $\frac{1}{5} + \left(+\frac{1}{7}\right)$ 18. $\frac{2}{3} + \left(-\frac{1}{4}\right)$ 19. $\left(-\frac{3}{4}\right) + \left(+\frac{2}{3}\right)$ 20. $\frac{3}{5} + \left(-\frac{2}{3}\right)$

In Exercises 21–32, simplify each expression and verify with your calculator. Copy your screen display to justify your work.

21. $5 + [4 + (-2)]$ 22. $-6 + [(-3) + 8]$ 23. $-2 + (-4 + 5)$ 24. $5 + [-4 + (-6)$

25. $[-4 + (-3)] + [2 + (-2)]$ 26. $[3 + (-1)] + [-2 + (-3)]$

27. $-4 + (-3 + 2) + (-3)$ 28. $5 + [2 + (-5)] + (-2)$

29. $-|-9 + (-3)| + (-6)$ 30. $-|8 + (-4)| + 7$

31. $-5.2 + |-2.5 + (-4)|$ 32. $6.8 + |8.6 + (-1.1)|$

In Exercises 33–38, find each difference and verify with your calculator. Copy your screen display to justify your work.

33. $\dfrac{5}{3} - \dfrac{7}{6}$ **34.** $-\dfrac{5}{9} - \dfrac{5}{3}$ **35.** $-5 - \left(-\dfrac{3}{5}\right)$ **36.** $\dfrac{7}{8} - (-3)$

37. $|-6.7 - (-2.5)|$ **38.** $|17.5 - 25.3|$

In Exercises 39–48, find each difference and verify with your calculator. Copy your screen display to justify your work.

39. $3 - [(-4) - 3]$ **40.** $-5 - [4 - (-2)]$ **41.** $(5 - 3) + (3 - 5)$

42. $(3 - 5) - [5 - (-3)]$ **43.** $5 - [4 + (-2) - 5]$ **44.** $3 - [-(-2) + 5]$

45. $[5 - (-34)] - [-2 + (-23)]$ **46.** $-5 + \{-3 - [-2 - (+4)]\}$

47. $\left(\dfrac{5}{2} - 3\right) - \left(\dfrac{3}{2} - 5\right)$ **48.** $\left(\dfrac{7}{3} - \dfrac{5}{6}\right) - \left[\dfrac{5}{6} - \left(-\dfrac{7}{3}\right)\right]$

In Exercises 49–60, let $x = -4$, $y = 5$, and $z = -6$ and evaluate each expression (use fractions where applicable). Verify your computation by using the STOre feature of the calculator. Copy your screen display to justify your work.

49. $(x + y) + 3$ **50.** $(y + 5) + x$ **51.** $|2x + y|$ **52.** $3|x + y + z|$

53. $x - y - z$ **54.** $x - (y - z)$ **55.** $z - (x - y) + 10$ **56.** $3 - [x + (-3)]$

57. $\dfrac{y - x}{3 - z}$ **58.** $\dfrac{y - z}{3y + x}$ **59.** $\dfrac{|y + x|}{y - x}$ **60.** $\dfrac{|y - x|}{|x - y|}$

In Exercises 61–64, let $x = -2.34$, $y = 3.47$, and $z = 0.72$. Use the STOre feature on your calculator to perform the computation and round each answer to the nearest tenth.

61. $x^3 - y + z^2$ **62.** $y - z^2 - x^2$ **63.** $x^2 - y^2 - z^2$ **64.** $z^3 - x^2 - y^3$

APPLICATIONS *In Exercises 65–77, use signed numbers to write an arithmetic expression to model the problem. Use your calculator to perform the computation. Copy your screen display to justify your work.*

65. Temperature The temperature fell from zero to $-14°$ in one night. By 5:00 P.M. the next day, the temperature had risen $10°$. What was the temperature at 5:00 P.M.?

66. History In 1897, Joseph Thompson discovered the electron. Fifty-four years later, the first fission reactor was built. Nineteen years before the reactor, James Chadwick discovered the neutron. In what year was the neutron discovered?

67. History The Greek mathematician Euclid was alive in 300 B.C. The English mathematician Sir Isaac Newton was alive in A.D. 1700. How many years apart did they live?

68. Military science An army retreated 2300 meters. After regrouping, it moved forward 1750 meters. The next day it gained another 1875 meters. What was the army's net gain?

69. Sports A football player gained and lost the following yardage on six consecutive plays: $+5$, $+7$, -5, $+1$, -2, -6. How many yards were gained or lost?

70. Temperature Find the difference between a temperature of $32°$ above zero and a temperature of $27°$ above zero.

71. Temperature Find the difference between a temperature of $3°$ below zero and a temperature of $21°$ below zero.

72. Business The owner of a small business has a gross income of \$97,345.32. However, he paid \$37,675.66 in expenses, plus \$7537.45 in taxes, \$3723.41 in health care premiums, and \$5767.99 in pension payments. Find his profit.

73. Real estate A woman sold her house for \$115,000. Her fees at closing were \$78 for preparing a deed, \$446 for title work, \$216 for revenue stamps, and a sales commission of \$7612.32. In addition, there was a deduction of \$23,445.11 to pay off her old mortgage. As a part of the deal, the buyer agreed to pay half the title work. How much money did the woman receive?

74. Carpentry John is assembling a set of metal shelves. The bolts that were packaged with the shelves are $\frac{3}{8}$ inches long. He finds that the bolts are $\frac{5}{16}$ inches too long. What length bolt is needed?

75. Farming The Mitchells harvested $28\frac{1}{2}$ tons of wheat, sold $19\frac{3}{4}$ tons, and lost $5\frac{1}{10}$ tons to spoilage. How much wheat was left?

76. Medical A premature baby weighs $3\frac{1}{2}$ lb at birth. From the chart below, determine the weight at the end of the week.

Day 1	2	3	4	5	6	7
loss: $\frac{1}{8}$ lb	loss: $\frac{1}{4}$ lb	gain: $\frac{1}{16}$ lb	gain: $\frac{1}{8}$ lb	gain: $\frac{5}{16}$ lb	gain: $\frac{1}{16}$ lb	gain: $\frac{3}{16}$ lb

77. Distance If the distance between two points, x and y, on the number line is defined to be $|x - y|$, find the distance between each pair of points.

x	y
$\frac{1}{2}$	$-\frac{2}{3}$
$-\frac{4}{5}$	$-\frac{1}{8}$

78. Enrollment Illustration 1 shows the enrollment in music, art, and theatre courses at a small liberal arts college over a two-year period.

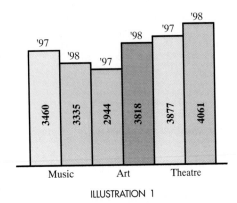

ILLUSTRATION 1

a. What is the difference in enrollment from 1997 to 1998 for each department?

b. Which department had the greatest increase of students? How many students were gained?

c. What is the total change in student enrollment for all three departments combined?

✍ **79.** Explain why, using a concrete everyday application, the sum of two negative numbers is always negative and the sum of two positive numbers is always positive.

✍ **80.** Explain why, using a concrete everyday application, the sum of a positive number and a negative number could be either positive or negative.

✍ **81.** Think of two numbers. First add the absolute values of the two numbers and write your answer. Second, add the two numbers, take the absolute value of that sum, and write that answer. Do your two answers agree? Try this several different times, using different combinations of positive and negative numbers. Can you find two numbers that produce different answers? When do you get answers that agree, and when don't you?

REVIEW *In Exercises 82–83, simplify each expression. Verify with your calculator, copying the screen display to justify your work.*

82. $(7 - 3)^2$

83. $2 + 5 \cdot 3^2$

In Exercises 84–85, evaluate each expression. Use the STOre feature of the calculator to justify your work. Let $x = 3$ and $y = 2$.

84. $(4x)^y$

85. $6x^2 - 2y^2$

86. Grades An algebra class has 24 students. One-third of the class made an A. How many students did not make an A?

87. Averages If the five highest grades on a biology test were 90, 89, 82, 76, and 71, find the average. Write the arithmetic expression necessary to determine the average and then use your calculator for the computation. Copy your screen display to justify your work.

1.4 Multiplying and Dividing Real Numbers

■ MULTIPLYING REAL NUMBERS ■ DIVIDING REAL NUMBERS

■ MULTIPLYING REAL NUMBERS

Because the times sign, $\times$, looks like the letter x, it is seldom used in algebra. Instead, a dot, parentheses, or no symbol at all is used to denote multiplication. Each of the following expressions indicates the **product** obtained when two real numbers, x and y, are multiplied.

$$x \cdot y \qquad (x)(y) \qquad x(y) \qquad (x)y \qquad xy$$

■ STUDY TIP

In order to be successful in algebra you must attend class regularly and be an active participant. Sit near the front of the class and participate in class discussions.

To develop rules for multiplying real numbers, we rely on the definition of multiplication. The expression $5 \cdot 4$ indicates that 4 is to be used as a term in a sum five times. That is

$$5(4) = 4 + 4 + 4 + 4 + 4 \qquad \text{Read } 5(4) \text{ as "5 times 4."}$$
$$= 20$$

Likewise, the expression $5(-4)$ indicates that -4 is to be used as a term in a sum five times. Thus,

$$5(-4) = (-4) + (-4) + (-4) + (-4) + (-4)$$
$$= -20 \qquad \text{Read } 5(-4) \text{ as "5 times negative 4."}$$

We realize that the order in which we multiply two whole numbers does not affect the outcome: $2 \cdot 3 = 3 \cdot 2 = 6$. This applies to all real numbers, not just whole numbers, and is called the **commutative property of multiplication.** Thus, if we are given the problem $(-5)(4)$, we can apply the commutative property to rewrite the product as $4(-5)$:

$$4(-5) = (-5) + (-5) + (-5) + (-5) = -20$$

Thus far we can conclude that

a. The product of two positive numbers is a positive number.

b. The product of a positive number and a negative number (or the product of a negative number and a positive number) is a negative number.

EXPLORING THE CONCEPT   Use your calculator to examine the pattern that is developed below to help you determine what happens when two negative numbers are multiplied together.

$$5(-4) = ?$$
$$4(-4) = ?$$
$$3(-4) = ?$$
$$2(-4) = ?$$
$$1(-4) = ? \qquad \text{Multiplicative identity: any number multiplied by 1 is that number itself.}$$
$$0(-4) = ? \qquad \text{The product of any number and 0 is 0.}$$
$$-1(-4) = ?$$

Conclusion   The pattern of numbers above is -20, -16, -12, -8, -4, 0, 4. Because $-1(-4) = 4$, the product of two negative numbers is a positive number.   ■

Rules for Multiplying Signed Numbers
1. The product of two real numbers with like signs is the product of the absolute values.
2. The product of two real numbers with unlike signs is the negative of the product of their absolute values.
3. Any number multiplied by 0 is 0: $a \cdot 0 = 0 \cdot a = 0$.
4. Any number multiplied by 1 is that number itself: $a \cdot 1 = 1 \cdot a = a$.

EXAMPLE 1   Find each product: **a.** $4(-7)$,   **b.** $(-5)(-4)$,   **c.** $(-7)(6)$,   **d.** $8(6)$,   **e.** $(-3)(5)(-4)$,   **f.** $(-4)(-2)(-3)$

Solution   **a.** $4(-7) = -(4 \cdot 7)$ **b.** $(-5)(-4) = +(5 \cdot 4)$
$= -28$ $= +20$

c. $(-7)(6) = -(7 \cdot 6)$ **d.** $8(6) = +(8 \cdot 6)$
$= -42$ $= +48$

e. $(-3)(5)(-4) = (-15)(-4)$ **f.** $(-4)(-2)(-3) = 8(-3)$
$= 60$ $= -24$

a.
```
4(-7)
           -28
(-5)(-4)
            20
(-7)(6)
           -42
```
b.
c.

d.
```
8(6)
            48
(-3)(5)(-4)
            60
(-4)(-2)(-3)
           -24
```
e.
f.

EXAMPLE 2 If $x = -3$, $y = 2$, and $z = 4$, evaluate   **a.** $y + xz$ and   **b.** $x(y - z)$

Solution   We substitute -3 for x, 2 for y, and 4 for z in each expression and simplify.

a. $y + xz = 2 + (-3)(4)$
$= 2 + (-12)$
$= -10$

b. $x(y - z) = -3[2 - 4]$
$= -3[2 + (-4)]$
$= -3(-2)$
$= 6$

a.
```
-3→X:2→Y:4→Z:Y+X
Z
           -10
X(Y-Z)
            6
```
b.

■ TECHNOLOGY TIP
Be careful! Graphing calculators DO NOT always recognize the implied multiplication xz as x "times" z. If your calculator gives an error message when xz is entered, try entering $x * z$ instead.

EXAMPLE 3 If $x = -2$ and $y = 3$, evaluate   **a.** $x^2 - y^2$ and   **b.** $-x^2$

Solution   **a.** We substitute -2 for x and 3 for y and simplify.

$x^2 - y^2 = (-2)^2 - 3^2$
$= 4 - 9$ Simplify the exponential expressions first.
$= -5$ Do the subtraction.

b. We substitute -2 for x and simplify.

$-x^2 = -(-2)^2$
$= -4$ Simplify $(-2)^2$.

a.
```
-2→X:3→Y:X²-Y²
            -5
-X²
            -4
```
b.

EXAMPLE 4 Find each product: **a.** $\left(-\dfrac{2}{3}\right)\left(-\dfrac{6}{5}\right)$ and   **b.** $\left(\dfrac{3}{10}\right)\left(-\dfrac{5}{9}\right)$

Solution   **a.** $\left(-\dfrac{2}{3}\right)\left(-\dfrac{6}{5}\right) = +\left(\dfrac{2}{3} \cdot \dfrac{6}{5}\right)$ **b.** $\left(\dfrac{3}{10}\right)\left(-\dfrac{5}{9}\right) = -\dfrac{3}{10} \cdot \dfrac{5}{9}$

$= +\dfrac{2 \cdot 6}{3 \cdot 5}$ $= -\dfrac{3 \cdot 5}{10 \cdot 9}$

$= +\dfrac{12}{15}$ $= -\dfrac{15}{90}$

$= +\dfrac{4}{5}$ $= -\dfrac{1}{6}$

a.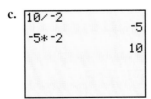

EXAMPLE 5   **Temperature**   If the temperature is dropping 4° each hour, how much warmer was it 3 hours ago?

Solution   A temperature drop of 4° per hour can be represented by $-4°$ per hour. Three hours ago can be represented by -3. The temperature 3 hours ago is the product $(-3)(-4)$.

$$(-3)(-4) = +12$$

The temperature was 12° warmer 3 hours ago.   ■

■ DIVIDING REAL NUMBERS

In general, the rule

$$\frac{a}{b} = c \quad \text{if and only if} \quad c \cdot b = a$$

is true for the division of any real number, a, by any nonzero real number, b. For example,

$$\frac{8}{4} = 2 \quad \text{because} \quad 2 \cdot 4 = 8$$

EXPLORING THE CONCEPT   Use your calculator to perform each division problem and verify the results by computing the corresponding multiplication problem.

a. $\dfrac{10}{2}$ **b.** $\dfrac{-10}{-2}$ **c.** $\dfrac{10}{-2}$ **d.** $\dfrac{-10}{2}$

e. Compare the rules for signed numbers to the operation of division.

Conclusion   a.   b.

c.   d.

e. Parts **a** and **b** indicate that the division of two numbers with like signs produces a positive result. Parts **c** and **d** indicate that the division of two numbers with different signs produces a negative result.   ■

> **Rules for Dividing Signed Numbers**
> 1. The quotient of two real numbers with like signs is the quotient of their absolute values.
> 2. The quotient of two real numbers with unlike signs is the negative of the quotient of their absolute values.
> 3. $\dfrac{a}{0}$ is undefined; $\dfrac{0}{0}$ is undefined (indeterminate).
> 4. If $a \neq 0$, then $\dfrac{0}{a} = 0$.

EXAMPLE 6   Find each quotient: **a.** $\dfrac{36}{18}$,   **b.** $\dfrac{-44}{11}$,   **c.** $\dfrac{27}{-9}$,   **d.** $\dfrac{-64}{-8}$,   and   **e.** $\dfrac{6}{0}$.

Solution   **a.** $\dfrac{36}{18} = +\dfrac{36}{18} = 2$ The quotient of real numbers with like signs is the quotient of their absolute values.

b. $\dfrac{-44}{11} = -\dfrac{44}{11} = -4$ The quotient of real numbers with unlike signs is the negative of the quotient of their absolute values.

■ **STUDY TIP**
Study in 30-minute segments of time with breaks of 5 to 10 minutes between each 30-minute session.

c. $\dfrac{27}{-9} = -\dfrac{27}{9} = -3$ The quotient of real numbers with unlike signs is the negative of the quotient of their absolute values.

d. $\dfrac{-64}{-8} = +\dfrac{64}{8} = 8$ The quotient of real numbers with like signs is the quotient of their absolute values.

e. $\dfrac{6}{0}$ is undefined.

a.
```
36/18
             2
```
b.
```
-44/11
            -4
```
c.
```
27/-9
            -3
```

d.
```
-64/-8
             8
```

e.
```
ERR:DIVIDE BY 0
1■Quit
2:Goto
```

EXAMPLE 7   If $x = -64$, $y = 16$, and $z = -4$, evaluate   **a.** $\dfrac{yz}{-x}$,   **b.** $\dfrac{z^3 y}{x}$,   and   **c.** $\dfrac{x + y}{-z^2}$.

Solution   We substitute -64 for x, 16 for y, and -4 for z in each expression and simplify.

a. $\dfrac{yz}{-x} = \dfrac{16(-4)}{-(-64)}$ **b.** $\dfrac{z^3 y}{x} = \dfrac{(-4)^3 (16)}{-64}$ **c.** $\dfrac{x + y}{-z^2} = \dfrac{-64 + 16}{-(-4)^2}$

$= \dfrac{-64}{+64}$ $= \dfrac{(-64)(16)}{(-64)}$ $= \dfrac{-48}{-16}$

$= -1$ $= 16$ $= 3$

■ **TECHNOLOGY TIP**
Parentheses were necessary in the numerator of Example 7**c** to indicate the addition occurs before the division. In **a** and **b**, no parentheses were necessary because the calculator follows the order of operations and computes multiplication before division.

a.
```
-64→X:16→Y: -4→Z:
YZ/-X
            -1
```
b.
```
Z^3Y/X
            16
```
c.
```
(X+Y)/-Z²
             3
■
```

EXAMPLE 8   If $x = -50$, $y = 10$, and $z = -5$, evaluate   **a.** $\dfrac{xyz}{x - 5z}$ and   **b.** $\dfrac{3xy + 2yz}{2(x + y)}$.

Solution   We substitute -50 for x, 10 for y, and -5 for z in each expression and simplify.

■ STUDY TIP
Place a piece of paper over the solution to an example in your text. Read the example and try to work the problem on your paper. If you get stuck, slide the paper down and jump start yourself.

a. $\dfrac{xyz}{x - 5z} = \dfrac{(-50)(10)(-5)}{-50 - 5(-5)}$

$= \dfrac{(-500)(-5)}{-50 + 25}$

$= \dfrac{2500}{-25}$

$= -100$

b. $\dfrac{3xy + 2yz}{2(x + y)} = \dfrac{3(-50)(10) + 2(10)(-5)}{2(-50 + 10)}$

$= \dfrac{-150(10) + (20)(-5)}{2(-40)}$

$= \dfrac{-1500 - 100}{-80}$

$= \dfrac{-1600}{-80}$

$= 20$

a.
b.
```
-50→X:10→Y:-5→Z:
XYZ/(X-5Z)
             -100
(3XY+2YZ)/(2(X+Y
))
              20
```

EXAMPLE 9   **Stock reports**   In its annual report, a publicly held corporation reports its performance on a per-share basis. When a company with 35 million shares outstanding loses \$2.3 million, what will be the per-share loss?

Solution   A loss of \$2.3 million can be represented by $-2,300,000$. Because there are 35 million shares, the per-share amount lost can be represented by the quotient $\frac{-2,300,000}{35,000,000}$.

$$\frac{-2,300,000}{35,000,000} \approx -0.065714285 \qquad \text{Use a calculator.}$$

The company lost about 6.6 cents per share.

Remember these facts about dividing real numbers.

Division

1. $\dfrac{a}{0}$ is undefined.

2. If $a \neq 0$, then $\dfrac{0}{a} = 0$.

3. $\dfrac{a}{1} = a$.

4. If $a \neq 0$, then $\dfrac{a}{a} = 1$.

EXERCISE 1.4

VOCABULARY   *Fill in the blank to make a true statement.*

1. A _____ is the result obtained when two real numbers are multiplied.

CONCEPTS

✍2. Explain the difference between products and factors.

✍3. Explain the difference between the divisors of a number and the factors of a number.

4. Find each product and verify with your calculator.
  a. $4(9)$ **b.** $(-4)(9)$
  c. $4(-9)$ **d.** $(-4)(-9)$

✍5. In your own words, explain the rule for multiplication of signed numbers.

6. Find each quotient and verify with your calculator.

  a. $\dfrac{10}{5}$ **b.** $-\dfrac{10}{5}$

  c. $\dfrac{10}{-5}$ **d.** $\dfrac{-10}{-5}$

✎ **7.** Explain the rule for dividing signed numbers.

✎ **8.** Explain how you would decide whether the product of several numbers is positive or negative.

9. If the quotient of two numbers is undefined, what would their product be?

✎**10.** If x^5 is a negative number, can you determine if x is negative? Explain.

PRACTICE   *In Exercises 11–22, simplify each expression. Confirm each result with the graphing calculator, making sure to record the screen display.*

11. $\left(\dfrac{1}{2}\right)(32)$ **12.** $\left(-\dfrac{3}{4}\right)(12)$ **13.** $\left(-\dfrac{3}{4}\right)\left(-\dfrac{8}{3}\right)$ **14.** $\left(-\dfrac{12}{5}\right)\left(\dfrac{15}{2}\right)$

15. $(3)(-4)(-6)$ **16.** $(-1)(-3)(-6)$ **17.** $(-2)(3)(4)$ **18.** $(5)(0)(-3)$

19. $\dfrac{8-12}{-2}$ **20.** $\dfrac{16-2}{2-9}$ **21.** $\dfrac{20-25}{7-12}$ **22.** $\dfrac{2(15)^2-2}{-2^3+1}$

In Exercises 23–44, let $x = -1$, $y = 2$, and $z = -3$. Evaluate each expression and confirm results by using the STOre feature of the graphing calculator.

23. x^2 **24.** $-z^2$ **25.** $-xz$ **26.** $y - xz$

27. $z - xy$ **28.** $(x + y)z$ **29.** $y(x - z)$ **30.** $xy + yz$

31. $zx - zy$ **32.** y^2z^2 **33.** z^3y **34.** $y(x - y)^2$

35. $z(y - x)^2$ **36.** $\dfrac{yz}{x}$ **37.** $\dfrac{zx}{y}$ **38.** $\dfrac{z + x}{y}$

39. $\dfrac{xyz}{y - 1}$ **40.** $\dfrac{z - xy}{x + y}$ **41.** $\dfrac{x^2y^3}{yz}$ **42.** $\dfrac{2x^2 + 2y}{x + y}$

43. $\dfrac{y^2 + z^2}{y + z}$ **44.** $\dfrac{2x^2 - 2z^2}{x + z}$

In Exercises 45–52, evaluate each expression if $x = \frac{1}{2}$, $y = -\frac{2}{3}$, and $z = -\frac{3}{4}$. Confirm results by using the STOre feature of the graphing calculator. Express all results as fractions (when applicable).

45. $x + y$ **46.** $y + z$ **47.** $x + y + z$ **48.** $y + x - z$

49. $(x + y)(x - y)$ **50.** $(x - z)(x + z)$ **51.** $(x + y + z)(xyz)$ **52.** $xyz(x - y - z)$

✎**53.** If the product of five numbers is negative, how many of the factors could be negative?

✎**54.** If x^6 is a positive number, can you determine if x is positive?

APPLICATIONS   *In Exercises 55–61, (a) write an arithmetic expression for each problem, and (b) use your graphing calculator to simplify each expression. Copy the screen display to verify results.*

55. Mowing lawns   Rafael worked all day mowing lawns and was paid $8.00 per hour. If he had $94 at the end of an 8-hour day, how much did he have before he started working?

56. Gambling   In Las Vegas, Robert lost $30 per hour playing the slot machines for 15 hours. What product of signed numbers represents the change in his financial condition?

57. Dieting   A man lost 37.5 pounds. If he lost 2.5 pounds each week, how long had he been dieting?

58. Saving for college   A student has saved $15,000 to attend graduate school. If she estimates that her expenses will be $613.50 a month while in school, does she have enough to complete an 18-month master's degree program?

59. Earnings per share   Over a 5-year period, a corporation reported profits of $18 million, $21 million, and $33 million in 3 of the years. In the other 2 years, it reported losses of $5 million and $71 million. Find the average gain (or loss) per year.

✍**60.** Describe two situations in which negative numbers are useful.

61. Stock market   A stock originally went on the market for $2.40. During the next 6-month period, the stock gained $\frac{3}{4}$ point, gained 2 points, dropped $1\frac{1}{2}$ points, dropped $\frac{7}{8}$ point, rose $\frac{1}{4}$ point, and dropped $\frac{3}{8}$ point, as indicated in Illustration 1. If each point represents $1.00, what is the stock's value at the end of the 6-month period?

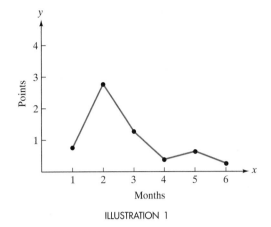

ILLUSTRATION 1

REVIEW   In Exercises 62–66, simplify and verify your results with your calculator.

62. -4^4

63. $\left(-\frac{2}{3}\right) + \left(\frac{4}{5}\right)$

64. $[-4 - (6 - 1)] + [3 - 7]$

65. Evaluate for $y = -2$ and $x = -\frac{1}{2}$
   a. $x + y$ **b.** $x - y$

66. A pattern for a Halloween costume requires $1\frac{1}{8}$ yards of blue material, $\frac{2}{3}$ yard of red material, $\frac{1}{2}$ yard of white material, and $\frac{3}{4}$ yard of yellow material. What is the total amount of material purchased?

1.5 Basic Applications

■ IRRATIONAL NUMBERS ■ GEOMETRY ■ COMMONLY USED FORMULAS IN MATHEMATICS
■ BUSINESS ■ ROUNDING DECIMALS

■ IRRATIONAL NUMBERS

In the first section, we talked about studying problem situations and developing our tools for solving these problems. Many application problems can be solved using various mathematical formulas: Determining the volume of a swimming pool is often necessary to compute the amount of chemicals needed to treat the water; converting Fahrenheit temperatures to Celsius is necessary for scientific applications; computing interest on loans or savings accounts is necessary for personal finances.

We will begin by examining some basic geometric formulas for perimeter, area, and volume. To do this, we need to discuss the set of irrational numbers in more detail than was done in Section 1.1.

Irrational Numbers
An **irrational number** is a number that cannot be expressed as a fraction with an integer in its numerator and a nonzero integer in its denominator.

Numbers such as $\sqrt{2}$ and π are not rational numbers because they cannot be written as fractions with an integer numerator and an integer denominator. Such numbers are called **irrational numbers.** We can find *decimal approximations* for the values of irrational numbers with a calculator.

EXPLORING THE CONCEPT

Enter each of the following on your graphing calculator. Determine which numbers are rational and which are irrational.

■ TECHNOLOGY TIP
To enter 17/20 on the calculator, use the division sign.

a. $\sqrt{2}$ **b.** π **c.** $\dfrac{17}{20}$

Conclusions

a.
b.
c.

```
√(2)
          1.414213562
π
          3.141592654
17/20
                  .85
■
```

You must be careful not to make generalizations about the decimal numbers displayed on your calculator. For any denominator of more than four digits, the calculator will always display a decimal approximation. For example, the fraction $\frac{1234}{1235}$ is a rational number because it is a quotient of two integers. However, if you enter the fraction $\frac{1234}{1235}$ on your calculator followed by the convert to fraction command, it will display the decimal equivalent because the denominator has four digits.

```
1234/1235▶Frac
          .9991902834
■
```

WARNING! You must know the definitions of rational and irrational numbers and not rely on calculator displays.

Recall that the set of real numbers is comprised of the sets of rational numbers and irrational numbers. Figure 1-6 shows how the various sets of numbers are interrelated. Examples of numbers within each set are displayed in the boxes.

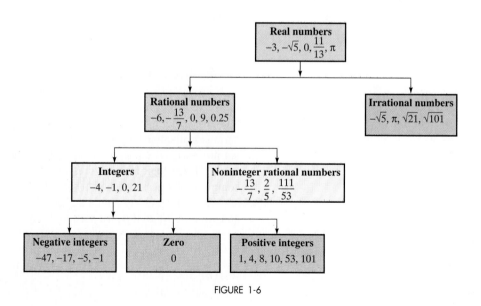

FIGURE 1-6

EXAMPLE 1   List the numbers in the set $\left\{-3, 0, \frac{1}{2}, 1.25, \sqrt{3}, 5\right\}$ that are

a. natural numbers   **b.** whole numbers   **c.** negative integers

d. rational numbers   **e.** irrational numbers   **f.** real numbers

Solution   **a.** The only natural number is 5.

b. The whole numbers are 0 and 5.

c. The only negative integer is -3.

d. The rational numbers are $-3, 0, \frac{1}{2}, 1.25$, and 5. (1.25 is rational because 1.25 can be written in the form $\frac{5}{4}$.)

e. The only irrational number is $\sqrt{3}$.

f. All of the numbers are real numbers.   ■

■ GEOMETRY

Substituting numbers for variables is often required when finding perimeters and areas of geometric figures. The **perimeter** of a geometric figure is the distance around it, and the **area** of the figure is the amount of surface that it encloses. The perimeter of a circle is called its **circumference.**

Table 1-1 shows the formulas for the perimeter and area of several geometric figures.

Figure	Name	Perimeter	Area
	Square	$P = 4s$	$A = s^2$
	Rectangle	$P = 2l + 2w$	$A = lw$
	Triangle	$P = a + b + c$	$A = \frac{1}{2}bh$
	Trapezoid	$P = a + b + c + d$	$A = \frac{1}{2}h(b + d)$
	Circle	$C = 2\pi r$	$A = \pi r^2$

TABLE 1-1

The **volume** of a three-dimensional geometric solid is the amount of space it encloses. Table 1-2 shows the formulas for the volume of several solids.

Figure	Name	Volume
	Rectangular solid	$V = lwh$
	Cylinder	$V = Bh$ where B is the area of the base
	Pyramid	$V = \frac{1}{3}Bh$ where B is the area of the base
	Cone	$V = \frac{1}{3}Bh$ where B is the area of the base (If the base is a circle, then $B = \pi r^2$)
	Sphere	$V = \frac{4}{3}\pi r^3$

TABLE 1-2

EXAMPLE 2   Prefab 4-foot-high fencing is $4.99 per linear foot. Find the cost of fencing the rectangular lot displayed:

Solution   The formula for the perimeter of a rectangle is $p = 2l + 2w$. Substitute in the known values of $l = 108.9$ and $w = 75.6$.

$$p = 2(108.9) + 2(75.6)$$
$$= 217.8 + 151.2$$
$$= 369$$

We need 369 linear feet of fencing. Because the fencing is $4.99 per linear foot, we have

$$369(4.99) = \$1841.31$$

The cost of fencing the lot is $1841.31.

75.6 ft

108.9 ft

EXAMPLE 3   **Winter driving**   Find the number of cubic feet of road salt in a conical pile that is 18.75 feet high and covers a circular area of 28.60 feet in diameter. Use a calculator and round the answer to two places.

Solution   We can find the area of the circular base of the cone shown in Figure 1-7 by substituting 14.3 for the radius in the formula for the area of a circle.

$$A = \pi r^2$$
$$\approx \pi (14.3)^2$$
$$\approx 642.4242817 \qquad \text{Use a calculator.}$$

We then substitute 642.4242817 for B and 18.75 for h in the formula for the volume of a cone.

$$V = \frac{1}{3}Bh$$

$$\approx \frac{1}{3}(642.4242817)(18.75)$$

$$\approx 4015.151761 \qquad \text{Use a calculator.}$$

To two decimal places, there are 4015.15 cubic feet of salt in the pile.   ■

FIGURE 1-7

■ COMMONLY USED FORMULAS IN MATHEMATICS

Distance (d) can be determined as it relates to rate of speed (r) and time (t). This is given by the formula $d = rt$. For example, if we were riding in a car traveling at a speed of 50 mph for 3 hours, we would have traveled 150 miles.

Suppose we want to determine the distance traveled by a train going 72 mph at 2-hour intervals for the first 12 hours of the excursion. We could display the information in a table like the one below.

Time (hours)	2	4	6	8	10	12	t
Distance (miles)	144	288	432	576	720	864	$72t$

This table can be created analytically (paper-and-pencil computation), via the calculator by merely entering the arithmetic expression as in Example 2, or via the STOre key, also illustrated in Example 2. The calculator itself will also generate a table.

First, go to the table setup menu. You must designate the start value (or minimum). The table above has a start value of 2 hours. The table increment must also be designated. The Greek letter delta, Δ, is used in mathematics to represent the change in value. Thus $\Delta t = 2$ would represent time intervals of 2-hour increments. Your calculator may have ΔTbl to designate the table increment.

To generate your table, go to the y= screen and enter the expression 72x. The calculator will only recognize x as a variable when generating a table. The screens below should match your screens. Access the TABLE screen to display the table seen below.

It displays the same information as the table created analytically above.

EXAMPLE 4   The formula for converting Fahrenheit temperature to Celsius is given as

$$C = \frac{5}{9}(F - 32).$$

a. Use the conversion formula to determine the equivalent Celsius temperature for 23° F.

b. Use the calculator to generate a conversion table for 23° F, 33° F, 43° F, and 53° F. Approximate the equivalent Celsius temperature to the nearest tenth of a degree for each of the given Fahrenheit temperatures.

Solution   **a.** $C = \frac{5}{9}(F - 32)$

Substitute 23 for F (the Fahrenheit temperature).

$$C = \frac{5}{9}(23 - 32)$$

$$C = \frac{5}{9}(9)$$

$$C = 5$$

Therefore, 23° Fahrenheit is equivalent to 5° Celsius.

b. The calculator will be evaluating the expression $\frac{5}{9}(F - 32)$ for the different values of F. In order to generate a table, the expression must be entered at the y1= prompt. Enter it now.

Access the table setup and set the table to start at 23 with increments of 10. Display the table:

X	Y1
23	-5
33	.55556
43	6.1111
53	11.667
63	17.222
73	22.778
83	28.333

X=23

We can conclude the following approximations of temperatures: 23° F = −5° C, 33° F = 0.6° C, 43° F = 6.1° C, and 53° F = 11.7° C.   ■

▧ BUSINESS

The basic formula for computing simple interest is $i = prt$, where i represents interest earned, p represents the principal (the amount invested or borrowed), r is the rate (percent), and t is the amount of time the money is invested. A **percent** represents parts out of 100. Thus, 6% would be $\frac{6}{100}$ or 0.06; $6\frac{1}{4}\%$ would be $\frac{6.25}{100}$ or 0.0625. All percents should be converted to fractions or the equivalent terminating decimal (DO NOT ROUND!) before performing computations.

EXAMPLE 5   **Paying tuition**   Juan signs a one-year note to borrow $8500 for tuition. If the rate of interest is $6\frac{1}{4}\%$, how much interest will he pay?

Solution   For the privilege of using the bank's money for one year, Juan must pay $6\frac{1}{4}\%$ of $8500. We calculate the interest, i, as follows:

$$i = 6\frac{1}{4}\% \text{ of } 8500$$

$$= 0.0625 \cdot 8500 \qquad \text{\small The word } of \text{ means } times.$$

$$= 531.25$$

Juan will pay $531.25 interest.   ■

■ ROUNDING DECIMALS

■ TECHNOLOGY TIP
Your graphing calculator displays a finite number of digits and will automatically round the number being displayed if its digits exceed the designated number of places. One graphing calculator displays the fraction 17/11,371 as the decimal 0.0014950312 while another displays 0.00149503122. Both decimal fractions are rounded *approximations* of 17/11,371.

When decimal fractions are long, we often **round** them to a specific number of decimal places. This gives an *approximate value* of the decimal fraction. For example, the decimal fraction 25.36124 rounded to one place (or to the nearest tenth) is 25.4. Rounded to two places (or to the nearest one-hundredth), the decimal is 25.36. To round decimals, we use the following rules.

> **Rounding Decimals**
> 1. Determine to how many decimal places you wish to round.
> 2. Look at the first digit to the right of that decimal place.
> 3. If that digit is 4 or less, drop it and all digits that follow. If it is 5 or greater, add 1 to the digit in the position in which you wish to round, and drop all the digits that follow.

EXERCISE 1.5

VOCABULARY In Exercises 1–6, fill in the blank to make a true statement.

1. An _____ number is a number that cannot be expressed as a quotient of integers.

2. The _____ of a geometric figure is the distance around it.

3. The ____ of a geometric figure is the amount of surface it encloses.

4. The perimeter of a circle is called the _____.

5. The _____ of a three-dimensional geometric solid is the amount of space it encloses.

6. A _____ represents the number of parts out of 100.

CONCEPTS

7. List the numbers in the set $\left\{ -3, -\dfrac{1}{2}, 0, \sqrt{2}, 3, 5.\overline{6} \right\}$ that are
   a. natural numbers
   b. whole numbers
   c. negative integers
   d. rational numbers
   e. irrational numbers

8. Determine whether perimeter, area, or volume would be used to determine the measure of the indicated quantity.
   a. water for a swimming pool
   b. wallpaper to cover two walls of a bedroom
   c. carpeting for a bedroom
   d. wallpaper border for a kitchen
   e. paint for the exterior of a house

   f. sand in an hour glass
   g. concrete edging for flower bed/garden
   h. compost for flower bed/garden
   i. weed killer for flower bed/garden

✍ 9. True or False: Two-thirds $\left(\dfrac{2}{3} \right)$ is equal to 0.67. Explain.

✍ 10. Which is more accurate when computing the area of a circle, 3.14 or the value displayed by the π key on your calculator? Explain.

11. Convert each of the following percentages to a decimal representation.
    a. 5%   b. $4\frac{1}{2}\%$ c. $8\frac{3}{4}\%$

PRACTICE In Exercises 12–15, find the perimeter of each figure. Perimeter is considered a linear measure, therefore your results should be expressed in the same units as the given measurement. If the calculator is used, make sure screens are copied exactly as they appear.

12. 3 m 5 m 7 m

13. 10 cm 3 cm 3 cm 10 cm

14. 6 cm 7 cm 9 cm 14 cm

15. 4 in. 4 in. 4 in. 4 in.

In Exercises 16–19, find the area of each figure. Area is always expressed in square units, or units². If the calculator is used, make sure screens are copied exactly as they appear.

16.
5 m
5 m

17.
5 cm
4 cm
8 cm

18.
6 ft
10 ft

19.
16 cm
12 cm
22 cm

In Exercises 20 and 21, find the circumference of each circle. Use the π key on your calculator. Circumference is considered a linear measurement. If the calculator is used, make sure screens are copied exactly as they appear.

20.
14 m

21.
21 cm

In Exercises 22 and 23, find the area of each circle. Use the π key on your calculator. Make sure results are expressed in square units. If the calculator is used, make sure screens are copied exactly as they appear.

22.
42 ft

23.
7 m

In Exercises 24–29, find the volume of each figure. Use the π key on your calculator. Make sure results are expressed in cubic units. If the calculator is used, make sure screens are copied exactly as they appear. Round results to the nearest thousandth.

24.
2 cm
3 cm
3 cm
3 cm
3 cm

25.
6 ft
2 ft
3 ft

26.
6 m

27.
14 in.
12 in.

28.
21 cm
14 cm
8 cm

29.
4 in.
4 in.
3 in.
6 in.

APPLICATIONS *In Exercises 30–56, use the π key on your calculator where applicable. Make sure solutions are expressed in the appropriate units. Copy calculator screens exactly as they appear and round results to two decimal places.*

30. Circumference Find the circumference of the bicycle wheel shown in Illustration 1.

ILLUSTRATION 1

31. Area A built-in kitchen eating area has the dimensions pictured in Illustration 2. Find the amount of floor area occupied.

ILLUSTRATION 2

32. Area Find the area of the 25-inch television screen in Illustration 3 if the width is 17 inches and the height is 18.5 inches.

ILLUSTRATION 3

33. Volume Find the volume, in cubic inches, of the lawn cart shown in Illustration 4.

ILLUSTRATION 4

34. Volume Find the volume of the hot water heater with the dimensions pictured in Illustration 5.

ILLUSTRATION 5

35. Volume Find the volume of the can of tomato sauce shown in Illustration 6.

ILLUSTRATION 6

36. Shipping Find the volume of the crate that was used to ship the stove shown in Illustration 7. Assume there is no extra space between the stove and the crate.

ILLUSTRATION 7

37. Construction A chicken coop is to be constructed with a tin roof and sides that are covered in poultry netting. If the coop has the dimensions indicated in Illustration 8, will one roll of the netting be sufficient? Justify your response.

ILLUSTRATION 8

38. Storing solvents   A hazardous solvent fills a rectangular tank with dimensions of 12 by 9.5 by 7.3 inches. For disposal, it must be transferred to a cylindrical canister 7.5 inches in diameter and 18 inches high. Find how much solvent will be left over.

39. Wallpapering   One roll of wallpaper covers about 33 square feet. At $27.50 per roll, how much would it cost to paper two walls 8.5 feet high and 17.3 feet long? (*Hint:* Wallpaper can only be purchased in full rolls.)

40. Sealing asphalt   A rectangular parking lot is 253.5 feet long and 178.5 feet wide. A 55-gallon drum of asphalt sealer covers 4000 square feet and costs $97.50. Find the cost to seal the parking lot. Sealer can only be purchased in full drums.

41. Installing carpet   What will it cost to carpet a 23-by-17.5-foot living room and a 17.5-by-14-foot dining room with carpet that costs $29.79 per square yard? One square yard is 9 square feet.

42. Temperature   Use your calculator to generate a table to compute the Celsius temperature for each Fahrenheit temperature to the nearest tenth of a degree.

Fahrenheit temp.	x	-20	-15	-10	-5	0
Celsius temp.	$\frac{5}{9}(x-32)$					

43. Scroll through the table constructed in Exercise 42 to answer the following questions:
   a. When it is 35° F, what is the Celsius temperature (to the nearest degree)?
   b. When it is 10° C, what is the Fahrenheit temperature?
   c. If the temperature is 72° C, why is it not possible to determine the Fahrenheit temperature?

44. Distance   The formula for finding the distance between two points a and b on a number line is $|a - b|$. For each pair of points specified below, find the distance between them.

   a. $\overset{a}{\underset{-14}{\vert}} \qquad \overset{b}{\underset{-8}{\vert}}$

   b. $\overset{a}{\underset{-8\frac{1}{2}}{\vert}} \qquad \overset{b}{\underset{2\frac{3}{5}}{\vert}}$

   c. $\overset{a}{\underset{\frac{2}{5}}{\vert}} \qquad \overset{b}{\underset{7\frac{5}{9}}{\vert}}$

45. Business   Blaire deposits her summer job earnings of $2300 in a savings account earning $3\frac{3}{4}\%$ simple interest. How much interest will she have
   a. at the end of 2 years?
   b. at the end of 6 months?

46. Ecology   Approximately 242,000,000 old tires are disposed of each year. If 77.7% are dumped illegally, how many old tires are dumped illegally?

47. Banking   If Clarksville, Tennessee, has 52,000 people with bank accounts, according to Illustration 9, how many people do their bank transactions by phone?

48. Tax rate   According to the 1998 tax rate schedule in Illustration 10, determine the amount of tax paid by a single individual with an income of $28,324.

OFF-SITE BANKING

Less than half of all banking transactions are done at bank branch offices. Automated teller machines (ATMs) are the most popular off-site option:

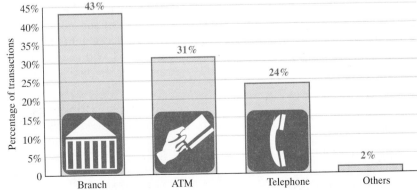

SOURCE: Bank Administration Institute

ILLUSTRATION 9

Tax Rate Schedules, 1998

Taxable income	What you pay
Single individual	
$ 0–$ 22,100	15.0% of sum over $0.00
$ 22,100–$ 53,500	$ 3,315 + 28.0% of sum over $ 22,100
$ 53,500–$115,000	$12,107 + 31.0% of sum over $ 53,500
$115,000–$250,000	$31,172 + 36.0% of sum over $115,000
over $250,000	$79,772 + 39.6% 0f sum over $250,000
Heads of households	
$ 0–$ 28,750	15.0% of sum over $0.00
$ 28,750–$ 76,400	$ 4,312.50 + 28.0% of sum over $ 28,750
$ 76,400–$127,500	$17,024.50 + 31.0% of sum over $ 76,400
$127,500–$250,000	$33,385.00 + 36.0% of sum over $127,500
over $250,000	$77,485.00 + 39.6% 0f sum over $250,000
Married individuals filing jointly, or qualifying widow(er)	
$ 0–$ 36,900	15.0% of sum over $0.00
$ 36,900–$ 89,150	$ 5,535.00 + 28.0% of sum over $ 36,900
$ 89,150–$140,000	$20,165.00 + 31.0% of sum over $ 89,150
$140,000–$250,000	$35,928.50 + 36.0% of sum over $140,000
over $250,000	$75,528.50 + 39.6% 0f sum over $250,000
Married, filing separate return	
$ 0–$ 18,450	15.0% of sum over $0.00
$ 18,450–$ 44,575	$ 2,767.50 + 28.0% of sum over $ 18,450
$ 44,575–$70,000	$10,082.50 + 31.0% of sum over $ 44,575
$ 70,000–$125,000	$17,964.25 + 36.0% of sum over $ 70,000
over $125,000	$37,764.25 + 39.6% 0f sum over $125,000

SOURCE: Internal Revenue Service, 1997

ILLUSTRATION 10

49. Statistics An elementary school has an enrollment of 871 students. Based on Illustration 11, how many of these children do not have snacks?

WHO CHOOSES SNACKS FOR KIDS?

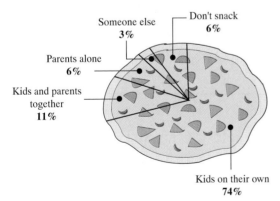

Someone else 3% Don't snack 6%
Parents alone 6%
Kids and parents together 11%
Kids on their own 74%

SOURCE: Statistics based on *USA Today*

ILLUSTRATION 11

50. Speed skating In tryouts for the Olympics, a speed skater had times of 44.47, 43.24, 42.77, and 42.05 seconds. Find the average time.

51. Cost of gasoline Diego drove his car 15,675.2 miles last year, averaging 25.5 miles per gallon of gasoline. The average cost of gasoline was $1.27 per gallon. Find the fuel cost to drive the car.

52. Paying taxes A businesswoman earns $48,712.32 in taxable income. She must pay 15% tax on the first $23,000 and 28% on the rest. In addition, she must pay a Social Security tax of 15.4% on the total amount. How much tax will she need to pay?

53. Dairy production If a Holstein cow produces 7600 gallons of milk with a $3\frac{1}{2}$% butterfat content, and a Guernsey cow produces 6500 gallons of milk that is 5% butterfat, which cow produces more butterfat?

54. Feeding dairy cows Each year a typical dairy cow will eat 12,000 pounds of food, which is 57% silage. To feed a herd of 30 cows, how much silage will a farmer use in a year?

55. Comparing bids Two contractors bid on a home remodeling project. The first bids $9350 for the entire job. The second contractor will work for $27.50 per hour, plus $4500 for materials. He estimates that the job will take 150 hours. Which contractor has the lower bid?

56. Choosing a furnace A high-efficiency home heating system can be installed for $4170, with an average monthly heating bill of $57.50. A regular furnace can be installed for $1730, but monthly heating bills average $107.75. After three years, which system is more expensive?

REVIEW

57. Simplify each of the following. Verify the results with your calculator.
    a. -3^2 **b.** $(-3)^2$
    c. 3^2 **d.** $4 \cdot 3^2$

58. Evaluate each of the following for $x = -\dfrac{1}{2}$ and $y = -\dfrac{2}{3}$. Express as fractions.
    a. $-x^2$ **b.** xy **c.** xy^3

59. Area A college student purchased 4 gallons of paint to paint her efficiency apartment. If she uses wallpaper in the bathroom, she will only use $2\frac{7}{9}$ gallons of paint. How much paint will be left over?

60. Perimeter An irregularly shaped lot has sides of $14\frac{1}{2}$ yards, $11\frac{2}{3}$ yards, $17\frac{5}{8}$ yards, and $15\frac{1}{3}$ yards. How much chainlink fencing will be needed to enclose the lot?

CHAPTER SUMMARY

CONCEPTS

REVIEW EXERCISES

Section 1.1

Review of Fractions

All results that are fractions should be expressed in lowest terms.

A fraction is completely reduced when the numerator and the denominator have no common prime factors.

Fractions may be added or subtracted as long as they have a common denominator.

Fractions may be multiplied by finding the product of the numerator over the product of the denominator.

To divide one fraction by another, multiply the first fraction by the reciprocal of the second fraction.

Before performing arithmetic operations with mixed numbers, first convert them to improper fractions.

Fractions may be converted to their terminating decimal equivalent before performing arithmetic operations. Fractions that represent repeating decimals may not be converted to decimals before performing arithmetic operations.

1. Simplify each of the expressions below arithmetically. Confirm results with the calculator, copying screen displays exactly as they appear. Express as fractions.

    a. $\dfrac{1}{3} + \dfrac{1}{7}$ **b.** $\dfrac{2}{3} - \dfrac{1}{7}$

    c. $\dfrac{31}{15} \cdot \dfrac{10}{62}$ **d.** $\dfrac{18}{21} \div \dfrac{6}{7}$

    e. $3\dfrac{2}{3} + 4\dfrac{1}{2}$ **f.** $6\dfrac{4}{5} - 2\dfrac{1}{3}$

    g. $3\dfrac{1}{2} \cdot 4\dfrac{1}{3}$ **h.** $3\dfrac{3}{4} \div 1\dfrac{5}{8}$

2. Flu shots Faculty at a local university are required to get flu shots on one of 3 days. The first day, $\frac{1}{5}$ of the faculty receive shots and $\frac{2}{3}$ receive their shots on the second day.
    a. What fractional part of the faculty received flu shots on the third day?

    b. If the faculty totaled 600, how many received shots each day?

| Section 1.2 | Exponents and Order of Operations |

If n is a natural number,

$$n \text{ factors of } x$$
$$\text{then } x^n = \overbrace{x \cdot x \cdot \ldots \cdot x}$$

Order of Mathematical Operations

Within each pair of grouping symbols (working from the innermost pair to the outermost pair), do the following operations:
1. Find the values of any exponential expressions.
2. Do all multiplications and divisions as they are encountered while working from left to right.
3. Do all additions and subtractions as they are encountered while working from left to right.

When all of the grouping symbols have been removed, repeat the above rules to finish the calculation.

Only the parentheses keys may be used as grouping symbols on your graphing calculator. Replace all [ ] and { } with parentheses when entering expressions to be evaluated on the calculator.

In a fraction simplify the numerator and denominator separately. Then simplify the fraction, if possible.

3. Simplify each of the following expressions. Confirm results with the graphing calculator, copying screen displays exactly as they appear.

   a. 3^4
   b. $\left(\dfrac{2}{3}\right)^2$
   c. -5^2
   d. $(-5)^2$

4. Simplify each of the following expressions. Confirm results with the graphing calculator, copying screen displays exactly as they appear.

   a. $4 + (8 \div 4)$
   b. $(4 + 8) \div 4$
   c. $(5 - 2)^2 + 5^2 + 2^2$
   d. $\dfrac{4 \cdot 3 + 3^4}{31}$
   e. $\left(\dfrac{3}{2}\right)^2 - \dfrac{1}{3}$
   f. $\dfrac{4}{3} \cdot \dfrac{9}{2} + \dfrac{1}{2} \cdot 18$

5. For each of the following, write the arithmetic expression whose calculator entry is displayed below. Your expression should be written as it would be displayed in your text.

   a.

   b.

   c.

6. **Average study time**   Four students recorded time spent working on a take-home exam: 5.2, 4.7, 9.5, and 8 hours. Find the average time spent. Make sure you write an arithmetic expression to model the problem and then simplify the expression. Confirm your calculations with the graphing calculator, copying your screen display to justify the results.

| Section 1.3 | Adding and Subtracting Real Numbers |

The absolute value of a number x, denoted by $|x|$, is the distance between x and 0 on the number line,
$$|x| \geq 0.$$

To find the sum of two real numbers with the same sign, we add their absolute values and keep their common sign.

7. Simplify each of the given expressions. Confirm your calculations with the graphing calculator, copying the screen display exactly as it appears.

   a. $1 - [5 - (-3)]$
   b. $\left|\dfrac{3}{7} - \left(-\dfrac{4}{7}\right)\right|$
   c. $\dfrac{3}{7} - \left|-\dfrac{4}{7}\right|$
   d. $\dfrac{2}{3} - \left(\dfrac{1}{3} - \dfrac{2}{3}\right)$
   e. $-\dfrac{3}{14} \cdot \left(-\dfrac{7}{6}\right)$
   f. $-\dfrac{1}{2} \cdot \dfrac{4}{3}$

To add two real numbers that have unlike signs, we subtract their absolute values (the smaller from the larger) and use the sign of the number with the greatest absolute value.

If a and b are two real numbers, then

$$a - b = a + (-b).$$

On the graphing calculator, be careful to observe the difference between the operation key of subtraction and the key used for a negative sign.

8. Simplify each of the given expressions. Confirm your calculations with the graphing calculator, copying the screen display exactly as it appears.

   a. $\left(\dfrac{-3 + (-3)}{3} \right)\left(-\dfrac{15}{5} \right)$ b. $\dfrac{-2 - (-8)}{5 + (-1)}$

   c. $\dfrac{[-3 + (-4)]^2}{10 + (-3)}$ d. $\left(\dfrac{-10}{2} \right)^2 - (-1)^3$

Section 1.4

Multiplying and Dividing Real Numbers

The product of two real numbers with like signs is the product of their absolute values.

The product of two real numbers with unlike signs is the negative of the product of their absolute values.

The quotient of two real numbers with like signs is the quotient of their absolute values.

The quotient of two real numbers with unlike signs is the negative of the quotient of their absolute values.

Division by zero is undefined.

Some graphing calculators may not recognize xy as a product of x and y, and you will have to enter the expression as $x * y$.

In Review Exercises 9–24, let $x = 2$, $y = -3$, and $z = -1$. Evaluate each expression below both analytically (paper-and-pencil computation) and using the STOre feature of the graphing calculator. Be sure to copy your screen display exactly as it appears.

9. $y + z$ 10. $x + y$ 11. $x + (y + z)$

12. $x - y$ 13. $x - (y - z)$ 14. $(x - y) - z$

15. xy 16. yz 17. $x(x + z)$

18. xyz 19. $y^2z + x$ 20. $yz^3 + (xy)^2$

21. $\dfrac{xy}{z}$ 22. $\dfrac{|xy|}{3z}$

23. $\dfrac{3y^2 - x^2 + 1}{y|z|}$ 24. $\dfrac{2y^2 - xyz}{x^2|yz|}$

Section 1.5

Basic Applications

Mathematical formulas can be used to solve many application problems.

If using π in a formula, be sure to use the π key on your calculator instead of a decimal or fractional approximation.

25. **Packaging**   Four steel bands surround the shipping crate in Illustration 1. Find the total length of strapping needed.

4.2 ft

2.7 ft

1.2 ft

ILLUSTRATION 1

The graphing calculator can be used for more than just computation. It can be used to generate tables to display basic information relating to a specific formula.

Percentages should always be converted to the decimal or fractional equivalent before performing the operations of multiplication or division.

26. Petroleum Storage   Find the volume of the cylindrical storage in Illustration 2. Round the result to the nearest thousandth.

ILLUSTRATION 2

27. Temperature   Use the formula to convert Fahrenheit degrees to Celsius to determine the Celsius temperature when the temperature is 95° F.

28. Purchasing   A blouse that costs $42.95 is on sale for 20% off. If the customer will pay 6% tax on the final sale, how much will the blouse cost?

■ Chapter Test

In Problems 1–8, simplify each expression (use fractions where applicable). Verify your results with the calculator, copying the screen display exactly as it appears.

1. $3(4 - 2) - 2(2 - 6)$

2. $6 + 4 \cdot 3 - 2(-7)$

3. $-13.7 - |-13.7|$

4. $\dfrac{7}{8} \cdot \dfrac{24}{21}$

5. $\dfrac{18}{35} \div \dfrac{9}{14}$

6. $\dfrac{24}{16} + 3$

7. $\dfrac{17 - 5}{36} - \dfrac{2(13 - 5)}{12}$

8. $\dfrac{|-7 - (-6)|}{-7 - |-6|}$

In Problems 9–14, write an arithmetic expression to model the given data. Simplify the expression both analytically (paper and pencil) and by using the calculator, copying the screen display exactly as it appears.

9. Find the area of a rectangle 12.8 feet wide and 23.56 feet long. Round the result to the nearest hundredth.

10. Find the area of the figure in Illustration 3.

ILLUSTRATION 3

11. Find the volume of the solid in Illustration 4. Round results to the nearest thousandth.

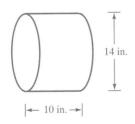

ILLUSTRATION 4

12. A student has $5000 in a certificate of deposit that earns $5\frac{1}{2}\%$ simple interest and $1500 in a savings account that earns 3% simple interest. What is the total amount of interest earned?

13. If a small city has 150,000 homes (or apartment units), according to Illustration 5, how many are heated by natural gas?

Source: Energy Information Administration

ILLUSTRATION 5

14. The ottoman in Illustration 6 is $14\frac{1}{2}$ inches long, $14\frac{1}{2}$ inches wide, and $18\frac{1}{2}$ inches high. What is the volume of stuffing required to fill a single ottoman?

ILLUSTRATION 6

In Problems 15–20, let $x = -2$, $y = 3$, and $z = 4$. Evaluate each expression both analytically and by using the calculator. Be sure to copy screen displays exactly as they appear to justify your results.

15. $xy + z$

16. $x(y + z)$

17. $\dfrac{z + 4y}{2x}$

18. $|x^y - z|$

19. $x^3 + y^2 + z$

20. $|x| - 3|y| - 4|z|$

Computer Systems Analyst   Computer systems analysts help businesses and scientific research organizations develop analytical systems to process and interpret data. Using such techniques as cost accounting, sampling, and mathematical model building, they analyze information and often present results graphically by charts and diagrams. They may also prepare cost–benefit analyses to help management decide whether proposed solutions are satisfactory.

Once the system is accepted, systems analysts adapt its logical requirements to the capabilities of computer machinery. They work closely with programmers to rid the system of errors.

Systems analysts solve a wide range of problems in many different industries. Because the work is so varied and complex, analysts usually specialize in either business or scientific and engineering applications.

Qualifications   Businesses prefer a college degree in accounting, business management, or economics. Scientific organizations prefer a college degree in the physical sciences, mathematics, or engineering. Advanced degrees and degrees in computer science and information science are becoming more important in both employment areas.

Job Outlook   Through the year 2005, the demand for systems analysts is expected to grow more rapidly than the average for all other occupations. As competition increases worldwide, systems analysts will help organizations take advantage of technological advances in automation, telecommunications, and scientific research. Because personal computers are now found on almost every desktop, systems analysts will be needed to design computer networks that facilitate the sharing of data, while providing adequate security for sensitive information.

Example Application   The process of arranging records into a sequential order, called **sorting,** is a common and important task in electronic data processing. In any sorting operation, records must be compared with other records to determine which one should precede the other. One sorting technique, called a **selection sort,** requires C comparisons to sort N records into their proper order, where C and N are related by the following formula:

$$C = \frac{N(N-1)}{2}$$

How many comparisons are required to

a. sort 20 records?

b. sort 10,000 records?

Solution   **a.** Substitute 20 for N in the formula

$$C = \frac{N(N-1)}{2}$$

61

and calculate C:

$$C = \frac{20(20 - 1)}{2}$$

$$C = \frac{20 \cdot 19}{2}$$

$$C = 190$$

Sorting 20 records requires 190 comparisons.

b. Substitute 10,000 for N in the formula

$$C = \frac{N(N - 1)}{2}$$

and calculate C:

$$C = \frac{10,000(10,000 - 1)}{2}$$

$$C = 49,995,000$$

Sorting 10,000 records requires almost 50 million comparisons. A selection sort is not efficient for large lists.

EXERCISES
1. How many comparisons are required to sort 500 records?

2. If the computer time required to sort 20 records were 0.5 second, how long would it take to sort 10,000 records?

3. Another important data-processing task is that of finding a particular entry in a large list. In a **sequential search,** an average of C comparisons is required to find an entry in a list of N items. C and N are related by the formula

$$C = \frac{(N + 1)}{2}$$

How many comparisons are needed (on the average) to search a list of 25 items?

4. How many comparisons are needed (on the average) to search 10,000 items?

Exponents
and Polynomials

2

A Father of Algebra

One important figure in the history of mathematics, François Vieta (1540–1603) was one of the first mathematicians to use notation close to that which we use today. Trained as a lawyer, Vieta served in the parliament of Brittany and as the personal lawyer of Henry of Navarre. If he had continued as a successful lawyer, Vieta might now be forgotten. Fortunately, he lost his job.

When political opposition forced him out of office in 1584, Vieta had time to devote himself entirely to his hobby, mathematics. He used the time to study the writings of earlier mathematicians, and adapted and improved their ideas. Vieta was the first to use letters to represent unknown numbers, but he did not use our modern notation for exponents. To us, his notation seems awkward. For example, what we would write as

$$(x + 1)^3 = x^3 + 3x^2 + 3x + 1$$

Vieta would have written as

$$\overline{x + 1} \text{ cubus aequalis x cubus } + \text{ x quad. } 3 + \text{ x in } 3 + 1$$

In this chapter, we shall develop rules for finding products, powers, and quotients of certain exponential expressions and then discuss how to add, subtract, multiply, and divide polynomials.

2.1 Whole-number Exponents

■ PROPERTIES OF NATURAL-NUMBER EXPONENTS ■ ZERO EXPONENT

We have used natural number exponents to indicate repeated multiplication. For example,

$$2^5 = 2 \cdot 2 \cdot 2 \cdot 2 \cdot 2 = 32 \qquad (-7)^3 = (-7)(-7)(-7) = -343$$
$$x^4 = x \cdot x \cdot x \cdot x \qquad -y^5 = -y \cdot y \cdot y \cdot y \cdot y$$

These examples suggest a definition for x^n, where n is a natural number.

Natural-Number Exponents

If n is a natural number, then

$$x^n = \overbrace{x \cdot x \cdot x \cdot \cdots \cdot x}^{n \text{ factors of } x}$$

In the exponential expression x^n, x is called the **base,** and n is called the **exponent.** The entire expression is called a **power of x.** If an exponent is a natural number, it tells how many times its base is to be used as a factor in a product. An exponent of 1 indicates that its base is to be used as a factor one time, an exponent of 2 indicates that its base is to be used as a factor two times, and so on.

$$3^1 = 3 \qquad (-y)^1 = -y \qquad (-4z)^2 = (-4z)(-4z) \qquad \text{and} \qquad (t^2)^3 = t^2 \cdot t^2 \cdot t^2$$

EXAMPLE 1   Show that -2^4, read "the opposite of 2 to the fourth power," and $(-2)^4$, read "negative 2 quantity to the fourth power," are different numbers.

Solution   We write each expression without exponents and observe different results.

$$-2^4 = -(2^4) \qquad\qquad (-2)^4 = (-2)(-2)(-2)(-2)$$
$$= -(2 \cdot 2 \cdot 2 \cdot 2) \qquad\qquad = 16$$
$$= -16$$

```
-2^4
              -16
(-2)^4
               16
```

Since $-16 \neq 16$, it follows that $-2^4 \neq (-2)^4$.   ■

EXAMPLE 2   Write each expression as repeated multiplication using the definition of natural-number exponent.

a. r^3,   **b.** $(-2s)^4$,   and   **c.** $\left(\dfrac{1}{3}ab\right)^5$.

Solution   **a.** $r^3 = r \cdot r \cdot r$

b. $(-2s)^4 = (-2s)(-2s)(-2s)(-2s)$

c. $\left(\dfrac{1}{3}ab\right)^5 = \left(\dfrac{1}{3}ab\right)\left(\dfrac{1}{3}ab\right)\left(\dfrac{1}{3}ab\right)\left(\dfrac{1}{3}ab\right)\left(\dfrac{1}{3}ab\right)$

■

■ PROPERTIES OF NATURAL-NUMBER EXPONENTS

■ **TECHNOLOGY TIP**
Remember, many graphing calculators cannot perform symbolic operations. If you enter $(x^2)\wedge 3$, the calculator will evaluate the expression for the value that is stored in x.

Because of the definition of natural-number exponents, we can develop a rule for multiplying exponential expressions. To multiply x^3 by x^2, for example, we note that the expression x^2 means that x is to be used as a factor two times and the expression x^3 means that x is to be used as a factor three times. Thus,

$$x^2x^3 = \overbrace{x \cdot x}^{2 \text{ factors of } x} \cdot \overbrace{x \cdot x \cdot x}^{3 \text{ factors of } x}$$

$$= \overbrace{x \cdot x \cdot x \cdot x \cdot x}^{5 \text{ factors of } x}$$

$$= x^5$$

In general,

$$x^m \cdot x^n = \overbrace{x \cdot x \cdot x \cdot \ldots \cdot x}^{m \text{ factors of } x} \cdot \overbrace{x \cdot x \cdot x \cdot \ldots \cdot x}^{n \text{ factors of } x}$$

$$= \overbrace{x \cdot x \cdot x \cdot x \cdot x \cdot x \cdot \ldots \cdot x \cdot x \cdot x}^{m + n \text{ factors of } x}$$

$$= x^{m+n}$$

This discussion justifies the rule for multiplying exponential expressions: *To multiply two exponential expressions with the same base, we keep the same base and add the exponents.*

Product Rule for Exponents
If m and n are natural numbers, then
$$x^m x^n = x^{m+n}$$

◎◎   a, c, e

EXAMPLE 3

a. $x^3 x^4 = x^{3+4}$
    $= x^7$

b. $y^2 y^4 = y^{2+4}$
    $= y^6$

c. $zz^3 = z^1 z^3$
    $= z^{1+3}$
    $= z^4$

d. $x^2 x^3 x^6 = (x^2 x^3)x^6$
    $= (x^{2+3})x^6$
    $= x^5 x^6$
    $= x^{5+6}$
    $= x^{11}$

e. $(2y^3)(3y^2) = 2(3)y^3 y^2$
    $= 6y^{3+2}$
    $= 6y^5$

f. $(4x)(-3x^2) = 4(-3)xx^2$
    $= -12x^{1+2}$
    $= -12x^3$

■

■ **STUDY TIP**
Make note cards (4 × 6 are the most convenient) for 10 problems from the section. On the front of the card, write the directions and a problem. On the back of the card, provide a complete solution to the problem. Finally, put the section number and problem number in the upper right corner of the card. Go over cards periodically during the day—while waiting for your pizza to be delivered or the microwave to buzz.

> **WARNING!** The product rule for exponents applies only to exponential expressions with the same base. An expression such as $x^2 y^3$ cannot be simplified, because x^2 and y^3 have different bases.

To find another rule of exponents, we consider the expression $(x^3)^4$, which can be written as $x^3 \cdot x^3 \cdot x^3 \cdot x^3$. Because each of the four factors of x^3 contains three factors of x, there are $4 \cdot 3$, or 12, factors of x. Thus, the product can be written as x^{12}.

$$(x^3)^4 = x^3 \cdot x^3 \cdot x^3 \cdot x^3$$

$$= \overbrace{\underbrace{x \cdot x \cdot x}_{x^3} \cdot \underbrace{x \cdot x \cdot x}_{x^3} \cdot \underbrace{x \cdot x \cdot x}_{x^3} \cdot \underbrace{x \cdot x \cdot x}_{x^3}}^{12 \text{ factors of } x}$$

$$= x^{12}$$

In general,

$$(x^m)^n = \overbrace{x^m \cdot x^m \cdot x^m \cdot \ldots \cdot x^m}^{n \text{ factors of } x^m}$$

$$= \overbrace{x \cdot x \cdot x \cdot x \cdot x \cdot x \cdot x \cdot \ldots \cdot x}^{mn \text{ factors of } x}$$

$$= x^{mn}$$

This discussion justifies a rule for raising an exponential expression to a power: *To raise an exponential expression to a power, we keep the same base and multiply the exponents.*

Power to a Power Rule

If m and n are natural numbers, then

$$(x^m)^n = x^{mn}$$

EXPLORING THE CONCEPT   Use your graphing calculator to explore the relationship between each pair of numbers.

a. $(2^4)^5$; $(2^5)^4$

b. $(4^2)^4$; $(4^4)^2$

c. Which property justifies the results of each pair of exponential expressions?

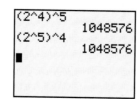

```
(2^4)^5
          1048576
(2^5)^4
          1048576
■
```

```
(4²)^4
          65536
(4^4)²
          65536
```

Since $(2^4)^5 = 2^{4 \cdot 5}$ and $(2^5)^4 = 2^{5 \cdot 4}$, it is the commutative property of multiplication that justifies the results. Thus, $(x^m)^n = (x^n)^m$. ■

EXAMPLE 4   **a.** $(2^3)^7 = 2^{3 \cdot 7}$ **b.** $(y^5)^2 = y^{5 \cdot 2}$
$= 2^{21}$ $= y^{10}$

c. $(z^7)^7 = z^{7 \cdot 7}$ **d.** $(u^x)^y = u^{x \cdot y}$
$= z^{49}$ $= u^{xy}$

In Example 5, both the product and power rules of exponents are applied.

EXAMPLE 5   **a.** $(x^2 x^5)^2 = (x^7)^2$ **b.** $(y y^6 y^2)^3 = (y^9)^3$
$= x^{14}$ $= y^{27}$

c. $(z^2)^4 (z^3)^3 = z^8 z^9$ **d.** $(x^3)^2 (x^5 x^2)^3 = x^6 (x^7)^3$
$= z^{17}$ $= x^6 x^{21}$
$= x^{27}$

To find two more power rules for exponents, we consider the expressions $(2x)^3$ and $\left(\frac{2}{x}\right)^3$.

$$(2x)^3 = (2x)(2x)(2x)$$
$$= 2 \cdot x \cdot 2 \cdot x \cdot 2 \cdot x$$
$$= 2 \cdot 2 \cdot 2 \cdot x \cdot x \cdot x \qquad \text{Commutative property of multiplication.}$$
$$= 2^3 \cdot x^3$$
$$= 8x^3$$

$$\left(\frac{2}{x}\right)^3 = \left(\frac{2}{x}\right)\left(\frac{2}{x}\right)\left(\frac{2}{x}\right) \qquad (x \neq 0)$$
$$= \frac{2 \cdot 2 \cdot 2}{x \cdot x \cdot x}$$
$$= \frac{2^3}{x^3}$$
$$= \frac{8}{x^3}$$

These examples suggest that *to raise a product to a power, we raise each factor of the product to that power,* and *to raise a fraction to a power, we raise both the numerator and denominator to that power.*

Product to a Power and Quotient to a Power
If n is a natural number, then
$$(xy)^n = x^n y^n \qquad \text{and if } y \neq 0, \text{ then} \qquad \left(\frac{x}{y}\right)^n = \frac{x^n}{y^n}$$

EXAMPLE 6   **a.** $(ab)^4 = a^4 b^4$ **b.** $(3c)^3 = 3^3 c^3$
$= 27c^3$

c. $(x^2 y^3)^5 = (x^2)^5 (y^3)^5$ **d.** $(-2x^3 y)^2 = (-2)^2 (x^3)^2 y^2$
$= x^{10} y^{15}$ $= 4x^6 y^2$

e. $\left(\frac{4}{k}\right)^3 = \frac{4^3}{k^3}$ **f.** $\left(\frac{3x^2}{2y^3}\right)^5 = \frac{3^5 (x^2)^5}{2^5 (y^3)^5}$
$= \frac{64}{k^3}$ $= \frac{243x^{10}}{32y^{15}}$

To find a rule for dividing exponential expressions, we consider the fraction

$$\frac{4^5}{4^2}$$

where the exponent in the numerator is greater than the exponent in the denominator. We can simplify the fraction as follows:

$$\frac{4^5}{4^2} = \frac{4 \cdot 4 \cdot 4 \cdot 4 \cdot 4}{4 \cdot 4}$$

$$= \frac{\overset{1}{\cancel{4}} \cdot \overset{1}{\cancel{4}} \cdot 4 \cdot 4 \cdot 4}{\underset{1}{\cancel{4}} \cdot \underset{1}{\cancel{4}}} \qquad \frac{4}{4} = 1.$$

$$= 4^3$$

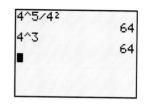

The result of 4^3 has a base of 4 and an exponent of $5 - 2$, or 3. This suggests that *to divide exponential expressions with the same base, we keep the base and subtract the exponents.*

Quotient Rule for Exponents

If m and n are natural numbers, $m > n$ and $x \neq 0$, then

$$\frac{x^m}{x^n} = x^{m-n}$$

EXAMPLE 7   If there are no divisions by 0, then

a. $\dfrac{x^4}{x^3} = x^{4-3}$

$\qquad\quad = x^1$

$\qquad\quad = x$

b. $\dfrac{8y^2 y^6}{4y^3} = \dfrac{8y^8}{4y^3}$

$\qquad\qquad = \dfrac{8}{4} y^{8-3}$

$\qquad\qquad = 2y^5$

c. $\dfrac{a^3 a^5 a^7}{a^4 a} = \dfrac{a^{15}}{a^5}$

$\qquad\qquad = a^{15-5}$

$\qquad\qquad = a^{10}$

d. $\dfrac{(a^3 b^4)^2}{ab^5} = \dfrac{a^6 b^8}{ab^5}$

$\qquad\qquad = a^{6-1} b^{8-5}$

$\qquad\qquad = a^5 b^3$

■ ZERO EXPONENT

If we apply the quotient rule to the fraction $\frac{5^3}{5^3}$, where the exponents in numerator and denominator are equal, we obtain

$$\frac{5^3}{5^3} = 5^{3-3} = 5^0$$

However, because any nonzero number divided by itself is equal to 1, we have

$$\frac{5^3}{5^3} = 1$$

To make the results of 5^0 and 1 consistent, we shall define 5^0 to be equal to 1. In general, we have

> ### Zero Exponent
> If x is any nonzero real number, then
> $$x^0 = 1$$
> Since $x \neq 0$, 0^0 is undefined.

EXAMPLE 8

a. $\left(\dfrac{1}{13}\right)^0 = 1$ **b.** $(-0.115)^0 = 1$

```
a. (1/13)^0
                    1
b. (-0.115)^0
                    1
c. 4²/4²
                    1
```

c. $\dfrac{4^2}{4^2} = 4^{2-2}$ **d.** $\dfrac{x^5}{x^5} = x^{5-5}$ $(x \neq 0)$

   $\quad = 4^0$ $\quad = x^0$

   $\quad = 1$ $\quad = 1$

e. $3x^0 = 3(1)$ **f.** $(3x)^0 = 1$

   $\quad = 3$

g. $\dfrac{6^n}{6^n} = 6^{n-n}$ **h.** $\dfrac{y^m}{y^m} = y^{m-m}$ $(y \neq 0)$

   $\quad = 6^0$ $\quad = y^0$

   $\quad = 1$ $\quad = 1$

It is clear from parts **e** and **f** that $3x^0 \neq (3x)^0$.

The rules for natural number and zero exponents are summarized as follows:

> ### Properties of Exponents
> If n is a natural number, then
> $$x^n = \overbrace{x \cdot x \cdot x \cdot \ldots \cdot x}^{n \text{ factors of } x}$$
> If m and n are natural numbers and there are no divisions by 0, then
> $$x^m x^n = x^{m+n} \qquad (x^m)^n = x^{mn} \qquad (xy)^n = x^n y^n \qquad \left(\frac{x}{y}\right)^n = \frac{x^n}{y^n}$$
> $$\frac{x^m}{x^n} = x^{m-n} \quad \text{provided } m > n \qquad x^0 = 1 \text{ provided } x \neq 0.$$

STUDY TIP

How do you cope with getting stuck? Whether you are at the beginning or middle of an exercise, getting stuck can be a frustrating experience. However, do not let this defeat you. Answer the following questions:

1. Have you thoroughly reviewed both the text and the notes before attempting the problem?

2. Have you spent adequate time attempting the problem or did you just "throw in the towel" when you could not figure it out at first glance? This is NOT getting stuck—it is giving up before you get started!

If, after carefully analyzing a similar problem, you are still unable to complete the problem, move on to something else. Come back later and tackle the problem again. If you are still unsuccessful, seek the help of your instructor or a tutor—preferably BEFORE the class meets again.

EXERCISE 2.1

VOCABULARY   *In Exercises 1–2, fill in the blank to make a true statement.*

1. In the exponential expression x^n, x is called the _____ and n is called the _____.

2. If m and n are natural numbers, then

a. $x^m x^n = $ _____   **b.** $(x^m)^n = $ ____

c. $(xy)^n = $ ____   **d.** $\left(\dfrac{x}{y}\right)^n = $ ____ $(y \neq 0)$

e. $\dfrac{x^m}{x^n} = $ _____, provided $m > n$ and $x \neq 0$

f. $x^0 = $ __, provided $x \neq 0$

CONCEPTS *In Exercises 3–10, identify the base and the exponent in each expression.*

3. 4^3

4. $(-5)^2$

5. $(-x)^2$

6. $-x^4$

7. x

8. (xy)

9. $2x^3$

10. $-3y^6$

In Exercises 11–16, write each expression using exponents.

11. $2 \cdot 2 \cdot 2$

12. $5 \cdot 5$

13. $x \cdot x \cdot x \cdot x \cdot x$

14. $y \cdot y \cdot y \cdot y \cdot y \cdot y$

15. $(2x)(2x)(2x)$

16. $(-4y)(-4y)$

PRACTICE *In Exercises 17–20, let $x = -2$ and $y = -3$. Evaluate each expression both analytically and with the calculator, copying screen displays exactly as they appear.*

17. $3x^3$

18. $-4x^5$

19. $(-2y)^4$

20. $(3y)^5$

In Exercises 21–30, write each expression as an expression involving only one exponent.

21. x^4x^3

22. y^5y^2

23. x^5x^5

24. yy^3

25. $y^3(y^2y^4)$

26. $(y^4y)y^6$

27. $4x^2(3x^5)$

28. $-2y(y^3)$

29. $(-y^2)(4y^3)$

30. $(-4x^3)(-5x)$

In Exercises 31–40, write each expression as an expression involving only one exponent.

31. $(3^2)^4$

32. $(4^3)^3$

33. $(y^5)^3$

34. $(b^3)^6$

35. $(a^3)^7$

36. $(b^2)^3$

37. $(x^2x^3)^5$

38. $(y^3y^4)^4$

39. $(2x^5)^2(x^7)^3$

40. $(2y^3y)^2(y^2)^2$

In Exercises 41–56, write each expression without using parentheses.

41. $(xy)^3$

42. $(uv^2)^4$

43. $(r^3s^2)^2$

44. $(a^3b^2)^3$

45. $(4ab^2)^2$

46. $(3x^2y)^3$

47. $(-2r^3s^3t)^3$

48. $(-3x^2y^4z)^2$

49. $\left(\dfrac{a}{b}\right)^3$

50. $\left(\dfrac{r^2}{s}\right)^4$

51. $\left(\dfrac{x^2}{y^3}\right)^5$

52. $\left(\dfrac{u^4}{v^2}\right)^6$

53. $\left(\dfrac{-2a}{b}\right)^5$

54. $\left(\dfrac{2t}{3}\right)^4$

55. $\left(\dfrac{b^2}{3a}\right)^3$

56. $\left(\dfrac{a^3b}{c^4}\right)^5$

In Exercises 57–66, simplify each expression.

57. $\dfrac{x^5}{x^3}$

58. $\dfrac{a^6}{a^3}$

59. $\dfrac{x^3y^4}{yy^2}$

60. $\dfrac{b^4b^5}{b^2b^3}$

61. $\dfrac{(ab^2)^3}{(ab)^2}$

62. $\dfrac{(m^3n^4)^3}{(mn^2)^3}$

63. $\dfrac{20(r^4s^3)^4}{6(rs^3)^3}$

64. $\dfrac{35(r^3s^2)^2}{49r^2s^4}$

65. $\left(\dfrac{y^3y}{2yy^2}\right)^3$

66. $\left(\dfrac{3t^3t^4t^5}{4t^2t^6}\right)^3$

In Exercises 67–70, simplify each expression.

67. $2x^0$

68. $(2x)^0$

69. $\dfrac{x^0 - 5x^0}{2x^0}$

70. $\dfrac{4a^0 + 2a^0}{3a^0}$

✎**71.** Is the operation of raising to a power commutative? That is, is $a^b = b^a$? Explain. Use your calculator to explore this idea, letting a and b assume various positive integer values.

✎**72.** Is the operation of raising to a power associative? That is, is $(a^b)^c = a^{(bc)}$? Explain. Use your calculator to explore this idea, letting a, b, and c assume various positive integer values.

APPLICATIONS *In Exercises 73–76, write an expression involving only one exponent to represent the area of the figure.*

73.

74.

75.

76.

REVIEW

77. Mental arithmetic: Perform the following computations without using pencil and paper or calculator.

   a. $-4 + -5$ **b.** $-5(-6)$

   c. $-8 - (-5)$ **d.** $12 - 15$

   e. $\frac{1}{2}(-6)$ **f.** $-\frac{1}{4}(-12)$

78. For the set of numbers $\left\{-3, -2, -1.5, -\frac{1}{4}, 0, \sqrt{3}, \sqrt{4}, 5, 7.2\right\}$, identify the following groups of numbers:

   a. whole numbers

   b. irrational numbers

   c. rational numbers

   d. negative integers

   e. positive integers

   f. natural numbers

79. Convert each percentage to a decimal number.

   a. 45%   **b.** 3%   **c.** 2.5%

80. Use your calculator to convert each of the following fractions to a decimal and round the result to the nearest thousandth:

   a. $\frac{17}{19}$ **b.** $\frac{2}{3}$ **c.** $\frac{5}{7}$

2.2 Algebraic Expressions and Polynomials

■ TRANSLATING EXPRESSIONS ■ ALGEBRAIC TERMS ■ POLYNOMIALS ■ *P(x)* NOTATION

■ TRANSLATING EXPRESSIONS

Variables and numbers can be combined with the operations of arithmetic to produce **algebraic expressions.** For example, if x and y are variables, the algebraic expression $x + y$ represents the **sum** of x and y, and the algebraic expression $x - y$ represents their **difference.**

There are many ways to read the **sum** $x + y$. Some of them are:

• the **sum of** x and y

• x **increased by** y

• x **plus** y

• y **more than** x

• y **added to** x

There are many ways to read the **difference** $x - y$. Some of them are:

• the result obtained when y is **subtracted from** x

• the result of **subtracting** y **from** x

• x **less** y

• y **less than** x

• x **decreased by** y

• x **minus** y

EXAMPLE 1   Let x represent a certain number. Write an expression that represents   **a.** the number that is 5 more than x and   **b.** the number 12 decreased by x.

Solution   **a.** The number "5 more than x" is the number found by adding 5 to x. It is represented by $x + 5$.

b. The number "12 decreased by x" is the number found by subtracting x from 12. It is represented by $12 - x$.   ∎

EXAMPLE 2   **Income taxes**   Luciano worked x hours preparing his income taxes. He worked 3 hours less than that preparing his son's return. Write an expression that represents **a.** the number of hours he spent preparing his son's return   and   **b.** the total number of hours he worked.

Solution   **a.** Because Luciano worked x hours on his return and 3 hours less on his son's return, he worked $(x - 3)$ hours on his son's return.

b. Because he worked x hours on his return and $(x - 3)$ hours on his son's return, the total time he spent on taxes was $[x + (x - 3)]$ hours.   ∎

There are several ways to indicate the **product** xy in words. Some of them are:

- x **multiplied by** y
- the **product of** x and y
- x **times** y

EXAMPLE 3   Let x represent a certain number. Denote a number that is   **a.** twice as large as x, **b.** 5 more than 3 times x,   and   **c.** 4 less than $\frac{1}{2}$ of x.

Solution   **a.** The number "twice as large as x" is found by multiplying x by 2. It is represented by $2x$.

b. The number "5 more than 3 times x" is found by adding 5 to the product of 3 and x. It is represented by $3x + 5$.

c. The number "4 less than $\frac{1}{2}$ of x" is found by subtracting 4 from the product of $\frac{1}{2}$ and x. It is represented by $\frac{1}{2}x - 4$.   ∎

EXAMPLE 4   **Stock valuation**   Jim owns x shares of Transitronic stock, valued at $29 a share, y shares of Positone stock, valued at $32 a share, and 300 shares of Baby Bell, valued at $42 a share.   **a.** How many shares of stock does he own?   **b.** What is the value of his stock?

Solution   **a.** Because there are x shares of Transitronic, y shares of Positone, and 300 shares of Baby Bell, his total number of shares is $x + y + 300$.

b. The value of x shares of Transitronic is $\$29x$, the value of y shares of Positone is $\$32y$, and the value of 300 shares of Baby Bell is $\$42(300)$. The total value of the stock is $\$(29x + 32y + 12{,}600)$.   ∎

If x and y represent two numbers and $y \neq 0$, the **quotient** obtained when x is divided by y is denoted by each of the following expressions:

$$x \div y, \qquad x/y, \qquad \text{and} \qquad \frac{x}{y}$$

EXAMPLE 5   Let x and y represent two numbers. Write an algebraic expression that represents the sum obtained when 3 times the first number is added to the quotient obtained when the second number is divided by 6.

Solution   Three times the first number, x, is denoted as $3x$. The quotient obtained when the second number, y, is divided by 6 is the fraction $\frac{y}{6}$. Their sum is expressed as $3x + \frac{y}{6}$.

EXAMPLE 6   **Cutting a rope**   A 5-foot section is cut from the end of a rope that is l feet long. The remaining rope is then divided into three equal parts. Find the length of each of the equal pieces.

Solution   After a 5-foot section is cut from one end of l feet of rope, the rope that remains is $(l - 5)$ feet long. When that remaining rope is cut into 3 equal pieces, each piece will be $\frac{l-5}{3}$ feet long. See Figure 2-1.

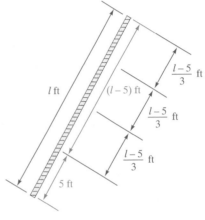

FIGURE 2-1

■ ALGEBRAIC TERMS

Numbers without variables, such as 7, 21, and 23, are called **constants.** Expressions such as 37, xyz, or $32t$, which are constants, variables, or products of constants and variables, are called **algebraic terms.** Because the expression $3x + 5y$ denotes the sum of the two algebraic terms $3x$ and $5y$, the expression $3x + 5y$ contains two terms. Likewise, the expression $xy - 7$ contains two terms. The expression $3 + x + 2y$ contains three terms. Its first term is 3, its second term is x, and its third term is $2y$.

Numbers that are part of an indicated product are called **factors** of that product. For example, the product $7x$ has two factors, 7 and x. Either factor in this product is called the **coefficient** of the other factor. When we speak of the coefficient in a term such as $7x$, however, we generally mean the **numerical coefficient,** which in this case is 7. For example, the numerical coefficient of the term $12xyz$ is 12. The coefficient of such terms as x, ab, or rst is understood to be 1. Thus,

$$x = 1x, \qquad ab = 1ab, \qquad \text{and} \qquad rst = 1rst$$

EXAMPLE 7   **a.** The expression $5x + y$ has two terms. The numerical coefficient of its first term is 5. The numerical coefficient of its second term is 1.

b. The expression $-17wxyz$ has one term, which contains the five factors -17, w, x, y, and z. Its numerical coefficient is -17.

c. The expression 37 has one term, the constant 37. Its numerical coefficient is 37.

■ POLYNOMIALS

An algebraic expression that is the sum of one or more terms containing whole-number exponents on its variables is called a **polynomial.** The expressions

$$8xy^2t \qquad 3x + 2 \qquad 4y^2 - 2y + 3 \qquad \text{and} \qquad 3a - 4b - 4c + 8d$$

are examples of polynomials.

> **WARNING!** An expression such as $2x^3 - 3y^{-2}$ is not a polynomial, because the second term contains a negative exponent on a variable base.

A polynomial with exactly one term is called a **monomial.** A polynomial with exactly two terms is called a **binomial.** A polynomial with exactly three terms is called a **trinomial.** Here are some examples of each:

Monomials	Binomials	Trinomials
$5x^2y$	$3u^3 - 4u^2$	$-5t^2 + 4t + 3$
$-6x$	$18a^2b + 4ab$	$27x^3 - 6x - 2$
29	$-29z^{17} - 1$	$-32r^6 + 7y^3 - z$

The monomial $7x^6$ is called a **monomial of sixth degree** or a **monomial of degree 6** because the variable x occurs as a factor six times. The monomial $3x^3y^4$ is a monomial of the seventh degree because the variables x and y occur as factors a total of seven times. Other examples are:

$-2x^3$ is a monomial of degree 3

$47x^2y^3$ is a monomial of degree 5

$18x^4y^2z^8$ is a monomial of degree 14

8 is a monomial of degree 0 because $8 = 8x^0$

These examples illustrate the following definition.

> **Degree of a Monomial**
> If a is a nonzero constant, the **degree of the monomial** ax^n is n. The degree of a monomial with several variables is the sum of the exponents of those variables.

We define the degree of a polynomial by considering the degree of each of its terms.

> **Degree of a Polynomial**
> The **degree of a polynomial** is the same as the degree of its term with the highest degree.

For example,

- $x^2 + 2x$ is a binomial of degree 2 because the degree of its first term is 2 and the degree of its other term is less than 2.
- $3x^3y^2 + 4x^4y^4 - 3x^3$ is a trinomial of degree 8 because the degree of its second term is 8 and the degree of each of its other terms is less than 8.
- $25x^4y^3z^7 - 15xy^8z^{10} - 32x^8y^8z^3 + 4$ is a polynomial of degree 19 because its second and third terms are of degree 19. Its other terms have degrees less than 19.

■ *P(x)* NOTATION

Polynomials that contain a single variable are often denoted by symbols such as

$P(x)$ Read as "*P* of *x*."

$Q(t)$ Read as "*Q* of *t*."

$R(z)$ Read as "*R* of *z*."

where the letter within the parentheses represents the variable of the polynomial.

WARNING! The symbol $P(x)$ does not indicate the product of *P* and *x*. Instead, it represents a polynomial with the variable *x*.

■ **STUDY TIP**
Polynomials are typically labeled as $P(x)$, $Q(x)$, or $R(x)$. However, any letter of the alphabet may be used to designate the polynomial. Likewise, we are not restricted to the variable *x*. For example, $M(a)$ would designate the polynomial *M* which is in terms of the variable *a*.

The symbols $P(x)$, $Q(t)$, and $R(z)$ could represent the polynomials

$$P(x) = 3x + 4, \qquad Q(t) = 3t^2 + 4t - 5, \qquad \text{or} \qquad R(z) = -z^3 - 2z + 3$$

The symbol $P(x)$ is convenient, because it provides a way to indicate the value of a polynomial in *x* at different values of *x*. If $P(x) = 3x + 4$, for example, then $P(1)$ represents the value of the polynomial $P(x) = 3x + 4$ when $x = 1$.

$$P(x) = 3x + 4$$
$$P(1) = 3(1) + 4$$
$$= 7$$

Likewise, if $Q(t) = 3t^2 + 4t - 5$, then $Q(-2)$ represents the value of the polynomial $Q(t) = 3t^2 + 4t - 5$ when $t = -2$.

$$Q(t) = 3t^2 + 4t - 5$$
$$Q(-2) = 3(-2)^2 + 4(-2) - 5$$
$$= 3(4) - 8 - 5$$
$$= 12 - 8 - 5$$
$$= -1$$

```
-2→T:3T²+4T-5
                -1
■
```

EXAMPLE 8   Consider the polynomial $P(z)$ where $P(z) = 3z^2 + 2$. Find   **a.** $P(0)$,   **b.** $P(2)$,   **c.** $P(-3)$,   and   **d.** $P(s)$.

Solution   **a.** $P(z) = 3z^2 + 2$ **b.** $P(z) = 3z^2 + 2$

$\qquad\qquad P(0) = 3(0)^2 + 2 \qquad\qquad\qquad P(2) = 3(2)^2 + 2$

$\qquad\qquad\qquad\quad = 2 \qquad\qquad\qquad\qquad\qquad\quad = 3(4) + 2$

$\qquad\qquad\qquad\qquad\qquad\qquad\qquad\qquad\qquad\qquad = 12 + 2$

$\qquad\qquad\qquad\qquad\qquad\qquad\qquad\qquad\qquad\qquad = 14$

```
0→Z:3Z²+2
              2
2→Z:3Z²+2
              14
-3→Z:3Z²+2
              29
■
```

   c. $P(z) = 3z^2 + 2$ **d.** $P(z) = 3z^2 + 2$

$\qquad\qquad P(-3) = 3(-3)^2 + 2 \qquad\qquad\quad P(s) = 3s^2 + 2$

$\qquad\qquad\qquad\quad = 3(9) + 2$

$\qquad\qquad\qquad\quad = 27 + 2$

$\qquad\qquad\qquad\quad = 29$

EXAMPLE 9   **Trajectory**   The polynomial $h(t) = -16t^2 + 28t + 8$ gives the height (in feet) of an object *t* seconds after it has been thrown upward. Find the height of the object at   **a.** 1 second   and   **b.** 2 seconds.

Solution **a.** To find the height at 1 second, we find $h(1)$.

$$h(t) = -16t^2 + 28t + 8$$
$$h(1) = -16(1)^2 + 28(1) + 8$$
$$= -16 + 28 + 8$$
$$= 20$$

At 1 second, the object is 20 feet above the ground.

b. To find the height at 2 seconds, we find $h(2)$.

$$h(t) = -16t^2 + 28t + 8$$
$$h(2) = -16(2)^2 + 28(2) + 8$$
$$= -16(4) + 28(2) + 8$$
$$= -64 + 56 + 8$$
$$= 0$$

At 2 seconds, the height is 0 feet. That is when the object strikes the ground.

EXAMPLE 10 Consider $P(x)$ where $P(x) = x^3 + 1$. Find **a.** $P(2t)$, **b.** $P(-3y)$, **c.** $P(s^4)$, and **d.** $P(x) + P(a)$.

Solution **a.** $P(x) = x^3 + 1$
$$P(2t) = (2t)^3 + 1$$
$$= 8t^3 + 1$$

b. $P(x) = x^3 + 1$
$$P(-3y) = (-3y)^3 + 1$$
$$= -27y^3 + 1$$

c. $P(x) = x^3 + 1$
$$P(s^4) = (s^4)^3 + 1$$
$$= s^{12} + 1$$

d. $P(x) + P(a) = x^3 + 1 + a^3 + 1$
$$= x^3 + a^3 + 2$$

EXERCISE 2.2

VOCABULARY *In Exercises 1–7 fill in the blank to make a true statement.*

1. Variables and numbers combined with arithmetic operations produce _____.

2. Numbers without variables are _____ because they never change in value.

3. Expressions that are composed of factors of numbers and variables are called _____.

4. In the term $4x$, 4 is called the _____ of the term.

5. An algebraic expression that is the sum of one or more terms containing whole-number exponents on the variables is called a _____.

6. A polynomial consisting of one term is called a _____, of two terms a _____, and of three terms a _____.

7. The _____ of a polynomial is the same as the _____ of its term with the highest _____.

CONCEPTS

8. Identify each of the following as a sum, product, quotient, or difference.
   a. xy
   b. $x - y$
   c. $x + y$
   d. $\dfrac{x}{y}$

9. Which of the following pairs of algebraic expressions is *not* equivalent?
   a. $\dfrac{1}{2}x, \dfrac{x}{2}$
   b. $3x, x^3$
   c. $x + 2, 2 + x$
   d. $x - 2, 2 - x$

10. If a board is 12 ft long and you cut off x ft, what algebraic expression represents the remaining length of board?

11. Using the polynomial $3x^2 - 5x + 6$,
    a. determine if the polynomial is a monomial, binomial, or trinomial.

b. determine the degree of the polynomial.
c. determine the numerical coefficient of the second term.

12. State the number of terms in each algebraic expression and also state the numerical coefficient of the second term.
   a. $-4c + 3d$
   b. $3ab + bc - cd - ef$
   c. $-4xyz + 7xy - z$
   d. $5uvw - 4uv + 8uw$

13. If $P(x) = 3x + 5$, without performing any computation circle the line in the TABLE that allows you to determine $P(-2)$.

X	P(X)
-3	-4
-2.5	-2.5
-2	-1
-1.5	.5
-1	2
-.5	3.5
0	5

X= -3

PRACTICE *In Exercises 14–21, let x, y, and z represent three real numbers. Write an algebraic expression to denote each quantity.*

14. The sum of x and y.
15. The product of x and y.
16. The product of x and twice y.
17. The sum of twice x and twice y.
18. The difference obtained when x is subtracted from y.
19. The quotient obtained when y is divided by x.
20. The quotient obtained when the sum of x and y is divided by z.
21. y decreased by x.

In Exercises 22–27, write an algebraic expression in an appropriate English phrase.

22. $x + 3$
23. $y - 2$
24. $\frac{x}{y}$
25. xz
26. $\frac{5}{x + y}$
27. $\frac{3x}{y + z}$

In Exercises 28–35, classify each polynomial as a monomial, a binomial, or a trinomial, if possible.

28. $3x + 7$
29. $3y - 5$
30. $3y^2 + 4y + 3$
31. $3xy$
32. $3z^2$
33. $3x^4 - 2x^3 + 3x - 1$
34. $5t - 32$
35. $9x^2y^3z^4$

In Exercises 36–47, state the degree of each polynomial.

36. $3x^4$
37. $3x^5 - 4x^2$
38. $-2x^2 + 3x^3$
39. $-5x^5 + 3x^2 - 3x$
40. $3x^2y^3 + 5x^3y^5$
41. $-2x^2y^3 + 4x^3y^2z$
42. $-5r^2s^2t - 3r^3st^2$
43. $4r^2s^3t^3 - 5r^2s^8 + 3$
44. $x^{12} + 3x^2y^3z^4$
45. 17^2x
46. 38
47. -25

In Exercises 48–55, let $P(x) = 5x - 3$. Find each value analytically (by hand), and verify with your calculator.

48. $P(2)$
49. $P(0)$
50. $P(-1)$
51. $P(-2)$
52. $P(w)$
53. $P(t)$
54. $P(-y)$
55. $P(2t)$

In Exercises 56–67, let $P(x) = 5x - 2$. Find each value analytically (by hand), and verify with your calculator.

56. $P\left(\frac{1}{5}\right)$
57. $P\left(\frac{1}{10}\right)$
58. $P(u^2)$
59. $P(-v^4)$
60. $P(-4z^6)$
61. $P(10x^7)$
62. $P(x + h)$
63. $P(x - h)$
64. $P(x) + P(h)$
65. $P(x) - P(h)$
66. $P(2y + z)$
67. $P(-3r + 2s)$

APPLICATIONS

68. **Course load** A man enrolls in college for c hours of credit, and his sister enrolls for 4 more hours than her brother. Write an expression that represents the number of hours the sister is taking. Could the sister be taking 3 hours of courses? Why or why not?

69. **Mileage** An antique Ford has 25,000 more miles on its odometer than a new car. If the new car has traveled m miles, find an expression that represents the mileage on the Ford.

70. **T-bills** Write an expression that represents the value of t T-bills, each worth $9987.

71. Real estate   Write an expression that represents the value of *a* vacant lots if each lot is worth $35,000.

72. Cutting rope   A rope *x* ft long is cut into 5 equal pieces. Write an expression to represent the length of each piece.

73. Plumbing   A plumber cuts a pipe that is 12 ft long into *x* equal pieces. Find the length of each piece.

74. Comparing assets   A girl had *d* dollars, and her brother had $5 less than three times that amount. Write an expression to represent the brother's amount of money.

What is the smallest dollar amount (integer value) that the girl could have?
What is the smallest dollar amount the brother could have?

75. Comparing investments   Wendy has *x* shares of stock. Her sister has 8 fewer shares than twice Wendy's shares. Write an expression to represent the sister's stock shares.
Is it possible for Wendy to own only 3 shares of stock? Why or why not?

In Exercises 76–79, assume that the height in feet of an object t seconds after it is thrown is given by $h(t) = -16t^2 + 120t + 16$. *Find each value.*

76. The height from which the object was thrown: find $h(0)$.

77. The height at 2 seconds.

78. The height at 6 seconds.

79. The height at 7 seconds.

REVIEW

80. Complete each table.

a.

x	0	1	2	3
2^x				

b.

x	0	1	2	3
3^x				

81. Based on each table, determine the missing base value.

a.

x	0	1	2	3
$\square^x$	1	9	81	243

b.

x	0	1	2	3
$\square^x$	−1	−16	−256	−4096

82. Simplify. Verify your results with your calculator.
a. $25 + 2[18 - 4(5 - 2)]$
b. $(6 - 4)^2 + (8 + -6)^3$
c. $-2(3 - 4) - 5(2^2 - 3^2)$

2.3 Adding and Subtracting Polynomials

■ DISTRIBUTIVE PROPERTY ■ ADDING AND SUBTRACTING LIKE TERMS ■ ADDING AND SUBTRACTING MULTIPLES OF POLYNOMIALS

■ DISTRIBUTIVE PROPERTY

In order to perform the polynomial operations of addition and subtraction, we will need the distributive property. The **distributive property of multiplication over addition** (often simply referred to as the **distributive property**) allows us to multiply the sum of two numbers by a third number. Because of this property, we can often add first and then multiply, or multiply first and then add.

For example, $2(3 + 7)$ can be calculated in two different ways. We can add and then multiply, or we can multiply each number within the parentheses by 2 and then add.

$$2(3 + 7) = 2(10) \quad \text{and} \quad 2(3 + 7) = 2 \cdot 3 + 2 \cdot 7$$
$$= 20 \qquad\qquad = 6 + 14$$
$$= 20$$

Either way, the result is 20. In general, we have the following property:

> **Distributive Property of Multiplication over Addition**
> If a, b, and c are real numbers, then $a(b + c) = ab + ac$.

Because multiplication is commutative, the distributive property can be written in the form

$$(b + c)a = ba + ca$$

EXAMPLE 1   Use the distributive property to write $3(x + 2)$ without using parentheses.

Solution $$3(x + 2) = 3x + 3 \cdot 2$$
$$= 3x + 6$$

The distributive property can be extended to three or more terms. For example, if a, b, c, and d are real numbers, then

$$a(b + c + d) = ab + ac + ad$$

■ ADDING AND SUBTRACTING LIKE TERMS

Terms with identical variable factors are called **like terms.** For example

Like terms	*Unlike terms*
$4x$ and $5x$	$4x$ and $5x^2$
$3xyz^2$ and $-2xyz^2$	$\dfrac{1}{2}ab^2c$ and $\dfrac{1}{3}a^2bc$

Because of the distributive property, we can combine like terms by adding their coefficients and using the same variables and exponents. For example,

$$2y + 5y = (2 + 5)y \qquad \text{and} \qquad -3x^2 + 7x^2 = (-3 + 7)x^2$$
$$= 7y \qquad \qquad \qquad \qquad = 4x^2$$

Likewise,

$$4x^3y^2 + 9x^3y^2 = 13x^3y^2 \qquad \text{and} \qquad 4r^2s^3t^4 + 7r^2s^3t^4 = 11r^2s^3t^4$$

Thus, to add like monomials, we simply combine like terms.

EXAMPLE 2   **a.** $5xy^3 + 7xy^3 = (5 + 7)xy^3 = 12xy^3$

b. $-7x^2y^2 + 6x^2y^2 + 3x^2y^2 = (-7 + 6 + 3)x^2y^2$
$$= (-1 + 3)x^2y^2$$
$$= 2x^2y^2$$

c. $(2x^2)^2 + (3x)^4 = 4x^4 + 81x^4$
$$= (4 + 81)x^4$$
$$= 85x^4$$

d. $2(x + y) + 3(x + y) = 5(x + y)$
$$= 5x + 5y$$

e. $8x^2 - 3x^2 = (8 - 3)x^2 = 5x^2$

f. $6x^3y^2 - 9x^3y^2 = (6 - 9)x^3y^2 = (6 + -9)x^3y^2 = -3x^3y^2$

g. $-3r^2st^3 - 5r^2st^3 = (-3 - 5)r^2st^3 = -8r^2st^3$

■ **STUDY TIP**
Be sure to do work neatly. Get into the neatness habit by being precise when doing your homework exercises. Do not hesitate to write yourself notes and tips as you work through or review exercises. Organization and precision go a long way towards making you a successful math student.

Because of the distributive property, we can also remove parentheses enclosing several terms when the sign preceding the parentheses is a − sign. The negative sign indicates that we want the negative, or opposite, of the polynomial. Because a polynomial is a sum of monomials, we must get the opposite of each *term* of the monomial. It is comparable to multiplying through the polynomial by negative 1.

$$-(3x^2 + 3x - 2) = -1(3x^2 + 3x - 2)$$
$$= -1(3x^2) + (-1)(3x) + (-1)(-2)$$
$$= -3x^2 + (-3x) + 2$$
$$= -3x^2 - 3x + 2$$

EXAMPLE 3   **a.** $(3x - 4) - (5x + 7) = 3x - 4 - 5x - 7$
$$= 3x + -4 + -5x + -7 \qquad a - b = a + -b$$
$$= 3x + -5x + -4 + -7 \qquad \text{Commutative property of addition.}$$
$$= -2x + -11$$
$$= -2x - 11 \qquad a - b = a + -b$$

b. $(3x^2 - 4x - 6) - (2x^2 - 6x + 12)$
$$= 3x^2 - 4x - 6 - 2x^2 + 6x - 12$$
$$= 3x^2 + -4x + -6 + -2x^2 + 6x + -12 \qquad a - b = a + -b$$
$$= 3x^2 + -2x^2 + -4x + 6x + -6 + -12 \qquad \text{Commutative property of addition.}$$
$$= x^2 + 2x + -18$$
$$= x^2 + 2x - 18 \qquad a - b = a + -b$$

c. $(-4rt^3 + 2r^2t^2) - (-3rt^3 + 2r^2t^2)$
$$= -4rt^3 + 2r^2t^2 + 3rt^3 - 2r^2t^2$$
$$= -4rt^3 + 2r^2t^2 + 3rt^3 + -2r^2t^2 \qquad a - b = a + -b$$
$$= -4rt^3 + 3rt^3 + 2r^2t^2 + -2r^2t^2 \qquad \text{Commutative property of addition.}$$
$$= -rt^3$$

ADDING AND SUBTRACTING MULTIPLES OF POLYNOMIALS

Because of the distributive property, we can remove parentheses enclosing several terms when a monomial precedes the parentheses. We simply multiply every term within the parentheses by that monomial. For example, to add $3(2x + 5)$ and $2(4x - 3)$, we proceed as follows:

$$3(2x + 5) + 2(4x - 3) = 6x + 15 + 8x - 6$$
$$= 6x + 8x + 15 - 6$$
$$= 14x + 9$$

This is consistent with the order of operations. We cannot simplify what is grouped because the parentheses contain expressions that are not like terms and cannot be combined. We then must multiply.

EXAMPLE 4   **a.** $3(x^2 + 4x) + 2(x^2 - 4) = 3x^2 + 12x + 2x^2 - 8$
$$= 5x^2 + 12x - 8$$

Before tackling homework exercises, go back over your notes and review the examples. Then attempt the homework without looking back for an example problem to mimic. Students often become dependent on looking at examples for ideas on how to get started on a problem. Remember, there will be no examples to follow on a test!

b. $8(y^2 - 2y + 3) - 4(2y^2 + y - 3) = 8y^2 - 16y + 24 - 8y^2 - 4y + 12$

$$= 8y^2 + -16y + 24 + -8y^2 + -4y + 12$$

$$= 8y^2 + -8y^2 + -16y + -4y + 24 + 12$$

$$= 0y^2 + -20y + 36$$

$$= -20y + 36$$

c. $-4x(xy^2 - xy + 3) - x(xy^2 - 2) + 3(x^2y^2 + 2x^2y)$

$$= -4x^2y^2 + 4x^2y - 12x - x^2y^2 + 2x + 3x^2y^2 + 6x^2y$$

$$= -4x^2y^2 + 4x^2y + -12x + -x^2y^2 + 2x + 3x^2y^2 + 6x^2y$$

$$= -4x^2y^2 + -x^2y^2 + 3x^2y^2 + 4x^2y + 6x^2y + -12x + 2x$$

$$= -2x^2y^2 + 10x^2y + -10x$$

$$= -2x^2y^2 + 10x^2y - 10x$$

EXERCISE 2.3

VOCABULARY   In Exercises 1–2, fill in the blank to make a true statement.

1. The _____ property allows the expression $3(x + y)$ to be calculated as $3x + 3y$.

2. _____ terms have identical variable factors.

CONCEPTS   In Exercises 3–10 tell whether the terms are like or unlike terms. If they are like terms, find their sum.

3. $3y, 4y$

4. $3x^2, 5x^2$

5. $3x, 3y$

6. $3x^2, 6x$

7. $3x^3, 4x^3, 6x^3$

8. $-2y^4, -6y^4, 10y^4$

9. $32x^5y^3, -21x^5y^3, -11x^5y^3$

10. $-x^2y, xy, 3xy^2$

In Exercises 11–18, simplify each expression, if possible. To simplify means to combine like terms.

11. $4y + 5y$

12. $-2x + 3x$

13. $-8t^2 - 4t^2$

14. $15x^2 + 10x^2$

15. $32u^3 - 16u^3$

16. $25xy^2 - 7xy^2$

17. $18x^5y^2 - 11x^5y^2$

18. $17x^6y - 22x^6y$

PRACTICE   In Exercises 19–24, simplify each expression, if possible.

19. $3rst + 4rst + 7rst$

20. $-2ab + 7ab - 3ab$

21. $-4a^2bc + 5a^2bc - 7a^2bc$

22. $3x^2y - 5xy^2 + 7x^2y$

23. $3mn^3 - 5mn^3 + 3m^3n$

24. $6abc + 6bca + 7cba$

In Exercises 25–28, find the polynomial in simplified form that represents the perimeter of each figure.

25.

26.

27.

28.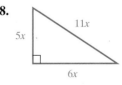

In Exercises 29–43, do the indicated operations and simplify.

29. $(3x + 7) + (4x - 3)$

30. $(2y - 3) + (4y + 7)$

31. $(4a + 3) - (2a - 4)$

32. $(5b - 7) - (3b + 5)$

33. $(3x^2 - 3x - 2) + (3x^2 + 4x - 3)$

34. $(2b^2 + 3b - 5) - (2b^2 - 4b - 9)$

35. $(3a^2 - 2a + 4) - (a^2 - 3a + 7)$

36. $2(x + 3) + 3(x + 3)$

37. $5(x + y) + 7(x + y)$

38. $-8(x - y) + 11(x - y)$

39. $-4(a - b) - 5(a - b)$

40. $2(x^2 - 5x - 4) - 3(x^2 - 5x - 4) + 6(x^2 - 5x - 4)$

41. $-5(y^2 - 2y - 6) + 6(2y^2 + 2y - 5)$

42. $2a(ab^2 - b) - 3b(a + 2ab) + b(b - a + a^2b)$

43. $3y(xy + y) - 2y^2(x - 4 + y) + 2(y^3 + y^2)$

In Exercises 44–47, find the polynomial that represents the perimeter of each figure.

44.

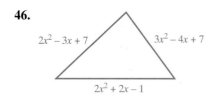

$3x^2 + 4x + 2$

$4x^2 + 4x - 2$

45.

$2x - 5$ $2x - 5$

x x

$2x - 5$ $2x - 5$

$2x - 5$

$2(3x^2 - 4x + 1)$

46.

$2x^2 - 3x + 7$ $3x^2 - 4x + 7$

$2x^2 + 2x - 1$

47.

$3x + 1$

$x - 2$ $x + 1$

$2x^2 - 3x - 8$

48. Use the STOre feature of your calculator to determine the following:
  a. the value of the perimeter in Exercise 44 based on the polynomial expression you determined to represent the perimeter if $x = 4$.
  b. the value of the length and width of the rectangle if $x = 4$.
  c. Conclusion: Is $x = 4$ a valid value for the given rectangle? Why or why not?

49. Use the STOre feature of your calculator to determine the following:
  a. the value of the perimeter in Exercise 45 based on the polynomial expression you determined to represent the perimeter if $x = 3$.
  b. the values of the sides of the figure if $x = 3$.
  c. Conclusion: Is $x = 3$ a valid value for the given figure? Why or why not?

50. Use the STOre feature of your calculator to determine the following:
  a. the value of the perimeter of the figure in Exercise 46 based on the polynomial you determined to represent the perimeter if $x = 2$.
  b. the values of each side if $x = 2$.
  c. Conclusion: Is $x = 2$ a valid value for the given figure? Why or why not?

51. Use the STOre feature of your calculator to determine the following:
  a. the value of the perimeter in Exercise 47 based on the polynomial you determined to represent the perimeter if $x = 2$.
  b. the values of the sides of the trapezoid if $x = 2$.
  c. Conclusion: Is $x = 2$ a valid value for the given trapezoid? Why or why not?

52. Find the sum when $3y^2 - 5y + 7$ is added to the sum of $-3y^2 - 7y + 4$ and $5y^2 + 5y - 7$.

53. Find the sum when $x^2 + x - 3$ is added to the sum of $2x^2 - 3x + 4$ and $3x^2 - 2$.

54. Find the difference when $-3z^3 - 4z + 7$ is subtracted from the sum of $2z^2 + 3z - 7$ and $-4z^3 - 2z - 3$.

55. Find the difference when $t^3 - 2t^2 + 2$ is subtracted from the sum of $3t^3 + t^2$ and $-t^3 + 6t - 3$.

56. Find the difference when $32x^2 - 17x + 45$ is subtracted from the sum of $23x^2 - 12x - 7$ and $-11x^2 + 12x + 7$.

57. Find the sum when $3x^2 + 4x - 7$ is added to the sum of $-2x^2 - 7x + 1$ and $-4x^2 + 8x - 1$.

APPLICATIONS

58. Rental cars A rental car agency charges a flat fee of $37 per day plus $0.21 per mile. Write a polynomial expression to model the cost for one day's rental if the car is driven x miles.

59. Portrait studio A photographer charges a $75 sitting fee and then $10 per unit purchased for studio photographs. (A unit consists of one 8 × 10, or a sheet of two 5 × 7s, or a sheet of 8 wallets.) Write a polynomial expression when x represents the number of units purchased to model the cost of having pictures made.

60. River travel   A boat traveling x mph upstream against a 3 mph current travels a distance of $5(x - 3)$ miles in 5 hours ($d = rt$). When the boat returns downstream in 4 hours, the distance traveled is $4(x + 3)$ miles. Write a polynomial expression and simplify it to model the total distance traveled.

61. Investments   George has $5000 to invest in two simple interest-bearing accounts. He plans to invest x dollars in a certificate of deposit that earns 6% interest and the remaining money, $5000 - x$ dollars, in a money market account that earns $3\frac{1}{2}\%$. Write a simplified polynomial expression to represent the total interest earned in the two accounts. (*Hint: i = prt*; time is 1 year in this problem.)

62. Geometry   The length of a rectangle is 4 more than the width. If the width is x and the length is $4 + x$, write a polynomial expression to model the perimeter and simplify this expression.

63. Travel   In order to reach her destination, a woman drives x hours at 45 mph to the train station and then rides the train for $x + 2$ hours traveling at a speed of 72 mph. Write a simplified polynomial expression to model the total distance traveled.

64. Distance   Two cars start at City A and drive in opposite directions for x hours. One vehicle is

traveling 55 mph and the other, 65 mph. Write a polynomial expression to model the total distance between the two cars and simplify your expression.

65. Pages in a manual   A calculator manual is being assembled as indicated below. Find an expression to model the total number of pages (write the expression in simplified form).

preface	x pages
table of contents	2 fewer pages than the preface
body of text	40 times as many pages as the preface
index	4 more pages than the preface

66. Geometry   Find the circumference of the given circle, expressing the circumference in terms of π and x.

$4x + 5$

REVIEW

67. What repeated multiplication does 4^3 represent?

68. Insert parentheses to make the statement true.
   a. $14 + 3 \cdot 7 - 5 = 20$ **b.** $33 - 8 - 10 = 15$
   c. $18 - 5 + 6 - 9 = 10$ **d.** $4 + 2 \cdot 3^4 = 486$

69. Tax   The sales tax for Clarksville, TN is $8\frac{1}{4}\%$.
   a. Represent this percentage as a decimal.

b. Find the amount of sales tax on a purchase of $299.99.

✎ **c.** Explain how you determine if it is appropriate to round the results in **b.**

70. If the polynomial $P(x) = 2x^2 + 11x + 5$ represents the area of a rectangle whose sides measure $2x + 1$ and $x + 5$, find the area if $x = 2$.

2.4 Multiplying Polynomials

■ MULTIPLYING POLYNOMIALS   ■ THE FOIL METHOD

In Section 2.1, we multiplied certain monomials by other monomials. To multiply $4x^2$ by $-2x^3$, for example, we use the commutative property of multiplication to group the numerical factors and the variable factors together and multiply.

$$4x^2(-2x^3) = 4 \cdot x^2(-2) \cdot x^3 = -8x^5$$

EXAMPLE 1   **a.** $3x^5(2x^5) = 3 \cdot x^5 \cdot 2 \cdot x^5$
       $= 3 \cdot 2 \cdot x^5 \cdot x^5$
       $= 6x^{10}$

b. $-2a^2b^3(5ab^2) = -2 \cdot a^2 \cdot b^3 \cdot 5 \cdot a \cdot b^2$
       $= -2 \cdot 5 \cdot a^2 \cdot a \cdot b^3 \cdot b^2$
       $= -10a^3b^5$

c. $-4y^5z^2(2y^3z^3)(3yz) = -4 \cdot y^5 \cdot z^2 \cdot 2 \cdot y^3 \cdot z^3 \cdot 3 \cdot y \cdot z$
       $= -4 \cdot 2 \cdot 3 \cdot y^5 \cdot y^3 \cdot y \cdot z^2 \cdot z^3 \cdot z$
       $= -24y^9z^6$ ■

The previous example suggests the following rule:

> **Multiplying Two Monomials**
> To multiply two monomials, first multiply the numerical factors and then multiply the variable factors.

To find the product of a monomial and a polynomial with more than one term, we use the distributive property. To multiply $x + 4$ by $3x$, for example, we proceed as follows:

$$3x(x + 4) = 3x \cdot x + 3x \cdot 4$$
$$= 3x^2 + 12x$$

EXAMPLE 2   **a.** $2a^2(3a^2 - 4a) = 2a^2 \cdot 3a^2 - 2a^2 \cdot 4a$
$$= 6a^4 - 8a^3$$

b. $-2xz^2(2x - 3z + 2x^2z^2) = -2xz^2(2x) - (-2xz^2)(3z) + (-2xz^2)(2x^2z^2)$
$$= -4x^2z^2 + 6xz^3 - 4x^3z^4$$

The results of Example 2 suggest the following rule:

> **Multiplying Polynomials by Monomials**
> To multiply a polynomial with more than one term by a monomial, use the distributive property to remove parentheses and simplify.

We must use the Distributive Property more than once to multiply a polynomial by a binomial. For example, to multiply $3x^2 + 3x - 5$ by $2x + 3$, we proceed as follows:

$$(2x + 3)(3x^2 + 3x - 5) = (2x + 3)3x^2 + (2x + 3)3x - (2x + 3)5$$
$$= 3x^2(2x + 3) + 3x(2x + 3) - 5(2x + 3)$$
$$= 6x^3 + 9x^2 + 6x^2 + 9x - 10x - 15$$
$$= 6x^3 + 15x^2 - x - 15$$

■ MULTIPLYING POLYNOMIALS

EXAMPLE 3   **a.** $(2x - 4)(3x + 5) = (2x - 4)3x + (2x - 4)5$
$$= 2x \cdot 3x - 4 \cdot 3x + 2x \cdot 5 - 4 \cdot 5$$
$$= 6x^2 - 12x + 10x - 20$$
$$= 6x^2 - 2x - 20$$

b. $(3x - 2y)(2x + 3y) = (3x - 2y)2x + (3x - 2y)3y$
$$= 3x \cdot 2x - 2y \cdot 2x + 3x \cdot 3y - 2y \cdot 3y$$
$$= 6x^2 - 4xy + 9xy - 6y^2$$
$$= 6x^2 + 5xy - 6y^2$$

c. $(3y + 1)(3y^2 + 2y + 2) = (3y + 1)3y^2 + (3y + 1)2y + (3y + 1)2$
$$= 3y \cdot 3y^2 + 1 \cdot 3y^2 + 3y \cdot 2y + 1 \cdot 2y + 3y \cdot 2 + 1 \cdot 2$$
$$= 9y^3 + 3y^2 + 6y^2 + 2y + 6y + 2$$
$$= 9y^3 + 9y^2 + 8y + 2$$

The results of Example 3 suggest the following rule:

> **Multiplying Polynomials**
> To multiply one polynomial by another, multiply each term of one polynomial by each term of the other polynomial and combine like terms.

It is often convenient to organize the work vertically.

EXAMPLE 4   **a.** Multiply:

$$
\begin{array}{r}
2x - 4 \\
3x + 2 \\
\hline
\end{array}
$$

$3x(2x - 4) \rightarrow \quad 6x^2 - 12x$

$2(2x - 4) \rightarrow \quad \underline{ + 4x - 8}$

$\qquad\qquad\qquad\quad 6x^2 - 8x - 8$

b. Multiply:

$$3a^2 - 4a + 7$$
$$\underline{\ 2a + 5}$$

$2a(3a^2 - 4a + 7) \rightarrow \quad 6a^3 - 8a^2 + 14a$

$5(3a^2 - 4a + 7) \rightarrow \quad \underline{\ + 15a^2 - 20a + 35}$

$\qquad\qquad\qquad\qquad\quad 6a^3 + 7a^2 - 6a + 35$

c. Multiply:

$$3y^2 - 5y + 4$$
$$\underline{\ -\ 4y^2 - 3}$$

$-4y^2(3y^2 - 5y + 4) \rightarrow -12y^4 + 20y^3 - 16y^2$

$-3(3y^2 - 5y + 4) \rightarrow \underline{\ -\ 9y^2 + 15y - 12}$

$\qquad\qquad\qquad\qquad -12y^4 + 20y^3 - 25y^2 + 15y - 12$

■ THE FOIL METHOD

The mnemonic FOIL (First Outer Inner Last) is often used to quickly multiply two *binomials*. Remember, when multiplying two polynomials each term of one polynomial must multiply each term of the other polynomial. To use the FOIL method to multiply $2a - 4$ by $3a + 5$, we

1. multiply the **F**irst terms $2a$ and $3a$ to obtain $6a^2$,
2. multiply the **O**uter terms $2a$ and 5 to obtain $10a$,
3. multiply the **I**nner terms -4 and $3a$ to obtain $-12a$, and
4. multiply the **L**ast terms -4 and 5 to obtain -20.

Then we simplify the resulting polynomial, if possible.

$$(2a - 4)(3a + 5) = 2a(3a) + 2a(5) + (-4)(3a) + (-4)(5)$$
$$= 6a^2 + 10a - 12a - 20 \qquad \text{Simplify.}$$
$$= 6a^2 - 2a - 20 \qquad \text{Combine like terms.}$$

EXAMPLE 5   Use the FOIL method to find each product.

a. $(3x + 4)(2x - 3) = 3x(2x) + 3x(-3) + 4(2x) + 4(-3)$
$= 6x^2 - 9x + 8x - 12$
$= 6x^2 - x - 12$

b. $(2y - 7)(5y - 4) = 2y(5y) + 2y(-4) + (-7)(5y) + (-7)(-4)$
$= 10y^2 - 8y - 35y + 28$
$= 10y^2 - 43y + 28$

c. $(2r - 3s)(2r + t) = 2r(2r) + 2r(t) - 3s(2r) - 3s(t)$
$= 4r^2 + 2rt - 6rs - 3st$

EXAMPLE 6   Simplify each expression.

a. $3(2x - 3)(x + 1) = 3(2x^2 + 2x - 3x - 3)$ Use FOIL to multiply the binomials.

$= 3(2x^2 - x - 3)$ Combine like terms.
$= 6x^2 - 3x - 9$ Use the distributive property to remove parentheses.

b. $(x + 1)(x - 2) - 3x(x + 3) = x^2 - 2x + x - 2 - 3x^2 - 9x$
$= -2x^2 - 10x - 2$ Combine like terms.

The products discussed in Example 7 are called **special products.**

EXAMPLE 7   Use the FOIL method to find each special product.

a. $(x + y)^2 = (x + y)(x + y)$

$= x^2 + xy + xy + y^2$
$= x^2 + 2xy + y^2$

The square of the sum of two quantities such as $x + y$ has three terms: the square of the first quantity, plus twice the product of the quantities, and the square of the second quantity.

b. $(x - y)^2 = (x - y)(x - y)$

$= x^2 - xy - xy + y^2$
$= x^2 - 2xy + y^2$

The square of the difference of two quantities such as $x - y$ has three terms: the square of the first quantity, minus twice the product of the quantities, and the square of the second quantity.

c. $(x + y)(x - y) = x^2 - xy + xy - y^2$
$= x^2 - y^2$

■ STUDY TIP

Go back and work a few problems from each section previously covered in the chapter after completing your homework. Frequently reviewing will help you see how the concepts in the chapter relate to one another. This also distributes your study time for the next test over a long period of time. This will be more beneficial than studying for large blocks of time a day or two before the exam.

The product of a sum and a difference of two quantities such as $x + y$ and $x - y$ is a binomial. It is the product of the first quantities minus the product of the second quantities.

Binomials that have the same terms but different signs between them are often called **conjugate binomials.** For example, the conjugate of $x + y$ is $x - y$, and the conjugate of $ab - c$ is $ab + c$.

Because the special products discussed in Example 7 occur so often, it is wise to learn their forms.

> ### Special Products
> $$(x + y)^2 = x^2 + 2xy + y^2$$
> $$(x - y)^2 = x^2 - 2xy + y^2$$
> $$(x + y)(x - y) = x^2 - y^2$$

 WARNING!   Note that $(x + y)^2 \neq x^2 + y^2$ and $(x - y)^2 \neq x^2 - y^2$.

EXPLORING THE CONCEPT

A frequent student error is to simplify $(x + y)^2$ as $x^2 + y^2$ and $(x - y)^2$ as $x^2 - y^2$. Enter the following expressions on your calculator to numerically justify that $(x + y)^2 \neq x^2 + y^2$ and $(x - y)^2 \neq x^2 - y^2$.

a. $(5 + 7)^2$; $5^2 + 7^2$ **b.** $(-3 + -8)^2$; $(-3)^2 + (-8)^2$

c. $(8 - 2)^2$; $8^2 - 2^2$ **d.** $(4 - 11)^2$; $4^2 - 11^2$

Conclusion   **a.**   **b.**

c.   **d.**

EXERCISE 2.4

VOCABULARY   *In Exercises 1–3, fill in the blank to make a true statement.*

1. Binomials that have the same terms but different signs between the terms are called _____ binomials.

3. Two binomials such as $(2x + 3)$ and $(3x - 4)$ can be multiplied by using the mnemonic device called _____.

2. To multiply $2a^2(3a^2 - 4a)$ we need the _____ property.

CONCEPTS

4. Which of the following is equal to $(x + y)^2$?
   a. $x^2 + y^2$ **b.** $(x + y)(x + y)$

5. Find each product.
   a. $(2x)(3x)$ **b.** $(-4x^2)(-x)$
   c. $\left(\dfrac{1}{2}x\right)\left(\dfrac{4}{5}x\right)$

6. Find each product.
   a. $3(x + 40)$ **b.** $-3(a - 2)$
   c. $-4(t + 7)$ **d.** $6(x^2 - 3)$

PRACTICE *In Exercises 7–14, find each product.*

7. $(3x^2)(4x^3)$ **8.** $(-2a^3)(3a^2)$ **9.** $(3b^2)(-2b)(4b^3)$ **10.** $(3y)(2y^2)(-y^4)$

11. $(x^2y^3)^5$ **12.** $(a^3b^2c)^4$ **13.** $(a^3b^2c)(abc^3)^2$ **14.** $(xyz^3)(xy^2z^2)^3$

In Exercises 15–22, find each product.

15. $3x(x - 2)$ **16.** $4y(y + 5)$

17. $-2x^2(3x^2 - x)$ **18.** $-4x^2(3x^2 - x)$

19. $2x^2(3x^2 + 4x - 7)$ **20.** $3y^3(2y^2 - 7y - 8)$

21. $\frac{1}{4}x^2(8x^5 - 4)$ **22.** $-\frac{2}{3}r^2t^2(9r - 3t)$

In Exercises 23–32, use the FOIL method to find each product.

23. $(a + 4)(a + 5)$ **24.** $(y - 3)(y + 5)$ **25.** $(3x - 2)(x + 4)$

26. $(t + 4)(2t - 3)$ **27.** $(2a + 4)(3a - 5)$ **28.** $(2b - 1)(3b + 4)$

29. $(3x - 5)(2x + 1)$ **30.** $(2y - 5)(3y + 7)$ **31.** $(2t + 3s)(3t - s)$

32. $(3a - 2b)(4a + b)$

In Exercises 33–42, find the special products.

33. $(x + 4)(x + 4)$ **34.** $(a + 3)(a + 3)$ **35.** $(x + 5)^2$

36. $(y - 6)^2$ **37.** $(r + 4)(r - 4)$ **38.** $(b + 2)(b - 2)$

39. $(4x + 5y)(4x - 5y)$ **40.** $(6p + 5q)(6p - 5q)$ **41.** $(2a - 3b)^2$

42. $(3a + 2b)^2$

In Exercises 43–44, find the volume of each figure.

43.

44.

In Exercises 45–48, find the area of each figure.

45.

46.

47.

48.

In Exercises 49–58, simplify each expression.

49. $2t(t + 2) + 3t(t - 5)$ **50.** $3y(y + 2) + (y + 1)(y - 1)$

51. $3xy(x + y) - 2x(xy - x)$ **52.** $(a + b)(a - b) - (a + b)(a + b)$

53. $(x + y)(x - y) + x(x + y)$ **54.** $(2x - 1)(2x + 1) + x(2x + 1)$

55. $(x + 2)^2 - (x - 2)^2$ **56.** $(x - 3)^2 - (x + 3)^2$

57. $(2s - 3)(s + 2) + (3s + 1)(s - 3)$ **58.** $(3x + 4)(2x - 2) - (2x + 1)(x + 3)$

APPLICATIONS *In Exercises 59–68, write a simplified expression to model each problem.*

59. The length of a rectangle is 5 more than the width. If the width is x, then the length is $5 + x$. Find the area of the rectangle.

60. The side of a square is $2x - 3$. Find the area.

61. The sides of a rectangular prism are x, $x + 2$, and $x - 2$ in length. Find the volume.

62. Find the product of three consecutive integers represented by x, $x + 1$, and $x + 2$.

63. Find the product of three consecutive even integers represented by x, $x + 2$, and $x + 4$.

64. If two consecutive even integers are x and $x + 2$, find the difference of their squares.

65. If two consecutive odd integers are x and $x + 2$, find the sum of their squares.

66. Area   Find the polynomial to represent the area of the desk calendar in Illustration 1.

ILLUSTRATION 1

67. Area   If the top of a desk has a width of $2x^2 - 5x$ and a depth of $3x^2 + 2x$, find the polynomial that models the amount of the desktop that is *not* occupied by the calendar in Exercise 66.

68. Volume   Find the polynomial that models the volume of the space occupied by the microwave in Illustration 2.

ILLUSTRATION 2

REVIEW

69. Distance   If a car is traveling 65 mph, then the formula $d = 65t$ can be used to determine the distance traveled at any given time. Use your calculator to generate a table of values relating time to distance that corresponds to the following table. (Replace the variable t, time, with the variable x when using the calculator.)

x	.5	1	1.5	2
$65x$				

Record the process with a step-by-step description.

70. Cutting a board   The board below has been cut into two pieces. What was the length of the board (in feet) before it was cut?

71. Distance   A car traveling from Nashville, TN to Louisville, KY must travel at two different speeds: 70 mph in Tennessee and 60 mph in Kentucky. Write the expression that models the total distance traveled.

72. Perimeter   Find the perimeter of the quadrilateral.

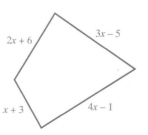

2.5 Dividing Polynomials by Monomials

■ DIVIDING A MONOMIAL BY A MONOMIAL ■ DIVIDING A POLYNOMIAL BY A MONOMIAL

■ DIVIDING A MONOMIAL BY A MONOMIAL

Dividing a number is equivalent to multiplying by its reciprocal. For example, dividing the number 8 by 2 gives the same answer as multiplying 8 by $\frac{1}{2}$.

$$\frac{8}{2} = 4 \quad \text{and} \quad 8 \cdot \frac{1}{2} = 4 \quad \text{or} \quad \frac{1}{2} \cdot 8 = 4$$

In general, the following is true.

> **Division**
>
> $$\frac{a}{b} = a \cdot \frac{1}{b} \quad \text{or} \quad \frac{a}{b} = \frac{1}{b} \cdot a \quad (b \neq 0)$$

■ **STUDY TIP**
In the few minutes before class begins, review your notes, have your list of questions ready, and make sure you have all the tools necessary for learning with you—notebook, text, calculator, and pencil(s).

Recall that to simplify a fraction, we write both its numerator and denominator as the product of several factors and then divide out all common factors. For example, to simplify $\frac{4}{6}$, we can write

$$\frac{4}{6} = \frac{2 \cdot 2}{2 \cdot 3} = \frac{\overset{1}{\cancel{2}} \cdot 2}{\underset{1}{\cancel{2}} \cdot 3} = \frac{2}{3} \qquad \frac{2}{2} = 1.$$

To simplify the fraction $\frac{20}{25}$, we can write

$$\frac{20}{25} = \frac{4 \cdot 5}{5 \cdot 5} = \frac{4 \cdot \overset{1}{\cancel{5}}}{\underset{1}{\cancel{5}} \cdot 5} = \frac{4}{5} \qquad \frac{5}{5} = 1.$$

To simplify algebraic fractions, we can either use the method just used for simplifying arithmetic fractions or use the rules of exponents.

EXAMPLE 1   Simplify   **a.** $\dfrac{x^2 y}{x y^2}$ and   **b.** $\dfrac{-8a^3 b^2}{4ab^3}$.

Solution

Method for arithmetic fractions

a. $\dfrac{x^2 y}{x y^2} = \dfrac{xxy}{xyy}$

$$= \frac{\overset{1}{\cancel{x}}\overset{}{x}\overset{1}{\cancel{y}}}{\underset{1}{\cancel{x}}y\underset{1}{\cancel{y}}}$$

$$= \frac{x}{y}$$

b. $\dfrac{-8a^3 b^2}{4ab^3} = \dfrac{-2 \cdot 4aaabb}{4abbb}$

$$= \frac{-2 \cdot \overset{1\,\,1}{\cancel{4}}\overset{}{\cancel{a}}aa\overset{1\,\,1}{\cancel{b}\cancel{b}}}{\underset{1\,1\,1\,1}{\cancel{4}\cancel{a}\cancel{b}\cancel{b}}b}$$

$$= \frac{-2a^2}{b}$$

Using the rules of exponents

a. $\dfrac{x^2 y}{x y^2} = x^{2-1} y^{1-2}$

$$= x^1 y^{-1} \qquad y^{-1} = \frac{1}{y}. \text{ Negative}$$
exponents will be discussed in depth in Chapter 7.

$$= \frac{x}{y}$$

b. $\dfrac{-8a^3 b^2}{4ab^3} = \dfrac{-2^3 a^3 b^2}{2^2 ab^3}$

$$= -2^{3-2} a^{3-1} b^{2-3}$$

$$= -2^1 a^2 b^{-1} \qquad b^{-1} = \frac{1}{b}$$

$$= \frac{-2a^2}{b}$$

■

◼ DIVIDING A POLYNOMIAL BY A MONOMIAL

To divide a polynomial with more than one term by a monomial, we rewrite the division as a product, use the distributive law to remove parentheses, and simplify each resulting fraction.

EXAMPLE 2   Simplify $\dfrac{9x + 6y}{3xy}$.

Solution $\dfrac{9x + 6y}{3xy} = \dfrac{1}{3xy}(9x + 6y)$

$= \dfrac{9x}{3xy} + \dfrac{6y}{3xy}$ Remove parentheses.

$= \dfrac{3}{y} + \dfrac{2}{x}$ Simplify each fraction.   ◼

EXAMPLE 3   Simplify $\dfrac{6x^2y^2 + 4x^2y - 2xy}{2xy}$.

Solution $\dfrac{6x^2y^2 + 4x^2y - 2xy}{2xy} = \dfrac{1}{2xy}(6x^2y^2 + 4x^2y - 2xy)$

$= \dfrac{6x^2y^2}{2xy} + \dfrac{4x^2y}{2xy} - \dfrac{2xy}{2xy}$ Remove parentheses.

$= 3xy + 2x - 1$ Simplify each fraction.   ◼

EXAMPLE 4   Simplify $\dfrac{12a^3b^2 - 4a^2b + a}{6a^2b^2}$.

Solution $\dfrac{12a^3b^2 - 4a^2b + a}{6a^2b^2} = \dfrac{1}{6a^2b^2}(12a^3b^2 - 4a^2b + a)$

$= \dfrac{12a^3b^2}{6a^2b^2} - \dfrac{4a^2b}{6a^2b^2} + \dfrac{a}{6a^2b^2}$ Remove parentheses.

$= 2a - \dfrac{2}{3b} + \dfrac{1}{6ab^2}$ Simplify each fraction.   ◼

EXAMPLE 5   Simplify $\dfrac{(x - y)^2 - (x + y)^2}{xy}$.

Solution $\dfrac{(x - y)^2 - (x + y)^2}{xy} = \dfrac{x^2 - 2xy + y^2 - (x^2 + 2xy + y^2)}{xy}$ Multiply the binomials in the numerator.

$= \dfrac{x^2 - 2xy + y^2 - x^2 - 2xy - y^2}{xy}$ Remove parentheses.

$= \dfrac{-4xy}{xy}$ Combine like terms.

$= -4$ Divide out xy.   ◼

◼ STUDY TIP
As you work through an exercise section, write down questions for your instructor as they occur to you. Keep the list on a separate piece of paper and write down the answers to *all* of your questions.

EXERCISE 2.5

CONCEPTS *In Exercises 1–4, perform each division by simplifying each fraction.*

1. $\dfrac{8x}{2}$

2. $\dfrac{4x}{12}$

3. $\dfrac{xy}{yx}$

4. $\dfrac{ab}{bc}$

In Exercises 5–8, perform each division using the distributive property.

5. $\dfrac{4x + 12y}{4}$

6. $\dfrac{6x - 3y}{3}$

7. $\dfrac{7x + 9xy}{x}$

8. $\dfrac{5ab - 6bc}{b}$

PRACTICE *In Exercises 9–34, do each division by simplifying each fraction. Write all answers without using negative or zero exponents.*

9. $\dfrac{r^3 s^2}{rs^3}$

10. $\dfrac{y^4 z^3}{y^2 z^2}$

11. $\dfrac{8x^3 y^2}{4xy^3}$

12. $\dfrac{-3y^3 z}{6yz^2}$

13. $\dfrac{12u^5 v}{-4u^2 v^3}$

14. $\dfrac{16rst^2}{-8rst^3}$

15. $\dfrac{-16r^3 y^2}{-4r^2 y^4}$

16. $\dfrac{35xyz^2}{-7x^2 yz}$

17. $\dfrac{-65rs^2 t}{15r^2 s^3 t}$

18. $\dfrac{112u^3 z^6}{-42u^3 z^6}$

19. $\dfrac{x^2 x^3}{xy}$

20. $\dfrac{(xy)^2}{x^2 y^3}$

21. $\dfrac{(a^3 b^4)^3}{ab^4}$

22. $\dfrac{(a^2 b^3)^3}{a^6 b^6}$

23. $\dfrac{15(r^2 s^3)^2}{-5(rs^2)^3}$

24. $\dfrac{-5(a^2 b)^3}{10(ab)^3}$

25. $\dfrac{-32(x^3 y)^3}{128(xy^2)^3}$

26. $\dfrac{68(a^6 b^7)^2}{-96(abc^2)^3}$

27. $\dfrac{(5a^2 b)^3}{(2a^2 b)^3}$

28. $\dfrac{-(4x^3 y^3)^2}{x^2 y^4}$

29. $\dfrac{-(3x^3 y^4)^3}{-(9x^4 y^5)^2}$

30. $\dfrac{(2r^3 s^2 t)^2}{-(4r^2 s^2 t)^2}$

31. $\dfrac{(a^2 a^3)^4}{(a^4)^3}$

32. $\dfrac{(b^3 b^4)^5}{(bb^2)^2}$

33. $\dfrac{(z^3 z^4)^3}{(z^3)^2}$

34. $\dfrac{(t^3 t^5)}{(t^2)^3}$

In Exercises 35–48, perform the division.

35. $\dfrac{6x + 9y}{3xy}$

36. $\dfrac{8x + 12y}{4xy}$

37. $\dfrac{5x - 10y}{25xy}$

38. $\dfrac{2x - 32}{16x}$

39. $\dfrac{3x^2 + 6y^3}{3x^2 y^2}$

40. $\dfrac{4a - 9b^2}{12ab}$

41. $\dfrac{15a^3 b^2 - 10a^2 b^3}{5a^2 b^2}$

42. $\dfrac{9a^4 b^3 - 16a^3 b^4}{12a^2 b}$

43. $\dfrac{4x - 2y + 8z}{4xy}$

44. $\dfrac{5a^2 + 10b^2 - 15ab}{5ab}$

45. $\dfrac{12x^3 y^2 - 8x^2 y - 4x}{4xy}$

46. $\dfrac{12a^2 b^2 - 8a^2 b - 4ab}{4ab}$

47. $\dfrac{-25x^2 y + 30xy^2 - 5xy}{-5xy}$

48. $\dfrac{-30a^2 b^2 - 15a^2 b - 10ab^2}{-10ab}$

In Exercises 49–58, simplify each numerator and perform the division.

49. $\dfrac{5x(4x - 2y)}{2y}$

50. $\dfrac{9y^2(x^2 - 3xy)}{3x^2}$

51. $\dfrac{(-2x)^2 + (3x^2)^2}{6x^2}$

52. $\dfrac{(-3x^2 y)^3 + (3xy^2)^3}{27x^3 y^4}$

53. $\dfrac{4x^2 y^2 - 2(x^2 y^2 + xy)}{2xy}$

54. $\dfrac{-5a^3 b - 5a(ab^2 - a^2 b)}{10a^2 b^2}$

55. $\dfrac{(3x - y)(2x - 3y)}{6xy}$

56. $\dfrac{(2m - n)(3m - 2n)}{-3m^2n^2}$

57. $\dfrac{(a + b)^2 - (a - b)^2}{2ab}$

58. $\dfrac{(x - y)^2 + (x + y)^2}{2x^2y^2}$

APPLICATIONS   *In Exercises 59–64, write an expression involving division to model each problem. Simplify your expression.*

59. Perimeter   The perimeter of a square is $8x + 12$. Find the length of one side.

60. Perimeter   The perimeter of an equilateral triangle is $6x^2 - 3x + 9$. Find the length of one side.

61. Area   Find the width of a rectangle whose length is $4x^2$ if the area is $8x^3 + 2x^2 - 12$. (*Hint:* If $A = lw$, then $w = \frac{A}{l}$.)

62. Area   Find the length of a rectangle whose width is $3x^3$ if the area is $6x^4 - 6x^3$.

63. Volume   The volume of a swimming pool is $18x^4 - 12x^2 + 8$. If the length is $2x$ and the width is $3x^2$, find the depth. $\left(Hint:\ V = lwh;\ h = \frac{V}{lw}.\right)$

64. Volume   The volume of a box of cereal is $6x^3 + 4x^2 + 3$. If the width is x and the depth is 2, find the height.

In Exercises 65–67, use the STOre feature of your calculator for the arithmetic computation.

65. The perimeter of a square is $x^3 - 12x^2$. Is $x = 2$ a valid value for x? Why or why not? Justify your response with appropriate calculator or algebraic work.

66. The area of a rectangle is $4x^2 - 2x - 6$. If the length is $2x$, determine if $x = 1$ is a valid value of x for the width.

Why or why not? Justify your response with appropriate calculator or algebraic work.

67. The perimeter of a square is $8x^3 - 12x^2$. Is $x = 3$ a valid value for x? Why or why not? Justify your response with appropriate calculator or algebraic work.

REVIEW

68. Translate each phrase to an algebraic expression.
  a. 4 less than x
  b. 2 more than three times x
  c. half x increased by 5

69. The polynomial $P(x)$ has been evaluated for $x = 3$ by using the calculator. Based on the calculator screen,   **a.** what was the polynomial   and   **b.** what was the value of the polynomial when $x = 3$?

70. If $P(x) = 3x^2 + 5x - 1$ and $Q(x) = 2x^2 - 6x - 3$, find
  a. $P(x) - Q(x)$
  b. $Q(x) - P(x)$

2.6 Dividing Polynomials by Polynomials

■ DIVIDING A POLYNOMIAL BY A POLYNOMIAL

■ DIVIDING A POLYNOMIAL BY A POLYNOMIAL

To divide one polynomial by another, we can use a process similar to long division. We have already divided a polynomial by a monomial using the Distributive Property. This approach can be compared to "short division." We will begin the examples using division by a monomial, with the arithmetic division paralleled alongside.

EXAMPLE 1   Divide $6x^3 + 8x^2 - 4x$ by $2x$

Solution

Arithmetic division
Divide 684 by 2.

$$\begin{array}{r} 3 \\ 2\overline{)684} \end{array}$$

How many times does 2 divide 6? $\frac{6}{2} = 3$

$$\begin{array}{r} 3 \\ 2\overline{)684} \\ 6 \end{array}$$

Multiply the divisor, 2, by the partial quotient, 3.

$$\begin{array}{r} 3 \\ 2\overline{)684} \\ -6 \\ \hline 8 \end{array}$$

Subtract; bring down the next term.

$$\begin{array}{r} 34 \\ 2\overline{)684} \\ -6 \\ \hline 8 \end{array}$$

How many times does 2 divide 8? $\frac{8}{2} = 4$. Place a 4 above the 8.

$$\begin{array}{r} 34 \\ 2\overline{)684} \\ -6 \\ \hline 8 \\ 8 \end{array}$$

Multiply the divisor of 2 by the 4 in the quotient.

$$\begin{array}{r} 34 \\ 2\overline{)684} \\ -6 \\ \hline 8 \\ -8 \\ \hline 4 \end{array}$$

Subtract; bring down the next term.

$$\begin{array}{r} 34 \\ 2\overline{)684} \\ -6 \\ \hline 8 \\ -8 \\ \hline 4 \end{array}$$

How many times does 2 divide 4? $\frac{4}{2} = 2$. Place a 2 above the 4.

$$\begin{array}{r} 342 \\ 2\overline{)684} \\ -6 \\ \hline 8 \\ -8 \\ \hline 4 \\ 4 \end{array}$$

Multiply 2 by 2.

$$\begin{array}{r} 342 \\ 2\overline{)684} \\ -6 \\ \hline 8 \\ -8 \\ \hline 4 \\ -4 \\ \hline 0 \end{array}$$

Subtract.

Polynomial division
Divide $6x^3 + 8x^2 - 4x$ by $2x$

$$\begin{array}{r} 3x^2 \\ 2x\overline{)6x^3 + 8x^2 - 4x} \end{array}$$

How many times does $2x$ divide $6x^3$?
$\dfrac{6x^3}{2x} = 3x^2$

$$\begin{array}{r} 3x^2 \\ 2x\overline{)6x^3 + 8x^2 - 4x} \\ 6x^3 \end{array}$$

Multiply the divisor, $2x$, by the partial quotient, $3x^2$.

$$\begin{array}{r} 3x^2 \\ 2x\overline{)6x^3 + 8x^2 - 4x} \\ -6x^3 \\ \hline 8x^2 \end{array}$$

Subtract; bring down the next term.

$$\begin{array}{r} 3x^2 + 4x \\ 2x\overline{)6x^3 + 8x^2 - 4x} \\ -6x^3 \\ \hline 8x^2 \end{array}$$

How many times does $2x$ divide $8x^2$?
$\dfrac{8x^2}{2x} = 4x$. Place $4x$ above the $8x^2$.

Remember: We are dividing *terms;* the appropriate sign must be placed between the terms.

$$\begin{array}{r} 3x^2 + 4x \\ 2x\overline{)6x^3 + 8x^2 - 4x} \\ -6x^3 \\ \hline 8x^2 \\ 8x^2 \end{array}$$

Multiply the divisor of $2x$ by the $4x$ in the quotient.

$$\begin{array}{r} 3x^2 + 4x \\ 2x\overline{)6x^3 + 8x^2 - 4x} \\ -6x^3 \\ \hline 8x^2 \\ -8x^2 \\ \hline -4x \end{array}$$

Subtract; bring down the next term.

$$\begin{array}{r} 3x^2 + 4x \\ 2x\overline{)6x^3 + 8x^2 - 4x} \\ -6x^3 \\ \hline 8x^2 \\ -8x^2 \\ \hline -4x \end{array}$$

How many times does $2x$ divide $-4x$?
$\dfrac{-4x}{2x} = -2$. Place (-2) above the $-4x$.

$$\begin{array}{r} 3x^2 + 4x - 2 \\ 2x\overline{)6x^3 + 8x^2 - 4x} \\ -6x^3 \\ \hline 8x^2 \\ -8x^2 \\ \hline -4x \\ -4x \end{array}$$

Multiply $2x$ by (-2).

$$\begin{array}{r} 3x^2 + 4x - 2 \\ 2x\overline{)6x^3 + 8x^2 - 4x} \\ -6x^3 \\ \hline 8x^2 \\ -8x^2 \\ \hline -4x \\ -(-4x) \\ \hline 0 \end{array}$$

Subtract.

$$684 \div 2 = 342$$
dividend divisor quotient

The remainder is 0.

Check by multiplication:

The solution checks.

$$(6x^3 + 8x^2 - 4x) \div (2x) = 3x^2 + 4x - 2$$
dividend   divisor   quotient

The remainder is 0.

Check by multiplication:

$$2x(3x^2 + 4x - 2) = 2x(3x^2) + 2x(4x) - 2x(2)$$
$$= 6x^3 + 8x^2 - 4x$$

The solution checks. ∎

You should make note of the pattern that occurs: divide, multiply, subtract, bring down the next term. We will now extend the concept to the division of a polynomial by a polynomial.

EXAMPLE 2   Divide $x^2 + 5x + 6$ by $x + 2$

Solution   *Step 1:*

$$\begin{array}{r} x \phantom{{}+5x+6} \\ x + 2 \overline{)x^2 + 5x + 6} \end{array}$$

How many times does x divide x^2? $\frac{x^2}{x} = x$. Place the x above the division symbol.

Step 2:

$$\begin{array}{r} x \phantom{{}+5x+6} \\ x + 2 \overline{)x^2 + 5x + 6} \\ x^2 + 2x \phantom{{}+6} \end{array}$$

Multiply each term in the divisor by x. Place the product under $x^2 + 5x$ as indicated and draw a line.

Step 3:

$$\begin{array}{r} x \phantom{{}+5x+6} \\ x + 2 \overline{)x^2 + 5x + 6} \\ x^2 + 2x \phantom{{}+6} \\ \hline 3x + 6 \end{array}$$

Subtract $x^2 + 2x$ from $x^2 + 5x$ by adding the negative of $x^2 + 2x$ to $x^2 + 5x$.

Bring down the next term.

Step 4:

$$\begin{array}{r} x + 3 \\ x + 2 \overline{)x^2 + 5x + 6} \\ x^2 + 2x \phantom{{}+6} \\ \hline 3x + 6 \end{array}$$

How many times does x divide $3x$? $\frac{3x}{x} = +3$. Place the $+3$ above the division symbol.

Step 5:

$$\begin{array}{r} x + 3 \\ x + 2 \overline{)x^2 + 5x + 6} \\ x^2 + 2x \phantom{{}+6} \\ \hline 3x + 6 \\ 3x + 6 \end{array}$$

Multiply each term in the divisor by 3. Place the product under the $3x + 6$ as indicated and draw a line.

Step 6:

$$\begin{array}{r} x + 3 \\ x + 2 \overline{)x^2 + 5x + 6} \\ x^2 + 2x \phantom{{}+6} \\ \hline 3x + 6 \\ 3x + 6 \\ \hline 0 \end{array}$$

Subtract $3x + 6$ from $3x + 6$ by adding the negative of $3x + 6$.

The quotient is $x + 3$ and the remainder is 0.

Step 7: Check the work by verifying that the product of $x + 2$ and $x + 3$ is $x^2 + 5x + 6$.

$$(x + 2)(x + 3) = x^2 + 3x + 2x + 6$$
$$= x^2 + 5x + 6$$

The answer checks.   ∎

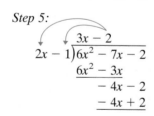

EXAMPLE 3   Divide: $\dfrac{6x^2 - 7x - 2}{2x - 1}$

Solution   *Step 1:*

$$\begin{array}{r} 3x \\ 2x - 1\overline{\smash{)}6x^2 - 7x - 2} \end{array}$$

How many times does $2x$ divide $6x^2$? $\frac{6x^2}{2x} = 3x$. Place the $3x$ above the division symbol.

Step 2:

$$\begin{array}{r} 3x \\ 2x - 1\overline{\smash{)}6x^2 - 7x - 2} \\ 6x^2 - 3x \end{array}$$

Multiply each term in the divisor by $3x$. Place the product under $6x^2 - 7x$ as indicated and draw a line.

Step 3:

$$\begin{array}{r} 3x \\ 2x - 1\overline{\smash{)}6x^2 - 7x - 2} \\ 6x^2 - 3x \\ \hline -4x - 2 \end{array}$$

Subtract $6x^2 - 3x$ from $6x^2 - 7x$ by adding the negative of $6x^2 - 3x$ to $6x^2 - 7x$.

Bring down the next term.

Step 4:

$$\begin{array}{r} 3x - 2 \\ 2x - 1\overline{\smash{)}6x^2 - 7x - 2} \\ 6x^2 - 3x \\ \hline -4x - 2 \end{array}$$

How many times does $2x$ divide $-4x$? $\frac{-4x}{2x} = -2$. Place the -2 above the division symbol.

Step 5:

$$\begin{array}{r} 3x - 2 \\ 2x - 1\overline{\smash{)}6x^2 - 7x - 2} \\ 6x^2 - 3x \\ \hline -4x - 2 \\ -4x + 2 \end{array}$$

Multiply each term in the divisor by -2. Place the product under the $-4x - 2$ as indicated and draw a line.

Step 6:

$$\begin{array}{r} 3x - 2 \\ 2x - 1\overline{\smash{)}6x^2 - 7x - 2} \\ 6x^2 - 3x \\ \hline -4x - 2 \\ -4x + 2 \\ \hline -4 \end{array}$$

Subtract $-4x + 2$ from $-4x - 2$ by adding the negative of $-4x + 2$.

In this example, the quotient is $3x - 2$ and the remainder is -4. It is common to write the answer in quotient + $\frac{\text{remainder}}{\text{divisor}}$ form:

$$3x - 2 + \frac{-4}{2x - 1}$$

where the fraction $\dfrac{-4}{2x - 1}$ is formed by dividing the remainder by the divisor.

Step 7: To check the answer, we multiply $3x - 2 + \dfrac{-4}{2x - 1}$ by $2x - 1$. The product should be the dividend.

$$(2x - 1)\left(3x - 2 + \frac{-4}{2x - 1}\right) = (2x - 1)(3x - 2) + (2x - 1)\left(\frac{-4}{2x - 1}\right)$$
$$= (2x - 1)(3x - 2) - 4$$
$$= 6x^2 - 4x - 3x + 2 - 4$$
$$= 6x^2 - 7x - 2$$

Because the result is the dividend, the answer checks.

EXAMPLE 4   Divide $4x^2 + 2x^3 + 12 - 2x$ by $x + 3$

Solution   The division process works most efficiently if the terms of both the divisor and the dividend are written in descending order. This means that the term involving the highest power of x appears first, the term involving the second-highest power of x appears second, and so on. We can use the commutative property to rearrange the terms of the dividend in descending order and divide as follows:

$$
\begin{array}{r}
2x^2 - 2x + 4 \\
x + 3 \overline{)2x^3 + 4x^2 - 2x + 12} \\
\underline{2x^3 + 6x^2} \\
-2x^2 - 2x \\
\underline{-2x^2 - 6x} \\
+4x + 12 \\
\underline{+4x + 12}
\end{array}
$$

Check: $(x + 3)(2x^2 - 2x + 4) = 2x^3 - 2x^2 + 4x + 6x^2 - 6x + 12$
$$= 2x^3 + 4x^2 - 2x + 12$$

EXAMPLE 5   Divide $\dfrac{x^2 - 4}{x + 2}$

Solution   Since the binomial $x^2 - 4$ does not have a term involving x, we must either include the term $0x$ or leave a space for it. After this adjustment, the division is routine.

$$
\begin{array}{r}
x - 2 \\
x + 2 \overline{)x^2 + 0x - 4} \\
\underline{x^2 + 2x} \\
-2x - 4 \\
\underline{-2x - 4}
\end{array}
$$

Check: $(x + 2)(x - 2) = x^2 - 2x + 2x - 4$
$$= x^2 - 4$$

EXAMPLE 6   Divide $x^3 + y^3$ by $x + y$

Solution   We write $x^3 + y^3$ leaving spaces for the missing terms and proceed as follows:

$$
\begin{array}{r}
x^2 - xy + y^2 \\
x + y \overline{)x^3 \qquad\qquad + y^3} \\
\underline{x^3 + x^2y} \\
-x^2y \\
\underline{-x^2y - xy^2} \\
+xy^2 + y^3 \\
\underline{xy^2 + y^3}
\end{array}
$$

Check: $(x + y)(x^2 - xy + y^2) = x^3 - x^2y + xy^2 + x^2y - xy^2 + y^3$
$$= x^3 + y^3$$

EXERCISE 2.6

CONCEPTS

1. Divide each of the following:

   a. $\dfrac{6x^3}{2x}$

   b. $\dfrac{8x^2}{2x}$

   c. $\dfrac{-4x}{2x}$

   d. $\dfrac{6x^3 + 8x^2 - 4x}{2x}$

2. Arrange the terms in descending order by the powers of x.

   a. $11x + 10x^2 + 3$ **b.** $-x - 21 + 2x^2$

   c. $xy - 2y^2 + 6x^2$ **d.** $-10y^2 + 13xy + 3x^2$

3. Fill in the missing terms using 0 coefficients.

   a. $x^2 - 9$ **b.** $x^3 - 8$

PRACTICE

4. Divide $x^2 + 4x + 4$ by $x + 2$

5. Divide $x^2 - 5x + 6$ by $x - 2$

6. Divide $y^2 + 13y + 12$ by $y + 1$

7. Divide $z^2 - 7z + 12$ by $z - 3$

8. Divide $a^2 + 2ab + b^2$ by $a + b$

9. Divide $a^2 - 2ab + b^2$ by $a - b$

In Exercises 10–15, do each division.

10. $\dfrac{6a^2 + 5a - 6}{2a + 3}$

11. $\dfrac{8a^2 + 2a - 3}{2a - 1}$

12. $\dfrac{3b^2 + 11b + 6}{3b + 2}$

13. $\dfrac{3b^2 - 5b + 2}{3b - 2}$

14. $\dfrac{2x^2 - 7xy + 3y^2}{2x - y}$

15. $\dfrac{3x^2 + 5xy - 2y^2}{x + 2y}$

In Exercises 16–24, rearrange the terms so that the powers of x are in descending order. Then do each division.

16. $5x + 3 \overline{)\, 11x + 10x^2 + 3}$

17. $2x - 7 \overline{)\, -x - 21 + 2x^2}$

18. $4 + 2x \overline{)\, -10x - 28 + 2x^2}$

19. $1 + 3x \overline{)\, 9x^2 + 1 + 6x}$

20. $2x - y \overline{)\, xy - 2y^2 + 6x^2}$

21. $2y + x \overline{)\, 3xy + 2x^2 - 2y^2}$

22. $3x - 2y \overline{)\, -10y^2 + 13xy + 3x^2}$

23. $2x + 3y \overline{)\, -12y^2 + 10x^2 + 7xy}$

24. $4x + y \overline{)\, -19xy + 4x^2 - 5y^2}$

In Exercises 25–30, do each division.

25. $2x + 3 \overline{)\, 2x^3 + 7x^2 + 4x - 3}$

26. $2x - 1 \overline{)\, 2x^3 - 3x^2 + 5x - 2}$

27. $3x + 2 \overline{)\, 6x^3 + 10x^2 + 7x + 2}$

28. $4x + 3 \overline{)\, 4x^3 - 5x^2 - 2x + 3}$

29. $2x + y \overline{)\, 2x^3 + 3x^2y + 3xy^2 + y^3}$

30. $3x - 2y \overline{)\, 6x^3 - x^2y + 4xy^2 - 4y^3}$

In Exercises 31–40, do each division. If there is a remainder, leave the answer in quotient $+ \frac{remainder}{divisor}$ form.

31. $\dfrac{2x^2 + 5x + 2}{2x + 3}$

32. $\dfrac{3x^2 - 8x + 3}{3x - 2}$

33. $\dfrac{4x^2 + 6x - 1}{2x + 1}$

34. $\dfrac{6x^2 - 11x + 2}{3x - 1}$

35. $\dfrac{x^3 + 3x^2 + 3x + 1}{x + 1}$

36. $\dfrac{x^3 + 6x^2 + 12x + 8}{x + 2}$

37. $\dfrac{2x^3 + 7x^2 + 4x + 3}{2x + 3}$

38. $\dfrac{6x^3 + x^2 + 2x + 1}{3x - 1}$

39. $\dfrac{2x^3 + 4x^2 - 2x + 3}{x - 2}$

40. $\dfrac{3y^3 - 4y^2 + 2y + 3}{y + 3}$

In Exercises 41–50, do each division.

41. $\dfrac{x^2 - 1}{x - 1}$

42. $\dfrac{x^2 - 9}{x + 3}$

43. $\dfrac{4x^2 - 9}{2x + 3}$

44. $\dfrac{25x^2 - 16}{5x - 4}$

45. $\dfrac{x^3 + 1}{x + 1}$

46. $\dfrac{x^3 - 8}{x - 2}$

47. $\dfrac{a^3 + a}{a + 3}$

48. $\dfrac{y^3 - 50}{y - 5}$

49. $3x - 4\overline{)15x^3 - 23x^2 + 16x}$

50. $2y + 3\overline{)21y^2 + 6y^3 - 20}$

APPLICATIONS   *In Exercises 51–58, write an expression that includes polynomial division to model each of the problems below. Compute the division to respond to the question.*

51. Perimeter   The perimeter of a square is $4x^2 - 20x + 16$. What is the length of each side?

52. Using the STOre feature of your calculator, determine if $x = 2$ is a valid value for the variable in Exercise 51.

53. Perimeter   The perimeter of an equilateral triangle is $27x^3 - 6x^2 + 9$. What is the length of a side?

54. Using the STOre feature of your calculator, determine if $x = 2$ is a valid value for the variable in Exercise 53.

55. Perimeter   a. What is the width of a rectangle whose length is $x + 5$ and whose area is $x^2 + 12x + 35$? (*Hint:* If $A = lw$, then $w = \frac{A}{l}$.)   **b.** What is the perimeter?

56. Perimeter   a. Find the length of a rectangular garden whose area is $3x^2 - x - 10$ and whose width is $x - 2$.   **b.** What is the perimeter?

57. Volume   The volume occupied by a swimming pool is $12x^6 + 54x^5 + 60x^4$ cu. ft. Find the depth if the width is $6x^4$ and the length is $2x + 5$.

58. Volume   The volume of cereal occupied by the box in Illustration 1 is $12x^3 + 2x^2 - 2x$. Find the width of the box.

ILLUSTRATION 1

REVIEW

59. If the dimensions of a crate are $2x - 3$ units high by $x + 2$ units wide by $3x$ units long, find the polynomial expression that represents the volume.

60. Use the STOre feature of your calculator to compute the volume of the crate in Exercise 59 if $x = 2.5$ ft.

61. Simplify:

  a. $(4x^2)^3$ **b.** $\dfrac{4x^3y}{8x^2y^2}$ **c.** $\dfrac{(2x)^2}{(6x)^3}$

62. The dimensions of the crate in Exercise 59 are $2x - 3$, $x + 2$, and $3x$. Translate each of these dimensions to an English phrase.

63. Translate each English phrase to an algebraic expression and then find the sum of the expressions: 4 more than x, twice x decreased by 5, 4 times x.

CHAPTER SUMMARY

CONCEPTS

REVIEW EXERCISES

Section 2.1

Properties of exponents. If n is a natural number,

$$\text{then } x^n = \overbrace{x \cdot x \cdot x \cdot \ldots \cdot x}^{n \text{ factors of } x}$$

If m and n are integers, then

$x^m x^n = x^{m+n}$

$(x^m)^n = x^{mn}$

$(xy)^n = x^n y^n$

$\left(\dfrac{x}{y}\right)^n = \dfrac{x^n}{y^n}$ provided $y \neq 0$

$\dfrac{x^m}{x^n} = x^{m-n}$ provided $x \neq 0$

$x^0 = 1$ provided $x \neq 0$

Remember that most graphing calculators cannot perform symbolic operations. Expressions entered with variables will be evaluated, not simplified, at whatever value is stored for that specific variable.

Whole Number Exponents

1. Simplify each expression. Confirm results with the calculator, copying the screen display to justify your work.
   - **a.** 5^3
   - **b.** 3^5
   - **c.** $(-8)^2$
   - **d.** -8^2
   - **e.** $3^2 + 2^2$
   - **f.** $(3 + 2)^2$
   - **g.** $3(3^3 + 3^3)$
   - **h.** $1^{17} + 17^1$

2. Simplify each expression.
   - **a.** $x^3 x^2$
   - **b.** $x(x^2 y)$
   - **c.** $y^7 y^3$
   - **d.** $x^0 y^5$
   - **e.** $2b^3 b^4 b^5$
   - **f.** $(-z^2)(z^3 y^2)$
   - **g.** $(4^4 x)s^2$
   - **h.** $-3y(y^5)$

3. Simplify each expression.
   - **a.** $(x^2 x^3)^3$
   - **b.** $(2x^2 y)^2$
   - **c.** $(3x^0)^2$
   - **d.** $(3x^2 y^2)^0$
   - **e.** $\dfrac{x^7}{x^3}$
   - **f.** $\left(\dfrac{x^2 y^3}{xy^2}\right)^2$
   - **g.** $\dfrac{8(y^2 x)^2}{2^3 (yx)^2}$
   - **h.** $\dfrac{(5x^0 y^2 z^3)^3}{25(yz)^5}$

Section 2.2

The degree of a monomial with several variables is the sum of the exponents of those variables.

The degree of a polynomial is the same as the degree of its term with the highest degree.

If $P(x)$ is a polynomial in x, then $P(r)$ is the value of the polynomial when r is substituted for x, i.e., $x = r$.

Algebraic Expressions and Polynomials

4. Give the degree of each polynomial below and classify it as a monomial, a binomial, or a trinomial.
   - **a.** $13x^7$
   - **b.** $5^3 x + x^2$
   - **c.** $-3x^5 + x - 1$
   - **d.** $9x + 21x^3$

5. Let $P(x) = 5x^4 - x$. Find each value. Confirm your result with the calculator, copying the screen display exactly as it appears to justify your results.
   - **a.** $P(3)$
   - **b.** $P(0)$
   - **c.** $P(-2)$
   - **d.** $P(2t)$

6. Let a, b, and c represent three real numbers. Write an algebraic expression to represent each quantity.
   - **a.** a decreased by c
   - **b.** the sum of b and c
   - **c.** the product of a and b

7. **Investing** George has $10,000 to invest. If he invests b dollars in a money market account, how much money is left to invest in stocks?

8. **Mileage** A car averages 55 miles an hour on a trip. Write expressions to represent the number of miles traveled in 2 hours, 5 hours, and b hours.

Section 2.3	**Adding and Subtracting Polynomials**

When adding polynomials, remove parentheses and combine like terms by adding or subtracting the numerical coefficients and using the same variables and the same exponents.

When subtracting polynomials, be careful to distribute the operation of subtraction when removing the parentheses.

9. Simplify each expression.
  a. $3x + 5x - x$ **b.** $3x + 2y$
  c. $(xy)^2 + 3x^2y^2$ **d.** $-2x^2yz + 3yx^2z$
  e. $3x^2y^0 + 2x^2$ **f.** $2(x + 7) + 3(x + 7)$
  g. $(7a^2 + 2a - 5) - (3a^2 - 2a + 1)$
  h. $4(4x^3 + 2x^2 - 3x - 8) - 5(2x^3 - 3x + 8)$

10. Find the polynomial that represents the perimeter of the pictured figure.

11. Find the difference when $4x^2 - 5x$ is subtracted from the sum of $3x^2 + 2x - 9$ and $5x^2 + 6x + 1$.

Section 2.4	**Multiplying Polynomials**

To multiply two monomials, first multiply the numerical factors and then multiply the variable factors.

To multiply a polynomial with more than one term by a monomial, multiply each term of the polynomial by the monomial and simplify.

To multiply one polynomial by another, multiply each term of one polynomial by each term of the other polynomial and simplify.

To multiply two binomials, use the FOIL method.

Special products:
$(x + y)^2 = x^2 + 2xy + y^2$
$(x - y)^2 = x^2 - 2xy + y^2$
$(x + y)(x - y) = x^2 - y^2$

12. Find each product.
  a. $(2x^2y^3)(5xy^2)$ **b.** $5(x + 3)$
  c. $x^2(3x^2 - 5)$ **d.** $-3xy(xy - x)$
  e. $(x + 3)(x + 2)$

13. Find each product.
  a. $(3a - 3)(2a + 2)$ **b.** $(a - 1)(a + 1)$
  c. $6(y - 2)(y + 2)$ **d.** $(x + 4)^2$
  e. $(x - 3)^2$

14. Find the polynomial that represents the area of the pictured rectangle.

15. Find the product of $y^2 + 1$ and its conjugate.

Section 2.5	**Dividing Polynomials by Monomials**

To simplify a fraction, divide out all factors common to the numerator and the denominator of the fraction.

To divide a polynomial by a monomial, rewrite the division as a product, use the Distributive Law to remove parentheses, and simplify each resulting fraction.

16. Divide each polynomial by the given monomial.
  a. $\dfrac{3x + 6y}{2xy}$ **b.** $\dfrac{14xy - 21x}{7xy}$
  c. $\dfrac{15a^2bc + 20ab^2c - 25abc^2}{-5abc}$
  d. $\dfrac{(x + y)^2 + (x - y)^2}{-2xy}$

Section 2.6	Dividing Polynomials by Polynomials

Use long division to divide one polynomial by another.	**17.** Perform the divisions.

a. $x - 1\overline{)x^2 - 6x + 5}$ **b.** $x + 2\overline{)x^2 + 3x + 5}$

c. $3x - 1\overline{)3x^2 + 14x - 2}$ **d.** $3x + 1\overline{)-13x - 4 + 9x^3}$

◼ Chapter Test

1. Use exponents to rewrite $2\,x\,x\,x\,y\,y\,y\,y$

In Problems 2–4, write each expression as an expression containing only one exponent.

2. $y^2(y\,y^3)$ **3.** $(2x^3)^5(x^2)^3$ **4.** $(2r\,r^2\,r^3)^3$

5. Simplify: $3x^0$.

6. Identify $3x^2 + 2$ as a monomial, a binomial, or a trinomial.

7. Find the degree of the polynomial $3x^2y^3z^4 + 2x^3y^2z - 5x^2y^3z^5$.

8. If $P(x) = x^2 + x - 2$, find $P(-2)$.

9. Simplify: $(xy)^2 + 5x^2y^2 - (3x)^2y^2$.

10. Add: $(3x^3 + 4x^2 - x - 7) + (2x^3 - 2x^2 + 3x + 2)$.

11. Subtract: $(2x^2 - 7x + 3) - (3x^2 - 2x - 1)$.

12. Simplify: $-6(x - y) + 2(x + y) - 3(x + 2y)$.

13. Simplify: $-2(x^2 + 3x - 1) - 3(x^2 - x + 2) + 5(x^2 + 2)$.

In Problems 14–17, find each product.

14. $(-2x^3)(2x^2y)$ **15.** $3y^2(y^2 - 2y + 3)$ **16.** $(2x - 5)(3x + 4)$ **17.** $(2x - 3)(x^2 - 2x + 4)$

18. Simplify the fraction: $\dfrac{8x^2y^3z^4}{16x^3y^2z^4}$.

19. Do the division: $\dfrac{6a^2 - 12b^2}{24ab}$.

20. Divide: $2x + 3\overline{)2x^2 - x - 6}$.

◼ Cumulative Review: Chapters 1 and 2

VOCABULARY / CONCEPTS

1. A number greater than one, whose only factors are one and itself is called a _____ number.

✍ **2.** Explain, in a step-by-step format, how to find the least common denominator for two fractions.

Following your step-by-step directions, find the least common denominator for $\frac{5}{24}$ and $\frac{7}{18}$.

✍ **3.** Explain the difference between the expressions $4x$ and x^4.

✍ **4.** Explain the difference between *evaluate* and *simplify* when applied to the expression $3x + 5x$.

5. Determine the absolute value of each expression and verify with your calculator.
   a. $|-3|$ **b.** $-|-3|$ **c.** $-|3|$

6. What rule supports the statement $-(-5) = 5$?

✍ **7.** Explain why $\frac{x}{0}$ is undefined and $\frac{0}{x} = 0$ provided $x \neq 0$.

8. Which of the following represent a product?
   a. $(-4) - (4)$ **b.** $5(-3)$
   c. $-4(-3)$ **d.** $6 - (-6)$

9. List the numbers in the set $\left\{-5, -\frac{7}{2}, -1, 0, \sqrt{2}, 4.2, 6\right\}$ that are
   a. whole numbers
   b. integers
   c. rational numbers
   d. irrational numbers
   e. nonnegative integers

10. Determine whether perimeter, area, or volume would be used to determine the measure of the indicated quantity.
    a. molding for a window frame
    b. water in a hot tub
    c. sand on a barge
    d. carpeting for a living room

11. Complete the Laws of Exponents. Assume m and n are natural numbers.
    a. $x^m \cdot x^n = $ _____
    b. $(x^m)^n = $ _____
    c. $(xy)^n = $ _____
    d. $\left(\dfrac{x}{y}\right)^n = $ _____ $(y \neq 0)$
    e. $\dfrac{x^m}{x^n} = $ _____ provided $m > n$ and $x \neq 0$.
    f. $x^0 = $ __ provided $x \neq 0$.

12. In the expression $3x^5$, identify the base and the exponent.

13. In the expression $3x^5$, 3 is called the _____ of the term.

14. Polynomials consisting of two terms are called _____.

15. If George lives 15 blocks from his office in downtown Chicago and Kevin lives x blocks further from the same office building, write an expression to describe how far Kevin lives from the office building.

16. If $4(x + y) = 4x + 4y$, then the _____ property has been applied.

17. Which of the following is equal to $(x + 2)^3$?
    a. $x^3 + 8$ b. $(x + 2)(x + 2)(x + 2)$

18. The binomials $(4 + x)$ and $(4 - x)$ are called _____ binomials and have a product of _____.

19. Divide by using the Distributive Property: $\dfrac{4x + 12}{3}$.

20. Arrange in descending order by the powers of x: $4 - 3x - x^2$

PRACTICE *In Exercises 21–28, perform the indicated operation and verify with your calculator.*

21. $\dfrac{5}{6} \cdot \dfrac{3}{20}$

22. $\dfrac{2}{3}\left(\dfrac{4}{5} \div \dfrac{28}{35}\right)$

23. $5^3 - (5 - 1)^3$

24. $4 + 2 \cdot 3 - [6 - (5 - 1)]$

25. $-|8 - 12| - 5$

26. $-\dfrac{7}{9} - \dfrac{7}{3}$

27. $\dfrac{-4 - 11}{4 - 7}$

28. $\dfrac{3(2^2) - 3}{-2^3 - 1}$

29. Arithmetically evaluate $zx + x^2 - z$ for $x = 3$ and $z = 2$ and confirm with the STOre feature of your calculator.

30. Arithmetically evaluate $\dfrac{x^2 - y}{x + y}$ for $x = -2$ and $y = -3$ and confirm with the STOre feature of your calculator.

In Exercises 31–32, write an expression involving only one exponent.

31. $-3x(x^4)$

32. $\left(\dfrac{x^8}{x^2}\right)^3$

33. Translate each English phrase to an algebraic expression.
    a. the product of x and y
    b. twice x increased by 4

34. If $P(x) = 3x^2 - 4x + 2$, find
    a. the degree of $P(x)$, and
    b. $P(-2)$. Verify with your calculator.

35. Simplify: $(4x^2 - 2x + 6) - 2(x^2 - 2x - 3)$.

36. Find the difference when $2x + 4y$ is subtracted from the sum of $3x + 6y$ and $x - 4y$.

37. Multiply: $(3x + 5)(x^2 - 3x + 1)$.

38. Simplify: $(x + 2)^2 - (x + 3)^2$

39. Divide: $\dfrac{-30x^4 + 25xy^2 - 5xy}{-5xy}$.

40. Divide: $\dfrac{8x^3 - 6x^2 + 3x - 2}{2x + 1}$. If there is a remainder, express the answer in the form: $quotient + \frac{remainder}{divisor}$.

41. Use the STOre feature to evaluate $(3x^2y)^2$ when $x = -3$ and $y = \frac{1}{3}$.

42. If $P(x) = -4x^2 - 5x + 2$, circle the line in the TABLE that indicates the value of $P(-1)$.

43. You are performing the computation $\dfrac{4 + 8}{3}$. Make the appropriate corrections on the displayed screen to ensure the correct result.

44. Simplify with your calculator. Copy your complete screen display to justify your result.
$10^2 + [10^3 \cdot (18 \div 2) - 3]$

45. Simplify with your calculator. Copy your complete screen display to justify your result.
$$\frac{|4.02 - 0.001|}{|0.003 - 0.01|}$$

46. An arithmetic expression is displayed on the screen at the right. What would this expression look like if written in this textbook?

APPLICATIONS

47. Course enrollment   In the fall semester of 1998, 24 students were enrolled in the 12:30 P.M. Tuesday-Thursday Intermediate Algebra Course. Five-sixths of these students graduated from high school in the spring of 1998.
   a. Write an arithmetic expression to determine how many students in the course graduated from high school in the spring of 1998. Simplify your expression.
   b. Of this group of recent high school graduates, $\frac{3}{4}$ had taken Algebra I and Algebra II in high school. Write an arithmetic expression to determine how many students had not taken Algebra I and Algebra II. Simplify your expression.

48. Telephone solicitor   A telephone solicitor earns $2.50 per hour plus 10 cents per call. Write an arithmetic expression to determine how much money is earned by the solicitor in one hour if 20 calls are made. Simplify your expression.

49. Room dimensions   Find the length of a room whose width is $x + 4$ and whose area is $x^3 + 5x^2 + 6x + 8$.

50. Grade distribution   At the end of the semester an instructor's grade distributions are as indicated in the chart below. Find an expression to model the total number of grades and thus the total number of students. Express your result as a simplified polynomial expression.

As	x grades
Bs	double the number of As
Cs	3 times as many As
Ds	4 more than the As
Fs	1 more than the As
Ws	2 less than the As

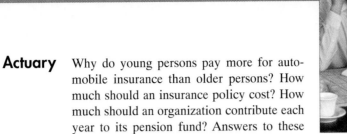

Actuary

Why do young persons pay more for automobile insurance than older persons? How much should an insurance policy cost? How much should an organization contribute each year to its pension fund? Answers to these and similar questions are provided by actuaries, who design insurance and pension plans and follow their performance to make sure that the plans are maintained on a sound financial basis.

Qualifications

A good educational background for a beginning job in a large life insurance or casualty insurance company is a bachelor's degree in mathematics or statistics; a degree in actuarial science is even better. Companies prefer well-rounded individuals with a liberal arts background, including social science and communications courses, as well as a good technical background.

Job Outlook

Shifts in the age distribution of the population will result in large increases in the number of people with established careers and family responsibilities. This group will account for increases in sales of insurance and pension plans. The outlook for employment is very good through the year 2005. Job opportunities will be best for new college graduates who have passed at least two actuarial examinations while still in school and have a strong mathematical and statistical background.

Example Application 1

An **annuity** is a sequence of equal payments made periodically over a length of time. The sum of the payments and the interest earned during the **term** of the annuity is called the **amount** of the annuity.

After a sales clerk works 6 months, her employer will begin an annuity for her and will contribute $500 semiannually to a fund that pays 8% annual interest. After she has been employed two years, what will be the amount of her annuity?

Solution

Because the payments are to be made semiannually, there will be four payments of $500, each earning a rate of 4% per 6-month period. These payments will occur at the end of 6 months, 12 months, 18 months, and 24 months. The first payment, to be made after 6 months, will earn interest for three interest periods. Thus, the amount of the first payment after two years is $500 $(1.04)^3$. The amounts of each of the four payments after two years are shown in Table 1.

Payment (at the end of period)	Amount of Payment at the End of 2 Years
1	$500(1.04)^3 = \$ 562.43$
2	$500(1.04)^2 = \$ 540.80$
3	$500(1.04)^1 = \$ 520.00$
4	$500 = \$ 500.00$
	$A_n = \$2123.23$

TABLE 1

The amount of the annuity is the sum of the amounts of the individual payments. This sum is $2123.23.

Example Application 2   The **present value** of the above annuity is the lump-sum principal that must be invested on the first day of her employment at 8% annual interest, compounded semiannually, to grow to $2123.23 at the end of two years. Find the present value of the annuity.

Solution   A sum of $500(1.04)^{-1}$ invested at 4% interest per 6-month period will grow to be the first payment of $500, to be made after 6 months. A sum of $500(1.04)^{-2}$ will grow to be the second payment of $500, to be made after 12 months. The present values of each of the four payments are shown in Table 2.

Payment	Present Value of Each Payment
1	$500(1.04)^{-1} = \$\ 480.77$
2	$500(1.04)^{-2} = \$\ 462.28$
3	$500(1.04)^{-3} = \$\ 444.50$
4	$500(1.04)^{-4} = \$\ 427.40$
	Present value = $1814.95

TABLE 2

The present value of the annuity is $1814.95. Verify that $1814.95 will grow to $2123.23 in two years at 8% annual interest, compounded semiannually.

EXERCISES

1. Find the amount of an annuity if $1000 is paid semiannually for two years at 6% annual interest. Assume that the first of the four payments is made immediately.

2. Find the present value of the annuity of Exercise 1.

3. Note that the amounts for the payments in Table 1 form a geometric progression. Verify the answer for the first example by using the formula for the sum of a geometric progression.

4. Note that the present values for the payments in Table 2 form a geometric progression. Verify the answer for the second example by using the formula for the sum of a geometric progression.

Modeling Linear Equations in One and Two Variables

3

■ PERSPECTIVE

The concept of equations has a long history, and the techniques we will study in this chapter have been developed over many centuries.

The mathematical notation that we use today is the result of thousands of years of development. The ancient Egyptians used a word for variables, best translated as *heap*. Others used the word *res*, which is Latin for

thing. In the 15th century, the letters *p*: and *m*: were used for *plus* and *minus*. What we would now write as $2x + 3 = 5$ might have appeared to those early mathematicians as

2 *res p*: 3 *aequalis* 5

To explore the relationships between different quantities, equations are often used. This chapter investigates solutions of linear equations (both algebraically and graphically) and applications of linear equations. Equations in both one and two variables are discussed.

3.1 Introduction to Linear Equations in One Variable

■ BASIC DEFINITIONS ■ PROPERTIES OF EQUALITY ■ ALGEBRAIC SOLUTIONS OF LINEAR EQUATIONS IN ONE VARIABLE ■ TYPES OF EQUATIONS AND THEIR SOLUTIONS ■ INTRODUCTION TO PROBLEM SOLVING

■ BASIC DEFINITIONS

An **equation** is a statement indicating that two quantities are equal. Below are some examples of equations.

$$x + 5 = 21 \qquad 3(3x + 4) = 3 + x \qquad 3x^2 - 4x + 5 = 0$$

In the equation $x + 5 = 21$, $x + 5$ is called the **left-hand side,** and 21 is called the **right-hand side.** The letter x is called a **variable.** The initial discussion will focus on **linear equations in one variable.** A linear equation in one variable, x, is any equation that *can be* written in the form $ax + c = 0$, where a and c are real numbers and $a \neq 0$.

An equation can be true or false. It is true when the left side is equivalent to the right side. The equation $16 + 5 = 21$ is true, whereas the equation $10 + 5 = 21$ is false. An equation that contains a variable may be true or false, depending upon the value of that variable. Consider the equation $2x - 5 = 11$. If $x = 8$, the equation is true because

$$
\begin{aligned}
2x - 5 &= 11 \\
2(8) - 5 &= 11 \\
16 - 5 &= 11 \\
11 &= 11
\end{aligned}
$$

However, this equation is false for all other values of x. Any number that makes an equation true when substituted for its variable is said to satisfy the equation. All numbers that satisfy an equation are called its **solution(s)** or **roots.** Because the number 8 is the only number that satisfies the equation $2x - 5 = 11$, it is its only solution.

Often solutions to equations are expressed as **solution sets.** In the example above the solution set of the equation is {8}. A solution set, therefore, is simply a set that contains the solution(s) of an equation.

EXAMPLE 1   Is 5 a solution of $3x - 5 = 2x$? The substitution property of equality allows us to replace x with any value and simplify the resulting expressions. Only the solution will yield a true equation.

Solution   We substitute 5 for x in the equation and simplify.

$$3x - 5 = 2x$$
$$3 \cdot 5 - 5 = 2 \cdot 5 \qquad \text{Replace } x \text{ with } 5$$
$$15 - 5 = 10$$
$$10 = 10$$

Because the left side is equivalent to the right side when x is replaced by 5, 5 is the solution or root of the equation.

Store the value of 5 for the variable x and confirm that the left side of the equation is equivalent to the right side.

EXAMPLE 2   Which of the following numbers, 2, 4, or $-3/2$, are solutions of $2x^2 = 5x + 12$?

Solution
$$2(2)^2 = 5(2) + 12$$
$$2(4) = 10 + 12$$
$$8 \neq 22$$

Because $8 \neq 22$, 2 is not a solution.

$$2(4)^2 = 5(4) + 12$$
$$2(16) = 20 + 12$$
$$32 = 32$$

Because $32 = 32$, 4 is a solution.

$$2\left(-\frac{3}{2}\right)^2 = 5\left(-\frac{3}{2}\right) + 12$$
$$\frac{9}{2} = \left(-\frac{15}{2}\right) + 12$$
$$\frac{9}{2} = -\frac{15}{2} + \frac{24}{2}$$
$$\frac{9}{2} = \frac{9}{2}$$

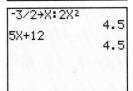

■ **TECHNOLOGY TIP**
If desired, the FRAC option of the graphing calculator can be used to perform arithmetic operations with fractions.

Because $\frac{9}{2} = \frac{9}{2}$, $-\frac{3}{2}$ is a solution.

The equation in Example 2 is a second-degree or quadratic equation. We will begin our discussion of solving equations with first-degree linear equations; second-degree equations will be solved in Chapter 5.

■ PROPERTIES OF EQUALITY

The four properties of equality defined and illustrated below are used to algebraically solve linear equations. Recall that an equation has a left side and a right side and is a statement that these two sides are equivalent. Thus, relationships between constants and variables must be preserved. The properties of equality are the rules that must be followed to preserve equality.

Addition Property of Equality

Suppose that a, b, and c are real numbers. It follows that

If $a = b$, then $a + c = b + c$.

Subtraction Property of Equality

Suppose that a, b, and c are real numbers. It follows that

If $a = b$, then $a - c = b - c$.

Multiplication Property of Equality

Suppose that a, b, and c are real numbers and $c \neq 0$. It follows that

If $a = b$, then $a \cdot c = b \cdot c$.

Division Property of Equality

Suppose that a, b, and c are real numbers and $c \neq 0$. It follows that

If $a = b$, then $\dfrac{a}{c} = \dfrac{b}{c}$.

Stated informally, the properties allow you to add or subtract the same quantity to both *sides* of an equation and to multiply or divide both *sides* of an equation by the same *nonzero* quantity. Careful application of these properties is used to algebraically solve linear equations.

Linear equations are solved by writing a series of **equivalent equations.** Equivalent equations are equations that have the same solution set.

EXPLORING THE CONCEPT

Consider the mathematical statement $x = 12$. This is an equation because it contains an equal sign and is true because replacement of x by 12 yields the statement $12 = 12$. Adding the number 2 to both sides of the equation results in a new equation, $x + 2 = 12 + 2$ or $x + 2 = 14$. The new equation is equivalent to the original equation because, again, when the variable is replaced by 12 the new equation is true. The original equation ($x = 12$) and the new equation ($x + 2 = 14$) have the same solution. Therefore, they are **equivalent equations.**

Create three equivalent equations by

a. Subtracting 4 from each side of the original equation ($x = 12$).

b. Multiplying each side of the original equation ($x = 12$) by 4.

c. Dividing each side of the original equation ($x = 12$) by 4.

Conclusion

a. $x - 4 = 8$; this is an equivalent equation to $x = 12$ because replacing the variable x with 12 results in $12 - 4 = 8$.

b. $4x = 48$; this is an equivalent equation to $x = 12$ because replacing the variable x with 12 results in $4 \cdot 12 = 48$.

c. $\frac{x}{4} = 3$; this is an equivalent equation to $x = 12$ because replacing the variable x with 12 results in $\frac{12}{4} = 3$. ∎

Adding, subtracting, multiplying, or dividing both sides of an equation by the same constant (excluding zero for the operations of multiplication and division) results in a series of equivalent equations.

We will now examine the use of the properties of equality in solving an equation by transformation.

■ ALGEBRAIC SOLUTIONS OF LINEAR EQUATIONS IN ONE VARIABLE

To **solve** an equation means to find the value(s) of the variable(s) that will make the equation a true statement. The equations in this chapter will be solved by writing a series of equivalent equations.

When solving equations it is best to first simplify both sides of the equation and then begin transforming the equation by using the properties of equality to write a series of equivalent equations. To solve any equation, we must isolate the variable on one side. This can be a multistep process that may require combining like terms. As we solve equations, we will follow the steps below.

> **Steps for Solving a Linear Equation in One Variable**
> 1. Use the distributive property to remove grouping symbols.
> 2. Combine like terms on each **side** of the equation.
> 3. Clear fractions, if desired, by multiplying each **term** on both sides of the equation by the LCD.
> 4. Use the addition or subtraction property of equality to collect all terms containing the variable on one side of the equation and all constants on the other side.
> 5. Use the multiplication or division property of equality to isolate the variable. The equation should be in the form:
>   **variable = constant**
> 6. Check by ensuring that the proposed solution produces a true statement when the variable is replaced by the constant. Always use the original equation when checking.

EXAMPLE 3   Solve $\dfrac{5x}{4} = 9$.

Solution

$4 \cdot \dfrac{5x}{4} = 4 \cdot 9$ Use the multiplication property of equality to clear the fraction.

$5x = 36$ Simplify both sides.

$\dfrac{5x}{5} = \dfrac{36}{5}$ Use the division property of equality to isolate the variable.

$x = \dfrac{36}{5}$ Simplify.

Therefore, $\frac{36}{5}$ is the solution of the original equation. The check is shown below, both analytically (paper-and-pencil computation) and by using the STOre feature of the calculator:

$\dfrac{5 \cdot \dfrac{36}{5}}{4} = 9$

$\dfrac{36}{4} = 9$

$9 = 9$

EXAMPLE 4 Solve $\dfrac{x}{3} + 4 = 5$.

Solution

$$3 \cdot \frac{x}{3} + 3 \cdot 4 = 3 \cdot 5$$ Use the multiplication property of equality to clear the equation of fractions. Each **term** of both **sides** of the equation must be multiplied by the LCD.

$$x + 12 = 15$$ Simplify both sides of the equation.

$$x + 12 - 12 = 15 - 12$$ Subtract 12 from both **sides** of the equation to place all constants on the right (notice the variable is on the left).

$$x + 0 = 3$$ Because zero is the *additive identity*, $x + 0$ is equivalent to x.

$$x = 3$$

The solution, or root, is 3. Verify with the calculator. ∎

```
3→X:X/3+4
            5
■
```

EXAMPLE 5 Solve $-4x - 8x + 5 = 17$.

Solution

$$-12x + 5 = 17$$ Combine like terms.

$$-12x + 5 - 5 = 17 - 5$$ Use the subtraction property of equality to place constants together on one side of the equation.

$$-12x = 12$$ Simplify.

$$\frac{-12x}{-12} = \frac{12}{-12}$$ Use the division property of equality to isolate the variable.

$$x = -1$$

The solution set is $\{-1\}$. Verify with the calculator. ∎

```
-1→X: -4X-8X+5
             17
■
```

EXAMPLE 6 Solve $2x - 5 = x + 4$.

Solution

$$2x - 5 = x + 4$$

$$2x - 5 - x = x + 4 - x$$ The subtraction property of equality allows us to subtract x from both sides of the equation.

$$x - 5 = 4$$ Simplify both sides.

$$x - 5 + 5 = 4 + 5$$ The addition property of equality allows us to add 5 to both sides of the equation.

$$x = 9$$ Simplify both sides.

The equation $2x - 5 = x + 4$ is true when x has a value of 9. The solution (root) is 9; the solution set is $\{9\}$. To solve the equation, a series of equivalent equations was written using the properties of equality, which guarantee preservation of the relationship of equality. Verify with the calculator. ∎

```
9→X:2X-5
            13
X+4
            13
```

EXAMPLE 7 Solve $3(x + 2) - 2x = 4x$.

Solution

$$3x + 6 - 2x = 4x$$ Use the distributive property to remove grouping.

$$x + 6 = 4x$$ Combine like terms.

$$x + 6 - x = 4x - x$$ Use the subtraction property of equality to get all variable terms together.

$$6 = 3x$$ Simplify each side.

$$\frac{6}{3} = \frac{3x}{3}$$ Use the division property of equality to isolate the variable.

$$2 = x$$ Simplify each side. ∎

Some students do not like having the variable on the right. However, the relationship of equality is **symmetric.** This property ensures that entire *sides* of equations may be transposed. Therefore, $2 = x$ is an equivalent mathematical statement to $x = 2$. The variable terms could have been placed on the left side of the equation. This is illustrated below, beginning with the step in bold print above. The steps prior to that would remain the same regardless of where you prefer the variable to be.

$$x + 6 = 4x$$

$$x + 6 - \mathbf{4x} = 4x - \mathbf{4x}$$ Use the subtraction property of equality.

$$-3x + 6 = 0$$ Simplify both sides.

$$-3x + 6 - \mathbf{6} = 6 - \mathbf{6}$$ Use the subtraction property of equality to place the constant on the right side of the equation.

$$-3x = -6$$ Simplify.

$$\frac{-3x}{-3} = \frac{-6}{-3}$$ Use the division property of equality to isolate the variable.

$$x = 2$$

Because the properties of equality have been carefully applied, placement of the variable does not affect the solution to the equation. Verify with the calculator.

■ TYPES OF EQUATIONS AND THEIR SOLUTIONS

The equations in Examples 1–7 are true for some real numbers. These equations are called **conditional equations** because they are true for some, but not all, real numbers.

An equation that is true for all real numbers is called an **identity.** Regardless of the value used to replace the variable, the equation will be true. For example, the equation $x + x = 2x$ is an identity because it is true regardless of the real number substituted in for the variable x. Its solution is **all real numbers.** This can be expressed with the symbol $\mathbb{R}$, which means "the set of all real numbers."

An equation that has no solution is called a **contradiction.** The equation $x = x + 1$ is a contradiction because no number is equal to itself plus 1. Its solution is **no solution,** which can be represented with the symbol $\varnothing$ representing the **null set** or the set that contains no elements.

> **WARNING!** Do not confuse the **type** of equation with its **solution.** The terms *conditional, identity,* and *contradiction* refer to the type. Remember the solution or root is all real numbers that make the equation a true statement.

Examples 8 and 9 illustrate the solutions of equations that represent a contradiction and an identity.

EXAMPLE 8   Solve $6(x + 1) - 4x = 2(x + 1) + 1$.

Solution

$$6x + 6 - 4x = 2x + 2 + 1$$ Use the distributive property to clear grouping.

$$2x + 6 = 2x + 3$$ Combine like terms on each side of the equation.

$$2x + 6 - 2x = 2x + 3 - 2x$$ Use the subtraction property of equality to put all the variables on one side of the equation.

$$6 = 3$$ The result is a false statement.

■ STUDY TIP
When designating an *empty set* use either the notation ϕ (the Greek letter phi) or { }. Do not use the notation {ϕ} as this set is not empty—it contains one element.

The equation is a contradiction and has **no solution.** This is often denoted with the symbol $\varnothing$ or { } to denote an empty set.   ■

Think about the equation printed in bold in Example 8. At that point in the solving process it is obvious that the equation has no solution. There is no real number that you can double and add six to that will equal itself doubled plus three.

EXAMPLE 9   Solve $x + 6 = 2\left(3 + \dfrac{x}{2}\right)$.

Solution   Use the distributive property to clear grouping symbols and yield

$$x + 6 = 6 + x$$

■ STUDY TIP

When learning to solve equations students often check solutions only by referring to the back of the text or comparing answers with a study partner. It is important to check all solutions so that the process becomes a natural part of problem solving.

Because the left side of the equation is equivalent to the right side, the equation is an identity. There is no need to proceed further. Therefore the solution is **all real numbers,** which is often symbolized as $\mathbb{R}$. ■

Carefully look back over the examples in this section. As you work through the exercise for this section, keep in mind that NOT all the equations you will be asked to solve will be conditional equations. Therefore, be very careful.

■ INTRODUCTION TO PROBLEM SOLVING

Application problems illustrate the power of algebra in solving practical problems, or ones that occur every day. Several types of application problems will be considered in the rest of this section and throughout the text. However, rarely do real-life applications fall into these categories. The emphasis, therefore, will be on developing a repertoire of problem-solving *tools*. The strategies below will help you begin. It is suggested that these be copied on a note card for quick reference when working application problems—not just in this section, but throughout the text.

Problem Solving Strategies

1. Read the problem several times and organize the facts. What is known? What are you asked to find? It may help you to organize the facts under the two headings, **known** and **unknown.**

2. **a.** If there is only one unknown quantity, choose a variable to represent that quantity. Clearly define what the variable is and what it represents with a "Let" statement.

   b. When there is more than one unknown, choose a variable to represent one of the quantities and express all other important quantities in terms of the chosen variable.

3. Translate the information expressed in the words of the problem to an algebraic equation.

4. Solve the equation.

5. Answer the question posed in the problem.

6. Check your results in terms of both the equation and with the words of the problem. Make sure your solution makes sense within the context of the problem.

EXAMPLE 10   Five more than double a number is 19. What is the number?

Solution   Because there is only one unknown, x will represent the unknown number. This is expressed by using a "Let" statement.

Let x = the number   Clearly define the variable and what it represents.

Five **more than double a number** is 19:

$$5 \quad + \quad 2\cdot \quad x \quad = 19$$

The words of the problem are translated to an algebraic equation: $5 + 2x = 19$.

$$5 + 2x = 19 \qquad \text{Solve the equation.}$$
$$5 + 2x - 5 = 19 - 5$$
$$2x = 14$$
$$\frac{2x}{2} = \frac{14}{2}$$
$$x = 7$$

The number is 7.   Answer the question.

Make sure your result makes sense in **both** the equation and in the context of the problem. Checking the equation can be done by the tools previously introduced: analytically (substitution) or by using the STOre feature of the calculator. ■

EXAMPLE 11   A rental car agency charges a flat fee of $37 a day plus $0.21 per mile driven. If the rental charge for one day is $61.15, how many miles were driven?

Solution   Let x = number of miles driven.   Clearly identify a variable and what it represents.

$37 a day **plus $0.21 per mile** translates to the expression

$$37 \qquad + \quad .21 \quad \cdot \quad x$$

The *total* rental charge is $61.15, so the resulting equation is

$$37 + .21x = 61.15$$

Now solve the equation:

$$37 + .21x = 61.15$$
$$37 + .21x - 37 = 61.15 - 37$$
$$.21x = 24.15$$
$$\frac{.21x}{.21} = \frac{24.15}{.21}$$
$$x = 115$$

The car was driven 115 miles.   Answer the question. ■

EXERCISE 3.1

VOCABULARY In Exercises 1–8, fill in the blank to make a true statement.

1. A _____ equation in one variable is any equation in one variable, x, that can be written in the form $ax + b = 0$.

2. All numbers that satisfy an equation are called _____ or _____.

3. Equations with the same solution are called _____ equations.

4. To _____ an equation means to find the value(s) for the variable(s) that will make the equation a true statement.

5. The statement that $x = 2$ is equivalent to the statement $2 = x$ is an example of the _____ property.

6. An equation that is true for *all* acceptable real numbers is called an _____.

7. A contradiction is an equation that is true for ___ real numbers.

8. An equation that is true for _____ real numbers is called a _____.

CONCEPTS

9. If an equation has no solution, then the solution set is represented by the symbol ___.

10. When the solution to an equation is the set of all real numbers, we represent it by the symbol ___.

11. In Exercises a–f, the calculator screens pictured show the verification of a solution to an equation. Write the equation that was being checked and express its solution as a solution set.

   a.

   b.

   c.

   d.

   e.

   f.

12. If x is the first integer, write expressions for each of the next two consecutive integers.

13. If x is the first even integer, write expressions for each of the next two consecutive even integers.

14. If x is the first odd integer, write expressions for each of the next two consecutive odd integers.

15. If x is the number, write an expression for each of the following:
   a. a number increased by 4
   b. a number decreased by 7

16. If x is the number, write an expression for each of the following:
   a. twice a number is decreased by 3
   b. half a number increased by 7

PRACTICE   *In Exercises 17–28, state whether the given number is a solution of the equation. Justify your response.*

17. $x + 2 = 3$; 1

18. $x - 2 = 4$; 6

19. $2x = 4$; 2

20. $3x = 6$; 3

21. $\dfrac{x}{5} = x$; 0

22. $\dfrac{x}{7} = 7x$; 0

23. $3k + 5 = 5k - 1$; 3

24. $2s - 1 = s + 7$; 6

25. $2(t + 5) - 3t = -2t + 1$; -9

26. $3(a - 5) + 2 = 3a - 13$; 6

27. $\dfrac{5 + x}{10} - x = \dfrac{1}{2}$; 0

28. $\dfrac{x - 5}{6} = 12 - x$; 11

In Exercises 29–38, solve each equation algebraically, showing ALL steps in the solution process. Solutions should be expressed as sets. Check all problems, either analytically or by using the STOre feature of the calculator, to verify roots.

29. $\dfrac{b}{3} = \dfrac{1}{3}$ **30.** $\dfrac{a}{13} = \dfrac{1}{26}$ **31.** $2v - \dfrac{7}{3} = -\dfrac{5}{6}$ **32.** $z + \dfrac{7}{9} = \dfrac{2}{9}$

33. $5(b + 7) = 6b - 2$ **34.** $8(b + 2) = 9b$

35. $2 + 3(x - 5) = 4(x - 1)$ **36.** $10x + 3(2 - x) = 5(x + 2) - 4$

37. $11x + 6(3 - x) = 3 + x$ **38.** $9(t - 1) = 6(t + 2) - t$

In Exercises 39–54, solve each equation algebraically, showing ALL steps in the solution process. Solutions should be expressed as sets. Check all problems, either analytically or by using the STOre feature of the calculator, to verify roots.

39. $9(x + 11) + 5(13 - x) = 2x.$ **40.** $3(x + 15) + 4(11 - x) = 5x$

41. $\dfrac{3(t - 7)}{2} = t - 6$ **42.** $\dfrac{2(t + 9)}{3} = t - 8$

43. $8x + 3(2 - x) = 5(x + 2) - 4$ **44.** $4(a + 2) = 2(a - 4) + 16$

45. $3(b - 1) + 18 = 4(b + 5)$ **46.** $x + 7 = \dfrac{2x + 6}{2} + 4$

47. $\dfrac{2(t - 1)}{6} - 2 = \dfrac{t + 2}{6}$ **48.** $2(3z + 4) = 2(3z - 2) + 13$

Recall that the decimal, 0.1, is equivalent to the fraction $\frac{1}{10}$. Therefore, if a problem contains decimals, you may clear them in exactly the same way as you clear fractions—by multiplying each term on both sides by a number that will clear the decimal.

49. $4.8x + 6 = 0.8x - 2$ **50.** $6.5x - 5.2 = 2.5x + 6.8$

51. $3.1(x - 2) = 1.3x + 2.8$ **52.** $0.6x - 0.8 = 0.8(2x - 1) - 0.7$

53. $3.2x + 5.64 = 5.6x + x - 1.16$ **54.** $0.65x + 2.7 = 0.05x + 2(x + 1)$

APPLICATIONS

55. Perimeter The perimeter of a square is 28 inches and each side is $(2x + 3)$ inches. Find the length of each side.
 a. Write an equation to model the perimeter.

 b. Simplify the equation.
 c. Solve the equation.
 d. Determine the length of each side.

56. Perimeter The perimeter of a square is 32 inches and each side is $(3x - 1)$ inches in length. Find the length of the sides of the square.
 a. Write an equation to model the perimeter.

 b. Simplify the equation.
 c. Solve the equation.
 d. Determine the length of the sides.

57. Perimeter The perimeter of a rectangle is 20 feet. The length is $(x - 4)$ feet and the width is x feet. Find the dimensions.
 a. Write an equation to model the perimeter.

 b. Simplify the equation.
 c. Solve the equation.
 d. Determine the dimensions of the rectangle.

58. Perimeter The perimeter of a rectangle is 250 feet. The length is $(2x + 5)$ feet and the width is $4x$ feet. Find the dimensions of the rectangle.
 a. Write an equation to model the perimeter.

 b. Simplify the equation.
 c. Solve the equation.
 d. Determine the length of each side.

In Exercises 59–66, (a) write an equation to model the problem, (b) simplify the equation, (c) solve the equation, and (d) answer the question posed.

59. Integer The sum of two consecutive even integers is 54. Find the integers.

60. Integer The sum of two consecutive odd integers is 88. Find the integers.

61. Number Twice a number increased by 5 is 15. Find the number.

62. Number Four less than three times a number is 17, find the number.

63. Cable charges   The basic cable rate is $29.95 a month, plus $3.75 for each pay-per-view movie. If a subscriber's total bill is $52.45, how many movies were viewed?

64. Catering charges   A wedding consultant charges $1250 plus $12.75 per guest for a wedding reception. How many guests attended the reception if the bill was $3940.25?

65. Kennel charges   A sportsman boarded his dog at the kennel for $16, plus $12 a day. If the stay cost $100, how many days was the owner gone?

66. Phone rates   A call to Tucson from a pay phone in Chicago cost $0.85 for the first minute and $0.27 for each additional minute or portion of a minute. If a student has $8.50 in change, how long can she talk?

REVIEW

67. Simplify each of the following:

    a. $\dfrac{4}{5} - \dfrac{1}{3}$ **b.** $\left(3\dfrac{1}{2}\right) \cdot \left(-\dfrac{3}{14}\right)$

    c. $-\dfrac{4}{5} \div \left(1\dfrac{1}{2}\right)$

69. Simplify: $(xx^7y^3)^2$

71. Simplify: $2x^2(x^3 - 4xy)$

68. Evaluate each of the following for $x = -1$ and $y = 2$:

    a. $(xx^7y^3)^2$ **b.** $2x^2(x^3 - 4xy)$

    c. $(2x - y)^2$

70. If you write $a(b + c) = ab + ac$, then the _____ property has been applied.

72. Simplify: $(2x - y)^2$

3.2   Literal Equations and Formulas

■ SOLVING LITERAL EQUATIONS   ■ FORMULAS

■ SOLVING LITERAL EQUATIONS

Equations with several variables are called **literal equations.** Often these equations are **formulas,** such as $A = lw$, the formula for finding the area of a rectangle. Suppose that we wish to find the lengths of several rectangles whose areas and widths are known. It would be tedious to substitute values for A and w into the formula and then repeatedly solve the formula for l. It would be better to solve the formula $A = lw$ for l first and then substitute values for A and w and compute l directly.

EXAMPLE 1   Solve $A = lw$ for l.

Solution   To isolate l on the right-hand side, we undo the multiplication by w by dividing both sides of the equation by w.

■ **STUDY TIP**
As much as possible, study when you are most alert. Beginning a two-hour study session at 11:00 P.M. is often nonproductive.

$$A = lw$$

$$\frac{A}{w} = \frac{lw}{w} \qquad \text{Divide both sides by } w.$$

$$\frac{A}{w} = l \qquad \text{Simplify.}$$

$$l = \frac{A}{w} \qquad \text{Symmetric property of equality.}$$

EXAMPLE 2   The formula $A = \frac{1}{2}bh$ gives the area of a triangle with base b and height h. Solve this formula for b.

Solution

$$A = \frac{1}{2}bh$$

$$2A = 2 \cdot \frac{1}{2}bh \qquad \text{Multiply both sides by 2.}$$

$$2A = bh \qquad \text{Simplify.}$$

$$\frac{2a}{h} = \frac{bh}{h} \qquad \text{Divide both sides by } h.$$

$$\frac{2A}{h} = b \qquad \text{Simplify.}$$

$$b = \frac{2A}{h} \qquad \text{Symmetric property of equality.}$$

If the area A and the height h of a triangle are known, the base b is given by the formula $b = \frac{2A}{h}$. ■

EXAMPLE 3   The formula $C = \frac{5}{9}(F - 32)$ is used to convert Fahrenheit temperature readings into their Celsius equivalents. Solve the formula for F.

Solution

$$C = \frac{5}{9}(F - 32)$$

$$\frac{9}{5}C = \frac{9}{5} \cdot \frac{5}{9}(F - 32) \qquad \text{Multiply both sides by } \frac{9}{5}.$$

$$\frac{9}{5}C = 1(F - 32) \qquad \text{Simplify.}$$

$$\frac{9}{5}C = F - 32 \qquad \text{Remove parentheses.}$$

$$\frac{9}{5}C + 32 = F - 32 + 32 \qquad \text{Add 32 to both sides.}$$

$$\frac{9}{5}C + 32 = F \qquad \text{Combine like terms.}$$

$$F = \frac{9}{5}C + 32 \qquad \text{Symmetric property of equality.}$$

The formula $F = \frac{9}{5}C + 32$ is used to convert degrees Celsius to degrees Fahrenheit.

■

EXAMPLE 4   The area A of the trapezoid shown in Figure 3-1 is given by the formula

$$A = \frac{1}{2}(B + b)h$$

where B and b are its bases and h is its height. Solve the formula for b.

FIGURE 3-1

Solution   *Method 1:* $A = \dfrac{1}{2}(B + b)h$

$$2A = 2 \cdot \dfrac{1}{2}(B + b)h \qquad \text{Multiply both sides by 2.}$$

$$2A = Bh + bh \qquad \text{Simplify and remove parentheses.}$$

$$2A - Bh = Bh + bh - Bh \qquad \text{Subtract } Bh \text{ from both sides.}$$

$$2A - Bh = bh \qquad \text{Combine like terms. .}$$

$$\dfrac{2A - Bh}{h} = \dfrac{bh}{h} \qquad \text{Divide both sides by } h.$$

$$\dfrac{2A - Bh}{h} = b \qquad \text{Simplify.}$$

Method 2: $A = \dfrac{1}{2}(B + b)h$

$$2A = 2 \cdot \dfrac{1}{2}(B + b)h \qquad \text{Multiply both sides by 2.}$$

$$2A = (B + b)h \qquad \text{Simplify.}$$

$$\dfrac{2A}{h} = \dfrac{(B + b)h}{h} \qquad \text{Divide both sides by } h.$$

$$\dfrac{2A}{h} = B + b \qquad \text{Simplify.}$$

$$\dfrac{2A}{h} - B = B + b - B \qquad \text{Subtract } B \text{ from both sides.}$$

$$\dfrac{2A}{h} - B = b \qquad \text{Combine like terms.}$$

Although they look different, the results of Methods 1 and 2 are equivalent. ■

EXAMPLE 5   Solve $3x + 2y = 8$, for y.

Solution
$$3x + 2y - 3x = -3x + 8 \qquad \text{Subtract } 3x \text{ from both sides.}$$
$$2y = -3x + 8 \qquad \text{Simplify both sides.}$$
$$\dfrac{2y}{2} = \dfrac{-3x + 8}{2} \quad \text{or} \quad \dfrac{2y}{2} = \dfrac{-3x}{2} + \dfrac{8}{2} \qquad \text{Divide both sides by 2.}$$
$$y = \dfrac{-3x + 8}{2} \quad \text{or} \quad y = -\dfrac{3}{2}x + 4 \qquad \text{Simplify.}$$
■

■ FORMULAS

In Sections 1.5, many formulas were given. It is very important that before using a formula you understand what each variable represents. Example 6 illustrates this idea.

FIGURE 3-2

EXAMPLE 6   The volume V of the right-circular cone shown in Figure 3-2 is given by the formula

$$V = \dfrac{1}{3}Bh$$

where B is the area of its circular base and h is its height. Solve the formula for h and find the height of a right-circular cone with a volume of 64 cubic centimeters and a base area of 16 square centimeters.

Solution   We first solve the formula for h.

$$V = \frac{1}{3}Bh$$

$$3V = 3 \cdot \frac{1}{3}Bh \qquad \text{Multiply both sides by 3.}$$

$$3V = Bh \qquad \text{Simplify.}$$

$$\frac{3V}{B} = \frac{Bh}{B} \qquad \text{Divide both sides by } B.$$

$$\frac{3V}{B} = h \qquad \text{Simplify.}$$

$$h = \frac{3V}{b} \qquad \text{Symmetric property of equality.}$$

■ **STUDY TIP**
As you complete a section of exercises, go back to the previous sections and review each section's title and objectives. Look for connections between sections and for the flow of the mathematics.

We then substitute 64 for V and 16 for B and simplify.

$$h = \frac{3V}{B}$$

$$h = \frac{3(64)}{16}$$

$$= 3(4)$$

$$= 12$$

The height of the right-circular cone is 12 centimeters.   ■

EXERCISE 3.2

VOCABULARY   *Fill in the blank to make a true statement.*

1. Equations with several variables are called _____.

CONCEPTS   *In Exercises 2–10, solve for the indicated variable. Each exercise requires the use of exactly one property of equality from page 110 of Section 3.1. Identify the property used.*

2. $K = A + 32$; for A

3. $P = a + b + c$; for c

4. $P = 4s$; for s

5. $d = rt$; for t

6. $E = IR$; for I

7. $i = prt$; for r

8. $V = lwh$; for h

9. $C = 2\pi r$; for r

10. $P = I^2R$; for R

PRACTICE   *In Exercises 11–24, solve each formula for the variable indicated.*

11. $P = 2l + 2w$; for w

12. $A = P + Prt$; for t

13. $a = \frac{1}{2}(B + b)h$; for h

14. $K = \frac{wv^2}{2g}$; for w

15. $I = \frac{E}{R}$; for R

16. $V = \frac{1}{3}\pi r^2 h$; for h

17. $\frac{1}{2}x + 3y = 10$; for y

18. $\frac{2}{3}x + 4y = 12$; for y

19. $\frac{1}{2}x + \frac{2}{3}y = 5$; for y

20. $x + 4y = \frac{1}{2}$; for y

21. $\frac{2}{3}(x + 4) + y = -1$; for y

22. $\frac{1}{2}(x - 5) + y = 6$; for y

23. $2(x - 4) + 3(y + 2) = 6$; for y

24. $3(x + 2) - 4(y - 1) = 5$; for y

In Exercises 25–32, solve each formula for the variable indicated. Then substitute numbers to find the variable's value.

25. $d = rt$; find t if $d = 135$ and $r = 45$.

26. $d = rt$; find r if $d = 275$ and $t = 5$.

27. $i = prt$; find t if $i = 12$, $p = 100$, and $r = 0.06$.

28. $i = prt$; find r if $i = 120$, $p = 500$, and $t = 6$.

29. $P = a + b + c$; find c if $P = 37$, $a = 15$, and $b = 19$.

30. $y = mx + b$; find x if $y = 30$, $m = 3$, and $b = 5$.

31. $K = \frac{1}{2}h(a + b)$; find h if $K = 48$, $a = 7$, and $b = 5$.

32. $\frac{x}{2} + y = z^2$; find x if $y = 3$ and $z = 3$.

APPLICATIONS

33. Ohm's law   The formula $E = IR$, called **Ohm's law,** is used in electronics. Solve for I and then calculate the current I if the voltage E is 48 volts and the resistance R is 12 ohms. Current has units of *amperes.*

34. Volume of a cone   The volume V of a cone is given by the formula $V = \frac{1}{3}\pi r^2 h$. Solve the formula for h and then calculate the height h if V is 36π cubic inches and the radius r is 6 inches.

35. Circumference of a circle   The circumference C of a circle is given by $C = 2\pi r$, where r is the radius of the circle. Solve the formula for r and then calculate the radius of a circle with a circumference of 14.32 feet. Round your answer to the nearest hundredth of a foot.

36. Growth of money   At a simple interest rate r, an amount of money P grows to an amount A in t years according to the formula $A = P(1 + rt)$. Solve the formula for P. After $t = 3$ years, a girl has an amount $A = \$4357$ on deposit. What amount P did she start with. Assume an interest rate of 6%.

37. Power loss   The power P lost when an electric current I passes through a resistance R is given by the formula $P = I^2R$. Solve for R. If P is 2700 watts and I is 14 amperes, calculate R to the nearest hundredth of an ohm.

38. Geometry   The perimeter P of a rectangle with length l and width w is given by the formula $P = 2l + 2w$. Solve this formula for w. If the perimeter of a certain rectangle is 58.37 meters and its length is 17.23 meters, find its width. Round the answer to two decimal places.

39. Force of gravity   The masses of the two objects in Illustration 1 are m and M. The force of gravitation, F, between the masses is given by

$$F = \frac{GmM}{d^2}$$

where G is a constant and d is the distance between them. Solve for m.

ILLUSTRATION 1

40. Thermodynamics   In thermodynamics, the Gibbs free-energy function is given by

$$G = U - TS + pV$$

Solve this equation for the pressure, p.

41. Pulleys   The approximate length L of a belt joining two pulleys of radii r and R feet with centers D feet apart is given by the formula $L = 2D + 3.25(r + R)$. (See Illustration 2.) Solve the formula for D. If a 25-foot belt joins pulleys of radius 1 foot and 3 feet, how far apart are the centers of the pulleys?

ILLUSTRATION 2

42. Geometry   The measure a of an interior angle of a regular polygon with n sides is given by the formula $a = 180°\left(1 - \frac{2}{n}\right)$. (See Illustration 3.) Solve the formula for n. How many sides does a regular polygon have if an interior angle is 108°? (*Hint:* Distribute first.)

ILLUSTRATION 3

REVIEW

43. Evaluate the expression $2(x - 1) + 2$ for $x = -3$.

44. Find $P(4)$ when $P(x) = 2(x - 1) + 2$.

45. Simplify the expression $2(x - 1) + 2$.

46. Solve for x: $2(x - 1) + 2 = 3x - 5$.

3.3 Introduction to Equations in Two Variables

■ VERIFYING SOLUTIONS OF EQUATIONS IN TWO VARIABLES ■ SOLUTIONS TO EQUATIONS IN TWO VARIABLES ■ FUNCTIONS

C°   F°

■ VERIFYING SOLUTIONS OF EQUATIONS IN TWO VARIABLES

Section 1.5 addressed several types of everyday applications that involved two related quantities. For example, the relationship of temperature measurement in Fahrenheit versus Celsius can be expressed as $F = \frac{5}{9}(C - 32)$. There are other similar relationships that can be modeled using two variables. In order to represent these relationships graphically with our calculator, we will use the variables x and y. Thus $y = \frac{5}{9}(x - 32)$, where $x =$ Celsius temperature and $y =$ Fahrenheit temperature. Using the distributive property, we can simplify this equation to $y = \frac{5}{9}x - \frac{160}{9}$.

Solutions to equations in two unknowns must be expressed as **ordered pairs.** The pair is called an ordered pair because the x value is always written first while the y value is always written second: (x, y). Ordered pairs are *always* enclosed in parentheses.

EXAMPLE 1   Verify that the following ordered pairs are solutions of the equation $4x - 5y = 2$.

a. $(-7, -6)$ **b.** $(3, 2)$

Solution   To verify that each ordered pair is a solution, we will substitute the given values for x and y into the equation.

■ **STUDY TIP**

Read directions carefully to be sure you understand what is being asked. Concentrate on the process and then check your work.

a.
$$4x - 5y = 2$$
$$4(-7) - 5(-6) \; ? \; 2$$
$$-28 + 30 \; ? \; 2$$
$$2 = 2$$

The ordered pair $(-7, -6)$ is a solution.

b.
$$4x - 5y = 2$$
$$4(3) - 5(+2) \; ? \; 2$$
$$12 - 10 \; ? \; 2$$
$$2 = 2$$

The ordered pair $(3, 2)$ is a solution.

EXAMPLE 2   Verify that the following ordered pairs are solutions of the equation $y = x^2 + 3$.

a. $(-3, 12)$ **b.** $(5, 28)$

Solution   To verify that each ordered pair is a solution, we will substitute the given values for x and y into the equation.

a.
$$y = x^2 + 3$$
$$12 \; ? \; (-3)^2 + 3$$
$$12 \; ? \; 9 + 3$$
$$12 = 12$$

The ordered pair $(-3, 12)$ is a solution.

b. $y = x^2 + 3$

$28 \ ? \ (5)^2 + 3$

$28 \ ? \ 25 + 3$

$28 = 28$

The ordered pair (5, 28) is a solution.

■ SOLUTIONS TO EQUATIONS IN TWO VARIABLES

When finding solutions to equations in two variables, we are trying to find x and y values that relate in such a way as to make the equation a true statement when the variables are replaced with the given values. The examples above demonstrated this relationship. We will now determine the value of one variable when given the value of the other variable.

EXAMPLE 3   Find the ordered pair solution to $y = 5x - 7$ when $x = 3$.

Solution   Substitute **3** for x and evaluate for y.

$y = 5(3) - 7$

$= 15 - 7$

$= 8$

The ordered pair solution is (3, 8).

EXAMPLE 4   Find the ordered pair solution to $y = 5x - 7$ when $y = -4$.

Solution   Substitute **−4** for y and solve for x.

$$-4 = 5x - 7$$

$$-4 + 7 = 5x - 7 + 7 \qquad \text{Add 7 to both sides.}$$

$$3 = 5x$$

$$\frac{3}{5} = \frac{5x}{5} \qquad \text{Divide both sides by 5.}$$

$$\frac{3}{5} = x$$

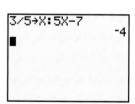

The ordered pair solution is $\left(\frac{3}{5}, -4\right)$.

When solving a linear equation in x and y, you may select any value for x and solve for the related y value, or select any value of y and solve for the related x value.

Many different ordered pair solutions can be viewed in the TABLE of your calculator. The equation $y = 5x - 7$ must be entered on the Y= screen. Set the TABLE (TBLSET) to begin at $x = -3$ (any value can be selected for the table start value) and to be incremented by 1 (ΔTbl = 1).

From the TABLE we can see that the equation $y = 5x - 7$ has the ordered pairs (−3, −22), (−2, −17), (−1, −12), (0, −7), (1, −2), (2, 3), (3, 8) as solutions. In fact, the equation has infinitely many solutions, as you can see if you scroll through the table. Because the table was incremented by units of 1, we will not be able to see ordered pair solutions that involve rational numbers such as 3/5 unless we change the table increment or the table start value. The screens below indicate a table increment of 1/5.

Observe that the table increment of 1/5 is converted to .2 when displayed in the table. The ordered pair solution of $(3/5, -4)$ occurs in the table as $(.6, -4)$.

EXAMPLE 5   The following table of values gives selected values for x and y. Complete the table by substituting the given value into the equation and solving for the remaining variable.

$3x - 2y = 6$

x	y
4	3
-2	-6
2	0
$\frac{8}{3}$	1
0	-3

Solution   If $x = 4$, then $3(4) - 2y = 6$
$$12 - 2y = 6$$
$$-2y = -6$$
$$y = 3 \qquad \text{The ordered pair } (4, 3) \text{ is a solution.}$$

If $x = -2$, then $3(-2) - 2y = 6$
$$-6 - 2y = 6$$
$$-2y = 12$$
$$y = -6 \quad \text{The ordered pair } (-2, -6) \text{ is a solution.}$$

If $y = 0$, then $3x - 2(0) = 6$
$$3x = 6$$
$$x = 2 \qquad \text{The ordered pair } (2, 0) \text{ is a solution.}$$

If $y = 1$, then $3x - 2(1) = 6$
$$3x - 2 = 6$$
$$3x = 8$$
$$x = \frac{8}{3} \qquad \text{The ordered pair } \left(\frac{8}{3}, 1\right) \text{ is a solution.}$$

If $x = 0$, then $3(0) - 2y = 6$
$$-2y = 6$$
$$y = -3 \qquad \text{The ordered pair } (0, -3) \text{ is a solution.}$$

The completed chart looks like this:

x	y
4	3
-2	-6
2	0
$\frac{8}{3}$	1
0	-3

The ordered pairs that result from this chart are $(4, 3)$, $(-2, -6)$, $(2, 0)$, $\left(\frac{8}{3}, 1\right)$ and $(0, -3)$.

To verify with the TABLE on your calculator, you will need to solve the equation $3x - 2y = 6$ for the variable y in order to enter the equation on the Y= screen.

$$3x - 2y = 6$$

$$3x - 3x - 2y = 6 - 3x \qquad \text{Subtract } 3x \text{ from both sides of the equation.}$$

$$-2y = 6 - 3x \qquad \text{Simplify.}$$

$$\frac{-2y}{-2} = \frac{6 - 3x}{-2} \qquad \text{Divide both sides of the equation by } -2.$$

$$y = \frac{6 - 3x}{-2} \qquad \text{Simplify.}$$

```
Plot1 Plot2 Plot3
\Y1⊟(6-3X)/-2
\Y2=
\Y3=
\Y4=
\Y5=
\Y6=
\Y7=
```

```
  X   │ Y1
 -2   │ -6
 -1   │ -4.5
  0   │ -3
  1   │ -1.5
  2   │ 0
  3   │ 1.5
  4   │ 3
X=-2
```

In order to verify the ordered pair $\left(\frac{8}{3}, 1\right)$, we could either set the table increment to 1/3 or merely set the table to start at 8/3 and verify the single ordered pair. ■

EXAMPLE 6   The following table of values gives selected values for x. Complete the table by substituting the given value into the equation and solving for the remaining variable.

$$y = x^3 - 2$$

x	y
-4	
0	
2	

Solution   If $x = -4$, then $y = (-4)^3 - 2$ The ordered pair $(-4, 66)$ is a solution.

$$y = -64 - 2$$

$$y = -66$$

If $x = 0$, then $y = (0)^3 - 2$ The ordered pair $(0, -2)$ is a solution.

$$y = 0 - 2$$

$$y = -2$$

If $x = 2$, then $y = (2)^3 - 2$ The ordered pair $(2, 6)$ is a solution.

$$y = 8 - 2$$

$$y = 6$$

The completed chart looks like this:

x	y
-4	-66
0	-2
2	6

The ordered pairs that result from this chart are $(-4, -66)$, $(0, -2)$, and $(2, 6)$. Verified with the TABLE we have the following:

```
Plot1 Plot2 Plot3
\Y1◻X^3-2
\Y2=
\Y3=
\Y4=
\Y5=
\Y6=
\Y7=
```

```
 X  │ Y1
-4  │ -66
-3  │ -29
-2  │ -10
-1  │ -3
 0  │ -2
 1  │ -1
 2  │ 6
X=-4
```

This view of the table provides several additional ordered pair solutions to the equation $y = x^3 - 2$. ∎

▦ FUNCTIONS

All of the equations we have examined have formed a relationship between a specific value for x and a specific value for y. We call the equations *relations*. A **relation** is merely a set of ordered pairs. When each x value is related to exactly one y value, we call the relation a *function*.

> **Function**
> A **function** is a collection of ordered pairs in which each x value of the ordered pair is paired with a unique y value.

EXAMPLE 7 The table of values represents the conversion of the measurement of cups to pints. Does this relationship represent a function?

x = number of cups	$\frac{1}{2}$	1	$1\frac{1}{2}$	2	$2\frac{1}{2}$
y = number of pints	$\frac{1}{4}$	$\frac{1}{2}$	$\frac{3}{4}$	1	$1\frac{1}{4}$

Solution Because each x value is paired with a unique (one and only one) y value, we have a function. In other words, $\frac{1}{2}$ cup is always equal to $\frac{1}{4}$ of a pint. We do not have to worry about $\frac{1}{2}$ cup sometimes equaling a different amount. The ordered pairs for this function would be $\left\{\left(\frac{1}{2}, \frac{1}{4}\right), \left(1, \frac{1}{2}\right), \left(1\frac{1}{2}, \frac{3}{4}\right), (2, 1), \left(2\frac{1}{2}, 1\frac{1}{4}\right)\right\}$. ∎

EXAMPLE 8 Is the formula for converting Fahrenheit temperature to Celsius a function? The formula is

$$y = \frac{5}{9}(x - 32)$$

where y = degrees Celsius and x = degrees Fahrenheit.

Solution We need to answer the question "does each x produce a unique y?" Select a value for x, say 41°. Then

$$y = \frac{5}{9}(41 - 32)$$

$$y = \frac{5}{9}(9) \qquad \text{Subtraction produces a unique result.}$$

$$y = 5 \qquad \text{Multiplication produces a unique result.}$$

Because the operations of addition, subtraction, multiplication and division produce unique results, any equation that applies these operations to x will produce unique results for y and is thus a function. ∎

When discussing functions and their applications we are often interested in just the x values of the function or just the y values. These classifications are called the **domain** and **range** of the function, respectively.

Domain of a Function
The domain of a function is the set of all x values of the ordered pairs of the function.

Range of a Function
The range of a function is the set of all y values of the ordered pairs of the function.

■ STUDY TIP

Think of a function as a machine. The domain is the data that you put into the machine, and the range is the results that the machine puts back out.

EXAMPLE 9 In the conversion chart, identify the domain and range.

x = number of cups	$\frac{1}{2}$	1	$1\frac{1}{2}$	2	$2\frac{1}{2}$
y = number of pints	$\frac{1}{4}$	$\frac{1}{2}$	$\frac{3}{4}$	1	$1\frac{1}{4}$

Solution The domain is $\left\{\frac{1}{2}, 1, 1\frac{1}{2}, 2, 2\frac{1}{2}\right\}$. The braces indicate we have a **set,** a collection of objects. The range is $\left\{\frac{1}{4}, \frac{1}{2}, \frac{3}{4}, 1, 1\frac{1}{4}\right\}$. The ordered pairs for this function would be $\left\{\left(\frac{1}{2}, \frac{1}{4}\right), \left(1, \frac{1}{2}\right), \left(1\frac{1}{2}, \frac{3}{4}\right), (2, 1), \left(2\frac{1}{2}, 1\frac{1}{4}\right)\right\}$ ∎

EXAMPLE 10 An arching stream of water in a botanical garden fountain can be modeled by the function $y = -0.05x^2 + 2x$, where x represents the horizontal distance and y the vertical height. Complete the table of values and determine the domain and range.

x = horizontal distance	10	15	20
y = vertical height			

Solution

$y = -0.05x^2 + 2x$
$y = -0.05(10)^2 + 2(10)$
$y = -0.05(100) + 20$
$y = -5 + 20$
$y = 15$

$y = -0.05x^2 + 2x$
$y = -0.05(15)^2 + 2(15)$
$y = -0.05(225) + 30$
$y = -11.25 + 30$
$y = 18.75$

$y = -0.05x^2 + 2x$
$y = -0.05(20)^2 + 2(20)$
$y = -0.05(400) + 40$
$y = -20 + 40$
$y = 20$

Set the TABLE to begin at $x = 10$ and to be incremented by 5 to verify your results.

```
TABLE SETUP
 TblStart=10
 ΔTbl=5
Indent: Auto Ask
Depend: Auto Ask
```

```
 X    Y₁
10    15
15    18.75
20    20
25    18.75
30    15
35    8.75
40    0
X=10
```

x = horizontal distance	10	15	20
y = vertical height	15	18.75	20

This table translates to the ordered pairs (10, 15), (15, 18.75), and (20, 20), which means that the arch of water reaches a height of 15 feet when the arch travels 10 feet, a height of 18.75 feet when the arch travels 15 feet, and a height of 20 feet when the arch travels 20 feet.

The domain of the function, as indicated by the TABLE, is {10, 15, 20} and the range, as indicated by the TABLE, is {15, 18.75, 20}. Remember, the domain and range contain many other values than those that are represented in our table. ■

EXAMPLE 11 A parking garage charges the building tenants $5.00 for valet service and $1.50 for each hour (or part of an hour) that they are parked.

a. Write an equation where y is the total cost and x is the number of hours to model this relationship.

b. Is this relationship a function?

c. During any one 24-hour period, what is the domain and what is the range?

VALET SERVICE $5

PARKING $1.50 per hour

Solution **a.** The unknown amount is the number of hours, thus

x = number of hours parked

We know that $1.50 is charged for each hour, thus

$1.50x$ = cost for parking

and a flat fee of $5 is charged for valet service. The equation that models the total cost of parking is

$$y = 1.5x + 5$$

b. This equation is a function because the operations of multiplication (1.5 times x) and addition (plus 5) yield unique results.

c. Build a table of values to help you picture the domain and range. You may use your calculator TABLE to help speed up the process.

x = no. of hours	1	2	3	4	5	6	7	. . .	23	24
y = total cost	6.5	8	9.5	11	12.5	14	15.5	. . .	39.5	41

```
 X    Y₁
1    6.5
2    8
3    9.5
4    11
5    12.5
6    14
7    15.5
X=1
```

```
 X    Y₁
8    17
9    18.5
10   20
11   21.5
12   23
13   24.5
14   26
X=8
```

```
 X    Y₁
15   27.5
16   29
17   30.5
18   32
19   33.5
20   35
21   36.5
X=15
```

```
 X    Y₁
22   38
23   39.5
24   41
25   42.5
26   44
27   45.5
28   47
X=22
```

The domain is $\{1, 2, 3, 4, \ldots, 23, 24\}$ because there are 24 hours in a day. The range is $\{6.5, 8, 9.5, 11, 12.5, 14, 15.5, 17, 18.5, 20, 21.5, 23, 24.5, 26, 27.5, 29, 30.5, 32, 33.5, 35, 36.5, 38, 39.5, 41\}$ because you will pay \$5 plus \$1.50 per hour for any given number of hours parked during this 24-hour period.   ■

The equation in Example 10, $y = -0.05x^2 + 2x$, is a second-degree (also called *quadratic*) equation/function, and the equation in Example 6, $y = x^3 - 2$, is a third-degree equation/function. Throughout this text we will examine functions of different degrees and their behavior. Initially, however, our focus will be on first-degree (also called *linear*) equations/functions. The last example demonstrated a linear function.

EXERCISE 3.3

VOCABULARY   In Exercises 1–7, fill in the blanks to make a true statement.

1. The equation $y = \frac{5}{9}(x - 32)$ is an equation in ____ variables.

2. Solutions to equations in two unknowns must be expressed as _____.

3. ΔTbl determines how the TABLE of the calculator is to be _____.

4. A ___ is a collection of objects.

5. A _____ is a collection of ordered pairs where each x value is paired with a unique y value.

6. The _____ is the set of all x coordinates in a collection of ordered pairs.

7. The _____ is the set of all y coordinates in a collection of ordered pairs.

CONCEPTS

8. Is the ordered pair $(-2, -2)$ a solution to the equation $y = 2x^3 - 5$?

9. Is the ordered pair $(2, 2)$ a solution to the equation $3x - 4y = 2$?

✍**10.** If $(2, 0)$ is a solution of $2x + y = 4$, will $(0, 2)$ also be a solution? Justify your answer.

✍**11.** In general, if (a, b) is a solution to a linear equation in two variables, will (b, a) be a solution? Justify your answer.

12. If an equation produces the table of values displayed at the right, write all the ordered pairs that are solutions to the equation.

X	Y1	
-2	-14	
-1.5	-13	
-1	-12	
-.5	-11	
0	-10	
.5	-9	
1	-8	
X= -2		

✍**13.** Does the table of values in Exercise 12 represent a function? Explain in complete sentences.

14. What is the pictured domain of the set of ordered pairs in Exercise 12?

15. What is the pictured range of the set of ordered pairs in Exercise 12?

PRACTICE   In Exercises 16–31, find the corresponding y value for the given x value. Write the result as an ordered pair.

16. $x + y = 2$; $x = 2$, $x = -3$

17. $x - y = 3$; $x = -1$, $x = 5$

18. $2x + y = -1$; $x = -3$, $x = 4$

19. $x + 3y = -4$; $x = -5$, $x = -8$

20. $y = 3x + 2$; $x = \frac{2}{3}$, $x = -\frac{5}{3}$

21. $y = -4x + 1$; $x = \frac{1}{4}$, $x = -\frac{3}{4}$

22. $x^2 = y$; $x = -3$, $x = 9$

23. $x^2 - 20 = y$; $x = 10$, $x = -5$

24. $y = 2(x - 1)^2$; $x = -3$, $x = 4$

25. $y = 3(x + 2)^2$; $x = 2$, $x = -6$

26. $y = \frac{1}{4}(x + 2)^2$; $x = 0$, $x = 3$

27. $y = \frac{2}{3}(x + 3)^2 - 2$; $x = -1$, $x = 2$

28. $y = x^3$; $x = 4$, $x = -5$

29. $y = -x^3$; $x = -2$, $x = 10$

30. $y = -\dfrac{2}{3}x^3 + 5$; $x = 3$, $x = -6$

31. $y = \dfrac{4}{5}x^3 - 1$; $x = 1$, $x = -1$

In Exercises 32–35, complete the table of values for the given equation.

32. $5x + y = 7$

33. $5x + y = 7$

34. $\dfrac{1}{2}x + \dfrac{2}{3}y = 7$

35. $\dfrac{1}{2}x + \dfrac{2}{3}y = 7$

x	y
-2	
0	
2	

x	y
	-3
	0
	3

x	y
-4	
2	
6	

x	y
	-5
	0
	3

In Exercises 36–43, determine if each relationship or equation defines a function and then determine the domain and range.

36. $\{(2, 1), (3, -2), (4, 5)\}$

37. $\{(-4, 5), (5, -4), (3, -4)\}$

38.

X	Y₁
-8	-26
-7	-24
-6	-22
-5	-20
-4	-18
-3	-16
-2	-14

X=-8

39.

X	Y₁
-1	-8
0	-5
1	-2
2	1
3	4
4	7
5	10

X=-1

40. $y = 2x + 3$

42. $y = x^2 + 5$

41. $2x + y = 1$

43. $y = x^3 - 1$

APPLICATIONS

44. Rental car   A rental company charges $25 plus $0.21 per mile to rent a car. The linear equation $y = .21x + 25$ models the problem, where y represents the total cost and x is the number of miles traveled. Fill in the table of values to determine the cost, y, for each of the given mileages.

x	5	25	100	150
y				

Answer each of the following questions in a complete sentence:

a. What is the cost of traveling 25 miles?
b. What is the cost of traveling 150 miles?
c. If the company increased the cost per mile to $0.25, what would the new equation be?

d. What would the cost be for traveling 150 miles with the new price?

45. Entertainment   The county fair charges $5 gate admission and $0.75 per carnival ride. The linear equation that models the relationship between the total cost, y, and the number of rides, x, is $y = .75x + 5$. Fill in the table of values to determine the cost of attending the fair if riding the indicated number of rides.

x	0	5	10	15
y				

Answer each of the following questions in a complete sentence:

a. What is the cost of riding 5 rides?
b. What is the cost of riding 15 rides?
c. If the carnival decreased the cost per ride to $0.50, what would the new equation be?

d. What would the cost be for riding 15 rides with the new price?

46. Depreciation   A car that originally costs $18,000 depreciates $1,200 per year. The linear equation that models this depreciation is $y = -1200x + 18000$, where x represents the number of years and y represents the value of the car.
a. Find the value of the car after 8 years.
b. After how many years will the car have no value?

47. Appreciation   A house purchased in 1987 for $82,000 increases in value at a rate of $1,250 each year. If y represents the value of the house and x represents the number of years, the linear equation that models this appreciation is $y = 1250x + 82000$.
a. Find the value of the house in 12 years.
b. When will the house have a value of $92,000?

In Exercises 48–53, write a linear equation in the form $y = mx + b$ to model the problem. Use your model to complete the given tables.

48. Tutoring   A tutor charges $15 for a study guide and $12 per hour for tutoring. After determining the equation to model this relationship, fill in the table of values to determine the tutoring costs for the indicated hours.

hours	2	6	$8\frac{1}{2}$
cost			

49. Auto repairs   A repair service charges a $20 "diagnostic fee" and $37.50 per hour for repairs. After determining the equation to model this relationship, fill in the table of values to determine the total repair costs.

hours	1 hour	1.5 hours	5 hours	5.75 hours
cost				

50. Appreciation   An antique gold necklace purchased in 1950 originally cost $12. Its current appraised value indicates it has been increasing in value at the rate of $8.20 per year.
a. Write an equation to model this appreciation.

b. What is the necklace's value in 50 years?

c. If it continues to appreciate at its current rate, after how many years will it have a value of $627?

51. Depreciation   A computer system purchased for $4000 depreciates at a rate of $800 per year. Write an equation to model this relation and determine the value of the computer after 4 years.

52. Distance   The distance, y, that a vehicle travels can be determined by multiplying the rate of speed times the time, x.
a. Write an equation in x and y to model the distance traveled by a vehicle that averages 55 mph.

b. Fill in the table to determine the designated distance.

x	2	3.5	$4\frac{3}{4}$
y			

53. Child care   If the campus day care charges $5.50 per hour for "drop-ins," write an equation to model the cost, y for x number of hours.
Use your model to determine the cost for 3.5 hours of drop-in day care.

REVIEW

54. Evaluate $x^4 y^3$ for $x = -2$ and $y = 3$.
55. Let $P(x) = 3x^2 - 5x + 2$.
a. State the degree of the polynomial.
b. Find the value of $P(-3)$.
56. Simplify: $(2x^2 - 3x + 5) - (3x^2 - 5x + 4)$.

57. Find the volume of the figure in Illustration 1.

58. Perimeter   The perimeter of a square is $12 + 16x$. Write an expression involving division to determine the length of a side. Simplify your expression.

3x

Legal size
3x + 2

Letter size
x + 1

ILLUSTRATION 1

59. Class grades   An algebra class has 36 students. One-third of the class made an A. How many students did not make an A?

60. Interest   Jorge signs a one-year note to borrow $9000 for tuition. If the rate of interest is $7\frac{1}{4}\%$, how much interest must he pay?

3.4 The Rectangular Coordinate System

■ REVIEW OF ORDERED PAIR SOLUTIONS ■ THE RECTANGULAR COORDINATE SYSTEM ■ GRAPHING LINEAR EQUATIONS ■ GRAPHING WITH A GRAPHING CALCULATOR ■ GRAPHING LINEAR FUNCTIONS ■ *x*- AND *y*-INTERCEPTS

■ REVIEW OF ORDERED PAIR SOLUTIONS

In Chapter 1 we drew one-dimensional pictures on a number line. We now want to draw two-dimensional pictures in a number plane. Just as each point on a number line is associated with a real number, each point in a number plane is associated with an ordered pair of real numbers.

The equation $3x + 2 = 5$ has the single variable x, and its only solution is 1. This solution can be graphed (or plotted) on the number line, as in the figure below.

The equation $x + 2y = 5$ contains the two variables x and y. The solutions of such equations are ordered pairs of numbers. For example, the pair $x = 1$ and $y = 2$, $(1, 2)$, is a solution, because the equation is satisfied when $x = 1$ and $y = 2$.

$$x + 2y = 5$$
$$1 + 2(2) = 5 \qquad \text{Substitute 1 for } x \text{ and 2 for } y.$$
$$1 + 4 = 5$$
$$5 = 5$$

The pair $x = 5$ and $y = 0$ $(5, 0)$, is also a solution, because the pair satisfies the equation:

$$x + 2y = 5$$
$$5 + 2(0) = 5 \qquad \text{Substitute 5 for } x \text{ and 0 for } y.$$
$$5 + 0 = 5$$
$$5 = 5$$

■ THE RECTANGULAR COORDINATE SYSTEM

Solutions of equations with two variables can be plotted on a **rectangular coordinate system,** sometimes called a **Cartesian coordinate system** after the 17th-century French mathematician René Descartes. To design the McDonald's arches a mathematical model had to be created on just such a coordinate system.

The rectangular coordinate system consists of two number lines, called the ***x*-axis** and the ***y*-axis,** drawn at right angles to each other, as shown in Figure 3-3. The two axes intersect at a point called the **origin,** which is the 0 point on each axis. The positive direction on the *x*-axis is to the right, and the positive direction on the *y*-axis is upward. The two axes divide the coordinate system into four regions, called **quadrants,** which are numbered as shown in Figure 3-3.

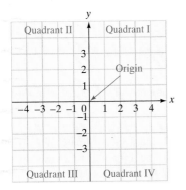

FIGURE 3-3

Each point P is associated with a unique ordered pair (x, y) of real numbers called the **coordinates of the point.**

EXPLORING THE CONCEPT   Use your graphing calculator to determine the sign of the x- and y-coordinates in each quadrant. Display the coordinate axes on your calculator graph screen by using the GRAPH feature. The directional arrow keys, ▶ ◀ ▲ ▼, will allow you to move the cursor (cross hairs) around the various quadrants. Observe the x- and y-coordinates as they are displayed at the bottom of the screen and record the sign of the coordinates in the table below.

Quadrant	Sign of x-coordinate	Sign of y-coordinate
I	+	+
II	−	+
III	−	−
IV	+	−

Conclusion   Quadrant I: $(+, +)$ Quadrant III: $(-, -)$

Quadrant II: $(-, +)$ Quadrant IV: $(+, -)$

To plot the ordered pair $(3, 2)$, where $x = 3$ and $y = 2$, we start at the origin and move 3 units to the right along the x-axis and then 2 units up in the positive y direction, as in Figure 3-4. This locates point A. Point A has an **x-coordinate** or **abscissa** of 3 and a **y-coordinate** or **ordinate** of 2. This information is denoted by the pair of numbers $(3, 2)$, called the **coordinates** of point A. Point A is the **graph** of the pair $(3, 2)$.

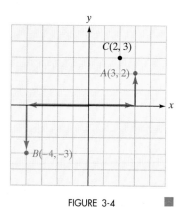

To plot the pair $(-4, -3)$, we start at the origin, move 4 units to the left along the x-axis, and then move 3 units down to point B in the figure. Point C has coordinates of $(2, 3)$.

FIGURE 3-4

 WARNING!   The order of the coordinates of a point is important. Point A with coordinates of $(3, 2)$ is not the same as point C with coordinates $(2, 3)$. For this reason, the pair of coordinates (x, y) of a point is called an **ordered pair.** The x-coordinate is the first number in the pair, and the y-coordinate is the second number.

EXAMPLE 1   On a rectangular coordinate system, plot the points   **a.** $A(-1, 2)$,   **b.** $B(0, 0)$,   **c.** $C(5, 0)$,   **d.** $D\left(-\frac{5}{2}, -3\right)$,   and   **e.** $E(3, -2)$.

Solution   **a.** The point $A(-1, 2)$ has an x-coordinate of -1 and a y-coordinate of 2. To plot point A, we start at the origin and move 1 unit to the left and then 2 units up. (See Figure 3-5.) Point A lies in quadrant II.

b. To plot point $B(0, 0)$, we start at the origin and move 0 units to the right and 0 units up. Because there was no movement, point B is the origin.

c. To plot point $C(5, 0)$, we start at the origin and move 5 units to the right and 0 units up. Point C lies on the x-axis, 5 units to the right of the origin.

d. To plot point $D\left(-\frac{5}{2}, -3\right)$, we start at the origin and move $\frac{5}{2}$ units $\left(\frac{5}{2}\text{ units} = 2\frac{1}{2}\text{ units}\right)$ to the left and then 3 units down. Point D lies in quadrant III.

e. To plot point $E(3, -2)$, we start at the origin and move 3 units to the right and then 2 units down. Point E lies in quadrant IV.

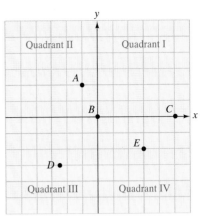

FIGURE 3-5 ■

EXAMPLE 2   **Braking distances**   Approximate braking distances for a car in good condition on dry pavement are given in Table 3-1.

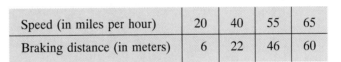

Speed (in miles per hour)	20	40	55	65
Braking distance (in meters)	6	22	46	60

TABLE 3-1

Plot these points on a rectangular coordinate system.

Solution   If we plot the speeds along the x-axis and the braking distances along the y-axis, we obtain the graph shown in Figure 3-6.

From the position of the points plotted in the figure, we can see that the braking distances are much longer at greater speeds. ■

FIGURE 3-6

■ GRAPHING LINEAR EQUATIONS

We have seen that solutions to equations in x and y are ordered pairs of numbers. To find some ordered pairs (x, y) that satisfy $y = 5 - x$, we pick numbers at random, substitute them for x, and find the corresponding values of y. If we pick $x = 1$, we have

$y = 5 - x$
$y = 5 - \mathbf{1}$ Substitute 1 for x.
$y = 4$

The ordered pair $(1, 4)$ satisfies the equation. If we let $x = 2$, we have

$y = 5 - x$
$y = 5 - \mathbf{2}$ Substitute 2 for x.
$y = 3$

A second solution of the equation is $(2, 3)$. If we let $x = 5$, we have

$y = 5 - x$
$y = 5 - \mathbf{5}$ Substitute 5 for x.
$y = 0$

A third solution of the equation is $(5, 0)$. If we let $x = -1$, we have

$y = 5 - x$
$y = 5 - (\mathbf{-1})$ Substitute -1 for x.
$y = 6$

A fourth solution is $(-1, 6)$. If we let $x = 6$, we have

$y = 5 - x$

$y = 5 - 6$ Substitute 6 for x.

$y = -1$

A fifth solution is $(6, -1)$.

We list the ordered pairs $(1, 4)$, $(2, 3)$, $(5, 0)$, $(-1, 6)$, and $(6, -1)$ in the table shown in Figure 3-7. Their graphs lie on a line, called the **graph** of the equation.

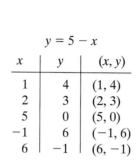

$y = 5 - x$

x	y	(x, y)
1	4	$(1, 4)$
2	3	$(2, 3)$
5	0	$(5, 0)$
-1	6	$(-1, 6)$
6	-1	$(6, -1)$

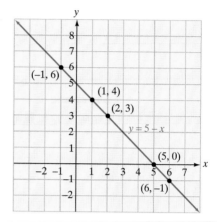

FIGURE 3-7

Any equation of the form, $y = mx + b$ whose graph is a line is called a *linear equation in two variables.* Any point on the line has coordinates that satisfy the equation, and the graph of any pair (x, y) that satisfies the equation is a point on the line.

Linear Function

An equation that can be written in the form $y = mx + b$, where m and b are real numbers, is called a *linear function.*

Although only two points are needed to graph a linear equation, we often plot a third point as a check. If the three points do not lie on a line, at least one of them is in error.

Procedure for Graphing Linear Equations in x and y

1. Find two pairs (x, y) that satisfy the equation by picking arbitrary numbers for x and solving for the corresponding values of y. A third point provides a check.

2. Plot each resulting pair (x, y) on a rectangular coordinate system. If they do not lie on a line, check your calculations.

3. Draw the line passing through the points.

EXAMPLE 3   Graph the ordered pair solutions to the equation $y = 3x - 4$.

Solution   If we substitute 1 for x, we get $y = -1$, and the ordered pair $(1, -1)$ is a solution. If we substitute 2 for x, we get $y = 2$, and the ordered pair $(2, 2)$ is a solution. If we substitute 3 for x, we get $y = 5$, and the ordered pair $(3, 5)$ is a solution. These ordered pairs appear in the table accompanying Figure 3-8. We plot these points and join them with a line, which is the graph of the solutions of the equation $y = 3x - 4$.

Every point that lies on this line is a solution to the equation $y = 3x - 4$, and every ordered pair whose coordinates satisfy the equation $y = 3x - 4$ lies on the line.

$y = 3x - 4$

x	y	(x, y)
1	−1	$(1, -1)$
2	2	$(2, 2)$
3	5	$(3, 5)$

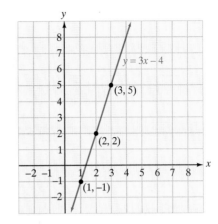

FIGURE 3-8

Scroll through the TABLE on your calculator to view other ordered pairs that are solutions to the equation. A few of the possible pairs are displayed below.

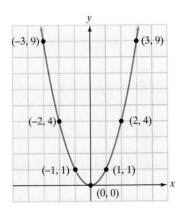

EXAMPLE 4 Graph the ordered pair solutions to $y = x^2$. Use the specified x values to determine your graph.

x	y
−3	
−2	
−1	
0	
1	
2	
3	

Solution The equation $y = x^2$ is a second degree equation and is therefore not linear. Complete the table of values by substituting each value for x into the equation and solving for y.

x	y
−3	9
−2	4
−1	1
0	0
1	1
2	4
3	9

Verify your table of values by using the TABLE feature of your calculator.

Plot these points and connect them with a smooth curve.

EXAMPLE 5   Graph the ordered pair solutions to $y = x^3$ by using the selected x values to determine your graph.

x	y
-2	
-1	
0	
1	
2	

Solution   The equation $y = x^3$ is a third degree equation and is therefore not linear. Complete the table of values by substituting each value for x into the equation and solving for y.

x	y
-2	-8
-1	-1
0	0
1	1
2	8

Verify your table of values by using the TABLE feature of your calculator.

Plot these points and connect them with a smooth curve.

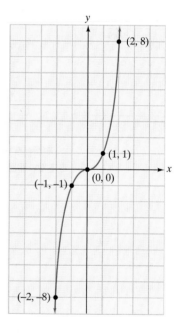

■ GRAPHING WITH A GRAPHING CALCULATOR

Unlike traditional algebra texts, this text employs full integration of a graphics calculator. All references to a calculator will imply a graphics calculator, not a scientific calculator. In Chapter 1 we drew one-dimensional pictures on a number line. We now want to draw two-dimensional pictures in a number plane. Just as each point on a number line is associated with a real number, each point in a number plane is associated with an ordered pair of real numbers.

When you sketch a graph on your graph paper, the size of your graph is limited by the size of your paper space. Theoretically, each axis extends indefinitely, which we represent by arrows at the end of each axis. However, the number of tic marks you place on each axis limits its visual size. Similar limitations occur when using a graphing calculator to display a coordinate system. You can limit the visual size of your calculator graph by designating a minimum and maximum value for the *x*-axis and a minimum and maximum value for the *y*-axis.

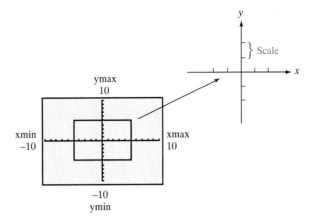

To condense this information, we will use the following notation:

xmin xmax   ymin ymax

$[-10, 10]$ by $[-10, 10]$

Observe that there are 10 tic marks to the left, right, above, and below the origin on the calculator graph. The distance between the tic marks is called the *scale*. On this graph the scale for both the *x* and *y* axes is 1. We would write this as xscl = 1 and yscl = 1. Unless otherwise noted, it should be assumed xscl = 1 and yscl = 1 on all graphs displayed in this text.

In order to set the limitations on your calculator graph, you should be familiar with the key that accesses this feature. It may be denoted as "WINDOW," "RANGE," or "WIND" on your calculator. The viewing window above $[-10, 10]$ by $[-10, 10]$, is called the **standard viewing window.** It can be set manually by entering the information or automatically using the ZOOM feature.

WARNING! What you "see" may be deceiving. Each of the calculator graphs displayed below look the same, and yet each represents a different-size coordinate system based on the limitations placed on xmin, xmax, xscl, ymin, ymax, and yscl.

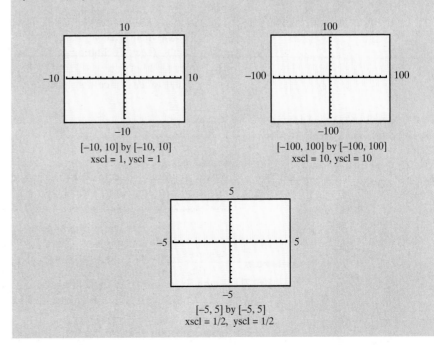

Before proceeding to the examples, set your calculator viewing window to Zstandard by either manually entering the window values of $[-10, 10]$ by $[-10, 10]$ or automatically by using the ZOOM menu of your calculator.

■ GRAPHING LINEAR FUNCTIONS

EXAMPLE 6   To use the graphing calculator to graph the function $y = 3x - 4$ the expression $3x - 4$ should be entered at the y1= prompt on the y= screen. Press GRAPH to display the graph of the equation.

Solution

$y = 3x - 4$

x	y
1	−1
2	2
3	5

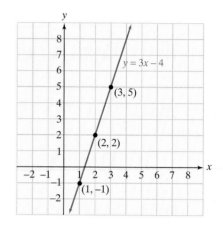

FIGURE 3-9

You may now access the TABLE feature of the calculator to compare your table of values, as shown in Figure 3-9, to those of the calculator.

If you access the TRACE feature you can use the left and right cursor keys to trace along the line. You will notice that the coordinate values are cumbersome decimal values.

For this reason, we will *not* use the standard viewing window when we want to trace. The calculator has a preset window called ZInteger that yields integer values when tracing (see "Technology 3.4: Graphing Basics"). To set these values manually, enter $[-47, 47]$ by $[-31, 31]$ on your WINDOW screen now, with x and y scales equal to 10. Trace and observe the x and y coordinates.

EXAMPLE 7   Graph $3x + 2y = 6$ in the ZInteger window: $[-47, 47]$ by $[-31, 31]$, with x and y scales equal to 10. Trace to confirm the values displayed in the table in Figure 3-10.

Solution   First, solve the equation for y in order to enter it on the y= screen.

$$3x + 2y = 6$$
$$3x - 3x + 2y = 6 - 3x$$
$$2y = 6 - 3x$$
$$\frac{2y}{2} = \frac{6 - 3x}{2}$$
$$y = 3 - \frac{3}{2}x$$
$$y = -\frac{3}{2}x + 3$$

$$3x + 2y = 6$$

x	y	(x, y)
0	3	$(0, 3)$
2	0	$(2, 0)$
4	-3	$(4, -3)$

FIGURE 3-10

■ **STUDY TIP**
When studying for a test, focus on directions as well as problems. Words such as *evaluate, solve, simplify,* and *graph* should be the triggers that move you toward successful problem solving on tests.

■ **x- AND y-INTERCEPTS**

The points where a graph intersects the *x*- and *y*-axes are called **intercepts** of the graph.

> **x- and y-Intercepts**
> An *x*-intercept of a graph is a point $(a, 0)$ where the graph intersects the *x*-axis. To find *a*, substitute 0 for *y* in the equation of the graph and solve for *x*.
>
> The **y-intercept** of a graph is the point $(0, b)$ where the graph intersects the *y*-axis. To find *b*, substitute 0 for *x* in the equation of the graph and solve for *y*.

Plotting the *y*- and *x*-intercepts to graph a first-degree equation is called the **intercept method of graphing.** This method is useful for graphing equations written in **standard form.**

> **Standard Form of the Equation of a Line**
> If *A*, *B*, and *C* are real numbers and *A* and *B* are not both 0, then the equation
> $$Ax + By = C$$
> is called the **standard form** of the equation of a line.

Whenever possible, we will write the standard form $Ax + By = C$ so that *A*, *B*, and *C* are integers and $A \geq 0$.

EXAMPLE 8   Graph the solution to the equation $3x + 2y = 6$ by plotting the *x*- and *y*-intercepts.

Solution   To find the *y*-intercept, we let $x = 0$ and solve for *y*.

$$3x + 2y = 6$$
$$3(0) + 2y = 6 \qquad \text{Substitute 0 for } x.$$
$$2y = 6 \qquad \text{Simplify.}$$
$$y = 6 \qquad \text{Divide both sides by 2.}$$

The y-intercept is the ordered pair $(0, 3)$.

To find the x-intercept, we let $y = 0$ and solve for x.

$$3x + 2y = 6$$
$$3x + 2(0) = 6 \qquad \text{Substitute 0 for } y.$$
$$3x = 6 \qquad \text{Simplify.}$$
$$x = 2 \qquad \text{Divide both sides by 3.}$$

The x-intercept is the pair $(2, 0)$.

As a check, we plot one more point. If $x = 4$, then

$$3x + 2y = 6$$
$$3(4) + 2y = 6 \qquad \text{Substitute 4 for } x.$$
$$12 + 2y = 6 \qquad \text{Simplify}$$
$$2y = -6 \qquad \text{Add } -12 \text{ to both sides.}$$
$$y = -3 \qquad \text{Divide both sides by 2.}$$

The point $(4, -3)$ lies on the graph. We plot these three points and join them with a line. The graph of $3x + 2y = 6$ is shown in Figure 3-11.

$3x + 2y = 6$

x	y
0	3
2	0
4	-3

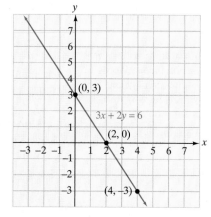

FIGURE 3-11

Verify the solutions and the graph with your calculator.

The graphs of nonhorizontal and nonvertical lines will have one x-intercept and one y-intercept. Graphs of equations that are nonlinear may have varying numbers of x-intercepts. For example, $y = x^2 - 4$, shown in Figure 3-12, has two x-intercepts and one y-intercept, yet $y = 0.2x^3 - 2.6x + 2.4$ has three x-intercepts.

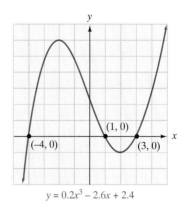

$y = x^2 - 4$ $y = 0.2x^3 - 2.6x + 2.4$

FIGURE 3-12

EXAMPLE 9   Graph the equations   **a.** $y = 3$ and   **b.** $x = -2$.

Solution   **a.** We can write the equation $y = 3$ in standard form as $0x + y = 3$. Since the coefficient of x is 0, the numbers assigned to x have no effect on y. The value of y is always 3. For example, if we replace x with -3, we get

$$0x + y = 3$$
$$0(-3) + y = 3$$
$$0 + y = 3$$
$$y = 3$$

The table in Figure 3-13 gives several ordered pairs that satisfy the equation $y = 3$. After plotting these pairs (x, y) and joining them with a line, we see that the graph of $y = 3$ is a horizontal line, parallel to the x-axis and intersecting the y-axis at 3. The y-intercept is $(0, 3)$, and there is NO x-intercept.

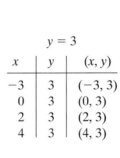

$y = 3$

x	y	(x, y)
-3	3	$(-3, 3)$
0	3	$(0, 3)$
2	3	$(2, 3)$
4	3	$(4, 3)$

FIGURE 3-13

$[-10, 10]$ by $[-10, 10]$

b. We can write the equation $x = -2$ in standard form as $x + 0y = -2$. Since the coefficient of y is 0, the values of y have no effect on x. The number x is always negative 2. A table of values and the graph are shown in Figure 3-14.

The graph of $x = -2$ is a vertical line, parallel to the y-axis and intersecting the x-axis at negative 2. The x-intercept is $(-2, 0)$, and there is no y-intercept.

$$x = -2$$

x	y	(x, y)
-2	-2	$(-2, -2)$
-2	0	$(-2, 0)$
-2	2	$(-2, 2)$
-2	3	$(-2, 3)$

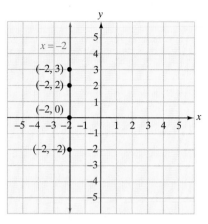

FIGURE 3-14

The graphing calculator will graph linear functions in the form $y = mx + b$ and linear functions that are horizontal lines, $y = b$, as in Example 9a. However, the calculator cannot graph the equation of vertical lines, $x = a$, as in Example 9b, because these lines are not functions. If you examine the table of ordered pairs for the graph of the line $x = -2$, you will discover that each x value DOES NOT have a *unique* y value. Thus, $x = -2$ is *not* a function and cannot be graphed with your calculator.

From the results of Example 9, we have the following facts.

Equations of Lines Parallel to the Coordinate Axes

The equation $y = b$ represents a horizontal line that intersects the y-axis at $(0, b)$. If $b = 0$, the line is the x-axis.

The equation $x = a$ represents a vertical line that intersects the x-axis at $(a, 0)$. If $a = 0$, the line is the y-axis.

EXERCISE 3.4

VOCABULARY *In Exercises 1–12, fill in the blanks to make a true statement.*

1. Solutions of equations in two variables can be plotted on a _____.

2. The __ axis is the horizontal axis and the __ axis is the vertical axis.

3. The two axes intersect at the _____.

4. The two axes divide the coordinate system into four regions called _____.

5. Another name for the x-coordinate of the ordered pair (x, y) is the _____.

6. Another name for the y-coordinate of the ordered pair (x, y) is the _____.

7. The point $(0, b)$ of a line represents the __-intercept.

8. The point $(a, 0)$ of a line represents an __-intercept.

9. An equation in the form $Ax + By = C$ is called the _____ form for the equation of a line.

10. The calculator viewing window indicated at the right is called the _____ viewing window.

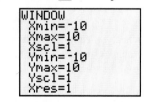

11. The space between the tic marks on the axes is called the _____.

12. The TI-83 viewing window that yields integer values when tracing has xmin = _____, xmax = ___, ymin = _____ and ymax = ___. It also has both scales set to 10.

CONCEPTS

13. In a–d below, plot each point on a rectangular coordinate system. Indicate in which quadrant each point lies.
   a. E(2, 5)   **b.** F(−3, 1)
   c. G(−2, −3)   **d.** H(3, −2)

14. In a–f below, plot each point on a rectangular coordinate system. Indicate on which axis each point lies.
   a. F(0, 5)   **b.** E(0, −2)
   c. D(2, 0)   **d.** C(−4, 0)
   e. B(0, 0)   **f.** A(−5, 0)

15. In a–h below, refer to Illustration 1 and find the coordinates of each point.
   a. *H*   **b.** *G*
   c. *F*   **d.** *E*
   e. *B*   **f.** *C*
   g. *D*   **h.** *A*

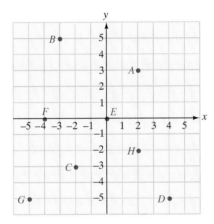

ILLUSTRATION 1

16. **Data analysis**   Based on the graph in Illustration 2, express the information in the boxes, 1988 and 1996, as ordered pairs. 51.2 77.0
Explain the meaning of the ordered pairs as they relate to the graph.

Rise in teen smoking

The incidents of teens becoming regular smokers jumped 50 percent between 1988 and 1996. A look at the rate of teens who started smoking daily, age 12-17:

considered nonsmokers at start of year

SOURCE: Centers for Disease Control and Prevention (AP)

ILLUSTRATION 2

17. **Data analysis**   Write an ordered pair that relates the information that "there were 12.6 deaths per 100,000 people in 1992" based on the graph in Illustration 3.

Deaths decrease

Powerful new AIDS drugs helped dramatically reduce the number of AIDS deaths in the United States last year. The numbers dropped to their lowest rate since 1987. A look:

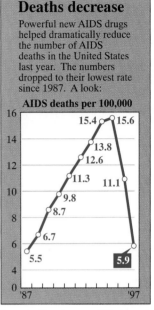

SOURCE: Department of Health and Human Services (AP/Justin Gilbert)

ILLUSTRATION 3

18. In Exercises a–d, which of the following ordered pairs would be displayed in the viewing window that is indicated at the right?
   a. (2, 5)
   b. (−3.7, 2.9)
   c. (4.8, −3.2)
   d. (−3.1, −4.7)

[−4.7, 4.7] by [−3.1, 3.1]

19. In a–d below, determine the viewing window and indicate it in the form [xmin, xmax] by [ymin, ymax]. Both the *x* and *y* scales are equal to 1.

a.   **b.**

c.   **d.**

20. In a–d below, sketch the set of coordinate axes indicated by the given viewing window.

    a. $[-6, 6]$ by $[-6, 6]$ **b.** $[-4, 4]$ by $[-4, 4]$
        xscl=1, yscl=1   xscl=1, yscl=1

    c. $[-8, 8]$ by $[-8, 8]$ **d.** $[-8, 8]$ by $[-8, 8]$
        xscl=2, yscl=1   xscl=4, yscl=4

21. In order to more fully understand why lines and curves on the graphing calculator appear jagged, shade every box that the curve on the graph in Illustration 4 passes through. This simulates the calculator action of lighting up the pixel squares that display a graph.

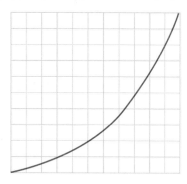

ILLUSTRATION 4

✐22. Explain why the ZInteger viewing window is a better choice of viewing window than ZStandard when using the TRACE feature.

✐23. Discuss the disadvantages to displaying a graph in the ZInteger viewing window.

24. Give the **coordinates** of the x- and y-intercepts of the graph in Illustration 5.

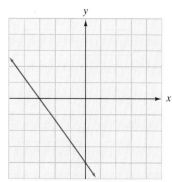

ILLUSTRATION 5

25. Give the **coordinates** of the x- and y-intercepts of the graph of $y = |x| - 4$ in Illustration 6.

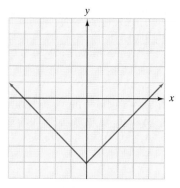

ILLUSTRATION 6

✐26. Which of the following are linear functions? In your own words, justify your answer.

    a. $y = x^2 - 5x$
    b. $2x - y = 6$
    c. $y = x^4 + 2$
    d. $y = x + 2$

✐27. Explain why you would want to graph three ordered pairs when graphing the equation of a line when it only takes two points to determine a line.

✐28. You have been asked to determine the x- and y-intercepts of the equation $2x - 4y = 7$. Explain the process in detail.

✐29. What is the equation of the vertical line pictured in Illustration 7? Is this the equation of a function? Explain.

ILLUSTRATION 7

✐30. What is the equation of the horizontal line pictured in Illustration 8? Is this the equation of a function? Explain.

ILLUSTRATION 8

31. Find the equation of the line that is horizontal and passes through $(3, -2)$.

32. Find the equation of the x-axis.

33. Find the equation of the y-axis.

PRACTICE *In Exercises 34–47, complete the table of solutions for each equation and graph the equation on graph paper. Confirm both the table and the graph with your calculator. (Hint: Be sure the calculator is in the ZInteger viewing window so you can TRACE.)*

34. $y = x + 2$

x	y
3	
1	
-2	

35. $y = x - 4$

x	y
5	
4	
-1	

36. $y = -2x$

x	y
2	
1	
-3	

37. $y = \dfrac{x}{2}$

x	y
1	
-1	
-4	

38. $y = 2x - 1$

x	y
3	
-1	
-2	

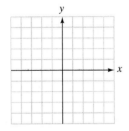

39. $y = 3x + 1$

x	y
-2	
0	
1	

40. $y = \dfrac{x}{2} - 2$

x	y
8	
0	
-2	

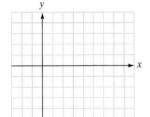

41. $y = \dfrac{x}{3} - 3$

x	y
6	
0	
-3	

42. $y = x^2 + 2$

x	y
-2	
-1	
0	
1	
2	

43. $y = x^2 - 3$

x	y
-2	
-1	
0	
1	
2	

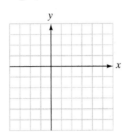

44. $y = (x - 5)^2$

x	y
3	
4	
5	
6	
7	

45. $y = (x + 9)^2 - 7$

x	y
-11	
-10	
-9	
-8	
-7	

46. $y = x^3 + 4$

x	y
-2	
-1	
0	
1	
2	

47. $y = x^3 - 4$

x	y
-2	
-1	
0	
1	
2	

In Exercises 48–59, graph the solution to each equation by using the intercept method. Verify the intercepts and the graph with your calculator. Remember, to graph with the calculator you will need to solve each equation for y.

48. $x + y = 7$

49. $x + y = -2$

50. $x - y = 7$

51. $x - y = -2$

52. $2x + y = 5$

53. $3x + y = -1$

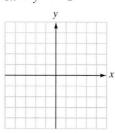

54. $2x + 3y = 12$

55. $3x - 2y = 6$

56. $3x + 12 = 4y$

57. $2x + 12 = 9y$

58. $2(x + 2) - y = 4$

59. $3(y + 1) - x = 4$

In Exercises 60–71, graph each equation on graph paper and verify your solution graphically when possible.

60. $y = -5$

61. $x = 4$

62. $x = 5$

63. $y = 4$

64. $y = 0$

65. $x = 0$

66. $2x = 5$

67. $3y = 7$

68. $3(x + 2) + x = 4$

69. $2(y + 3) - y = 5$

70. $3(y - 2) + 2 = y$

71. $4(x + 2) - 2(x - 1) = 6$

In Exercises 72–79, graph each function in the ZInteger viewing window and use TRACE to determine the x and y intercepts.

72. $y = 2x + 8$

73. $y = -3x - 12$

74. $2x - 3y = 10$

75. $4x - y = 8$

76. $y = x^2 - 3x - 28$

77. $y = -x^2 + 2x - 15$

78. $y = -0.01(x - 10)^3$

79. $y = (x - 3)^3$

APPLICATIONS

80. Stopping distances Approximate stopping distances for a car in good condition on dry pavement, including the normal $\frac{3}{4}$-second reaction time, are as shown in Table 1. Plot the points and draw any appropriate conclusions.

Speed (in miles per hour)	20	40	55	65
Stopping distance (in meters)	12	35	64	80

TABLE 1

81. Tutoring A tutor charges \$12 an hour for tutoring mathematics. Complete Table 2 showing the tutor's fee for working 1, 2, 3, and 5 hours.

Hours worked	1	2	3	5
Tutor's fee	12	24	36	60

TABLE 2

82. Tutoring Let y represent the tutor's fee for working x hours. Plot the pairs found in Exercise 81. Do the points lie on a line?

83. Stopping distances   Consuming two or three drinks can easily increase the reaction time to put on the brakes by 1 second. Approximate stopping distances for a car in good condition on dry pavement, including $1\frac{3}{4}$ seconds for reaction time, are as shown in Table 3. Plot the points and draw any appropriate conclusions.

Speed (in miles per hour)	20	40	55	65
Stopping distance (in meters)	22	53	87	109

TABLE 3

84. Maximum heart rate   A rule of thumb for a person's maximum heart rate for safe aerobic exercise is 220 minus the person's age. Complete Table 4 showing the maximum heart rate for persons 20, 30, 40, and 60 years old.

Age	20	30	40	60
Heart rate	200	190	180	160

TABLE 4

85. Let y represent the maximum heart rate for a person x years old. (See Exercise 84.) Write an equation that relates x and y and graph it.

86. Installing computer systems   A computer system installer charges $10 for materials and $40 per hour. Complete Table 5 showing the charge for jobs taking 1, 2, 3, and $5\frac{1}{2}$ hours.

Hours worked	1	2	3	$5\frac{1}{2}$
Charge	50	90	130	230

TABLE 5

87. Let y represent the charge for working x hours. Plot the pairs found in Exercise 86. Do the points lie on a line?

REVIEW

88. The area of a kitchen floor is $12x^2 + 5x - 2$. If the length of one side of the room is $3x + 2$ feet, find the polynomial that represents the length of the adjacent side.

89. A circle has a radius of $2x^2$. Find the area, leaving π as part of your answer.

90. The calculator screen indicates that an equation with a supposed solution of $x = 4$ is being checked. Does the display verify that 4 is a solution or does it dispute it? Explain.

91. Verify that $F = 95$ is a solution to $35 = \frac{5}{9}(F - 32)$.

Use your calculator for the verification.

92. Classify these equations as an identity, contradiction, or conditional equation.
   a. $2x - 1 = 5$
   b. $2x + 4 = 2(x - 2)$
   c. $2(x + 1) = 2(x + 3) - 4$

Read the box headings carefully to be sure that you are following the directions for your specific calculator model. All the TI-86 information begins on page 154.

Setting Up the Graph Display: TI-83

The calculator's display is controlled through the **MODE** and **WINDOW FORMAT** screens.

1. Press the [MODE] key. **MODE** controls how numbers and graphs are displayed and interpreted. The current settings on each row should be highlighted as displayed. The blinking rectangle can be moved using the four **cursor** (arrow) keys. To change the setting on a particular row, move the blinking rectangle to the desired setting and press [ENTER].

Note: Items must be highlighted to be activated.

Normal vs. scientific notation

Floating decimal vs. fixed to 9 places

Type of angle measurement

Type of graphing: function, parametric, polar, sequence

Graphed points connected or dotted

Functions graphed one by one

Results displayed with real or complex numbers

Full screen, horizontally split screen, or a vertically split screen that displays a graph and table simultaneously

2. To return to the home screen at this point, press [CLEAR] or [2nd] [QUIT].

3. Press [y=]. The calculator can graph up to ten different equations at the same time. Because **MODE** is in the sequential setting, the graphs will be displayed sequentially. Note that cursoring down accesses additional **y=** prompts. The display of the equations entered on the **y=** screen is controlled by the size of the viewing window. The dimensions of the viewing window are determined by the values entered on the **WINDOW** screen.

This calculator has a feature called the *graph style icon* that allows you to distinguish between the graphs of equations. Press [y=] and observe the "\" in front of each y. Use the left arrow to cursor over to this "\". The diagonal should now be moving up and down. Pressing [ENTER] once changes the diagonal from thin to thick. The graph of y1 will now be displayed as a thick line. Repeatedly pressing [ENTER] displays the following:

◥: shades above y1

◣: shades below y1

−o: traces the leading edge of the graph followed by the graph

o: traces the path but does not plot

⋰ : displays graph in dot, not connected, MODE

It is important to remember that the graphing icon takes precedence over the **MODE** screen. If the icon is set for a solid line and the **MODE** screen is set for DOT and not solid, the graphing icon will determine how the graph is displayed. Pressing [CLEAR] to delete an entry at the **y=** prompt will automatically reset the graphing icon to default, a solid line.

4. Press **[ZOOM] [6:ZStandard] [WINDOW].** This is called the standard viewing window. The information on this screen indicates that in a rectangular coordinate system the x-values will range from -10 to 10 and the y-values will range from -10 to 10. The interval notation for this is $[-10, 10]$ by $[-10, 10]$. The xscl$=1$ and yscl$=1$ settings indicate that the tic marks

on the axes are one unit apart. The values entered on this screen may be changed by using the cursor arrows to move to the desired line and typing over the existing entry. When drawing a graph, you may set the desired viewing rectangle on the calculator as well as scale the x-axis and y-axis. This calculator has a row labeled Xres$=$. This determines the screen resolution. It should be set equal to 1, which means that each pixel on the x-axis will be evaluated and graphed.

5. Access **FORMAT** (located above the ZOOM key) by pressing **[2nd] [FORMAT].** The following settings should be highlighted:

Graphs on a rectangular coordinate system

Cursor location is displayed on screen

Graphing grid is not displayed

Axes are visible

Axes are not labeled with an x and y

If **ExprOn** is highlighted, the equation is displayed on the graph screen when the TRACE feature is activated.

Altering the Viewing Window: TI-83

The calculator viewing window is 95 pixel points wide by 63 pixel points high; there are 94 horizontal spaces and 62 vertical spaces to light up. When tracing on a graph, the readout changes according to the size of the space. The size of the space can be controlled by the following formulas:

$$\frac{\text{Xmax} - \text{Xmin}}{94} = \text{horizontal space width}, \quad \frac{\text{Ymax} - \text{Ymin}}{62} = \text{vertical space height}$$

We will now examine some preset viewing windows and how they affect the pixel space size.

6. Press **[ZOOM].** There are 10 entries on this screen. The down arrow key can be used to view remaining entries.

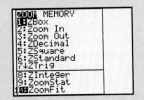

1: Boxes in and enlarges a designated area.

2: Acts like a telephoto lens and "zooms in."

3: Acts like a wide-angle lens and "zooms out."

4: Moves cursor $\frac{1}{10}$ of a unit per move.

5: "Squares up" the previously used viewing window.

6: Sets axes to $[-10, 10]$ by $[-10, 10]$.

7: Graphs trigonometric functions.

8: Moves cursor ONE integer unit per move.

9: Used when graphing statistics.

10: Adjusts the viewing window to fit the graph.

7. a. ZDecimal is useful for graphs that require the use of the calculator's TRACE feature. Applying the horizontal space width formula,

$$\frac{\text{Xmax} - \text{Xmin}}{94} = \frac{4.7 - (-4.7)}{94} = 0.1$$

changes the x-values by $\frac{1}{10}$ of a unit each time the cursor is moved. This is why this screen yields "friendly" values when TRACING. (In general, Xmax − Xmin needs to be a multiple of 94 to produce a "friendly" screen.) Enter Y1 = −2X + 3 and Y2 = 1/2X − 2 on the **Y=** screen. Press **[ZOOM] [4:ZDecimal]** to display the graph of these two lines in the ZDecimal viewing window. TRACE along the graph of one of the lines and observe the changes in x-values. The change should be $\frac{1}{10}$ of a unit. (Remember, the y-values are dependent on the values selected for x.)

 b. ZInteger is useful for application problems where the x-value is valid only if represented as an integer (such as when x equals the number of tickets sold, number of passengers in a vehicle, etc.). Application of the horizontal space width formula,

$$\frac{\text{Xmax} - \text{Xmin}}{94} = \frac{47 - (-47)}{94} = 1$$

changes the x-values by one unit each time the cursor is moved. (ZInteger is the only preset viewing window that allows you to set the cursor to recenter the axes and the only preset window that *requires* you to press **[ENTER]** to activate the viewing window.)

 Press **[ZOOM] [8:ZInteger]**, move the cursor to the origin of the graph ($x = 0$ and $y = 0$), and press **[ENTER]** to display the graph of these two lines, as indicated at the right.

 c. ZStandard provides a good visual comparison between hand-sketched graphs (or textbook graphs) that are approximately $[-10, 10]$ by $[-10, 10]$. Applying the horizontal space width formula,

$$\frac{\text{Xmax} - \text{Xmin}}{94} = \frac{10 - (-10)}{94} \approx 0.212765974$$

the x-values will change by .212765974 each time the TRACE cursor is moved. If you are using the TRACE feature, you will usually want a screen with "friendlier" x-values than this one provides.

 Press **[ZOOM] [6:ZStandard]** to display the graph of these two lines as indicated at the right. TRACE along the graph of one of the lines and observe the changes in x-values. If you select two consecutive x-values and find the difference, it should be 0.212765974.

 8. ZDecimal frequently does not provide a large-enough viewing window. When this is the case, you may multiply the Xmin and Xmax by the same constant and the Ymin and Ymax by the same constant to produce a larger viewing rectangle that still provides cursor moves in tenths of units. Multiplying Xmin and Xmax by 2 would mean cursor moves of $\frac{2}{10}$ of a unit:

$$\frac{\text{Xmax} - \text{Xmin}}{94} = \frac{2(4.7) - 2(-4.7)}{94} = 0.2$$

whereas multiplying by 3 would mean cursor moves of $\frac{3}{10}$ of a unit. The screen above is the **ZDecimal** screen with the max and min values multiplied by 2. This WINDOW will be referred to in the future as ZDecimal × 2.

Setting Up the Graph Display: TI-86

The calculator's display is controlled through the **MODE** and **GRAPH/FORMAT** screens.

1. Press **[2nd] [MODE].** The current settings on each row should be highlighted as displayed. The blinking rectangle can be moved using the four **cursor** (arrow) keys. To change the setting on a particular row, move the blinking rectangle to the desired setting and press **[ENTER].**

Note: Items must be highlighted to be activated.

Normal vs. scientific notation

Floating decimal vs. fixed to 11 places

Type of angle measurement

Complex number display

Type of graphing: function, polar, parametric, differential equation

Computations performed in bases other than base 10

Format of vector display

Type of differentiation

2. The **FORMAT** screen is accessed by pressing **[GRAPH] [MORE] [F3](FORMAT).** The following settings should be highlighted:

Graphs on both rectangular and polar coordinate system

Cursor location is displayed on the screen

Graphed points are connected or discrete

Functions displayed sequentially or simultaneously

Graphing grid is not displayed

Axes are visible

Axes are not labeled with an x and y

3. To return to the home screen at this point, press **[CLEAR]** or **[EXIT].**

4. Press **[GRAPH] [F1](y(x)=).** The calculator can graph up to 99 different equations at the same time. Because **MODE** is in the sequential setting, the graphs will be displayed sequentially. The display of the equation(s) entered on the **y(x)=** screen is controlled by the size of the viewing window. The dimensions of the viewing window are determined by the values entered on the **WIND** screen, which is referred to as the **WINDOW** screen in the text.

This calculator has a feature called the "graph style icon" that allows you to distinguish between the graphs of equations. Press **[GRAPH] [F1](y(x)=)** and observe the "\" in front of each y. Press **[MORE]** followed by **[F3](STYLE).** The diagonal should now have changed from thin to thick. The graph of y1 will now be displayed as a thick line. Repeatedly pressing **[STYLE]** displays the following:

◥: shades above y1

◣: shades below y1

‾o: traces the leading edge of the graph followed by the graph

o: traces the path but does not plot

⋰: displays graph in dot, not connected, MODE

It is important to remember that the graphing icon takes precedence over the **MODE** screen. If the icon is set for a solid line and the **MODE** screen is set for DOT and not solid, the graphing icon will determine how the graph is displayed. Pressing **[CLEAR]** to delete an entry at the **y(x)=** prompt will automatically reset the graphing icon to default, a solid line.

5. Press [GRAPH] [F3](ZOOM) [F4](ZSTD) [2nd] [M2](WIND). This is called the standard viewing window. The information on this screen indicates that in a rectangular coordinate system the x-values will range from -10 to 10 and the y-values will range from -10 to 10. The interval notation for this is $[-10, 10]$ by $[-10, 10]$. The xScl$=1$ and yScl$=1$ set-

tings indicate that the tic marks on the axes are one unit apart. The values entered on this screen may be changed by using the cursor arrows to move to the desired line and typing over the existing entry. When drawing a graph, you may set the desired viewing window on the calculator as well as scale the x-axis and y-axis.

Remember: Every time the text indicates "press [WINDOW]" you will need to press [GRAPH] [F2](WIND). When the text indicates "press [GRAPH]", you should press [F5](GRAPH) from the **GRAPH** MENU.

Altering the Viewing Window: TI-86

The calculator viewing window is 127 pixel points wide by 63 pixel points high; there are 126 horizontal spaces and 62 vertical spaces to light up. When tracing on a graph, the readout changes according to the size of the space. The size of the space can be controlled by the following formulas:

$$\frac{xMax - xMin}{126} = \text{horizontal space width}, \frac{yMax - yMin}{62} = \text{vertical space height}$$

We will now examine some preset viewing windows and how they affect the pixel space size. Press [GRAPH] [F3](ZOOM) to display a partial zoom menu.

6. a. ZDecimal (ZDECM) is useful for graphs that require the use of the calculator's TRACE feature. Zdecimal can be located on the ZOOM menu by pressing [MORE]. Applying the horizontal space width formula

$$\frac{xMax - xMin}{126} = \frac{6.3 - (-6.3)}{126} = 0.1$$

changes the x-values by $\frac{1}{10}$ of a unit each time the cursor is moved. This is why this screen yields "friendly" values when TRACING. (In general, xMax $-$ xMin needs to be a multiple of 126 to produce a "friendly" screen.) Enter y1 $= 2x + 3$ and y2 $= (1/2)x - 2$ on the y(x)$=$ screen. Press [ZOOM] [MORE] [F4] (ZDECM) to display the graph of these two lines in the ZDecimal viewing window. TRACE along the graph of one of the lines and observe the changes in x-values. The change should be $\frac{1}{10}$ of a unit. (Remember, the y-values are dependent on the values selected for x.)

b. ZInteger is useful for application problems where the x-value is valid only if represented as an integer (such as when x equals the number of tickets sold, number of passengers in a vehicle, etc.). Application of the horizontal space width formula,

$$\frac{xMax - xMin}{126} = \frac{63 - (-63)}{126} = 1$$

changes the x-values by one unit each time the cursor is moved. Press [ZOOM] [MORE] [MORE] [F5] (ZINT), move the cursor to the origin of the graph ($x = 0$ and $y = 0$), and press [ENTER] to display the graph of these two lines as indicated at the right. TRACE along the graph of one of the lines and observe the changes in x-values. The change should be one unit.

c. ZStandard provides a good visual comparison between hand-sketched graphs (or textbook graphs) that are approximately $[-10, 10]$ by $[-10, 10]$. Applying the horizontal space width formula,

$$\frac{xMax - xMin}{126} = \frac{10 - (-10)}{126} \approx 0.15873015873$$

the x-values will change by 0.15873015873 each time the TRACE cursor is moved. If you are using the TRACE feature, you will usually want a screen with "friendlier" x-values than this one provides.

Press [ZOOM] [F4](ZSTD) to display the graph of these two lines as indicated at the right. TRACE along the graph of one of the lines and observe the changes in x-values. If you select two consecutive x-values and find the difference, it should be 0.15873015873.

7. ZDecimal frequently does not provide a large-enough viewing window. When this is the case, you may multiply the xMin and xMax by the same constant and the yMin and yMax by the same constant to produce a larger viewing rectangle that still provides cursor moves in tenths of units. Multiplying xMin and xMax by 2 would mean cursor moves of $\frac{2}{10}$ of a unit:

$$\frac{xMax - xMin}{126} = \frac{2(6.3) - 2(-6.3)}{126} = 0.2$$

whereas multiplying by 3 would mean cursor moves of $\frac{3}{10}$ of a unit. The screen at the left is the ZDecimal screen with the max and min values multiplied by 2. This WINDOW will be referred to in the future as ZDecimal $\times$ 2.

8. Press [GRAPH] [F3](ZOOM). Pressing [MORE] repeatedly will display the remaining menu selections and then return the display to the original screen.

ZBOX	Boxes in and enlarges a designated area.
ZIN	Acts like a telephoto lens and "zooms in."
ZOUT	Acts like a wide-angle lens and "zooms out."
ZSTD	Automatically sets standard viewing window to $[-10, 10]$ by $[-10, 10]$.
ZPREV	Resets **WIND** values to values used prior to the previous ZOOM operation.
ZFIT	Resets yMin and yMax on the WIND screen to include the minimum and maximum y-values that occur between the current xMin and xMax settings.
ZSQR	"Squares up" the previously used viewing window.
ZTRIG	Sets the WIND to built-in trig values.
ZDECM	Sets cursor moves to $\frac{1}{10}$ of a unit per move.
ZDATA	Automatically sets the viewing window to accommodate statistical data.
ZRCL	Sets **WIND** values to those stored by the user (see ZSTO).
ZFACT	Sets the zoom factors used in **ZIN** and **ZOUT.**
ZOOMX	Bases graph display on xFact only when zooming in or out.
ZOOMY	Bases graph display on yFact only when zooming in or out.
ZINT	Moves cursor ONE integer unit per move.
ZSTO	Stores current WIND values for future use; values are recalled by **ZRCL.**

3.5 Linear Applications and Models

■ SETTING GOOD VIEWING WINDOWS ■ LINEAR APPLICATIONS ■ LINEAR MODELS

■ SETTING GOOD VIEWING WINDOWS

In Section 3.4 we examined the construction of hand-drawn graphs and calculator-generated graphs. The standard viewing window and the integer viewing window were used to generate most of the calculator graphs. ("Technology 3.4" also discusses other preset viewing windows.) These programmed viewing windows provide a quick-and-easy approach to adjusting the size of your viewing window. However, in application models you will frequently discover the need to structure your own viewing window.

We want to focus on what constitutes an appropriate viewing window. It is important that all the distinguishing features of the graph be displayed. You should be able to see all the x-intercepts and the y-intercepts as well as any high points or low points (peaks or valleys) of the graph. For example, the graph pictured has interesting features that lie outside of the standard viewing window. This portion of the graph needs to be included. To do this, you will have to individually adjust the max (maximum) and min (minimum) values for both/either the x- and y-axis. These values are found on the Window screen. When large adjustments are made, you may want to experiment with setting the x- and y-scales to 0 so as to not display tic marks.

EXAMPLE 1   Graph the cubic function $y = x^3 - 5x^2 + 2x - 1$ in an appropriate viewing window.

Solution   Begin by displaying the graph in the standard viewing window. Now TRACE left to right along the graph of the function. We discover that the y-minimum on the graph is approximately -12 and yet our y-minimum in the standard viewing window is only -10. Reset the y-minimum to a value *smaller* than -12 to ensure a clear view of the lower portion of the graph.

We now have a *good* view of the graph. Tracing allows you to locate the approximate *highs* and *lows* of the graph when adjusting the viewing window. The window pictured is $[-10, 10]$ by $[-20, 10]$.

EXAMPLE 2   Suppose \$12,000 is invested in two accounts, a money market account earning $3\frac{1}{2}\%$ interest and a six-month certificate of deposit earning 5% interest. If x dollars is invested in the money market account then there will be \$12,000 $- x$ left to invest in the CD. The equation $y = .035x + .05(12,000 - x)$ models the total interest earned by the two accounts. Graph this model in an appropriate viewing window.

Solution   Upon entering the equation on the y= edit screen and displaying the graph in the standard viewing window, you discover that none of the graph is visible.

Pressing TRACE reveals that the y-intercept is located at $(0, 600)$. This means that the y-axis must be adjusted to display $(0, 600)$. We will set the ymax to 700.

We know that $y = .035x + .05(12,000 − x)$ *is not* a horizontal line because it is not in the form $y = b$. It is, however, linear because the equation $y = .035x + .05(12000 − x)$ simplifies to $y = −.015x + 600$. For application models it is important to take into consideration what your variables represent as you are adjusting the viewing window. In this problem, x represents the amount invested in the $3\frac{1}{2}\%$ account. The *total* investment is $12,000, so it is safe to assume that x will not exceed $12,000. Thus we set the xmax to $12,000. The pictured graph has a viewing window of $[−10, 12000]$ by $[−10, 700]$. The x- and y-scales of the pictured graph are set equal to 1. The axes appear thick-ened because of the larger number of tic marks on each axis. Setting the x- and y-scales equal to zero returns them to a more normal appearance.

You can now TRACE to determine the approximate x-intercept and complete the graph as displayed below.

$[−10, 40000]$ by $[−10, 700]$

■ LINEAR APPLICATIONS

Let us once again consider the investment of $12,000 in two different interest-bearing accounts.

EXAMPLE 3   Suppose $12,000 is invested in two accounts, a money market account earning $3\frac{1}{2}\%$ interest and a six-month certificate of deposit earning 5% interest. If x dollars is invested in the money market account then there will be $12,000 − x$ left to invest in the CD. The equation $y = .035x + .05(12000 − x)$ models the total interest earned by the two accounts. Use the graphical display to determine the amount of interest in the account when $4246 is deposited in the $3\frac{1}{2}\%$ account.

Solution   **a.** We could TRACE to *approximately* $x = 4246$ and determine that the amount of interest, y, is approximately $536.

The STOre feature indicates that the exact amount of interest when $4246 is invested in the $3\frac{1}{2}\%$ account is $536.31.

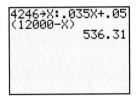

The TABLE feature supports the STOre feature.

Thus, if you are asked to graphically determine a specific value, TRACE is not the most accurate approach.

b. The use of your calculator's VALUE feature (or eval feature on some calculators) provides an accurate value *and* at the same time displays the information graphically. This feature is found under the CALC menu.

■ **TECHNOLOGY TIP**
When using the value or eval x feature, your x-value must be between the values defined for xmin and xmax on the Window screen.

It is important to remember that in most viewing windows, TRACE will only yield approximate values. Unless you can TRACE to the exact value required, you will need to use the value/eval feature.

The next example uses a preset viewing window (ZInteger) because the unknown quantity, x, represents the number of rides at a fair.

EXAMPLE 4

The first day of the fair, admission is $5, and it costs $0.75 to ride each ride. We can represent this algebraically with the equation $y = 5 + .75x$ where x represents the number of rides and y represents the total amount of money spent.

a. Graph the equation in the ZInteger viewing window $[-47, 47]$ by $[-31, 31]$ with both x and y scales equal to 10. Trace along the graph of the line. How would you interpret the information displayed at the bottom of the screen in the context of the state fair?

b. Continuing to trace, answer each question: How much would you spend if you rode 5 rides?, 15 rides? If you cursor left past the y-axis, you will get negative values of x and y. These points are meaningless in the context of our problem. Why?

Solution **a.** The x-value represents the number of tickets purchased and the y-value represents the total amount of money spent. In fact, the ZInteger screen must be used in order for the x-value displayed to be an integer because x represents the quantity of tickets purchased.

b. The total cost for 5 rides would be $8.75 (the interpretation of the graph pictured at the right). Trace to $x = 15$. The interpretation of your graphical display is that to ride 15 rides, the cost would be $16.25.

The x values that are less than zero are meaningless because x represents the quantity of tickets purchased. ■

EXAMPLE 5

■ **TECHNOLOGY TIP**
To deselect an equation so it is not displayed on the graph, place the cursor over the equal sign and press ENTER. To select the equation to be graphed, place the cursor over the equal sign and press ENTER to highlight the equal sign.

The normal price of the fair is $1 admission and $1.25 per ride. The second night that the fair is in town, it will run its normal rates. The equation that would represent cost in terms of number of rides would be $y = 1 + 1.25x$. Graph the equation of this line at the y2 prompt. To view *only* the new graph, deselect the equation entered at y1.

a. Trace along the graph of the line to determine how much you would spend if you rode 5 rides, 15 rides.

b. Reselect the graph of $y = 5 + .75x$ and display both graphs, representing the cost and number of rides for both nights. How many rides would you have to ride for the cost to be the same no matter which night you attended the fair?

Solution

a. The total cost for 5 rides would be $7.25 and for 15 rides would be $19.75.

b. Trace to the point of intersection of the two graphs. You may use the up and down arrow keys to move from the graph of one equation to the graph of the other equation. Be sure that the point of intersection is the **one** point that the two lines have in common. The common ordered pair is (8, 11). Translated to the context of the problem, the ordered pair means that when 8 tickets have been purchased, the total cost will be $11 regardless of which night you attend the fair. ■

EXAMPLE 6

The third night that the fair is in town, the managers decide to run a special: $10 admission and no additional cost per ride. We would represent this with the equation $y = 10$. Enter this equation at the y3 = prompt and display the graph of all three equations.

How many rides would you have to ride for the $10 price to be a better deal than the first night?, second night? (*Hint:* You might want to turn the graphs ON and OFF to make comparisons.)

Solution

After six rides the cost of attending the first night exceeds the cost of attending the third night.

After seven rides the cost of attending the second night exceeds the cost of attending the third night.

EXAMPLE 7

Intelix Home Video produces videotapes of weddings, baptisms, etc. For each tape produced there is a basic fixed cost of $100 and a variable cost of $2 per tape. The tapes then sell for $20 each. Write an equation to represent the production costs y for x tapes and an equation to represent the revenues y for x tapes. Explore these graphs and their meaning in the context of the problem.

a. What is the cost of producing 23 tapes?

b. What is the revenue for producing 23 tapes?

c. Does the company sustain a loss or a profit when 23 tapes are sold?

d. How many tapes must be produced and sold for the company to break even (cost = revenue)?

Solution   **a.** The cost equation would be $y = 100 + 2x$. Graph this equation in the viewing window $[-20, 74]$ by $[-5, 181]$ with x- and y-scales equal to 10. Trace to find the production cost for 23 tapes. At the point $(23, 146)$ we can see that the cost for producing 23 tapes is $146.

■ **TECHNOLOGY TIP**
The calculator will trace to a point not visible in your designated viewing window. The coordinates at the bottom of the screen indicate the cursor position.

b. The revenue equation would be $y = 20x$. Graph this equation on the same screen as the cost equation and trace on the **revenue** equation to determine the revenues produced for 23 tapes. The point $(23, 460)$ indicates that 23 tapes produce a revenue of $460.

c. Does the company sustain a loss or make a profit when 23 tapes are sold? Because the production cost is $146 and the revenues are $460, the company makes a profit of $314.

■ **TECHNOLOGY TIP**
See "Technology 3.5: Intersect Method for Solving Equations" on page 165 for specific help in using the INTERSECT feature of your calculator.

d. How many tapes must be produced and sold for the company to break even (cost = revenue)? Attempting to trace to the point of intersection is futile because the x value at the point of intersection is not an integer. (The window you were given was designed to produce integer values for x when tracing.) The x-value of the point of intersection of the two equations represents the solution to the equation cost = revenue (i.e., $100 + 2x = 20x$). Use the INTERSECT feature of your calculator to locate the point of intersection displayed at the right. The company would need to produce approximately 5.6 tapes to break even.

■ LINEAR MODELS

Many times the data that we are given to work with do not produce a straight-line graph. When this happens, we use a **line of best fit** to predict trends in the data. We will not discuss the methods used to determine the equation as this is a topic for statistics; however, you may want to refer to your calculator manual to learn how your calculator can be used to determine these types of equations.

EXAMPLE 8   The following table gives the winning time (in seconds) for the men's 400-meter freestyle swimming event in the Olympics from 1904 to 1992. The Olympics were not held during the years 1916, 1940, and 1944 due to the two world wars.

YEAR	1904	1908	1912	1920	1924	1928	1932	1936	1948	1952
TIME	376.2	336.8	324.4	326.8	304.2	301.6	288.4	284.5	281	270.7

1956	1960	1964	1968	1972	1976	1980	1984	1988	1992	1996
267.3	258.3	252.2	249	240.3	231.9	231.3	231.2	227	225	227.97

SOURCE: *The New York Times 1998 Almanac.*

The linear equation that is the best fit for the set of data points is

$$y = -1.42x + 3042.79$$

Use the graph to predict the winning times for the missing years during the two world wars.

Solution   The data points are graphed as (year, time) at the right in the viewing window [1895, 2000] by [199, 401] with both x- and y-scales set to zero.

The graph at the right demonstrates the fit between the graph of the equation and the data points.

■ TECHNOLOGY TIP

When using the value option, the value chosen for x **must** be between the xMax and xMin as defined in the window.

Access the value option on your calculator to enter specific x-values and have the calculator plot the location of the point and give the coordinates. Entering 1914 at the $x=$ prompt computes a winning time of 324.91 seconds, as displayed on the graph at the right. Repeating the procedure for the years 1940 and 1944 produces winning times of 287.99 seconds and 282.31 seconds, respectively.

■

EXERCISE 3.5

VOCABULARY   *In Exercise 1, fill in the blank to make a true statement.*

1. A line of _____ is used to predict trends in data when the data are not linear.

CONCEPTS

✍ 2. Explain what constitutes a "good" or "appropriate" graphical display.

✍ 3. Allison's mother pays her babysitter, Sherah, $5.00 per hour to babysit Allison and an additional $5 for gas used in driving to and from her home. If $y = 5x + 5$ models the cost of babysitting for x hours, interpret the meaning of the graphical display.

Quick-Copy is $20 plus 5¢ a copy for bulk printing. If y = cost and x = number of copies, then $y1 = 10 + .09x$ would model the cost for using Fast-Copy and $y2 = 20 + .05x$ would model the cost at Quick-Copy. Interpret the meaning of the graphical display.

✍ 4. The cost for using Fast-Copy printers is $10 plus 9¢ a copy for bulk printing, and the cost for using

✍ 5. In Exercise 4, if more than 250 copies are printed, which company offers a "better deal"?

PRACTICE   *In Exercises 6–15, graph each function in an appropriate viewing window. Begin each graph in the standard viewing window and adjust the window so the x- and y-intercepts are visible, as well as any interesting features of the graph.*

6. $y = 3x + 15$

7. $y = 5x + 18$

8. $y = -\frac{1}{2}x - 8$

9. $y = -\frac{3}{5}x - 9$

10. $y = |x| - 14$

11. $y = |x| + 16$

12. $y = x^2 - 11$

13. $y = x^2 - 15$

14. $y = x^2 + 20$

15. $y = x^2 + 14$

APPLICATIONS

16. Stopping distance The approximate stopping distance of a car in good conditions on dry pavement includes a $\frac{3}{4}$-second reaction time. Two or three drinks can increase the reaction time to $1\frac{3}{4}$ seconds. The equation that models this is $y = \frac{87}{45x} - \frac{50}{3}$, where x is the speed of the vehicle and y is the stopping distance in meters. Graph this model in an appropriate viewing window.

17. College fees If a local community college charges an activity fee of $50 per student and $25 per credit hour for a technology fee, the equation $y = 25x + 50$ models the cost y for x credit hours taken. Graph this equation in an appropriate viewing window.

18. Gratuity George and Annette eat out every Friday night at their favorite restaurant and leave the waitress a 15% tip.
   a. Write an equation where y is the total cost of the meal plus tip and x is the cost of their meal.

   b. Graph the equation in the ZInteger viewing window, $[-47, 47]$ by $[-31, 31]$, x- and y-scale = 10.
   c. Trace to determine the amount of money they can spend on food if they only have $20 to spend tonight.
   d. Use the value feature to determine the total bill if they order the $35.25 steak dinner for two.

19. Sales tax The combined state and local sales tax for Clarksville, TN is $8\frac{3}{4}\%$.
   a. Write an equation where y is the amount of tax and x is the amount of the purchase.
   b. Graph the equation in the ZInteger viewing window.
   c. Trace to determine the amount of the purchase if the tax is $2.54.

20. Interest Jean received gifts of money from her grandparents for graduation. The money she received in May was deposited in a certificate of deposit earning 4% simple interest. She received twice as much money from her parents and deposited this money in a CD earning 5% simple interest.
   a. Write an equation in x and y where y represents the total amount of interest earned in one year on the two accounts and x represents the amount invested.

   b. Graph this equation in the viewing window $[0, 141]$ by $[0, 30]$ where the x- and y-scales are 1.
   c. Trace to determine the amount of interest earned in the two accounts if her grandparents gave her $108.

   d. How much money would be in the 5% account if her total interest for the year is $31.50?

21. Ticket sales The local elementary school sells tickets to a play at a cost of $0.50 per ticket for faculty and students (to cover the cost of printing the programs) and $2.50 for everyone else.

   a. Write an equation to model the income y from the sale of 25 total tickets.
   b. Graph this equation in the viewing window $[0, 94]$ by $[0, 62]$ where the x- and y-scales equal 10.
   c. Trace to determine the number of faculty/student tickets sold if the total income for the event is $36.50.
   d. Trace to determine the number of $2.50 tickets sold if the income is $20.50.
   e. Is it possible to obtain an income of $66.50 from the sale of tickets? Explain.

22. Grade averages A student makes a 75 and a 65 on the first two tests.
   a. Write an equation where y represents the average grade after three tests.

   b. Graph in the viewing window designated in Exercise 21.
   c. Trace to the point on the graph that indicates the student would have an average of 72 if he made a 76 on the third test.
   d. Is it possible for the student to have an average of 81 after the third test? If so, what grade would he have to make on the third test?

23. Markup The campus bookstore purchases books at the retail price from a textbook dealer and marks the cost up by 40% before selling to the student.
   a. Write an equation to represent the cost y (to the student) of a text purchased for x dollars.
   b. Graph this equation in the viewing window $[0, 94]$ by $[0, 75]$, with x- and y-scales equal to 10.
   c. Trace to determine the retail cost of a textbook that sells for $46.20.
   d. Use the value feature to determine the cost of a textbook to the student if the bookstore purchases the text for $48.

24. Printing costs The Mathematics Department uses the campus media center to print flyers advertising the sale of $3\frac{1}{2}$-inch computer disks. The cost of printing the flyers is $15 and the disks are purchased from a wholesale catalogue for $0.75 each. The disks are then sold to the student at a cost of $1.10 each.
   a. Write an equation to represent the cost y to the department.
   b. Graph the cost equation in the ZInteger viewing window.
   c. Trace to determine the number of disks purchased by the department if the total cost is $81.
   d. Write an equation to represent the revenue y produced by the sale of x disks.
   e. Reset the viewing window to ZInteger and display both the cost equation and the revenue equation.

f. Adjust the size of the viewing window to display the point of intersection and use the INTERSECT feature to determine how many disks must be sold to break even.

g. How many disks must be sold in order for the department to make a profit?

25. Data analysis   A survey (see Table 1) was taken of notably tall buildings on the east coast, midwest, and west cost. The survey compares height, in feet from sidewalk to roof, to number of stories (beginning at street level). The linear equation that models this data is $y = 10.69x + 191.11$. Graph this equation in the viewing window [26, 120] by [348, 1615] with both x- and y-scales set to zero. If the Empire State building has 102 stories, use the given information to predict the height of the building.

Building	Stories	Height (ft)
Baltimore U.S. Fidelity and Guaranty Co.	40	529
Maryland National Bank (Baltimore, MD)	34	509
Sears Tower (Chicago, IL)	110	1454
John Hancock Building (Chicago, IL)	100	1127
Transamerica Pyramid (San Francisco, CA)	48	853
Bank of America (San Francisco, CA)	52	778

SOURCE: *The World Alamanac and Book of Facts 1995;* actual height of the Empire State building is 1250 ft.

TABLE 1

REVIEW

27. Determine the dimensions of the viewing window that could simultaneously display the following groups of points.

a. (2, 3), (−11, 3), (−15, −4), and (9, −20)

b. (−18, 13.4), (16.5, −4) and (25, 30)

c. (50, −50), (−75, 15), and (25, −80)

28. Graph the linear equation $y = x - 15$ in both the standard (ZStandard) viewing window and in the integer (ZInteger) viewing window and explain the difference in what you see.

26. Statistics   The statistics in Table 2 represent dollars per 100 pounds for cattle. The data is plotted in the viewing window [1934, 2010] by [−5, 100] and displayed below. The linear equation that models this data is $y = 1.30x - 2515.32$.

a. Graph the linear equation in the given viewing window and predict the price per 100 pounds of cattle for the years 2000–2005.

b. Are these predictions a "sure thing"? Discuss the factors that might affect your predictions.

Year	1940	1950	1960	1970	1975
Price	7.56	23.30	20.40	27.10	32.20

1979	1980	1984	1985	1986	1987
66.10	62.40	57.30	53.70	52.60	61.10

1988	1989	1990	1991	1992	1993
66.60	69.50	74.60	72.70	71.30	72.60

SOURCE: *The Word Almanac and Book of Facts 1995.*

TABLE 2

29. Graph each of the following functions in the integer viewing window and TRACE to determine the coordinates of the x-intercepts.

a. $y = x^2 - 14x + 45$ **b.** $y = x^2 - x - 56$

c. $y = x^2 + 11x + 24$

30. Trace to find the y-intercept of each of the functions defined in Exercise 29. Compare each of the y-values of the intercepts with the constant value of the function. What do you observe?

The INTERSECT features on the TI-86 and the TI-83 work similarly. If using the TI-86, watch for parenthetical information that specifies the differences for your calculator.

To graphically solve an equation, the graphical representation of the algebraic expression on each side of the equation will be examined. Consider the equation $3x + 4 = 2 - x$. The left side of the equation, $3x + 4$, will be designated as y1. Enter $3x + 4$ after **y1=** (remember to use the [X,T,θ,n] key on the TI-83 and the [x-VAR] key on the TI-86). The right side of the equation, $2 - x$, will be designated as y2. Enter $2 - x$ after **y2=.** We want to determine graphically *where* y1 = y2.

Press [ZOOM] [ZStandard] [GRAPH] (TI-86 users press [2nd] [ZOOM] [F4](ZSTD) to view the graphical representation of each of the algebraic expressions graphed in the standard viewing window). The graphical solution to the equation is the intersection of y1 and y2. At the intersection point, y1 is equal to y2. Circle the intersection point.

To interpret this graph, we want the *x-value* that is the solution to the equation. The *x*-value can be found by using the intersect feature (denoted as ISECT on the TI-86). The only requirement of the intersect feature is that the point of intersection be visible on the screen.

To access the INTERSECT option, press [2nd] [CALC] [5:intersect] to select INTERSECT (TI-86 users press [MORE] [MATH] [MORE] [F3] (ISECT)). Move the cursor along the first curve to the approximate point of intersection and press [ENTER]. At the *second curve* prompt, press [ENTER] again because the cursor will still be close to the point of intersection. At the *guess* prompt, press [ENTER]. This is instructing the calculator that our *guess* is the approximate point of intersection that was designated at the *first curve* prompt. The solution for *x* is -0.5. Your screen should look like the one pictured.

Check the solution. Recall the two options that have been introduced thus far for checking solutions, the **STO**re key and TABLE feature of the calculator. Comparing the results of your check work to the information displayed on the INTERSECT screen shows that when -0.5 is substituted in for *x*, both sides of the equation evaluate to 2.5. This is the *y*-value displayed on the intersection screen. Although the ordered pair $(-0.5, 2.5)$ is the coordinate of the intersection of the two lines displayed, ONLY -0.5 is the solution to the one variable equation, $3x + 4 = 2 - x$.

3.6 Algebraic and Graphical Solutions to One-Variable Linear Equations

■ THE THEORY BEHIND EQUATION SOLVING ■ USING THE TABLE TO CONFIRM ALGEBRAIC SOLUTIONS ■ SOLVING LINEAR EQUATIONS WITH THE GRAPHING CALCULATOR ■ PROBLEM SOLVING REVISITED

■ THE THEORY BEHIND EQUATION SOLVING

The steps for equation solving presented in the first section can be used to solve any linear equation in one variable.

> **Linear Equation**
>
> A linear equation in one variable, x, is any equation that *can be* written in the form
>
> $$ax + c = 0$$
>
> where a and c are real numbers, $a \neq 0$.

We suggest that the steps be carefully followed until you are comfortable with the process.

The theory behind solving linear equations by transformation (writing a series of equivalent equations) is simple: Using opposite operations, arithmetic processes are reversed or "undone." This is usually done in the reverse order of the order of operations. For example, additions and subtractions of terms are done prior to multiplications and divisions.

When the solving procedure is examined, it becomes apparent that the first two steps involve the simplification of the expressions on each side of the equal sign. The third step suggests multiplying each term on both sides of the equation by the LCD to eliminate fractions. This step is a convenience rather than a necessity. The example below illustrates two valid algebraic solutions of the same equation. Study them carefully.

EXAMPLE 1   Solve $\dfrac{2}{3}x + 5 = \dfrac{1}{2}x - 2$.

■ TECHNOLOGY TIP

The graphing calculator can be used to simplify $\frac{2}{3}x - \frac{1}{2}x$. Recall that the distributive property allows you to rewrite the expression as $\left(\frac{2}{3} - \frac{1}{2}\right)x$. The fraction option returns the correct result of $\frac{1}{6}$.

$$\frac{2}{3}x + 5 = \frac{1}{2}x - 2$$

$$6 \cdot \frac{2}{3}x + 6 \cdot 5 = 6 \cdot \frac{1}{2}x - 6 \cdot 2$$

$$4x + 30 = 3x - 12$$

$$4x + 30 - 3x = 3x - 12 - 3x$$

$$x + 30 = -12$$

$$x + 30 - 30 = -12 - 30$$

$$x = -42$$

$$\frac{2}{3}x + 5 = \frac{1}{2}x - 2$$

$$\frac{2}{3}x + 5 - \frac{1}{2}x = \frac{1}{2}x - 2 - \frac{1}{2}x$$

$$\frac{1}{6}x + 5 = -2$$

$$\frac{1}{6}x + 5 - 5 = -2 - 5$$

$$\frac{1}{6}x = -7$$

$$x = -42 \quad ■$$

The equation on the left was solved by clearing the fraction first, whereas the equation on the right kept the fraction throughout the solution process. Make sure you can justify each step in the solving process for **both** the solutions in Example 1.

Example 1 illustrates the fact that clearing an equation of fractions is a solving *option* rather than a *requirement*. Most students choose to eliminate fractions before beginning the solving process. Care must be taken, however, if you choose to eliminate the fraction(s), that each *term* on both sides of the equation be multiplied by the LCD. Consider the two solutions of the given equation below. In the solution on the left, the fraction is eliminated before anything else is done. The solution on the right retains the fraction. Both solutions are correct.

EXAMPLE 2   Solve $\frac{1}{2}(x + 4) + 5 = x - 3$.

$$\frac{1}{2}(x + 4) + 5 = x - 3$$

$$2 \cdot \frac{1}{2}(x + 4) + 2 \cdot 5 = 2 \cdot x - 2 \cdot 3$$

$$(x + 4) + 10 = 2x - 6$$

$$x + 14 = 2x - 6$$

$$x + 14 - 2x = 2x - 6 - 2x$$

$$-x + 14 = -6$$

$$-x + 14 - 14 = -6 - 14$$

$$-x = -20$$

$$\frac{-x}{-1} = \frac{-20}{-1}$$

$$x = 20$$

$$\frac{1}{2}(x + 4) + 5 = x - 3$$

$$\frac{1}{2}x + \frac{1}{2} \cdot 4 + 5 = x - 3$$

$$\frac{1}{2}x + 2 + 5 = x - 3$$

$$\frac{1}{2}x + 7 - x = x - 3 - x$$

$$-\frac{1}{2}x + 7 = -3$$

$$-\frac{1}{2}x + 7 - 7 = -3 - 7$$

$$-\frac{1}{2}x = -10$$

$$(-2)\left(-\frac{1}{2}x\right) = (-2)(-10)$$

$$x = 20 \quad \blacksquare$$

Care must be taken to justify steps when solving. THINK about and mentally justify each step in the solving process.

■ USING THE TABLE TO CONFIRM ALGEBRAIC SOLUTIONS

The STOre feature of the graphing calculator was used to check solutions in Section 3.1. Another way to verify solutions is by using the TABLE feature of the calculator. Consider the equation $5x + 4 = 2x + 10$ and the following chart in which different values of x were substituted into the left and right sides of the equation. Each side was simplified and then compared. Notice that the number 2 is the solution to the equation because both sides evaluate to 14 when x has a value of 2.

x	$5x + 4$	$2x + 10$
1	$5(1) + 4 = 9$	$2(1) + 10 = 12$
2	$5(2) + 4 = 14$	$2(2) + 10 = 14$
3	$5(3) + 4 = 19$	$2(3) + 10 = 16$
4	$5(4) + 4 = 24$	$2(4) + 10 = 18$

The TABLE feature of the graphing calculator can be used to generate a table similar to the preceding one. This is illustrated in the checking process of the following example.

EXAMPLE 3   Solve $3(x - 5) = 4(x + 9)$.

Solution

$$3x - 15 = 4x + 36$$ Distributive property.

$$3x - 15 - \mathbf{4x} = 4x + 36 - \mathbf{4x}$$ Subtraction property of equality.

$$-x - 15 = 36$$ Simplify each side.

$$-x - 15 + \mathbf{15} = 36 + \mathbf{15}$$ Addition property of equality.

$$-x = 51$$ Simplify each side.

$$\frac{-x}{\mathbf{-1}} = \frac{51}{\mathbf{-1}}$$ Division property of equality.

$$x = -51$$ Simplify both sides.

■ **STUDY TIP**

AS you learn various techniques on the graphing calculator, think about what algebraic processes are supported with the technology. For example, use of the STOre feature is comparable to the algebraic process of evaluation.

Check:

a. Using the TABLE feature of the graphing calculator: To use the calculator's TABLE feature to check, first enter the left side of the equation at the Y1= prompt and the right side at the Y2= prompt. Access the TABLE setup and have the TABLE start at the value being checked. Access the TABLE. If the solution checks, the value in the Y1 column will be the same as the value in the Y2 column.

b. Analytically.

$$3x - 15 = 4x + 36$$
$$3(-\mathbf{51}) - 15 \ ? \ 4(-\mathbf{51}) + 36$$
$$-153 - 15 \ ? \ -204 + 36$$
$$-168 = -168$$

c. Using the STOre feature of the graphing calculator.   ■

EXAMPLE 4   Solve $\dfrac{3x + 11}{5} = x + 3$.

■ **TECHNOLOGY TIP**

Recall the entire numerator, $3x + 11$, MUST be enclosed in parentheses when entering the fraction into the calculator.

$$5 \cdot \left(\frac{3x + 11}{5} \right) = 5 \cdot x + 5 \cdot 3$$

$$3x + 11 = 5x + 15$$

$$3x + 11 - \mathbf{3x} = 5x + 15 - \mathbf{3x}$$

$$11 = 2x + 15$$

$$11 - \mathbf{15} = 2x + 15 - \mathbf{15}$$

$$-4 = 2x$$

$$\frac{-4}{2} = \frac{2x}{2}$$

$$-2 = x$$

■

Study the algebraic solution and make sure you can justify each step in the solution process.

■ SOLVING LINEAR EQUATIONS WITH THE GRAPHING CALCULATOR

In each of the examples below, the equations are solved both algebraically and with the graphing calculator. For complete keystroking information, see "Technology 3.5: INTERSECT Method for Solving" on page 165.

EXAMPLE 5   Solve $2(x + 2) - x = 10 - x$.

Solution   *Algebraic solution:* $2x + 4 - x = 10 - x$

$$x + 4 = 10 - x$$
$$2x + 4 = 10$$
$$2x = 6$$
$$x = 3$$

Graphic solution: $y1 = 2(x + 2) - x$

$$y2 = 10 - x$$

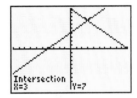

Notice the display at the bottom of the screen. This means that when the variable x has a value of 3, the expressions entered at the $y =$ prompts evaluate to 7.

Check: $2(3) + 4 - 3 \, ? \, 10 - 3$

$$6 + 4 - 3 \, ? \, 7$$
$$7 = 7$$

■

EXAMPLE 6   Solve $-\dfrac{3}{5}x + \dfrac{21}{5} = -\dfrac{4}{3}x + \dfrac{17}{3}$.

Solution   *Algebraic solution:* $15 \cdot -\dfrac{3}{5}x + 15 \cdot \dfrac{21}{5} = 15 \cdot -\dfrac{4}{3}x + 15 \cdot \dfrac{17}{3}$

$$-9x + 63 = -20x + 85$$
$$11x = 22$$
$$x = 2$$

Graphic solution: $y1 = (-3/5)x + 21/5$

$$y2 = (-4/3)x + 17/3$$

The check is left to the reader. We suggest that the check be done analytically, using the TABLE feature and the STOre feature of the calculator.

■

■ PROBLEM SOLVING REVISITED

The strategies below were introduced in Section 3.1. These should have been copied on a note card for quick reference. This card will be used throughout the text when working application problems.

Problem-Solving Strategies

1. Read the problem several times and organize the facts. What is known? What are you asked to find? It may help you to organize the facts under the two headings, **known** and **unknown.**

2. **a.** If there is only one unknown quantity, choose a variable to represent that quantity. Clearly define what the variable is and what it represents with a "Let" statement.

   b. When there is more than one unknown, choose a variable to represent one of the quantities and express all other important quantities in terms of the chosen variable.

3. Translate the information expressed in the words of the problem into an algebraic equation.

4. Solve the equation.

5. Answer the question posed in the problem.

6. Check your results both in terms of the equation and the words of the problem. Make sure your solution makes sense within the context of the problem.

EXAMPLE 7   Three more than double a number is 3 times the number plus 15. What is the number?

Solution   Because there is only one unknown, x will represent the unknown number. This is expressed by using a "Let" statement.

■ STUDY TIP
Read this text with paper and pencil in hand. Work out examples as you read them to be sure you understand each step of the work.

Let x = the number.   Clearly define the variable and what it represents.

The words of the problem are translated to an algebraic equation:

$$3 + 2x = 3x + 15$$

Solve the equation

$$3 + 2x = 3x + 15$$
$$3 + 2x - 3x = 3x + 15 - 3x$$
$$3 - x = 15$$
$$3 - x - 3 = 15 - 3$$
$$-x = 12$$
$$x = -12$$

The number is -12.   Answer the question.   ■

Make sure your result makes sense BOTH in the equation and in the context of the problem. Checking the equation can be done by the tools previously introduced: analytically (substitution), using the STOre feature of the calculator, or using the TABLE feature of the calculator. Another helpful strategy is to draw a picture or diagram whenever possible. This is illustrated in Example 8.

EXAMPLE 8   **Plumbing**   A plumber wants to cut a 17-foot pipe into three sections. The longest section is to be 3 times as long as the shortest, and the middle-sized section is to be 2 feet longer than the shortest. How long should each section be?

Solution

Known:

a 17-foot pipe will be cut into three sections
the longest section is 3 times as long as the **shortest**
the middle section is 2 feet longer than the **shortest**

Unknown: the length of each section.

17 ft

Because there is more than one unknown, we must first determine which piece of the pipe will be represented by the variable x. Because the lengths of the longest section and the middle section are given *in terms of* the length of the shortest section,

Let $x =$ the length of the shortest section.

We can now translate our known phrases into algebraic statements:

longest section is 3 times as long as the shortest

longest section = 3   · x

longest section = $3x$

middle section is 2 feet longer than the shortest

middle section = 2   + x

middle section = $2 + x$

Couple this with a labeled drawing:

17 ft = total length

x	$x + 2$	$3x$
Length of shortest section	Length of middle–sized section	Length of longest section

Because the pipe is 17 feet long, the sum of the 3 pieces will equal the total length. This translates to the equation below.

$$x + (x + 2) + 3x = 17$$
$$x + x + 2 + 3x = 17 \qquad \text{Remove grouping.}$$
$$5x + 2 = 17 \qquad \text{Combine like terms.}$$
$$5x = 15 \qquad \text{Subtract 2 from both sides.}$$
$$x = 3 \qquad \text{Divide both sides by 5.}$$

The equation could have been solved with the calculator.

Accessing the graph in the ZStandard viewing window does not display the inter-section of the y1 and y2 equations. Because y2 = 17 is the horizontal line whose *y*-intercept is 17, the ymax on the WINDOW screen must be adjusted. Let the ymax equal 20. The point of intersection is clearly displayed and the INTERSECT option under the CALC menu is used to find the intersection of the two graphs.

This confirms our analytic solution that *x* must have a value of 3 for the equation to be a true statement. We then substitute in 3 for *x* and find the length of our pieces of pipe. ∎

EXAMPLE 9   **Road construction**   A truck made five trips hauling asphalt to a road construction site. A larger truck made eight trips. When fully loaded, the larger truck carries 2 tons more asphalt than the smaller truck. If the two trucks hauled a total of 55 tons, how much asphalt can the smaller truck carry in one load?

Solution   Known:

there are two trucks, one large and one small
the small truck made five trips hauling asphalt
the larger truck made eight trips hauling asphalt
the large truck carries 2 tons more asphalt than the smaller
the two trucks hauled a total of 55 tons

Unknown: how much asphalt can the smaller truck carry in one load?

Because we're only asked for the amount of asphalt carried by one truck (the smaller), our let statement becomes

Let *x* = amount of asphalt carried by the smaller truck.

We know there are two trucks, so we look in our known information for a phrase that relates the amounts hauled by each truck. Once we find the phrase, we translate it.

the large truck carries 2 tons **more** asphalt than **the smaller**

large truck   =   2   +   *x*

large truck = 2 + *x*

Our variable expressions, *x* and *x* + 2, represent the amounts carried by each truck per trip. We know the small truck made five trips and the larger truck made eight trips. Therefore, the amount hauled by each truck is

small truck   large truck
$5x$ $8(x + 2)$
no. of trips   am't per trip   no. of trips   am't per trip

Because the two trucks hauled a total amount of 55 tons of asphalt, the key is the word *total*, which implies the arithmetic operation of addition. Again, we translate to an equation, solve, and check:

Total tonnage carried by small truck	plus	total tonnage carried by large truck	equals	total tonnage carried by both trucks
$5x$	$+$	$8(x + 2)$	$=$	55

$5x + 8(x + 2) = 55$ The equation to solve.

$5x + 8x + 16 = 55$ Remove parentheses.

$13x + 16 = 55$ Combine like terms.

$13x = 39$ Subtract 16 from both sides.

$x = 3$ Divide both sides by 13.

The smaller truck can carry 3 tons of asphalt.
This equation could have been solved with the calculator.

Because the point of intersection is not displayed, the WINDOW screen is accessed and the ymax is adjusted. Because the number 55 was entered at the y2= prompt, the ymax is set equal to 60 to accommodate the value. The INTERSECT option under the CALC menu is used to find the intersection. The intersect screen on the right confirms our algebraic value of *x*.

EXERCISE 3.6

VOCABULARY In Exercises 1–2, fill in the blanks to make a true statement.

1. A linear equation in **one** variable, *x*, is any equation that can be written in the form _____, where *a* and *c* are real numbers and $a \neq 0$.

2. Two features of the calculator that can be used to verify solutions are the _____ and _____ features.

CONCEPTS

3. The graphical solution to an equation whose left side has been graphed at y1 and whose right side has been graphed at y2 is displayed at the right. What is the solution?

4. The solution to an equation whose left side has been entered at y1 and whose right side has been entered at y2 is verified in the table at the right. What is the solution?

In Exercises 5–14, the left side of each equation was entered at the y1= prompt and the right side of the equation was entered at the y2= prompt. The pictured display is in the ZStandard viewing window, and the point of intersection is not visible. To display the point of intersection, a new viewing window must be set. Specify an appropriate viewing window for each of the following.

5. $3(x - 4) - 6 = 16$

6. $5(x - 2) + 9 = 20$

7. $8x - 2 = 3(x - 8)$

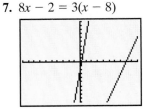

8. $\dfrac{8x + 3}{5} + 6 = 22$

9. $\dfrac{5x-4}{2} - 8 = 15 + x$

10. $\dfrac{1}{2}(x+4) - \dfrac{1}{2} = -\dfrac{23}{2} - \dfrac{1}{2}x$

11. $[3(x+1) - 2] - 5 = -16$

12. $\dfrac{3x+10}{5} + 7 = 12$

13. $\dfrac{x+3}{5} + 12 = 13$

14. $2[4(x-1) + 3] - 6 = -12$

PRACTICE *In Exercises 15–30, solve each equation algebraically, showing all steps of the solving process. Check solutions using the STOre and/or TABLE feature(s) of the graphing calculator.*

15. $\dfrac{5x+10}{7} = 0$

16. $\dfrac{5k-8}{9} = 1$

17. $\dfrac{17k-28}{21} + \dfrac{4}{3} = 0$

18. $\dfrac{3t-5}{5} + \dfrac{1}{2} = -\dfrac{19}{2}$

19. $\dfrac{11(x-12)}{2} = 9 - 2x$

20. $\dfrac{8(5-s)}{5} = -2s$

21. $\dfrac{5z}{3} + 3 = -2$

22. $\dfrac{3y}{2} + 3 = -2$

23. $19.1x - 4(x + 0.3) = -46.5$

24. $14.3(x + 2) + 13.7(x - 3) = 15.5$

25. $\dfrac{2}{3}(x+6) + \dfrac{5}{6}(x-6) = x + 3$

26. $\dfrac{1}{8}(x-8) + \dfrac{1}{4}(x-4) = x + 2$

27. $\dfrac{2x+3}{5} = \dfrac{3x-5}{3}$

28. $\dfrac{6a-1}{8} = \dfrac{5a+2}{3}$

29. $-[2x - (5x - 1) + 3] = -4(x + 3)$

30. $3[x - (3x - 1)] = 3(x + 4)$

In Exercises 31–40, solve each equation using the graphing calculator. Copy the graphing calculator screen display exactly as it appears, recording ALL displayed information. If the WINDOW values were adjusted from ZStandard, specify the viewing window used to obtain a graph that clearly displayed the point of intersection.

31. $3(x + 4) - 8 = 5(x - 1) + 16$

32. $4(x - 1) + 2 = 3x - 1$

33. $\dfrac{2x+4}{2} = \dfrac{5x-2}{8}$

34. $\dfrac{3x+1}{5} = \dfrac{4x-1}{6}$

35. $\dfrac{1}{5}(x - 10) - 8 = -11 + 2x$

36. $3x + 2(x - 4) = -2x - 22$

37. $2(x + 6) - 10 = 20$

38. $3(x - 6) = 2(x - 5)$

39. $\dfrac{x}{5} - 3 = 15$

40. $\dfrac{x}{3} - 2 = -10$

In Exercises 41–50, translate each statement below into an algebraic equation and solve. Let x represent the variable.

41. Three more than twice a number is 5 less than the number. What is the number?

42. Six less than 3 times a number is 8 more than the number. What is the number?

43. Twice the sum of a number and 3 is 5 times the number. What is the number?

44. Triple the difference of a number and 2 to get an expression equal to twice the number plus 3. What is the number?

45. A number and three times its additive inverse is equal to 4. What is the number?

46. The additive inverse and 5 times a number is equal to 6 less than the number. What is the number?

47. Four times the quantity of a number and 6 is 3 less than the number. What is the number?

48. Eight times the quantity of a number decreased by 2 is equal to 8. What is the number?

49. The quotient of a number and 2 is equivalent to the number plus 3. What is the number?

50. The quotient of a number and 3 is 14 less than 5 times the number. What is the number?

APPLICATIONS *Solve each of the application problems below. Make sure that all four component parts of a complete solution are present. This includes the appropriate "Let" statement(s), an equation, the complete solution to the equation, and an answer to the question.*

51. Shares of stock At a recent stockholders' meeting, 4.5 million shares were voted in favor of a proposal for a mandatory retirement age for the board of directors. This represented 75% of the total number of shares, so the proposal passed. How many shares were there?

52. Buying real estate The cost of a condominium is $57,595 less than the cost of a house. If the house costs $102,744, find the cost of the condominium.

53. Buying paint After reading the ad in Illustration 1, a decorator bought one gallon of primer, one gallon of paint, and a brush. The total cost was $30.44. Find the cost of the brush.

ILLUSTRATION 1

54. Customer satisfaction Two-thirds of the movie audience left the theater in disgust. If 78 angry patrons walked out, how many were there originally?

55. Stock split After a three-for-two stock split, a shareholder will own 1.5 times as many shares as before. If 555 shares are owned after the split, how many were owned before?

56. Off-campus housing Four-sevenths of the senior class is living in off-campus housing. If 868 students live off campus, how large is the senior class?

57. Union membership The 2484 union members represent 90% of a factory's workforce. How many employees are there?

58. Hospital occupancy Thirty-six percent of hospital patients stay for less than 2 days. If 1008 patients in January stayed for less than 2 days, what total number of patients did the hospital treat in January?

59. Advertising The manager of a supermarket hires a woman to distribute advertising circulars door to door. She will be paid $5 a day plus $0.05 for every advertisement distributed. How many circulars must she distribute to earn $42.50?

60. Integer problem Six less than 3 times a certain integer is 9. Find the integer.

61. Integer problem If a certain integer is increased by 7 and that result is divided by 2, the integer 5 is obtained. Find the original integer.

62. Apartment rental A student moves into a bigger apartment, which rents for $400 per month. That rent is $100 less than twice what she is now paying. Find the current rate.

63. Auto repair A mechanic charged $20 an hour to repair the water pump on a car, plus $95 for parts. If the total bill was $155, how many hours did the repair take?

J.D's Automotive		
Labor	Hrs	Cost
Parts		
Total		

64. Water billing The city's water department charges $7 per month plus 42 cents for every 100 gallons of water used. Last month, one homeowner used 1900 gallons and received a bill for $17.98. Was the billing correct? Justify your response.

65. Monthly sales A clerk's sales in February were $2000 less than three times her sales in January. If her February sales were $7000, by what amount did her sales increase?

66. Ticket sales A music group charges $1500 for each performance plus 20% of the total ticket sales. After a concert, the group received $2980. How much money did the ticket sales raise?

67. Carpentry The 12-foot board in Illustration 2 has been cut into two sections, one twice as long as the other. How long is each section?

x $2x$

12 ft

ILLUSTRATION 2

68. Plumbing The 20-foot pipe in Illustration 3 has been cut into two sections, one 3 times as long as the other. How long is each section?

x $3x$

20 ft

ILLUSTRATION 3

69. Discount shopping A stereo system discounted 20% is selling for $969.20. What was its original price?

STEREO $969.20 SAVE $$$

70. Installing solar heating One solar panel in Illustration 4 is 3.4 feet wider than the other. Find the width of each.

|← 18 ft →|

ILLUSTRATION 4

REVIEW

78. Dosing guidelines The dosing guidelines for synthetic thyroid replacement hormone are given in Illustration 5. What is the **minimum** dosage for a child of 7 years of age weighing 55 pounds (25 kilograms)?

71. Waste disposal Two tanks hold a total of 45 gallons of a toxic solvent. One tank holds 6 gallons more than twice the amount in the other. Can the smaller tank be emptied into a 10-gallon waste disposal canister?

72. Buying vitamins If you buy one bottle of vitamins, you can get a second bottle for half price. Two bottles cost $2.25. Find the usual price for a single bottle of vitamins.

SALE VITAMINS BUY ONE GET THE SECOND ONE HALF PRICE!

73. Atomic weights Water is made up of hydrogen and oxygen. In water, the mass of the oxygen is 8 times the mass of the hydrogen. How much hydrogen is in 2700 grams of water?

74. Choosing salary options Carrie has her choice of two salary options. The first plan pays $600 per month plus a 2% commission on sales. The second plan is straight commission—5% of sales. To decide which plan to choose, Carrie needs to know the monthly sales that will produce equal income. Find the monthly sales.

75. Sales quotas The salesforce is hoping that last month's sales will exceed their one-million-dollar quota. Jennifer was last month's top sales representative, responsible for 45% of total sales. Jim came in second at 23%, with the remaining $317,920 credited to others. Did the sales team make their quota?

76. Publisher's inventories A novel can be purchased in a hardcover edition for $15.95 or in paperback for $4.95. The publisher printed 11 times as many paperbacks as hardcover books, a total of 114,000 copies. How many hardcover books were printed?

77. Manufacturing concrete Concrete contains 3 times as much gravel as cement. How much cement is in 500 pounds of dry concrete mix?

Dosing Guidelines for Pediatric Hypothyroidism	
Age	Daily dose per kg body weight*
0-3 mos	10-15 mcg
3-6 mos	8-10 mcg
6-12 mos	6-8 mcg
1-5 yrs	5-6 mcg
6-12 yrs	4-5 mcg
>12 years	2-3 mcg
Growth & puberty complete	1.6 mcg

*To be adjusted on the basis of clinical response and laboratory tests (see **Laboratory Tests**).

SOURCE: Knoll Pharmaceutical Company

ILLUSTRATION 5

79. Sports A baseball diamond is 90 feet on each side as shown in Illustration 6. Find the area.

ILLUSTRATION 6

80. Cricket chirps The number of cricket chirps, N, per minute is related to the temperature, T, by the formula

$$T = 50 + \frac{N - 92}{4.7}$$

a. Fill in the following table to determine the relationship between temperature and chirps. Round all results to the nearest whole number.

N		235		0
T	90°		40°	

✎**b.** What is the relationship between the number of chirps and the temperature?

✎**c.** Is this a linear relationship? (That is, if T were replaced with y and N with x, would the equation define a linear function and would its graph be a straight line?)

d. Graph the equation in an appropriate viewing window.

81. Write as an algebraic phrase: "the product of x and y divided by 3."

82. Write as an English phrase: $2x - 5$.

83. In order to get to work every day, a college student walks $x + 2$ miles to the bus stop, rides the bus $2x - 1$ miles, and then walks another $x - 1$ miles to campus. What is the total distance walked?

CHAPTER SUMMARY

CONCEPTS

REVIEW EXERCISES

Section 3.1

Any real number can be added to (or subtracted from) both sides of an equation to form another equation with the same solution(s) as the original equation.

Both sides of an equation can be multiplied by (or divided by) any *nonzero* real number to form another equation with the same solution(s) as the original equation.

Conditional equations are true for some, but not all, real numbers.

Identities are equations that are true for ALL real numbers. Their solutions are represented by $\mathbb{R}$, a symbol for "the set of all real numbers."

Contradictions are equations that are NEVER true for any real number. Their solution is represented by $\varnothing$, the null set.

This section includes "Steps for Solving a Linear Equation in One Variable" as well as "Problem-Solving Strategies."

Introduction to Linear Equations in One Variable

1. Tell whether the indicated number is a solution of the given equation.
  a. $3x + 7 = 1$; -2 **b.** $5 - 2x = 3$; -1
  c. $2(x + 3) = x$; -3 **d.** $5(3 - x) = 2 - 4x$; 13

2. Solve each equation algebraically, showing all steps of your work. Check all results in the original equation.
  a. $3x + 7 = 1$ **b.** $7 - 9x = 8$
  c. $\dfrac{x + 3}{4} = 2$ **d.** $\dfrac{x - 7}{2} = -2$
  e. $\dfrac{x}{2} + \dfrac{7}{2} = 3$ **f.** $\dfrac{x}{3} - 3 = 2$

3. Solve each equation algebraically, showing all steps of your work. Check all results in the original equation.
  a. $2x - 19 = 2 - x$ **b.** $5x - 19 = 20 - 2x$
  c. $10(x - 3) = 3(x + 11)$ **d.** $2(5x - 7) = 2(x - 16)$
  e. $\dfrac{2}{3}(5x - 3) = 38$ **f.** $\dfrac{3}{2}(3x - 12) = 9$

4. Write an equation in one variable to model the problem and solve the equation algebraically. Make sure your result makes sense within the context of the problem.
  a. If twice an integer is decreased by 7, the result is 9. Find the original integer.

  b. Four less than 3 times a number is 77. Find the number.

  c. A contractor charges $35 for the installation of rain gutters plus $1.50 per foot. One installation costs $162.50. How many feet of gutter is required?

  d. The electric company charges $17.50 per month plus 18 cents for every kilowatt hour of energy used. One resident's bill was $43.96. How many kilowatt hours were used?

  e. A 45-foot rope is to be cut into three sections. One section is to be 15-feet long. Of the remaining sections, one must be 2 feet less than 3 times the length of the other. Find the length of the shortest section.

Section 3.2

A literal equation, or formula, can often be solved for any of its variables.

Literal Equations and Formulas

5. Solve each equation for the indicated variable.
  a. $3x + y = 12$; y **b.** $4x - 3y = 8$; y **c.** $d = rt$; t

  d. $P = 2l + 2w$; w **e.** $A = 2\pi rh$; h **f.** $A = P + Prt$; r

  g. $P = \dfrac{RT}{V}$; V

| Section 3.3 | Introduction to Equations in Two Variables |

Linear equations in two variables will have their solutions expressed as an ordered pair (x, y).

Linear equations in two variables have infinite solutions in the form (x, y).

A function is a collection of ordered pairs in which each x-value is paired with a unique y-value.

The domain of a function is the set of all x-values of the ordered pairs of the function.

The range of a function is the set of all y-values of the ordered pairs of the function.

A linear equation in Standard Form is of the form $Ax + By = C$, where A, B, and C are real numbers.

6. Tell whether the indicated ordered pair is a solution to the linear equation in two unknowns. Do this both analytically (by substitution and simplifying) and using the STOre feature of the calculator.
   a. $4x + y = 7$; $(2, -1)$ **b.** $x - 3y = 5$; $(11, 2)$
   c. $3x - 2y = -1$; $(3, -4)$ **d.** $2x + 3y = 3$; $(-2, -3)$

7. Find the corresponding y-value for each given x-value. Write the result as an ordered pair.
   a. $x - 2y = 6$; $x = 2$ **b.** $5x + 2y = 9$; $x = 3$
   c. $x^2 + 3x + 2 = y$; $x = -1$ **d.** $x^2 - 9 = y$; $x = 3$

8. Complete the table of values for the equation $3x + y = 6$.

x	y
-2	
-1	
0	
1	
2	

9. Determine if each relation or equation specified below defines a function. Determine both the domain and the range of the relation or function.
   a. $\{(2, 3), (4, 6), (2, -3)\}$ **b.** $\{(1, 5), (2, 8), (3, 11)\}$

   c. $y = 4x + 5$ **d.** $y = x^2 - 1$

| Section 3.4 | The Rectangular Coordinate System |

Ordered pairs of numbers (x, y) are associated with points in a rectangular coordinate system.

To graph a linear equation in the variables x and y, choose at least three values for x, use substitution to find the corresponding values of y, plot the points (x, y), and finally draw the line that passes through the points.

10. Graph the equation by hand on graph paper.
    a. $y = 3x - 1$ **b.** $y = -4x + 2$ **c.** $3x + y = 8$

    d. $x - 4y = 6$ **e.** $y = 3$ **f.** $x = -5$

To find the *x*-intercept of a line, let $y = 0$ and solve for the value of *x*.

To find the *y*-intercept of a line, let $x = 0$ and solve for the value of *y*.

Equations of nonvertical and nonhorizontal lines can be written either in the form $y = mx + b$ or $Ax + By = C$.

The equation $x = a$ is the equation of a vertical line. If $a = 0$ then $x = a$ represents the *y*-axis. If $a \neq 0$ then $x = a$ represents a line parallel to the *y*-axis.

The equation $y = b$ is the equation of a horizontal line. If $b = 0$ then $y = b$ represents the *x*-axis. If $b \neq 0$ then $y = b$ represents a line parallel to the *x*-axis.

The Standard (ZStandard) viewing window for a graphing calculator is $[-10, 10]$ by $[-10, 10]$ with *x*- and *y*-scales set equal to 1.

The Integer (ZInteger) viewing window for most graphing calculators is $[-47, 47]$ by $[-31, 31]$ with *x*- and *y*-scales set equal to 10.

Graphing Basics are explained in "Technology 3.4."

11. Algebraically, find the *x*- and *y*-intercepts of the equations in Exercise 10 (if they exist). Identify those equations that are horizontal and vertical lines.

12. Graph each equation in the ZInteger viewing window and TRACE to determine *x*- and *y*-intercepts.
   a. $y = 3x - 6$
   b. $y = x^2 - 4x - 5$
   c. $y = (x - 4)^3$
   d. $4x - y = 12$

Section 3.5 Linear Applications and Models

A good viewing window is one that displays *x*-intercepts, *y*-intercepts, and any high or low points of the graph.

Both graphs and tables of the graphing calculator can be used to explore application problems in two unknowns.

Solutions to linear equations in one variable can be confirmed via the STOre feature or by using the TABLE feature of the graphing calculator.

13. Determine an appropriate viewing window for each function specified below. Begin in ZStandard.
   a. $x^2 - 16x + 48 = y$
   b. $y = 5x - 100$
   c. $30x^2 - 155x + 140 = y$

In Review Exercises 14–16, use only your graphing calculator to respond to the questions.

14. A machine shop buys a lathe for \$1970. Its expected life is 10 years. If the linear equation that models this is $y = -170x + 1970$, then
   a. graph the equation in the viewing window of $[0, 11.75]$ by $[0, 1974]$.
   b. trace to determine the value of the lathe in 8 years.
   c. use the VALUE feature to determine the value of the lathe in 6 years 9 months.

Equations may be solved graphically by graphing each side of the equation and then using the INTERSECT feature to determine the point(s) of intersection. See "Technology 3.5."

15. An antique clock was purchased for $125. It has been increasing in value in a linear pattern at a rate of $35 per year.
    a. Write an equation to model its value y as related to years x.
    b. Graph this equation in the viewing window of $[0, 9.4]$ by $[-10, 500]$.
    c. TRACE to determine the value in 3 years.
    d. TRACE to determine how many years it will take to have a value of $405.

16. A community theater has receipts of $1069 after selling adult tickets for $12 and student/child tickets for $5. Let x represent the number of adult tickets sold and y the number of student tickets sold.
    a. Write an equation in x and y to model the total sales.
    b. Solve the equation for y and graph it in a $[0, 94]$ by $[0, 215]$ viewing window.
    c. Using TRACE or VALUE, determine the number of student tickets if 42 adult tickets are sold.
    d. Is it possible for 60 adult tickets to be sold? Based on the graph, why or why not?
    e. Is it possible for 90 adult tickets to be sold? Why or why not?

Section 3.6	Algebraic and Graphical Solutions to One-Variable Linear Equations

The INTERSECT feature of the calculator, the TABLE, and the STOre feature can be used to confirm algebraic solutions to equations.

17. Using the equations in Review Exercises 2 and 3,
    a. solve each equation graphically using the INTERSECT feature.
    b. confirm solutions in the TABLE.
    c. confirm solutions using the STOre feature.

18. Using Review Exercise 4,
    a. solve each equation graphically.
    b. confirm solutions in the TABLE.
    c. confirm solutions using the STOre feature.

■ Chapter Test

In Problems 1 and 2, tell whether the indicated number is a solution of the given equation.

1. $-3(2 - x) = 0$; -2

2. $x(x + 1) = x^2 + 1$; -1

In Problems 3 and 4, solve each equation for the indicated variable.

3. $4x + 3y = 8$; solve for y

4. $\frac{1}{2}x + 2y = \frac{3}{5}$; solve for y

In Problems 5 and 6, tell whether the indicated ordered pair is a solution of the given equation.

5. $2x - 5y = 7$; $(-1, 2)$

6. $x + 3y = -2$; $(1, -1)$

In Problems 7–14, solve for x both algebraically and graphically.

7. $\frac{x}{7} = 1$

8. $8x + 2 = -14$

9. $23 - 5(x + 10) = -12$

10. $x(x + 5) = x^2 + 3x - 8$

11. $\frac{3(x - 6)}{2} = 6x$

12. $\frac{5}{3}(x - 7) = 15(x + 1)$

13. $x(x + 1) = x^2 + x + 1$ **14.** $x + 5(x - 3) = 2(x - 1) + 3(x - 5) + 2$

In Problems 15–18, graph the equation on graph paper and with your graphing calculator. Indicate the x and y intercepts on your hand-sketched graph. Copy the screen display from your graphing calculator and specify the window values.

15. $y = 4x - 1$ **16.** $y = -\dfrac{1}{2}x + 5$ **17.** $2x + y = 1$ **18.** $x + 3y = -2$

19. The number of television sets that consumers buy depends on the price. The higher the price, the fewer TVs people will buy. The equation that relates the price, y, to the number, x, of TVs sold at that price is the demand equation. The demand equation for a 13-inch TV is

$$y = -\frac{1}{10}x + 170$$

    a. Graph the demand equation in a [0, 188] by [0, 200] viewing window and sketch the graph.

    b. Use the graph to determine how many televisions will be sold at a price of $150.

In Problems 20–22, (a) write an equation to model the problem, (b) solve the equation, (c) answer the question in a complete sentence.

20. The sum of two consecutive odd integers is 36. Find the integers.

21. A Turkish rug was purchased for $560. It has increased 96% in value. Find its current value.

22. A real estate agent earns $600 plus a 6% commission for the sale of each house. If she earns $5700 on the sale of a house, what was the value of the house?

CAREERS & MATHEMATICS

Pilot Pilots are highly trained professionals who fly airplanes and helicopters to carry out a wide variety of tasks. Most pilots transport passengers and cargo, but some are involved in crop dusting, spreading seed, testing aircraft, fire-fighting, tracking criminals, traffic reporting, and rescue and evacuation.

Qualifications All pilots paid to transport passengers or cargo must have a commercial pilot's license with an instrument rating. They must also have good health, vision correctable to 20/20, good hearing, and no physical handicaps that would interfere with safety. Most employers also require at least 2 years of college, and many prefer college graduates. Requirements for a private pilot's license are less restrictive.

Job Outlook Because pilots like their jobs and few change careers, competition for employment is stiff. Over the next decade, however, retirements will provide job openings for qualified applicants.

Example Application The air passing over the curved top of the wing in Illustration 1 moves faster than the air beneath, because it has a greater distance to travel. This faster-moving air exerts less pressure on the wing than the air beneath, causing *lift.*

Lift

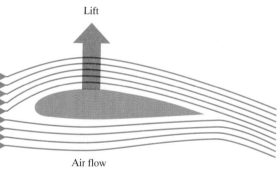

Air flow

ILLUSTRATION 1

Two factors that determine lift are controlled by the pilot. One is the speed or *velocity* of the plane. Another, the *angle of attack,* is the angle between the direction the plane is aimed and the direction it is actually moving, as shown in Illustration 2.

For one particular light aircraft, the lift, velocity, and angle of attack are related by the equation

$$L = (0.017a + 0.023)V^2$$

where L is the lift in pounds, a is the angle of attack in degrees, and V is the velocity in feet per second. To support this aircraft, the lift must equal the plane's weight, 2050 pounds. If the velocity is 130 feet per second, find the correct angle of attack.

Solution First, the pilot substitutes the values for lift and velocity into the given equation and solves for a, the angle of attack.

Direction plane is aimed

Angle of attack

Path of plane

Horizontal

ILLUSTRATION 2

$$L = (0.017a + 0.023)V^2$$

$$2050 = (0.017a + 0.023)(130)^2$$ Substitute 2050 for L and 130 for V.

$$\frac{2050}{130^2} = 0.017a + 0.023$$ Undo the multiplication on the right by dividing both sides by 130^2.

$$\frac{2050}{130^2} - 0.023 = 0.017a$$ Undo the addition on the right by subtracting 0.023 from both sides.

$$0.098301775 = 0.017a$$ Use a calculator to simplify the left-hand side.

$$\frac{0.098301775}{0.017} = a$$ Undo the multiplication on the right by dividing both sides by 0.017.

$$5.782457362 = a$$ Use a calculator.

To provide proper lift, the pilot must adjust the climb to maintain an angle of attack of about $5\frac{3}{4}^\circ$.

EXERCISES   As the angle of attack approaches about 17°, the plane begins to stall. With more cargo on the return trip, the same plane weighs 2325 pounds. If the pilot allows the velocity to drop to 80 feet per second (about 55 miles per hour), will the plane stall?

(*Answer:* The calculated angle of attack is 20°. The plane will stall unless the pilot increases the speed.)

An Extended Look
at Applications and
Absolute Value Equations

4

■ PERSPECTIVE

How to Solve It

As a young student, George Polya (1888–1985) enjoyed his studies of mathematics and understood the solutions presented by his teachers. However, Polya had questions still asked by mathematics students today: "Yes, the solution works, but how is it possible to come up with such a solution? How could *I* discover such things by myself?" These questions still concerned him years later when, as Professor of Mathematics at Stanford University, he developed an approach to teaching mathematics that was very popular with faculty and students. His book, *How to Solve It*, became a bestseller.

Polya's problem-solving approach involves four steps.

- *Understand the problem.* What is the unknown? What information is known? What are the conditions?
- *Devise a plan.* Have you seen anything like it before? Do you know any related problems you have solved before? If you can't solve the proposed problem, can you solve a similar but easier problem?

- *Carry out the plan.* Check each step. Can you explain why each step is correct?
- *Look back.* Examine the solution. Can you check the result? Can you use the result, or the method, to solve any other problem?

George Polya (1888–1985)

4.1 Applications—A Numeric Approach

■ CONSTRUCTING TABLES ■ GEOMETRY PROBLEMS ■ RATIO AND PROPORTION ■ SIMILAR TRIANGLES

■ CONSTRUCTING TABLES

When investigating problem-solving strategies, it becomes critical to first organize the given data. In the previous chapter, the strategy of writing down the known information and the unknown information provided a method for organization. Tables are useful for both organizing data and for helping generalize from the specific to the general. Consider the following example.

EXAMPLE 1   Blaire is offered a summer job for 3 months at a popular teen clothing store in the local mall. She has a choice of salary—either a set salary of $1040 per month or $500 per month plus 10% of all sales. How much would she have to sell each month for the salaried offer to equal the commission offer?

Solution   She constructs the following table to organize her information. She begins computing sales of $400 per month and moves up by increments of $200.

Sales	$400	$600	$800	x
Pay	.10(400) + 500 = 540	.10(600) + 500 = 560	.10(800) + 500 = 580	.10x + 500

The table of specific values illustrates a pattern. She can now move from the specifics of a set amount of sales to a generalized expression for the amount of money earned from sales. This allows her to set up an equation that can be solved and she can compute the amount of sales necessary to equal the proposed salary of $1040.

Let x = amount of sales per month

$.10x$ = amount of commission

$$.10x + 500 = 1040$$
$$.10x + 500 - 500 = 1040 - 500$$
$$.10x = 540$$
$$x = 5400$$

She would have to sell $5400 worth of clothing each month to equal the fixed salary of $1040.

Calculator solution: Once Blaire constructed her equation, she entered the expression $.10x + 500$ at the y1= prompt. She set the table to start at 400 (to represent $400 in clothing sales) and incremented the table by 200. Scrolling down the table allows her to compare the amount of clothing sales with her income from those sales.

The next example illustrates the use of the table with a consecutive integer problem. Recall the set of integers is {. . . , -5, -4, -3, -2, -1, 0, 1, 2, 3, 4, 5, . . .}. This example illustrates a consecutive integer problem where x is the first integer and $x + 1$ represents the second integer. If x represents an *odd* or an *even* integer, the expression $x + 2$ represents the *next* odd (or even) integer.

EXAMPLE 2   Bonnie is a mathematics teacher who loves to read. She's currently reading *War and Peace*. While placing her bookmark in the book, she notices that the sum of the page numbers is 2065. Between which two pages is the bookmark?

Solution   Again, a table can help us organize the information. In this problem, we begin with the pages 1000 and 1001 (this is simply a guess). The table allows us to generalize from the specific to the general and write an algebraic equation that we can solve to answer the question posed.

Left page number	1000	1002	x
Right page number	1001	1003	$x + 1$
Sum	2001	2005	$x + (x + 1)$

Let x = left page number

$x + 1$ = right page number

$$x + (x + 1) = 2065$$
$$2x + 1 = 2065$$
$$2x + 1 - 1 = 2065 - 1$$
$$2x = 2064$$
$$x = 1032$$

The bookmark is between pages 1032 and 1033.

Calculator solution: Once the equation is constructed, enter the left side at the y1= prompt. Set the table to start at 1000 and to increment by 1. Enter 2065 at the y2= prompt. Scroll down until the value in the Y1 column equals the value in the Y2 column.

EXAMPLE 3 Marie wants to attend the state fair, yet cannot decide on which night she should go. On Friday night, the price of admission is $3.00 and it costs $0.75 per ride. Saturday night is armband night and the admission price of $15 allows you to purchase an armband and ride all the rides you desire. Help her decide which is the best night to go.

Number of rides	1	2	x
Cost for Friday	3 + 1(.75)	3 + 2(.75)	3 + x(.75)

Again, the table allows Marie to move from the specific to the general and construct the following equation:

Let x = number of rides

$3 + x(.75)$ = cost on Friday night

$$3 + x(.75) = 15.00$$
$$3 + x(.75) - 3 = 15.00 - 3$$
$$.75x = 12.00$$
$$x = 16$$

Marie concludes she would have to ride 16 rides on Friday night for the cost of attendance to equal the cost on Saturday.

The calculator solution is left up to the student. ■

■ GEOMETRY PROBLEMS

The technique of using a table introduced previously in this section coupled with the problem-solving strategies introduced in the previous chapter can be used as you work through the remaining examples in the section. The solutions provided illustrate only one algebraic method of obtaining the equation to be solved. Keep in mind there are other methods available.

EXAMPLE 4 The length of a rectangle is 4 meters longer than twice its width. If the perimeter of the rectangle is 26 meters, find its dimensions.

Solution   We can draw a sketch as illustrated. The facts are organized into known facts and those that are unknown.

> Known:
>
> the length is 4 meters longer than twice the width
>
> the perimeter is 26 meters
>
> The formula for the perimeter of a rectangle is $p = 2l + 2w$
>
> Unknown:
>
> the dimensions (length and width)

Solution   *Algebraic solution:* Let w represent the width of the rectangle. Then $4 + 2w$ represents the length of the rectangle.

We can form the equation

| 2 · | the length | plus 2 · | the width | equals | the perimeter |

$2(4 + 2w) + 2w = 26$	The equation to solve.
$8 + 4w + 2w = 26$	Remove parentheses.
$6w + 8 = 26$	Combine like terms.
$6w = 18$	Subtract 8 from both sides.
$w = 3$	Divide both sides by 6.

The width of the rectangle is 3 meters, and the length, $4 + 2w$, is 10 meters.

Graphic solution:

$$y1 = 2(4 + 2x) + 2x$$
$$y2 = 26$$

$[-10, 10]$ by $[-10, 30]$

Check: If a rectangle has a width of 3 meters and a length of 10 meters, then the length is 4 meters longer than twice the width ($4 + 2 \cdot 3 = 10$). Furthermore, the perimeter is $2 \cdot 10 + 2 \cdot 3$, or 26 meters. The solution checks.

■ RATIO AND PROPORTION

An indicated quotient of two numbers is often called a **ratio.**

> **Ratio**
> A **ratio** is the comparison of two numbers by their indicated quotient.

The definition implies that a ratio is a fraction. Some examples of ratios are

$$\frac{7}{8}, \qquad \frac{21}{24}, \qquad \text{and} \qquad \frac{117}{223}$$

Recall, from Chapter 1, that fractions can represent parts of a whole or an indicated division. Another interpretation of a fraction is that it represents the **ratio** of one number to another. The fraction $\frac{7}{8}$ can be read as "the ratio of 7 to 8," the fraction $\frac{21}{24}$ can be read as "the ratio of 21 to 24," and the fraction $\frac{117}{223}$ can be read as "the ratio of 117 to 223." Because the fractions $\frac{7}{8}$ and $\frac{21}{24}$ represent equal numbers, they are called **equal ratios.** Ratios should be expressed in lowest terms.

Proportion
A **proportion** is a statement indicating that two ratios are equal.

Some examples of proportions are

$$\frac{1}{2} = \frac{3}{6}, \qquad \frac{3}{7} = \frac{9}{21}, \qquad \text{and} \qquad \frac{8}{1} = \frac{40}{5}$$

The proportion $\frac{1}{2} = \frac{3}{6}$ can be read as "1 is to 2 as 3 is to 6," the proportion $\frac{3}{7} = \frac{9}{21}$ can be read as "3 is to 7 as 9 is to 21," and the proportion $\frac{8}{1} = \frac{40}{5}$ can be read as "8 is to 1 as 40 is to 5."

In the proportion $\frac{1}{2} = \frac{3}{6}$, the numbers 1 and 6 are called the **extremes** of the proportion, and the numbers 2 and 3 are called the **means.** If we find the product of the extremes and the product of the means in this proportion, we see that the products are equal:

$$1 \cdot 6 = 6 \qquad \text{and} \qquad 3 \cdot 2 = 6$$

Theorem
In any proportion, the product of the extremes is equal to the product of the means.

EXAMPLE 5   Determine if the equation $\dfrac{x}{3y} = \dfrac{xy + 3x}{3y^2 + 9y}$ is a proportion.

Solution   We check to see if the product of the means is equal to the product of the extremes.

$$3y(xy + 3x) = 3xy^2 + 9xy \qquad \text{The product of the means.}$$
$$x(3y^2 + 9y) = 3xy^2 + 9xy \qquad \text{The product of the extremes.}$$

Because the products are equal, the equation is a proportion.   ■

■ SOLVING PROPORTIONS

EXAMPLE 6   Solve the proportion $\dfrac{12}{18} = \dfrac{3}{x}$ for x.

Solution   We proceed as follows:

■ TECHNOLOGY TIP
When a variable appears in the denominator of a fraction, the student should be advised to *not* solve graphically at this point in time. Use the STOre or TABLE feature to check solutions.

$$\frac{12}{18} = \frac{3}{x}$$

$$12x = 3 \cdot 18 \qquad \text{The product of the extremes equals the product of the means.}$$

$$x = \frac{54}{12} \qquad \text{Divide both sides by 12.}$$

$$x = \frac{9}{2} \qquad \text{Simplify.}$$

Thus, x represents the fraction $\frac{9}{2}$.   ■

EXAMPLE 7   Solve the proportion $\dfrac{y+1}{y} = \dfrac{y}{y+2}$ for y.

Solution

$$\dfrac{y+1}{y} = \dfrac{y}{y+2}$$

$y^2 = (y+1)(y+2)$ The product of the means is equal to the product of the extremes.

$y^2 = y^2 + 3y + 2$ Remove parentheses.

$0 = 3y + 2$ Add $-y^2$ to both sides.

$-3y = 2$ Add $-3y$ to both sides.

$y = -\dfrac{2}{3}$ Divide both sides by -3. ∎

EXAMPLE 8   **Grocery shopping**   If 5 tomatoes cost \$1.15, how much will 16 tomatoes cost?

Solution   We can let c represent the cost of 16 tomatoes. Because the ratio of the number of tomatoes is the same as the ratio of their costs, we can express this relationship as a proportion and find c.

$$\dfrac{5}{16} = \dfrac{1.15}{c}$$

$5c = 1.15(16)$ The product of the extremes is equal to the product of the means.

$5c = 18.4$ Do the multiplication.

$c = \dfrac{18.4}{5}$ Divide both sides by 5.

$c = 3.68$ Simplify.

Sixteen tomatoes will cost \$3.68. ∎

■ SIMILAR TRIANGLES

Two triangles are **similar** when corresponding angles have the same measure and the lengths of corresponding sides are in proportion. The following theorem enables us to measure sides of triangles indirectly.

> **Theorem**
> If two triangles are similar, then all pairs of corresponding sides are in proportion.

Using this theorem, we can find the height of a tree on a sunny day and stay safely on the ground.

EXAMPLE 9   A tree casts a shadow 18 feet long at the same time a woman 5 feet tall casts a shadow that is 1.5 feet long. Find the height of the tree.

Solution   Figure 4-1 shows the triangles determined by the tree and its shadow and the woman and her shadow.

FIGURE 4-1

Because the triangles have the same shape, they are similar, and the lengths of their corresponding sides are in proportion. If we let h represent the height of the tree, we can find h by solving the following proportion.

$$\frac{h}{5} = \frac{18}{1.5}$$ $\dfrac{\text{Height of the tree}}{\text{Height of the woman}} = \dfrac{\text{Shadow of the tree}}{\text{Shadow of the woman}}$.

$$1.5h = 5(18)$$ In a proportion, the product of the extremes is equal to the product of the means.

$$h = 60$$ To undo the multiplication by 1.5, divide both sides by 1.5 and simplify.

The tree is 60 feet tall.   ∎

EXERCISE 4.1

VOCABULARY   *In Exercises 1–4, fill in the blanks to make a true statement.*

1. A _____ is a comparison of two numbers by their indicated quotient.

2. _____ ratios represent _____ numbers.

3. In any proportion, the product of the _____ is equal to the product of the _____.

4. Two triangles with the same shape are called _____ triangles.

CONCEPTS

5. Express each phrase as a ratio in lowest terms.
   a. 5 to 7   **b.** 3 to 5   **c.** 17 to 34
   d. 19 to 38   **e.** 22 to 33   **f.** 14 to 21

6. Tell whether each statement is a proportion.

   a. $-\dfrac{7}{3} = \dfrac{14}{-6}$ **b.** $\dfrac{13}{-19} = \dfrac{-65}{95}$ **c.** $\dfrac{x^2}{y} = \dfrac{x}{y^2}$

   d. $\dfrac{x^2y}{x^2z} = \dfrac{y}{z}$ **e.** $\dfrac{x+2}{x(x+2)} = \dfrac{1}{x}$ **f.** $\dfrac{x+y}{x-2} = \dfrac{y}{2}$

7. Identify the pairs that represent similar triangles.

   a.

b.

c.

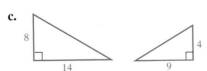

8. In two similar triangles, all pairs of corresponding sides are in proportion. Set up a proportion for each set of similar triangles.

a.

b.

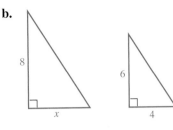

PRACTICE   *In Exercises 9–14, solve for the variable in each proportion.*

9. $\dfrac{x+3}{12} = \dfrac{-7}{6}$

10. $\dfrac{x+7}{-4} = \dfrac{3}{12}$

11. $\dfrac{3x-2}{7} = \dfrac{x}{28}$

12. $\dfrac{2x+1}{9} = \dfrac{x}{27}$

13. $\dfrac{7(x+6)}{6} = \dfrac{6(x+3)}{5}$

14. $\dfrac{3(x+5)}{2} = \dfrac{5(x-2)}{3}$

APPLICATIONS   *In Exercises 15–30, use the problem-solving strategies you have learned thus far to solve each of the given problems. Experiment with setting up tables and separating the known information from what is unknown. For each problem, make sure you clearly define your variable and what it represents, write an equation, provide a complete solution to that equation, and answer the question(s) posed. Be sure and check all results both in the equation as well as in the context of the problem.*

15. Finding consecutive even integers   The sum of two consecutive even integers is 54. Find the integers.

16. Finding consecutive odd integers   The sum of two consecutive odd integers is 88. Find the integers.

17. Finding consecutive integers   The sum of three consecutive integers is 120. Find the integers.

18. Finding consecutive even integers   The sum of three consecutive even integers is 72. Find the three even integers.

19. Finding integers   The sum of an integer and twice the next integer is 23. Find the smaller integer.

20. Finding integers   If 4 times the smallest of three consecutive integers is added to the largest, the result is 112. Find the three integers.

21. Finding integers   The larger of two integers is 10 greater than the smaller. The larger is 3 less than twice the smaller. Find the smaller integer.

22. Finding integers   The smaller of two integers is one-half of the larger, and 11 greater than one-third of the larger. Find the smaller integer.

23. Triangular bracing   The outside perimeter of the triangular brace in Illustration 1 is 57 feet. If all three sides are equal, find the length of each side.

ILLUSTRATION 1

24. Circuit boards   The perimeter of the circuit board in Illustration 2 is 90 centimeters. Find the dimensions of the board.

ILLUSTRATION 2

25. Swimming pools   The width of a rectangular swimming pool is 11 meters less than the length. The perimeter is 94 meters. Find the dimensions of the pool.

26. Wooden truss   The truss in Illustration 3 is in the form of an isosceles triangle. Each of the two equal sides is 4 feet less than the third side. If the perimeter is 25 feet, find the lengths of the sides.

ILLUSTRATION 3

27. Framing pictures   The length of a rectangular picture frame is 5 inches greater than twice the width. If the perimeter is 112 inches, find the dimensions of the frame.

28. Guy wires The two guy wires in Illustration 4 form an isosceles triangle. One of the two equal angles of the triangle is 4 times the third angle (the vertex angle). Find the measure of the vertex angle.

ILLUSTRATION 4

In Exercises 31–42, set up and solve the required proportion.

31. Gardening Garden seeds are on sale at 3 packets for 50 cents. How much will 39 packets cost?

32. Increasing a recipe A recipe for spaghetti sauce requires four 16-ounce bottles of ketchup to make 2 gallons of sauce. How many bottles of ketchup are needed to make 10 gallons of sauce?

33. Model railroading An HO-scale model railroad engine is 9 inches long. The HO scale is 87 feet to 1 foot. How long is a real engine?

34. Model railroading An N-scale model railroad caboose is 3.5 inches long. The N scale is 169 feet to 1 foot. How long is a real caboose to the nearest tenth of a foot?

35. Mixing fuel The instructions on a can of oil intended to be added to lawn mower gasoline read as follows:

Recommended	Gasoline	Oil
50 to 1	6 gal	16 oz

Are these instructions correct? (*Hint:* There are 128 ounces in 1 gallon.)

36. Height of a tree A tree casts a shadow of 26 feet at the same time as a 6-foot man casts a shadow of 4 feet. (See Illustration 5.) Find the height of the tree.

ILLUSTRATION 5

37. Height of a flagpole A man places a mirror on the ground and sees the reflection of the top of a flagpole, as in Illustration 6. The two triangles in the illustration

are similar. Find the height, *h*, of the flagpole rounded to the nearest tenth.

29. Equilateral triangles Find the measure of each angle of an equilateral triangle. (*Hint:* The three angles of an equilateral triangle are all equal.)

30. Land areas The perimeter of a square parcel of land is twice the perimeter of an equilateral (equal-sided) triangular plot. If one side of the square is 60 meters, find the length of a side of the triangle.

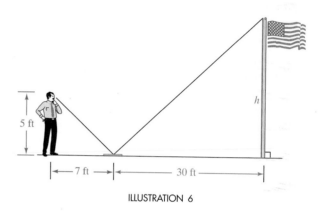

ILLUSTRATION 6

38. Width of a river Use the dimensions in Illustration 7 to find *w*, the width of the river. The two triangles in the illustration are similar.

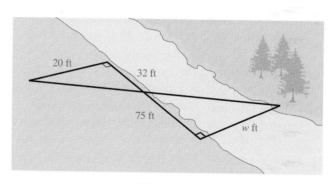

ILLUSTRATION 7

39. Flight path An airplane ascends 100 feet as it flies a horizontal distance of 1000 feet. How much altitude will it gain as it flies a horizontal distance of 1 mile? (See Illustration 8.) (*Hint:* 5280 feet = 1 mile.)

ILLUSTRATION 8

40. **Flight path**   An airplane descends 1350 feet as it flies a horizontal distance of 1 mile. How much altitude is lost as it flies a horizontal distance of 5 miles?

41. **Ski runs**   A half-mile long ski course falls 100 feet in every 300 feet of horizontal run. Find the height of the hill.

42. **Mountain travel**   A mountain road ascends 375 feet in every 2500 feet of travel. By how much will the road rise in a trip of 10 miles?

REVIEW

43. Simplify each expression by performing the indicated operation.
    a. $(4x^2 - 5x + 2) - (6x^2 - 3x - 4)$
    b. $(2x + 5)^3$
    c. $(x^3 + 27)$ divided by $(x + 3)$

44. If $P(x) = 4x^5 - 5x^2 + 6x$, find $P(-1)$ and verify with your calculator.

45. The left side of an equation has been graphed at y1 and the right side at y2.
    a. Based on the calculator display, what is the solution of the equation y1 = y2?

    b. Based on the table, what is the solution of the equation y1 = y2?

46. Algebraically solve the equation $\frac{2}{3}(x - 6) + 4 = 2x + 1$.

    Determine the **three** ways that you can use the graphing calculator to check your results and then apply one of these methods as your check.

4.2   Business Applications

■ BREAK-EVEN ANALYSIS ■ QUANTITY AND VALUE PROBLEMS ■ INVESTMENT PROBLEMS

■ BREAK-EVEN ANALYSIS

In manufacturing, there are two types of costs—**fixed costs** and **unit costs.** Fixed costs do not depend on the amount of product manufactured. Fixed costs would include the cost of plant rental, insurance, and machinery. Unit costs depend on the amount of product manufactured. Unit costs would include the cost of raw materials and labor.

Break-even analysis is used to find the production level where revenue will just offset the cost of production. When production exceeds the break-even point, the company will make a profit.

EXAMPLE 1   An electronics company has fixed costs of $6405 a week and a unit cost of $75 for each compact disk player manufactured. If the company can sell all the CD players it can make at a wholesale price of $90, find the company's break-even point.

Solution   Known:

fixed cost of $6405

unit cost of $75 for each CD player produced

selling price of $90 per CD player

Unknown:

break-even point (the number of CD players produced that would make cost equal revenue).

Suppose the company manufactures x CD players each week. The cost of manufacturing these players is the sum of the fixed and unit costs. We are given that the fixed costs are $6405 each week. The unit cost is the product of x, the number of CD players manufactured each week, and $75, the cost of manufacturing a single player. Thus, the weekly cost is

Cost = $75x + 6405$

The company can sell all the machines it can make, so the weekly revenue is the product of x, the number of players manufactured (and sold) each week, and $90, the wholesale price of each CD player. Thus, the weekly revenue is

Revenue = $90x$

The break-even point is the value of x for which the weekly revenue is equal to the weekly cost.

Let x represent the number of CD players manufactured each week. Then

$90x$ represents the weekly revenue, and

$75x + 6405$ represents the weekly cost.

Because the break-even point occurs when revenue equals cost, we set up and solve the following equation:

The weekly revenue	=	the weekly cost
$90x$	=	$75x + 6405$

$15x = 6405$ Subtract $75x$ from both sides.

$x = 427$ Divide both sides by 15.

If the company manufactures 427 CD players, the revenue will equal the cost, and the company will break even.

Graphic solution:

y1 = $90x$

y2 = $75x + 6405$

$[-10, 500]$ by $[-10, 45000]$ ■

■ QUANTITY AND VALUE PROBLEMS

EXAMPLE 2   An investor has stock worth $240,000 in three segments of the economy. He owns five times as many shares in transportation as he owns in utilities, and 2000 more shares in pharmaceuticals than in transportation. Each share in transportation is worth $55, each utility share is worth $75, and each share in pharmaceuticals is worth $30. How many shares does he own in each segment?

Solution   Known:

the total worth of the stock is $240,000

he owns five times as many shares in transportation as in utilities

he owns 2000 more shares in pharmaceuticals than in transportation

each share in transportation is worth $55

each share in utilities is worth $75

each share in pharmaceuticals is worth $30

Unknown:

the number of shares in each segment

Upon examination of the known information, it becomes apparent that the number of shares in transportation is given in terms of the number of shares in utilities, therefore,

Let n = number of shares in utilities

Translating, we see that

$5n$ = number of shares in transportation and

$5n + 2000$ = number of shares in pharmaceuticals

It is important to distinguish between the *number* of shares and the *value* of those shares. Because the investor owns n shares of utility stock worth $75 each, the value of the stock is $75n$. Because he owns $5n$ shares of transportation stock worth $55 each, their value is $55(5n)$. Finally, his $(5n + 2000)$ shares of pharmaceutical stock, at $30 each, are worth $30(5n + 2000)$. The total value can be expressed in two ways: as

$$\$[75n + 55(5n) + 30(5n + 2000)] \quad \text{and as} \quad \$240{,}000$$

Solution

The value of the utility stock	+	the value of the transportation stock	+	the value of the pharmaceutical stock	=	the total value
$75n$	+	$55(5n)$	+	$30(5n + 2000)$	=	$240{,}000$

$75n + 275n + 150n + 60{,}000 = 240{,}000$	Remove parentheses.
$500n + 60{,}000 = 240{,}000$	Combine like terms.
$500n = 180{,}000$	Subtract 60,000 from both sides.
$n = 360$	Divide both sides by 500.

Graphic solution:

$$y1 = 75x + 55(5x) + 30(5x + 2000)$$
$$y2 = 240000$$

Intersection
X=360   Y=240000

$[-10, 500]$ by $[-10, 300000]$

Summarizing the results in a table is a helpful way to check:

Segment	Shares	Value
Utilities	$n = 360$	$27,000
Transportation	$5n = 1800$	$99,000
Pharmaceuticals	$5n + 2000 = 3800$	$114,000
Total		$240,000

■ **INVESTMENT PROBLEMS**

EXAMPLE 3   **Investing money**   A retired teacher invested part of $12,000 at 6% annual interest and the rest at 9%. If the annual income from these investments was $945, how much did the teacher invest at each rate?

Analysis   The interest i earned by an amount p invested at an annual rate r for t years is given by the formula $i = prt$. In this example, $t = 1$ year. Hence, if x dollars were invested at 6%, the interest earned would be $0.06x$. If x dollars were invested at 6%, then the rest of the money, $(12,000 - x)$, would be invested at 9%. The interest earned on that money would be $.09(12,000 - x)$ dollars. The total interest earned in dollars can be expressed in two ways: as 945 and as the sum $.06x + .09(12,000 - x)$.

Solution   Let x represent the amount of money invested at 6%. Then $12,000 - x$ represents the amount of money invested at 9%.
We can form an equation as follows:

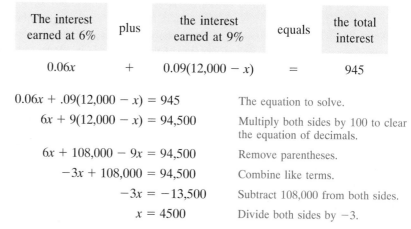

The interest earned at 6%	plus	the interest earned at 9%	equals	the total interest
$0.06x$	$+$	$0.09(12,000 - x)$	$=$	945

$0.06x + .09(12,000 - x) = 945$	The equation to solve.
$6x + 9(12,000 - x) = 94,500$	Multiply both sides by 100 to clear the equation of decimals.
$6x + 108,000 - 9x = 94,500$	Remove parentheses.
$-3x + 108,000 = 94,500$	Combine like terms.
$-3x = -13,500$	Subtract 108,000 from both sides.
$x = 4500$	Divide both sides by -3.

The teacher invested $4500 at 6% and $12,000 - $4500, or $7500, at 9%.

Graphic solution:

$$y1 = .06x + .09(12000 - x)$$
$$y2 = 945$$

Intersection
X=4500 ——— Y=945

$[-10, 10000]$ by $[-10, 1000]$

Check: The first investment yielded 6% of $4500, or $270. The second investment yielded 9% of $7500, or $675. Because the total return was $270 + $675, or $945, the answers check. ■

EXERCISE 4.2

VOCABULARY   *In Exercises 1–3, fill in the blanks to make a true statement.*

1. Costs that do not depend on the quantity of a product manufactured are called _____ costs.

2. Costs that depend on the quantity of a product manufactured are called _____ costs.

3. We use _____ analysis to find the production level where the revenue will just offset the cost of production.

CONCEPTS

4. If paint sells for $21 per gallon and x represents the number of gallons sold, is $21x$ a representation of cost of production or revenue?

5. If the cost of production is represented by $C(x) = \$4000 + \$25x$, identify the unit cost and the fixed cost.

APPLICATIONS   *In Exercises 6–28, pick a variable to represent the unknown quantity, set up an equation involving the variable, solve the equation, and check it.*

6. A paint manufacturer can choose between two processes for manufacturing house paint, with monthly costs shown in Table 1. The paint can be sold for $21 per gallon.

   a. Find the break-even point for process A.
   b. Find the break-even point for process B.

Process	Fixed costs	Unit cost (per gallon)
A	$ 75,000	$11
B	$128,000	$ 5

TABLE 1

c. If expected sales will be 8800 gallons per month, which process should the company choose?

d. If expected sales will be 9000 gallons per month, which process should the company choose?

7. **Performance bond** A construction company has agreed to finish a highway repair in 60 days, and pay a $1000 fine for each day the work takes beyond 60 days. To finish by the deadline, management expects to spend $12,000 per day. By cutting the work crew, expenses would be $9400 per day, but the job would take longer. How many days beyond deadline can the project last to make expenditures equal?

8. **Tax-free investments** An executive has $1000 to invest. One investment pays 11% interest, but the income is fully taxable. Income from a municipal bond fund is not taxable. If the executive is in a 32% tax bracket, what tax-free interest rate makes both investments equally attractive?

9. **Manufacturing shoes** A shoe company has fixed costs of $9600 per month and a unit cost of $20 per pair of shoes. The company can sell all the shoes it can make at a wholesale price of $30 per pair. Find the break-even point.

10. **Manufacturing insulators** A manufacturer of high-voltage insulators has fixed costs of $5400 per month and a unit cost of $12 per insulator. It can sell all the units it can make at a wholesale price of $15. Find the break-even point.

11. **Scheduling machine usage** A machine shop has two machines that can mill a certain brass plate. One machine has a setup cost of $500 and a cost of $2 per plate, while the other machine has a setup cost of $800 and a cost of $1 per plate. How many plates should be manufactured if the cost is to be the same using either machine?

12. **Oriental rugs** A rug manufacturer has two looms for weaving Oriental-style rugs. One loom has a setup cost of $750 and can produce a rug for $115. The other loom has a setup cost of $950 and can produce a rug for $95. How many rugs can be manufactured if the cost is to be the same using either loom?

13. **Software sales** The three best-selling software applications at one computer store are priced as in Table 2.

Spreadsheet	$150
Database	$195
Word processing	$210

TABLE 2

Spreadsheet and database programs sold in equal numbers, but 15 more word processing applications were sold than the other two combined. The three applications generated sales of $72,000. How many spreadsheets were sold?

14. **Appliance store inventory** Stoves and refrigerators are expected to sell in equal numbers. With summer approaching, the number of air conditioners sold is expected to be double that of stoves and refrigerators combined. Stoves sell for $350, refrigerators for $450, and air conditioners for $500, and sales of $56,000 are expected. How many of each should be stocked?

15. **Sorting hardware** Liz has an assortment of bolts, washers, and locknuts—99 pieces in all. She has twice as many washers as bolts and one less locknut than washers. How many bolts does she have?

16. **CD collections** A musician has a collection of 178 CDs. He has twice as many jazz as rock and 23 more classical than jazz. How many of each does he have?

17. **Stock performance** An investor owns twice as many shares of stock with average performance as he owns of shares that have far outpaced the market. He owns 200 shares of a stock that is now worthless. If he owns 3500 shares in all, how many shares have performed better than average?

18. **Raising capital** 30% of a corporation's stock offering is preferred stock worth $25 per share. The rest is common stock, selling at $15 per share. If the entire offering raised $306 million, how many common shares were sold?

19. **Warehousing costs** Monthly storage costs are $1.50 for each portable television, $4.00 for each console, and $7.50 for each wide-screen entertainment center. An appliance store warehouses 40 more portables than wide-screen sets, and 15 fewer consoles than portables. [*more* ... *wide screens*] Television storage costs $276 for the month. How many wide-screen sets are in stock?

20. **Apartment rental** The designers of a new apartment complex will include one-, two-, and three-bedroom apartments. They feel they can rent equal numbers of each, with the monthly rents given in Table 3.

One bedroom	$550
Two bedroom	$700
Three bedroom	$900

TABLE 3

If the total monthly rental income will be $36,550, how many units of each kind are planned?

21. **Investment problem** A broker has invested $24,000 in two accounts, one earning 9% annual interest and the other earning 8%. After 1 year, his combined interest is $1965. How much was invested at each rate?

22. Investment problem A woman's rollover IRA of $18,750 is invested in two accounts, one earning 7% interest and the other earning 10%. After 1 year, the combined interest income is $1512. How much has she invested at each rate?

23. Investment problem One investment pays 8% and another pays 11%. If equal amounts are invested in each, the combined interest income for 1 year is $712.50. How much is invested at each rate?

24. Investment problem When equal amounts are invested in each of three accounts paying 5%, 6%, and 7%, one year's combined interest income is $882. How much is invested in each account?

25. Investment problem A college professor wants to supplement her retirement income with investment interest. If she invests $15,000 at 6% annual interest, how much more would she have to invest at 7% to achieve a goal of $1250 in supplemental income?

26. Investment problem A retired teacher has a choice of two investment plans: an insured fund that pays 11% interest or a riskier investment that promises a 13% return. The same amount invested at the higher rate would generate an extra $150 per year. How much does the teacher have to invest?

27. Investment problem A financial adviser recommends investing twice as much in CDs as in a bond fund. A client follows his advice and invests $21,000 in CDs paying 1% more interest than the fund. The CDs generate $840 more interest than the fund. Find the two rates. (*Hint:* 1% = 0.01)

28. Investment problem The amount of annual interest earned by $8000 invested at a certain rate is $200 less than $12,000 would earn at a 1% lower rate. At what rate is the $8000 invested?

REVIEW

29. Simplify.
a. $(4x)^2$ **b.** $(4x + 3)^2$

30. Evaluate for $x = -2$ and verify with your calculator.
a. $(4x)^2$ **b.** $(4x + 3)^2$

31. Algebraically solve $4\left(\dfrac{2}{3}x - 1\right) = (2x + 5) - (x - 1)$ and verify your results with
a. the STOre feature
b. a graphical solution
c. the TABLE

32. Algebraically solve $\dfrac{5x - 1}{3} = \dfrac{6x + 2}{4}$

4.3 More Applications of Equations

■ LIQUID MIXTURE PROBLEMS ■ DRY MIXTURE PROBLEMS ■ UNIFORM MOTION PROBLEMS

This section examines several different types of problems that are commonly encountered in algebra courses. An analysis of each problem is provided that offers another problem-solving strategy. Remember the tools available for problem solving: writing the known and the unknown facts, translating from English words to algebraic symbols, using a table to organize information, and moving from the specific to the general. These are valuable tools that you must experiment with in order to become a successful problem solver. You will probably discover that you use a combination of tools rather than any single problem-solving strategy. As you study and work through the examples in the section, think about ways to organize the information other than those illustrated. You are encouraged to form study groups and discuss the problems together. Verbalizing the mathematics is a valuable tool that students often neglect to capitalize on.

■ LIQUID MIXTURE PROBLEMS

EXAMPLE 1   **Mixing acids** A chemistry instructor has one solution that is 50% sulfuric acid and another that is 20% sulfuric acid. How much of each should she use to make 12 liters of a solution that is 30% acid?

Analysis   The sulfuric acid present in the final mixture comes from the two solutions to be mixed. If x represents the numbers of liters of the 50% solution required for the mixture, then the rest of the mixture ($(12 - x)$ liters) must be the 20% solution. (See Figure 4-2.) Only 50% of the x liters and only 20% of the $(12 - x)$ liters are pure sulfuric acid. The total of these amounts is also the amount of acid in the final mixture, which is 30% of 12 liters.

x liters   (12 − x) liters   12 liters

50%   20%   30%

FIGURE 4-2

Solution   Let x represent the required number of liters of the 50% solution. Then $12 - x$ represents the required number of liters of the 20% solution.
We can form the equation

The acid in the 50% solution	plus	the acid in the 20% solution	equals	the acid in the final mixture
50% of x	+	20% of $(12 - x)$	=	30% of 12

$0.50x + 0.20(12 - x) = 0.30(12)$	The equation to solve.
$5x + 2(12 - x) = 3(12)$	Multiply both sides by 100 to clear the equation of decimals.
$5x + 24 - 2x = 36$	Remove parentheses.
$3x + 24 = 36$	Combine like terms.
$3x = 12$	Subtract 24 from both sides.
$x = 4$	Divide both sides by 3.

The chemist must mix 4 liters of the 50% solution and 8 liters ($(12 - 4)$ liters) of the 20% solution.

Graphic solution:

$y1 = .50x + .20(12 - x)$

$y2 = .30(12)$

$[-10, 10]$ by $[-10, 10]$ ■

■ DRY MIXTURE PROBLEMS

EXAMPLE 2   **Mixing nuts** Fancy cashews are not selling at $9 per pound because they are too expensive. Filberts are selling at $6 per pound. How many pounds of filberts should be combined with 50 pounds of cashews to obtain a mixture that can be sold at $7 per pound?

Analysis   Dry mixture problems are based on the formula $v = pn$, where v is the value of the mixture, p is the price per pound, and n is the number of pounds. Suppose x pounds of filberts are used in the mixture. At $6 per pound, they are worth $$6x$. At $9 per pound, the 50 pounds of cashews are worth $9 \cdot 50$, or $450. The mixture will weigh $(50 + x)$ pounds, and at $7 per pound, it will be worth $$7(50 + x)$. The value of the ingredients, $\$(6x + 450)$, is equal to the value of the mixture, $\$7(50 + x)$. (See Figure 4-3.)

Solution   Let x represent the number of pounds of filberts in the mixture.

	v	$=$	p	$\cdot$	n
Filberts	$6x$		6		x
Cashews	$9(50)$		9		50
Mixture	$7(50 + x)$		7		$50 + x$

FIGURE 4-3

We can form the equation

The value of the filberts	plus	the value of the cashews	equals	the value of the mixture
$6x$	$+$	$9 \cdot 50$	$=$	$7(50 + x)$

$$6x + 9 \cdot 50 = 7(50 + x) \qquad \text{The equation to solve.}$$
$$6x + 450 = 350 + 7x \qquad \text{Remove parentheses and simplify.}$$
$$100 = x \qquad \text{Subtract } 6x \text{ and } 350 \text{ from both sides.}$$

The storekeeper should use 100 pounds of filberts in the mixture.

Check:

The value of 100 pounds of filberts at \$6 per pound is	\$600
The value of 50 pounds of cashews at \$9 per pound is	450
The value of the mixture is	\$1050

The value of 150 pounds of mixture at \$7 per pound is also \$1050.

Graphic solution:

$$y1 = 6x + 9(50)$$
$$y2 = 7(50 + x)$$

Intersection
X=100   Y=1050
$[-10, 200]$ by $[-10, 1500]$

■ UNIFORM MOTION PROBLEMS

EXAMPLE 3   **Driving times**   Chicago, Illinois, and Green Bay, Wisconsin, are about 200 miles apart. A car leaves Chicago traveling toward Green Bay at 55 miles per hour. At the same time, a truck leaves Green Bay bound for Chicago at 45 miles per hour. How long will it take them to meet?

Analysis   Uniform motion problems are based on the formula $d = rt$, where d is the distance traveled, r is the rate, and t is the time. We can organize the information of this problem in chart form, as in Figure 4-4.

	r	$\cdot$	r	$=$	d
Car	55		t		$55t$
Truck	45		t		$45t$

200 mi

Chicago   Green Bay

55 mph   45 mph

FIGURE 4-4

We know that the two vehicles travel for the same amount of time; say, t hours. The faster car travels $55t$ miles, and the slower truck travels $45t$ miles. The total distance can be expressed in two ways: as the sum $55t + 45t$ and as 200 miles.

Solution   Let t represent the time that each vehicle travels until they meet. Then

$55t$ represents the distance traveled by the car, and

$45t$ represents the distance traveled by the truck.

After referring to Figure 4-4, we form the equation

The distance the car goes	plus	the distance the truck goes	equals	the total distance
$55t$	$+$	$45t$	$=$	200

$[-10, 10]$ by $[-10, 250]$

$55t + 45t = 200$ The equation to solve.

$100t = 200$ Combine like terms.

$t = 2$ Divide both sides by 100.

The vehicles will meet after 2 hours.

Graphic solution:

$y1 = 55x + 45x$

$y2 = 200$

Check: During those 2 hours, the car travels $55 \cdot 2$, or 110 miles, while the truck travels $45 \cdot 2$, or 90 miles. The total distance traveled is $110 + 90$, or 200 miles. This is the total distance between Chicago and Green Bay. The answer checks. ■

EXERCISE 4.3

CONCEPTS

1. Express each percentage as a decimal number *and* as a fraction.
   a. 3%   **b.** 30%   **c.** $3\frac{1}{2}$%   **d.** 3.5%

2. If a 20-ounce bottle of saline solution for contacts is 5% salt, how much salt is in the bottle?

3. If the total liquid mixture is 30 grams and the liquid was originally divided between two bottles, how much was in each of the original bottles? (*Hint:* Your response will be in terms of polynomials, not specific quantities.)

4. If one beaker contains 25 ounces of liquid and a second beaker contains an unknown amount, x, how much liquid will there be if the liquid of the two beakers is combined into one larger beaker?

APPLICATIONS   *In Exercises 5–24, use the problem-solving strategies you have learned thus far to solve each of the given problems. For each problem, make sure you clearly define your variable and what it represents, write an equation, provide a complete solution to that equation, and answer the question(s) posed. Be sure and check all results both in the equation as well as in the context of the problem.*

5. **Mixing fuels**   How many gallons of fuel costing \$1.15 per gallon must be mixed with 20 gallons of a fuel costing \$0.85 per gallon to obtain a mixture costing \$1 per gallon? (See Illustration 1.)

6. **Mixing paint**   Paint costing \$19 per gallon is to be mixed with 5 gallons of a \$3-per-gallon thinner to make a paint that can be sold for \$14 per gallon. How much paint will be produced?

7. **Brine solution**   How many gallons of a 3% salt solution must be mixed with 50 gallons of a 7% solution to obtain a 5% solution?

x gal

\$1.15 per gal

20 gal

$x + 20$ gal

\$.85 per gal   \$1.00 per gal

ILLUSTRATION 1

8. Making cottage cheese To make low-fat cottage cheese, milk containing 4% butterfat is mixed with 10 gallons of milk containing 1% butterfat to obtain a mixture containing 2% butterfat. How many gallons of the richer milk must be used?

9. Antiseptic solutions A nurse wishes to add water to 30 ounces of a 10% solution of benzalkonium chloride to dilute it to an 8% solution. How much water must she add?

10. Mixing photographic chemicals A photographer wishes to mix 2 liters of a 5% acetic acid solution with a 10% solution to get a 7% solution. How many liters of 10% solution must be added?

11. Mixing candy Lemon drops worth $1.90 per pound are to be mixed with jelly beans that cost $1.20 per pound to make 100 pounds of a mixture worth $1.48 per pound. How many pounds of each candy should be used?

12. Blending gourmet tea One grade of tea worth $3.20 per pound, is to be mixed with another grade worth $2 per pound to make 20 pounds that will sell for $2.72 per pound. How much of each grade of tea must be used?

13. Mixing nuts A pound of peanuts is worth $0.30 less than a pound of cashews. Equal amounts of peanuts and cashews are used to make 40 pounds of a mixture that sells for $1.05 per pound. How much is a pound of cashews worth?

14. Mixing candy Twenty pounds of lemon drops are to be mixed with cherry chews to make a mixture that will sell for $1.80 per pound. How much of the more expensive candy should be used? (See Table 1.)

	Price per pound
Peppermint patties	$1.35
Lemon drops	$1.70
Licorice lumps	$1.95
Cherry chews	$2.00

TABLE 1

15. Coffee blends A store sells regular coffee for $4 a pound and a gourmet coffee for $7 a pound. To get rid of 40 pounds of the gourmet coffee, the shopkeeper makes a gourmet blend that he will put on sale for $5 a pound. How many pounds of regular coffee should be used?

16. Lawn seed blends A garden store sells Kentucky bluegrass seed for $6 per pound and ryegrass seed for $3 per pound. How much rye must be mixed with 100 pounds of bluegrass to obtain a blend that will sell for $5 per pound?

17. Travel time Ashford and Bartlett are 315 miles apart. A car leaves Ashford bound for Bartlett at 50 miles per hour. At the same time, another car leaves Bartlett and heads toward Ashford at 55 miles per hour. In how many hours will the two cars meet?

18. Travel time Granville and Preston are 535 miles apart. A car leaves Preston bound for Granville at 47 miles per hour. At the same time, another car leaves Granville and heads toward Preston at 60 miles per hour. How long will it take them to meet?

19. Travel time Two cars leave Peoria at the same time, one heading east at 60 miles per hour and the other west at 50 miles per hour. (See Illustration 2.) How long will it take them to be 715 miles apart?

ILLUSTRATION 2

20. Boating Two boats leave port at the same time, one heading north at 35 knots (nautical miles per hour), the other south at 47 knots. How long will it take them to be 738 nautical miles apart?

21. Travel time Two cars start together and head east, one at 42 miles per hour and the other at 53 miles per hour. (See Illustration 3.) In how many hours will the cars be 82.5 miles apart?

ILLUSTRATION 3

22. Speed of trains Two trains are 330 miles apart, and their speeds differ by 20 miles per hour. They travel toward each other and meet in 3 hours. Find the speed of each train.

23. Speed of an airplane Two planes are 6000 miles apart, and their speeds differ by 200 miles per hour. They travel toward each other and meet in 5 hours. Find the speed of the slower plane.

24. Average speed An automobile averaged 40 miles per hour for part of a trip and 50 miles per hour for the remainder. If the 5-hour trip covered 210 miles, for how long did the car average 40 miles per hour?

REVIEW

25. Simplify:
   a. $[4 - (2 - 3)^2 + 5] - (6 - 2)^3$
   b. $(3x + 5)(2x^2 - x + 1)$
   c. $\dfrac{4x^2y^3 - 6xy^2 + 8x^3}{4xy}$

26. Evaluate $(3x + 5)(2x^2 - x + 1)$ for $x = -2$. Verify with your calculator.

27. Solve graphically $6x + 2 = 4 - 3x$ and express the solution as a fraction.

28. **Finding consecutive even integers** The sum of two consecutive even integers is 206. Find the numbers.

29. **Finding integers** Four more than 3 times the quantity of an integer plus 2 is equal to 13. Find the integer.

4.4 Equations Containing Absolute Value

■ ABSOLUTE VALUE ■ EQUATIONS WITH ONE ABSOLUTE VALUE ■ EQUATIONS WITH TWO ABSOLUTE VALUES

■ ABSOLUTE VALUE

■ **STUDY TIP**
Remember to read numbers like "$-x$" as "the opposite of x." You do not want to confuse yourself into thinking that $-x$ always represents a negative number. For example, if $x = -5$ then $-x = -(-5) = 5$. Thus $-x = 5$.

We begin this section by defining the absolute value of x.

> **Absolute Value**
> If $x \geq 0$, then $|x| = x$
> If $x < 0$, then $|x| = -x$

This definition provides a way to associate a nonnegative real number with any real number:

   If $x \geq 0$, then x is its own absolute value.

   If $x < 0$, then $-x$ (which is positive) is the absolute value.

Either way, $|x|$ is positive or 0:

> $|x| \geq 0$ **for all real numbers** x

EXAMPLE 1   Find   **a.** $|9|$,   **b.** $|-5|$,   **c.** $|0|$.

Solution   **a.** Since $9 \geq 0$, 9 is its own absolute value: $|9| = 9$.

b. Since $-5 < 0$, the negative of -5 is the absolute value: $|-5| = -(-5) = 5$.

c. Since $0 \geq 0$, 0 is its own absolute value: $|0| = 0$.

Because absolute value is always positive and distance is always positive, absolute value may also be thought of as a number's distance from 0 on the number line. Reconsidering the three parts of Example 1 in this context reveals

a. $|9| = 9$ because the number 9 is located 9 units from zero on the number line:

b. $|-5| = 5$ because the number -5 is located 5 units from zero on the number line.

5 units

c. $|0| = 0$ because the number 0 is 0 units from zero on the number line.

0

■ EQUATIONS WITH ONE ABSOLUTE VALUE

Consider the absolute value equation $|x| = 5$. The variable x can have a value of either 5 or -5 because both of these numbers are a distance of 5 units from 0 on the number line.

5 units   5 units

Therefore, the solutions to the equation are $x = 5$ or $x = -5$.

This leads to the following generalization:

> **Absolute Value Equations**
> If $k > 0$, then $|x| = k$ is equivalent to $x = k$ or $x = -k$

It is important to note two things about the given algorithm:

1. Absolute value is isolated on one side of the equation.

2. k is a positive constant.

EXAMPLE 2   Solve the equation $|x - 3| = 7$.

Solution   The equation $|x - 3| = 7$ indicates that a point on the number line with a coordinate of $x - 3$ is 7 units from the origin. Thus, $x - 3$ can be either 7 or -7.

$$x - 3 = 7 \quad \text{or} \quad x - 3 = -7$$
$$x = 10 \quad | \quad x = -4$$

The solutions of the equation $|x - 3| = 7$ are 10 and -4. (See Figure 4-5.) If either of these numbers is substituted for x in the equation $|x - 3| = 7$, the equation is satisfied:

$$|x - 3| = 7$$
$$|10 - 3| \stackrel{?}{=} 7$$
$$|7| \stackrel{?}{=} 7$$
$$7 = 7$$

$$|x - 3| = 7$$
$$|-4 - 3| \stackrel{?}{=} 7$$
$$|-7| \stackrel{?}{=} 7$$
$$7 = 7$$

-4   10

FIGURE 4-5

Calculator solution:

Notice on the second screen that the point of intersection is right at the edge of the screen. It would be clearer had the xmax been increased to provide a better window. The third screen has an adjusted xmax value of 15. ■

EXAMPLE 3   Solve the equation $|3x - 2| = 5$.

Solution   We can write $|3x - 2| = 5$ as

$$3x - 2 = 5 \quad \text{or} \quad 3x - 2 = -5$$

and solve each equation for x:

$$
\begin{array}{c|c}
3x - 2 = 5 \quad \text{or} & 3x - 2 = -5 \\
3x = 7 & 3x = -3 \\
x = \dfrac{7}{3} & x = -1
\end{array}
$$

Verify that both solutions check.

Calculator solution:

■

EXAMPLE 4   Solve the equation $\left| \dfrac{2}{3}x + 3 \right| + 4 = 10$

Solution   Careful examination of the algorithm reveals that it is valid if and only if the absolute value is isolated on one side and a constant is on the other side.
    We first isolate the absolute value on the left-hand side.

$$\left| \frac{2}{3}x + 3 \right| + 4 = 10$$

1. $\left| \dfrac{2}{3}x + 3 \right| = 6$ Subtract 4 from both sides.

We can now write Equation 1 as

$$\frac{2}{3}x + 3 = 6 \quad \text{or} \quad \frac{2}{3}x + 3 = -6$$

and solve each equation for x:

$$
\begin{array}{c|c}
\dfrac{2}{3}x + 3 = 6 \quad \text{or} & \dfrac{2}{3}x + 3 = -6 \\[2mm]
\dfrac{2}{3}x = 3 & \dfrac{2}{3}x = -9 \\[2mm]
2x = 9 & 2x = -27 \\[2mm]
x = \dfrac{9}{2} & x = \dfrac{-27}{2}
\end{array}
$$

Verify that both solutions check.

Calculator solution:

$$[-15, 10] \text{ by } [-10, 20]$$

EXAMPLE 5   Solve the equation $\left| 7x + \dfrac{1}{2} \right| = -4$.

Solution   Again, careful examination of the algorithm reveals that our constant, k, must be positive in order to apply the algorithm. In this example, $k = -4$. Therefore, the algorithm cannot be applied. Relating absolute value to distance on the number line translates to "what number(s) are a *negative* four units from 0 on the number line?" This does not make sense because we know that distance is always positive. We also know that absolute value is always equal to 0 or some positive number. We therefore conclude that the given equation is a contradiction, and has no solution.

The calculator display confirms our conclusion because there are no points of intersection.

EXAMPLE 6   Solve the equation $\left| \dfrac{1}{2}x - 5 \right| - 4 = -4$.

Solution   We first isolate the absolute value on the left-hand side.

$$\left| \frac{1}{2}x - 5 \right| - 4 = -4$$

$$\left| \frac{1}{2}x - 5 \right| = 0 \qquad \text{Add 4 to both sides.}$$

■ **TECHNOLOGY TIP**
Notice that when confirming our algebraic solution with the graphing calculator we enter the original expressions on each side of the equation at the appropriate prompts. This helps ensure that our conclusion is not based on a careless arithmetic error.

Because 0 is the only number whose absolute value is 0, the binomial $\frac{1}{2}x - 5$ must be equal to 0, and we have

$$\frac{1}{2}x - 5 = 0$$

$$\frac{1}{2}x = 5 \qquad \text{Add 5 to both sides.}$$

$$x = 10 \qquad \text{Multiply both sides by 2.}$$

$$[-10, 20] \text{ by } [-10, 10]$$

The calculator confirms our algebraic solution.

■ **EQUATIONS WITH TWO ABSOLUTE VALUES**

The equation $|a| = |b|$ is true when $a = b$ or when $a = -b$. For example,

$$|3| = |3| \qquad \text{or} \qquad |3| = |-3|$$
$$3 = 3 \qquad | \qquad 3 = 3$$

Thus, we have the following result.

Equations with Two Absolute Values

If a and b represent algebraic expressions, the equation $|a| = |b|$ is equivalent to the pair of equations

$$a = b \quad \text{or} \quad a = -b$$

EXAMPLE 7   Solve the equation $|5x + 3| = |3x + 25|$.

Solution   This equation is true when $5x + 3 = 3x + 25$, or when $5x + 3 = -(3x + 25)$. We solve each equation for x.

$$
\begin{array}{ll}
5x + 3 = 3x + 25 \quad \text{or} & 5x + 3 = -(3x + 25) \\
\qquad 2x = 22 & 5x + 3 = -3x - 25 \\
\qquad\ x = 11 & \qquad 8x = -28 \\
& \qquad\ x = -\dfrac{28}{8} \\
& \qquad\ x = -\dfrac{7}{2}
\end{array}
$$

The graphing calculator can be used to confirm the solutions.

$[-20, 20]$ by $[-10, 70]$

EXAMPLE 8   Solve the equation $|x + 4| = |x - 4|$.

Solution   This equation is true when $x + 4 = x - 4$ or when $x + 4 = -(x - 4)$. We solve each equation for x.

$$
\begin{array}{ll}
x + 4 = x - 4 & x + 4 = -(x - 4) \\
\qquad 4 = -4 & x + 4 = -x + 4 \\
\qquad \varnothing & 2x + 4 = 4 \\
& \qquad 2x = 0 \\
& \qquad\ x = 0
\end{array}
$$

The graphing calculator can be used to confirm the solutions.

EXERCISE 4.4

VOCABULARY   In Exercises 1–2, fill in the blanks to make a true statement.

1. If $x \geq 0$, then $|x| =$ __. If $x < 0$, then $|x| =$ ___.

2. Absolute value can be thought of as a number's _____ from zero on the number line.

CONCEPTS

3. Determine the indicated absolute value.
    a. $|4|$ **b.** $|-4|$ **c.** $-|-4|$
    d. $|a|$ **e.** $|3 - \pi|$ (Do not approximate!)

4. If $|x| = 8$, then $x =$ __ or $x =$ ___.

⚖5. Explain why the statement $|x| = -2$ makes no sense.

⚖6. Is it possible to have a solution for $|x| + 7 = 2$? Explain why or why not.

PRACTICE   *In Exercises 7–40, solve each equation algebraically. Confirm your solutions graphically.*

7. $|x| = 8$

8. $|x| = 9$

9. $|x - 3| = 6$

10. $|x + 4| = 8$

11. $|2x - 3| = 5$

12. $|4x - 4| = 20$

13. $|3x + 2| = 16$

14. $|5x - 3| = 22$

15. $\left|\dfrac{7}{2}x + 3\right| = -5$

16. $|2x + 10| = 0$

17. $\left|\dfrac{x}{2} - 1\right| = 3$

18. $\left|\dfrac{4x - 64}{4}\right| = 32$

19. $|3 - 4x| = 5$

20. $|8 - 5x| = 18$

21. $|3x + 24| = 0$

22. $|x - 21| = -8$

23. $\left|\dfrac{3x + 48}{3}\right| = 12$

24. $\left|\dfrac{x}{2} + 2\right| = 4$

25. $|x + 3| + 7 = 10$

26. $|2 - x| + 3 = 5$

27. $\left|\dfrac{3}{5}x - 4\right| - 2 = -2$

28. $\left|\dfrac{3}{4}x + 2\right| + 4 = 4$

29. $|2x + 1| = |3x + 3|$

30. $|5x - 7| = |4x + 1|$

31. $|3x - 1| = |x + 5|$

32. $|3x + 1| = |x - 5|$

33. $|2 - x| = |3x + 2|$

34. $|4x + 3| = |9 - 2x|$

35. $\left|\dfrac{x}{2} + 2\right| = \left|\dfrac{x}{2} - 2\right|$

36. $|7x + 12| = |x - 6|$

37. $\left|x + \dfrac{1}{3}\right| = |x - 3|$

38. $\left|x - \dfrac{1}{4}\right| = |x + 4|$

39. $|3x + 7| = -|8x - 2|$

40. $-|17x + 13| = |3x - 14|$

41. Construct several examples to show that $|a \cdot b| = |a| \cdot |b|$

42. Construct several examples to show that $\left|\dfrac{a}{b}\right| = \dfrac{|a|}{|b|}$

43. Construct several examples to show that $|a + b| \neq |a| + |b|$

44. Construct several examples to show that $|a - b| \neq |a| - |b|$

REVIEW

45. Simplify and verify with your calculator.

   a. $\dfrac{4}{5} + \dfrac{-8}{9}$

   b. $\left(-\dfrac{7}{8}\right)\left(-\dfrac{3}{5}\right) \div \left(\dfrac{-9}{20}\right)$

46. Identify the property that is exemplified below:

   a. $-(-x) = x$

   b. $a(b + c) = ab + ac$

   c. $xy = yx$

47. Find $P(-1)$ if $P(x) = -2x^2 - x + 2$. Verify with your calculator.

48. If each tic mark is equal to 1 unit, give the dimensions of each viewing window.

   a.   **b.**

49. Graph $y = \frac{1}{2}x - 12$ in a good viewing window and record the viewing window's dimensions in the following form: [xmin, xmax] by [ymin, ymax].

CHAPTER SUMMARY

CONCEPTS

REVIEW EXERCISES

Constructing tables is a valuable way to organize information in application problems.

A ratio is a comparison of two numbers by their indicated quotient.

A proportion is a statement that two ratios are equal.

In any proportion, the product of the extremes is equal to the product of the means.

If two triangles are similar, then the corresponding sides are in proportion.

Applications—A Numeric Approach

1. In each of the problems below, determine if each statement is a proportion.

   a. $-\dfrac{4}{5} = \dfrac{28}{-35}$

   b. $\dfrac{14}{19} = \dfrac{-42}{57}$

   c. $\dfrac{x+4}{x-2} = \dfrac{3x}{x+1}$

   d. $\dfrac{x+1}{x^2+x} = \dfrac{2}{2x}$

2. Solve each proportion. Check your results both graphically (using the INTERSECT feature) and by using the STOre feature of the calculator.

   a. $\dfrac{x-4}{4} = \dfrac{x+4}{8}$

   b. $\dfrac{x+3}{2} = \dfrac{x-3}{3}$

   c. $\dfrac{x-5}{3} = \dfrac{2x+1}{2}$

   d. $\dfrac{3x-1}{2} = \dfrac{4x+5}{3}$

In Review Exercises 3–6, (a) clearly identify the variable and what it represents, (b) write an equation that models the problem, (c) solve the equation either algebraically or graphically, and (d) answer the question posed in the problem with a complete sentence.

3. If the length of the rectangular painting in Illustration 1 is 3 inches more than twice the width, how wide is the rectangle?

84 in.

ILLUSTRATION 1

4. If 5 tons of iron ore yields 3 tons of pig iron, how much iron ore is needed to make 18 tons of pig iron?

5. A pharmacist mixes 3 grams of medicine with 300 milliliters of sugar syrup. Find out how much medicine is in a single dose of 5 milliliters.

6. How tall is the building if a 5-foot-tall woman casts a shadow of 2 feet (Illustration 2)?

ILLUSTRATION 2

Section 4.2	Business Applications

The break-even point occurs when the costs equal the revenue.

In Review Exercises 7–10, (a) clearly identify the variable and what it represents, (b) write an equation that models the problem, (c) solve the equation either algebraically or graphically, and (d) answer the question posed in the problem with a complete sentence.

It is important to distinguish between *number* and *value* when translating written expressions to algebraic equations.

7. Costs for maintaining automotive parts run $668.50 per week plus variable costs of $1.25 per unit. The company can sell as many as it can make for $3 each. Find the break-even point.

With investment problems remember that $i = prt$, where i represents interest earned, p represents amount invested, r is annual rate, and t is time. Rate must be expressed as a decimal or fraction for computational purposes.

8. A company can manufacture baseball caps on either of two machines, with costs shown in Table 1. At the projected sales level, they find the costs of the two machines equal. Find the expected sales.

Machine	Startup cost	Unit cost (per cap)
1	$ 85	$3
2	$105	$2.50

TABLE 1

9. A woman has $27,000. Part is invested for one year in a certificate of deposit paying 7% interest and the remaining amount in a cash management fund paying 9%. The total interest on the two investments is $2110. How much did she invest at each rate?

10. Part of $13,750 is invested at $4\frac{1}{2}$% and the rest at 5%. After one year, the account paid $647.50. How much was invested at the lower rate?

Section 4.3	More Applications of Equations

When working liquid mixture problems it is important that the concentrations represented in equations are consistent in that they all represent the same material, whether it be salt, alcohol, etc.

In Review Exercises 11–13, (a) clearly identify the variable and what it represents, (b) write an equation that models the problem, (c) solve the equation either algebraically or graphically, and (d) answer the question posed in the problem with a complete sentence.

11. How many liters of water must be added to 30 liters of a 10% brine solution to dilute it to an 8% solution?

Dry mixtures problems are based on the formula $v = pn$, where v is the value of the mixture, p is the price per pound, and n is the number of pounds.

12. A store manager mixes candy worth $0.90 per pound with candy worth $1.65 per pound to make 20 lb of a mixture worth $1.20 per pound. How many pounds of each must be mixed?

13. A bicycle path is 5 miles long. A man walks from one end at the rate of 3 mph. At the same time a friend bicycles from the other end, traveling 12 mph. In how many minutes will they meet?

Uniform motion problems are based on the formula $d = rt$, where d is the distance traveled, r is the rate, and t is the time.

Section 4.4	Equations Containing Absolute Value

Absolute Value:
If $x \geq 0$ then $|x| = x$.

If $x < 0$ then $|x| = -x$.

If $k > 0$, then $|x| = k$ is equivalent to $x = k$ or $x = -k$.

If a and b represent algebraic expressions, then the equation $|a| = |b|$ is equivalent to the pair of equations $a = b$ or $a = -b$.

14. Solve each equation algebraically and confirm your solution graphically.

   a. $|3x + 1| = 10$ **b.** $\left|\dfrac{3}{2}x - 4\right| = 9$

   c. $|x - 5| = -3$ **d.** $|2x - 3| + 2 = 5$

   e. $|4x - 1| - 7 = -3$

15. Solve each equation algebraically and confirm your solution graphically.

   a. $|3x + 2| = |2x - 3|$ **b.** $|5x - 4| = |4x - 5|$

■ Chapter Test

1. Is $-\dfrac{42}{22} = \dfrac{14}{-11}$ a proportion?

2. Is $\dfrac{2x - 1}{3} = \dfrac{4x - 1}{6}$ a proportion?

3. Solve $\dfrac{4 - x}{13} = \dfrac{11}{26}$.

4. Solve $\dfrac{2x + 5}{4} = \dfrac{x - 1}{6}$.

5. A plane drops 575 feet as it flies a horizontal distance of $\frac{1}{2}$ mile. How much altitude will it lose as it flies a horizontal distance of 7 miles (Illustration 1)?

6. A car leaves Rockford, Illinois at a rate of 65 mph, bound for Madison, Wisconsin. At the same time, a truck leaves Madison at a rate of 55 mph, bound for Rockford. If the cities are 72 miles apart, how long will it take the truck to meet the car?

ILLUSTRATION 1

7. Rose invests part of a $50,000 inheritance in a certificate of deposit at a rate of 6%. The rest is invested in a high-interest savings account earning $5\frac{1}{2}$% to be accessible for emergencies that her $1140-a-month Social Security check cannot cover. In one year, the two accounts earn $2975. How much is in the certificate of deposit?

8. A leading juice brand advertises that its pink grapefruit juice cocktail is 35% juice. How many ounces of juice must be mixed with 13 ounces of other ingredients (e.g., filtered water, corn syrup, preservatives, etc.) to produce a bottle of pink grapefruit juice cocktail that is 35% juice?

9. One side of a triangle is 4 more than twice the shortest side. If the third side is twice the shortest side, find the short side if the perimeter is 24.

10. A high school student invests $100 in supplies for an "earn at home" envelope-stuffing job. It costs her $0.35 to mail each envelope, but the company pays her $0.45 per envelope mailed. How many envelopes will she need to stuff and mail to break even?

Solve each absolute value equation below both algebraically and graphically.

11. $\left|\dfrac{2}{3}x - 5\right| = 6$

12. $\left|4x + \dfrac{2}{3}\right| = -1$

13. $|x + 5| - 3 = 2$

14. $|3x - 2| + 4 = 2$

15. $|x - 5| = -|x + 3|$

16. $|4x + 1| = |2x - 3|$

■ Cumulative Review: Chapters 3 and 4

VOCABULARY / CONCEPTS

1. An identity is an equation that is true _____ values of the variable, and the symbol ___ represents the solution.

2. A contradiction is an equation that is true for ___ real numbers and the symbol _____ represents the solution.

3. Solutions to equations in two variables are expressed as _____.

4. The set of all x-coordinates in a collection of ordered pairs is called the _____.

5. The set of all y-coordinates in a collection of ordered pairs is called the _____.

6. Determine the quadrant location of each ordered pair.
   a. $(-2, 3)$ **b.** $(-4, -1)$
   c. $(3, -4)$ **d.** $(5, -2)$

7. Sketch the set of coordinate axes that is indicated by the given viewing window: $[-5, 5]$ by $[-10, 10]$ with xscl $= 1$ and yscl $= 2$.

8. Explain the difference between tracing on a graph in the standard (ZStandard) viewing window and tracing on a graph in an integer viewing window (ZInteger).

9. The display at the right is a graphical model of the linear equation $y = \$39 + \$0.21x$, which represents the cost, y, of renting a car for one day at a fixed rate of $39 for x miles at 21¢ a mile. Interpret the meaning of the graphical display.

X=50   Y=49.5

10. Two features of the calculator that can be used to verify solutions of linear equations in one variable are the _____ and _____ features.

11. A ratio is a _____ of two numbers by their indicated quotient.

12. Express the phrase "18 to 24" as a ratio in lowest terms.

13. Two triangles with the same shape are called _____ triangles, and their corresponding sides are in _____.

14. If a CD sells for $16.99 and x represents the quantity sold, write an expression to represent the revenue from the sale of x number of CDs.

15. Express each percentage as a decimal number.
   a. 6%   **b.** $7\frac{3}{4}\%$

16. If 15 liters of a solution is added to a quantity of alcohol to produce a total liquid solution of $15 + x$ liters, how much alcohol is there?

17. If $a \geq 0$ then $|a| =$ ___. If $a < 0$ then $|a| =$ ___.

18. Why is $|x| = -4$ an invalid statement?

19. Determine the indicated absolute value.
   a. $-|-3|$
   b. $|2 - \pi|$ (do not approximate)

20. Solve for p: $i = prt$.

PRACTICE

21. Algebraically solve $3x + 4(2 - x) = 3(x + 4) - 5$ and verify with the STOre feature of your calculator.

22. Algebraically solve $\dfrac{3(x - 1)}{6} - 4 = \dfrac{x + 3}{6}$ and verify with the STOre feature of your calculator.

23. Complete the table of values for $y = x^2 + 3$.

x	y
-3	
-2	
-1	
0	
1	
2	
3	

24. Which ordered pair(s) is *not* a solution of $2x + y = 5$?
  a. $(2, 1)$ **b.** $(-2, -1)$ **c.** $(-3, 11)$ **d.** $(5, -5)$

25. Which of the following are *not* functions?
  a. $\{(2, 4), (-2, 4), (3, 1), (-3, 1)\}$
  b. $\{(2, 4), (2, -4), (3, 1), (-3, -1)\}$
  c. $y = 2x - 5$

26. Give the coordinate of the *x*- and *y*-intercepts in Illustration 2.

ILLUSTRATION 2

27. Graph the equation of the line $y = 4x - 1$.

28. Graph $y = x^2 - 8x + 7$ in the ZInteger viewing window and use the TRACE feature to determine the *x*- and *y*-intercepts.

✍**29.** Explain why ZInteger is a better viewing window for Exercise 28 than the ZStandard viewing window would be.

✍**30.** Why is the standard viewing window not a good viewing window for solving the equation $3 + [x - (2 - x)] = -16$? What must be done to have a good viewing window?

31. Solve for x in the proportion $\dfrac{4(x + 1)}{3} = \dfrac{x - 2}{4}$.

In Exercises 32–34, algebraically solve each equation and verify by solving graphically.

32. $|6x - 5| = 2$

33. $\left|\dfrac{x}{2} + 1\right| - 5 = 3$

34. $|2x - 1| = |3x + 4|$

In Exercises 35–36, solve for the indicated variable.

35. $\dfrac{1}{2}x - \dfrac{1}{3}y = 6$; for y

36. $Ax + By = C$; for y

In Exercises 37–40, graphically solve each equation.

37. $\left|\dfrac{1}{2}x - 1\right| = 6$

38. $-|x + 2| = -3$

39. $|x + 2| = -3$

40. $|3x + 2| = |x - 4|$

APPLICATIONS

41. Mixing solutions A veterinarian recommends a solution of 3 parts vinegar to 1 part alcohol as ear drops for a dog that is prone to ear infections. If the solution is to be stored in a 5-ounce bottle, will 3.75 ounces of vinegar and 1.25 ounces of alcohol be the correct amounts?

42. Finding integers The sum of an integer and three times the next integer is 171. Find the smaller integer.

43. Investing money $3000 earns 7% simple interest in a savings account, and a $4000 investment earns 4% simple interest in another savings account. How long will it take for the accounts to have the same amount of money?

44. Buying stamps Sherah purchased 33-cent stamps for her Christmas cards and 29-cent stamps for postcards. If she spent $24.15 and purchased four times as many 33-cent stamps as 29-cent stamps, how many 33-cent stamps did she purchase?

45. Computing travel time A postal transport truck leaves town traveling at 55 miles per hour. Unfortunately, one hour after the truck left the main post office, it was discovered that a large bag of priority mail had not been loaded. A vehicle was immediately dispatched to carry the priority mail to the transport truck. If the vehicle can travel 70 miles per hour, how long will it take to overtake the transport?

46. Mixing solutions A fruit juice beverage containing 10% juice is mixed with 64 ounces of another fruit juice beverage containing 50% juice. How many ounces of the 10% juice beverage will be needed to produce a beverage that is 35% juice?

47. Volume The volume of a sphere is given by the formula $V = \frac{4}{3}\pi r^3$. How many cubic inches of air are in a beach ball having a radius of 9 inches (to the nearest tenth of a cubic inch)?

48. Investments At a simple interest rate r, an amount of money P grows to an amount A in t years according to the formula $A = P(1 + rt)$. If $2500 grows to $2937.50 when invested at a $3\frac{1}{2}\%$ interest rate, find the length of time the money was invested.

In Section 4.2, Example 1, the equation for the break-even point for an electronics company was determined to be $90x = 75x + 6405$. Review the example and then graph the left and right sides of the equation in the window $[-10, 930]$ by $[-10, 60{,}000]$ before completing Exercises 49–50.

49. Trace along the graph of y1 = 90x to complete the following.
   a. When the number of CD players is 230, the revenue is _____.
   b. When the revenue is $15,300, the number of CD players produced is ____.

50. Trace along the graph of y2 = 75x + 6405 to complete the following.
   a. When 170 CD players are produced the cost is _____.
   b. How does this compare to the revenue earned (Exercise 49) from this same quantity of CD players?
   c. If the cost is $52,905, how many CD players were produced? Is the company making a profit or loss at this point?

CAREERS & MATHEMATICS

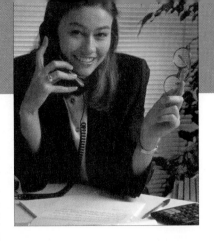

Banker Practically every bank has a group of officers who make decisions affecting bank operations: the president, who directs overall operations; one or more vice-presidents, who act as general managers or are in charge of bank departments, such as trust or credit; a comptroller, or cashier, who (as an executive officer) is generally responsible for all bank property; and treasurers and other senior officers as well as junior officers, who supervise sections within departments.

These officers make decisions within a framework of policy set by a board of directors and existing laws and regulations. They must have a broad knowledge of business activities to relate to the operations of their departments, because their customers will include a variety of individuals and businesses applying for loans, seeking investment advice, organizing trusts, setting up pensions, and so on. Besides supervising these financial services, officers advise individuals and businesses and participate in community projects.

Qualifications Bank officer and management positions are filled by management trainees and by promoting outstanding bank clerks or tellers. A college degree in finance or liberal arts, including accounting, economics, commercial law, political science, and statistics, is necessary. A Master of Business Administration degree (MBA) is preferred by some banks, although people with backgrounds as diverse as nuclear physics and forestry are hired by some banks to meet the needs of the complex, high-technology industries with which they deal.

Job Outlook Through the year 2005, employment of bank officers is expected to increase faster than the average for other occupations, due to expanding bank services, both domestic and international, and the increasing dependence on computers.

Example Application In banking, the calculation of compound interest uses exponential expressions. At the end of each period, a savings account is credited with a dividend, calculated as a fixed percentage of the savings balance. The money will grow to an amount FV, called the **future value,** according to the formula

$$FV = PV(1 + i)^n$$

where PV, called the **present value,** is the initial deposit. The percent of the current balance credited as interest is called the **periodic interest rate,** denoted by i. The **number of periods** that the money is left on deposit is n.

If $1000 is deposited in an account that earns 6% annual interest compounded quarterly, find the amount in the account after 3 years.

Solution Interest paid quarterly is paid four times each year. The periodic interest rate is one-fourth of 6%, or 1.5%. Express the periodic interest rate as a decimal: $i = 0.015$. In 3 years, there will be $3 \cdot 4$, or 12 periods. Thus, $n = 12$. The present value is the initial deposit: $PV = \$1000$. Substitute these values into the equation and calculate FV.

$$FV = PV(1 + i)^n$$
$$= 1000(1 + 0.015)^{12} \qquad \text{Substitute 1000 for } PV, 0.015 \text{ for } i, \text{ and 12 for } n.$$
$$= 1000(1.015)^2$$
$$= 1195.618171 \qquad \text{Use a calculator.}$$

After 3 years, the account will contain $1195.62.

217

EXERCISES

1. If $10,000 is deposited in an account earning 6% annual interest, compounded once a year, how much will be in the account at the end of 2 years?

2. If $10,000 is deposited in an account earning 6% annual interest, compounded twice a year, how much will be in the account at the end of 2 years?

3. If $10,000 is deposited in an account earning 6% annual interest, compounded four times a year, how much will be in the account at the end of 2 years?

4. Refer to Exercises 1–3. How much more is earned by compounding twice a year, instead of once a year? How much more is earned by compounding four times a year, instead of twice a year?

(Answers: 1. $11,236 2. $11,255.09 3. $11,264.93 4. $19.09; $9.84)

Factoring

5

■ PERSPECTIVE

No End of Primes

A composite number always has prime factors that are smaller than the number itself. The composite number 30, for example, has the prime factors 2, 3, and 5—all less than 30. We might think that the larger a number is, the more likely it is to be composite, because there are so many smaller prime numbers that could be possible factors. Maybe all numbers, if large enough, would be composite, and therefore there would be a largest prime number.

That is not what happens, however. In about 300 B.C. the Greek mathematician Euclid proved that the number of primes is unlimited—that there are infinitely many prime numbers. This is an important fact of a branch of mathematics called **number theory.** Although Euclid is best known for his study of geometry, many of his writings deal with number theory.

Who could possibly care that prime numbers can be very large? Cryptographers working for the federal government care, because several codes that are very difficult to crack rely on the properties of large primes.

Euclid (about 300 B.C.)

In this chapter we shall reverse the operation of multiplication and find the factors of a polynomial. In Chapter 2, polynomials were discussed. Polynomials are comprised of terms; ungrouped plus and minus signs are term indicators. The process of factoring a polynomial enables us to write polynomials as a product of **factors** rather than a sum/difference of terms. Factoring polynomials is important because it may be used to algebraically solve equations that are not first degree.

5.1 Using the Distributive Property to Factor

■ IDENTIFYING THE GREATEST COMMON FACTOR ■ FACTORING OUT THE GREATEST COMMON FACTOR ■ FACTORING OUT A NEGATIVE FACTOR ■ POLYNOMIALS AS GREATEST COMMON FACTORS ■ FACTORING BY GROUPING

■ IDENTIFYING THE GREATEST COMMON FACTOR

The **factors** of an expression (either a constant or a polynomial) are those expressions (a) that evenly divide the given expression or (b) whose product is equivalent to the given expression. To **factor** a number or a polynomial simply means to write the given quantity as a product of factors. Because the number 2 divides the number 6 evenly, 2 is a **factor** of 6. Three is also a factor of 6 because it divides 6 evenly. Therefore,

$$6 = \underset{\text{factor}}{2} \cdot \underset{\text{factor}}{3}$$

The numbers 2 and 3 are each factors of 6 because (a) they both evenly divide 6 and (b) their product, $2 \cdot 3$, is equivalent to 6. The expression $2 \cdot 3$ is the factored form of 6. To factor a natural number means to write the number as a product of prime numbers. Recall that a prime number is a natural number *greater than 1* whose only factors are one and the number itself.

■ STUDY TIP

The words *divisor* and *factor* mean the same thing. Thus 5 is both a factor *and* a divisor of 20. By the same token, $3xy$ is both the greatest common factor *and* the greatest common divisor of the expression $6xy + 3x^3y$.

Prime Factored Form

A natural number is said to be in **prime factored form** if it is written as the product of factors that are prime numbers.

The Fundamental Theorem of Arithmetic states that there is *exactly one* prime factorization for every natural number greater than 1. When asked to **factor** in algebra, it is implicitly stated that the prime factorization is required.

EXAMPLE 1   Factor each of the following: **a.** 42   **b.** 60   **c.** 90.

Solution   **a.** $42 = 2 \cdot 21$ **b.** $60 = 2 \cdot 30$ **c.** $90 = 2 \cdot 45$
$= 2 \cdot 3 \cdot 7$ $= 2 \cdot 2 \cdot 15$ $= 2 \cdot 3 \cdot 15$
$= 2 \cdot 2 \cdot 3 \cdot 5$ $= 2 \cdot 3 \cdot 3 \cdot 5$
$= 2^2 \cdot 3 \cdot 5$ $= 2 \cdot 3^2 \cdot 5$

It is a good habit to check factorization by multiplying the factors together to ensure the product of the factors is equivalent to the original number.

Examining the prime factorizations of 42, 60, and 90 reveals that the three numbers have several common factors. We are interested in the *greatest common factor* of the three numbers. This is defined as the largest natural number that divides each of several given numbers. In Example 1, 6 is the GCF because it is the largest of the common factors of 2, 3, and 6.

■ STUDY TIP
We suggest that you write the given steps on a notecard for quick reference.

Determining the Greatest Common Factor
1. Determine the prime factorization of each constant. Use exponents rather than repeated factors to write the number in prime factored form.
2. List each prime number that is a factor of all the numbers in the list. When exponents appear, choose the common factor with the *smallest* exponent.
3. Multiply the listed primes from the previous step. This is the greatest common factor, or GCF.

EXAMPLE 2   State the GCF for each list of numbers.

20, 40, 80

Solution $20 = 2 \cdot 10$ $40 = 4 \cdot 10$ $80 = 2 \cdot 40$
$= 2 \cdot 2 \cdot 5$ $= 2 \cdot 2 \cdot 2 \cdot 5$ $= 2 \cdot 2^3 \cdot 5$
$= 2^2 \cdot 5$ $= 2^3 \cdot 5$ $= 2^4 \cdot 5$

It is a good idea to double-check your factorization by confirming that the product of the factors is equivalent to the original number.

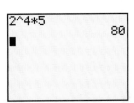

The common factors are 2^2 and 5 (notice that 2^2 was chosen because it has the smallest exponent of all factors with a base of 2). The GCF is the *product* of the common factors: $2^2 \cdot 5 = 20$. The greatest common factor of 20, 40, and 80 is 20. ∎

EXAMPLE 3   Find the GCF of $6x^2y^3$, $4x^3y^2$, and $18x^2y$.

Solution

$6x^2y^3 = 2 \cdot 3x^2 \cdot y^3$ Write each constant in prime factored form using exponents.

$4x^3y^2 = 2^2 \cdot x^3 \cdot y^2$

$18x^2y = 2 \cdot 3^2\, x^2 \cdot y$

$2, x^2, y$ List the common factors, specifying the factor with the smallest exponent (this applies to bases that are constants or variables).

The greatest common factor is $2x^2y$, the product of the common factors.

You can check your choice of GCF by ensuring that it evenly divides each monomial AND that the quotients have no common factors.

Check:

$$\frac{6x^2y^3}{2x^2y} = 3y^2 \qquad \frac{4x^3y^2}{2x^2y} = 2xy \qquad \frac{18x^2y}{2x^2y} = 9$$

The quotients $3y^2$, $2xy$, and 9 are **relatively prime,** meaning that they have no common factors. ∎

■ FACTORING OUT THE GREATEST COMMON FACTOR

Recall the distributive property: $a(b + c) = ab + ac$. This property allows multiplication over the operation of addition (and subtraction, because subtraction is defined as adding the opposite of a number). Consider the use of the distributive property in the following expression:

$$3x^2(2x - 3y) = 3x^2 \cdot 2x - 3x^2 \cdot 3y$$
$$= 6x^3 - 9x^2y$$

To factor the polynomial $6x^3 - 9x^2y$, we find the greatest common factor of each *term* and then rewrite each *term* as the product of the GCF and another factor.

The GCF of the two terms in the polynomial is $3x^2$. Rewriting each term of the polynomial $6x^3 - 9x^2y$ as the indicated product of the GCF and another factor gives us the expression

$$3x^2 \cdot 2x - 3x^2 \cdot 3y$$

We then use the distributive property to rewrite the polynomial as the *product of factors* rather than the *sum of terms:*

$$3x^2(2x - 3y)$$

The steps for factoring out a GCF are summarized as follows:

> **Factoring Out the Greatest Common Factor**
> 1. Determine the GCF of the terms of the polynomial.
> 2. Rewrite each *term* as the product of the GCF and another factor.
> 3. Use the distributive property to rewrite the polynomial as the *product* of the GCF and a polynomial.

EXAMPLE 4   Factor: $12a^3b^2 - 20a^2b^3$.

Solution

$$12a^3b^2 = 2^2 \cdot 3a^3b^2$$

Factor each term of the polynomial to determine the GCF.

$$-20a^2b^3 = -2^2 \cdot 5a^2b^3$$

$2^2a^2b^2$

The GCF is the product of the common factors.

$$12a^3b^2 - 20a^2b^3 = 4a^2b^2 \cdot 3a - 4a^2b^2 \cdot 5b$$

Rewrite each term as the product of the GCF and another factor.

$$= 4a^2b^2(3a - 5b)$$

Use the distributive property.

Check the result by verifying that $4a^2b^2(3a - 5b) = 12a^3b^2 - 20a^2b^3$.   ■

EXAMPLE 5   Factor: $a^2b^2 - ab$.

Solution   The GCF is ab.

$$a^2b^2 - ab = ab \cdot ab - ab \cdot 1$$

Rewrite each term as the product of the GCF and another factor.

$$= ab(ab - 1)$$

Use the distributive property.

Check the result by verifying that $ab(ab - 1) = a^2b^2 - ab$.   ■

WARNING!   It is important to include the number 1 in the binomial factor in Example 5. It serves as a place holder.

■ FACTORING OUT A NEGATIVE FACTOR

It is often convenient when working with polynomials for the leading coefficient to be a positive number. For this reason, a negative one (-1) is usually factored out of a polynomial when the leading coefficient is negative.

EXAMPLE 6   Factor: $-18a^2b + 6ab^2 - 12a^2b^2$.

Solution

$$-18a^2b = -2 \cdot 3^2a^2b$$

Factor each term of the polynomial to determine the GCF.

$$6ab^2 = 2 \cdot 3ab^2$$
$$-12a^2b^2 = -2^2 \cdot 3a^2b^2$$

The GCF is $-2 \cdot 3ab$, or $-6ab$

Because the leading coefficient is negative, we will use negative one as part of our greatest common factor.

$$-18a^2b + 6ab^2 - 12a^2b^2 = (-6ab)3a - (-6ab)b + (-6ab)2ab$$
$$= -6ab(3a - b + 2ab)$$

Check by verifying that $-6ab(3a - b + 2ab) = -18a^2b + 6ab^2 - 12a^2b^2$.   ■

As you progress through this chapter on factoring, ALWAYS look for and factor out the GCF first.

POLYNOMIALS AS GREATEST COMMON FACTORS

Sometimes the greatest common factor of several terms is a polynomial, and we can factor out that polynomial as a common factor. It is helpful to carefully identify the number of terms in a polynomial before beginning the process.

EXAMPLE 7   Factor $a + 3$ out of the expression $(a + 3) + (a + 3)^2$.

Solution   The polynomial has two terms, $(a + 3)$ and $(a + 3)^2$. We recall that $a + 3$ is equal to $(a + 3)1$ and that $(a + 3)^2$ is equal to $(a + 3)(a + 3)$. We can then factor out the GCF of $(a + 3)$ and simplify:

$$(a + 3) + (a + 3)^2 = (a + 3)1 + (a + 3)(a + 3)$$

$$= (a + 3)[1 + (a + 3)] \qquad \begin{array}{l}\text{Use the distributive}\\\text{property to factor out the}\\\text{quantity } (a + 3) \text{ from}\\\text{each term of the}\\\text{polynomial.}\end{array}$$

$$= (a + 3)(a + 4) \qquad \begin{array}{l}\text{Simplify the second}\\\text{factor.}\end{array}$$

EXAMPLE 8   Factor: $6a^2b^2(x + 2y) - 9ab(x + 2y)$.

Solution
$$6a^2b^2(x + 2y) - 9ab(x + 2y) = (x + 2y)(6a^2b^2 - 9ab) \qquad \begin{array}{l}\text{Factor out the}\\\text{common factor}\\(x + 2y).\end{array}$$

$$= (x + 2y)(3ab)(2ab - 3) \qquad \begin{array}{l}\text{Factor out the}\\\text{common factor}\\\text{of } (3ab) \text{ from}\\(6a^2b^2 - 9ab).\end{array}$$

$$= 3ab(x + 2y)(2ab - 3) \qquad \begin{array}{l}\text{Use the}\\\text{commutative}\\\text{property of}\\\text{multiplication.}\end{array}$$

It is important that you develop the habit of always checking each polynomial factor to ensure it is prime before concluding that the factoring process is complete.

FACTORING BY GROUPING

Suppose we wish to factor the expression

$$ax + ay + cx + cy$$

Although no factor is common to all four terms, there is a common factor of a in $ax + ay$ and a common factor of c in $cx + cy$. We can factor out the a and c as follows:

$$ax + ay + cx + cy = (ax + ay) + (cx + cy) \qquad \begin{array}{l}\text{Associative property of}\\\text{addition.}\end{array}$$

$$= a(x + y) + c(x + y) \qquad \begin{array}{l}\text{Distributive property (factoring}\\\text{the GCF of } a \text{ out of the first}\\\text{two terms and the GCF of } c\\\text{out of the last two terms).}\end{array}$$

$$= (x + y)(a + c) \qquad \begin{array}{l}\text{Distributive property (factoring}\\\text{the GCF of the quantity}\\(x + y) \text{ out of each term).}\end{array}$$

Therefore, $ax + ay + cx + cy = (x + y)(a + c)$. This type of factoring is called **factoring by grouping.**

EXAMPLE 9   Factor: $2c + 2d - cd - d^2$.

Solution $2c + 2d - cd - d^2 = (2c + 2d) + (-cd - d^2)$ Associative property and definition of subtraction.

$= 2(c + d) - d(c + d)$ Distributive property. Notice that after grouping $(-cd - d^2)$, the leading coefficient of this group was negative. Therefore, a negative one became part of the GCF.

$= (c + d)(2 - d)$ Distributive property. ■

 WARNING!   When factoring expressions by grouping make sure that the process of taking out a GCF is continued until the result is a *product of factors*. There should be no ungrouped plus or minus signs in the final result of any factoring exercise.

EXAMPLE 10   Factor: **a.** $a(c - d) + b(d - c)$ and   **b.** $ac + bd - ad - bc$.

Solution   **a.** $a(c - d) + b(d - c) = a(c - d) - b(-d + c)$ Use the distributive property to factor a negative one from $(d - c)$.

$= a(c - d) - b(c - d)$ The commutative property assures us that $-d + c = c - d$.

$= (c - d)(a - b)$ Factor out $(c - d)$ using the distributive property.

b. In this example, we cannot factor anything (other than 1) from the first two terms or the last two terms. However, if we use the commutative property to rearrange the terms, the factoring is routine.

$ac + bd - ad - bc = ac - ad + bd - bc$ By the commutative property, $bd - ad = -ad + bd$.

$= a(c - d) - b(-d + c)$ Factor a from $ac - ad$ and $-b$ from $bd - bc$ using the distributive property.

$= a(c - d) - b(c - d)$ $-d + c = c - d$

$= (c - d)(a - b)$ Factor out $(c - d)$ using the distributive property. ■

EXERCISE 5.1

VOCABULARY   *In Exercises 1–9, fill in each blank to make a true statement.*

1. The _____ of an expression evenly divides the expression.

2. Two or more expressions whose product is equivalent to a given expression are called _____ of the original expression.

3. A _____ number is a number whose *only* factors are 1 and itself.

4. The expression $3 + x$ is an indicated *sum* of _____.

5. The _____ property of multiplication over addition enables us to factor polynomials.

6. Factorization of a polynomial requires that a _____ or _____ of terms be rewritten as a _____ of factors.

7. The greatest common factor of a polynomial can be a monomial or a _____.

8. The Fundamental Theorem of Arithmetic states there is _____ prime factorization for every natural number greater than 1.

9. If a polynomial contains four terms, it may be possible to factor the polynomial by _____.

CONCEPTS *In Exercises 10–19, classify each of the statements as true or false.*

10. The factors of 12 are 1, 2, 3, 4, 6, and 12.

11. The prime factors of 12 are 1, 2, and 3.

12. The numbers 10 and 17 have no common factors (other than 1).

13. The greatest common factor of 12 and 16 is 4.

14. The polynomial $3x(x + 3) + 4(x - 5)$ is completely factored.

15. The polynomial $3x^2(x + 3)(x - 2)$ is completely factored.

16. $6(x^2 - y^3)$ is equivalent to the polynomial $6x^2 - 6y^3$.

17. $6x^2y - 3x^2 = 3x(2xy - x)$ illustrates a prime factorization of a polynomial.

18. The GCF of the polynomial $3x^2y^3 + 6xy^4 - 9x^3y^5$ is $3x$.

19. The GCF of the polynomial $15a^2b^3c - 20a^3b^2c^2 + 25a^2b^4c^2$ is $5a^2b^2c$.

In Exercises 20–25, determine the GCF of each polynomial.

20. $24 + 12x$

21. $36y + 6$

22. $2x^2y^4 + 6x^2y^5 - 2x^2y^3$

23. $4x^3y^5 - 2x^2y^4 + 6x^3y^3$

24. $16a^3b^2c - 4a^2bc^3 + 8a^3b^2c^2$

25. $20t^3v - 15t^2v^3 + 3tv$

In Exercises 26–31, complete the factorization.

26. $2x^2 - 6x = 2x(\quad)$

27. $6x^3 + 3x = 3x(\quad)$

28. $3x^2y + 2x^3y^2 = x^2y(\quad)$

29. $5a^3b^2 + 7a^2b = a^2b(\quad)$

30. $-6x^2y^3 - 3xy = -3xy(\quad)$

31. $-4x^3y - 2xy = -2xy(\quad)$

PRACTICE *In Exercises 32–53, factor each polynomial. When appropriate, factor out negative one as part of the GCF.*

32. $3x + 6$

33. $xy - xz$

34. $-a - b$

35. $-2x + 5y$

36. $t^3 + 2t^2$

37. $r^4 - r^2$

38. $-2a + 3b$

39. $-3m - 4n + 1$

40. $a^3b^3z^3 - a^2b^3z^2$

41. $24x^2y^3z^4 + 8xy^2z^3$

42. $12uvw^3 - 18uv^2w^2$

43. $3x + 3y - 6z$

44. $-3xy + 2z + 5w$

45. $-3ab - 5ac + 9bc$

46. $ab + ac - ad$

47. $4y^2 + 8y - 2xy$

48. $-3x^2y - 6xy^2$

49. $-4a^2b^3 + 12a^3b^2$

50. $-4a^2b^2c^2 + 14a^2b^2c - 10ab^2c^2$

51. $-14a^6b^6 + 49a^2b^3 - 21ab$

52. $abx - ab^2x + abx^2$

53. $4x^2y^2z^2 - 6xy^2z^2 + 12xyz^2$

In Exercises 54–61, factor each polynomial.

54. $3(r - 2s) - x(r - 2s)$

55. $x(a + 2b) + y(a + 2b)$

56. $(x - 3)^2 + (x - 3)$

57. $(3t + 5)^2 - (3t + 5)$

58. $3x^2(r + 3s) - 6y^2(r + 3s)$

59. $9a^2b^2(3x - 2y) - 6ab(3x - 2y)$

60. $(3x - y)(x^2 - 2) + (x^2 - 2)$

61. $(x - 5y)(a + 2) - (x - 5y)$

In Exercises 62–71, factor each polynomial.

62. $xr + xs + yr + ys$

63. $pm - pn + qm - qn$

64. $2ab + 2ac + 3b + 3c$

65. $3ac + a + 3bc + b$

66. $2x^2 + 2xy - 3x - 3y$

67. $3ab + 9a - 2b - 6$

68. $mp - np - m + n$

69. $6x^2u - 3x^2v + 2yu - yv$

70. $x(a - b) + y(b - a)$

71. $p(m - n) - q(n - m)$

In Exercises 72–77, factor each polynomial completely. You may have to rearrange some terms.

72. $2r - bs - 2s + br$

73. $5x + ry + rx + 5y$

74. $ax + by + bx + ay$

75. $mr + ns + ms + nr$

76. $ac + bd - ad - bc$

77. $sx - ry + rx - sy$

78. When we add $5x$ and $7x$, we combine like terms to simplify the expression: $5x + 7x = 12x$. Explain how this is related to factoring.

79. A student factors the polynomial $3x^2y + 6xy^2$ as follows: $3x^2y + 6xy^2 = 3xy(x + 2)$. To check his work, he used his graphing calculator, storing 8 for x and 6 for y. His screen display is

Is his factorization correct? Why or why not?

   Another student factors the same polynomial as follows: $3x^2y + 6xy^2 = 3y(x^2 + 2xy)$. To check his work, he used his graphing calculator, storing 5 for x and 6 for y. His screen display is

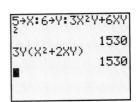

Is his factorization correct? Why or why not?

REVIEW

80. Use the distributive property to multiply the two polynomials $(3x + 4)$ and $(5x - 2)$. Show each step of the process.

82. a. Evaluate the expression $4x^2 + 5x^2y$ both analytically and by using the STOre feature of the graphing calculator when $x = -2$ and $y = 3$.

   b. Evaluate the expression $x^2(4 + 5y)$ both analytically and by using the STOre feature of the graphing calculator using the same values given for x and y in part (a).

   c. Should the results of the evaluations in part (a) and part (b) be the same? Why or why not?

81. Simplify the expression $3x + 4y - 2(x - 2y)$.

83. a. Solve the equation $3(x + 2) - 4x = \frac{1}{2}(x - 3) + 2$ analytically.

   b. Solve the same equation graphically.

   c. Is the solution a rational or an irrational number? Justify your conclusion.

5.2 Factoring Trinomials

■ FACTORING TRINOMIALS BY GROUPING

Recall that trinomials are polynomials that contain three terms. The method of factoring by grouping can be used to factor trinomials of the form $ax^2 + bx + c$, where $a > 0$. The following steps will be helpful:

Factoring a Trinomial by Grouping
1. Write the trinomial in descending powers of one variable.
2. Identify a, b, and c, where a is the coefficient of the first term ($a > 0$), b is the coefficient of the middle term, and c is the coefficient of the third term (or the constant term).
3. Find the product $a \cdot c$.
4. Find two numbers whose **product** is ac and whose **sum** is b.
5. Replace the middle term of the original trinomial with two terms, using the numbers found in step 4 as coefficients. You now have a four-term polynomial.
6. Factor by grouping.

EXAMPLE 1   Factor: $4x^2 - 4x - 3$.

Solution   The trinomial is in descending powers of the variable, x.

Identify a, b, and c: $a = 4$, $b = -4$, $c = -3$
The product $a \cdot c = 4 \cdot -3 = -12$.
The factors of -12 are $-1(12)$, $-2(6)$, $-3(4)$, and $1(-12)$, $2(-6)$, $3(-4)$.

Because the sum of the factors must equal -4 (the value of b), the only pair of factors that will work are 2 and -6: $2 + -6 = -4$.
Replace the middle term with two terms using 2 and -6 as coefficients:

$$4x^2 - 4x - 3 = 4x^2 + 2x - 6x - 3$$

Factor by grouping:

$$4x^2 + 2x - 6x - 3 = (4x^2 + 2x) + (-6x - 3)$$

Use the associative property and the definition of subtraction to group terms.

$$= 2x(2x + 1) - 3(2x + 1)$$

Use the distributive property to factor the GCF out of each grouped pair.

$$= (2x + 1)(2x - 3)$$

Use the distributive property to factor out the GCF of $(2x + 1)$.

Verify the results by multiplication:

$$(2x + 1)(2x - 3) = 2x(2x) + 2x(-3) + 1(2x) + 1(-3)$$

Use the distributive property.

$$= 4x^2 - 6x + 2x - 3$$
$$= 4x^2 - 4x - 3$$

Combine like terms.

STUDY TIP
Work assigned problems in pencil. If your result is not correct, circle the problem number with a colored pencil or pen. Leave space for a correct result to be worked. When you receive help in working the problem, write it in colored pencil/ink. Then, when reviewing exercises, any problems done in color need to be done by *you*. The colored pencil/ink indicates a result you received help on.

It is often helpful to chart your information when trying to find two numbers whose product is *ac* and sum is *b*. This is illustrated in the next example.

EXAMPLE 2   Factor: $6x^3 + 3x - 11x^2$

Solution
$$6x^3 + 3x - 11x^2 = 6x^3 - 11x^2 + 3x \qquad \text{Write the polynomial in descending powers of a variable.}$$

$$= x(6x^2 - 11x + 3) \qquad \text{Factor out the GCF of } x.$$

To factor the trinomial, identify *a*, *b*, and *c*: $a = 6$, $b = -11$, $c = 3$.
Find the product, $a \cdot c$: $a \cdot c = 6(3) = 18$.

Because the product of the two numbers is positive and the sum of the same two numbers must equal *b*, which is negative, the desired factors must both be negative.

Two numbers	Product is 18	Sum is −11
−1, −18	$(-1)(-18) = 18$	$(-1) + (-18) = -19$
−2, −9	$(-2)(-9) = 18$	$(-2) + (-9) = -11\checkmark$

The required numbers are -2 and -9.
Replace the middle term of the trinomial with two terms, using -2 and -9 as coefficients. Do not forget the GCF of *x* when working with the trinomial. Factor by grouping.

$$x(6x^2 - 11x + 3) =$$
$$x[6x^2 - 2x - 9x + 3] = x[2x(3x - 1) - 3(3x - 1)]$$
$$= x(3x - 1)(2x - 3)$$

Verify the result by multiplication.

EXAMPLE 3   Factor: $2x^2y - 8x^3 + 3xy^2$.

Solution
$$2x^2y - 8x^3 + 3xy^2 = -8x^3 + 2x^2y + 3xy^2 \qquad \text{Write the polynomial in descending powers of a variable.}$$

$$= -x(8x^2 - 2xy - 3y^2) \qquad \text{Factor out the GCF of } -x.$$

To factor the trinomial, identify *a*, *b*, and *c*: $a = 8$, $b = -2$, $c = -3$.
Find the product $a \cdot c$: $a \cdot c = 8(-3) = -24$.

Because the product of the two numbers is negative and the sum of the same two numbers must equal *b*, which is negative, the desired factors must have opposite signs. Moreover, the number with the larger absolute value will be negative.

Two numbers	Product is −24	Sum is −2
1, −24	$(1)(-24) = -24$	$1 + (-24) = -23$
2, −12	$(2)(-12) = -24$	$2 + (-12) = -10$
3, −8	$(3)(-8) = -24$	$3 + (-8) = -5$
4, −6	$(4)(-6) = -24$	$4 + (-6) = -2\checkmark$

The required numbers are 4 and -6.

Replace the middle term of the trinomial with two terms, using 4 and -6 as coefficients. Do not forget the GCF of $-x$ when working with the trinomial.

Factor by grouping:

$$
\begin{aligned}
-8x^3 + 2x^2y + 3xy^2 &= -x(8x^2 - 2xy - 3y^2) \\
&= -x(8x^2 + 4xy - 6xy - 3y^2) \\
&= -x[4x(2x + y) - 3y(2x + y)] \\
&= -x(2x + y)(4x - 3y)
\end{aligned}
$$

Verify the result by multiplication. ■

■ FACTORING TRINOMIALS BY TRIAL AND ERROR

The previous examples illustrate the fact that the product of two binomials is often a trinomial. Many polynomials, therefore, can be factored by simply testing reasonable products of binomial factors (trial and error). When the leading coefficient is 1, this is often not difficult. Consider factoring the trinomial

$$x^2 + 5x + 6$$

To find its binomial factors, note that the product of the first two terms must be x^2. Thus, the first term of each binomial must be x. Recalling the mnemonic FOIL (first, outer, inner, last) used for multiplying two binomials may be helpful when factoring by trial and error.

$$
\overset{\displaystyle x^2}{\overbrace{\qquad}}
$$
$$(x \qquad)(x \qquad)$$

Because the product of their last terms must be 6, and the sum of the products of the outer and inner terms must be $5x$, we must find two numbers whose product is 6 and whose sum is 5.

$$
\overset{\displaystyle 6}{\overbrace{\qquad}}
$$
$$(x + \text{?})(x + \text{?})$$
$$\underbrace{\qquad}_{O + I = 5x}$$

Two such numbers are $+3$ and $+2$. Thus, we have

$$x^2 + 5x + 6 = (x + 3)(x + 2)$$

We can verify this factorization by multiplying $x + 3$ and $x + 2$ and observing that the product is $x^2 + 5x + 6$.

$$
\begin{aligned}
(x + 3)(x + 2) &= x^2 + 2x + 3x + 6 \\
&= x^2 + 5x + 6 \qquad \text{Combine like terms.}
\end{aligned}
$$

Because of the commutative property of multiplication, the order of the factors is not important. Therefore,

$$x^2 + 5x + 6 = (x + 2)(x + 3)$$

EXAMPLE 4   Factor: $y^2 - 7y + 12$.

Solution   If this trinomial is to be the product of two binomials, the product of their first terms must be y^2. Thus, the first term of each binomial must be y.

$$(y \qquad)(y \qquad)$$

Because the product of the last terms must be $+12$, and the sum of the products of the outer and inner terms must be $-7y$, we must find two negative numbers whose product is $+12$ and whose sum is -7.

$$(y - ?)(y - ?) \overset{12}{\frown}$$
$$O + I = -7y$$

Two such numbers are -4 and -3. Thus,

$$y^2 - 7y + 12 = (y - 4)(y - 3)$$

We can check by verifying that the product of $y - 4$ and $y - 3$ is $y^2 - 7y + 12$.

$$(y - 4)(y - 3) = y^2 - 3y - 4y + 12$$
$$= y^2 - 7y + 12 \qquad \blacksquare$$

We must consider more combinations when we factor trinomials with lead coefficients other than 1. To factor $2x^2 - 7x + 3$, for example, we must find binomials of the form $ax + b$ and $cx + d$ such that

$$2x^2 - 7x + 3 = (ax + b)(cx + d)$$

Because the first term of the trinomial $2x^2 - 7x + 3$ is $2x^2$, the first terms of the binomial factors must be $2x$ and x.

$$(2x \quad)(x \quad) \overset{2x^2}{\frown}$$

Because the product of the last terms is 3, and the sum of the products of the outer terms and inner terms is $-7x$, we must find the two numbers with a product of 3 that will give a middle term of $-7x$.

$$(2x \quad ?)(x \quad ?) \overset{+3}{\frown}$$
$$O + I = -7x$$

Because both $(3)(1)$ and $(-3)(-1)$ give a product of 3, there are four possible combinations to consider:

$$(2x + 3)(x + 1) \qquad (2x + 1)(x + 3)$$
$$(2x - 3)(x - 1) \qquad (2x - 1)(x - 3)$$

Of these possibilities, only the last one gives the required middle term of $-7x$. Thus,

$$2x^2 - 7x + 3 = (2x - 1)(x - 3)$$

We can check this result by multiplication.

$$(2x - 1)(x - 3) = 2x^2 - 6x - x + 3$$
$$= 2x^2 - 7x + 3$$

EXAMPLE 5   Factor: $3y^2 - 4y - 4$.

Solution   Because the first term of this trinomial is $3y^2$, the first terms of the binomial factors must be $3y$ and y.

$$(3y \quad)(y \quad) \overset{3y^2}{\frown}$$

The product of the last terms must be -4, and the sum of the products of the outer terms and inner terms must be $-4y$.

$$(3y \quad ?)(y \quad ?)$$
$$-4$$
$$O + I = -4y$$

Because $(1)(-4)$, $(-1)(4)$, and $(-2)(2)$ all give a product of -4, there are six possible combinations to consider:

$$(3y + 1)(y - 4) \qquad (3y - 4)(y + 1)$$
$$(3y - 1)(y + 4) \qquad (3y + 4)(y - 1)$$
$$(3y - 2)(y + 2) \qquad (3y + 2)(y - 2)$$

Again, only the last possibility gives the required middle term of $-4y$. Thus,

$$3y^2 - 4y - 4 = (3y + 2)(y - 2)$$

We can check by multiplication.

$$(3y + 2)(y - 2) = 3y^2 - 6y + 2y - 4$$
$$= 3y^2 - 4y - 4 \qquad \blacksquare$$

EXAMPLE 6   Factor: $-8x^3 + 22x^2 - 12x$.

Solution   We factor out the common monomial factor of $-2x$.

$$-8x^3 + 22x^2 - 12x = -2x(4x^2 - 11x + 6)$$

Then we find the binomial factors of $4x^2 - 11x + 6$.

$$-8x^3 + 22x^2 - 12x = -2x(x - 2)(4x - 3)$$

We can check by multiplication.

$$-2x(x - 2)(4x - 3) = -2x(4x^2 - 3x - 8x + 6)$$
$$= -2x(4x^2 - 11x + 6)$$
$$= -8x^3 + 22x^2 - 12x \qquad \blacksquare$$

It is not easy to give specific rules for factoring trinomials by trial and error, because guesswork is often necessary. However, the following hints are often helpful:

Factoring a Trinomial by Trial and Error

1. Write the trinomial in descending powers of one variable.

2. Factor out any greatest common factor (including -1 if that is necessary to make the coefficient of the first term positive).

3. If the sign of the third term is positive, the signs between the terms of the binomial factors are the same as the sign of the middle term of the trinomial. If the sign of the third term is negative, the signs between the terms of the binomial factors are opposite.

4. Mentally try various combinations of first terms and last terms until you find one that works, or until you exhaust all the possibilities. In that case, the trinomial does not factor using only integer coefficients.

5. Check the factorization by multiplication.

■ PRIME POLYNOMIALS

Not all polynomials are factorable—some are prime. The trinomial $x^2 + 2x + 3$ is an example of a polynomial that is prime. It cannot be factored using only integer coefficients by either method presented in this section.

EXAMPLE 7 Factor: $x^2 + 2x + 3$.

Solution By grouping:
Identify a, b, and c: $a = 1$, $b = 2$, $c = 3$.
Find two numbers whose product is $a \cdot c = (1)(3) = 3$ AND whose sum is 2. The only factors of 3 are 1 and 3, and the sum of these two numbers is 4 (which does not equal the value of b).
Therefore, the polynomial is prime.

By trial and error:

Because the last term of the trinomial is 3 and the middle term is $2x$, we must find two factors of 3 whose sum is 2 so that

$$(x \quad ?)(x \quad ?)$$

Because 3 factors only as $(1)(3)$ and $(-1)(-3)$, it has no factors whose sum is 2. Thus, $x^2 + 2x + 3$ cannot be factored over the integers. It is a prime polynomial. ■

When factoring trinomials by grouping, the chart becomes a valuable tool for identifying prime polynomials. This is illustrated in the following example.

EXAMPLE 8 Factor: $6x^2 + 3x + 2$.

Solution Identify a, b, and c: $a = 6$, $b = 3$, $c = 2$.
The product, $a \cdot c$, is 12, therefore, we must find two numbers whose product is 12 and whose sum is 3.

Two numbers	Product is 12	Sum is 3
1, 12	$1 \cdot 12 = 12$	$1 + 12 = 13$
2, 6	$2 \cdot 6 = 12$	$2 + 6 = 8$
3, 4	$3 \cdot 4 = 12$	$3 + 4 = 7$
4, 3	$4 \cdot 3 = 12$	$4 + 3 = 7$
6, 2	$6 \cdot 2 = 12$	$6 + 2 = 8$
12, 1	$12 \cdot 1 = 12$	$12 + 1 = 13$

There are no combinations of factors of 12 whose sum is 3. Observe the pattern in the sum column: 13, 8, **7, 7**, 8, 13. A sum of 3 would have had to occur between the two 7s. This is impossible when using integer factors of 12. The rows in bold are actually a repeat of the first three rows. Because multiplication is commutative, they were not required. ■

EXERCISE 5.2

VOCABULARY *Fill in each blank to make a true statement.*

1. _____ are polynomials that contain three terms.

2. When factoring three-term polynomials, begin by writing the terms in _____ order of one of the variables.

3. A correct factorization can always be justified by _____.

4. Polynomials that are not factorable are called _____ polynomials.

5. A factorable trinomial can be expressed as the product of two (or more) _____.

CONCEPTS

6. To factor $10x^2 - 27 - 3x$, we must first rewrite the trinomial as _____.

7. Given the trinomial $x^2 + 5x + 6$,
   a. What is the leading coefficient?
   b. What is the coefficient of the second term?
   c. What is the coefficient of the first term?
   d. What is the constant term?

8. Complete each factorization:
   a. $2x^2 + x - 15 = (2x - 5)(x + \underline{?})$.
   b. $10x^2 + 41x - 18 = (5x - \underline{?})(2x + 9)$
   c. $3x^2 + x - 10 = (? - 5)(x + ?)$

In Exercises 9–14, decide which of the three lettered choices is the correct factored form of the given polynomial.

9. $x^2 - 2x - 15$ (a) $(x + 3)(x - 5)$ (b) $(x - 3)(x + 5)$ (c) $(x + 15)(x - 1)$
10. $x^2 + 4x - 12$ (a) $(x + 4)(x - 3)$ (b) $(x - 6)(x + 2)$ (c) $(x + 6)(x - 2)$
11. $6x^2 - x - 15$ (a) $(2x - 3)(3x + 5)$ (b) $(3x + 3)(2x - 5)$ (c) $(2x + 3)(3x - 5)$
12. $10x^2 - 19x + 6$ (a) $(2x - 6)(5x - 1)$ (b) $(5x - 2)(2x - 3)$ (c) $(2x + 3)(5x + 2)$
13. $2x^2 + 2x - 4$ (a) $(2x + 4)(x - 1)$ (b) $2(x + 2)(x - 1)$ (c) $(2x - 4)(x + 1)$
14. $3x^2 - 6x - 24$ (a) $3(x + 2)(x - 4)$ (b) $(3x + 6)(x + 4)$ (c) $(3x + 6)(x + 1)$

15. When factoring $x^2 + 5x + 6$, one student wrote $(x + 3)(x + 2)$, while another wrote $(x + 2)(x + 3)$. Are they both correct? Why or why not?

16. When factoring $x^2 + 10x + 9$, one student wrote $(x + 9)(x + 1)$, while another student wrote $(x + 1)(x + 9)$. Are they both correct? Why or why not?

PRACTICE *In Exercises 17–43, factor each polynomial completely. If it is prime, so state. Do not forget to factor out a -1 if the leading coefficient is negative and to **always** look for and factor out a GCF first.*

17. $x^2 - 9x - 12$ 18. $y^2 + 11y - 26$ 19. $4 - 5x + x^2$
20. $y^2 + 5 + 6y$ 21. $x^2 + 3x + 2$ 22. $y^2 + 4y + 3$
23. $2x^2 + 10x + 12$ 24. $3y^2 - 21y + 18$ 25. $-5t^2 + 25t - 30$
26. $-2x^2 + 20x - 18$ 27. $x^2 + 9x + 5$ 28. $y^2 - 8y + 6$
29. $u^2 + 2uv - 15v^2$ 30. $m^2 + 3mn - 10n^2$ 31. $-3rs + r^2 + 2s^2$
32. $-13yz + y^2 - 14z^2$ 33. $-4x^2y - 4x^3 + 24xy^2$ 34. $8x^2 - 12xy - 8y^2$
35. $24y^2 + 14xy + 2x^2$ 36. $6x^3 - 15x^2 - 9x$ 37. $9y^3 + 3y^2 - 6y$

38. $x^2 + 9x + 20$

39. $y^2 - 8y + 15$

40. $6x^2 - 7x - 1$

41. $2y^2 + 5y - 2$

42. $4r^2 - 4rs - 8s^2$

43. $6x^2 + 3xy - 18y^2$

In Exercises 44–46, a polynomial, P(x), has been entered on the calculator and evaluated for three values of x. For each problem below, (a) factor P(x), and (b) use your calculator to evaluate the factored form of the polynomial for the same three values of x that are displayed on the screen. If your results are the same, you can conclude that you have factored correctly. This provides a partial check of your factorizations. Copy both screen displays to justify your results. (Hint: Your factored form of the expression should be entered at the y1 prompt.)

44.

45.

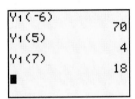

46.

In Exercises 47–52, factor each of the polynomials by grouping.

47. $6x^2 - 15x + 4x - 10$

48. $35x^2 - 63x + 10x - 18$

49. $2y^2 + 16y - 5y - 40$

50. $4x^2 + 12x - x - 3$

51. $42x^2 + 48x - 7x - 8$

52. $5y^2 - 5y - 2y + 2$

In Exercises 53–62, each of the trinomials is a **perfect square trinomial**. Factor each, using the methods presented in this section. Do you see why they are considered perfect squares?

53. $x^2 + 6x + 9$

54. $x^2 + 10x + 25$

55. $y^2 - 8y + 16$

56. $z^2 - 2z + 1$

57. $t^2 + 20t + 100$

58. $r^2 + 24r + 144$

59. $u^2 - 18u + 81$

60. $v^2 - 14v + 49$

61. $x^2 + 4xy + 4y^2$

62. $x^2 + 6xy + 9y^2$

The trinomials in Exercises 53–62 are called *perfect square trinomials* because each can be written as the square of a binomial. The special product formulas may be helpful:

Special Product Formulas

$$x^2 + 2xy + y^2 = (x + y)^2$$
$$x^2 - 2xy + y^2 = (x - y)^2$$

Consider the trinomial $z^2 + 12z + 36$.

Notice that the first term is a perfect square: $z^2 = (z)^2$.

The last term is a perfect square: $36 = (6)^2$.

The middle term is twice the product of the bases of the first and third terms: $12z = 2(z)(6)$.

Therefore, the trinomial can be factored as the sum of the bases, quantity squared:

$$z^2 + 12z + 36 = (z + 6)^2$$

Anytime you notice this pattern in a trinomial, you may use a special product formula to factor it if you desire. However, remember that factoring by grouping or trial and error could also be used, and the results would be the same.

APPLICATIONS

63. Volume The volume of a rectangular solid is represented by the polynomial $h^3 + 7h^2 + 12h$, where h represents its height. Recall that the volume of a rectangular solid is the product of its length, width, and height ($V = lwh$).

   a. How much longer are the length and width than the height? (assume the length is the longest dimension)

   b. State the length and the width if the height is 6 inches.

64. Volume The volume of a rectangular solid is represented by the polynomial $h^3 + 7h^2 + 10h$, where h represents its height. Recall that the volume of a

rectangular solid is the product of its length, width, and height ($V = lwh$).

   a. How much longer are the length and width than the height? (assume the length is the longest dimension)

   b. State the length and the width if the height is 4 inches.

65. For what values of b will the polynomial $6x^2 + bx + 6$ be factorable?

66. For what values of b will the polynomial $5x^2 + bx + 4$ be factorable?

REVIEW

67. Simplify: $(x + 3)^2$.

68. **a.** Simplify the expression $(5x^2y^3)^2$.
   b. Evaluate the original expression in part (a) if $x = -2$ and $y = -4$.
   c. Evaluate your simplified expression from part (a) when $x = -2$ and $y = -4$.
   d. Are the evaluations equal? Why or why not?

69. Rewrite $\dfrac{a^{-4}b^2}{c^{-5}}$ without negative exponents.

70. If $P(x) = 2x^2 + 3x - 4$, find each of the following:
   a. $P(-5)$ **b.** $P(6)$
   c. $P(-2.3)$ (use the VARS or STOre feature of your calculator)

71. Solve the equation
   $x^2 + 5x - 3 = x(x - 4) + 3(2x - 6)$
   a. algebraically.
   b. graphically (specify a window that shows a complete graph).

5.3 Factoring the Difference of Two Squares and a Summary of Factoring Techniques

■ FACTORING THE DIFFERENCE OF TWO SQUARES ■ SUMMARY OF FACTORING TECHNIQUES

This section will focus on the factoring of binomials that fit a pattern. Much of the section will be devoted to the discovery of those patterns.

■ FACTORING THE DIFFERENCE OF TWO SQUARES

EXPLORING THE CONCEPT   Using the distributive property, multiply each of the following binomials together.

a. $(x + 2)(x - 2) = ?$

b. $(x + 2)(x + 2) = ?$

c. $(ab - 1)(ab + 1) = ?$

d. $(ab - 1)(ab + 2) = ?$

e. $(2x + 3)(2x - 3) = ?$

f. $(2x + 3)(3x - 2) = ?$

Conclusion   ***a.** $x^2 - 4$

b. $x^2 + 4x + 4$

***c.** $a^2b^2 - 1$

d. $a^2b^2 + ab - 2$

***e.** $4x^2 - 9$

f. $6x^2 + 5x - 6$

*Whenever we multiply a binomial of the form $x + y$ by a binomial of the form $x - y$, we obtain another binomial:

$$(x + y)(x - y) = x^2 - y^2$$

The binomials $(x + y)$ and $(x - y)$ are called **conjugate binomials**.

The binomial $x^2 - y^2$ is called the **difference of two squares,** because x^2 is the square of x and y^2 is the square of y. The difference of the squares of two quantities such as x and y always factors into the *sum* of those two quantities multiplied by the *difference* of those two quantities. The product of conjugate binomials is *always* the difference of squares.

> **Factoring the Difference of Two Squares**
>
> $x^2 - y^2 = (x + y)(x - y)$

To factor $x^2 - 9$, for example, we note that $x^2 - 9$ can be written in the form $(x)^2 - (3)^2$ and that this is the *difference* between the square of x and the square of 3. Thus, it factors into the product of the sum of the bases (x plus 3) and the difference between the bases (x minus 3).

$$x^2 - 9 = (x)^2 - (3)^2$$
$$= (x + 3)(x - 3)$$

As in all of our factoring problems, we can check the result by verifying that $(x + 3)(x - 3) = x^2 - 9$.

Recall that the square root key, $\sqrt{}$, can be used to determine if a constant is a perfect square. Expressions containing variables such as x^4y^2 are also perfect squares because they can be written as the square of a quantity:

$$x^4y^2 = (x^2y)^2$$

EXAMPLE 1   Factor: $4y^4 - 25z^2$.

Solution   Because the binomial $4y^4 - 25z^2$ can be written in the form $(2y^2)^2 - (5z)^2$, it represents the difference of the squares of $2y^2$ and $5z$. Thus, it factors into the sum of these two quantities times the difference of these two quantities.

$$4y^4 - 25z^2 = (2y^2)^2 - (5z)^2$$
$$= (2y^2 + 5z)(2y^2 - 5z)$$

Check this result by multiplication.   ■

We can often factor out a greatest common factor before factoring the difference of two squares. For example, to factor $8x^2 - 32$, we begin by factoring out the greatest common factor of 8 and then factor the resulting difference of two squares.

$$8x^2 - 32 = 8(x^2 - 4) \qquad \text{Factor out 8.}$$
$$= 8(x^2 - 2^2) \qquad \text{Write 4 as } 2^2.$$
$$= 8(x + 2)(x - 2) \qquad \text{Factor the difference of two squares.}$$

We can verify this result by multiplication:

$$8(x + 2)(x - 2) = 8(x^2 - 4)$$
$$= 8x^2 - 32$$

EXAMPLE 2   Factor: $2a^2x^3y - 8b^2xy$.

Solution
$$2a^2x^3y - 8b^2xy = 2xy(a^2x^2 - 4b^2) \qquad \text{Factor out } 2xy.$$
$$= 2xy[(ax)^2 - (2b)^2]$$
$$= 2xy(ax + 2b)(ax - 2b) \qquad \text{Factor the difference of two squares.}$$

Check this result by multiplication.   ■

Sometimes we must factor a difference of two squares more than once to completely factor a polynomial. For example, the binomial $625a^4 - 81b^4$ can be written in the form $(25a^2)^2 - (9b^2)^2$. The difference of the squares of $25a^2$ and $9b^2$ factors as

$$625a^4 - 81b^4 = (25a^2)^2 - (9b^2)^2$$
$$= (25a^2 + 9b^2)(25a^2 - 9b^2).$$

The factor $25a^2 - 9b^2$, however, can be written in the form $(5a)^2 - (3b)^2$ and can be factored as $(5a + 3b)(5a - 3b)$. Thus, the complete factorization of $625a^4 - 81b^4$ is

$$625a^4 - 81b^4 = (25a^2 + 9b^2)(5a + 3b)(5a - 3b)$$

WARNING!   The binomial $25a^2 + 9b^2$ is called the **sum of two squares** because it can be written in the form $(5a)^2 + (3b)^2$. Such binomials cannot be factored if we are limited to integer coefficients. Polynomials that do not factor over the integers are called **irreducible** or **prime polynomials.**

✎**52.** A student entered a polynomial at the y1 prompt of his calculator and what he believed to be the correct factorization of that polynomial at the y2 prompt. His TABLE display is pictured below.

X	Y1	Y2
-3	-16	-16
-2	-21	-21
-1	-24	-24
0	-25	-25
1	-24	-24
2	-21	-21
3	-16	-16

X=-3

Does the display confirm that his factorization was correct? Why or why not?

✎**53.** A student entered a polynomial at the y1 prompt on her calculator and what she believed to be the correct factorization of that polynomial at the y2 prompt. Her TABLE display is pictured below.

X	Y1	Y2
-3	-40	-40
-2	-45	-45
-1	-48	-48
0	-49	-49
1	-48	-48
2	-45	-45
3	-40	-40

X=-3

Does the display confirm that her factorization was correct? Why or why not?

✎**54.** A student entered a polynomial at the y1 prompt of his calculator and what he believed to be the correct factorization of that polynomial at the y2 prompt. His TABLE display is pictured below.

X	Y1	Y2
-3	34	4
-2	29	9
-1	26	16
0	25	25
1	26	36
2	29	49
3	34	64

X=-3

Does the display confirm that his factorization was correct? Why or why not?

✎**55.** A student entered a polynomial at the y1 prompt of her calculator and what she believed to be the correct factorization of that polynomial at the y2 prompt. Her TABLE display is pictured below.

X	Y1	Y2
-3	-40	100
-2	-45	81
-1	-48	64
0	-49	49
1	-48	36
2	-45	25
3	-40	16

X=-3

Does the display confirm that her factorization was correct? Why or why not?

In Exercises 56–99, factor each polynomial completely. If a polynomial is prime, so state. Check by multiplication.

56. $6x + 3$

57. $x^2 - 9$

58. $x^3 - 27$

59. $x^3 + 8$

60. $6t^2 + 7t - 3$

61. $3rs^2 - 6r^2st$

62. $4x^2 - 25$

63. $ac + ad + bc + bd$

64. $t^2 - 2t + 1$

65. $6p^2 - 3p - 2$

66. $3a^2 - 12$

67. $2x^2 - 32$

68. $x^2y^2 - 2x^2 - y^2 + 2$

69. $2xy^2 + 8xy - 24x$

70. $-3x^2 + 3x + 60$

71. $-5x^2 - 10x + 40$

72. $t^4 - 16$

73. $-8p^3q^7 - 4p^2q^3$

74. $27x^3 - 8y^3$

75. $64y^3 - z^3$

76. $3rs + 6r^2 - 18s^2$

77. $x^2 + 7x + 1$

78. $3a^2 + 27b^2$

79. $25x^2 + 16y^2$

80. $16 - 40z + 25z^2$

81. $14t^3 - 40t^2 + 6t^4$

82. $6x^2 + 7x - 20$

83. $-6x^2 - 7x + 20$

84. $-20x^2 - 11x + 3$

85. $16x^3 - 54y^3$

86. $64a^3 - 4b^3$

87. $a^2(x - a) - b^2(x - a)$

88. $2x^2 - 5xy - 3y^2$

89. $8a^2x^3y - 2b^2xy$

90. $2ac + 4ad + bc + 2bd$

91. $81p^4 - 16q^4$

92. $6x^2 - x - 16$

93. $4x^2 + 9y^2$

94. $30x^4 + 5x^3 - 200x^2$

95. $4x^2y^2 - y^4$

96. $6y^3 + 35y^2 - 6y$

97. $10r^2 - 13r - 4$

98. $4x^2 + 4x + 1 - y^2$

99. $21t^3 - 10t^2 + t$

REVIEW

100. a. Divide $6x^2y + 9xy$ by $3xy$.
  b. Factor $6x^2y + 9xy$
  ✎**c.** What relationship(s) do you see between parts (a) and (b) above?

101. The *nth* term, *l*, of an arithmetic sequence is $l = f + (n - 1)d$, where *f* is the first term and *d* is

the common difference. Remove the parentheses and solve the equation for the variable *n*.

102. Geometry A rectangle is 13 times as long as it is wide, and its perimeter is 140 centimeters. Find the area of the rectangle.

5.4 Using Factoring to Solve Equations

■ REVIEW OF EQUATION SOLVING AND BASIC DEFINITIONS ■ SOLVING QUADRATIC EQUATIONS BY FACTORING ■ SOLVING HIGHER-DEGREE EQUATIONS BY FACTORING ■ GRAPHICAL SOLUTIONS ■ APPLICATIONS

■ REVIEW OF EQUATION SOLVING AND BASIC DEFINITIONS

Factoring can be used to solve equations that cannot be solved by methods introduced previously in Chapter 3. Recall that we solved equations by a process called *transformation*. After simplifying each side of the equation, we used the addition, subtraction, multiplication, and/or division properties of equality to rewrite (transform) the original equation into a series of equivalent equations until the variable was isolated on one side of the equation. This process is used to solve linear (first-degree) equations.

The degree of a polynomial equation (when simplified) is the same as the highest degree of any of its terms. Equations that are of degree two (second-degree equations) are called **quadratic equations.**

Quadratic Equation
A quadratic equation is any equation that can be written in the form $ax^2 + bx + c = 0$, where *a*, *b*, and *c* are real numbers, $a \neq 0$.

The solution (or root) of an equation is the value (or values) of the variable that make the equation a true statement. To find the solutions of an equation that is quadratic or of higher degree, the **zero factor property** is used.

Zero Factor Property
If $a \cdot b = 0$ and *a* and *b* are real numbers, it follows that $a = 0$ or $b = 0$.

This property simply states that if a product of factors is zero, *at least one* of the factors must be equal to zero. To apply this property, the polynomial equations must be set equal to zero. A polynomial equation is said to be in **standard form** when one side of the equation is zero.

■ SOLVING QUADRATIC EQUATIONS BY FACTORING

EXAMPLE 1 Solve: $(x - 4)(x + 5) = 0$.

Solution Because the product of two factors equals 0, the zero factor property assures us that one of the factors must be equivalent to zero.

Either $x - 4 = 0$ or $x + 5 = 0$

We can solve each of these linear equations to get

$x = 4$ or $x = -5$

Therefore, the equation $(x - 4)(x + 5) = 0$ has two solutions, 4 and -5. These could be expressed as a solution set: $\{4, -5\}$.

These solutions can be verified in any one of the following ways:

Analytically: By substituting the solutions into the original equation:

Using the STOre feature:

When $x = 4$ When $x = -5$

$$(x - 4)(x + 5) = 0 \qquad\qquad (x - 4)(x + 5) = 0$$

$$(4 - 4)(4 + 5) = 0 \qquad\qquad (-5 - 4)(-5 + 5) = 0$$

$$(0)(9) = 0 \qquad\qquad\qquad (-9)(0) = 0$$

$$0 = 0\surd \qquad\qquad\qquad\qquad 0 = 0\surd$$

```
4→X:(X-4)(X+5)
                 0
-5→X:(X-4)(X+5)
                 0
```

Using the TABLE feature: The left side of the equation is entered at the y1 prompt and the right side is entered at the y2 prompt. Initially, the table is set up to start with x having a value of -5. The second screen verifies the solution of -5 because the value in the y1 column is equal to the value in the y2 column. Scrolling down the table, the third screen shows the verification of the solution 4.

```
Plot1 Plot2 Plot3
\Y1冒(X-4)(X+5)
\Y2冒0
\Y3=
\Y4=
\Y5=
\Y6=
\Y7=
```

X	Y₁	Y₂
-5	0	0
-4	-8	0
-3	-14	0
-2	-18	0
-1	-20	0
0	-20	0
1	-18	0

X=-5

X	Y₁	Y₂
-2	-18	0
-1	-20	0
0	-20	0
1	-18	0
2	-14	0
3	-8	0
4	0	0

X=4

Solving Equations by Factoring

1. Simplify both sides of the equation by removing grouping symbols and combining like terms. Fractions may be eliminated by multiplying each term on both sides of the equation by the LCD.

2. Put the equation in standard form (set one side equal to zero).

3. Factor the polynomial.

4. Set each factor containing a variable equal to zero.

5. Solve the resulting equation.

6. Check solutions either analytically or with the graphing calculator.

EXAMPLE 2   Solve: $9x^2 - 6x = 0$.

Solution   Because the polynomial equation is in standard form, we begin by factoring the polynomial.

■ **TECHNOLOGY TIP**
Verify solutions to equations using the method you like best and with which you feel most comfortable. However, experiment with both methods presented in Example 1 and link the calculator verification to the analytical check.

$$9x^2 - 6x = 0$$

$$3x(3x - 2) = 0 \qquad \text{Factor out the GCF of } 3x.$$

$$3x = 0 \quad \text{or} \quad 3x - 2 = 0 \qquad \text{Use the zero factor property to set each factor equal to zero.}$$

$$\frac{3x}{3} = \frac{0}{3} \qquad\qquad 3x = 2 \qquad \text{Solve each linear equation for the variable.}$$

$$\frac{3x}{3} = \frac{2}{3}$$

$$x = 0 \qquad\qquad x = \frac{2}{3}$$

Solutions: $0, \frac{2}{3}$; Solution set: $\left\{0, \frac{2}{3}\right\}$.

The verification using the STOre feature is shown; however, remember that the TABLE feature or an analytical verification could also have been used. Also, remember that solutions should be verified using the original equation.

■ SOLVING HIGHER-DEGREE EQUATIONS BY FACTORING

EXAMPLE 3   Solve: $x^3 - 2x^2 = 63x$.

Solution   This equation is *not* a quadratic equation (it is a third-degree equation), but it can be solved by factoring.

$$x^3 - 2x^2 - 63x = 0$$ Put the equation in standard form.

$$x(x^2 - 2x - 63) = 0$$ Begin factoring the polynomial by factoring out the GCF of x.

$$x(x - 9)(x + 7) = 0$$ Factor the trinomial.

$$x = 0 \quad x - 9 = 0 \quad x + 7 = 0$$ Use the zero factor property to set each factor equal to 0.

$$x = 0 \quad x = 9 \quad x = -7$$ Solve each linear equation.

Solutions: 0, 9, −7; Solution set: {0, 9, −7}.

The verification, using the TABLE feature is illustrated below:

The left side of the original equation is entered at the y1 prompt and the right side is entered at the y2 prompt.

The TABLE is set up to begin with an x value of −7. Because the values in the y1 and y2 columns are equivalent, −7 is verified as a solution. Scrolling through the TABLE verifies the other roots of 0 and 9.

EXAMPLE 4   Solve: $\dfrac{3}{4}(6x^2 - x) = \dfrac{3}{2}$.

Solution

$$\frac{18}{4}x^2 - \frac{3}{4}x = \frac{3}{2}$$ Use the distributive property to eliminate grouping.

$$4 \cdot \frac{18}{4}x^2 - 4 \cdot \frac{3}{4}x = 4 \cdot \frac{3}{2}$$ Multiply through the equation by the LCD of 4 to eliminate fractions.

$$18x^2 - 3x = 6$$ Simplify.

$$18x^2 - 3x - 6 = 0$$ Put the equation in standard form.

$$3(6x^2 - x - 2) = 0$$ Factor the GCF of 3 from the polynomial.

$$3(3x - 2)(2x + 1) = 0$$ Factor the trinomial.

$$3x - 2 = 0 \qquad 2x + 1 = 0$$

Using the zero factor property, set each factor **containing a variable** equal to zero. It does not make sense to set the factor of 3 equal to 0.

$$3x = 2 \qquad 2x = -1$$

Solve each linear equation.

$$x = \frac{2}{3} \qquad x = -\frac{1}{2}$$

Solutions: $\frac{2}{3}$, $-\frac{1}{2}$; Solution set: $\left\{\frac{2}{3}, -\frac{1}{2}\right\}$.

The verification using the STOre feature is shown:

 WARNING! In the previous example, factoring the polynomial produced the equation $3(3x - 2)(2x + 1) = 0$. The product of three factors equaled zero. The zero could only have occurred in the factor $(3x - 2)$ or the factor $(2x + 1)$ because the third factor was the constant, 3.

■ GRAPHICAL SOLUTIONS

Quadratic and higher-degree equations can also be solved graphically, using the INTERSECT feature of the calculator. Each equation that has been solved algebraically in this section will now be solved graphically, using this feature.

EXAMPLE 5

Solve $(x - 4)(x + 5) = 0$ graphically.

Solution

Enter the left side of the equation at the y1 prompt and the right side at the y2 prompt and view the graph in the standard viewing window $[-10, 10]$ by $[-10, 10]$:

■ **TECHNOLOGY TIP**

The right side of the equation, 0, was entered at the y2 prompt. The graph of this line is on top of the x-axis and is not distinguishable as a separate line. Turning your axes off will enable you to view the line.

To get a complete graph, reset the ymin to -25. (This is an educated guess!)

■ **TECHNOLOGY TIP**

A complete graph includes both the x- and y-intercepts as well as any peaks and valleys. TRACING along the graph and observing x- and y-values allows you to set a good viewing window.

Using the INTERSECT feature twice gives the roots of the equation:

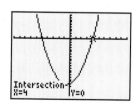

This verifies the solutions found in Example 1.

EXAMPLE 6   Solve $9x^2 - 6x = 0$ graphically.

Solution   Enter the left side of the equation at the y1 prompt and the right side at the y2 prompt and view the graph in the standard viewing window. To get a better look at the points of intersection, the window values are set to $[-2, 2]$ by $[-2, 2]$.

y= edit screen   ZStandard $[-2, 2]$ by $[-2, 2]$

Using the INTERSECT feature twice gives the roots of the equation:

Remember, the value for x that appears in the middle screen is automatically stored in the "x" slot in the calculator when the INTERSECT feature is used. Returning to the home screen allows us to convert this value to its equivalent fraction form. This verifies the solution found in Example 2.   ∎

EXAMPLE 7   Solve $x^3 - 2x^2 = 63x$ graphically.

Solution   Enter the left side of the equation at the y1 prompt and the right side at the y2 prompt and view the graph in the standard viewing window.

The right side of the equation ($63x$) has graphed almost on top of the y-axis. To get a better graph, we'll rewrite the original equation in standard form, $x^3 - 2x^2 - 63x = 0$, and graph it in the ZStandard viewing window.

Because this is not a good window, we TRACE to find the peaks and valleys of the graph so that we can reset our window values. The graph is pictured below in a window of $[-10, 10]$ by $[-250, 200]$, with all intersections displayed.

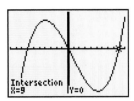

This verifies the solution found in Example 3. ∎

EXAMPLE 8   Solve $\dfrac{3}{4}(6x^2 - x) = \dfrac{3}{2}$ graphically.

Enter the left side of the equation at the y1 prompt and the right side at the y2 prompt and view the graph in the standard viewing window. To get a better view of the points of intersection, the window values are reset to $[-2, 2]$ by $[-3, 3]$. This was not necessary; the points of intersection could have been found when the graph was in the ZStandard viewing window; however, we suggest that you practice getting a good, clear graph of desired points.

y= edit screen   ZStandard $[-2, 2]$ by $[-3, 3]$

The solutions, or roots, are found using the INTERSECT feature.

This verifies the solutions found in Example 4. ∎

■ APPLICATIONS

EXAMPLE 9   **Integer problem**   One negative integer is 5 less than another, and their product is 84. Find the integers.

Solution   Let x represent the larger number. Then $x - 5$ represents the smaller number. Because their product is 84, we form the equation $x(x - 5) = 84$ and solve it.

$$x(x - 5) = 84$$
$$x^2 - 5x = 84 \qquad \text{Simplify by removing parentheses.}$$
$$x^2 - 5x - 84 = 0 \qquad \text{Put the equation in standard form.}$$
$$(x - 12)(x + 7) = 0 \qquad \text{Factor.}$$
$$x - 12 = 0 \quad \text{or} \quad x + 7 = 0 \qquad \text{Set each factor containing a variable equal to 0.}$$
$$x = 12 \quad \text{or} \quad x = -7 \qquad \text{Solve each linear equation.}$$

Because we need two negative numbers, we discard the result $x = 12$. The two integers are

$$x = -7 \qquad \text{and} \qquad x - 5 = -7 - 5$$
$$= -12$$

Check: The number -12 is 5 less than -7, and $(-12)(-7) = 84$.
The graphical solutions are pictured below.

$$[-10, 15] \text{ by } [-10, 100]$$

EXAMPLE 10   **Ballistics**   If an object is thrown straight up into the air with an initial velocity of 112 feet per second, its height after t seconds is given by the formula

$$h = 112t - 16t^2$$

where h represents the height of the object in feet. After this object has been thrown, in how many seconds will it hit the ground?

Solution   When the object hits the ground, its height will be 0. Thus, we set h equal to 0 and solve for t.

$$h = 112t - 16t^2$$
$$0 = 112t - 16t^2$$

$0 = \mathbf{16t(7 - t)}$ Factor.

$\mathbf{16t = 0}$ or $\mathbf{7 - t = 0}$ Set each factor equal to zero.

$t = 0$ or $t = 7$ Solve each linear equation.

When $t = 0$, the object's height above the ground is 0 feet, because it has not been released. When $t = 7$, the height is again 0 feet. The object has hit the ground. The solution is 7 seconds.
The graphical solutions are pictured below.

$$[-10, 10] \text{ by } [-50, 200]$$

Recall that the area of a rectangle is given by the formula $A = lw$, where A represents the area, l the length, and w the width of the rectangle. The perimeter of a rectangle is given by the formula $P = 2l + 2w$, where P represents the perimeter, l the length, and w the width of the rectangle.

EXAMPLE 11   **Perimeter of a rectangle**   Assume that the rectangle in Figure 5-1 has an area of 52 square centimeters and that its length is 1 centimeter more than 3 times its width. Find the perimeter of the rectangle.

FIGURE 5-1

Solution   Let w represent the width of the rectangle. Then $3w + 1$ represents its length. Because the area is 52 square centimeters, we substitute 52 for A and $3w + 1$ for l in the formula $A = lw$ and solve for w.

$$A = lw$$
$$52 = (3w + 1)w$$
$$52 = 3w^2 + w$$ Simplify by removing parentheses.
$$0 = 3w^2 + w - 52$$ Put the equation in standard form by subtracting 52 from both sides.
$$0 = (3w + 13)(w - 4)$$ Factor.

$3w + 13 = 0$ or $w - 4 = 0$ Set each factor equal to 0.

$3w = -13$ or $w = 4$ Solve each linear equation.

$$w = -\frac{13}{3}$$

Because the width of a rectangle cannot be negative, we discard the result $w = -\frac{13}{3}$. Thus, the width of the rectangle is 4, and the length is given by

$$3w + 1 = 3(\mathbf{4}) + 1$$
$$= 12 + 1$$
$$= 13$$

The dimensions of the rectangle are 4 centimeters by 13 centimeters. We find the perimeter by substituting 13 for l and 4 for w in the formula for the perimeter.

$$P = 2l + 2w$$
$$= 2(13) + 2(4)$$
$$= 26 + 8$$
$$= 34$$

The perimeter of the rectangle is 34 centimeters.

Check: A rectangle with dimensions of 13 centimeters by 4 centimeters does have an area of 52 square centimeters, and the length is 1 centimeter more than 3 times the width. A rectangle with these dimensions has a perimeter of 34 centimeters.
    The graphical solutions are pictured below.

$[-10, 10]$ by $[-10, 75]$

EXAMPLE 12   **Dimensions of a triangle**   The triangle in Figure 5-2 has an area of 10 square centimeters and a height that is 3 centimeters less than twice the length of its base. Find the length of the base and the height of the triangle.

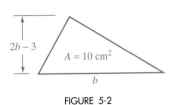

FIGURE 5-2

Solution   Let b represent the length of the base of the triangle. Then $2b - 3$ represents the height. Because the area is 10 square centimeters, we substitute 10 for A and $2b - 3$ for h in the formula $A = \frac{1}{2}bh$ and solve for b.

$$A = \frac{1}{2}bh$$

$$10 = \frac{1}{2}b(2b - 3)$$

$$20 = b(2b - 3) \qquad \text{Multiply both sides by 2 to clear fractions.}$$

$$20 = 2b^2 - 3b \qquad \text{Simplify by removing parentheses.}$$

$$0 = 2b^2 - 3b - 20 \qquad \text{Put the equation in standard form.}$$

$$0 = (2b + 5)(b - 4) \qquad \text{Factor.}$$

$$2b + 5 = 0 \qquad \text{or} \qquad b - 4 = 0 \qquad \text{Set each factor equal to 0.}$$

$$2b = -5 \qquad\qquad\qquad b = 4 \qquad \text{Solve each linear equation.}$$

$$b = -\frac{5}{2}$$

Because a triangle cannot have a negative number for the length of its base, we discard the result $b = -\frac{5}{2}$. The length of the base of the triangle is 4 centimeters. Its height is $2(4) - 3$, or 5 centimeters.

Check: If the base of the triangle has a length of 4 centimeters and the height of the triangle is 5 centimeters, its height is 3 centimeters less than twice the length of its base. Its area is 10 centimeters.

$$A = \frac{1}{2}bh$$

$$= \frac{1}{2}(4)(5)$$

$$= 2(5)$$

$$= 10$$

The graphical solution is pictured below.

$[-10, 10]$ by $[-10, 25]$

EXERCISE 5.4

VOCABULARY   *Fill in each blank to make a true statement.*

1. The process used to solve linear equations in one variable is called _____ because the addition, subtraction, multiplication, and division properties of equality are used to rewrite the original equation into a series of equivalent equations.

2. Quadratic equations are of degree ____.

3. Another word for the solution of an equation is its ____.

4. The _____ property is used to find solutions of quadratic and higher-degree equations when factoring is used.

5. The property referred to in Exercise 4 states that if a product of factors is zero, _____ of its factors must actually be zero.

6. A polynomial equation is said to be in _____ when one side of the equation is equal to zero.

7. The _____ feature of the graphing calculator can be used to solve equations graphically.

8. Solutions to equations can be checked by using the _____ or _____ features of the calculator.

9. A complete graphical display should show both __ and __ intercepts as well as any peaks and valleys of the graph.

CONCEPTS

10. Simplify $2x^2 - 10x + 3x - 15$.
11. Factor $2x^2 - 10x + 3x - 15$.
12. Evaluate $2x^2 - 10x + 3x - 15$ when x has a value of 6.
13. Solve $2x^2 - 7x - 15 = 0$.
✍14. Check the solutions obtained in Exercise 13. Does this relate to the process of simplifying, evaluating, or factoring? Explain.

15. Classify each of the following statements as an *equation* or an *expression*.
   a. $x^2 - 25$
   b. $x^2 = -3x + 10$
   c. $x^2 - 25 = 0$
   d. $x^2 + 3x - 10$

16. Classify each of the following equations as linear, quadratic, or higher degree.
   a. $3x(x + 4) = 9$
   b. $5x - 6 = 2x - 8(x - 3)$

   c. $x^2(x + 1) = 10 + 20\left(x - \dfrac{1}{2}\right)$

PRACTICE   *In Exercises 17–46, solve each equation by factoring. Verify your results by using either the STOre or TABLE feature of your calculator.*

17. $(x + 5)(x - 2) = 0$
18. $(x - 6)(x + 8) = 0$
19. $15x^2 - 20x = 0$
20. $10x^2 - 2x = 0$
21. $2x^2 - 5x + 2 = 0$
22. $2x^2 + x - 3 = 0$
23. $5x^2 - 6x + 1 = 0$
24. $6x^2 - 5x + 1 = 0$
25. $x(6x + 5) = 6$
26. $x(2x - 3) = 14$
27. $(x + 1)(8x + 1) = 18x$
28. $4x(3x + 2) = x + 12$
29. $6x^3 + x^2 = 12x$
30. $4x^2 - 1 = 0$
31. $9y^2 - 1 = 0$
32. $2x(3x^2 + 10x) = -6x$
33. $2x^3 = 2x(x + 2)$
34. $x^3 + 7x^2 = x^2 - 9x$
35. $x^2(x + 10) = 2x(x - 8)$
36. $\dfrac{1}{2}x^2 = 6 - 2x$
37. $\dfrac{3}{2}x^2 + \dfrac{27}{2}x = -30$
38. $6x^3 - 8x^2 = 8x$
39. $4x^2 = 81$
40. $9y^2 = 64$
41. $3x^2 - \dfrac{11}{2}x = 5$
42. $2x^2 = \dfrac{7}{3}x - \dfrac{2}{3}$
43. $(x - 2)^2 - 3 = -3$
44. $(x + 3)^2 - 4 = 0$
45. $\dfrac{1}{8}x^2 - \dfrac{1}{4}x - 6 = 0$
46. $\dfrac{1}{3}x^2 + x - 6 = 0$

In Exercises 47–54, (a) match each graphical display with its equation; (b) state the solutions of the equation; (c) write the equation in standard form (multiply the factors so the left side is a polynomial).

47. $(x - 2)(x + 3) = 0$

48. $x(x - 5) = 0$

49. $x(x - 3)(x + 3) = 0$

50. $x(x - 2)(x + 2) = 0$

51. $x(x - 3)(x + 2) = 0$

52. $x(x - 2)(x + 3) = 0$

53. $(x + 4)(x - 2) = 0$

54. $(x + 4)(x + 2) = 0$

A.

B.

C.

D.

E.

F.

G.

H.

In Exercises 55–60, solve each equation using the graphing calculator. Unless specified otherwise, the standard viewing window may be used.

55. Solve $x^2 + 3.44x - 0.21 = 0$.

56. Solve $x^2 - 3.88x - 1.806 = 0$.

57. Solve $x^2 + 4.9x - 1.56 = 0$.

58. Solve $x^2 + 4.08x - 2.392 = 0$.

59. Solve $x^3 + 1.2x^2 - 15x + 2.944 = 0$.
(Window: $[-10, 10]$ by $[-15, 35]$)

60. Solve $x^3 + x^2 - 7.79x - 2.4 = 0$.

APPLICATIONS   *In Exercises 61–66, an object has been thrown straight up into the air. The formula $h = vt - 16t^2$ gives the height, h, of the object above the ground after t seconds when it is thrown upward with an initial velocity, v.*

61. Time of flight   After how many seconds will an object hit the ground if it was thrown with a velocity of 144 feet per second?

62. Time of flight   After how many seconds will an object hit the ground if it was thrown with a velocity of 160 feet per second?

63. Ballistics   If a cannonball is fired with an upward velocity of 220 feet per second, at what time s will it be at a height of 600 feet?

64. Ballistics   A cannonball's initial upward velocity is 128 feet per second. At what times will it be 192 feet above the ground?

65. Exhibition diving   A popular tourist attraction consists of swimmers diving from a cliff to the water 64 feet below. A diver's height, h, above the water t seconds after diving is given by $h = -16t^2 + 64$. How long does a dive last?

66. Forensic medicine   The kinetic energy, E, of a moving object is given by $E = \frac{1}{2}mv^2$, where m is the mass of the object (in kilograms) and v is the object's velocity (in meters per second). Kinetic energy is measured in joules. By measuring the damage done to a victim who has been struck by a 3-kilogram club, a police pathologist finds that the energy at impact was 54 joules. Find the velocity of the club at impact.

In Exercises 67–86, write an equation to model each problem. Solve the equation and state the answer to the question posed in the problem with a complete sentence. Make sure you check results in the words of the problem.

67. Integer problem   One positive integer is 2 more than another. Their product is 35. Find the integers.

68. Integer problem   One positive integer is 5 less than 4 times another. Their product is 21. Find the integers.

69. Integer problem   If 4 is added to the square of a composite integer, the result is 5 less than 10 times that integer. Find the integer. (*Hint:* A composite number is one that is *not* prime nor equal to 1.)

70. Number problem   If 3 times the square of a certain natural number is added to the number itself, the result is 14. Find the number.

71. Insulation   The area of the rectangular slab of foam insulation in Illustration 1 is 36 square meters. Find the dimensions of the slab.

ILLUSTRATION 1

72. Shipping pallet   The length of a rectangular shipping pallet is 2 feet less than 3 times its width. Its area is 21 square feet. Find the dimensions of the pallet.

73. Carpentry   A room containing 143 square feet is 2 feet longer than it is wide. How long a crown molding is needed to trim the perimeter of the ceiling?

74. Designing solar panels   The length of a standard rectangular solar heat exchange panel is 2 meters longer than the width. If the length remained the same but the width were doubled, the area would be 48 square meters. Find the perimeter of a standard panel.

75. Designing a tent   The length of the base of the triangular sheet of canvas above the door of the tent shown in Illustration 2 is 2 feet more than twice its height. The area is 30 square feet. Find the height and the length of the base of the triangle.

ILLUSTRATION 2

76. Dimensions of a triangle   The height of a triangle is 2 inches less than 5 times the length of its base. The area is 36 square inches. Find the length of the base and the height of the triangle.

77. Area of a triangle   The base of a triangle is numerically 3 less than its area, and the height is numerically 6 less than its area. Find the area of the triangle.

78. Area of a triangle   The length of the base and the height of a triangle are numerically equal. Their sum is 6 less than the number of units in the area of the triangle. Find the area of the triangle.

79. Dimensions of a parallelogram   The formula for the area of a parallelogram is $A = bh$. The area of the parallelogram in Illustration 3 is 200 square centimeters. If its base is twice its height, how long is the base?

ILLUSTRATION 3

80. Swimming pool border   The owners of the rectangular swimming pool in Illustration 4 want to surround the pool with a crushed-stone border of uniform width. They have enough stone to cover 74 square meters. How wide should they make the border? (*Hint:* The area of the larger rectangle minus the area of the smaller is the area of the border.)

ILLUSTRATION 4

81. House construction   The formula for the area of a trapezoid is $A = \dfrac{h(B + h)}{2}$. The area of the trapezoidal truss in Illustration 5 is 24 square meters. Find the height of the trapezoid if one base is 8 meters and the other base is the same as the height.

ILLUSTRATION 5

82. **Volume**   The volume of a rectangular solid is given by the formula $V = lwh$, where l is the length, w is the width, and h is the height. The volume of the rectangular solid in Illustration 6 is 210 cubic centimeters. Find the width of the rectangular solid if its length is 10 centimeters and its height is 1 centimeter longer than twice its width.

ILLUSTRATION 6

83. **Volume of a pyramid**   The volume of a pyramid is given by the formula $V = \dfrac{Bh}{3}$, where B is the area of its base and h is its height. The volume of the pyramid in Illustration 7 is 192 cubic centimeters. Find the dimensions of its rectangular base if one edge of the base is 2 centimeters longer than the other and the height of the pyramid is 12 centimeters.

84. **Volume of a pyramid**   The volume of a pyramid is 84 cubic centimeters. Its height is 9 centimeters, and one side of its rectangular base is 3 centimeters shorter

ILLUSTRATION 7

than the other. Find the dimensions of its base. (See Exercise 83).

85. **Volume of a solid**   The volume of a rectangular solid is 72 cubic centimeters. Its height is 4 centimeters, and its width is 3 centimeters shorter than its length. Find the sum of its length and width. (See Exercise 82).

86. Here is an easy-sounding word problem: The length of a rectangle is 2 feet greater than the width, and the area is 18 square feet. Find the width of the rectangle.
    a. Write an equation to model the data.

    ✎**b.** Can you solve the given equation? How did you solve it? Explain fully.

In Exercises 87–91, write an equation that has the specified values as solutions. Your equation should contain only integral coefficients.

87. $3, -4$

88. $-2, 5$

89. $-3, 0, \dfrac{7}{3}$

90. $-5, -1, \dfrac{1}{2}$

91. $\dfrac{3}{4}, \dfrac{2}{3}$

REVIEW   *Solve each of the following equations.*

92. $\dfrac{2}{3}(5x - 3) = 38$

93. $2q^2 - 9 = q(q + 3) + q^2$

94. $\dfrac{4x + 1}{15} = \dfrac{3x - 3}{6} - 2$

95. $1 + \dfrac{2}{x - 3} = \dfrac{1}{3}$

96. Graph $y = 2x + 1$.

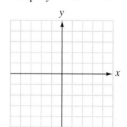

97. Is $(6, -2)$ a solution of $2x + 3y = 6$?

CHAPTER SUMMARY

CONCEPTS

REVIEW EXERCISES

Section 5.1

A natural number is in prime factored form when written as a product of factors that are prime numbers.

Identifying the Greatest Common Factor
1. Determine the prime factorization of each constant. Use exponents rather than repeated factors to write the number in prime factored form.
2. List each prime number that is a factor of all the numbers in the list. When exponents appear, choose the common factor with the *smallest* exponent.
3. Multiply the listed primes from the previous step. This is the greatest common factor, or GCF.

Factoring Out the Greatest Common Factor
1. Determine the GCF of the terms of the polynomial.
2. Rewrite each *term* as the product of the GCF and another factor.
3. Use the distributive property to rewrite the polynomial as the *product* of the GCF and a polynomial.
   When the leading coefficient is negative, a -1 is usually factored out of a polynomial.

Using the Distributive Property to Factor

In Review Exercises 1–20, factor each polynomial completely.

1. $3x + 9y$
2. $5ax^2 + 15a$
3. $7x^2 + 14x$

4. $3x^2 - 3x$
5. $2x^3 + 4x^2 - 8x$
6. $ax + ay + az$

7. $ax + ay - a$
8. $x^2yz + xy^2z$

9. $5a^2 + 6ab^2 + 10acd - 15a$
10. $7axy + 21x^2y - 35x^3y + 7xy^2$

11. $(x + y)a + (x + y)b$
12. $(x + y)^2 + (x + y)$

13. $2x^2(x + 2) + 6x(x + 2)$
14. $3x(y + z) - 9x(y + z)^2$

15. $3p + 9q + ap + 3aq$
16. $ar - 2as + 7r - 14s$

17. $x^2 + ax + bx + ab$
18. $xy + 2x - 2y - 4$

19. $3x^2y - xy^2 - 6xy + 2y^2$
20. $5x^2 + 10x - 15xy - 30y$

Section 5.2

Factoring a Trinomial by Grouping
1. Write the trinomial in decreasing powers of one variable.
2. Identify a, b, and c, where a is the coefficient of the first term $(a > 0)$, b is the coefficient of the middle term, and c is the coefficient of the third term (or the constant term).

Factoring Trinomials

In Review Exercises 21–30, factor each polynomial completely.

21. $x^2 + 10x + 21$
22. $x^2 + 4x - 21$
23. $x^2 + 2x - 24$

24. $x^2 - 4x - 12$
25. $2x^2 - 5x - 3$
26. $3x^2 - 14x - 5$

27. $6x^2 + 7x - 3$
28. $6x^2 + 3x - 3$

29. $6x^3 + 17x^2 - 3x$
30. $4x^3 - 5x^2 - 6x$

3. Find the product, $a \cdot c$.

4. Find two numbers whose **product** is ac and whose **sum** is b.

5. Replace the middle term of the original trinomial with two terms, using the numbers found in step 4 as coefficients. You now have a four-term polynomial.

6. Factor by grouping.

Factoring a Trinomial by Trial and Error

1. Write the trinomial in descending powers of one variable.

2. Factor out any greatest common factor (including -1 if that is necessary to make the coefficient of the first term positive).

3. If the sign of the third term is positive, the signs between the terms of the binomial factors are the same as the sign of the middle term of the trinomial. If the sign of the third term is negative, the signs between the terms of the binomial factors are opposite.

4. Mentally, try various combinations of first terms and last terms until you find one that works, or until you exhaust all the possibilities. In that case, the trinomial does not factor using only integer coefficients.

5. Check the factorization by multiplication.

| **Section 5.3** | **Factoring the Difference of Two Squares and a Summary of Factoring Techniques** |

To factor the difference of two squares, use the pattern
$x^2 - y^2 = (x - y)(x + y)$.

The binomials $(x + y)$ and $(x - y)$ are called *conjugate binomials*.

The sum of two squares is NOT factorable over the set of real numbers.

In Review Exercises 31–40, factor each polynomial completely.

31. $x^2 - 9$

32. $x^2y^2 - 16$

33. $(x + 2)^2 - y^2$

34. $z^2 - (x + y)^2$

35. $6x^2y - 24y^3$

36. $(x + y)^2 - z^2$

37. $64a^3 - 1$

38. $5 - 40y^3$

39. $x^3 + 125$

40. $3x + 81x^4$

To factor the difference of two cubes, use the pattern
$x^3 - y^3 = (x - y)(x^2 + xy + y^2)$.

To factor the sum of two cubes, use the pattern
$x^3 + y^3 = (x + y)(x^2 - xy + y^2)$.

Factoring Polynomials
1. Always check for and factor out the greatest common factor (GCF) first. If the leading coefficient is negative, factor out a -1 with/as the GCF.
2. Count the number of terms in the polynomial, and then follow the given procedures for that specific type of polynomial.

In Review Exercises 41–50, factor each polynomial completely.

41. $24xy^2 - 6xy$

42. $x^3 + 27$

43. $6x^2 + x - 2$

44. $9x^2 - 49$

45. $16x^2 - 25$

46. $2x^2 - 10x + 3x - 15$

47. $4x^2 + 8x - 12$

48. $4x^2 + 12x + 9$

49. $x^3 - 1$

50. $2x^3 - 16$

Section 5.4   Using Factoring to Solve Equations

The degree of a polynomial equation (when simplified) is the same as the highest degree of any of its terms.

A quadratic equation is an equation of degree two that can be written in the form $ax^2 + bx + c = 0$.

Zero factor property If $a \cdot b = 0$ and a and b are real numbers, it follows that $a = 0$ or $b = 0$.

Solve each equation below by factoring. Check roots using the STOre feature of the calculator.

51. $x^2 + 2x = 0$

52. $2x^2 - 6x = 0$

53. $x^2 - 9 = 0$

54. $x^2 - 25 = 0$

55. $a^2 - 7a + 12 = 0$

56. $x^2 - 2x - 15 = 0$

57. $2x - x^2 + 24 = 0$

58. $16 + x^2 - 10x = 0$

59. $2x^2 - 5x - 3 = 0$

60. $2x^2 + x - 3 = 0$

61. $4x^2 = 1$

62. $9x^2 = 4$

63. $x^3 - 7x^2 + 12x = 0$

64. $x^3 + 5x^2 + 6x = 0$

65. $2x^3 + 5x^2 = 3x$

66. $3x^3 - 2x = x^2$

Solving Equations by Factoring
1. Simplify both sides of the equation by removing grouping symbols and combining like terms. Fractions may be eliminated by multiplying each term on both sides of the equation by the LCD.
2. Put the equation in standard form (set one side equal to zero).

Solve each of the following problems. Make sure you clearly identify what the variable represents, write an equation, provide a complete solution to the equation (either algebraically or graphically), and answer the question posed. Be sure to check your solution in the words of the problem.

67. Integer problem   The sum of two integers is 12, and their product is 35. Find the integers.

68. Number problem   If 3 times the square of a positive number is added to 5 times the number, the result is 2. Find the number.

3. Factor the polynomial.
4. Set each factor containing a variable equal to zero.
5. Solve the resulting equation.
6. Check solutions either analytically or with the graphing calculator.

69. Construction   The base of the triangular preformed concrete panel in Illustration 1 is 3 feet longer than twice its height, and its area is 45 square feet. How long is the base?

ILLUSTRATION 1

70. Bombing run   A pilot releases a bomb from an altitude of 3000 feet. The bomb's height, h, above the target t seconds after its release is given by the formula $h = 3000 + 40t - 16t^2$. How long will it be before the bomb hits the target?

71. Gardening   A rectangular flower bed is 3 feet longer than twice its width, and its area is 27 square feet. Find its dimensions.

72. Geometry   A rectangle is 3 feet longer than it is wide. Its area is numerically equal to its perimeter. Find its dimensions.

■ Chapter Test

In Problems 1 and 2, factor out the greatest common factor.

1. $4xy^2 + 12x^2y^3 - 6xz$

2. $3x^2(a + b) - 6xy(a + b)$

In Problems 3–14, factor each expression.

3. $ax + ay + bx + by$

4. $x^2 - 25$

5. $3a^2 - 27b^2$

6. $16x^4 - 81y^4$

7. $x^2 + 4x + 3$

8. $x^2 - 9x - 22$

9. $27 - x^3$

10. $3x^2 + 13x + 4$

11. $2y^2 + 5y - 12$

12. $2x^2 + 3xy - 2y^2$

13. $12 - 25x + 12x^2$

14. $2y^4 - 16y$

In Problems 15–20, solve each equation.

15. $(x - 5)(x + 5) = 0$

16. $2x^2 + 5x + 3 = 0$

17. $10x^2 - 13x = 9$

18. $-3(y - 6) + 2 = y^2 + 2$

19. Cannon fire   A cannonball is fired into the air with a velocity of 192 feet per second. In how many seconds will it hit the ground? (Its height above the ground is given by the formula $h = vt - 16t^2$, where v is the velocity and t is the time.)

20. Base of a triangle   The base of a triangle with an area of 40 square meters is 2 meters longer than its height. Find the base of the triangle.

CAREERS & MATHEMATICS

Computer Programmer

Computers process vast quantities of information rapidly and accurately when they are given programs to follow. Computer programmers write those programs, which logically list the steps the machine must follow to organize data, solve a problem, or do some other task.

Programmers work from descriptions (prepared by systems analysts) of the task that the computer system should perform. The programmer then writes the specific program from the description, by breaking down each step into a series of coded instructions using one of several possible computer languages.

Application programmers are usually oriented toward business, engineering, or science. System programmers maintain the **software** that controls the computer system. Because of their knowledge of operating systems, system programmers often help application programmers determine the source of problems that may occur with their programs.

Qualifications

Most programmers have taken special courses in computer programming or have degrees in computer or information science, mathematics, engineering, or the physical sciences. Graduate degrees are required for some jobs.

Some employers who use computers for business applications do not require college degrees but prefer applicants who have had college courses in data processing.

Employers look for people who can think logically, who are capable of exacting, analytical work, and who demonstrate ingenuity and imagination when solving problems.

Job Outlook

The need for computer programmers will increase as business, government, schools, and scientific organizations develop new applications and require improvements in the software they already use. System programmers will be needed to develop and maintain the complex operating systems required by new, more powerful computer languages and by the networking of desktop computers. Job prospects are excellent into the next century for college graduates who have had computer-related courses—especially for those who have a major in computer science or a related field. Graduates of two-year programs in data-processing technologies also have good prospects, primarily in business applications.

Example Application

The polynomial

$$3x^4 + 2x^3 + 5x^2 + 7x + 1$$

can be written as

$$3 \cdot x \cdot x \cdot x \cdot x + 2 \cdot x \cdot x \cdot x + 5 \cdot x \cdot x + 7 \cdot x + 1$$

to illustrate that it involves ten multiplications and four additions. Multiplications are more time-consuming on a computer than are additions. Rewrite the polynomial to require fewer multiplications.

Solution

Factor the common factor of x from the first four terms of the given polynomial as follows.

$$3x^4 + 2x^3 + 5x^2 + 7x + 1$$
$$x(3x^3 + 2x^2 + 5x + 7) + 1$$

261

Now factor the common x from the first three terms of the polynomial appearing within the parentheses.

$$x[x(3x^2 + 2x + 5) + 7] + 1$$

Again, factor an x from the first two terms within the set of parentheses to get

$$x\{x[x(3x + 2) + 5] + 7\} + 1$$

To emphasize the number of multiplications, we rewrite the previous expression as

$$x \cdot \{x \cdot [x \cdot (3 \cdot x + 2) + 5] + 7\} + 1$$

Each colored dot represents one multiplication. There are now only four multiplications, although there are still four additions. If the polynomial were to be evaluated for many different values of x, the savings in computer time afforded by using the revised form of the polynomial would be substantial.

EXERCISES

1. Evaluate the polynomial in the example at $x = -1$. Do this twice—once by direct substitution into the original polynomial, and then by substitution into the revised form.

2. Write $5x^7 + 3x^4 + 9x + 2$ in revised form.

3. Evaluate the polynomial of Exercise 2 at $x = 1$ by using each method.

4. Evaluate the polynomial of Exercise 2 at $x = -2$ by using each method.

(*Answers:* **1.** 0 **2.** $x \cdot \{x \cdot [x \cdot x \cdot (5 \cdot x \cdot x \cdot x + 3)] + 9\} + 2$ **3.** 19 **4.** -608)

Rational Expressions and Equations

■ **PERSPECTIVE**

Some Complex Fractions in Architecture

Each of the complex fractions in the list

$$1 + \frac{1}{2}, \; 1 + \cfrac{1}{1 + \frac{1}{2}}, \; 1 + \cfrac{1}{1 + \cfrac{1}{1 + \frac{1}{2}}}, \; 1 + \cfrac{1}{1 + \cfrac{1}{1 + \cfrac{1}{1 + \frac{1}{2}}}}, \ldots$$

can be simplified by using the value of the expression preceding it. For example, to simplify the second expression in the list, replace $1 + \frac{1}{2}$ with $\frac{3}{2}$:

$$1 + \cfrac{1}{1 + \frac{1}{2}} = 1 + \cfrac{1}{\frac{3}{2}} = 1 + \frac{2}{3} = \frac{5}{3}$$

To simplify the third expression, replace

$$1 + \cfrac{1}{1 + \frac{1}{2}} \text{ with } \frac{5}{3}:$$

$$1 + \cfrac{1}{1 + \cfrac{1}{1 + \frac{1}{2}}} = 1 + \cfrac{1}{\frac{5}{3}} = 1 + \frac{3}{5} = \frac{8}{5}$$

Can you show that the expressions in the list simplify to the fractions $\frac{3}{2}, \frac{5}{3}, \frac{8}{5}, \frac{13}{8}, \frac{21}{13}, \frac{34}{21}, \ldots$? Do you see a pattern, and can you predict the next fraction?

Use a calculator to write each of these fractions as a decimal. The values produced get closer and closer to the irrational number $1.61803398875\ldots$, which is known as the **golden ratio.** This number often appears in the architecture of the ancient Greeks and Egyptians. The width of the stairs in front of the Greek Parthenon (Illustration 1), divided by the building's height, is the golden ratio. The height of the triangular face of the Great Pyramid of Cheops (Illustration 2), divided by the pyramid's width, is also the golden ratio.

ILLUSTRATION 1

ILLUSTRATION 2

In Chapter 2 we added, subtracted, multiplied, and divided expressions that contained polynomials. This chapter will explore the indicated quotient of two polynomials in more depth. The expression $\dfrac{3x^2 + 6x}{2x^2 + 4}$ is an indicated quotient of two polynomials; however, it is more often known as a rational expression.

6.1 Simplifying Rational Expressions

■ BASIC DEFINITIONS AND EXCLUDED VALUES ■ SIMPLIFYING RATIONAL EXPRESSIONS
■ EQUIVALENT RATIONAL EXPRESSIONS

■ BASIC DEFINITIONS AND EXCLUDED VALUES

Just as the quotient of two integers (where the denominator is not zero) is called a rational number, the quotient of two polynomials is called a **rational expression.** The denominator of this expression cannot equal zero.

Rational Expression

A rational expression is any expression that can be written as $\dfrac{P}{Q}$, where P and Q are polynomials, $Q \neq 0$.

Examples of rational expressions are $\dfrac{3}{x+5}$, $\dfrac{2x^2}{3y}$, $\dfrac{2x^2+3x+4}{x^2-4}$, and $\dfrac{x^2+3}{x^2+1}$. An **excluded value** of a rational expression is the value(s) of the variable that would make the denominator equal 0. When the denominator of a fraction is equal to 0, the fraction is undefined. This is also true of a rational expression. To find the excluded values, simply set the denominator equal to zero and solve the resulting equation for the variable. The solutions are those values that the variable *cannot* take on.

EXAMPLE 1 State the excluded values of **a.** $\dfrac{3}{x+5}$, **b.** $\dfrac{2x^2}{x^2+5x+6}$, **c.** $\dfrac{2x^2+3x+4}{x^2-4}$, and **d.** $\dfrac{x^2+3}{x^2+1}$.

Solution **a.** The fraction $\dfrac{3}{x+5}$ is undefined when its denominator equals zero.

$x + 5 = 0$ Set the denominator equal to zero.
$x = -5$ Solve for the variable.

The number -5 is an excluded value for the expression $\dfrac{3}{x+5}$. Another way to state this is, $x \neq -5$.

b. The fraction $\dfrac{2x^2}{x^2+5x+6}$ is undefined when its denominator equals zero.

$x^2 + 5x + 6 = 0$ Set the denominator equal to zero.
$(x+3)(x+2) = 0$ Because this is a quadratic (second-degree) equation, we factor the polynomial side.
$x+3=0$ or $x+2=0$ Use the zero factor property and set each factor equal to zero.
$x = -3$ or $x = -2$ Solve for the variable.

The numbers -3 and -2 are excluded values for the rational expression $\dfrac{2x^2}{x^2+5x+6}$. Another way to state this is $x \neq -3$ or -2.

c. The fraction $\dfrac{2x^2+3x+4}{x^2-4}$ is undefined when its denominator equals zero.

$x^2 - 4 = 0$ Set the denominator equal to zero.
$(x-2)(x+2) = 0$ Because this is a quadratic (second-degree) equation, factor the polynomial.
$x-2=0$ or $x+2=0$ Use the zero factor property and set each factor equal to zero.
$x = 2$ or $x = -2$ Solve for the variable.

The numbers 2 and -2 are excluded values for the expression $\dfrac{2x^2+3x+4}{x^2-4}$. Another way to state this is $x \neq 2$ or -2.

d. The fraction $\dfrac{x^2+3}{x^2+1}$ is undefined when its denominator is equal to zero.

$x^2 + 1 = 0$ Set the denominator equal to zero.

■ TECHNOLOGY TIP

Enter the expression $\dfrac{3}{x+5}$ at the y1 prompt. Set the TABLE to begin at -5 and then access the TABLE.

X	Y1
-5	ERROR
-4	3
-3	1.5
-2	1
-1	.75
0	.6
1	.5

X= -5

This shows us that when $x=-5$, the expression entered at the y1 prompt $\left(\dfrac{3}{x+5}\right)$ is undefined. This is the significance of the ERROR message that is displayed.

There are *no* excluded values for the given rational expression because squaring a number always yields a result that is either 0 or positive. Adding the constant 1 makes the denominator larger than 0. ■

■ SIMPLIFYING RATIONAL EXPRESSIONS

A rational number is in lowest terms (simplified) when the numerator and denominator have no common *factors* other than 1. The same is true of a rational expression. Recall the fundamental property of fractions from Chapter 1.

> **The Fundamental Property of Fractions**
> If a represents a real number, and b and x represent nonzero real numbers, then
> $$\frac{a \cdot x}{b \cdot x} = \frac{a}{b}.$$

This property also is true for rational expressions. It allows the process of dividing out common factors in the numerator and the denominator of a rational expression (often called canceling by students). It is imperative to remember that only common *factors* can be divided out and that no factors of the denominator are equal to zero.

EXAMPLE 2   Simplify: **a.** $\dfrac{18}{30}$ and   **b.** $\dfrac{21x^2y}{14xy^2}$.

Solution   The simplification of a rational number and a rational expression will be paralleled so that the similarities are apparent.

Begin by writing the numerator and the denominator as a product of factors:

a. $\dfrac{18}{30} = \dfrac{2 \cdot 3 \cdot 3}{2 \cdot 3 \cdot 5}$ **b.** $\dfrac{21x^2y}{14xy^2} = \dfrac{3 \cdot 7 \cdot x \cdot x \cdot y}{2 \cdot 7 \cdot x \cdot y \cdot y}$

Divide out common factors (apply the fundamental property of fractions):

$$= \frac{\overset{1}{\cancel{2}} \cdot \overset{1}{\cancel{3}} \cdot 3}{\underset{1}{\cancel{2}} \cdot \underset{1}{\cancel{3}} \cdot 5} \qquad\qquad = \frac{3 \cdot \overset{1}{\cancel{7}} \cdot \overset{1}{\cancel{x}} \cdot x \cdot \overset{1}{\cancel{y}}}{2 \cdot \underset{1}{\cancel{7}} \cdot \underset{1}{\cancel{x}} \cdot y \cdot \underset{1}{\cancel{y}}}$$

Express the numerator and the denominator as the product of the resulting factors:

$$= \frac{3}{5} \qquad\qquad = \frac{3x}{2y}$$

Because the numerator and denominator have no common factors (other than 1) the fractions are simplified.

Just as the fraction $\dfrac{18}{30}$ is equivalent to $\dfrac{3}{5}$, the rational expression $\dfrac{21x^2y}{14xy^2}$ is equivalent to $\dfrac{3x}{2y}$ for all values of x and y except 0. The number zero is an excluded value for both variables x and y.

The rational expression could also have been simplified by using the rules of exponents:

$$\frac{21x^2y}{14xy^2} = \frac{3 \cdot 7}{2 \cdot 7} \cdot x^{2-1} \cdot y^{1-2}$$

$$= \frac{3}{2}xy^{-1}$$

$$= \frac{3}{2} \cdot \frac{x}{1} \cdot \frac{1}{y} \qquad\qquad \text{Recall } y^{-1} = \tfrac{1}{y}.$$

$$= \frac{3x}{2y}$$

■

The following steps can be used when simplifying rational expressions:

Simplifying a Rational Expression
1. Factor the numerator and the denominator completely.

2. Divide out common factors.

EXAMPLE 3   Simplify the rational expression $\dfrac{x^2 + 3x}{3x + 9}$.

Solution $\dfrac{x(x + 3)}{3(x + 3)}$ Factor the numerator and the denominator.

$= \dfrac{x}{3}$ Divide out the common factor of $(x + 3)$.

The rational expression $\dfrac{x^2 + 3x}{3x + 9}$ is equivalent to $\dfrac{x}{3}$ for all values of x *except* -3. Notice that the excluded value of -3 must be found by setting the denominator of the *original* expression equal to zero rather the denominator of the reduced expression. ■

From now on we will assume that our simplified rational expression is equivalent to the original rational expression for all values of the variable that *do not* make a denominator equal to zero.

EXAMPLE 4   Simplify: **a.** $\dfrac{x - y}{y - x}$ and   **b.** $\dfrac{2a - 1}{1 - 2a}$.

Solution   We can rearrange terms in each numerator, factor out -1, and proceed as follows:

a. $\dfrac{x - y}{y - x} = \dfrac{-y + x}{y - x}$ 　　　**b.** $\dfrac{2a - 1}{1 - 2a} = \dfrac{-1 + 2a}{1 - 2a}$

$= \dfrac{-(y - x)}{y - x}$ 　　　　　　　$= \dfrac{-(1 - 2a)}{1 - 2a}$

$= \dfrac{-\overset{1}{\cancel{(y - x)}}}{\underset{1}{\cancel{y - x}}}$ 　　　　　　$= \dfrac{-\overset{1}{\cancel{(1 - 2a)}}}{\underset{1}{\cancel{1 - 2a}}}$

$= -1$ 　　　　　　　　　$= -1$ 　　■

The results of Example 4 illustrate this important fact.

Division of Negatives
The quotient of any nonzero expression and its negative is -1.

WARNING! It is important to remember that only *factors* that are common to the *entire numerator* and the *entire denominator* can be divided out. Consider the arithmetic fraction $\dfrac{5 + 8}{5}$. It simplifies to $\dfrac{13}{5}$. It would be incorrect to simplify it by dividing out the common *term* of 5.

EXAMPLE 5 Simplify: $\dfrac{5(x + 3) - 5}{7(x + 3) - 7}$.

Solution Resist the urge to divide out the binomial $(x + 3)$. It is *not* a common factor of the *entire numerator* and the *entire denominator*. Instead, we simplify both the numerator and denominator, factor, and divide out any common factors.

$$\frac{5(x + 3) - 5}{7(x + 3) - 7} = \frac{5x + 15 - 5}{7x + 21 - 7} \qquad \text{Remove parentheses.}$$

$$= \frac{5x + 10}{7x + 14} \qquad \text{Combine like terms.}$$

$$= \frac{5(x + 2)}{7(x + 2)} \qquad \text{Factor the numerator and denominator.}$$

$$= \frac{5\overset{1}{\cancel{(x + 2)}}}{7\underset{1}{\cancel{(x + 2)}}} \qquad \text{Divide out the common factor of } x + 2.$$

$$= \frac{5}{7}$$

EXAMPLE 6 Simplify: $\dfrac{xy + 2x + 3y + 6}{x^2 + x - 6}$.

Solution We can factor the numerator by grouping and factor the denominator as a general trinomial. Then we proceed as follows:

$$\frac{xy + 2x + 3y + 6}{x^2 + x - 6} = \frac{x(y + 2) + 3(y + 2)}{(x - 2)(x + 3)} \qquad \begin{array}{l}\text{Begin to factor the numerator}\\\text{and factor the denominator.}\end{array}$$

$$= \frac{(y + 2)(x + 3)}{(x - 2)(x + 3)} \qquad \text{Factor the numerator.}$$

$$= \frac{(y + 2)\overset{1}{\cancel{(x + 3)}}}{(x - 2)\underset{1}{\cancel{(x + 3)}}} \qquad \begin{array}{l}\text{Divide out the common factor}\\\text{of } x + 3.\end{array}$$

$$= \frac{y + 2}{x - 2}$$

Do not assume every rational expression will simplify. Some are already in lowest terms.

EXAMPLE 7 Simplify: $\dfrac{x^2 + x - 2}{x^2 + x}$.

Solution $\dfrac{x^2 + x - 2}{x^2 + x} = \dfrac{(x + 2)(x - 1)}{x(x + 1)} \qquad \text{Factor the numerator and the denominator.}$

Because there are no factors common to the numerator and denominator, this fraction is already simplified.

The following property holds for both rational numbers (fractions) and for rational expressions:

Properties of Fractions

$$\frac{a}{b} = \frac{-a}{-b} = -\frac{a}{-b} = -\frac{-a}{b}, \qquad (b \neq 0)$$

and

$$-\frac{a}{b} = \frac{-a}{b} = \frac{a}{-b} = -\frac{-a}{-b}, \qquad (b \neq 0)$$

The second property stated above is particularly useful when a result involves a negative. That negative can be placed with the numerator, with the denominator, or in front of the entire rational expression. The rational expression below can be written in several different ways, all of which are equivalent:

$$-\frac{3x+4}{2x-1} = \frac{-(3x+4)}{2x-1} = \frac{-3x-4}{2x-1} = \frac{3x+4}{-(2x-1)} = \frac{3x+4}{-2x+1} = \frac{3x+4}{1-2x}$$

■ EQUIVALENT RATIONAL EXPRESSIONS

When adding and subtracting fractions, it is often necessary to rewrite a given fraction as an equivalent fraction with a different denominator. The fundamental property of fractions allows us to divide out common *factors* AND to multiply the numerator and denominator of a given fraction by the same quantity. This is equivalent to multiplying by the number 1, which is the identity for multiplication.

EXAMPLE 8   **a.** Rewrite the fraction $\dfrac{2}{3}$ with a denominator of 6.

b. Rewrite the fraction $\dfrac{3}{x-2}$ with a denominator of $x^2 - 4$.

Solution   **a.** $\dfrac{2}{3} = \dfrac{?}{6}$ **b.** $\dfrac{3}{x-2} = \dfrac{?}{x^2-4}$

Begin by factoring denominators:

$$\frac{2}{3} = \frac{?}{2 \cdot 3} \qquad\qquad \frac{3}{x-2} = \frac{?}{(x-2)(x+2)}$$

Comparing denominators (old to new), we see that in part (a) the new denominator has the extra factor of 3, and in part (b) the new denominator has the extra factor of $(x + 2)$. We then multiply the original fraction by a fraction that has the appropriate "missing factor" as its numerator and denominator.

$$\frac{2}{3} \cdot \frac{2}{2} = \frac{4}{6} \qquad \frac{3}{x-2} \cdot \frac{x+2}{x+2} = \frac{3(x+2)}{(x-2)(x+2)} = \frac{3x+6}{(x-2)(x+2)} \qquad ■$$

For now, simplify numerators but leave denominators in factored form. The reason for this will become apparent when we add and subtract fractions in the next section.

EXAMPLE 9   Rewrite the fraction $\dfrac{4x + 1}{2x + 3}$ with a denominator of $2x^2 - 7x - 15$.

Solution

$$\frac{4x + 1}{2x + 3} = \frac{?}{2x^2 - 7x - 15}$$

■ STUDY TIP

As you work through this chapter you will realize that there is very little new information contained within each section. You are simply being asked to put concepts that you have previously learned together and extend your knowledge of rational numbers to rational expressions. Many students find it helpful to use unlined paper when working problems that involve rational expressions.

$$\frac{4x + 1}{(2x + 3)} = \frac{?}{(2x + 3)(x - 5)}$$

Factor denominators so that a comparison of factors can be made between the original denominator and the desired denominator. Notice that the original denominator is a prime polynomial. It is enclosed in parentheses so that it is treated as a single factor.

The missing factor is the quantity $(x - 5)$.

$$\frac{4x + 1}{2x + 3} \cdot \frac{x - 5}{x - 5}$$

Multiply the numerator and denominator of the original rational expression by the missing factor, $(x - 5)$.

$$= \frac{(4x + 1)(x - 5)}{(2x + 3)(x - 5)}$$

$$= \frac{4x^2 - 20x + x - 5}{(2x + 3)(x - 5)}$$

Simplify the numerator.

$$= \frac{4x^2 - 19x - 5}{(2x + 3)(x - 5)}$$

Therefore, $\dfrac{4x + 1}{2x + 3} = \dfrac{4x^2 - 19x - 5}{(2x + 3)(x - 5)}$.

EXERCISE 6.1

VOCABULARY   *In Exercises 1–2, fill in the blanks to make a true statement.*

1. The quotient of two polynomials is called a _____ expression.

2. The _____ is the value(s) of the variable that will make the denominator equal to 0.

CONCEPTS

3. State the excluded values of each rational expression.

a. $\dfrac{3 + x}{x}$ **b.** $\dfrac{5 + x}{2x}$

c. $\dfrac{x + 2}{x - 5}$ **d.** $\dfrac{2(x + 4)}{x + 6}$

✏️**4.** One by one, enter each of the rational expression in Exercises 3(a) and 3(d) at the y1 prompt and record the y1 value displayed in the TABLE for each restricted value. Explain why the "values" are the same and their meaning.

5. For each TABLE below, a rational expression has been entered at the y1 prompt. Based on the TABLE, determine the excluded value(s) for each rational expression.

a.

b.

X	Y₁
-3	-.5
-2	ERROR
-1	.25
0	0
1	-.1667
2	-.5
3	ERROR

X=-3

c.

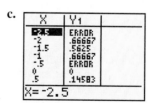

6. Simplify the following. Assume that no variable has a value that would make the denominator 0.

a. $\dfrac{x + y}{y + x}$ **b.** $\dfrac{x - y}{y - x}$

c. $\dfrac{a - 5}{5 - a}$ **d.** $\dfrac{a + 5}{5 + a}$

7. Simplify each rational expression. Assume that no variable has a value that would make the denominator 0.

a. $\dfrac{8}{52}$

b. $\dfrac{15}{21}$

c. $\dfrac{-6x}{18}$

d. $\dfrac{-25y}{5}$

e. $\dfrac{x+3}{3(x+3)}$

f. $\dfrac{2(x+7)}{x+7}$

8. Which of the following expressions is/are equivalent to $-\dfrac{x-y}{x+y}$?

a. $\dfrac{-x-y}{x+y}$

b. $\dfrac{y-x}{x+y}$

c. $\dfrac{x-y}{-x+y}$

d. $\dfrac{-x+y}{-x-y}$

9. Which of the following expressions is/are equivalent to $-\dfrac{x+y}{x-y}$?

a. $\dfrac{-(x+y)}{-(x-y)}$

b. $\dfrac{-x-y}{x-y}$

c. 1

d. $\dfrac{-x-y}{x+y}$

10. Rewrite each fraction as an equivalent fraction with the given denominator.

a. $\dfrac{25}{4}$; 20

b. $\dfrac{19}{21}$; 42

c. $\dfrac{8}{x}$; x^2y

d. $\dfrac{7}{y}$; xy^2

PRACTICE *In Exercises 11–14, state the excluded value(s) of each rational expression (if any exist).*

11. $\dfrac{x^2+9}{x^2+5}$

12. $\dfrac{x-4}{x^2+4}$

13. $\dfrac{x-3}{x^2-2x-24}$

14. $\dfrac{x+5}{x^2-x-30}$

In Exercises 15–48, simplify each rational expression. If a fraction is already simplified, so state. Assume that no variable has a value that would make a denominator equal to zero.

15. $\dfrac{15x^2y^2}{5xy^2}$

16. $\dfrac{12xz^3}{4xz^2}$

17. $\dfrac{5x+35}{x+7}$

18. $\dfrac{x-9}{3x-27}$

19. $\dfrac{x^2+3x}{2x+6}$

20. $\dfrac{xz-2x}{yz-2y}$

21. $\dfrac{15x-3x^2}{25y-5xy}$

22. $\dfrac{3y+xy}{3x+xy}$

23. $\dfrac{x-7}{7-x}$

24. $\dfrac{d-c}{c-d}$

25. $\dfrac{6x-3y}{3y-6x}$

26. $\dfrac{3c-4d}{4d-3c}$

27. $\dfrac{a+b-c}{c-a-b}$

28. $\dfrac{x-y-z}{z+y-x}$

29. $\dfrac{x^2+3x+2}{x^2+x-2}$

30. $\dfrac{x^2+x-6}{x^2-x-2}$

31. $\dfrac{x^2-8x+15}{x^2-x-6}$

32. $\dfrac{x^2-6x-7}{x^2+8x+7}$

33. $\dfrac{2x^2-8x}{x^2-6x+8}$

34. $\dfrac{3y^2-15y}{y^2-3y-10}$

35. $\dfrac{xy+2x^2}{2xy+y^2}$

36. $\dfrac{3x+3y}{x^2+xy}$

37. $\dfrac{x^2+3x+2}{x^3+x^2}$

38. $\dfrac{6x^2-13x+6}{3x^2+x-2}$

39. $\dfrac{x^2-8x+16}{x^2-16}$

40. $\dfrac{3x+15}{x^2-25}$

41. $\dfrac{x^2-2x-15}{x^2+2x-15}$

42. $\dfrac{x^2+4x-77}{x^2-4x-21}$

43. $\dfrac{x^2-3(2x-3)}{9-x^2}$

44. $\dfrac{x(x-8)+16}{16-x^2}$

45. $\dfrac{4(x+3)+4}{3(x+2)+6}$

46. $\dfrac{4+2(x-5)}{3x-5(x-2)}$

47. $\dfrac{x^2-9}{(2x+3)-(x+6)}$

48. $\dfrac{x^2+5x+4}{2(x+3)-(x+2)}$

In Exercises 49–54, (a) Using the STOre feature of your calculator, evaluate the given rational expressions when $x = 2$, $y = -3$, and $z = 5$. Copy the screen displays exactly as they appear. (b) Algebraically simplify the given expression. (c) Evaluate the reduced form of the rational expression for the given values of the variable. The results from parts (a) and (b) should be equal.

49. $\dfrac{28x}{32y}$

50. $\dfrac{14xz^2}{7x^2z^2}$

51. $\dfrac{6x-6y+6z}{9x-9y+9z}$

52. $\dfrac{3x-3y-6}{2x-2y-4}$

53. $\dfrac{2x^2-8}{x^2-3x+2}$

54. $\dfrac{3x^2-27}{x^2+3x-18}$

In Exercises 55–64, rewrite each fraction as an equivalent fraction with the given denominator.

55. $\dfrac{3x}{x+1}$; $(x+1)^2$

56. $\dfrac{5y}{y-2}$; $(y-2)^2$

57. $\dfrac{2y}{x}$; $x^2 + x$

58. $\dfrac{3x}{y}$; $y^2 - y$

59. $\dfrac{z}{z-1}$; $z^2 - 1$

60. $\dfrac{y}{y+2}$; $y^2 - 4$

61. $\dfrac{x+2}{x-2}$; $x^2 - 4$

62. $\dfrac{x-3}{x+3}$; $x^2 - 9$

63. $\dfrac{2}{x+1}$; $x^2 + 3x + 2$

64. $\dfrac{3}{x-1}$; $x^2 + x - 2$

65. Graph the rational expression $\dfrac{x+3}{x-3}$ on your graphing calculator.
  a. Sketch the graph (in the standard viewing window).
  b. Reset your calculator in dot mode and graph the same expression. Sketch the graph.
  ✍ **c.** Explain (reconcile) the differences between the two graphs. (*Hint:* Think about the excluded values of the rational expression.)

66. Graph the rational expression $\dfrac{x-4}{x+4}$ on your graphing calculator.
  a. Sketch the graph (in the standard viewing window).
  b. Reset your calculator in dot mode and graph the same expression. Sketch the graph.
  ✍ **c.** Explain (reconcile) the differences between the two graphs. (*Hint:* Think about the excluded values of the rational expression.)

67. a. Simplify the rational expression
  $\dfrac{x(x+3) - 3(x-1)}{x^2 + 3}$ algebraically.
  ✍ **b.** Enter the rational expression at the y1 prompt on your calculator; press graph. Reconcile the graph with the results of your simplification in part (a).

  ✍ **c.** Access the TABLE. Reconcile what you see with your simplification in part (a).

68. a. Simplify the rational expression
  $\dfrac{x(x-5) + 2(x-5)}{x^2 - 3x - 10}$ algebraically.
  ✍ **b.** Enter the rational expression at the y1 prompt on your calculator; press graph. Reconcile the graph with the results of your simplification in part (a).

  ✍ **c.** Access the TABLE. Reconcile what you see with your simplification in part (a).

REVIEW

69. Factor completely:
  a. $4x^3 - 10x^2 + 6x$
  b. $y^2 + 2xy - 63x^2$
  c. $15 + 3b - 5a^2 - a^2 b$

70. The perimeter of a triangle is 18 cm. If the third side is 3 cm less than the first side and the second side is 5 cm more than twice the first side, find the measure of the first side.

71. Graph $y = \dfrac{2}{3}x - 5$ by plotting the x- and y-intercepts.

72. Graph $y = \dfrac{2}{3}x - 5$ in the standard viewing window and sketch the graph.

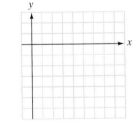

73. Solve $\dfrac{1}{2}(x-1) + 2 = 3 - 4x$ graphically. Copy your screen display to justify your work.

6.2 Adding and Subtracting Rational Expressions

■ ADDING AND SUBTRACTING RATIONAL EXPRESSIONS WITH LIKE DENOMINATORS ■ FINDING THE LCD ■ ADDING AND SUBTRACTING RATIONAL EXPRESSIONS WITH UNLIKE DENOMINATORS

■ ADDING AND SUBTRACTING RATIONAL EXPRESSIONS WITH LIKE DENOMINATORS

Rational expressions can be added and subtracted in exactly the same manner in which rational numbers are added or subtracted. As long as the denominators are the same, keep the common denominator, add (or subtract) the numerators, and then simplify the rational expression.

> ### Adding or Subtracting Rational Expressions with Common Denominators
>
> If $\dfrac{P}{Q}$ and $\dfrac{R}{Q}$ are rational expressions, then $\dfrac{P}{Q} + \dfrac{R}{Q} = \dfrac{P + R}{Q}$ and $\dfrac{P}{Q} - \dfrac{R}{Q} = \dfrac{P - R}{Q}$.

EXAMPLE 1   Add or subtract, as the signs indicate. Simplify results, if possible.

a. $\dfrac{x}{7} + \dfrac{3x}{7}$ **b.** $\dfrac{5x}{3} - \dfrac{2x}{3}$ **c.** $\dfrac{3x + y}{5x} + \dfrac{x + y}{5x}$

Solution   **a.** $\dfrac{x}{7} + \dfrac{3x}{7} = \dfrac{x + 3x}{7}$ **b.** $\dfrac{5x}{3} - \dfrac{2x}{3} = \dfrac{5x - 2x}{3}$

$= \dfrac{4x}{7}$ $= \dfrac{3x}{3}$ Combine like terms.

$= \dfrac{x}{1}$ Simplify.

$= x$ Denominators of 1 need not be written.

c. $\dfrac{3x + y}{5x} + \dfrac{x + y}{5x} = \dfrac{3x + y + x + y}{5x}$

$= \dfrac{4x + 2y}{5x}$ Combine like terms.

Factoring the numerator, $\dfrac{2(2x + y)}{5x}$, shows that the rational expression does not simplify. The final result could be left in factored form, if desired.   ■

> **WARNING!**   When subtracting fractions with polynomial numerators, care must be taken to use the distributive property to subtract the entire numerator rather than just a single term from the numerator.

EXAMPLE 2   Subtract and simplify: $\dfrac{5x + 1}{x - 3} - \dfrac{4x - 2}{x - 3}$.

Solution $\dfrac{5x + 1}{x - 3} - \dfrac{4x - 2}{x - 3} = \dfrac{(5x + 1) - (4x - 2)}{x - 3}$ Express the subtraction of the numerators over the common denominator of $(x - 3)$.

$= \dfrac{5x + 1 - 4x + 2}{x - 3}$ Use the distributive property to remove grouping in the numerator.

$= \dfrac{x + 3}{x - 3}$ Combine like terms.   ■

EXAMPLE 3   Perform the indicated operations: $\dfrac{3xy}{x-y} - \dfrac{x(3y-x)}{x-y} - \dfrac{x(x-y)}{x-y}$.

Solution $\dfrac{3xy}{x-y} - \dfrac{x(3y-x)}{x-y} - \dfrac{x(x-y)}{x-y} = \dfrac{3xy - x(3y-x) - x(x-y)}{x-y}$

Express the subtraction of the numerators over the common denominator.

$= \dfrac{3xy - 3xy + x^2 - x^2 + xy}{x-y}$ Use the distributive property to remove parentheses.

$= \dfrac{xy}{x-y}$ Combine like terms.   ■

■ FINDING THE LCD

To add rational numbers, they must first be converted to fractions with equivalent denominators (if they do not have like denominators originally). The same is true for adding and subtracting rational expressions. To find the least (lowest) common denominator, use the following steps:

> **Finding the Least Common Denominator (LCD)**
> **1.** Write down each of the different denominators that appear in the given fractions.
> **2.** Factor each of these denominators completely.
> **3.** Form a product using each of the different factors obtained in step 2. Use each different factor the *greatest* number of times it appears in any *single* factorization. This product is the least common denominator.

EXAMPLE 4   Find the LCD of the given pairs of rational expressions:

a. $\dfrac{5}{4x^4y^2}; \dfrac{3}{6x^3y^3}$ **b.** $\dfrac{3x}{x+2}; \dfrac{5x}{x^2-4}$

c. $\dfrac{x-2}{x^2+2x-15}; \dfrac{x+2}{x^2-6x+9}$

Solution   **a.** $4x^4y^2; 6x^3y^3$ Write down each denominator.

$2^2 \cdot x^4 \cdot y^2$ Factor each denominator.

$2 \cdot 3 \cdot x^3 \cdot y^3$

$\text{LCD} = 2^2 \cdot 3 \cdot x^4 \cdot y^3$ The LCD is the product of the different factors. The highest power of any common factor is chosen.

$= 12x^4y^3$

b. $x+2; x^2-4$ Write down each denominator.

$(x+2)(x-2)$ Factor each denominator completely.

$\text{LCD} = (x+2)(x-2)$ The LCD is the product of the different factors.

c. $x^2+2x-15; x^2-6x+9$ Write down each denominator.

$(x+5)(x-3)$ Factor each denominator completely.

$(x-3)(x-3) = (x-3)^2$

$\text{LCD} = (x+5)(x-3)^2$ The LCD is the product of the different factors. The highest power of any common factor is chosen.

■

■ ADDING AND SUBTRACTING RATIONAL EXPRESSIONS WITH UNLIKE DENOMINATORS

To add or subtract rational expressions with unlike denominators, use the following five-step process:

Adding/Subtracting with Unlike Denominators

1. Identify the LCD of the rational expressions. Use the process outlined earlier in this section.
2. Convert each rational expression to an equivalent rational expression having the LCD as denominator. This process was demonstrated in the previous section (Section 6.1).
3. Add or subtract the rational expressions by *expressing* the addition or subtraction over the common denominator.
4. Simplify the numerators, but leave the denominators in factored form.
5. Simplify the rational expression. Recall that this requires you to divide out common factors in the numerator and the denominator. Because the denominator is in factored form, you need only factor the numerator.

EXAMPLE 5   Add or subtract as the signs indicate: $\dfrac{3}{x^2y} + \dfrac{2}{xy} - \dfrac{1}{xy^2}$.

Solution   Find the least common denominator.

$$\left.\begin{array}{l} x^2y = x \cdot y \cdot y \\ xy = x \cdot y \\ xy^2 = x \cdot y \cdot y \end{array}\right\} \quad \text{Factor each denominator.}$$

In any one of these denominators, the factor x occurs at most twice, and the factor y occurs at most twice. Thus, the LCD is

$$\begin{aligned} \text{LCD} &= x \cdot x \cdot y \cdot y \\ &= x^2y^2 \end{aligned}$$

We build each given fraction into a new fraction with a denominator of x^2y^2.

$$\begin{aligned} &\frac{3}{x^2y} + \frac{2}{xy} - \frac{1}{xy^2} \\ &= \frac{3 \cdot y}{x \cdot x \cdot y \cdot y} + \frac{2 \cdot x \cdot y}{x \cdot y \cdot x \cdot y} - \frac{1 \cdot x}{x \cdot y \cdot y \cdot x} \qquad \text{Factor each denominator and build each fraction.} \\ &= \frac{3y + 2xy - x}{x^2y^2} \qquad \text{Do the multiplications and add the fractions.} \end{aligned}$$

EXAMPLE 6   Subtract: $\dfrac{x}{x+1} - \dfrac{3}{x}$.

Solution   The least common denominator is $(x+1)x$.

$$\begin{aligned} \frac{x}{x+1} - \frac{3}{x} &= \frac{x(x)}{(x+1)x} - \frac{3(x+1)}{x(x+1)} \qquad \text{Build the fractions to get the common denominator.} \\ &= \frac{x^2}{(x+1)x} - \frac{3x+3}{x(x+1)} \qquad \text{Do the multiplication in the numerator.} \\ &= \frac{x^2 - 3x - 3}{x(x+1)} \qquad \text{Subtract the numerators and keep the common denominator.} \end{aligned}$$

EXAMPLE 7 Do the indicated operations: $\dfrac{3}{x^2 - y^2} + \dfrac{2}{x - y} - \dfrac{1}{x + y}$.

Solution Find the least common denominator.

$$
\left.\begin{array}{l}
x^2 - y^2 = (x - y)(x + y) \\
x - y = x - y \\
x + y = x + y
\end{array}\right\} \text{Factor each denominator where possible.}
$$

Because the least common denominator is $(x - y)(x + y)$, we build each given fraction into a new fraction with that common denominator.

$$
\begin{aligned}
\frac{3}{x^2 - y^2} + \frac{2}{x - y} - \frac{1}{x + y} &= \frac{3}{(x - y)(x + y)} + \frac{2}{x - y} - \frac{1}{x + y} && \text{Factor } x^2 - y^2. \\
&= \frac{3}{(x - y)(x + y)} + \frac{2(x + y)}{(x - y)(x + y)} - \frac{1(x - y)}{(x + y)(x - y)} && \text{Build the fractions to get a common denominator.} \\
&= \frac{3}{(x - y)(x + y)} + \frac{2x + 2y}{(x - y)(x + y)} - \frac{x - y}{(x + y)(x - y)} && \text{Remove parentheses.} \\
&= \frac{3 + 2x + 2y - x + y}{(x - y)(x + y)} && \text{Add the fractions.} \\
&= \frac{3 + x + 3y}{(x - y)(x + y)} && \text{Combine like terms.} \quad\blacksquare
\end{aligned}
$$

EXAMPLE 8 Do the subtraction, $\dfrac{a}{a - 1} - \dfrac{2}{a^2 - 1}$, and simplify.

Solution Factor $a^2 - 1$ and write each fraction as a fraction with a denominator of $(a + 1)(a - 1)$.

$$
\begin{aligned}
\frac{a}{a - 1} - \frac{2}{a^2 - 1} &= \frac{a(a + 1)}{(a - 1)(a + 1)} - \frac{2}{(a + 1)(a - 1)} && \text{Factor } a^2 - 1 \text{ and build the first fraction.} \\
&= \frac{a^2 + a}{(a - 1)(a + 1)} - \frac{2}{(a + 1)(a - 1)} && \text{Remove parentheses in the numerator.} \\
&= \frac{a^2 + a - 2}{(a - 1)(a + 1)} && \text{Subtract the fractions.} \\
&= \frac{(a + 2)\overset{1}{\cancel{(a - 1)}}}{\underset{1}{\cancel{(a - 1)}}(a + 1)} && \text{Factor.} \\
&= \frac{a + 2}{a + 1} && \text{Divide out the common factor of } a - 1. \quad\blacksquare
\end{aligned}
$$

EXAMPLE 9 Do the addition: $\dfrac{m + 1}{2m + 6} + \dfrac{4 - m^2}{2m^2 + 2m - 12}$.

Solution

Factor $2m + 6$ and $2m^2 + 2m - 12$ and write each fraction with a denominator of $2(m + 3)(m - 2)$:

$$\frac{m+1}{2m+6} + \frac{4-m^2}{2m^2+2m-12} = \frac{m+1}{2(m+3)} + \frac{4-m^2}{2(m+3)(m-2)}$$
$$= \frac{(m+1)(m-2)}{2(m+3)(m-2)} + \frac{4-m^2}{2(m+3)(m-2)}$$

Now add the fractions by adding the numerators and keeping the common denominator, and simplify.

$$\frac{m+1}{2m+6} + \frac{4-m^2}{2m^2+2m-12} = \frac{(m+1)(m-2)+4-m^2}{2(m+3)(m-2)}$$
$$= \frac{m^2-m-2+4-m^2}{2(m+3)(m-2)} \quad \text{Multiply } (m+1)(m-2).$$
$$= \frac{-m+2}{2(m+3)(m-2)} \quad \text{Combine like terms.}$$
$$= \frac{-(m-2)}{2(m+3)(m-2)} \quad \text{Factor } -1 \text{ from } -m+2.$$
$$= \frac{-1}{2(m+3)} \quad \text{Divide out the common factor of } m-2.$$

EXERCISE 6.2

CONCEPTS *In Exercises 1–6, add or subtract the given rational expressions as the sign indicates. All results should be expressed as simplified rational expressions.*

1. $\frac{8}{7} + \frac{6}{7}$
2. $\frac{9}{11} + \frac{2}{11}$
3. $\frac{9y}{3x} - \frac{6y}{3x}$
4. $\frac{24}{7y} - \frac{10}{7y}$
5. $\frac{15z}{22} - \frac{-15z}{22}$
6. $\frac{-30rs}{21} + \frac{30rs}{21}$

In Exercises 7–10, find the least common denominator for the pair of fractions.

7. $\frac{2}{21}; \frac{5}{42}$
8. $\frac{4}{15}; \frac{3}{10}$
9. $\frac{4}{3x}; \frac{5}{9xy}$
10. $\frac{3}{x^2}; \frac{8}{3xy}$

In Exercises 11–16, simplify the sum or difference of the rational expressions. Results should be expressed in simplified form.

11. $\frac{1}{2} + \frac{2}{3}$
12. $\frac{2}{3} - \frac{5}{6}$
13. $\frac{2y}{9} + \frac{y}{3}$
14. $\frac{8a}{15} - \frac{5a}{12}$
15. $\frac{21x}{14} - \frac{5x}{21}$
16. $\frac{4x}{3} + \frac{2x}{y}$

PRACTICE *In Exercises 17–28, add or subtract the given rational expressions as the sign indicates. All results should be expressed as simplified rational expressions.*

17. $\frac{y+2}{2z} - \frac{y-4}{2z}$
18. $\frac{2x-3}{x^2} - \frac{x-3}{x^2}$
19. $\frac{y+2}{5z} + \frac{y+4}{5z}$
20. $\frac{x+3}{x^2} + \frac{x+5}{x^2}$
21. $\frac{3x-5}{x-2} + \frac{6x-13}{x-2}$
22. $\frac{8x-7}{x+3} + \frac{2x+37}{x+3}$

23. $\dfrac{3y-2}{y+3} - \dfrac{2y-5}{y+3}$

24. $\dfrac{5x+8}{x+2} - \dfrac{3x-2}{x+2}$

25. $\dfrac{3x}{y+2} - \dfrac{3y}{y+2} + \dfrac{x+y}{y+2}$

26. $\dfrac{3y}{x-5} - \dfrac{x}{x-5} - \dfrac{y-x}{x-5}$

27. $\dfrac{x+1}{x-2} - \dfrac{2(x-3)}{x-2} + \dfrac{3(x+1)}{x-2}$

28. $\dfrac{x^2-4}{x+2} + \dfrac{2(x^2-4)}{x+2} - \dfrac{3(x^2-5)}{x+2}$

In Exercises 29–36, find the LCD for the given pairs of fractions.

29. $\dfrac{x+3}{x^2-1}; \dfrac{x-2}{x+1}$

30. $\dfrac{y-2}{y^2-9}; \dfrac{y+3}{y-3}$

31. $\dfrac{6}{x^2+6x}; \dfrac{9}{x+6}$

32. $\dfrac{x+4}{xy^2-xy}; \dfrac{x-2}{xy}$

33. $\dfrac{x+5}{x^2-x-2}; \dfrac{x+6}{(x-2)^2}$

34. $\dfrac{x+4}{x^2+2x-3}; \dfrac{x+5}{(x+3)^2}$

35. $\dfrac{x+2}{x^2-4x-5}; \dfrac{x+3}{x^2-25}$

36. $\dfrac{x+4}{x^2-x-6}; \dfrac{x+2}{x^2-9}$

In Exercises 37–52, simplify the sum and/or difference of the given rational expressions. All results should be expressed in simplified form.

37. $\dfrac{y+2}{5y} + \dfrac{y+4}{15y}$

38. $\dfrac{x+3}{x^2} + \dfrac{x+5}{2x}$

39. $\dfrac{x+5}{xy} - \dfrac{x-1}{x^2y}$

40. $\dfrac{y-7}{y^2} - \dfrac{y+7}{2y}$

41. $\dfrac{3}{x-2} - (x-1)$

42. $a+1 - \dfrac{3}{a+3}$

43. $\dfrac{2x+2}{x-2} - \dfrac{2x}{x+2}$

44. $\dfrac{y+3}{y-1} - \dfrac{y+4}{y+1}$

45. $\dfrac{x}{(x-2)^2} + \dfrac{x-4}{(x+2)(x-2)}$

46. $\dfrac{a-2}{(a+3)^2} - \dfrac{a}{a-3}$

47. $\dfrac{2x}{x^2-3x+2} + \dfrac{2x}{x-1} - \dfrac{x}{x-2}$

48. $\dfrac{4x}{x-2} - \dfrac{3x}{x-3} + \dfrac{4x}{x^2-5x+6}$

49. $\dfrac{x+1}{2x+4} - \dfrac{x^2}{2x^2-8}$

50. $\dfrac{x+1}{x+2} - \dfrac{x^2+1}{x^2-x-6}$

51. $\dfrac{x-1}{x+2} + \dfrac{x}{3-x} + \dfrac{9x+3}{x^2-x-6}$

52. $\dfrac{x+1}{x-3} - \dfrac{x^2+9x}{x^2-2x-3} - \dfrac{5}{3-x}$

53. Perimeter

  a. Find a rational expression that models the perimeter of the pictured square.

$\dfrac{x}{x-6}$ ft

  b. If x has a value of 8, what would be the perimeter of the square?

  ✎**c.** Could x have a value of 6 feet? Why or why not?

54. Perimeter

  a. Find a rational expression that models the perimeter of the pictured square.

$\dfrac{x}{x-5}$ ft

  b. If x has a value of 10, what would be the perimeter of the square?

  ✎**c.** Could x have a value of 5 feet? Why or why not?

REVIEW

55. Factor $2x^3 - 7x^2 - 15x$.

56. If $P(x) = 2x^3 - 7x^2 - 15x$, find
   a. $P(5)$
   b. $P\left(-\dfrac{3}{2}\right)$
   c. $P(0)$

57. Solve $2x^3 - 7x^2 - 15x = 0$ graphically. Copy your screen display to justify your work.

58. Solve $2x^3 - 7x^2 - 15x = 0$ analytically and verify your solution with the calculator using the STOre feature.

6.3 Multiplying and Dividing Rational Expressions

■ MULTIPLYING RATIONAL EXPRESSIONS ■ DIVIDING RATIONAL EXPRESSIONS ■ COMBINED OPERATIONS

■ MULTIPLYING RATIONAL EXPRESSIONS

Recall that to multiply rational numbers (fractions) we multiply the numerators and multiply the denominators. The same is true for rational expressions.

Multiplying Rational Expressions

If $\dfrac{P}{Q}$ and $\dfrac{R}{S}$ are rational expressions, then $\dfrac{P}{Q} \cdot \dfrac{R}{S} = \dfrac{PR}{QS}$.

Because all rational expressions should be ultimately expressed in simplified form, it is suggested that you *express* the product of the numerators over the product of the denominators rather than actually multiplying the polynomials in the numerators and the denominators. The reason for this is because simplification occurs when common factors are divided.

EXAMPLE 1 Multiply: $\dfrac{35x^2y}{7y^2z} \cdot \dfrac{z}{5xy}$.

Solution
$$\dfrac{35x^2y}{7y^2z} \cdot \dfrac{z}{5xy} = \dfrac{35x^2y \cdot z}{7y^2z \cdot 5xy} = \dfrac{7 \cdot 5 \cdot x \cdot x \cdot y \cdot z}{7 \cdot y \cdot y \cdot z \cdot 5 \cdot x \cdot y}$$

$$= \dfrac{7 \cdot 5 \cdot x \cdot x \cdot y \cdot z}{7 \cdot y \cdot y \cdot z \cdot 5 \cdot x \cdot y} \quad \text{Divide out common factors to simplify.}$$

$$= \dfrac{x}{y^2}$$

EXAMPLE 2 Find the product: $\dfrac{x^2 - x}{2x + 4} \cdot \dfrac{x + 2}{x}$.

Solution
$$\dfrac{x^2 - x}{2x + 4} \cdot \dfrac{x + 2}{x} = \dfrac{(x^2 - x)(x + 2)}{(2x + 4)(x)} \quad \text{Multiply rational expressions by } \textit{expressing} \text{ the product of the numerators over the product of the denominators.}$$

$$= \dfrac{x(x - 1)(x + 2)}{2(x + 2)(x)} \quad \text{Factor the numerator and denominator completely.}$$

$$= \dfrac{x(x - 1)(x + 2)}{2(x + 2)(x)} \quad \text{Divide out common factors to simplify.}$$

$$= \dfrac{x - 1}{2}$$

■ STUDY TIP
Place a piece of paper over the solution to an example in your text. Read the example and try to work the problem on your paper. If you get stuck, slide the paper down and jump-start yourself.

EXAMPLE 3 Find the product of $\dfrac{x^2 - 3x}{x^2 - x - 6}$ and $\dfrac{x^2 + x - 2}{x^2 - x}$.

Solution

$\dfrac{x^2 - 3x}{x^2 - x - 6} \cdot \dfrac{x^2 + x - 2}{x^2 - x} = \dfrac{(x^2 - 3x)(x^2 + x - 2)}{(x^2 - x - 6)(x^2 - x)}$

> Express the product of the numerators over the product of the denominators.

$= \dfrac{x(x - 3)(x + 2)(x - 1)}{(x - 3)(x + 2)x(x - 1)}$

> Factor the numerator and the denominator completely.

$= \dfrac{\overset{1}{\cancel{x}}\overset{1}{\cancel{(x - 3)}}\overset{1}{\cancel{(x + 2)}}\overset{1}{\cancel{(x - 1)}}}{\underset{1}{\cancel{(x - 3)}}\underset{1}{\cancel{(x + 2)}}\underset{1}{\cancel{x}}\underset{1}{\cancel{(x - 1)}}}$

> Divide out common factors.

$= 1$

No matter what value is substituted in for the variable x, the product, $\dfrac{x^2 - 3x}{x^2 - x - 6} \cdot \dfrac{x^2 + x - 2}{x^2 - x}$, is equal to the constant 1. This is verified next using different options available on the graphing calculator.

First, store any value for x and evaluate the original expression:

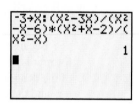

Another option is to enter the expression at y1 and access the TABLE. The advantage of this method is that it allows you to evaluate the expression for many values of the variable at one time.

  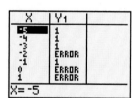

Notice that -2, 0, and 1 yield an error message because they are excluded values. ■

EXAMPLE 4 Multiply: $\dfrac{x^2 + 2x}{xy - 2y} \cdot \dfrac{x + 1}{x^2 - 4} \cdot \dfrac{x - 2}{x^2 + x}$.

Solution

$\dfrac{x^2 + 2x}{xy - 2y} \cdot \dfrac{x + 1}{x^2 - 4} \cdot \dfrac{x - 2}{x^2 + x} = \dfrac{x(x + 2)(x + 1)(x - 2)}{y(x - 2)(x + 2)(x - 2)x(x + 1)}$

> Multiply the fractions and factor where possible.

$= \dfrac{\overset{1}{\cancel{x}}\overset{1}{\cancel{(x + 2)}}\overset{1}{\cancel{(x + 1)}}\overset{1}{\cancel{(x - 2)}}}{y\underset{1}{\cancel{(x - 2)}}\underset{1}{\cancel{(x + 2)}}(x - 2)\underset{1}{\cancel{x}}\underset{1}{\cancel{(x + 1)}}}$

> Divide out common factors.

$= \dfrac{1}{y(x - 2)}$

■

■ DIVIDING RATIONAL EXPRESSIONS

Recall that when dividing rational numbers we multiply by the *reciprocal* (multiplicative inverse) of the divisor. This is stated formally below:

Dividing Rational Expressions

If $\dfrac{P}{Q}$ and $\dfrac{R}{S}$ are rational expressions, then $\dfrac{P}{Q} \div \dfrac{R}{S} = \dfrac{P}{Q} \cdot \dfrac{S}{R} = \dfrac{PS}{QR}.$

EXAMPLE 5   Divide: $\dfrac{x^2 + x}{3x - 15} \div \dfrac{x^2 + 2x + 1}{6x - 30}.$

Solution

$$\dfrac{x^2 + x}{3x - 15} \div \dfrac{x^2 + 2x - 1}{6x - 30}$$

$$= \dfrac{x^2 + x}{3x - 15} \cdot \dfrac{6x - 30}{x^2 + 2x + 1}$$ Invert the divisor and multiply.

$$= \dfrac{x(x + 1) \cdot 2 \cdot 3(x - 5)}{3(x - 5)(x + 1)(x + 1)}$$ Multiply the fractions and factor.

$$= \dfrac{x\cancel{(x + 1)} \cdot 2 \cdot 3\cancel{(x - 5)}}{\cancel{3}\cancel{(x - 5)}\cancel{(x + 1)}(x + 1)}$$ Divide out common factors.

$$= \dfrac{2x}{x + 1}$$ ■

EXAMPLE 6   Divide: $\dfrac{2x^2 - 3x - 2}{2x + 1} \div (4 - x^2).$

Solution

$$\dfrac{2x^2 - 3x - 2}{2x + 1} \div (4 - x^2)$$

$$= \dfrac{2x^2 - 3x - 2}{2x + 1} \cdot \dfrac{1}{(4 - x^2)}$$ Invert the divisor and multiply.

$$= \dfrac{(2x + 1)(x - 2) \cdot 1}{(2x + 1)(2 + x)(2 - x)}$$ Multiply the fractions and factor where possible.

$$= \dfrac{\cancel{(2x + 1)}\overset{-1}{\cancel{(x - 2)}} \cdot 1}{\cancel{(2x + 1)}(2 + x)\cancel{(2 - x)}}$$ Divide out common factors. Because $(x - 2)$ and $(2 - x)$ are opposites, their quotient is -1.

$$= \dfrac{-1}{2 + x}$$

which could be written as $-\dfrac{1}{2 + x}$ or $\dfrac{1}{-(2 + x)}$ or $\dfrac{1}{-2 - x}.$ ■

■ COMBINED OPERATIONS

The order of operations holds for the addition, subtraction, multiplication, and division of rational expressions. Unless parentheses indicate otherwise, perform multiplications and divisions in order from left to right.

EXAMPLE 7   Simplify the expression $\dfrac{x^2 - x - 6}{x - 2} \div \dfrac{x^2 - 4x}{x^2 - x - 2} \cdot \dfrac{x - 4}{x^2 + x}$.

Solution

$$\dfrac{x^2 - x - 6}{x - 2} \div \dfrac{x^2 - 4x}{x^2 - x - 2} \cdot \dfrac{x - 4}{x^2 + x}$$

$$= \dfrac{x^2 - x - 6}{x - 2} \cdot \dfrac{x^2 - x - 2}{x^2 - 4x} \cdot \dfrac{x - 4}{x^2 + x}$$ Invert the divisor and multiply.

$$= \dfrac{(x + 2)(x - 3)(x + 1)(x - 2)(x - 4)}{(x - 2)x(x - 4)x(x + 1)}$$ Multiply the fractions and factor.

$$= \dfrac{(x + 2)(x - 3)\overset{1}{\cancel{(x + 1)}}\overset{1}{\cancel{(x - 2)}}\overset{1}{\cancel{(x - 4)}}}{\underset{1}{\cancel{(x - 2)}}x\underset{1}{\cancel{(x - 4)}}x\underset{1}{\cancel{(x + 1)}}}$$ Divide out common factors.

$$= \dfrac{(x + 2)(x - 3)}{x^2}$$

EXERCISE 6.3

CONCEPTS

1. Factor each of the following into a product of prime numbers:
   a. 35   **b.** 24   **c.** 56   **d.** 27

2. Write each exponential expression in expanded form.
   a. x^2 **b.** y^5 **c.** $x^3 y^4$

In Exercises 3–12, perform the indicated multiplications and divisions. Simplify answers if possible.

3. $\dfrac{25}{35} \cdot \dfrac{21}{55}$

4. $\dfrac{27}{24} \cdot \dfrac{56}{35}$

5. $\dfrac{2}{3} \cdot \dfrac{15}{2} \cdot \dfrac{1}{7}$

6. $\dfrac{2}{5} \cdot \dfrac{10}{9} \cdot \dfrac{3}{2}$

7. $\dfrac{5y}{7} \cdot \dfrac{7x}{5z}$

8. $\dfrac{4x}{3y} \cdot \dfrac{3y}{7x}$

9. $\dfrac{35}{2} \div \dfrac{15}{2}$

10. $\dfrac{6}{14} \div \dfrac{10}{35}$

11. $\dfrac{x^2}{3} \div \dfrac{2x}{4}$

12. $\dfrac{z^2}{z} \div \dfrac{z}{3z}$

PRACTICE   *In Exercises 13–32, perform the indicated multiplications. Simplify answers if possible.*

13. $\dfrac{8x^2 y^2}{4x^2} \cdot \dfrac{2xy}{2y}$

14. $\dfrac{9x^2 y}{3x} \cdot \dfrac{3xy}{3y}$

15. $\dfrac{-2xy}{x^2} \cdot \dfrac{3xy}{2}$

16. $\dfrac{-3x}{x^2} \cdot \dfrac{2xz}{3}$

17. $\dfrac{z + 7}{7} \cdot \dfrac{z + 2}{z}$

18. $\dfrac{a - 3}{a} \cdot \dfrac{a + 3}{5}$

19. $\dfrac{x - 2}{2} \cdot \dfrac{2x}{x - 2}$

20. $\dfrac{y + 3}{y} \cdot \dfrac{3y}{y + 3}$

21. $\dfrac{(x + 1)^2}{x + 1} \cdot \dfrac{x + 2}{x + 1}$

22. $\dfrac{(y - 3)^2}{y - 3} \cdot \dfrac{y - 3}{y - 3}$

23. $\dfrac{2x + 6}{x + 3} \cdot \dfrac{3}{4x}$

24. $\dfrac{3y - 9}{y - 3} \cdot \dfrac{y}{3y^2}$

25. $\dfrac{x^2 + x - 6}{5x} \cdot \dfrac{5x - 10}{x + 3}$

26. $\dfrac{z^2 + 4z - 5}{5z - 5} \cdot \dfrac{5z}{z + 5}$

27. $\dfrac{x^2 + 7xy + 12y^2}{x^2 + 2xy - 8y^2} \cdot \dfrac{x^2 - xy - 2y^2}{x^2 + 4xy + 3y^2}$

28. $\dfrac{m^2 + 9mn + 20n^2}{m^2 - 25n^2} \cdot \dfrac{m^2 - 9mn + 20n^2}{m^2 - 16n^2}$

29. $\dfrac{3r^2 + 15rs + 18s^2}{6r^2 - 24s^2} \cdot \dfrac{2r - 4s}{3r + 9s}$

30. $\dfrac{2u^2 + 8u}{2u + 8} \cdot \dfrac{4u^2 + 8uv + 4v^2}{u^2 + 5uv + 4v^2}$

31. $\dfrac{x^2 + 5x + 6}{x^2} \cdot \dfrac{x^2 - 2x}{x^2 - 9} \cdot \dfrac{x^2 - 3x}{x^2 - 4}$

32. $\dfrac{x^2 - 1}{1 - x} \cdot \dfrac{4x}{x + x^2} \cdot \dfrac{x^2 + 2x + 1}{2x + 2}$

In Exercises 33–38, perform the indicated divisions. Simplify answers when possible.

33. $\dfrac{x+2}{3x} \div \dfrac{x+2}{2}$

34. $\dfrac{z-3}{3z} \div \dfrac{z+3}{z}$

35. $\dfrac{(z-2)^2}{3z^2} \div \dfrac{z-2}{6z}$

36. $\dfrac{(x+7)^2}{x+7} \div \dfrac{(x-3)^2}{x+7}$

37. $\dfrac{5x^2+13x-6}{x+3} \div \dfrac{5x^2-17x+6}{x-2}$

38. $\dfrac{x^2-x-6}{2x^2+9x+10} \div \dfrac{x^2-25}{2x^2+15x+25}$

In Exercises 39–48, perform the indicated operations. The order of operations is valid for rational expressions and must be followed.

39. $\dfrac{2}{3} \cdot \dfrac{15}{5} \div \dfrac{10}{5}$

40. $\dfrac{6}{5} \div \dfrac{3}{5} \cdot \dfrac{5}{15}$

41. $\dfrac{x^2}{18} \div \dfrac{x^3}{6} \div \dfrac{12}{x^2}$

42. $\dfrac{y^3}{3y} \cdot \dfrac{3y^2}{4} \div \dfrac{15}{20}$

43. $\dfrac{x^2-1}{x^2-9} \cdot \dfrac{x+3}{x+2} \div \dfrac{5}{x+2}$

44. $\dfrac{2}{3x-3} \div \dfrac{2x+2}{x-1} \cdot \dfrac{5}{x+1}$

45. $\dfrac{x-x^2}{x^2-4}\left(\dfrac{2x+4}{x+2} \div \dfrac{5}{x+2}\right)$

46. $\dfrac{2}{3x-3} \div \left(\dfrac{2x+2}{x-1} \cdot \dfrac{5}{x+1}\right)$

47. $\dfrac{y^2}{x+1} \cdot \dfrac{x^2+2x+1}{x^2-1} \div \dfrac{3y}{xy-y}$

48. $\dfrac{x^2-y^2}{x^4-x^3} \div \dfrac{x-y}{x^2} \div \dfrac{x^2+2xy+y^2}{x+y}$

REVIEW

49. Simplify: $\dfrac{x^3+2x^2-25x-50}{x^2-10x+25}$.

50. Evaluate $\dfrac{x^3+2x^2-25x-50}{x^2-10x+25}$ when

   a. $x=3$

   b. $x=5$

   c. Explain the significance of the result of part (b).

51. Solve $4x - 3y = 7$ for y.

52. Solve:

   a. $|3x-1| = 2$

   b. $|x+5| - 4 = -1$

   c. $|2x+1| = |x-5|$

6.4   Mixed Operations and Complex Fractions

■ COMPLEX FRACTIONS AND ORDER OF OPERATIONS

■ COMPLEX FRACTIONS AND ORDER OF OPERATIONS

Consider the arithmetic expression $2\dfrac{2}{3} \div 3\dfrac{3}{4}$. Because a fraction bar also can be used to indicate division, the expressions could have been written as $\dfrac{2\frac{2}{3}}{3\frac{3}{4}}$ or

$\dfrac{\frac{8}{3}}{\frac{15}{4}}$. This is an example of a **complex fraction. A complex fraction** is a fraction that contains a fraction in the numerator, the denominator, or both the numerator and denominator. One way to simplify a complex fraction is to replace the *main* fraction bar with a division sign. The expression is then simplified following the normal order of operations.

To simplify the complex fraction $\dfrac{\frac{3}{5}+1}{2-\frac{1}{5}}$ we can rewrite it as a division

problem: $\left(\dfrac{3}{5}+1\right) \div \left(2-\dfrac{1}{5}\right)$. The parentheses are required because the numerator of the complex fraction contains more than one term, as does the denominator. Continuing the simplification process:

$$\left(\frac{3}{5}+\frac{5}{5}\right) \div \left(\frac{10}{5}-\frac{1}{5}\right)$$

Simplify within grouping first by finding the common denominator and adding/subtracting the fractions as the signs indicate.

$$=\frac{8}{5} \div \frac{9}{5}$$

Continue to simplify grouping.

$$=\frac{8}{5} \cdot \frac{5}{9}$$

Invert the divisor and change the operation to multiplication.

$$=\frac{8 \cdot 5}{5 \cdot 9}$$

Express the multiplication of the fractions.

$$=\frac{8 \cdot \cancel{5}}{\cancel{5} \cdot 9}$$

Divide out common factors to simplify.

$$=\frac{8}{9}$$

Our analytical work is confirmed with the graphing calculator.

A second method for simplifying complex fractions uses the fundamental property of fractions. Because any expression can be multiplied by 1 to obtain an equivalent expression, we may multiply the numerator and the denominator of the complex fraction by the LCD of its component fractions. This makes the complex fraction a simple fraction.

Consider $\dfrac{\frac{3}{5}+1}{2-\frac{1}{5}}$. We may multiply the numerator and denominator of the

complex fraction by the LCD of its component fractions $\left(\frac{3}{5} \text{ and } \frac{1}{5}\right)$. We are actually multiplying the fraction by the multiplicative identity of 1, yet it is expressed as $\frac{5}{5}$.

$$\frac{\frac{3}{5}+1}{2-\frac{1}{5}} \cdot \frac{5}{5}$$

$$=\frac{5 \cdot \left(\frac{3}{5}+1\right)}{5 \cdot \left(2-\frac{1}{5}\right)}$$

Express the multiplication.

$$=\frac{5 \cdot \frac{3}{5}+5 \cdot 1}{5 \cdot 2-5 \cdot \frac{1}{5}}$$

Use the distributive property to remove parentheses.

$$=\frac{3+5}{10-1}$$

Use the order of operations to simplify.

$$=\frac{8}{9}$$

```
(3/5+1)/(2-1/5)▸
Frac
              8/9
```

This process is verified with the graphing calculator, as displayed by the following screen:

Either of the two methods illustrated and summarized here can be used to simplify a complex fraction.

Method 1: Simplifying a Complex Fraction Using the Order of Operations

1. Replace the main fraction bar with a division sign, grouping the numerator and denominator appropriately.

2. Simplify, using the order of operations.

Method 2: Simplifying a Complex Fraction by Multiplying by the LCD

1. Determine the LCD of the component fractions of the complex fraction.

2. Multiply the numerator and denominator of the complex fraction by the LCD. Be sure to distribute when the numerator and/or denominator contain more than one term.

The examples that follow illustrate both methods of simplification.

EXAMPLE 1   Simplify: $\dfrac{\frac{x}{3}}{\frac{y}{3}}$.

Solution

Method 1

$$\dfrac{\frac{x}{3}}{\frac{y}{3}} = \frac{x}{3} \div \frac{y}{3}$$

$$= \frac{x}{3} \cdot \frac{3}{y}$$

$$= \frac{3x}{3y}$$

$$= \frac{x}{y}$$

Method 2

$$\dfrac{\frac{x}{3}}{\frac{y}{3}} = \dfrac{\left(\frac{x}{3}\right)3}{\left(\frac{y}{3}\right)3}$$

$$= \dfrac{\frac{x}{1}}{\frac{y}{1}}$$

$$= \frac{x}{y}$$

EXAMPLE 2   Simplify: $\dfrac{\frac{x}{x+1}}{\frac{y}{x}}$.

Solution

Method 1

$$\frac{\dfrac{x}{x+1}}{\dfrac{y}{x}} = \frac{x}{x+1} \div \frac{y}{x}$$

$$= \frac{x}{x+1} \cdot \frac{x}{y}$$

$$= \frac{x^2}{y(x+1)}$$

Method 2

$$\frac{\dfrac{x}{x+1}}{\dfrac{y}{x}} = \frac{\left(\dfrac{x}{x+1}\right)x(x+1)}{\left(\dfrac{y}{x}\right)x(x+1)}$$

$$= \frac{\dfrac{x^2}{1}}{\dfrac{y(x+1)}{1}}$$

$$= \frac{x^2}{y(x+1)}$$

EXAMPLE 3 Simplify: $\dfrac{1 + \dfrac{1}{x}}{1 - \dfrac{1}{x}}$.

Solution

Method 1

$$\frac{1 + \dfrac{1}{x}}{1 - \dfrac{1}{x}} = \left(1 + \frac{1}{x}\right) \div \left(1 - \frac{1}{x}\right)$$

$$= \left(\frac{x}{x} + \frac{1}{x}\right) \div \left(\frac{x}{x} - \frac{1}{x}\right)$$

$$= \frac{x+1}{x} \div \frac{x-1}{x}$$

$$= \frac{x+1}{x} \cdot \frac{x}{x-1}$$

$$= \frac{(x+1) \cdot x}{x \cdot (x-1)}$$

$$= \frac{x+1}{x-1}$$

Method 2

$$\frac{1 + \dfrac{1}{x}}{1 - \dfrac{1}{x}} = \frac{\left(1 + \dfrac{1}{x}\right)x}{\left(1 - \dfrac{1}{x}\right)x}$$

$$= \frac{1 \cdot x + \dfrac{1}{x} \cdot x}{1 \cdot x - \dfrac{1}{x} \cdot x}$$

$$= \frac{x+1}{x-1}$$

EXAMPLE 4 Simplify the complex fraction $\dfrac{1}{1 + \dfrac{1}{x+1}}$.

Solution We use Method 2.

STUDY TIP

When simplifying complex fractions, we suggest that you experiment with both methods. Some students will use one method exclusively, which is fine. However, others will use Method 1 when the numerator and denominator of the complex fraction contain a single term and use Method 2 when the numerator and/or denominator contain more than one term. That is why only Method 2 was used in Example 4. It is, however, personal choice.

$$\frac{1}{1 + \dfrac{1}{x+1}} = \frac{1(x+1)}{\left(1 + \dfrac{1}{x+1}\right)(x+1)}$$

Multiply numerator and denominator by $x + 1$.

$$= \frac{x+1}{1(x+1) + 1}$$

Remove parentheses and simplify.

$$= \frac{x+1}{x+2}$$

Simplify.

EXERCISE 6.4

VOCABULARY *Fill in the blank to make a true statement.*

1. A _____ fraction is a fraction that contains a fraction in the numerator, the denominator, or both.

CONCEPTS

2. Is the expression $\dfrac{\frac{1}{2}}{4}$ equal to the expression $\dfrac{1}{\frac{2}{4}}$?

Why or why not?

3. Is the expression $\dfrac{\frac{3}{2}}{5}$ equal to the expression $\dfrac{3}{\frac{2}{5}}$?

Why or why not?

4. a. Perform the indicated division: $\dfrac{x^2}{3} \div \dfrac{2x}{4}$.

b. Rewrite the expression in part (a) as a complex fraction.

c. Would the complex fraction written in part (b) (when simplified) be equivalent to the quotient in part (a)? Why or why not?

5. a. Perform the indicated division: $\dfrac{x+2}{3x} \div \dfrac{x+2}{2}$.

b. Rewrite the expression in part (a) as a complex fraction.

c. Would the complex fraction written in part (b) (when simplified) be equivalent to the quotient in part (a)? Why or why not?

6. a. Evaluate the expression $\dfrac{\frac{1}{2}+\frac{2}{3}}{4}$ using the graphing calculator. Copy your screen display exactly as it appears.

b. Rewrite the complex fraction in part (a) as a division problem.

c. Evaluate the expression you wrote in part (b). Is the result equal to the result from part (a)? Should it be? Why or why not?

7. a. Evaluate the expression $\dfrac{\frac{3}{4}+\frac{1}{5}}{6}$ using the graphing calculator. Copy your screen display exactly as it appears.

b. Rewrite the complex fraction in part (a) as a division problem.

c. Evaluate the expression you wrote in part (b). Is the result equal to the result from part (a)? Should it be? Why or why not?

In Exercises 8–41, simplify each complex fraction. Check results of 8–15 using the graphing calculator.

8. $\dfrac{\frac{2}{3}}{\frac{3}{4}}$

9. $\dfrac{\frac{3}{5}}{\frac{2}{7}}$

10. $\dfrac{\frac{4}{5}}{\frac{32}{15}}$

11. $\dfrac{\frac{7}{8}}{\frac{49}{4}}$

12. $\dfrac{\frac{2}{3}+1}{\frac{1}{3}+1}$

13. $\dfrac{\frac{3}{5}-2}{\frac{2}{5}-2}$

14. $\dfrac{\frac{1}{2}+\frac{3}{4}}{\frac{3}{2}+\frac{1}{4}}$

15. $\dfrac{\frac{2}{3}-\frac{5}{2}}{\frac{2}{3}-\frac{3}{2}}$

16. $\dfrac{\frac{x}{y}}{\frac{1}{x}}$

17. $\dfrac{\frac{y}{x}}{\frac{x}{xy}}$

18. $\dfrac{\frac{5t^2}{9x^2}}{\frac{3t}{x^2t}}$

19. $\dfrac{\frac{5w^2}{4tz}}{\frac{15wt}{z^2}}$

20. $\dfrac{\dfrac{1}{x} - 3}{\dfrac{5}{x} + 2}$

21. $\dfrac{\dfrac{1}{y} + 3}{\dfrac{3}{y} - 2}$

22. $\dfrac{\dfrac{2}{x} + 2}{\dfrac{4}{x} + 2}$

23. $\dfrac{\dfrac{3}{x} - 3}{\dfrac{9}{x} - 3}$

24. $\dfrac{\dfrac{3y}{x} - y}{y - \dfrac{y}{x}}$

25. $\dfrac{\dfrac{y}{x} + 3y}{y + \dfrac{2y}{x}}$

26. $\dfrac{\dfrac{1}{x + 1}}{1 + \dfrac{1}{x + 1}}$

27. $\dfrac{\dfrac{1}{x - 1}}{1 - \dfrac{1}{x - 1}}$

28. $\dfrac{\dfrac{x}{x + 2}}{\dfrac{x}{x + 2} + x}$

29. $\dfrac{\dfrac{2}{x - 2}}{\dfrac{2}{x - 2} - 1}$

30. $\dfrac{1}{\dfrac{1}{x} + \dfrac{1}{y}}$

31. $\dfrac{1}{\dfrac{b}{a} - \dfrac{a}{b}}$

32. $\dfrac{\dfrac{2}{x}}{\dfrac{2}{y} - \dfrac{4}{x}}$

33. $\dfrac{\dfrac{2y}{3}}{\dfrac{2y}{3} - \dfrac{8}{y}}$

34. $\dfrac{3 + \dfrac{3}{x - 1}}{3 - \dfrac{3}{x}}$

35. $\dfrac{2 - \dfrac{2}{x + 1}}{2 + \dfrac{2}{x}}$

36. $\dfrac{\dfrac{3}{x} + \dfrac{4}{x + 1}}{\dfrac{2}{x + 1} - \dfrac{3}{x}}$

37. $\dfrac{\dfrac{5}{y - 3} - \dfrac{2}{y}}{\dfrac{1}{y} + \dfrac{2}{y - 3}}$

38. $\dfrac{\dfrac{2}{x} - \dfrac{3}{x + 1}}{\dfrac{2}{x + 1} - \dfrac{3}{x}}$

39. $\dfrac{\dfrac{5}{y} + \dfrac{4}{y + 1}}{\dfrac{4}{y} - \dfrac{5}{y + 1}}$

40. $\dfrac{\dfrac{1}{y^2 + y} - \dfrac{1}{xy + x}}{\dfrac{1}{xy + x} - \dfrac{1}{y^2 + y}}$

41. $\dfrac{\dfrac{2}{b^2 - 1} - \dfrac{3}{ab - a}}{\dfrac{3}{ab - a} - \dfrac{2}{b^2 - 1}}$

42. **a.** Simplify the four complex fractions below:

$$\dfrac{1}{1 + 1}, \quad \dfrac{1}{1 + \dfrac{1}{2}}, \quad \dfrac{1}{1 + \dfrac{1}{1 + \dfrac{1}{2}}}, \quad \text{and} \quad \dfrac{1}{1 + \dfrac{1}{1 + \dfrac{1}{1 + \dfrac{1}{2}}}}.$$

(*Hint:* Use Method 2, and work from the bottom up.)

✍ **a.** What is the pattern in the numerators? What would be the next expression and result?

REVIEW

43. Solve: $\dfrac{x + 5}{x - 3} = \dfrac{4}{5}$.

✍ 44. Explain why $x = 3$ would not be an acceptable solution to the equation in Exercise 43.

45. Samuel holds a yardstick perpendicular to the sidewalk and measures a 4-foot shadow. He then measures the shadow of the bell tower on campus and determines it to be 50 feet long. Using what he knows about similar triangles, he will be able to determine the height of the bell tower. What height does he find?

46. Simplify each of the following. Assume that all variables represent positive numbers.

   a. $\dfrac{(x^2 y)^3}{x^2}$

   b. $(2x^4)^3$

   c. $(2x + 3)^2$

6.5 Algebraic and Graphical Solutions of Rational Equations

■ EXTRANEOUS SOLUTIONS ■ ALGEBRAIC SOLUTIONS OF RATIONAL EQUATIONS ■ GRAPHICAL SOLUTIONS OF RATIONAL EQUATIONS

■ EXTRANEOUS SOLUTIONS

Sections 6.1 through 6.4 of this chapter focused on *simplifying* rational expressions. We were required to perform the arithmetic operations of addition, subtraction, multiplication and division. We now turn our attention to *solving* equations that contain rational expressions. The differences between the *simplification of expressions* and the *solving of equations* are reiterated below.

When simplifying an *expression,* follow the order of operations to remove groupings, raise to powers, and perform all arithmetic operations to rewrite the original *expression* as an equivalent *expression* in which all like terms have been combined.

When solving an *equation* that is linear, we use the addition, subtraction, multiplication, and division properties of equality to rewrite the original *equation* as a series of equivalent *equations* until the variable is isolated on one side of the equation. For equations of degree two or higher, the zero factor property is applied. The solutions, or roots, of the equation are the value(s) of the variable that make the equation a true statement.

Care must be taken when solving equations containing rational expressions with variables in the denominator. This is illustrated below.

EXAMPLE 1　　Solve the equation $\dfrac{x+3}{x-1} = \dfrac{4}{x-1}$.

Solution　　Generalizing the methods learned previously, begin by multiplying through the equation by the LCD, which is the quantity $(x-1)$.

$$\frac{x+3}{x-1} = \frac{4}{x-1}$$

$$(x-1) \cdot \frac{x+3}{x-1} = (x-1) \cdot \frac{4}{x-1}$$　　Multiply both sides of the equation by the LCD.

$$\frac{(x-1)\cdot(x+3)}{x-1} = \frac{(x-1)\cdot 4}{x-1}$$　　Express the multiplication.

$$\frac{\cancel{(x-1)}\cdot(x+3)}{\cancel{x-1}} = \frac{\cancel{(x-1)}\cdot 4}{\cancel{x-1}}$$　　Simplify, by dividing out common factors.

$$x+3 = 4$$　　Solve the resulting linear equation by subtracting 3 from both sides.

$$x = 1$$

Check: To check analytically, replace the variable, x, with its value, 1, in the original equation and simplify.

$$\frac{x+3}{x-1} = \frac{4}{x-1}$$

$$\frac{1+3}{1-1} = \frac{4}{1-1}$$　　Simplify.

$$\frac{4}{0} = \frac{4}{0}$$

Because zero appears in the denominator, the fractions are undefined. Therefore, the number 1 *cannot* be a solution of the equation. The original equation was a contradiction and has *no solution.*

The check can also be done with the graphing calculator using either the STOre feature or the TABLE. This is illustrated in the screens below.

 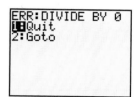

The number 1 is stored for the variable x and the left side of the equation is entered for evaluation. When ENTER is pressed we immediately see that the fraction is undefined. There is no need to repeat the process with the right side of the equation.

The left side of the equation is entered at the y1 prompt and the right side is entered at the y2 prompt. The TABLE is set to begin at 0 and to be incremented by 1. The error messages confirm the fact that 1 cannot be a solution to the equation. ∎

Looking back at the solution of the equation it does not appear that we have used any algebraic algorithms or properties incorrectly. The fallacy lies in the fact that the LCD used $(x - 1)$ contained a variable. Recall that the multiplication property of equality states multiplying through an equation by a *nonzero* quantity produces an equivalent equation. Because a variable appeared in our LCD, we were *not* assured that the multiplication produced an equivalent equation. In this case, it did not. The equation $x + 3 = 4$ is NOT equivalent to the original equation, $\frac{x + 3}{x - 1} = \frac{4}{x - 1}$.

Sometimes when solving rational equations apparent false solutions result. These are called **extraneous solutions** and are rejected. Any apparent solution that makes a denominator equal to zero is an extraneous solution.

ALGEBRAIC SOLUTIONS OF RATIONAL EQUATIONS

When arithmetically solving equations that contain variables in the denominators, solutions **must** be verified through a checking process. Extraneous solutions **must** be rejected. For this reason, it is suggested that excluded values be determined first. The following steps can be used:

Solving Rational Equations Algebraically

1. Determine any and all excluded values by setting each denominator containing a variable equal to zero and solving for the variable.
2. Determine the LCD.
3. Multiply each *term* on both sides of the equation by the LCD to clear fractions.
4. Simplify both sides of the equation.
5. Solve the equation. Remember, linear (first-degree) equations are solved by transformation, whereas quadratic (second-degree) and higher-order equations require the additional process of applying the zero factor property after the equation is put in standard form and the polynomial is factored.
6. Reject any extraneous solutions (those values that equal the excluded values found in step 1).
7. Check solutions in the original equation to verify solutions.

EXAMPLE 2   Solve the equation $\dfrac{3x + 1}{x + 1} - 2 = \dfrac{3(x - 3)}{x + 1}$.

Solution   Determine the excluded values by setting each denominator that contains a variable equal to zero and solving.

$$x + 1 = 0 \qquad \text{Set the denominator equal to zero. Solve by subtracting 1 from both sides of the equation.}$$

$$x = -1$$

The variable, x, **cannot** have a value of -1.

We begin solving by multiplying both sides of the equation by $x + 1$, the least common denominator of the fractions contained in the equation.

$$(x + 1) \cdot \left(\frac{3x + 1}{x + 1} - 2 \right) = (x + 1) \left[\frac{3(x - 3)}{x + 1} \right]$$

$$3x + 1 - 2(x + 1) = 3(x - 3) \qquad \text{Use the distributive property to remove groupings.}$$

$$3x + 1 - 2x - 2 = 3x - 9 \qquad \text{Use the distributive property to remove parentheses.}$$

$$x - 1 = 3x - 9 \qquad \text{Combine like terms.}$$

$$-2x - 1 = -9 \qquad \text{Subtract } 3x \text{ from both sides.}$$

$$-2x = -8 \qquad \text{Add 1 to both sides.}$$

$$x = 4 \qquad \text{Divide both sides by } -2.$$

Because our solution of 4 is not equal to the excluded value of -1, 4 is a valid (rather than an extraneous) solution.

Check:

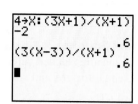

The STOre feature confirms the left side of the equation equals the right side when x is replaced by 4.

The pictured TABLE confirms both the solution and the excluded value.

EXAMPLE 3   Solve the equation $\dfrac{x+2}{x+3} + \dfrac{1}{x^2+2x-3} = 1$.

Solution   Determine the excluded values by setting each denominator equal to zero and solving for the variable.

$$x + 3 = 0 \qquad \text{and} \qquad x^2 + 2x - 3 = 0$$
$$x = -3 \qquad\qquad\qquad (x+3)(x-1) = 0$$

Because $x^2 + 2x - 3 = 0$ is a quadratic equation, we factor the polynomial side and use the zero factor property to find the solutions.

$$x + 3 = 0 \qquad \text{or} \qquad x - 1 = 0$$
$$x = -3 \qquad \text{or} \qquad x = 1$$

The excluded values are -3 and 1.

Now that the excluded values have been found, we begin solving the equation.

$$\frac{x+2}{x+3} + \frac{1}{x^2+2x-3} = 1$$

$$\frac{x+2}{x+3} + \frac{1}{(x+3)(x-1)} = 1 \qquad \text{To determine the LCD, factor the denominators.}$$

The LCD is $(x+3)(x-1)$. We multiply each term on both sides of the equation to eliminate fractions.

$$(x+3)(x-1) \cdot \frac{x+2}{x+3} + (x+3)(x-1) \cdot \frac{1}{(x+3)(x-1)} = (x+3)(x-1) \cdot 1$$

$$(x-1)(x+2) + 1 = (x+3)(x-1) \qquad \text{Simplify by dividing out common factors.}$$

$$x^2 + 2x - x - 2 + 1 = x^2 - x + 3x - 3 \qquad \text{Remove parentheses.}$$

$$x^2 + x - 1 = x^2 + 2x - 3 \qquad \text{Combine like terms.}$$

$$x - 1 = 2x - 3 \qquad \text{Subtract } x^2 \text{ from both sides.}$$

$$-x - 1 = -3 \qquad \text{Subtract } 2x \text{ from both sides.}$$

$$-x = -2 \qquad \text{Add 1 to both sides.}$$

$$x = 2 \qquad \text{Divide both sides by } -1.$$

Because 2 is *not* an excluded value, it *is* a solution to the original equation.

The verification using the STOre feature of the calculator is pictured below:

EXAMPLE 4   Solve the equation $\dfrac{2}{x-1} + 4 = \dfrac{4}{x^2-1} + 3$.

Solution   Determine the excluded values by setting each denominator equal to zero and solving for the variable.

$$x - 1 = 0 \qquad \text{and} \qquad x^2 - 1 = 0$$
$$x = 1 \qquad\qquad (x-1)(x+1) = 0$$

Because $x^2 - 1 = 0$ is a quadratic equation, we factor the polynomial side and use the zero factor property to find the solutions.

$$x - 1 = 0 \quad \text{or} \quad x + 1 = 0$$
$$x = 1 \quad \text{or} \quad x = -1$$

The excluded values are 1 and -1.

We now begin solving the rational equation. To determine the LCD, we factor denominators that are not prime.

$$\frac{2}{x-1} + 4 = \frac{4}{x^2-1} + 3$$

$$\frac{2}{x-1} + 4 = \frac{4}{(x-1)(x+1)} + 3$$

The LCD is $(x-1)(x+1)$.

$$(x-1)(x+1) \cdot \frac{2}{x-1} + (x-1)(x+1) \cdot 4 = (x-1)(x+1) \cdot \frac{4}{(x-1)(x+1)} + (x-1)(x+1) \cdot 3$$

$$2(x+1) + 4(x^2-1) = 4 + 3(x^2-1)$$

Simplify by dividing out common factors.

$$2x + 2 + 4x^2 - 4 = 4 + 3x^2 - 3$$

Simplify by removing grouping.

$$4x^2 + 2x - 2 = 3x^2 + 1$$

Combine like terms.

$$x^2 + 2x - 3 = 0$$

Because the equation is quadratic, use the subtraction property of equality to put it in standard form.

$$(x+3)(x-1) = 0$$

Factor the polynomial side.

$$x + 3 = 0 \quad \text{or} \quad x - 1 = 0$$

Use the zero factor property to set each factor equal to 0 and solve for the variable.

$$x = -3 \quad \text{or} \quad x = 1$$

Because the number 1 is an excluded value, it is rejected as a root. The solution is -3.

The solution of -3 is verified below:

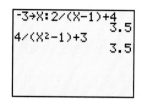

The fact that 1 is an extraneous root is also verified below:

    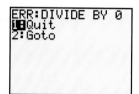

Before ENTER is   After ENTER is
pressed.   pressed.

The TABLE feature demonstrates that -3 is a solution because the expressions entered at the y1 and y2 prompts evaluate to the same constant (3.5) for $x = -3$. It also confirms the fact that 1 cannot be a solution. The ERROR message displayed indicates that YOU have made an error. In this case, you are "asking" the calculator to evaluate expressions for a value of the variable that makes those expressions undefined.

■ GRAPHICAL SOLUTIONS OF RATIONAL EQUATIONS

The INTERSECT option of the graphing calculator can also be used to solve rational equations. The INTERSECT feature is used in the examples that follow.

EXAMPLE 5   Solve the equation $\dfrac{3x}{2} + \dfrac{5}{4} = \dfrac{5x}{2}$.

Solution   Enter the left side of the equation at the y1 prompt and the right side at the y2 prompt. Use the INTERSECT feature to solve the equation.

The solution is 1.25. Because the original equation contained fractions, it is best to express the solution as a fraction. This may be done by returning to the home screen, accessing the variable x, and using the convert to fraction (▸Frac) option to express the solution as 5/4.

Looking at the graphical display of the two sides of the equation we see two straight lines. This means that each side of the original equation was linear. Notice that the original equation did not contain variables in the denominator, therefore there are no excluded values. The next example examines an equation in which a variable *does* appear in the denominator.   ∎

EXAMPLE 6   Solve $\dfrac{2}{x} + 6 = 8$ graphically.

Solution   Enter the left side of the equation at the y1 prompt and the right side at the y2 prompt. Access the graph.

When we access the INTERSECT feature, we see the first curve prompt. Notice that no value is displayed for y at the bottom of the screen. This is because when $x = 0$, the fraction $2/x$ is undefined. Therefore, cursor either right or left until a value is displayed for both x and y at the bottom of the screen. You may then respond to the prompts as usual.

The solution is 1. This can be verified algebraically or by using the STOre or TABLE feature of the calculator.   ∎

When we analyzed the last graphical display, we concluded that each side of the equation was linear. The same conclusion cannot be drawn from the previous example. To analyze this graph, return to the y= screen and turn off the graph of y1 by placing the cursor over the equal sign and pressing ENTER. The graph of y2 (y2 = 8) will be the only graph displayed. This is the right side of the equation and is a horizontal line.

If we turn the graph of the expression entered at y1 *ON* and the graph of y2 *OFF,* we have the following screen display:

Setting the TABLE to begin at −7 and to be incremented by 1, we can examine the values in the y1 column. Remember, those values represent the evaluation of the expression 2/x + 6 as *x* is replaced by the values displayed in the x-column. We see that as *x increases,* the value of 2/x + 6 *decreases.* This is consistent with the left side of the graph.

Scrolling down the TABLE verifies the excluded value of 0. As *x* continues to increase, we see that the value of the expression jumps to 8 when x = 1, and then decreases. This is consistent with the right side of the graph.

EXAMPLE 7   Solve $\dfrac{x^2 - 2x - 8}{x + 2} - 2x = x - 5$ both graphically and algebraically.

Solution   *Graphical solution:* Enter the left side of the equation at the y1 prompt and the right side at the y2 prompt. Use the INTERSECT feature to find the solution to the equation.

 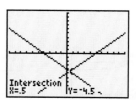

The solution is 1/2 (remember, because the original equation contained rational expressions, the solution should be expressed as a fraction rather than in its decimal form).

Algebraic solution: The graphical display shows us that *both* the left and the right sides of the equation are linear. However, the left side of the original equation does not appear linear. Therefore, as we begin solving we will start by simplifying the left side rather than multiplying through the entire equation by the LCD.

$$\frac{x^2 - 2x - 8}{x + 2} - 2x = x - 5$$

$$\frac{(x - 4)(x + 2)}{x + 2} - 2x = x - 5 \qquad \text{Simplify the rational expression by factoring and dividing out common factors.}$$

$$x - 4 - 2x = x - 5$$

$$-x - 4 = x - 5 \qquad \text{Combine like terms.}$$

$$-4 = 2x - 5 \qquad \text{Add } x \text{ to both sides of the equation.}$$

$$1 = 2x \qquad \text{Add 5 to both sides of the equation.}$$

$$\frac{1}{2} = x \qquad \text{Divide both sides of the equation by 2.}$$

This confirms the graphical solution.   ■

Notice that looking at the graphical display first helped us in our algebraic solution. Had we solved the equation by first multiplying through by the LCD of $(x + 2)$ we would have obtained the extraneous root of -2 as well as the actual root of $\frac{1}{2}$. The graphical displays only show actual roots, of course, because extraneous roots result from the process of multiplying through by an LCD and *not* obtaining an equation equivalent to the original.

EXAMPLE 8   Solve $\dfrac{5}{x} = -3x + 4$ both graphically and algebraically.

Solution   *Graphical solution:* Enter the left side of the equation at the y1 prompt and the right side at the y2 prompt. The display is pictured below.

Because there are no points of intersection, the equation is a contradiction and has *no solution.*

Algebraic solution:

$$\frac{5}{x} = -3x + 4$$
The excluded value is 0.

$$x \cdot \frac{5}{x} = -x \cdot 3x + x \cdot 4$$
Multiply through the equation by the LCD of x.

$$5 = -3x^2 + 4x$$
Simplify by performing multiplications.

$$3x^2 - 4x + 5 = 0$$
Because the equation is quadratic, put it into standard form.

Attempts to factor the quadratic polynomial $(3x^2 - 4x + 5)$ are unsuccessful. It is prime. Therefore, the equation has no rational solutions.   ■

EXAMPLE 9   Solve $\dfrac{1}{x - 1} + \dfrac{3}{x - 1} = 5$ graphically.

Solution   When we first look at the graphical display, we see a vertical line that we have not seen before. However if we change our calculator mode from CONNECTED to DOT (or change the graph style icon), this vertical line disappears.

The reason the line disappears is because it is not a part of the graph of the expression $\dfrac{1}{x - 1} + \dfrac{3}{x - 1}$. When in connected mode, the grapher has connected the two branches that are the graph of the side of the equation containing rational expressions. In dot mode this does not happen.

Using the INTERSECT feature we find the solution is 1.8, or, as a fraction, 9/5.

Equations containing rational expressions are sometimes difficult to solve graphically. However, the graphical display can often provide insights into the solving process. The following suggestions may prove helpful. Remember, explore the options available to you with the graphing calculator.

■ STUDY TIP

Before beginning an exercise set, read over your notes *and* read the section out of your textbook.

Graphical Displays of Rational Expressions

1. View the graph in *both* connected and dot mode.

2. Use the TABLE to help you see, numerically, what is happening. Reconcile the numerical information with the graphical display.

3. TRACE along the graph watching for patterns.

4. Use the INTERSECT feature to find/confirm solutions graphically.

5. For clarity, you may want to turn the x and y axes off to unclutter the screen. To do this, access the FORMAT menu highlight AxesOff and press ENTER.

6. Even when solving equations algebraically, look at the graphical display and practice reconciling your algebraic solution with the graphical display.

EXERCISE 6.5

VOCABULARY   *Fill in the blank to make a true statement.*

1. False solutions are called _____ solutions.

CONCEPTS

2. Determine the restricted values for each rational expression.

   a. $\dfrac{x}{x^2 - 1}$ **b.** $\dfrac{2}{4x}$ **c.** $\dfrac{3}{x^2 - 2x - 8}$

3. Identify each item as an expression or an equation. If it is an expression, simplify it. If it is an equation, solve it.

   a. $\dfrac{x}{3} + \dfrac{x}{2}$

   b. $\dfrac{x}{3} + \dfrac{x}{2} = 6$

   c. $\dfrac{3x}{x - 1} - \dfrac{x}{x^2 - 1}$

   d. $\dfrac{3}{x} + \dfrac{2}{4x}$

   e. $\dfrac{3x}{x - 1} - \dfrac{x}{x^2 - 1} = 3$

   f. $\dfrac{3}{x} + \dfrac{2}{4x} = 7$

PRACTICE   *In Exercises 4–35, solve each rational equation algebraically. Check solutions using either the STOre or TABLE feature of the calculator.*

4. $\dfrac{x}{2} + 4 = \dfrac{3x}{2}$

5. $\dfrac{y}{3} + 6 = \dfrac{4y}{3}$

6. $\dfrac{x}{3} + 1 = \dfrac{x}{2}$

7. $\dfrac{x}{2} - 3 = \dfrac{x}{5}$

8. $\dfrac{x - 3}{3} + 2x = -1$

9. $\dfrac{x + 2}{2} - 3x = x + 8$

10. $\dfrac{z - 3}{2} = z + 2$

11. $\dfrac{b + 2}{3} = b - 2$

12. $\dfrac{5(x+1)}{8} = x + 1$ **13.** $\dfrac{3(x-1)}{2} + 2 = x$ **14.** $\dfrac{5}{x} - \dfrac{4}{x} = 8 + \dfrac{1}{x}$ **15.** $\dfrac{11}{x} + \dfrac{13}{x} = 12$

16. $\dfrac{1}{x-1} + \dfrac{3}{x-1} = 1$ **17.** $\dfrac{3}{p+6} - 2 = \dfrac{7}{p+6}$ **18.** $\dfrac{x^2}{x+2} - \dfrac{4}{x+2} = x$ **19.** $\dfrac{z^2}{z+1} + 2 = \dfrac{1}{z+1}$

20. $\dfrac{x}{x-5} - \dfrac{5}{x-5} = 3$ **21.** $\dfrac{3}{y-2} + 1 = \dfrac{3}{y-2}$ **22.** $\dfrac{3r}{2} - \dfrac{3}{r} = \dfrac{3r}{2} + 3$ **23.** $\dfrac{2p}{3} - \dfrac{1}{p} = \dfrac{2p-1}{3}$

24. $\dfrac{1}{3} + \dfrac{2}{x-3} = 1$ **25.** $\dfrac{3}{5} + \dfrac{7}{x+2} = 2$ **26.** $\dfrac{7}{q^2-q-2} + \dfrac{1}{q+1} = \dfrac{3}{q-2}$

27. $\dfrac{-5}{x^2+x-2} + \dfrac{3}{x+2} = \dfrac{1}{x-1}$ **28.** $\dfrac{3x}{3x-6} + \dfrac{8}{x^2-4} = \dfrac{2x}{2x+4}$ **29.** $\dfrac{x-3}{4x-4} + \dfrac{1}{9} = \dfrac{x-5}{6x-6}$

30. $\dfrac{5}{4y+12} - \dfrac{3}{4} = \dfrac{5}{4y+12} - \dfrac{y}{4}$ **31.** $\dfrac{3}{5x-20} - \dfrac{4}{5} = \dfrac{3}{5x-20} - \dfrac{x}{5}$ **32.** $\dfrac{x}{x-1} - \dfrac{12}{x^2-x} = \dfrac{-1}{x-1}$

33. $1 - \dfrac{3}{x} = \dfrac{-8x}{x^2+3x}$ **34.** $\dfrac{x}{x^2-9} + \dfrac{x+8}{x+3} = \dfrac{x-8}{x-3}$ **35.** $\dfrac{x-3}{x-2} = \dfrac{1}{x} + \dfrac{x-3}{x}$

In Exercises 36–41, solve each rational equation graphically. Copy the INTERSECT screen exactly as it appears, including all displayed information, to justify your work. If a window other than Zstandard is used, specify it in the form [xmin, xmax] by [ymin, ymax].

36. $\dfrac{2x}{5} - 8 = \dfrac{4x}{5}$ **37.** $\dfrac{3x}{4} - 6 = \dfrac{x}{4}$ **38.** $\dfrac{2}{x+1} - \dfrac{12}{x+1} = -5$ **39.** $\dfrac{1}{x-3} + \dfrac{2}{x-3} = 1$

40. $x + \dfrac{2}{3} = \dfrac{2x-12}{3x-9}$ (*Hint:* Use a window in which you can clearly see both roots.)

41. $x + \dfrac{3}{4} = \dfrac{3x-50}{4x-24}$ (*Hint:* Use a window in which you can clearly see both roots.)

42. Consider the equation $\dfrac{x-4}{x-3} + \dfrac{x-2}{x-3} = x - 3$.

    a. Solve the given equation by first multiplying through by the LCD.

    b. Solve the given equation by first simplifying the left side.

    c. Solve the given equation graphically.

    d. Which solution method did you like best? Why?

    e. Would looking at the graphical display *first* have helped you in your algebraic solution? Why or why not?

43. Consider the equation, $\dfrac{x-5}{x-4} + \dfrac{x-3}{x-4} = x - 4$.

    a. Solve the given equation by first multiplying through by the LCD.

    b. Solve the given equation by first simplifying the left side.

    c. Solve the given equation graphically.

    d. Which solution method did you like best? Why?

    e. Would looking at the graphical display *first* have helped you in your algebraic solution? Why or why not?

REVIEW

44. Simplify: $2(3x^2 - 4x + 5) - (x^2 + 25)$

45. Divide: $\dfrac{4x^2 + 6xy - 12x^3}{4x}$

46. Uniform motion Two cars leave from Ashland City driving in opposite directions on Highway 41a. If one car is traveling at 55 mph and the other at 65 mph, how long before they are 60 miles apart?

47. Auto repair The service department at the local car dealership charges a flat rate of $75 for diagnosing problems and then $15 per hour for labor.
  a. Write an equation to model the cost where x represents the number of hours and y the total cost.
  b. Graph the equation in an appropriate viewing window on your calculator.

✎ **c.** Explain how to use the graph on the calculator (and the appropriate feature(s) of the calculator) to determine the cost for 3 hours of work.

6.6 Applications of Rational Equations

■ NUMBER PROBLEMS ■ UNIFORM MOTION PROBLEMS ■ WORK PROBLEMS ■ INVESTMENT PROBLEMS ■ AVERAGE COST

■ NUMBER PROBLEMS

The ability to solve equations containing rational expressions allows us to expand our knowledge of applications. We will begin with number problems. You may want to review the problem-solving strategies introduced in Section 3.1.

EXAMPLE 1   If the same number is added to both the numerator and the denominator of the fraction $\frac{3}{5}$, the result is $\frac{4}{5}$. Find the number.

Solution   Because there is only one unknown, we will let $x =$ the number. Translate the words of the problem to an algebraic equation: "The *same number* is added to both the numerator and denominator . . . the result is $\frac{4}{5}$":

$$\frac{3 + x}{5 + x} = \frac{4}{5}$$

To solve the equation, you may multiply through by the LCD of $5(5 + x)$ *OR*, because the equation is a proportion, cross-multiply, as shown here:

$5(3 + x) = 4(5 + x)$	Cross-multiply (the product of the means equals the product of the extremes).
$15 + x = 20 + 4x$	Use the distributive property to remove parentheses.
$15 + x = 20$	Subtract $4x$ from both sides.
$x = 5$	Subtract 15 from both sides.

The number is 5.
   The check is most easily done analytically, substituting 5 in for x in the expression $\dfrac{3 + x}{5 + x}$ to obtain the result of $\frac{4}{5}$:

$$\frac{3 + 5}{5 + 5} = \frac{8}{10} = \frac{4}{5}$$

■

■ UNIFORM MOTION PROBLEMS

Earlier, when solving uniform motion problems, we discovered two important tools for solving these types of problems, the formula $d = r \cdot t$ and organizing given information into a chart. This is particularly helpful when considering the movement of more than one object.

EXAMPLE 2 A coach can run 10 miles in the same amount of time that his best student-athlete can run 12 miles. If the student can run 1 mile per hour faster than the coach, how fast can the student run?

Solution Because the student's rate is given in terms of the coach's rate, we begin by writing our "Let" statements.

Let x = the coach's rate

$x + 1$ = the student's rate

Organize the given information into a chart:

	d	r	t
Coach	10	x	
Student	12	$x + 1$	

We do not know the times of either the coach or the student. However, we *do* know that if $d = r \cdot t$, then $t = \frac{d}{r}$. We can represent the times for each of the runners by dividing *their* distance by *their* rate. The last *column* of the table is determined by the information in the given row.

Completing the table, we have

	d	r	t
Coach	10	x	$\frac{10}{x}$
Student	12	$x + 1$	$\frac{12}{x + 1}$

Now that the information is organized in a table, return to the words of the problem to write the equation.

The first sentence tells us that their times are equal. Our equation can now be written and solved.

$\frac{10}{x} = \frac{12}{x + 1}$ Equate the runners' times.

$10(x + 1) = 12x$ Cross-multiply.

$10x + 10 = 12x$ Distribute.

$10 = 2x$ Subtract $10x$ from both sides of the equation.

$5 = x$ Divide both sides by 2.

Make sure you answer the question posed in the problem with a complete sentence: The coach runs 5 miles per hour and the student runs 6 miles per hour.

■ WORK PROBLEMS

Work problems often involve people or machines completing a job together. These problems are closely related to uniform motion problems, as will be illustrated in the following example. In each problem you must consider the *rate* at which each individual or machine works, the *time* actually worked, and the *part of the task* that is *completed* by that individual or machine. A table often proves helpful in organizing the information.

EXAMPLE 3   Mary can clean the house completely in 4 hours, whereas it takes her husband 6 hours to do the same job. How long will it take them to completely clean the house if they work together?

Solution   Let x = the time to complete the task when working together. Organize the information into a chart. Since it takes Mary 4 hours to complete the task, she is doing $\frac{1}{4}$ of the task per hour. Thus her rate of work is $\frac{1}{4}$. Multiplying the rate of work by the time worked yields the part of the task completed.

	Rate of work	Time worked	Part of task completed
Mary	$\frac{1}{4}$	x	$\frac{1}{4} \cdot x$ or $\frac{x}{4}$
Husband	$\frac{1}{6}$	x	$\frac{1}{6} \cdot x$ or $\frac{x}{6}$

Fractions are used in the "rate of work" column because Mary does $\frac{1}{4}$ of the total job in 1 hour. Her husband completes $\frac{1}{6}$ of the total job in 1 hour.

Because they are working together to complete the same job, we add the part of the task each completes and set this equal to 1 (they are completing one job of cleaning the house). We may then solve the equation.

$$\frac{x}{4} + \frac{x}{6} = 1$$

$6x + 4x = 24$ Multiply through the equation by the LCD of 24.

$10x = 24$ Combine like terms.

$x = \dfrac{24}{10}$ Divide both sides by 10.

They can complete the task together in $\frac{12}{5}$ hours. This is equivalent to 2.4 hours, which is better expressed as 2 hours and 24 minutes (four-tenths of an hour is $0.4 \cdot 60 = 24$ minutes).

The equation can be solved quite easily graphically, as illustrated below:

■ INVESTMENT PROBLEMS

Recall the formula $I = prt$, where I represents the interest, P the principal, r the rate (expressed as a decimal rather than a percentage), and t the time. Previously, we worked problems in which the principal amount was usually the unknown quantity. The example below illustrates a type of investment problem that results in a rational equation. Again, a chart is often helpful.

EXAMPLE 4   At a bank, a sum of money invested for one year will earn $96 interest. If invested in bonds, that same money would earn $108, because the interest rate paid by the bonds is 1% greater than that paid by the bank. Find the bank's rate.

Solution   Let x = the bank's rate.

$x + .01$ = the bond's rate

Organize the known information in a table:

	Interest	Principal	Rate	Time
Bank	96		x	1
Bonds	108		$x + .01$	1

The principal for each investment can be expressed in terms of the interest, rate, and time. $I = prt$; therefore, $P = \frac{I}{rt}$. In each row we divide the interest earned by the product of the rate and time. We can now complete the table.

	Interest	Principal	Rate	Time
Bank	96	$\dfrac{96}{x}$	x	1
Bonds	108	$\dfrac{108}{x + .01}$	$x + .01$	1

We must now write an equation. Referring back to the original problem, we notice the phrase ". . . that same money." This means that the principal values are equal. We can now write an equation and solve it.

$$\frac{96}{x} = \frac{108}{x + .01}$$ Equate the principal amounts from the two investments.

$$96(x + .01) = 108x$$ Because the equation is a proportion, cross-multiply.

$$96x + .96 = 108x$$ Distribute.

$$.96 = 12x$$ Subtract $96x$ from both sides of the equation.

$$.08 = x$$ Divide both sides by 12 to isolate the variable.

The bank's interest rate is 0.08, which is expressed as 8%. The bonds pay 9%, a rate that is 1% greater than that paid by the bank.   ∎

■ AVERAGE COST

Average cost is just what the term implies, the average cost to the consumer for the production of goods. Often, manufacturers must determine an average cost for each unit produced. The example below illustrates this concept.

EXAMPLE 5   A CD manufacturer can produce CDs at a cost of $8 per unit. It has operating costs of $24,000 per month. How many CDs must be produced monthly to have an average cost of $12 per unit?

Solution   Let x = number of CDs produced.

Average cost = Total cost ÷ Number of units

We substitute our information into this formula and solve the resulting equation:

$$12 = \frac{24000 + 8x}{x}$$

$$12x = 24000 + 8x \qquad \text{Multiply through the equation by the LCD of } x.$$

$$4x = 24000 \qquad \text{Subtract } 8x \text{ from both sides of the equation.}$$

$$x = 6000 \qquad \text{Divide both sides by 4.}$$

The company must produce 6000 CDs monthly to keep the average cost of the CD at \$12. ■

EXERCISE 6.6

CONCEPTS

1. If x and $x + 1$ are two consecutive integers, find their reciprocals.

2. What fraction do you have if the denominator of $\frac{7}{8}$ is doubled and the numerator is decreased by 2?

3. Write the rational expression determined by decreasing the numerator of $\frac{x + 2}{x}$ by 3 and increasing the denominator by 3.

APPLICATIONS
For each problem below, clearly identify the variable and what it represents, write an equation, provide a complete solution to that equation, and answer the question posed in a complete sentence.

4. **Number problem** If the denominator of the fraction $\frac{3}{4}$ is increased by a number and the numerator of the fraction is doubled, the result is 1. Find the number.

5. **Number problem** If a number is added to the numerator of the fraction $\frac{7}{8}$ and the same number is subtracted from the denominator, the result is 2. Find the number.

6. **Number problem** If a number is added to the numerator of the fraction $\frac{3}{4}$ and twice as much is added to the denominator, the result is $\frac{4}{7}$. Find the number.

7. **Number problem** If a number is added to the numerator of the fraction $\frac{5}{7}$ and twice as much is subtracted from the denominator, the result is 8. Find the number.

8. **Number problem** The sum of a number and its reciprocal is $\frac{13}{6}$. Find the numbers.

9. **Number problem** The sum of the reciprocals of two consecutive even integers is $\frac{7}{24}$. Find the integers.

10. **Sightseeing** A tourist can bicycle 28 miles in the same time he can walk 8 miles. If he can ride 10 miles per hour faster than he can walk, how much time should he allow to walk a 30-mile trail (see Illustration 1)? (*Hint:* How fast can he walk?)

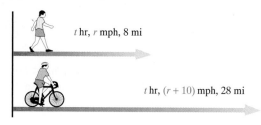

t hr, r mph, 8 mi

t hr, (r + 10) mph, 28 mi

ILLUSTRATION 1

11. **Comparing travel** A plane can fly 300 miles in the same time it takes a car to go 120 miles. If the car travels 90 miles per hour slower than the plane, how fast is the plane?

12. **Boating in a river** A boat that can travel 18 miles per hour in still water can travel 22 miles downstream in the same amount of time that it can travel 14 miles upstream. Find the speed of the current in the river. (See Illustration 2.)

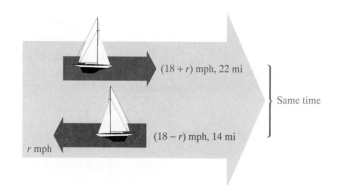

(18 + r) mph, 22 mi

(18 − r) mph, 14 mi

Same time

r mph

ILLUSTRATION 2

13. Wind speed A plane can fly 300 miles downwind in the same amount of time it can travel 210 miles upwind. Find the velocity of the wind if the plane can fly at 255 miles per hour in still air.

14. River tours A riverboat tour begins by going 60 miles upstream against a 5 mph current. There, the boat turns around and returns with the current. What still-water speed should the captain use to complete the tour in 5 hours?

15. Travel time The company president flew 680 miles in the corporate jet and returned in a smaller plane that could fly only half as fast. If the total travel time was 6 hours, find the speeds of the planes.

16. Roofing a house A homeowner estimates that it will take 7 days to roof his house. A professional roofer estimates that he could roof the house in 4 days. How long will it take if the homeowner helps the roofer?

17. Filling a pool An inlet pipe can fill an empty pool in 5 hours, and another inlet pipe can fill the pool in 4 hours. How long will it take both pipes to fill the pool?

18. Filling a pool One inlet pipe can fill an empty pool in 4 hours, and a drain can empty the pool in 8 hours. How long will it take the pipe to fill the pool if the drain is left open?

19. Mowing lawns Blaine can mow his yard in 3 hours. When his teen-aged son helps, they can complete the lawn in 2 hours. How long would it take the son to mow the entire yard himself?

20. Painting a room It takes Blaire 5 hours to paint her bedroom by herself. Her brother, Trey, can paint the entire room in 3 hours. How long will it take them to paint the room if they work together?

21. Processing forms Becky, a secretary, can process 500 university forms in $2\frac{1}{2}$ days. Sherry can process the same 500 forms in 2 days. If they work together, how long would it take them to process 500 of the university's forms?

22. Comparing investments Two certificates of deposit pay interest at rates that differ by 1%. Money invested for 1 year in the first CD earns $175 interest. The same principal invested in the other CD earns $200. Find the two rates of interest.

23. Comparing interest rates Two bond funds pay interest at rates that differ by 2%. Money invested for 1 year in the first fund earns $315 interest. The same amount invested in the other fund earns $385. Find the lower rate of interest.

24. Average cost A calculator manufacturing company has a yearly overhead of $120,000. It costs $84 to produce a single graphing calculator.
  a. How many graphing calculators must be produced annually for the average production cost of a calculator to be $100?

  ✎**b.** The left side of the equation is entered at the y1 prompt and the right side of the equation is entered at the y2 prompt (as shown below). The window values shown below were entered. The third graphical display shows x and y values. Interpret those values within the context of the problem.

  ✎**c.** Interpret the given display within the context of the problem.

25. Average cost A textbook company wants to keep the costs of its texts as low as possible.
  a. If it costs them $55 to produce a single mathematics text with overhead costs of $50,000 annually, how many texts must be produced for the average production cost per text to be held at $75?

  ✎**b.** The left side of the equation is entered at the y1 prompt and the right side of the equation is entered at the y2 prompt (as shown below). The window values shown below were entered. The third graphical display shows x and y values. Interpret those values within the context of the problem.

✐ **c.** Interpret the given display within the context of the problem.

is the left branch of the graph of the expression entered at the y2 prompt. Interpret the displayed information, as best you can. Is it applicable to this problem? Why or why not?

✐ **d.** Typically, the graphs of rational equations have two branches. In parts (b) and (c) above, only the right branch has been displayed for interpretation. Below

CHAPTER SUMMARY

CONCEPTS

REVIEW EXERCISES

Section 6.1

A rational expression is simplified when the numerator and the denominator have no common factors other than 1.

Simplifying (Reducing) a Rational Expression
1. Factor the numerator and the denominator completely.
2. Divide out common factors.

The fundamental property of fractions allows us to divide out common factors and to multiply the numerator and denominator of a given fraction by the same quantity. These processes produce equivalent fractions.

Simplifying Rational Expressions

In Review Exercises 1–3, simplify each rational expression. If a fraction is already in lowest terms, so indicate. State any excluded values of the variable.

1. **a.** $-\dfrac{51}{153}$ **b.** $-\dfrac{105}{45}$

    c. $\dfrac{3x^2}{6x^3}$ **d.** $\dfrac{5xy^2}{2x^2y^2}$

2. **a.** $\dfrac{x^2}{x^2+x}$ **b.** $\dfrac{x+2}{x^2+2x}$

    c. $\dfrac{x^2+4x+3}{x^2-4x-5}$ **d.** $\dfrac{x^2-x-56}{x^2-5x-24}$

3. **a.** $\dfrac{2x^2-16x}{2x^2-18x+16}$ **b.** $\dfrac{x^2+x-2}{x^2-x-2}$

Section 6.2

To add rational expressions with common denominators, keep the common denominator, add (or subtract) the numerators as the signs indicate, and then reduce the rational expression.

To add or subtract fractions with unlike denominators, you must first find the LCD. Then, convert each fraction to an equivalent fraction with the LCD as denominator.

Adding and Subtracting Rational Expressions

In Review Exercises 4–5, simplify each rational expression by adding or subtracting as the sign indicates. All results should be expressed in simplified form.

4. **a.** $\dfrac{x}{x+y}+\dfrac{y}{x+y}$ **b.** $\dfrac{3x}{x-7}-\dfrac{x-2}{x-7}$

    c. $\dfrac{x}{x-1}+\dfrac{1}{x}$ **d.** $\dfrac{1}{7}-\dfrac{1}{x}$

5. **a.** $\dfrac{3}{x+1}-\dfrac{2}{x}$ **b.** $\dfrac{x+2}{2x}-\dfrac{2-x}{x^2}$

    c. $\dfrac{x}{x+2}+\dfrac{3}{x}-\dfrac{4}{x^2+2x}$ **d.** $\dfrac{2}{x-1}-\dfrac{3}{x+1}+\dfrac{x-5}{x^2-1}$

Finding the Least Common Denominator (LCD)
1. Write down each of the different denominators that appear in the given fractions.
2. Factor each of these denominators completely.
3. Form a product using each of the different factors obtained in step 2. Use each different factor the *greatest* number of times it appears in any *single* factorization. This product is the least common denominator.

Adding/Subtracting with Unlike Denominators
1. Identify the LCD of the rational expressions. Use the process outlined earlier in this section.
2. Convert each rational expression to an equivalent rational expression having the LCD as denominator. This process was demonstrated in Section 6.1.
3. Add or subtract the rational expressions by *expressing* the addition or subtraction over the common denominator.
4. Simplify the numerators, but leave the denominators in factored form.
5. Simplify by reducing the rational expression to lowest terms. Recall that this requires you to divide out common factors in the numerator and the denominator. Because the denominator is in factored form, you need only factor the numerator.

Section 6.3 — Multiplying and Dividing Rational Expressions

To multiply rational expressions, multiply the numerators and multiply the denominators. Divide out common factors to simplify (reduce).

When dividing rational expressions, multiply the dividend by the reciprocal of the divisor.

In Review Exercises 6–7, perform the indicated operations. Make sure all results are simplified.

6. a. $\dfrac{3xy}{2x} \cdot \dfrac{4x}{2y^2}$ **b.** $\dfrac{3x}{x^2 - x} \cdot \dfrac{2x - 2}{x^2}$

c. $\dfrac{3x^2}{5x^2y} \div \dfrac{6x}{15xy^2}$ **d.** $\dfrac{x^2 + 5x}{x^2 + 4x - 5} \div \dfrac{x^2}{x - 1}$

7. a. $\dfrac{x^2 + 3x + 2}{x^2 + 2x} \cdot \dfrac{x}{x + 1}$ **b.** $\dfrac{x^2 + 4x + 4}{x^2 - x - 6} \cdot \dfrac{x^2 - 9}{x^2 + 5x + 6}$

c. $\dfrac{x^2 - x - 6}{2x - 1} \div \dfrac{x^2 - 2x - 3}{2x^2 + x - 1}$ **d.** $\dfrac{3n^2 + 5n - 2}{12n^2 - 13n + 3} \div \dfrac{n^2 + 3n + 2}{4n^2 + 5n - 6}$

Section 6.4

Mixed Operations and Complex Fractions

Complex fractions may be simplified by using either of the methods specified below. All complex fractions can be simplified using either method.

Method 1: Simplifying a Complex Fraction Using the Order of Operations
1. Replace the main fraction bar with a division sign, grouping the numerator and denominator appropriately.
2. Simiplify, using the order of operations.

Method 2: Simplifying a Complex Fraction by Multiplying by the LCD
1. Determine the LCD of the component fractions of the complex fraction.
2. Multiply the numerator and denominator of the complex fraction by the LCD. Be sure to distribute when the numerator and/or denominator contain more than one term.

8. Simplify each complex fraction.

a. $\dfrac{\dfrac{1}{x} + 1}{\dfrac{1}{x} - 1}$

b. $\dfrac{1 + \dfrac{3}{x}}{2 - \dfrac{1}{x^2}}$

c. $\dfrac{\dfrac{2}{x-1} + \dfrac{x-1}{x+1}}{\dfrac{1}{x^2-1}}$

d. $\dfrac{\dfrac{a}{b} + c}{\dfrac{b}{a} + c}$

Section 6.5

Algebraic and Graphical Solutions of Rational Equations

Solving Rational Equations Algebraically
1. Determine any and all excluded values by setting each denominator containing a variable equal to zero and solving for the variable.
2. Determine the LCD.
3. Multiply each *term* on both sides of the equation by the LCD to clear fractions.
4. Simplify both sides of the equation.
5. Solve the equation.
6. Reject any extraneous solutions (those values that equal the excluded values found in step 1).
7. Check solutions in the original equation to verify solutions.

9. Solve each equation algebraically. Check all results using either the STOre or TABLE features of the graphing calculator.

a. $\dfrac{3}{x} = \dfrac{2}{x-1}$

b. $\dfrac{5}{x+4} = \dfrac{3}{x+2}$

c. $\dfrac{2}{3x} + \dfrac{1}{x} = \dfrac{5}{9}$

d. $\dfrac{2x}{x+4} = \dfrac{3}{x-1}$

e. $\dfrac{2}{x-1} + \dfrac{3}{x+4} = \dfrac{-5}{x^2+3x-4}$

f. $\dfrac{4}{x+2} - \dfrac{3}{x+3} = \dfrac{6}{x^2+5x+6}$

Graphical Displays of Rational Expressions

1. View the graph in *both* connected and dot mode.
2. Use the TABLE to help you see, numerically, what is happening. Reconcile the numerical information with the graphical display.
3. TRACE along the graph watching for patterns.
4. Use the INTERSECT feature to find/confirm solutions graphically.
5. For clarity, you may want to turn the *x*- and *y*-axes off to unclutter the screen. To do this, access the FORMAT menu, highlight AxesOff, and press ENTER.
6. Even when solving equations algebraically, look at the graphical display and practice reconciling your algebraic solution with the graphical display.

10. Solve the rational equation $\dfrac{1}{x-1} + \dfrac{3}{x-1} = 1$ graphically. Sketch your screen display and copy all displayed information.

Section 6.6

Applications of Rational Equations

Be familiar with the problem-solving strategies introduced in Section 3.1.

The rate of work can be determined by the fraction $\dfrac{1}{\text{time worked}}$.

In Review Exercises 11–16, solve each problem. Be sure to clearly identify what the variable represents, to write an equation and include a complete solution, and to answer the question posed with a sentence.

11. **Pumping a basement**   If one pump can empty a flooded basement in 18 hours and a second pump can empty the basement in 20 hours, how long will it take to empty the basement if both pumps are used?

12. **Painting houses**   If a homeowner can paint a house in 14 days and a professional painter can paint it in 10 days, how long will it take if they work together?

13. **Jogging**   A jogger can bicycle 30 miles in the same time as he can jog 10 miles. If he can ride 10 miles per hour faster than he can jog, how fast can he jog?

14. **Wind speed**   A plane can fly 400 miles downwind in the same amount of time as it can travel 320 miles upwind. If the plane can fly at 360 miles per hour in still air, find the velocity of the wind.

15. Comparing interest rates Two savings accounts at a local bank pay interest rates that differ by $\frac{1}{2}\%$. Money invested for 1 year in one of the accounts earns \$108 in interest, whereas the same amount of money invested in the other account earns \$121.50. Find the lower rate of interest.

16. Average cost A new day care center is opening for small children. The overhead is \$6000 monthly. Expenditures per child (for supplies and lunches) will be approximately \$150 per month. How many students must be enrolled to keep the average cost per month for each child at \$400?

■ Chapter Test

1. Simplify: $\dfrac{48x^2y}{54xy^2}$

2. Simplify $\dfrac{2x^2 - x - 3}{4x^2 - 9}$ and state excluded values for the variable.

3. Simplify $\dfrac{3(x+2)-3}{2x-4-(x-5)}$ and state the excluded values for the variable.

4. Multiply and simplify: $\dfrac{12x^2y}{15xyz} \cdot \dfrac{25y^2z}{16xt}$.

5. Multiply and simplify: $\dfrac{x^2+3x+2}{3x+9} \cdot \dfrac{x+3}{x^2-4}$.

6. Divide and simplify: $\dfrac{8x^2y}{25xt} \div \dfrac{16x^2y^3}{30xyt^3}$.

7. Divide and simplify: $\dfrac{x^2-x}{3x^2+6x} \div \dfrac{3x-3}{3x^3+6x^2}$.

8. Add: $\dfrac{5x-4}{x-1} + \dfrac{5x+3}{x-1}$.

9. Subtract: $\dfrac{3y+7}{2y+3} - \dfrac{3(y-2)}{2y+3}$.

10. Add: $\dfrac{x+1}{x} + \dfrac{x-1}{x+1}$.

11. Simplify: $\dfrac{\frac{8x^2}{xy^3}}{\frac{4y^3}{x^2y^3}}$.

12. Simplify: $\dfrac{1+\frac{y}{x}}{\frac{y}{x}-1}$.

13. Solve: $\dfrac{x}{10} - \dfrac{1}{2} = \dfrac{x}{5}$.

14. Solve: $3x - \dfrac{2(x+3)}{3} = 16 - \dfrac{x+2}{2}$.

15. Solve: $\dfrac{7}{x+4} - \dfrac{1}{2} = \dfrac{3}{x+4}$.

16. Solve: $\dfrac{y}{y-1} = \dfrac{y-2}{y}$.

17. Cleaning highways One highway worker could pick up all the trash on a strip of highway in 7 hours, and his helper could pick up the trash in 9 hours. How long will it take them if they work together?

18. Boating A boat can motor 28 miles downstream in the same amount of time as it can motor 18 miles upstream. Find the speed of the current if the boat can motor at 23 miles per hour in still water.

19. Cycling A veteran cyclist can make a 35-mile trip in the same time that a novice cyclist can make a 20-mile trip. If the veteran is cycling 10 mph faster than the novice, how long will it take the novice to travel 20 miles?

20. Jogging Betsy can jog the 5 miles to the softball field in the same amount of time that she can cycle 20 miles to work. If she is cycling 10 mph faster than she is jogging, how fast is she jogging?

■ Cumulative Review: Chapters 5 and 6

VOCABULARY / CONCEPTS

1. True or False: The polynomial $4(x - 2) + x(x - 2)$ is completely factored.

2. True or False: The greatest common factor of the polynomial $3x^2(x - 1)^2 + 9x^3(x - 1)$ is $3x^2(x - 1)$.

3. A polynomial that contains three terms is called a _____.

4. What is the leading coefficient of the polynomial $-x^2 - 5x + 6$?

5. Why is $x^3 + x^2 + 1$ not a factorable trinomial?

6. A binomial of the form _____ is called the difference of two squares.

7. In order for a polynomial to be completely factored, each polynomial factor must be _____.

8. Quadratic equations are equation of degree ____.

✍ 9. Explain the difference between factoring $x^2 + 5x + 6$ and solving $x^2 + 5x + 6 = 0$.

10. Classify each equation below as either linear or quadratic.
   a. $4x(x - 5) = 2x(x + 1) - 3$
   b. $x^2 - 5x = 2x(x + 1) - x^2$
   c. $(x + 2)(x - 5) = (2x + 1)(x - 4)$

11. The quotient of two polynomials is called a _____ expression.

12. State the excluded value for each expression.
   a. $\dfrac{x + 5}{x - 2}$ **b.** $\dfrac{3x}{4x + 1}$ **c.** $\dfrac{5}{6x}$

13. Perform the indicated operation and simplify:
$\dfrac{14xy}{15} - \dfrac{-6xy}{15}$.

14. Determine the least common denominator for $\dfrac{5z}{6x^2y}$ and $\dfrac{3}{9xy^2}$.

15. Perform the indicated operation and simplify the result:
   a. $\dfrac{3xy}{5x^2} \cdot \dfrac{10}{6xy^2}$
   b. $\dfrac{-8y^2}{6x} \div \dfrac{2}{5y^2x}$

16. A fraction is defined to be a quotient of integers, and a _____ fraction is a quotient of fractions.

✍ 17. Is $\dfrac{\frac{1}{2} + \frac{2}{3}}{\frac{3}{4}}$ equivalent to $\dfrac{1}{2} + \dfrac{2}{3} \div \dfrac{3}{4}$?

18. Apparent solutions that are not valid are called _____ solutions.

19. Identify the excluded values and either simplify the expression or solve the equation.
   a. $\dfrac{1}{x - 3} + \dfrac{2}{x - 3} = 1$
   b. $\dfrac{1}{x - 3} + \dfrac{2}{x - 3}$

20. If you have two consecutive odd integers, using the variable x, write the two rational expressions that represent their reciprocals.

PRACTICE   *In Cumulative Review Exercises 21–28, factor the polynomials completely. If prime, so state.*

21. $x^2 - 49$

22. $x^2 - 8x - 9$

23. $18x^2 - 27x - 5$

24. $x^3 + 2x^2$

25. $x^2 - 18x + 81$

26. $5x^3 - 125x$

27. $ac - ad + bc - bd$

28. $a^3 - 8y^3$

In Cumulative Review Exercises 29–30, solve each equation by factoring. Verify your results using either the STOre or TABLE feature of the calculator.

29. $x(x + 8) - 5x = 5(1 - x) - 12$

30. $x^3 + x^2 - 12x = 0$

31. Solve graphically using the INTERSECT feature. Copy your complete screen display to justify your work. Solutions should be approximated to the nearest hundredth. $-0.65x^2 + 1.41x + 4 = 0$

32. The pictured screens display the solution to a quadratic equation. Write an equation in $ax^2 + bx + c = 0$ form that has these solutions.

33. A student entered a polynomial at y1 and at y2 entered what he *believed* to be the correct factorization of the polynomial entered at y1. He then displayed the TABLE at the right to verify his factorization. What does the TABLE reveal about the factorization and why.

X	Y1	Y2
3	-10	-2
4	-6	-2
5	0	0
6	8	4
7	18	10
8	30	18
9	44	28

X=3

34. Simplify by reducing to lowest terms: $\dfrac{4a + 4b}{a^2 + ab}$.

35. Rewrite the fraction as an equivalent fraction with the given denominator: $\dfrac{x}{x + 5}$; $x^2 - 25$.

36. Find the least common denominator and then subtract the two fractions: $\dfrac{3}{y + 2} - \dfrac{2}{y - 2}$.

37. Explain why $\dfrac{a - b}{b - a}$ is equal to -1.

38. Perform the indicated multiplication and simplify: $\dfrac{3 - a}{b^2 - b} \cdot \dfrac{b^2 - 2b + 1}{a^2 - 9}$.

39. Perform the indicated operations and simplify: $\dfrac{(x + 3)^5}{x^2 + 4x + 3} \div \dfrac{x^2 + 6x + 9}{x^2 - 3x} \cdot \dfrac{2x^2 + 8x + 6}{x^2 - 9}$.

40. Simplify $\dfrac{\dfrac{2}{3} + \dfrac{4}{5}}{\dfrac{2}{3} - \dfrac{4}{5}}$ and verify with your calculator.

41. Simplify: $\dfrac{\dfrac{a}{b} - \dfrac{b}{a}}{\dfrac{a}{b^2} - \dfrac{b}{a^2}}$.

42. Algebraically solve $\dfrac{5}{a - 3} + \dfrac{30}{9 - a^2} = 1$.

43. Graphically solve $\dfrac{6}{5x} + \dfrac{4}{5} - 2x = 0$.

44. Explain the meaning of the indicated "line" in the graphical solution of the equation $\dfrac{x-1}{x+1} = 3$ that is displayed at the right.

45. The TABLE at the right indicates that the left side of an equation has been entered at y1 and the right side at y2. Determine the excluded values, if any, and the solution of the equation.

APPLICATIONS

46. Time of flight   An object has been thrown straight up into the air. The formula $h = vt - 16t^2$ gives the height, h, of the object above the ground after t seconds with an initial velocity of v. After how many seconds will an object hit the ground if the initial velocity was 80 feet per second?

47. Fencing   A rancher wants to enclose a rectangular region for a garden because deer are eating the vegetables. If the length is to be 8 yards more than twice the width and the area is 10 square yards, find the measure of the width.

48. Data processing   Becky can enter admissions data on her computer terminal in 4 hours. Sherry can do the same job from her computer terminal in 6 hours because she is frequently interrupted by the telephone. If they work together, how long will it take to enter the admissions data?

49. Travel speed   The *Queen of the Cumberland* riverboat can travel 12 miles downstream in the same amount of time that it takes it to travel 8 miles upstream. If the current of the river is 1.5 miles per hour, what is the boat's rate in still water?

CAREERS & MATHEMATICS

Mechanical Engineer Mechanical engineers design and develop power-producing machines such as internal combustion engines, steam and gas turbines, and jet and rocket engines, as well as power-using machines such as refrigeration and air-conditioning equipment, machine tools, printing presses, and steel rolling mills. Many mechanical engineers do research, test, and design work; others work in maintenance, technical sales, and production operations. Some teach in colleges and universities or work as consultants.

Qualifications A bachelor's degree in engineering is generally required for beginning engineering jobs. College graduates with a degree in a natural science or mathematics also may qualify for some jobs.

Job Outlook Employment of mechanical engineers is expected to increase through the year 2005 due to the growing demand for complex industrial machinery and processes, as well as the need for new energy systems and solutions to environmental pollution problems.

Example Application Mechanical power is often transmitted from one location to another by means of a shaft—the drive shaft in an automobile is a good example. It is one concern of the mechanical engineer that the shaft be strong enough that it will not twist, bend, or snap. One critical value in the required calculations is the shaft's **stiffness,** denoted here by the letter k. The stiffness of a shaft is measured in units of inch pounds per radian.

If a single shaft consists of two sections (see Illustration 1), then the shaft's overall stiffness k is given by the formula

$$k = \frac{1}{\dfrac{1}{k_1} + \dfrac{1}{k_2}}$$

Section 1   Section 2

ILLUSTRATION 1

where k_1 and k_2 are the individual stiffnesses of Sections 1 and 2, respectively.

If the stiffness, k_2, of Section 2 of the shaft in Illustration 1 is 4,200,000 in. lb/rad, and the design specifies that the overall stiffness, k, of the entire shaft be 1,900,000 in. lb/rad, what must k_1, the stiffness of Section 1, be?

Solution We begin by solving the given equation for k_1.

$$k = \frac{1}{\dfrac{1}{k_1} + \dfrac{1}{k_2}}$$

$$k = \frac{1 k_1 k_2}{\left(\dfrac{1}{k_1} + \dfrac{1}{k_2}\right) k_1 k_2}$$

Simplify the complex fraction by multiplying numerator and denominator by $k_1 k_2$.

$$k = \frac{k_1 k_2}{k_2 + k_1}$$

314

$$k(k_2 + k_1) = \frac{k_1 k_2}{k_2 + k_1}(k_2 + k_1)$$ Clear the equation of fractions.

$$kk_2 + kk_1 = k_1 k_2$$

$$kk_2 = k_1 k_2 - kk_1$$ Prepare to solve for k_1 by subtracting kk_1 from both sides.

$$kk_2 = k_1(k_2 - k)$$ Factor out k_1 from the right-hand side.

$$\frac{kk_2}{k_2 - k} = k_1$$ Solve for k_1 by dividing both sides by $k_2 - k$.

To find the stiffness required for Section 1, substitute 4,200,000 for k_2 and 1,900,000 for k into the equation and find k_1.

$$k_1 = \frac{kk_2}{k_2 - k}$$

$$k_1 = \frac{(1,900,000)(4,200,000)}{4,200,000 - 1,900,000}$$

$$k_1 \approx 3,500,000$$ Use a calculator.

The stiffness of Section 1 must be approximately 3,500,000 in. lb/rad.

EXERCISES

1. Find k_1, the stiffness of Section 1, if $k_2 = 3,700,000$ in. lb/rad and $k = 1,530,000$ in. lb/rad.

2. What is the stiffness of a shaft whose two sections have the same stiffness, 4,000,000 in. lb/rad?

3. Show that the formula of this example may also be written as

$$k = \frac{k_1 k_2}{k_1 + k_2}.$$

4. The **compliance** of a shaft is the reciprocal of the stiffness. Show that the compliance of the entire shaft is the sum of the compliances of each of its sections.

Making the Transition from Elementary to Intermediate Algebra

7

■ PERSPECTIVE

The Metric System

A common metric unit of length is the kilometer, which is 1000 meters. Because 1000 is 10^3, we can write 1 km = 10^3 m. Similarly, 1 centimeter is one-hundredth of a meter: 1cm = 10^{-2} m. In the metric system, prefixes such as *kilo* and *centi* refer to powers of 10. Other prefixes are used in the metric system, as shown in the table.

Prefix	Symbol	Meaning	
peta	P	10^{15}	= 1,000,000,000,000,000.
tera	T	10^{12}	= 1,000,000,000,000.
giga	G	10^9	= 1,000,000,000.
mega	M	10^6	= 1,000,000.
kilo	k	10^3	= 1,000.
deci	d	10^{-1}	= 0.1
centi	c	10^{-2}	= 0.01
milli	m	10^{-3}	= 0.001
micro	μ	10^{-6}	= 0.000 001
nano	n	10^{-9}	= 0.000 000 001
pico	p	10^{-12}	= 0.000 000 000 001
femto	f	10^{-15}	= 0.000 000 000 000 001
atto	a	10^{-18}	= 0.000 000 000 000 000 001

To appreciate the magnitudes involved, consider these facts: Light, which travels 186,000 miles every second, will travel about 1 foot in one nanosecond. The distance to the nearest star is 43 petameters, and the diameter of an atom is about 10 nanometers. To measure some quantities, however, even these units are inadequate. The sun, for example, radiates 5×10^{26} watts. That's a lot of light bulbs!

In this chapter we will review many of the ideas covered in the first six chapters and then extend them to the intermediate algebra level. If you have trouble with any topic, review any section(s) of the text that first discusses the topic.

7.1 Negative Integral Exponents

■ REVIEW OF BASIC EXPONENT CONCEPTS ■ NEGATIVE INTEGRAL EXPONENTS ■ SCIENTIFIC NOTATION ■ USING SCIENTIFIC NOTATION TO SIMPLIFY COMPUTATIONS

■ REVIEW OF BASIC EXPONENT CONCEPTS

■ TECHNOLOGY TIP
If you are beginning this text at this chapter, it is assumed that you are familiar with the following features of the graphing calculator: basic arithmetic operations, raising to powers, extracting roots, simplifying absolute value, STOre feature, TABLE, graphing basics, and solving linear equations in one variable graphically. You should review Technology 1.1, 1.2, 3.4, and 3.5 to ensure proficiency in calculator usage.

In Section 2.1 we developed the rules for natural-number exponents. These rules are valid for both integer and rational number exponents. This section shall reexamine integer exponents, relying on your knowledge of positive integer (natural number) exponents and the zero exponent discussed in Section 2.1 to explore the area of negative integer exponents.

Points to remember

1. In the expression x^n, x is the base and n is the exponent.

2. When n is a natural number then $x^n = \overbrace{x \cdot x \cdot x \cdot x \cdot \ldots \cdot x}^{n \text{ factors of } x}$.

3. In -2^4, 2 is the base that is raised to the power of 4. The negative sign indicates we want the opposite of 2^4.

$$-2^4 = -2 \cdot 2 \cdot 2 \cdot 2 = -16$$

4. In $(-2)^4$, (-2) is the base that is raised to the power of 4.

$$(-2)^4 = (-2)(-2)(-2)(-2) = +16$$

5. If $x \neq 0$, then $x^0 = 1$. 0^0 is undefined.

■ NEGATIVE INTEGRAL EXPONENTS

When we first examined exponents in Section 2.1 we considered quotients such as $\frac{6^5}{6^2}$.

$$\frac{6^5}{6^2} = \frac{\overset{1}{\cancel{6}} \cdot \overset{1}{\cancel{6}} \cdot 6 \cdot 6 \cdot 6}{\underset{1}{\cancel{6}} \cdot \underset{1}{\cancel{6}}} = \frac{6^3}{1} \text{ or } 6^3 \qquad \text{Simplify using properties of fractions.}$$

or

$$\frac{6^5}{6^2} = 6^{5-2} = 6^3 \qquad \text{Use the quotient rule for exponents.}$$

In these problems, the exponent in the numerator was always larger than the exponent in the denominator. We will now investigate what happens when the exponent in the denominator is larger than the exponent in the numerator.

EXPLORING THE CONCEPT Use your calculator to simplify each pair of expressions below. Use the ►frac feature to display all results as fractions.

a. $\frac{6^2}{6^5}$, 6^{-3}

b. $\frac{4^4}{4^5}$, 4^{-1}

Conclusions **a.**   **b.**

If we apply the quotient rule to the fraction $\frac{6^2}{6^5}$, where the exponent in the numerator is less than the exponent in the denominator, we obtain

$$\frac{6^2}{6^5} = 6^{2-5} = 6^{-3}$$

However, we know that

$$\frac{6^2}{6^5} = \frac{\overset{1}{\cancel{6}} \cdot \overset{1}{\cancel{6}}}{\underset{1}{\cancel{6}} \cdot \underset{1}{\cancel{6}} \cdot 6 \cdot 6 \cdot 6} = \frac{1}{6^3}$$

To make the results of 6^{-3} and $\frac{1}{6^3}$ consistent, we define 6^{-3} to be equal to $\frac{1}{6^3}$. In general,

> **Negative Exponent**
>
> If x is any nonzero number and n is a natural number, then $x^{-n} = \dfrac{1}{x^n}$.

■

EXAMPLE 1   Express each quantity without using negative exponents. Assume that no denominators are zero.

a. $3^{-5} = \dfrac{1}{3^5}$

$\qquad\quad = \dfrac{1}{243}$

b. $x^{-4} = \dfrac{1}{x^4}$

■

EXPLORING THE CONCEPT   Use your calculator to simplify each pair of expressions below. Use the ▶frac feature to display all results as fractions.

a. $\dfrac{1}{3^{-1}}, \dfrac{3^{-1}}{1}$

b. $\dfrac{2}{3^{-1}}, \dfrac{3^{-1}}{2}$

Conclusions **a.**

```
1/3^-1▶Frac
                3
3^-1/1▶Frac
              1/3
```

b.

```
2/3^-1▶Frac
                6
3^-1/2▶Frac
              1/6
■
```

■

EXAMPLE 2   Use the definition of a negative exponent to simplify each quantity. Results should be expressed with no negative exponents; assume that no denominators are zero.

a. $\dfrac{2}{3^{-1}} = \dfrac{2}{\dfrac{1}{3^1}}$ Definition of negative exponent.

$\qquad\quad = 2 \div \dfrac{1}{3^1}$ Express the division.

$\qquad\quad = 2 \cdot \dfrac{3^1}{1}$ Dividing is the same as multiplying by the reciprocal.

$\qquad\quad = 2 \cdot 3$

$\qquad\quad = 6$

b. $\dfrac{3^{-1}}{2} = \dfrac{\frac{1}{3}}{2}$ Definition of a negative exponent.

$= \dfrac{1}{3} \div 2$ Express the division.

$= \dfrac{1}{3} \cdot \dfrac{1}{2}$ Dividing is the same as multiplying by the reciprocal.

$= \dfrac{1}{6}$

c. $\dfrac{2}{x^{-3}} = \dfrac{2}{\frac{1}{x^3}}$ Definition of a negative exponent.

$= 2 \div \dfrac{1}{x^3}$ Express the division.

$= 2 \cdot \dfrac{x^3}{1}$ Dividing is the same as multiplying by the reciprocal.

$= 2x^3$

Because of the definition of negative and zero exponents, the product, power, and quotient rules discussed in Section 2.1 are also true for all integral exponents. We restate the properties of exponents for integral exponents.

Properties of Exponents

If m and n are integers and there are no divisions by 0, then

$$x^m x^n = x^{m+n} \qquad (x^m)^n = x^{mn} \qquad (xy)^n = x^n y^n \qquad \left(\dfrac{x}{y}\right)^n = \dfrac{x^n}{y^n}$$

$$x^0 = 1 \ (x \neq 0) \qquad x^{-n} = \dfrac{1}{x^n} \qquad \dfrac{x^m}{x^n} = x^{m-n}$$

EXAMPLE 3   Use the properties of exponents to simplify each expression. Assume that no denominator is zero.

■ STUDY TIP

For an exponential expression to be simplified, each of the following must be true:

a. There should be no negative or zero exponents.

b. The expression should contain no parentheses.

c. Each base should appear only once.

a. $(x^{-3})^2 = x^{-6}$ $(x^m)^n = x^{mn}$

$= \dfrac{1}{x^6}$ $x^{-n} = \dfrac{1}{x^n}$

b. $\dfrac{x^3}{x^7} = x^{3-7}$ $\dfrac{x^m}{x^n} = x^{m-n}$

$= x^{-4}$

$= \dfrac{1}{x^4}$ $x^{-n} = \dfrac{1}{x^n}$

c. $\dfrac{y^{-4}y^{-3}}{y^{-20}} = \dfrac{y^{-7}}{y^{-20}}$ $x^m \cdot x^n = x^{m+n}$

$= y^{-7-(-20)}$ $\dfrac{x^m}{x^n} = x^{m-n}$

$= y^{-7+20}$

$= y^{13}$

d. $\dfrac{12a^3b^4}{4a^5b^2} = \dfrac{12}{4} \cdot \dfrac{a^3}{a^5} \cdot \dfrac{b^4}{b^2}$ Multiplication of fractions.

$= 3 \cdot a^{3-5} \cdot b^{4-2}$ Division of integers and $\dfrac{x^m}{x^n} = x^{m-n}$.

$= 3a^{-2}b^2$ Simplify.

$= \dfrac{3}{1} \cdot \dfrac{1}{a^2} \cdot \dfrac{b^2}{1}$ $x^{-n} = \dfrac{1}{x^n}$

$= \dfrac{3b^2}{a^2}$ Multiplication of fractions.

e. $\left(-\dfrac{x^3y^2}{xy^{-3}}\right)^{-2} = (-x^{3-1}y^{2-(-3)})^{-2}$ Order of operations: simplify inside grouping symbols first.

$\dfrac{x^m}{x^n} = x^{m-n}$

$= (-x^2y^5)^{-2}$ Simplify.

$= \dfrac{1}{(-x^2y^5)^2}$ $x^{-n} = \dfrac{1}{x^n}$

$= \dfrac{1}{x^4y^{10}}$ $(xy)^n = x^ny^n$

f. $\dfrac{x^{-1} + y^{-2}}{x^{-2} - y^{-1}} = \dfrac{\dfrac{1}{x} + \dfrac{1}{y^2}}{\dfrac{1}{x^2} - \dfrac{1}{y}}$ $x^{-n} = \dfrac{1}{x^n}$

$= \dfrac{x^2y^2\left(\dfrac{1}{x} + \dfrac{1}{y^2}\right)}{x^2y^2\left(\dfrac{1}{x^2} - \dfrac{1}{y}\right)}$ Multiply numerator and denominator by the LCD x^2y^2.

$= \dfrac{xy^2 + x^2}{y^2 - x^2y}$ Use the distributive property to clear parentheses.

$= \dfrac{x(y^2 + x)}{y(y - x^2)}$ Attempt to simplify the fraction by factoring the numerator and denominator.

The result cannot be simplified further. ◼

SCIENTIFIC NOTATION

Scientists often deal with extremely large and extremely small numbers. For example, the distance from the earth to the sun is approximately 150,000,000 kilometers, and ultraviolet light emitted from a mercury arc has a wavelength of approximately 0.000025 centimeters. The large number of zeros in these numbers makes them difficult to read and hard to remember. To make such numbers easier to work with, scientists use a compact form of notation called **scientific notation.**

Scientific Notation
A number is written in **scientific notation** if it is written as the product of a number between 1 (including 1) and 10 and an integer power of 10.

EXPLORING THE CONCEPT

Change your calculator mode from normal display to **SCI**entific notation display so that all results will be displayed in scientific notation. Enter each of the following on your calculator:

a. 30,000

b. 3×10^4

c. Compare the display to the definition of scientific notation and explain the calculator's approach to displaying numbers in scientific notation.

d. Display 30×10^3 on the calculator. Based on the calculator display, is 30×10^3 in scientific notation? Based on the definition of scientific notation, is it in the appropriate form?

Conclusion

a.
b.

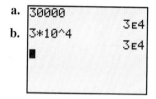

c. Instead of 3×10^4 the calculator displays 3 E 4. The notation E 4 indicates "times ten to the fourth power."

d. 30×10^3 is NOT in scientific notation because 30 is not between 1 and 10. ■

EXAMPLE 4

Change 150,000,000 to scientific notation.

Solution

To write the number 150,000,000 in scientific notation, we must write it as a product of a number between 1 and 10 and some power of 10. Note that the number 1.5 lies between 1 and 10.

To obtain the number 150,000,000, the decimal point in the number 1.5 must be moved eight places to the right. Because multiplying a number by 10 moves the decimal place one place to the right, we can accomplish this by multiplying 1.5 by 10 eight times. To see this, we count from the decimal point in 1.5 to where the decimal point should be in 150,000,000:

1 . 5 0 0 0 0 0 0 0.

8 places to the right

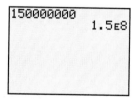

The number 150,000,000 written in scientific notation is 1.5×10^8. ■

EXAMPLE 5

Change 0.000025 to scientific notation.

Solution

To write the number 0.000025 in scientific notation, we write it as a product of a number between 1 and 10 and some power of 10. To obtain the number 0.000025, the decimal point in the number 2.5 must be moved five places to the left. We can accomplish this by dividing 2.5 by 10^5, which is equivalent to multiplying 2.5 by

$1/10^5$, or 10^{-5}. To do this, we count from the decimal point in 2.5 to where the decimal point should be in 0.000025:

.00002.5

5 places to the left

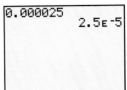

The number 0.000025 written in scientific notation is 2.5×10^{-5}. ■

We can change a number written in scientific notation to **standard notation.** For example, to write the number 9.3×10^7 in standard notation, we multiply 9.3×10^7.

$$9.3 \times 10^7 = 9.3 \times 10{,}000{,}000$$
$$= 93{,}000{,}000$$

EXAMPLE 6   Write   **a.** 3.4×10^5 and   **b.** 2.1×10^{-4} in standard notation.

Solution   **a.** $3.4 \times 10^5 = 3.4 \times 100{,}000$
$$= 340{,}000$$

b. $2.1 \times 10^{-4} = 2.1 \times \dfrac{1}{10^4}$

$$= 2.1 \times \dfrac{1}{10{,}000}$$

$$= 0.00021$$

■ **TECHNOLOGY TIP**
Be sure you have reset the calculator mode to "normal." To quickly enter an expression such as 3.4×10^5, you can use the calculator method of display. Locate the EE key on your calculator and enter 3.4×10^5 as 3.4 EE 5.

To summarize, each of the following numbers is written in both scientific and standard notation. In each case, the exponent gives the number of places that the decimal point moves, and the sign of the exponent indicates the direction that it moves.

$5.32 \times 10^5 = 5\,3\,2\,0\,0\,0.$ 5 places to the right.
$2.37 \times 10^6 = 2\,3\,7\,0\,0\,0\,0.$ 6 places to the right.
$8.95 \times 10^{-4} = 0.\,0\,0\,0\,8\,9\,5$ 4 places to the left.
$8.375 \times 10^{-3} = 0.\,0\,0\,8\,3\,7\,5$ 3 places to the left.
$9.77 \times 10^0 = 9.77$ No movement of the decimal point.

■

EXAMPLE 7   Write 432.0×10^5 in scientific notation.

Solution   The number 432.0×10^5 IS NOT written in scientific notation because 432.0 is not a number between 1 and 10. To write the number in scientific notation, we proceed as follows:

$$432.0 \times 10^5 = 4.32 \times 10^2 \times 10^5$$ Write 432.0 in scientific notation.

$$= 4.32 \times 10^7$$ $10^2 \times 10^5 = 10^7$.

■ **TECHNOLOGY TIP**
Be sure your calculator is set to the desired mode.

▥ USING SCIENTIFIC NOTATION TO SIMPLIFY COMPUTATIONS

Another advantage of scientific notation becomes apparent when simplifying fractions such as

$$\frac{(320)(25,000)}{0.000040}$$

that contain very large or very small numbers. Although we can simplify this fraction using ordinary arithmetic, scientific notation provides an easier way. First we write each number in scientific notation and then do the arithmetic on the numbers and the exponential expressions separately. Then we write the answer in standard form, if desired.

▥ **TECHNOLOGY TIP**
The calculator will automatically display very large or very small numbers in scientific notation regardless of the mode setting.

$$\frac{(320)(25,000)}{0.000040} = \frac{(3.2 \times 10^2)(2.5 \times 10^4)}{4.0 \times 10^{-5}}$$

$$= \frac{(3.2)(2.5)}{4.0} \times \frac{10^2 10^4}{10^{-5}}$$

$$= \frac{8.0}{4.0} \times 10^{2+4-(-5)}$$

$$= 2.0 \times 10^{11}$$

$$= 200,000,000,000$$

EXAMPLE 8   **Speed of light**   In a vacuum, light travels 1 meter in approximately 0.000000003 second. How long does it take for light to travel 500 kilometers?

Solution   Because 1 kilometer is equal to 1000 meters, the length of time for light to travel 500 kilometers (500 · 1000 meters) is given by

$$(0.000000003)(500)(1000) = (3 \times 10^{-9})(5 \times 10^2)(1 \times 10^3)$$

$$= 3(5) \times 10^{-9+2+3}$$

$$= 15 \times 10^{-4}$$

$$= \mathbf{1.5 \times 10^1} \times 10^{-4}$$

$$= 1.5 \times 10^{-3}$$

$$= 0.0015$$

Light travels 500 kilometers in approximately 0.0015 second.   ▪

EXERCISE 7.1

VOCABULARY   *Fill in the blank to make a true statement.*

1. A number written in _____ is the product of a number between 1 (including 1) and 10 and an integer power of 10.

CONCEPTS

2. In the expression x^n, __ is the base and __ is the exponent.

3. In the expression -3^2, __ is the base and the expanded form of the number is ___ · __.

4. In the expression $(-3)^2$, ____ is the base and the expanded form is ____ · ____.

5. If $x \neq 0$, then $x^0 =$ __.

6. 0^0 is said to be _____.

7. If $x \neq 0$ and n is a natural number, then $x^{-n} =$ ___.

8. Express each quantity without using negative exponents. Verify your results with your calculator.

   a. 2^{-3} **b.** 3^{-1} **c.** $\dfrac{1}{2^{-3}}$ **d.** $\dfrac{1}{3^{-1}}$

9. Review the properties of exponents that were introduced in Section 2.1 by simplifying each expression below.

   a. $x^4 \cdot x^7$ **b.** $(x^4)^7$ **c.** $\dfrac{x^7}{x^4}$

   d. $\left(\dfrac{x}{y}\right)^7$ **e.** $(xy)^3$ **f.** $(x^2y^3)^4$

10. Convert each pair of expressions written in scientific notation to standard notation using your calculator.

   a. 4.2×10^3 and 4.2×10^{-3}
   b. 1.513×10^2 and 1.513×10^{-2}

PRACTICE *In Exercises 11–16, simplify each expression using the laws of exponents. Answers should be expressed without parentheses or negative exponents and should be verified with your calculator.*

11. $2^5 \cdot 2^{-2}$

12. $10^2 \cdot 10^{-4}$

13. $\dfrac{4^{-5}}{4^{-3}}$

14. $\dfrac{2^{-3}}{2^3}$

15. $\dfrac{2^3 \cdot 2^{-1}}{2^2}$

16. $\dfrac{3^{-2} \cdot 3^4}{3^3}$

In Exercises 17–46, use the laws of exponents to simplify each expression. Answers should be expressed without using parentheses or negative exponents.

17. x^{-2}

18. y^{-3}

19. $(2y)^{-4}$

20. $(-3x)^{-1}$

21. $(ab^2)^{-3}$

22. $(m^2n^3)^{-2}$

23. $\dfrac{y^4y^3}{y^{-4}y^{-2}}$

24. $\dfrac{x^{12}x^{-7}}{x^3x^4}$

25. $(x^{-4}x^3)^3$

26. $(y^2y)^3$

27. $(-2x^3y^{-2})^{-5}$

28. $(3u^{-2}v)^{-3}$

29. $\left(\dfrac{a^3}{a^{-4}}\right)^2$

30. $\left(\dfrac{a^4}{a^{-3}}\right)^3$

31. $\left(\dfrac{x^5}{y^{-2}}\right)^{-2}$

32. $\left(\dfrac{x^{-2}}{y^3}\right)^{-3}$

33. $\left(\dfrac{4x^2}{3x^{-5}}\right)^4$

34. $\left(\dfrac{-3x^4x^{-3}}{x^{-3}x^7}\right)^3$

35. $\left(\dfrac{12y^3z^{-2}}{3y^{-4}z^3}\right)^2$

36. $\left(\dfrac{6xy^3}{3x^{-1}y}\right)^3$

37. $\left(\dfrac{9u^2v^3}{18u^{-3}v}\right)^4$

38. $\left(\dfrac{14u^{-2}v^3}{21u^{-3}v}\right)^4$

39. $\dfrac{16(x^{-2}yz)^{-2}}{(2x^{-3}z^0)^4}$

40. $\dfrac{(2x^{-2})^{-2}}{4(x^2y)^{-1}}$

41. $\dfrac{1+x^{-1}}{x^{-1}-1}$

42. $\dfrac{y^{-2}+1}{y^{-2}-1}$

43. $\dfrac{a^{-2}+a}{a+1}$

44. $\dfrac{x-x^{-2}}{1-x^{-2}}$

45. $\dfrac{2x^{-1}+4x^{-2}}{2x^{-2}+x^{-1}}$

46. $\dfrac{x^{-2}-3x^{-3}}{3x^{-2}-9x^{-3}}$

In Exercises 47–54, write each number in scientific notation and verify with your calculator.

47. 23,000

48. 4750

49. 0.062

50. 0.00073

51. 42.5×10^2

52. 0.3×10^3

53. 0.25×10^{-2}

54. 25.2×10^{-3}

In Exercises 55–62, write each number in standard notation and verify with your calculator.

55. 2.3×10^2

56. 3.75×10^4

57. 1.14×10^{-3}

58. 4.9×10^{-2}

59. 25×10^6

60. 0.07×10^3

61. 0.51×10^{-3}

62. 617×10^{-2}

In Exercises 63–65, convert each calculator display to the indicated notation.

63.
```
4.2E-5
```
a. scientific notation

b. standard notation

64.
```
-2.00001E7
```
a. scientific notation

b. standard notation

65.
```
-1.0000001E-11
```
a. scientific notation

b. standard notation

66. "Round-off" error can occur when performing calculator computations. When this happens, the calculator will yield a value that is very close to the anticipated value and yet not the exact value. For example, you may expect the result to be 5 but the calculator may display 5.0001 or 4.9999 instead. If the calculator displayed the value -1.0000001 E -11 as a result of a round-off error, what value is being represented?

In Exercises 67–72, use scientific notation to simplify each expression. Verify your work with your calculator. Give answers in both standard and scientific notation.

67. $(3.4 \times 10^2)(2.1 \times 10^3)$

68. $(4.1 \times 10^{-3})(3.4 \times 10^4)$

69. $\dfrac{9.3 \times 10^2}{3.1 \times 10^{-2}}$

70. $\dfrac{7.2 \times 10^6}{1.2 \times 10^8}$

71. $\dfrac{(12,000)(3600)}{0.0003}$

72. $\dfrac{(0.0004)(0.0012)}{80,000}$

APPLICATIONS

73. Distance to Alpha Centauri The distance from the earth to the nearest star outside our solar system is approximately 25,700,000,000,000 miles. Express this number in scientific notation.

74. Distance to Venus The distance from Venus to the sun is approximately 6.7×10^7 miles. Express this number in standard notation.

75. Speed of sound The speed of sound in air is 33,100 centimeters per second. Express this number in scientific notation.

76. Length of a meter One meter is approximately 0.00622 mile. Use scientific notation to express this number.

77. Distance to Mars The distance from Mars to the sun is approximately 1.14×10^8 miles. Express this number in standard notation.

78. Angstrom One angstrom is 1×10^{-7} millimeter. Express this number in standard notation.

79. Distance between Mercury and the sun The distance from Mercury to the sun is approximately 3.6×10^7 miles. Use scientific notation to express this distance in feet. (*Hint:* 5280 feet = 1 mile.)

80. Speed of sound The speed of sound in air is approximately 3.3×10^4 centimeters per second. Use scientific notation to express this speed in kilometers per second. (*Hint:* 100 centimeters = 1 meter, and 1000 meters = 1 kilometer.)

81. Mass of a proton The mass of one proton is approximately 1.7×10^{-24} gram. Use scientific notation to express the mass of 1 million protons.

82. Light year One light year is approximately 5.87×10^{12} miles. Use scientific notation to express this distance in feet. (*Hint:* 5280 feet = 1 mile.)

7.2 Review of Factoring

■ REVIEW OF FACTORING TECHNIQUES

Points to remember

1. To factor means to write as a product. The expression $4(x + 5) + y(x + 5)$ is NOT expressed as a product; it is the sum of two terms, $4(x + 5)$ and $y(x + 5)$.

2. Factoring out a negative 1 simply changes the sign of each term of the original polynomial.

3. The factors $(a + b)$ and $(b + a)$ are the same because the operation of addition is commutative.

4. The factors $(a - b)$ and $(b - a)$ are additive inverses (opposites). Thus $a - b = -(b - a)$.

5. The sum of squares, $x^2 + y^2$, does NOT factor over the set of integers. It is a prime polynomial.

■ REVIEW OF FACTORING TECHNIQUES

To completely factor a polynomial means to write the given polynomial as a *product* of prime polynomial factors. Factoring can always be checked by multiplying factors and ensuring that the product of factors is equivalent to the original polynomial. Remember to check that each polynomial factor is prime before stopping. Also, recall that there should be no *ungrouped* + or − signs in your final result. Below is a review of techniques that will help you in your factoring.

Factoring Polynomials

1. Always check for and factor out the greatest common factor (GCF) first. If the leading coefficient is negative, factor out a −1 with/as the GCF.

2. Count the number of terms in the polynomial, and then follow the given procedures for that specific type of polynomial.

Binomial

If it is the difference of two squares: $a^2 - b^2 = (a - b)(a + b)$.

If it is the difference of two cubes: $a^3 - b^3 = (a - b)(a^2 + ab + b^2)$.

If it is the sum of two cubes: $a^3 + b^3 = (a + b)(a^2 - ab + b^2)$.

If it is *not* one of these problem types it is prime. Refer to Section 5.3 if you need more practice in factoring binomials.

Trinomials

Make sure the trinomial is in descending powers of a variable and then attempt to factor it by either grouping or trial and error. The special product formulas below may be used if desired, but are not required:

$$x^2 + 2xy + y^2 = (x + y)^2$$
$$x^2 - 2xy + y^2 = (x - y)^2$$

Refer to Section 5.2 if you need more practice in factoring trinomials.

Four-Term Polynomials

Try to factor by grouping. Refer to Section 5.1 if you need more practice in factoring polynomials that contain four terms.

EXAMPLE 1   Factor: $-12y^2 + 20y$.

Solution   Recall the distributive property: $a(b + c) = ab + ac$. This property allows distribution of multiplication over both addition and subtraction. It will be used to factor out the **greatest common factor.**

$$-12y^2 = -1 \cdot 2^2 \cdot 3y^2$$
$$20y = 2^2 \cdot 5y$$
$$-1 \cdot 2^2 \cdot y = -4y$$

Determine the GCF of the terms of the polynomial.

The GCF, $-4y$, is the product of the common factors and the factor of -1.

$$-12y^2 + 20y = -4y \cdot 3y + (-4y)(-5)$$

Rewrite each term as the product of the GCF and another factor.

$$= -4y(3y - 5)$$

Use the distributive property.

Check the result by verifying that $-4y(3y - 5) = -12y^2 + 20y$.   ■

EXAMPLE 2   Factor: $(a + b)x + (a + b)y$.

Solution   The polynomial has two terms: $(a + b)x$ and $(a + b)y$. The greatest common factor of these two terms is the *quantity* $(a + b)$. Therefore, $(a + b)x + (a + b)y = (a + b)(x + y)$.

We can verify the result by multiplication:

$$(a + b)(x + y) = (a + b)x + (a + b)y$$ ■

EXAMPLE 3   Factor: $ax + ay + cx + cy$.

Solution   Although there is no factor common to all four terms, there is a common factor of a in $ax + ay$ and a common factor of c in $cx + cy$. We can factor out the a and c as follows:

$$ax + ay + cx + cy = (ax + ay) + (cx + cy)$$

Associative property of addition.

$$= a(x + y) + c(x + y)$$

Distributive property (factoring the GCF of a out of the first two terms and the GCF of c out of the last two terms).

$$= (x + y)(a + c)$$

Distributive property (factoring the GCF of the quantity $(x + y)$ out of each term).

Therefore, $ax + ay + cx + cy = (x + y)(a + c)$.   ■

This type of factoring is called **factoring by grouping.**

EXAMPLE 4   Factor: $6x^3 + 3x - 11x^2$.

Solution
$$6x^3 + 3x - 11x^2 = 6x^3 - 11x^2 + 3x$$

Put the polynomial in descending powers of a variable.

$$= x(6x^2 - 11x + 3)$$

Factor out the GCF of x.

The remaining trinomial, $6x^2 - 11x + 3$, will be factored by grouping. To do this, you will need the trinomial written in $ax^2 + bx + c$ format. If you prefer to factor by the trial-and-error method, you should refer back to Section 5.2.

To factor the trinomial, identify a, b, and c: $a = 6$, $b = -11$, $c = 3$; and find the product $a \cdot c : a \cdot c = 6(3) = 18$. Because the product of the two numbers is positive and the sum of the same two numbers must equal b, which is negative, the desired factors must both be negative.

Two numbers	Product is 18	Sum is -11
$-1, -18$	$(-1)(-18) = 18$	$-1 + (-18) = -19$
$-2, -9$	$(-2)(-9) = 18$	$-2 + (-9) = -11\checkmark$

The required numbers are -2 and -9.

Replace the middle term of the trinomial, $-11x$, using -2 and -9 as the coefficients. Do not forget the GCF of x when working with the trinomial. Factor by grouping.

$$x[6x^2 - 2x - 9x + 3] = x[2x(3x - 1) - 3(3x - 1)] \qquad \text{Factor by grouping.}$$
$$= x(3x - 1)(2x - 3) \qquad \text{Distributive property.}$$

■

EXAMPLE 5   Factor: $4y^4 - 25z^2$.

Solution   Because the binomial $4y^4 - 25z^2$ can be written in the form $(2y^2)^2 - (5z)^2$, it represents the difference of the squares of $2y^2$ and $5z$. Thus, it factors into the sum of these two quantities times the difference of these two quantities.

$$4y^4 - 25z^2 = (2y^2)^2 - (5z)^2$$
$$= (2y^2 + 5z)(2y^2 - 5z)$$

Verify this result by multiplication.

■

WARNING!   The binomial $25a^2 + 9b^2$ is called the **sum of two squares** because it can be written in the form $(5a)^2 + (3b)^2$. Such binomials cannot be factored if we are limited to integer coefficients. Polynomials that do not factor over the integers are called **irreducible** or **prime polynomials.**

EXAMPLE 6   Factor: $a^3 + 8$.

Solution   Since $a^3 + 8$ can be written as $a^3 + 2^3$, we have the sum of two cubes, which factors as follows:

$$F^3 + L^3 = (F + L)(F^2 - FL + L^2) \qquad \text{Recall that } F = \text{first term and } L = \text{last term.}$$
$$a^3 + 2^3 = (a + 2)(a^2 - a2 + 2^2)$$
$$= (a + 2)(a^2 - 2a + 4)$$

Thus, $a^3 + 8$ factors as the product of the sum of a and 2 and the trinomial $a^2 - 2a + 4$.

We can check by multiplication.

$$(a + 2)(a^2 - 2a + 4) = (a + 2)a^2 - (a + 2)2a + (a + 2)4$$
$$= a^3 + 2a^2 - 2a^2 - 4a + 4a + 8$$
$$= a^3 + 8$$

■

EXAMPLE 7 Factor: $a^3 - 64b^3$.

Solution Since $a^3 - 64b^3$ can be written as $a^3 - (4b)^3$, we have the difference of two cubes, which factors as follows:

$$F^3 - L^3 = (F - L)(F^2 + FL + L^2)$$
$$a^3 - (4b)^3 = (a - 4b)[a^2 + a(4b) + (4b)^2]$$
$$= (a - 4b)(a^2 + 4ab + 16b^2)$$

Thus, it factors as $(a - 4b)(a^2 + 4ab + 16b^2)$.
We can check by multiplication.

$$(a - 4b)(a^2 + 4ab + 16b^2) = (a - 4b)a^2 + (a - 4b)4ab + (a - 4b)16b^2$$
$$= a^3 - 4a^2b + 4a^2b - 16ab^2 + 16ab^2 - 64b^3$$
$$= a^3 - 64b^3$$

EXERCISE 7.2

VOCABULARY *In Exercises 1–3, fill in the blanks to make a true statement.*

1. To ___factor___ means to write as a product.

2. The GCF is the ___larg___ (smallest/largest) factor common to all terms.

3. ___Prime___ polynomials are polynomials that cannot be factored over the set of integers.

CONCEPTS

4. Multiply each of the following:
  a. $(x - 3)(x - 3)$
  b. $(x - 4)^2$
  c. $(x - 4)(x + 4)$
  d. $(2x + 3)(x - 5)$

5. Find the GCF of each pair of terms.
  a. $3y^3, 3y^2$
  b. $63x^3y^2, 81x^2y^4$

6. Factor out a factor of negative 1 from each expression. Verify by multiplication.
  a. $-5x - 6$ **b.** $-x^2 + 6x - 9$
  c. $a - b$ **d.** $b - a$

PRACTICE *In Exercises 7–14, factor each expression. Check your results by using the distributive property to multiply your final factorization.*

7. $2x + 8$
8. $3y - 9$
9. $2x^2 - 6x$
10. $3y^3 + 3y^2$
11. $15x^2y - 10x^2y^2$
12. $63x^3y^2 + 81x^2y^4$
13. $27z^3 + 12z^2 + 3z$
14. $25z^6 - 10z^3 + z^2$

In Exercises 15–18, factor out the negative of the greatest common factor. Check your results by using the distributive property to multiply your final factorization.

15. $-3a - 6$
16. $-6y + 12$
17. $-6x^2 - 3xy$
18. $-15y^3 + 25y^2$

In Exercises 19–24, factor by grouping. Factor out all common monomials first. Check your results by multiplication.

19. $ax + bx + ay + by$
20. $ar - br + as - bs$
21. $x^2 + yx + 2x + 2y$
22. $2c + 2d - cd - d^2$
23. $3c - cd + 3d - c^2$
24. $x^2 + 4y - xy - 4x$

In Exercises 25–44, factor each trinomial, if possible. If the coefficient of the first term is negative, begin by factoring out negative 1. Check your results by multiplication.

25. $x^2 + 5x + 6$
26. $y^2 + 7y + 6$
27. $3x^2 + 12x - 63$
28. $2y^2 + 4y - 48$

29. $-a^2 + 4a + 32$
30. $-x^2 - 2x + 15$
31. $6y^2 + 7y + 2$
32. $6x^2 - 11x + 3$

33. $8a^2 + 6a - 9$ **34.** $6z^2 + 17z + 12$ **35.** $a^2 - 3ab - 4b^2$ **36.** $b^2 + 2bc - 80c^2$

37. $-3a^2 + ab + 2b^2$ **38.** $-2x^2 + 3xy + 5y^2$ **39.** $3x^3 - 10x^2 + 3x$ **40.** $6y^3 + 7y^2 + 2y$

41. $x^4 + 8x^2 + 15$ **42.** $x^4 + 11x^2 + 24$ **43.** $y^4 - 13y^2 + 30$ **44.** $y^4 - 13y^2 + 42$

In Exercises 45–70, factor each expression. Be sure that your answer is completely factored and checks by multiplication.

45. $x^2 - 4$ **46.** $y^2 - 9$

47. $9y^2 - 64$ **48.** $16x^4 - 81y^2$

49. $81a^4 - 49b^2$ **50.** $64x^6 - 121y^2$

51. $(x + y)^2 - z^2$ **52.** $a^2 - (b - c)^2$

53. $x^4 - y^4$ **54.** $16a^4 - 81b^4$

55. $2x^2 - 288$ **56.** $8x^2 - 72$

57. $2x^3 - 32x$ **58.** $3x^3 - 243x$

59. $x^3 - 8$ **60.** $y^3 - 27$

61. $y^3 - 64$ **62.** $x^3y^3 - 125$

63. $125 + a^3$ **64.** $64 + y^3$

65. $x^3 + 8y^3$ **66.** $27x^3y^3 + 8$

67. $2x^3y - 16y$ **68.** $a^4b - ab^4$

69. $(a - b)^3 - 8$ **70.** $27 - (a - b)^3$

In Exercises 71–80, factor each expression. These problems represent a random mixture of the types of problems reviewed in this section. For more practice, return to Section 5.3. Check your results by multiplication.

71. $-3a^2 - 6ab - 3b^2$ **72.** $4x^2 + 4xy + y^2$ **73.** $-x^3 - 2x^2 - x$ **74.** $36a^2 - 1$

75. $x^2 - 4x + 3$ **76.** $x^2 - x - 6$ **77.** $y^3 - 216$ **78.** $4x^2 + 4x$

79. $-21x^7 + 14x^5 - 7x^4$ **80.** $125 + b^3$

REVIEW *In Exercises 81 and 82, simplify each expression using the properties of exponents. Leave no negative exponents in your answers.*

81. $4^{-5} \cdot 4^{-3}$ **82.** $\left(\dfrac{2x^{-3}}{y^{-2}} \right)^{-2}$

83. Write each expression in standard notation.   **84.** Write each number in scientific notation.
    a. 9.76×10^{-5} **b.** 8.12×10^5 **a.** $1{,}700{,}000$ **b.** 0.0000000051

7.3 Equation Review

■ ALGEBRAIC SOLUTIONS OF LINEAR EQUATIONS IN ONE VARIABLE ■ IDENTITIES AND IMPOSSIBLE EQUATIONS ■ LITERAL EQUATIONS AND FORMULAS ■ PROBLEM SOLVING ■ ABSOLUTE VALUE EQUATIONS ■ FACTORABLE POLYNOMIAL EQUATIONS ■ EQUATIONS WITH RATIONAL EXPRESSIONS

Points to remember

1. The addition and subtraction properties of equality allow you to add or subtract the same quantity from both sides of an equation.

2. The multiplication and division properties of equality allow you to multiply or divide both sides of an equation by the same nonzero quantity.

■ STUDY TIP
You should get in the habit of checking your work. Referring to the answer key in the back of the text is one approach. However, not all the answers appear in the back of the text and when taking an exam, you will not find the answers at the back of the test! Your method of checking should be a different approach from what you used on the problem to ensure that you do not repeat a previous error. Learning to find your own errors is as important as learning to solve the problem itself.

3. Equations involving absolute value will have

   a. Two solution parts if the absolute value is equal to a positive constant.

   b. One solution if the absolute value is equal to 0.

   c. No solution if the absolute value is equal to a negative number.

4. When the equation involves a factorable polynomial, the zero product property allows you to set the factored polynomial equal to zero and then solve by setting each factor equal to zero.

5. When solving equations involving a rational expression, multiply through the equation by the least common denominator to clear fractions. Solve the resulting equation.

■ ALGEBRAIC SOLUTIONS OF LINEAR EQUATIONS IN ONE VARIABLE

To **solve** an equation means to find the value(s) of the variable(s) that will make the equation a true statement. When solving linear equations in one variable it is best to first simplify both sides of the equation and then begin transforming the equation by using the properties of equality to write a series of equivalent equations. To solve any equation, we must isolate the variable on one side. This can be a multistep process that may require combining like terms. As we solve equations, we will use the following steps:

■ STUDY TIP
We suggest that you write the steps on a note card, prop it in front of you, and carefully follow it when solving equations.

Steps for Solving a Linear Equation in One Variable

1. Use the distributive property to remove grouping symbols.

2. Combine like terms on each **side** of the equation.

3. Clear fractions, if desired, by multiplying each **term** on both sides of the equation by the LCD.

4. Use the addition or subtraction property of equality to collect all terms containing the variable on one side of the equation and all constants on the other side.

5. Use the multiplication or division property of equality to isolate the variable. The equation should be in the form:

   variable = constant

6. Check by ensuring that the proposed solution produces a true statement when the variable is replaced by the constant. Always use the original equation when checking.

EXAMPLE 1   Solve: $3(x + 2) - 2x = 4x$.

Solution

$$3x + 6 - 2x = 4x \qquad \text{Use the distributive property to remove grouping.}$$

$$x + 6 = 4x \qquad \text{Combine like terms.}$$

$$x - x + 6 = 4x - x \qquad \text{Use the subtraction property of equality to get all variable terms together.}$$

$$6 = 3x \qquad \text{Simplify each side.}$$

$$\frac{6}{3} = \frac{3x}{3} \qquad \text{Use the division property of equality to isolate the variable.}$$

$$2 = x \qquad \text{Simplify each side.}$$

You can review the use of the STOre feature in "Technology 1.2" and the use of the TABLE function in Section 3.3.

Confirm your solution by storing the value of 2 for the variable x and using the STOre feature of your calculator to evaluate each side of the equation. Because each side of the equation evaluates to the same value for $x = 2$, we conclude that 2 is the solution of the equation.

You may check the solution(s) of your equation in this section either analytically (paper and pencil), by the STOre feature, or by the TABLE feature. If you have completed Chapters 1–6, you may also graphically solve your equations as a means of checking your analytical solution. The review of graphical solutions occurs in the next section of this chapter.

■ IDENTITIES AND IMPOSSIBLE EQUATIONS

The equation in Example 1 is called a **conditional equation** because it is true for some, but not all, real numbers. An equation that is true for all real numbers is called an **identity.** Regardless of the value used to replace the variable, the equation will be true. For example, the equation $x + x = 2x$ is an identity because it is true regardless of the real number substituted in for the variable x. Its solution is **all real numbers.** This can be expressed with the symbol $\mathbb{R}$, which means "the set of all real numbers."

An equation that has no solution is called a **contradiction.** The equation $x = x + 1$ is a contradiction because no number is equal to itself plus 1. Its solution is **no solution,** which can be represented by the symbol $\varnothing$ representing the **null set** or the set that contains no elements.

WARNING! Do not confuse the **type** of equation with its **solution.** The terms *conditional, identity,* and *contradiction* refer to the type. Remember the solution or root is all real numbers that make the equation a true statement.

■ LITERAL EQUATIONS AND FORMULAS

Equations with several variables are called **literal equations.** Often, these equations are **formulas** such as $A = lw$, the formula for finding the area of a rectangle.

EXAMPLE 2   Solve $A = lw$ for l.

Solution   To isolate l on the left-hand side, we undo the multiplication of w by dividing both sides of the equation by w.

■ S T U D Y   T I P
To help you focus on the variable that you are solving for, a good suggestion is to use a highlighter to emphasize the variable.

$$A = lw$$

$$\frac{A}{w} = \frac{lw}{w} \qquad \text{Divide both sides by } w.$$

$$\frac{A}{w} = l \qquad \text{Simplify.}$$

$$l = \frac{A}{w}$$

EXAMPLE 3   Solve $3x + 2y = 8$ for y.

Solution
$$2y = -3x + 8 \qquad \text{Subtract } 3x \text{ from both sides.}$$

$$\frac{2y}{2} = \frac{-3x + 8}{2} \quad \text{or} \quad \frac{2y}{2} = \frac{-3x}{2} + \frac{8}{2} \qquad \text{Divide both sides by 2.}$$

$$y = -\frac{3}{2}x + 4 \qquad \text{Simplify.}$$

STUDY TIP
Record the problem-solving strategies in the box below on a note card and keep it handy any time you have an application problem to tackle.

■ PROBLEM SOLVING

Application problems illustrate the power of algebra in solving practical, everyday problems. In order to solve any problem, you first need a game plan.

Problem-Solving Strategies

1. Read the problem several times and organize the facts.

   What is known?

   What are you asked to find?

   It may help you to organize the facts under the two headings, **known** and **unknown.**

2. **a.** If there is only one unknown quantity, choose a variable to represent that quantity. Clearly define what the variable is and what it represents with a "Let" statement.

   b. When there is more than one unknown, choose a variable to represent one of the quantities and express all other important quantities in terms of the chosen variable.

3. Translate the information expressed in the words of the problem to an algebraic equation.

4. Solve the equation.

5. Answer the question posed in the problem.

6. Check your results in terms of both the equation and the words of the problem. Make sure your solution makes sense within the context of the problem.

EXAMPLE 4   Bonnie is a mathematics teacher who loves to read. She's currently reading *War and Peace.* While placing her bookmark in the book she notices that the sum of the page numbers is 2065. Between which two pages is the bookmark?

Solution   A table can help us organize the information. In this problem, we begin with the pages 1000 and 1001 (this is simply a guess). The table allows us to generalize from the specific to the general and write an algebraic equation that we can solve to answer the question posed.

Left page no.	1000	1002	x
Right page no.	1001	1003	$x + 1$
Sum	2001	2005	$x + (x + 1)$

Let x = left page number

$x + 1$ = right page number

$$x + (x + 1) = 2065$$
$$2x + 1 = 2065$$
$$2x + 1 - 1 = 2065 - 1$$
$$2x = 2064$$
$$x = 1032$$

The bookmark is between pages 1032 and 1033.

Calculator solution: Once the equation is constructed, enter the left side at the y1= prompt. Set the table to start at 1000 and to increment by 1. Enter 2065 at the y2= prompt. Scroll down until the value in the Y1 column equals the value in the Y2 column.

Similar triangles are triangles in which corresponding angles are equal and lengths of corresponding sides are in proportion. Because all pairs of corresponding sides are in proportion in similar triangles, we can use proportions to solve application problems that involve similar triangles. This will enable us to measure sides of triangles indirectly, as seen in the following application.

EXAMPLE 5 **Measuring a tree**   On a sunny day, a large tree casts a shadow of 24 feet at the same time a vertical yardstick casts a shadow of 2 feet. Find the height of the tree.

Solution   Refer to Figure 7-1, which shows the triangles determined by the tree and its shadow and the yardstick and its shadow. Because the triangles have the same shape, they are similar, and the measures of their corresponding sides are in proportion. We can let h represent the height of the tree and find h by setting up and solving the following proportion.

$$\frac{h}{3} = \frac{24}{2}$$

$2h = 3(24)$ In a proportion, the product of the extremes is equal to the product of the means.

$2h = 72$ Simplify.

$h = 36$ Divide both sides by 2.

The tree is 36 feet tall.

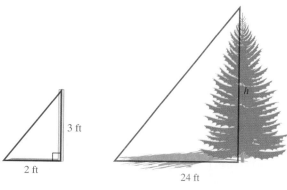

FIGURE 7-1

■ ABSOLUTE VALUE EQUATIONS

Consider the absolute value equation $|x| = 5$. The variable x can have a value of either 5 or -5 because both of these numbers are a distance of 5 units from 0 on the number line.

Therefore, the solutions to the equation are $x = 5$ or $x = -5$. This leads to the following generalization:

Absolute Value Equations
If $k > 0$, then $|x| = k$ is equivalent to $x = k$ or $x = -k$

It is important to note two things about the given algorithm:

1. Absolute value is isolated on one side of the equation.

2. k is a positive constant.

EXAMPLE 6   Solve the equation $|x - 3| + 5 = 12$.

Solution   Remember to first isolate the absolute value on the left-hand side.

$$|x - 3| = 7$$

The equation $|x - 3| = 7$ indicates that a point on the number line with a coordinate of $x - 3$ is 7 units from the origin. Thus, $x - 3$ can be either 7 or -7.

$$x - 3 = 7 \quad \text{or} \quad x - 3 = -7$$
$$x = 10 \quad \text{or} \quad x = -4$$

The solutions of the equation $|x - 3| + 5 = 12$ are 10 and -4. If either of these numbers is substituted in for x in the original equation $|x - 3| + 5 = 12$, the equation is satisfied.

$$
\begin{array}{c|c}
|x - 3| + 5 = 12 & |x - 3| + 5 = 12 \\
|x - 3| = 7 & |x - 3| = 7 \\
|10 - 3| \stackrel{?}{=} 7 & |-4 - 3| \stackrel{?}{=} 7 \\
|7| \stackrel{?}{=} 7 & |-7| \stackrel{?}{=} 7 \\
7 = 7 & 7 = 7
\end{array}
$$

Confirm the solution by using the TABLE feature:

```
Plot1 Plot2 Plot3
\Y1■abs(X-3)+5
\Y2■12
\Y3=
\Y4=
\Y5=
\Y6=
\Y7=
```

```
TABLE SETUP
TblStart=-4
△Tbl=1
Indpnt: Auto Ask
Depend: Auto Ask
```

X	Y1	Y2
-4	12	12
-3	11	12
-2	10	12
-1	9	12
0	8	12
1	7	12
2	6	12

X=-4

X	Y1	Y2
10	12	12
11	13	12
12	14	12
13	15	12
14	16	12
15	17	12
16	18	12

X=10

EXAMPLE 7   Solve the equation $\left| 7x + \dfrac{1}{2} \right| = -4$

Solution   Careful examination of the algorithm reveals that our constant k must be positive. In this example, $k = -4$, thus the algorithm cannot be applied. If we refer back to the definition of absolute value as it relates to the number line, the problem translates to "what number(s) are a *negative four* units from zero?" This does not make sense because not only is distance always positive but we also know from the definition that absolute value is always equal to 0 or some positive number.   ■

■ FACTORABLE POLYNOMIAL EQUATIONS

The solution (or root) of an equation is the value (or values) of the variable that make the equation a true statement. To find the solutions of an equation that is quadratic or of higher degree, the **zero factor property** is used.

> ### Zero Factor Property
> If $a \cdot b = 0$, a and b are real numbers, it follows that $a = 0$ or $b = 0$.

This property simply states that if a product of factors is zero, *at least one* of the factors must be equal to zero. To apply this property, the polynomial equations must be set equal to zero. A polynomial equation is said to be in **standard form** when one side of the equation is zero.

> ### Solving Equations by Factoring
> 1. Simplify both sides of the equation by removing grouping symbols and combining like terms. Fractions may be eliminated by multiplying each term on both sides of the equation by the LCD.
> 2. Put the equation in standard form (set one side equal to zero).
> 3. Factor the polynomial.
> 4. Set each factor containing a variable equal to zero.
> 5. Solve the resulting equations.
> 6. Check solutions either analytically or with the graphing calculator.

EXAMPLE 8   Solve: $x^3 - 2x^2 = 63x$.

Solution

$$x^3 - 2x^2 - 63x = 0$$ Put the equation in standard form.

$$x(x^2 - 2x - 63) = 0$$ Begin factoring the polynomial by factoring out the GCF of x.

$$x(x - 9)(x + 7) = 0$$ Factor the trinomial.

$$x = 0 \qquad x - 9 = 0 \qquad x + 7 = 0$$ Use the zero factor property to set each factor equal to 0.

$$x = 0 \qquad\qquad x = 9 \qquad\qquad x = -7$$ Solve each linear equation.

Solutions: 0, 9, −7; Solution set: $\{0, 9, -7\}$

The verification using the TABLE feature is illustrated.

The left side of the original equation is entered at the y1 prompt and the right side is entered at the y2 prompt.

The TABLE is set up to begin with an x value of -7. Because the values in the Y1 and Y2 columns are equivalent, -7 is verified as a solution. Scrolling through the TABLE verifies the other roots of 0 and 9.

■ EQUATIONS WITH RATIONAL EXPRESSIONS

When solving a linear equation, we use the addition, subtraction, multiplication, and division properties of equality to rewrite the original equation as a series of equivalent equations until the variable is isolated on one side of the equation. For equations of degree two or higher, the zero factor property is applied. The solutions, or roots, of the equation are the values of the variable that make the equation a true statement.

Care must be taken when solving equations containing rational expressions with variables in the denominator. Sometimes, when solving rational equations false solutions result. These are called **extraneous solutions** and are rejected. Any value of the variable that makes a denominator equal to zero is an extraneous solution.

WARNING! When arithmetically solving equations that contain variables in the denominators, solutions **must** be verified through a checking process. Extraneous solutions **must** be rejected. For this reason, it is suggested that excluded values be determined first.

The following steps can be used:

Solving Rational Equations Algebraically
1. Determine any and all excluded values by setting each denominator containing a variable equal to zero and solving for the variable.
2. Determine the LCD.
3. Multiply each *term* on both sides of the equation by the LCD to clear fractions.
4. Simplify both sides of the equation.
5. Solve the equation. Remember, linear (first-degree) equations are solved by transformation, whereas quadratic (second-degree) and higher-order equations require the additional process of applying the zero factor property after the equation is put in standard form and the polynomial is factored.
6. Reject any extraneous solutions (those values that equal the excluded values found in step 1).
7. Check solutions in the original equation to verify solutions.

EXAMPLE 9 Solve the equation $\dfrac{2}{x-1} + 4 = \dfrac{4}{x^2-1} + 3$.

Solution Determine the excluded values by setting each denominator equal to zero and solving for the variable.

$$x - 1 = 0 \quad \text{and} \quad x^2 - 1 = 0$$
$$x = 1 \quad\quad\quad\quad (x - 1)(x + 1) = 0$$
$$x - 1 = 0 \quad \text{or} \quad x + 1 = 0$$
$$x = 1 \quad \text{or} \quad x = -1$$

Because $x^2 - 1 = 0$ is a quadratic equation, we factor the polynomial side and use the zero factor property to find the solutions. The excluded values are 1 and -1.

We now begin solving the rational equation. To determine the LCD, we factor denominators that are not prime.

$$\frac{2}{x - 1} + 4 = \frac{4}{x^2 - 1} + 3$$

$$\frac{2}{x - 1} + 4 = \frac{4}{(x - 1)(x + 1)} + 3 \qquad \text{The LCD is } (x - 1)(x + 1).$$

We begin solving by multiplying both sides of the equation by the LCD $(x - 1)(x + 1)$.

$$(x - 1)(x + 1)\left(\frac{2}{x - 1} + 4\right) = (x - 1)(x + 1)\left(\frac{4}{(x - 1)(x + 1)} + 3\right)$$

Use the distributive property to remove groupings.

$$(x - 1)(x + 1) \cdot \frac{2}{x - 1} + (x - 1)(x + 1) \cdot 4 = (x - 1)(x + 1) \cdot \frac{4}{(x - 1)(x + 1)} + (x - 1)(x + 1) \cdot 3$$

$2(x + 1) + 4(x^2 - 1) = 4 + 3(x^2 - 1)$	Simplify by dividing out common factors.
$2x + 2 + 4x^2 - 4 = 4 + 3x^2 - 3$	Simplify by removing grouping.
$4x^2 + 2x - 2 = 3x^2 + 1$	Combine like terms.
$x^2 + 2x - 3 = 0$	Because the equation is quadratic, use the subtraction property of equality to put it in standard form.
$(x + 3)(x - 1) = 0$	Factor the polynomial side.
$x + 3 = 0 \quad \text{or} \quad x - 1 = 0$	Use the zero factor property to set each factor equal to 0 and solve for the variable.
$x = -3 \quad \text{or} \quad x = 1$	

Because the number 1 is an excluded value, it is rejected as a root. The solution is -3.

The solution of -3 is verified below:

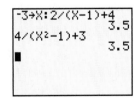

The fact that 1 is an extraneous root is also verified below:

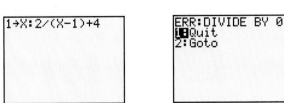

Before ENTER is   After ENTER is
pressed.   pressed.

■ **STUDY TIP**

Form a study group. Set up a time and place to meet regularly and do mathematics. Talk about the concepts with one another. The verbalization of concepts is often the key to success for students.

The TABLE feature show us -3 is a solution because the expressions entered at the y1 and y2 prompts evaluate to the same constant (3.5) for $x = -3$. It also confirms the fact that 1 cannot be a solution. The ERROR message displayed indicates that YOU have made an error—in this case, "asking" the calculator to evaluate expressions for a value of the variable that makes those expressions undefined.

EXERCISE 7.3

VOCABULARY   *In Exercises 1–7, fill in the blanks to make a true statement.*

1. To ___solve___ an equation means to find the value(s) of the variables(s) that will make the equation a true statement.

2. A ___conditional___ equation is true for some, but not all, real numbers.

3. An ___identity___ equation is an equation that is true for all acceptable values of the variable.

4. A ___contradiction___ is an equation that is never true for any value of the variable and thus has no solution.

5. A ___literal___ is an equation involving several variables.

6. The symbol ϕ denotes an empty set.

7. The symbol ___ℝ___ denotes the set of all real numbers.

CONCEPTS   *In Exercises 8–12, determine if the given values are or are not solutions to the equations. Use the STOre or TABLE features of your calculator.*

8. $4(x - 5) = \dfrac{1}{2}(x + 1) - 2$; 41

9. $|2x - 1| = 8$; $\dfrac{9}{2}, -\dfrac{7}{2}$

10. $-|x + 5| - 6 = 0$; -5

11. $6x^2 - 11x - 10 = 0$; $\dfrac{5}{2}, -\dfrac{2}{3}$

12. $\dfrac{6}{x} + \dfrac{5}{x} = 5$; 2

PRACTICE   *In Exercises 13–22, solve each linear equation and verify results with either the STOre or the TABLE features of your graphing calculator.*

13. $4(x - 1) = 28$

14. $3(x + 7) = 42$

15. $13(x - 9) - 2 = 7x - 5$

16. $\dfrac{8(x - 5)}{3} = 2(x - 4)$

17. $\dfrac{1}{3}x - 2 = \dfrac{1}{2}x + 2$

18. $\dfrac{7}{4}(x + 3) = \dfrac{3}{8}(x - 3)$

19. $6(x - 2) = 3(4 + x) + 3x$

20. $4(2x + 1) = 7x + 3 + x$

21. $6x - 5 = 2(x - 3) + 1 + 4x$

22. $-(2x + 5) - 3 = 4(1 - x) - 12 + 2x$

In Exercises 23–32, solve each literal equation or formula for the indicated variable.

23. $2x + y = 7$; y

24. $3x + y = 9$; y

25. $4x - y = 8$; y

26. $\dfrac{1}{2}x - y = 2$; y

27. $2x - 3y = 5$; y

28. $3x - 5y = 1$; y

29. $x + 2y = 6$; y

30. $2x + 2y = 8$; y

31. $V = \dfrac{1}{3}\pi r^2 h$; h

32. $A = P + Prt$; t

In Exercises 33–42, solve each absolute value equation and verify your results with the calculator.

33. $|x + 1| = 10$

34. $\left|\dfrac{3}{2}x - 4\right| = 9$

35. $|2x - 1| + 4 = 6$

36. $|3x + 5| - 2 = -1$

37. $|4x - 1| = -2$

38. $|5x + 7| + 5 = 1$

39. $2|x - 1| - 2 = 0$

40. $\dfrac{1}{2}|x + 2| - 1 = 0$

41. $|3x + 2| = |2x - 3|$

42. $|5x - 4| = |4x - 5|$

In Exercises 43–52, solve each polynomial equation by the zero product property. Verify your results with the calculator.

43. $(x + 1)(x - 2) = 0$

44. $(2x - 3)(x + 4) = 0$

45. $x^2 - 16 = 0$

46. $x^2 - 25 = 0$

47. $x^2 + 21x + 20 = 0$

48. $x^2 - 11x + 30 = 0$

49. $2x^3 - 10x^2 - 48x = 0$

50. $6x^3 + 5x^2 - 4x = 0$

51. $x(x - 12) = -20$

52. $x(x + 6) = -9$

In Exercises 53–62, solve each rational equation. Verify solutions using either the STOre or TABLE features of the calculator.

53. $\dfrac{11}{x} + \dfrac{13}{x} = 12$

54. $\dfrac{5}{x} - \dfrac{4}{x} = 8 + \dfrac{1}{x}$

55. $\dfrac{3}{p + 6} - 2 = \dfrac{7}{p + 6}$

56. $\dfrac{1}{x - 1} + \dfrac{3}{x - 1} = 1$

57. $\dfrac{z^2}{z + 1} + 2 = \dfrac{1}{z + 1}$

58. $\dfrac{x^2}{x + 2} - \dfrac{4}{x + 2} = x$

59. $\dfrac{3}{y - 2} + 1 = \dfrac{3}{y - 2}$

60. $\dfrac{x}{x - 5} - \dfrac{5}{x - 5} = 3$

61. $\dfrac{2p}{3} - \dfrac{1}{p} = \dfrac{2p - 1}{3}$

62. $\dfrac{3r}{2} - \dfrac{3}{r} = \dfrac{3r}{2} + 3$

APPLICATIONS   *In Exercises 63–72, solve each problem and verify results with your calculator. Remember, this section reviews problem solving. You may wish to return to Chapter 4 for specific examples not demonstrated in this section.*

63. Carpentry   A carpenter wants to cut a 20-foot rafter so that one piece is 3 times as long as the other. Where should he cut the board?

64. Geometry   A rectangle is 4 meters longer than it is wide. If the perimeter of the rectangle is 28 meters, find its area.

65. Rental car   Alamo car rental rents cars for $30 per day plus $0.24 a mile. Budget car rental rents the same car for $44 a day plus $0.16 a mile. How many miles would you have to drive in one day for the rental costs to be equal?

66. Height of a tree   A tree casts a shadow of 26 feet at the same time as a 6-foot man casts a shadow of 4 feet (see Illustration 1). Find the height of the tree.

67. Height of a flagpole   A man places a mirror on the ground and sees the reflection of the top of a flagpole, as in Illustration 2. The two triangles in the illustration are similar. Find the height, h, of the flagpole.

ILLUSTRATION 1

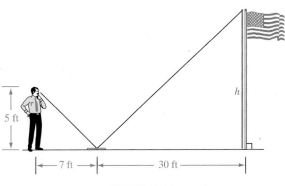

ILLUSTRATION 2

68. Elevation   A sign on a mountain highway near Cloudcroft, New Mexico, indicates a 5% downhill grade. How much decrease in elevation occurs per mile of horizontal distance? (*Hint:* 5280 feet = 1 mile)

69. Angles of a triangle   The sum of the angles of a triangle is 180°. Find the three angles if the base angles are the same and the third angle is four times one of the base angles.

70. Angles of a triangle   Find the angle measurements of a triangle if the first angle is 2 more than the second angle and the third angle is 3 times the second angle.

71. Investment problem   A retired teacher has a choice of two investment plans: an insured fund that pays 11% interest or a riskier investment that promises a 13% return. The same amount invested at the higher rate would generate an extra $150 per year. How much does the teacher have to invest?

72. Investment problem   A financial adviser recommends investing twice as much in CDs as in a bond fund. A client follows his advice and invests $21,000 in CDs paying 1% more interest than the fund. The CDs generate $840 more interest than the fund. Find the two rates.

REVIEW

73. Simplify each of the following. Leave no negative exponents in your answer.

    a. $\dfrac{x^{-2}y^{-4}}{(2x)^{-5}}$ **b.** $\left(\dfrac{3x^4y^{-2}}{2^{-2}x^{-3}y^2}\right)^{-1}$

74. Write in standard notation:
    a. 1.24×10^{-5} **b.** 3.005×10^2

75. Write in scientific notation:
    a. 3.14 E −3   **b.** 4,000,000,000

    c. 0.000 000 000 000 000 000 56

76. Factor completely:
    a. $9x^3 + 3x^5 - 30x$
    b. $x^3 - 343y^3$
    c. $y^2 + 2xy - 63x^2$

7.4   Graphing and Its Applications Revisited

■ THE RECTANGULAR COORDINATE SYSTEM ■ VIEWING WINDOWS ■ GRAPHING EQUATIONS IN TWO VARIABLES ■ GRAPHICAL SOLUTIONS OF ONE-VARIABLE EQUATIONS

■ **STUDY TIP**
Study in 30-minute segments of time with breaks of 5 to 10 minutes between each 30-minute session.

Points to remember

1. Solutions to one-variable equations are graphed on a number line. For example, $x = 1$ is the solution to $3x + 2 = 5$. Graphed on a number line this would be

2. Solutions to two-variable equations are graphed on a rectangular coordinate system. For example, the ordered pairs (3, 2), (2, 3), and (1, 4) are only three of an infinite number of solutions to the equation $x + y = 5$. These three ordered pairs are graphed on the coordinate system and connected to form a straight-line graph, as in Figure 7-2. ALL solutions to the equation $x + y = 5$ can be found on this straight line.

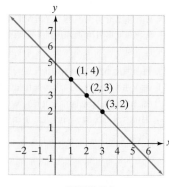

FIGURE 7-2

■ THE RECTANGULAR COORDINATE SYSTEM

Ordered pair solutions (x, y), of an equation in two variables can be graphed on a rectangular coordinate system. The system consists of two intersecting number lines called the x- and y-axes. The axes intersect at right angles and divide the coordinate system into four regions called *quadrants,* as shown in Figure 7-3.

Points on the graph are identified by **ordered pairs** of numbers. The order of the numbers, (x, y), is important. The **x-coordinate,** or **abscissa,** is first; and the **y-coordinate,** or **ordinate,** is second.

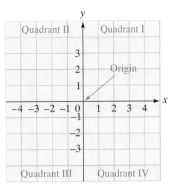

FIGURE 7-3

EXAMPLE 1   On a rectangular coordinate system, plot the points   **a.** $A(-1, 2)$,   **b.** $B(0, 0)$,   **c.** $C(5, 0)$,   **d.** $D\left(-\frac{5}{2}, -3\right)$,   and   **e.** $E(3, -2)$.

Solution   **a.** The point $A(-1, 2)$ has an x-coordinate of -1 and a y-coordinate of 2. To plot point A, we start at the origin and move 1 unit to the left and then 2 units up. (See Figure 7-4.) Point A lies in quadrant II.

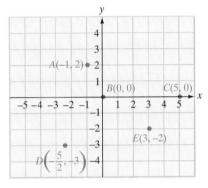

FIGURE 7-4

b. To plot point $B(0, 0)$, we start at the origin and move 0 units to the right and 0 units up. Because there was no movement, point B is the origin.

c. To plot point $C(5, 0)$, we start at the origin and move 5 units to the right and 0 units up. Point C lies on the x-axis, 5 units to the right of the origin.

d. To plot point $D\left(-\frac{5}{2}, -3\right)$, we start at the origin and move $\frac{5}{2}$ units $\left(\frac{5}{2} \text{ units} = 2\frac{1}{2} \text{ units}\right)$ to the left and then 3 units down. Point D lies in quadrant III.

e. To plot point $E(3, -2)$, we start at the origin and move 3 units to the right and then 2 units down. Point E lies in quadrant IV.   ■

■ **TECHNOLOGY TIP**
See "Technology 3.4: Graphing Basics" for more detail.

■ VIEWING WINDOWS

Unlike traditional algebra texts, this text employs full integration of a graphics calculator. All references to a calculator will imply a graphics calculator, not a scientific calculator. In Chapter 1 we drew one-dimensional pictures on a number line. We now want to draw two-dimensional pictures in a number plane. Just as each point on a number line is associated with a real number, each point in a number plane is associated with an ordered pair of real numbers.

When you sketch a graph on your graph paper, the size of your graph is limited by the size of your paper space. Theoretically, each axis extends indefinitely, which we represent by arrows at the end of each axis. However, the number of tic marks you place on each axis limits its visual size. Similar limitations occur when using a graphing calculator to display a coordinate system. You can limit the visual size of your calculator graph by designating a minimum and maximum value for the x-axis and a minimum and maximum value for the y-axis, as shown in Figure 7-5.

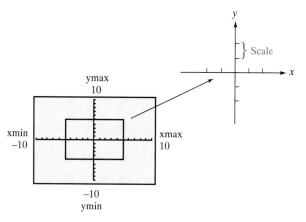

xmin: −10
xmax: 10
ymax: 10
−10 ymin

Scale

FIGURE 7-5

To condense this information, we will use the following notation:

xmin   xmax   ymin   ymax
$[-10, 10]$ by $[-10, 10]$

Observe that there are 10 tic marks to the left, right, above and below the origin on the calculator graph. The distance between the tic marks is called the *scale*. On this graph, the scale for both the *x*- and *y*-axes is 1. We would write this as xscl = 1 and yscl = 1. Unless otherwise noted, it should be assumed xscl = 1 and yscl = 1 on all graphs displayed in this text.

In order to set the limitations on your calculator graph, you should be familiar with the key that accesses this feature. It may be denoted as WINDOW, RANGE, or WIND on your calculator. The viewing window above, $[-10, 10]$ by $[-10, 10]$, is called the **standard viewing window.** It can be set manually by entering the information or automatically using the ZOOM feature.

WARNING! What you "see" may be deceiving. Each of the calculator graphs displayed below look the same and yet each represents a different-size coordinate system based on the limitations placed on xmin, xmax, xscl, ymin, ymax, and yscl.

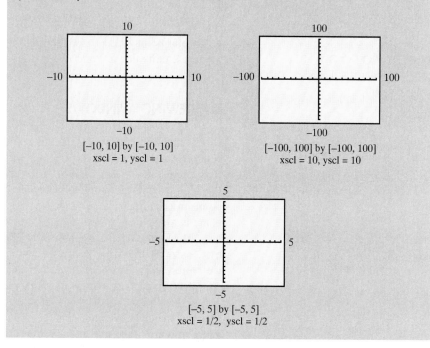

$[-10, 10]$ by $[-10, 10]$
xscl = 1, yscl = 1

$[-100, 100]$ by $[-100, 100]$
xscl = 10, yscl = 10

$[-5, 5]$ by $[-5, 5]$
xscl = 1/2, yscl = 1/2

■ GRAPHING EQUATIONS IN TWO VARIABLES

■ STUDY TIP
As much as possible, study when you are most alert. Beginning a two-hour study session at 11:00 P.M. is often nonproductive.

Equations may be graphed either with paper and pencil or with a graphing calculator. You should be proficient with both methods. We will review graphing **linear or straight-line equations** in two unknowns.

> **Standard Form**
> If A, B, and C are real numbers, where A and B are not both zero, then $Ax + By = C$ is called the **standard form of a linear equation.**
>
> **Linear Equation**
> Any equation that can be written in the form $y = mx + b$, where $y \neq 0$, is called a **linear equation.**

Whether you are graphing an equation of the form $Ax + By = C$ or one of the form $y = mx + b$, you will need at least two ordered pairs to produce a paper-and-pencil graph.

The points where the line intersects the x- and y-axes are called the *intercepts* of the line. Plotting these intercepts and drawing a line through them is called the *intercept method of graphing* and is a convenient method for graphing equations in standard form.

> **y- and x-Intercepts**
> The **y-intercept** of a line is the point $(0, b)$ where the line intersects the y-axis. To find b, substitute 0 for x in the equation of the line and solve for y.
>
> The **x-intercept** of a line is the point $(a, 0)$ where the line intersects the x-axis. To find a, substitute 0 for y in the equation of the line and solve for x.

EXAMPLE 2   Graph the equation $3x + 2y = 6$.

Solution   To find the y-intercept, we let $x = 0$ and solve for y.

$$3x + 2y = 6$$
$$3(0) + 2y = 6 \quad \text{Substitute 0 for } x.$$
$$2y = 6 \quad \text{Simplify.}$$
$$y = 3 \quad \text{Divide both sides by 2.}$$

The y-intercept is the pair $(0, 3)$.
To find the x-intercept, we let $y = 0$ and solve for x.

$$3x + 2y = 6$$
$$3x + 2(0) = 6 \quad \text{Substitute 0 for } y.$$
$$3x = 6 \quad \text{Simplify.}$$
$$x = 2 \quad \text{Divide both sides by 3.}$$

The x-intercept is the pair $(2, 0)$.
As a check, we plot one more point. If $x = 4$, then

$$3x + 2y = 6$$
$$3(4) + 2y = 6 \quad \text{Substitute 4 for } x.$$
$$12 + 2y = 6 \quad \text{Simplify.}$$
$$2y = -6 \quad \text{Add } -12 \text{ to both sides.}$$
$$y = -3 \quad \text{Divide both sides by 2.}$$

The point $(4, -3)$ lies on the graph. We plot these three points and join them with a line. The graph of $3x + 2y = 6$ is shown in Figure 7-6.

$$3x + 2y = 6$$

x	y
0	3
2	0
4	-3

FIGURE 7-6

EXAMPLE 3   To use the graphing calculator to graph the equation $3x + 2y = 6$ from Example 2, first solve the equation for y in order to enter it on the y= screen. Use the standard viewing window and press GRAPH to display the graph of the equation.

Solution   First, solve the equation for y in order to enter it on the y= screen.

$$3x - 3x + 2y = 6 - 3x$$

$$2y = 6 - 3x$$

$$\frac{2y}{2} = \frac{6 - 3x}{2}$$

$$y = 3 - \frac{3}{2}x$$

$$y = -\frac{3}{2}x + 3$$

You may now access the TABLE feature of the calculator to compare your table of values, as constructed in Example 2, to those of the calculator.

If you access the TRACE feature you can use the left and right cursor keys to trace along the line. You will notice that the coordinate values are cumbersome decimal values.

For this reason, we will *not* use the standard viewing window when we want to trace. The calculator has a preset window called ZInteger that yields integer values when tracing (see "Technology 3.4"). To set these values manually, enter $[-47, 47]$ by $[-31, 31]$ on your WINDOW screen now, with x- and y-scales equal to 10. Trace and observe the x- and y-coordinates.

■ TECHNOLOGY TIP
Both the standard and integer viewing windows can be automatically set by accessing them from the ZOOM menu. See "Technology 3.4: Graphing Basics" for details.

EXAMPLE 4   Graph the equation $y = 3$.

Solution   We can write the equation $y = 3$ in standard form as $0x + y = 3$. Because the coefficient of x is 0, the numbers assigned to x have no effect on y. The value of y is always 3. For example, if we replace x with -3, we get

$$0x + y = 3$$
$$0(-3) + y = 3$$
$$0 + y = 3$$
$$y = 3$$

The table in Figure 7-7 gives several ordered pairs that satisfy the equation $y = 3$. After plotting these pairs (x, y) and joining them with a line, we see that the graph of $y = 3$ is a horizontal line, parallel to the x-axis and intersecting the y-axis at 3. The y-intercept is $(0, 3)$, and there is no x-intercept.

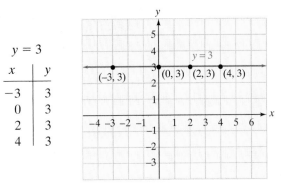

$y = 3$

x	y
-3	3
0	3
2	3
4	3

FIGURE 7-7

The horizontal line $y = 3$ is graphed below in the ZInteger viewing window with two of the four ordered pairs displayed by using the TRACE feature.

EXAMPLE 5   Graph the equation $x = -2$.

Solution   We can write the equation $x = -2$ in standard form as $x + 0y = -2$. Because the coefficient of y is 0, the values of y have no effect on x. The number x is always -2. A table of values and the graph are shown in Figure 7-8.

    The graph of $x = -2$ is a vertical line, parallel to the y-axis and intersecting the x-axis at -2. The x-intercept is $(-2, 0)$, and there is no y-intercept.

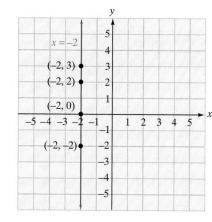

$x = -2$

x	y
-2	-2
-2	0
-2	2
-2	3

FIGURE 7-8

From the results of Examples 4 and 5, we have the following facts:

Equations of Lines Parallel to the Coordinate Axes

The equation $y = b$ represents a horizontal line that intersects the y-axis at $(0, b)$. If $b = 0$, the line is the x-axis.

The equation $x = a$ represents a vertical line that intersects the x-axis at $(a, 0)$. If $a = 0$, the line is the y-axis.

Example 5 was not graphed on the calculator because the calculator will graph nonvertical lines but will not graph vertical lines. Vertical lines are not **functions** and the calculator is designed to graph only functions.

Function

A function is a set of ordered pairs in which each x value in the ordered pair (x, y) is assigned a unique y value.

The ordered pairs used to graph the line $x = -2$ in Example 5 were $(-2, 3)$, $(-2, 2)$, $(-2, 0)$, and $(-2, -2)$. The x-value of -2 does not have a unique y-value; it has many y-values. Thus, vertical lines are not functions.

EXAMPLE 6   Is the formula for converting Fahrenheit temperature to Celsius a function? The formula is

$$y = \frac{5}{9}(x - 32)$$

where y = degrees Celsius and x = degrees Fahrenheit.

Solution   We need to answer the question "does each x produce a unique y?" Select a value for x, say $41°$. Then

$$y = \frac{5}{9}(41 - 32)$$

$$y = \frac{5}{9}(9) \qquad \text{Subtraction produces a unique result.}$$

$$y = 5 \qquad \text{Multiplication produces a unique result.}$$

Because the operations of addition, subtraction, multiplication, and division produce unique results, any equation that applies these operations to x will produce unique results for y, and thus a function.

Linear Function
Any linear equation that can be written in the form $y = mx + b$, where $y \neq 0$ and m and b are real numbers, is called a **linear function.**

EXAMPLE 7   Intelix Home Video produces videotapes of weddings, baptisms, etc. For each tape produced there is a basic fixed cost of $100 and a variable cost of $2 per tape. The tapes then sell for $20 each. Write an equation to represent the production costs y for x tapes and an equation to represent the revenues y for x tapes. Explore these graphs and their meaning in the context of the problem.

Solution   **a.** The cost equation would be $y = 100 + 2x$. Graph this equation in the viewing window $[-20, 74]$ by $[-5, 181]$ with x- and y-scales equal to 10. TRACE to find the production cost for 23 tapes. At the point $(23, 146)$, we can see that the cost for producing 23 tapes is $146.

■ TECHNOLOGY TIP
The calculator will trace to a point not visible in your designated viewing window. The coordinates at the bottom of the screen indicate the cursor position.

b. The revenue equation would be $y = 20x$. Graph this equation on the same screen as the cost equation and TRACE on the **revenue** equation to determine the revenues produced for 23 tapes. The point $(23, 460)$ indicates that the production of 23 tapes produces a revenue of $460.

c. Does the company sustain a loss or make a profit when 23 tapes are sold? Because the production cost is $146 and the revenues are $460, the company makes a profit of $314.

d. How many tapes must be produced and sold for the company to break even (cost = revenue)? Attempting to trace to the point of intersection is futile because the x-value at the point of intersection is not an integer. (The window you were given was designed to produce integer values for x when tracing.) The x-value of the point of intersection of the

■ TECHNOLOGY TIP
See "Technology 3.5: INTERSECT Method for Solving Equations" for specific help in using the INTERSECT feature of your calculator.

two equations represents the solution to the equation cost = revenue (i.e., $100 + 2x = 20x$). Use the INTERSECT feature of your calculator to locate the point of intersection displayed at the right. The company would need to produce approximately 5.6 tapes to break even.   ■

When discussing functions and their applications, we are often interested in just the x-values of the function or just the y-values. These classifications are called the **domain** and **range** of the function.

Domain of a Function
The **domain** of a function is the set of all x-values of the ordered pairs of the function.

Range of a Function
The **range** of a function is the set of all y-values of the ordered pairs of the function.

EXAMPLE 8   In the conversion chart, identify the domain and range.

x = number of cups	$\frac{1}{2}$	1	$1\frac{1}{2}$	2	$2\frac{1}{2}$
y = number of pints	$\frac{1}{4}$	$\frac{1}{2}$	$\frac{3}{4}$	1	$1\frac{1}{4}$

Solution   The domain is $\left\{\frac{1}{2}, 1, 1\frac{1}{2}, 2, 2\frac{1}{2}\right\}$. (The braces indicate we have a **set,** a collection of objects.) The range is $\left\{\frac{1}{4}, \frac{1}{2}, \frac{3}{4}, 1, 1\frac{1}{4}\right\}$.

GRAPHICAL SOLUTIONS OF ONE-VARIABLE EQUATIONS

It is important that you understand the difference between graphing linear equations in two variables and using a graphical display to interpret the solution of a linear equation in one variable.

To graphically solve an equation in one variable, we examine the graphical representation of the algebraic expression on each side of the equation. Each side of the equation is entered at the y1 and y2 prompts respectively. The solution occurs when y1 = y2, which is at the point of intersection of the two graphs. It is the **x-value** of the ordered pair that is the solution to the one-variable equation.

EXAMPLE 9   Graphically solve $2(x + 2) - x = 10 - x$.

Solution   Let y1 = $2(x + 2) - x$ and y2 = $10 - x$.

When solving graphically, begin in the standard viewing window and adjust the window as needed to display the point of intersection. Use the INTERSECT feature to display the graph at the right. The display at the bottom of the screen indicates that the ordered pair (3, 7) is common to both graphs. However, we have merely graphed the one-variable expressions of the left and right sides of an equation. It is the value $x = 3$ that is the solution to our equation. Checking analytically verifies this:

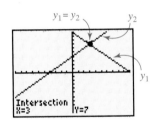

$$2(x + 2) - x = 10 - x$$
$$2(3 + 2) - 3 \ ? \ 10 - 3$$
$$2(5) - 3 \ ? \ 7$$
$$10 - 3 \ ? \ 7$$
$$7 = 7\sqrt{}$$

The y-value displayed on the calculator screen can be interpreted to mean that both sides of the equation will evaluate to 7 when x is replaced with 3.

Solution: 3; Solution set: {3}

EXAMPLE 10   Graphically solve $|3x + 1| = 4$.

Solution   Let y1 = abs $(3x + 1)$
                          y2 = 4

Use the INTERSECT feature **twice** to determine both solutions.

Remember, when the INTERSECT feature is used, the value computed for x is stored in x until you store another value there or until you use the INTERSECT feature again. You can return to the home screen and convert the decimal value displayed into a simplified fraction.

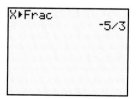

Solutions: 1, $-5/3$; Solution set: $\{1, -5/3\}$

EXAMPLE 11   Graphically solve $x^2 - x - 12 = 0$.

Solution   Let $y1 = x^2 - x - 12$
                    $y2 = 0$

Use the INTERSECT feature **twice** to determine both solutions.

Remember that although only one graph appears to be visible, the equation $y2 = 0$ is actually the x-axis.

Solution: -3, 4; Solution set: $\{-3, 4\}$

The two points of intersection in Example 11 occurred at $x = 4$, $y = 0$ and $x = -3$, $y = 0$. These ordered pairs $(4, 0)$ and $(-3, 0)$ represent the x-intercepts for the **two-variable** equation $y = x^2 - x - 12$. The x-intercept(s) represent the root(s) or zero(s) of the equation. The ZERO (or ROOT) feature of the graphing calculator offers not only another approach to graphical solutions of one-variable equations but also an accurate means of determining x-intercepts of any equation in two variables.

EXAMPLE 12   Graphically solve $x^2 - x - 12 = 0$ using the ZERO feature of the calculator.

Solution   In order to use this feature of the calculator, the equation MUST be set equal to zero. Enter the nonzero side as $y1$.

$$y1 = x^2 - x - 12$$

■ **TECHNOLOGY TIP**
See "Technology 7.4: ROOT/ZERO Method for Solving" at the end of this section for more details on using the ZERO feature. Some calculators may designate this feature as the ROOT feature.

Use the ZERO feature twice to determine both of the solutions (i.e., roots or zeros) of the equation as displayed on the two screens below.

It is now the *x*-intercepts that are displayed at the bottom of the screen, along with the word *zero*. Remember, it is the *x*-value that is the solution to your equation in one variable!

Solution: $-3, 4$; Solution set: $\{-3, 4\}$ ■

When solving graphically, it is best to let the format of your equation determine whether you use the INTERSECT or ZERO feature. For example, $x^2 - x - 12 = 0$ is in the required format for the ZERO feature, whereas $|3x + 1| = 4$ is not because neither side of the equation is zero. The less algebraic manipulation that occurs before you solve graphically, the higher your accuracy level.

Having two approaches to graphical solving provides two advantages. First, you can select the method that requires no algebraic manipulation on your part, thus decreasing the chance of human error when determining the solutions. Second, when the graphical display of one approach does not yield a good graphical display, you have an alternative approach.

EXAMPLE 13 Graphically solve $x^3 + x^2 = 4x + 4$

Solution Since neither side of the equation is zero, we will use the INTERSECT feature.

Let $y1 = x^3 + x^2$

$y2 = 4x + 4$

The graphical display does not clearly show all points of intersection. We would need to adjust the viewing window before using the INTERSECT feature. However, if we rewrite the equation to set one side equal to zero, we can produce a much improved graphical display.

$$x^3 + x^2 = 4x + 4$$
$$x^3 + x^2 - 4x = 4x - 4x + 4$$
$$x^3 + x^2 - 4x = 4$$
$$x^3 + x^2 - 4x - 4 = 4 - 4$$
$$x^3 + x^2 - 4x - 4 = 0$$

Now solve using the ZERO feature.

Let $y1 = x^3 + x^2 - 4x - 4$

solution: -2 solution: -1 solution: 2

Conclusion: The solutions are $-2, -1, 2$. The solution set is $\{-2, -1, 2\}$. ■

We can make several observations at this time.

1. Conditional linear equations reviewed in Example 9 will have one real root because linear equations are also first-degree equations.

2. Quadratic equations such as in Example 12 will have at most two real roots because quadratic equations are second-degree equations.

3. Third-degree polynomial equations such as in Example 13 will have at most three real roots.

    If all polynomial equations are set equal to zero, then the degree of the polynomial not only predicts the maximum number of real roots but it will also predict the maximum number of x-intercepts the graph of the polynomial will have.

Zeros of a Polynomial Equation
A polynomial equation of degree n will have at most n distinct roots/zeros.

EXERCISE 7.4

VOCABULARY   *In Exercises 1–7, fill in the blanks to make a true statement.*

1. Points on a graph are identified by ___order pair___ of numbers.

2. Another name for the x-coordinate is ___abscissa___ and for the y-coordinate is ___ordinate___.

3. The _____ window is defined to be $[-10, 10]$ by $[-10, 10]$.

4. A ___linear equation___ in the form $Ax + By = C$ is said to be in ___standard___ form.

5. The ___x-intersep___ of a line is the point $(a, 0)$ where the line intersects the x-axis.

6. The ___y-intercept___ of a line is the point $(0, b)$ where the line intersects the y-axis.

7. A _____ is a set of ordered pairs in which each x-value in the domain is assigned a unique y-value in the range.

CONCEPTS

8. Is the ordered pair $(4, 2)$ a solution to the equation $y = |x - 4| + 2$?

9. Plot the group of ordered pairs on a rectangular coordinate system:
   a. $(2, 5)$
   b. $(-3, 1)$
   c. $(-4, 2)$
   d. $(5, 0)$
   e. $(-5, 3)$
   f. $(0, -2)$

In Exercises 10–13, determine if each ordered pair would be visible if graphed on your calculator in the designated viewing window.

10. $[-15, 12]$ by $[-2, 2]$
   a. $(-12, 6)$
   b. $(4, 1)$

11. $[0, 150]$ by $[-10, 50]$
   a. $(4, 45)$
   b. $(5, 150)$

12. $[-25, 5]$ by $[25, 125]$
   a. $(-18, 27)$
   b. $(-26, 20)$

13. $[-20, 20]$ by $[-10, 5]$
   a. $(35, 10)$
   b. $(-15, -5)$

14. Each of the following represent solutions to an equation. Based on the given information, is the equation a function?
   a. $\{(4, 7), (2, 1), (-5, -7), (0, 3)\}$
   b.

15. A table of values for a function is displayed at the right.
   a. What is the domain?
   b. What is the range?

PRACTICE *In Exercises 16–23, use the x- and y-intercepts to construct a paper-and-pencil graph of the equation.*

16. $4x - 2y = 5$

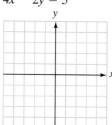

17. $x + 3y = 6$

18. $x + 2y = -1$

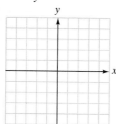

19. $2x - y = -2$

20. $3x + 1 = 4$

21. $4 - x = 5$

22. $y = -2$

23. $y = 4$

In Exercises 24–31, graph the given equations on your calculator and sketch the screen display. Use the ZInteger viewing window and TRACE to label the x- and y-intercepts that have integer coordinates.

24. $f(x) = 4x + 12$

25. $f(x) = -3x + 40$

26. $3x + y = 18$

27. $-2x + y = 24$

28. $4x + 6y = 36$

29. $5x - 8y = 80$

30. $6x - 7y = 35$

31. $5x + 9y = 55$

In Exercises 32–37, based on the displayed ordered pairs, determine if the graph displayed is a function.

32.

33.

34.

35.

36.

37.

In Exercises 38–41, determine the domain and range of each function.

38. $\{(2, 4), (3, 5), (-1, 2), (4, 2)\}$

39. $\left\{(-8, 5), (9, 2), (7, -3), \left(-\frac{1}{2}, \frac{2}{3}\right), (-4.5, 3)\right\}$

40.

X	Y₁	
-3	12	
-2	7	
-1	4	
0	3	
1	4	
2	7	
3	12	

X= -3

41.

X	Y₁	
-6	31	
-5.5	25.25	
-5	20	
-4.5	15.25	
-4	11	
-3.5	7.25	
-3	4	

X= -6

In Exercises 42–57, graphically solve each equation using the INTERSECT method. There are problems in this group that you have NOT yet learned how to solve analytically. Therefore, set your viewing window to ensure all existing points of intersection are visible. With the INTERSECT and ROOT features, you have the power to solve any equation.

42. $\dfrac{x-9}{5} = 4 - x$

43. $2 + \dfrac{x}{3} = \dfrac{5}{6} - x$

44. $4x - 5 = 7x + 4 - 3x$

45. $2(x-8) - 3 = 4x + 10 - 2x$

46. $x^2 - 24 = 2x$

47. $x^2 = -2x + 8$

48. $\dfrac{x+6}{x} = -7$

49. $\dfrac{x-3}{5} = \dfrac{8}{x}$

50. $|6x + 7| = 18$

51. $|3x - 5| = 10$

52. $-|4x + 7| = -11$

53. $-|2x - 8| = -12$

54. $|x + 3| + |x - 5| = 15$

55. $|x + 2| + |x - 3| = 12$

56. $\sqrt{4x - 3} = 7$

57. $5 = \sqrt{3x + 4}$

In Exercises 58–73, graphically solve each equation using the ZERO method. Begin in the standard viewing window and adjust your window as needed. (Hint: When solving equations in 66–73, the degree of the polynomial will predict the maximum number of possible roots if you set the equation equal to zero.)

58. $\dfrac{x-9}{5} - 4 + x = 0$

59. $2 + \dfrac{x}{3} - \dfrac{5}{6} + x = 0$

60. $\dfrac{x+6}{x} + 7 = 0$

61. $\dfrac{x-3}{5} - \dfrac{8}{x} = 0$

62. $|6x + 7| - 18 = 0$

63. $|3x - 5| - 10 = 0$

64. $-|4x + 7| + 11 = 0$

65. $-|2x - 8| + 12 = 0$

66. $x^2 - 2x - 24 = 0$

67. $x^2 + 2x - 8 = 0$

68. $15x^2 + 25x + 10 = 0$

69. $-6x^2 - 7x + 20 = 0$

70. $x^3 + x^2 - 30x = 0$

71. $x^4 - 1 = 0$

72. $6x^4 + x^3 - 40x^2 = 0$

73. $14x^3 - 40x^2 + 6x^4 = 0$

APPLICATIONS *In Exercises 74 and 75, graph the linear equation that models each problem and use your calculator to answer the questions posed.*

74. Mr. Hodge is part owner in a newly formed communications business. His monthly salary is $400 plus 10% of the business's net revenue for the month.
 a. Write a linear equation to model the problem where y represents Mr. Hodge's total salary for the month and x represents the amount of net revenue for the business.
 b. Graph the linear equation and state your viewing window.
 c. Use the VALUE feature to determine his salary if the net revenue is $9000.
 d. Resketch the graph in the following window: $[0, 94000]$ by $[-10,10000]$ (x- and y-scales equal 0).
 e. TRACE to determine the business net revenue if Mr. Hodge's earnings were $6500.

75. A car rental agency charges a flat fee of $45 plus 24 cents per mile.
 a. Write an equation to model the total cost, y, as related to miles driven, x.
 b. Graph on your calculator and indicate your viewing window.
 c. TRACE to determine the approximate cost (to the nearest cent) of traveling 131 miles.
 d. Use VALUE to determine the exact cost of traveling 131 miles.

REVIEW

76. Write an equation in one variable to model the problem and then determine the indicated value. What number must be added to the numerator of $\frac{6}{7}$ and subtracted from the denominator to produce the fraction $\frac{8}{5}$?

77. If a barrel contains 42 gallons of crude oil and a country has crude oil reserves of 2.61×10^{11} barrels, write an expression in scientific notation to indicate the number of gallons of crude oil reserves.

78. Simplify: $\dfrac{(2x^{-4}y^2)^3}{4x^{-2}y^{-1}}$.

79. Use the STOre feature of your calculator to evaluate $\dfrac{(2x^{-4}y^2)^3}{4x^{-2}y^{-1}}$ for $x = \dfrac{1}{2}$ and $y = \dfrac{1}{5}$.

80. Solve algebraically for x:
 a. $|3x + 5| - 6 = -2$ **b.** $2x^2 + x - 3 = 0$

The ROOT/ZERO feature on the TI-86 and the TI-83 work similarly. If using the TI-86, watch for parenthetical information that specifies the differences for your calculator.

If an equation is set equal to zero, when the nonzero side of the equation is graphed, the real roots or zeros are the x-values at the point(s) where the graph crosses the horizontal axis (x-axis). This method is *perfect* for equations of any type that are already set equal to zero.

1. To solve the equation $x^2 + 2x - 8 = 0$ using the root/zero method, enter the polynomial (in this case, the left side of the equation) at the **y1=** prompt. Nothing will be entered at the **y2=** prompt. Press **[ZOOM] [6:Zstandard]** (TI-86 users press **[2nd] [ZOOM] [F4] (ZSTD)**) to automatically set the standard viewing window. Your display should match the one given (the TI-86 screen will still have the GRAPH menu displayed at the bottom). Circle the points where the graph crosses the horizontal (x) axis. These are the roots or zeros.

2. To access the ROOT/ZERO option, press **[2nd] [CALC] [2:zero]** (TI-86 users press **[MORE] [MATH] [F1] (ROOT)**).

   a. **Set left bound:** The screen display asks for a left bound. This is an x-value smaller than the expected root. You may either enter a value for the variable x at the prompt or move the cursor to the left of the left-hand root and press **[ENTER]**. Because the roots are determined on the horizontal axis, a left bound is always determined by locating a value for x to the *left of the desired root*. At the top of the screen a ▶ marker will appear to designate the location of the left bound once ENTER has been pressed.

   b. **Set right bound:** Similarly, the right bound is always determined by moving the cursor to the *right of the root* or by entering a value for x that is larger than the desired root and pressing **[ENTER]**. Again, a ◀ marker is at the top of the screen to designate the location of the bound.

   Note: if the bound markers do not point toward each other, ▶ ◀, then you will get an "ERROR:bounds" message. If this happens, start the ROOT/ZERO calculation over. CARE MUST BE TAKEN TO ENSURE THAT ONLY ONE ROOT OCCURS BETWEEN THE SET BOUNDARIES.

   c. **Locate first root:** Move the cursor to the approximate location where the graph crosses the x-axis for your guess. When you press **[ENTER]** the calculator will search for the root within the area marked by ▶ and ◀. The root is $x = -4$.

   Note: The calculator should display $x = -4$ $y = 0$, which means that the expression entered at **y=** has a value of 0 when $x = -4$. **To verify that the y-value is actually zero, scroll through the TABLE to x = −4 (or use EVAL) and note that the y-value is 0 when $x = -4$.** In this particular problem, round-off error has occurred at the y-value. See Section 7.1 for a discussion of round-off error.

Troubleshooting Note: There will be times when the calculator will be very close to zero but will not display zero exactly. This occurred on the pictured screen. The *y*-value, y = 1.6 E −12 would be Y = −.0000000000016, which is, for all practical purposes a zero. Therefore, when using the graph screen to solve equations/inequalities, you should be aware that the display coordinate values approximate the actual mathematical coordinates. The accuracy of these display values is determined by the height and width of the pixel space being displayed. The space height/width formulas are discussed in detail in a later unit.

d. **Locate subsequent roots:** Repeat the entire process outlined above to determine the right-hand root. This root is x = 2. The screen displays below illustrate the process. Be aware that the bounds set below are arbitrary; the only requirement is that the bounds be to the left and right, respectively, of the desired root.

7.5 Solving Simple Linear Inequalities

■ GRAPHING INEQUALITIES ■ INTERVAL NOTATION ■ SOLVING LINEAR INEQUALITIES
■ GRAPHICAL SOLUTIONS ■ PROBLEM SOLVING

■ GRAPHING INEQUALITIES

The solution(s) to most of the equations in Section 7.3 were real numbers whose values could be graphed as points on a number line. For example, $x = 3$ graphed as

is the solution to $3(x + 3) − 2x = 4x$.

Suppose, however, that our solution had been all numbers *less than or equal to* three. First, we would need the following symbols to translate our English expression to the mathematical format of $x \leq 3$.

Inequality Symbols	
$<$ means	"is less than"
$>$ means	"is greater than"
$\leq$ means	"is less than or equal to"
$\geq$ means	"is greater than or equal to"

An **inequality** is a mathematical expression that indicates that two quantities are not necessarily equal. A **solution of an inequality** is any number that makes the inequality a true statement. The number 2 is a solution of the inequality

$$x \leq 3$$

because $2 \leq 3$.

The inequality $x \leq 3$ has many more solutions, because any real number that is less than or equal to 3 will satisfy the inequality. We can use a graph on the number line to exhibit the solutions of the inequality $x \leq 3$. The colored arrow in Figure 7-9 indicates all those points with coordinates that satisfy the inequality $x \leq 3$. The solid circle at the point with coordinate 3 indicates that the number 3 is a solution of the inequality $x \leq 3$. Because the value of 3 divides the number line into the set of points that ARE solutions and the set of points that ARE NOT solutions, it is called the **critical point.**

FIGURE 7-9

The graph of the inequality $x > 1$ appears in Figure 7-10. The colored arrow indicates all those points whose coordinates satisfy the inequality $x > 1$. The open circle at the point with coordinate 1 indicates that 1 is not a solution of the inequality $x > 1$; however, 1 is a critical point.

FIGURE 7-10

EXAMPLE 1   Graph each inequality on a number line:

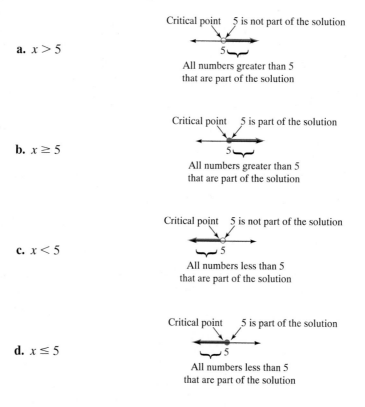

a. $x > 5$

Critical point   5 is not part of the solution

All numbers greater than 5 that are part of the solution

b. $x \geq 5$

Critical point   5 is part of the solution

All numbers greater than 5 that are part of the solution

c. $x < 5$

Critical point   5 is not part of the solution

All numbers less than 5 that are part of the solution

d. $x \leq 5$

Critical point   5 is part of the solution

All numbers less than 5 that are part of the solution

■ INTERVAL NOTATION

Inequalities are graphed as regions on a number line. The regions are called *intervals.* **Interval notation** is a concise way to describe the region that has been graphed on the number line.

EXAMPLE 2   Describe each inequality using interval notation.

        a. $x > 5$

        b. $x \geq 5$

        c. $x < 5$

        d. $x \leq 5$

Solution   It will be easier to express or describe each inequality using interval notation if you have a number line graph to look at. Your interval should correspond to the number line with the smaller value designating the left-hand end of the interval and the larger value designating the larger, right-hand end of the interval.

a. $x > 5$

Interval notation:

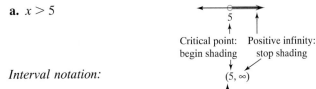

b. $x \geq 5$

Interval notation:

c. $x < 5$

Interval notation:

d. $x \leq 5$

Interval notation:

The infinity symbols (∞ and $-\infty$) indicate that the interval has no bound. For example, $x > 5$ and $x \geq 5$ have no right-hand boundary, whereas $x < 5$ and $x \leq 5$ have no left-hand boundary. When an interval has no left-hand boundary or no right-hand boundary, we say the graph is an **unbounded interval.** The symbols ∞ and $-\infty$ are not real numbers and therefore cannot be included in the interval. Thus, a parenthesis will always be used when an infinity symbol is involved.

Summary of unbounded intervals		
Inequality	**Graph**	**Interval notation**
$x > a$		(a, ∞)
$x \geq a$		$[a, \infty)$
$x < a$		$(-\infty, a)$
$x \leq a$		$(-\infty, a]$

■ SOLVING LINEAR INEQUALITIES

A **linear inequality** is any inequality that can be expressed in the form

$$ax + c < 0 \qquad ax + c > 0 \qquad ax + c \leq 0 \qquad ax + c \geq 0 \quad (a \neq 0)$$

To solve these inequalities we use the following properties of inequalities to transform our inequality into equivalent inequalities, much the same way that we used the property of equality to transform linear equations into equivalent linear equations.

> **Addition Property of Inequality**
> If a, b, and c are real numbers, and
> > if $a < b$, then $a + c < b + c$.
>
> Similar statements can be made for the symbols $>$, $\leq$, and $\geq$.

The **addition property of inequality** can be stated this way: *If any quantity is added to both sides of an inequality, the resulting inequality has the same direction as the original inequality.*

> **Subtraction Property of Inequality**
> If a, b, and c are real numbers, and
> > if $a < b$, then $a - c < b - c$.
>
> Similar statements can be made for the symbols $>$, $\leq$, and $\geq$.

The **subtraction property of inequality** can be stated this way: *If any quantity is subtracted from both sides of an inequality, the resulting inequality has the same direction as the original inequality.*

The subtraction property of inequality is included in the addition property: To *subtract* a number a from both sides of an inequality, we could instead *add* the *negative* of a to both sides.

EXAMPLE 3   Solve the inequality $2x + 5 > x - 4$. Graph the solution on the number line and translate to interval notation.

Solution   To isolate the x on the left-hand side of the $>$ sign, we proceed as we would when solving equations.

$$2x + 5 > x - 4$$
$$2x + 5 - 5 > x - 4 - 5 \qquad \text{Subtract 5 from both sides.}$$
$$2x > x - 9 \qquad \text{Combine like terms.}$$
$$2x - x > x - 9 - x \qquad \text{Subtract } x \text{ from both sides.}$$
$$x > -9 \qquad \text{Combine like terms.}$$

The graph of this solution (see Figure 7-11) includes all points to the right of -9 but does not include -9 itself. For that reason, we use an open circle at -9.

FIGURE 7-11

Because the critical point -9 is not included in the graph of the solution, we will use a parenthesis in the interval notation: $(-9, \infty)$. ∎

If both sides of the true inequality $2 < 5$ are multiplied by a *positive* number, such as 3, another true inequality results.

$$2 < 5$$
$$3 \cdot 2 < 3 \cdot 5 \qquad \text{Multiply both sides by 3.}$$
$$6 < 15$$

The inequality $6 < 15$ is a true inequality. However, if both sides of the inequality $2 < 5$ are multiplied by a negative number, such as -3, the direction of the inequality symbol must be reversed to produce another true inequality.

$$2 < 5$$
$$-3 \cdot 2 > -3 \cdot 5 \qquad \text{Multiply both sides by the \textit{negative} number } -3 \text{ and reverse the direction of the inequality.}$$
$$-6 > -15$$

The inequality $-6 > -15$ is a true inequality because -6 lies to the right of -15 on the number line.

Multiplication Property of Inequality

If a, b, and c are real numbers, and

if $a < b$	and	$c > 0$,	then	$ac < bc$.
If $a < b$	and	$c < 0$,	then	$ac > bc$.

There is a similar property for division.

Division Property of Inequality

If a, b, and c are real numbers, and

if $a < b$	and	$c > 0$,	then	$\dfrac{a}{c} < \dfrac{b}{c}$.
If $a < b$	and	$c < 0$,	then	$\dfrac{a}{c} > \dfrac{b}{c}$.

To *divide* both sides of an inequality by a nonzero number c, we could instead *multiply* both sides by $\frac{1}{c}$. The multiplication and division properties of inequality are also true for $\leq$, $>$, and $\geq$.

EXAMPLE 4 Solve $3x + 7 \leq -5$ and express the solution both as a number line graph and in interval notation.

Solution

$$3x + 7 \leq -5$$
$$3x + 7 - 7 \leq -5 - 7 \qquad \text{Subtract 7 from both sides.}$$
$$3x \leq -12 \qquad \text{Combine like terms.}$$
$$\frac{3x}{3} \leq \frac{-12}{3} \qquad \text{Divide both sides by 3.}$$
$$x \leq -4$$

The solution of $3x + 7 \leq -5$ consists of all real numbers less than, and also including, -4. The solid circle at -4 in the graph of the solution in Figure 7-12 indicates that -4 is one of the solutions of the given inequality.

FIGURE 7-12

A square bracket will be used in the interval notation to indicate that the critical point of -4 is part of the solution: $(-\infty, -4]$. ■

EXAMPLE 5   Solve $5 - 3x \leq 14$ and express the solution both as a number line graph and in interval notation.

Solution

$$5 - 3x \leq 14$$
$$5 - 3x - 5 \leq 14 - 5 \qquad \text{Subtract 5 from both sides.}$$
$$-3x \leq 9 \qquad \text{Combine like terms.}$$
$$\frac{-3x}{-3} \geq \frac{9}{-3} \qquad \text{Divide both sides by } -3 \text{ and reverse the direction of the } \leq \text{ symbol.}$$
$$x \geq -3$$

In the last step, both sides of the inequality were divided by -3. Because -3 is negative, the direction of the inequality was *reversed*. The graph of the solution appears in Figure 7-13. The solid circle at -3 indicates that -3 is one of the solutions.

FIGURE 7-13

The square bracket indicates that the critical point of -3 is one of the solutions to the inequality: $[-3, \infty)$. ■

WARNING!   Do not forget to reverse the direction of the inequality symbol when multiplying or dividing by a negative number.

EXPLORING THE CONCEPT   Solve each equation/inequality below and graph the solution set on a number line graph.

a. $4x - 5 = 2 - 3x$

b. $4x - 5 \leq 2 - 3x$

c. $4x - 5 < 2 - 3x$

d. $4x - 5 \geq 2 - 3x$

e. $4x - 5 > 2 - 3x$

f. What is the relationship between the critical points in **b–e** and the solution to the linear equation in **a?**

g. If $ax + b > cx + d$ has $x > -2$ as the solution, what is the solution to each of the following:

$$ax + b = cx + d \qquad ax + b \geq cx + d \qquad ax + b < cx + d$$

Solution **a.** $x = 1$

b. $x \leq 1$

c. $x < 1$

d. $x \geq 1$

e. $x > 1$

f. The critical points are all the same regardless of the direction or type of the inequality symbol, and they are the same as the solution to the equation.

g. $ax + b = cx + d$ would have $x = -2$ as its solution.
$ax + b \geq cx + d$ would have $x \geq -2$ as its solutions.
$ax + b < cx + d$ would have $x < -2$ as its solutions.

■ GRAPHICAL SOLUTIONS

Consider the inequality in Example 5, $5 - 3x \leq 14$. The solution of the inequality is the set of all values for x such that $x \geq -3$.

The calculator can be used to *confirm* that $x \geq -3$ is a solution. Enter the left side of the original inequality at y1 and the right side of the inequality at y2. Set the TABLE to begin at the critical point of $x = -3$ and scroll through the TABLE to examine the x-values that are less than -3, greater than -3, and equal to -3. Compare the y1 and y2 columns that represent the left and right sides of the inequality.

- At all values of $x < -3$, y1 > y2 (the expression $5 - 3x$ has values greater than 14).
- At $x = -3$, y1 = y2 (the expression $5 - 3x$ is equal to 14).
- At all values of $x > -3$, y1 < y2 (the expression $5 - 3x$ has values less than 14).

The inequality is true for values of x that satisfy y1 ≤ y2, so our solution of $x \geq -3$ is confirmed.

To graphically *solve* an inequality, we will use both the TABLE *and* a graphical display to interpret the solution. If you are graphically solving an inequality, you should graph both the left and right sides of the inequality and determine the point of intersection just as you do when graphically solving equations. (See Figure 7-14.)

The critical point of the inequality occurs at the point of intersection. As we discovered in the Exploring the Concept section, the critical point of the inequality and the solution to the corresponding equation are the same. Thus, we can now conclude that our number line will be divided by the value -3 (Figure 7-15).

The graph screen should be read from left to right. It is a visual representation of the numerical information that was previously displayed in the TABLE. For what x-values is y1 less than y2? Remember, the INTERSECT feature has already shown us the value for which y1 = y2. It should be the section that

FIGURE 7-14

FIGURE 7-15

is highlighted on the graph displayed. In general, y1 < y2 where y1 is **below** the graph of y2. (See Figure 7-16.)

$[-10, 10]$ by $[-10, 20]$

FIGURE 7-16

The x-values greater than -3 satisfy the requirement that y1 < y2. We can combine this information with the critical point on the number line to produce the final solution graphed on a number line (Figure 7-17).

Translated to interval notation: $[-3, \infty)$.

FIGURE 7-17

EXAMPLE 6   Graphically solve $10 - 3x < 2x + 5$.

Solution

$$y1 = 10 - 3x$$
$$y2 = 2x + 5$$

■ STUDY TIP

As you work through an exercise section, write down questions for your instructor as they occur to you. Keep the list on a separate piece of paper and write down the answers to *all* of your questions.

From the INTERSECT screen (Figure 7-18) we can determine that the critical point of the inequality is 1. The value of 1 will not be included as part of the solution because the original inequality is strictly less than.

$[10, 10]$ by $[-10, 15]$

FIGURE 7-18

Now determine where y1 < y2 by determining which x-values on the graph will occur where y1 < y2. Confirm this with the TABLE. (See Figure 7-19.)

FIGURE 7-19

Combine the information determined from interpreting the graph with the critical point determined through the INTERSECT feature to produce a number line graph of the solution, as shown in Figure 7-20.

FIGURE 7-20

■ PROBLEM SOLVING

EXAMPLE 7   **Averaging grades**   A student has scores of 72%, 74%, and 78% on three mathematics examinations. What score does he need on the last exam to earn a grade of at least a B (80% or better)?

Solution   We can let x represent the score on the fourth (and last) exam. To find the average grade, we add the four scores and divide by 4. To earn a B, this average must be greater than or equal to 80%.

The averages of the four grades	$\geq$	80

$$\frac{72 + 74 + 78 + x}{4} \geq 80$$

$y1 = (72 + 74 + 78 + x)/4$
$y2 = 80$

$[0, 150]$ by $[0, 100]$

We can solve this inequality for x.

$$\frac{224 + x}{4} \geq 80 \qquad 72 + 74 + 78 = 224.$$

$$224 + x \geq 320 \qquad \text{Multiply both sides by 4.}$$

$$x \geq 96 \qquad \text{Subtract 224 from both sides.}$$

To earn a B, the student must score 96% or better on the last exam. The graph of this solution appears in Figure 7-21.

FIGURE 7-21

EXAMPLE 8   **Geometry**   If the perimeter of an equilateral triangle is greater than 15 feet, how long could a side be?

Solution   Recall that each side of an equilateral triangle is the same length and that the perimeter of a triangle is the sum of the lengths of its three sides.

     Let x represent the length of one side of the triangle.

     Then $x + x + x$ represents the perimeter.

The perimeter is to be greater than 15 feet. We indicate this fact with the following inequality.

$$x + x + x > 15$$

$$3x > 15 \qquad \text{Combine like terms.}$$

$$x > 5 \qquad \text{Divide both sides by 3.}$$

Each side of the triangle must be more than 5 feet long.

$y1 = x + x + x$
$y2 = 15$

$[-10, 10]$ by $[-10, 20]$

EXERCISE 7.5

VOCABULARY   *In Exercises 1–4, fill in the blanks to make a true statement.*

1. An ~~inequality~~ is a mathematical expression that indicates that two quantities are not necessarily equal.

2. The ~~critical point~~ divides the number line into the set of points that are the solution and the set of points that are not the solution.

3. ~~Interval~~ notation is a concise method used to describe the region that has been graphed on the number line.

4. An ~~unbounded~~ interval has no left- or right-hand boundary.

CONCEPTS

5. The left side of an equation has been stored at y1 and the right side at y2. Based on the TABLE shown, determine the *x*-value that is the solution to y1 = y2.

X	Y₁	Y₂
-12	-32	-29
-11	-29	-27
-10	-26	-25
-9	-23	-23
-8	-20	-21
-7	-17	-19
-6	-14	-17

X= -12

6. The solution to y1 = y2 can be determined from the TABLE shown.
Based on the information displayed in the TABLE, determine a number line graph of the solution to the following inequalities.

X	Y₁	Y₂
-4	1	4
-3	3	5
-2	5	6
-1	7	7
0	9	8
1	11	9
2	13	10

X= -1

a. y1 < y2

b. y1 ≤ y2

c. y1 > y2

d. y1 ≥ y2

7. The solution to y1 = y2 can be determined from the TABLE shown. Based on the information displayed in the TABLE, determine a number line graph of the solution to the following inequalities.

X	Y₁	Y₂
-5	-29	-20
-4	-22	-16
-3	-15	-12
-2	-8	-8
-1	-1	-4
0	6	0
1	13	4

X= -2

a. y1 < y2

b. y1 ≤ y2

c. y1 > y2

d. y1 ≥ y2

8. The solution to y1 = y2 can be determined from the accompanying graph. Based on the information displayed, determine a number line graph of the solution to the following inequalities.

a. y1 < y2

b. y1 ≤ y2

c. y1 > y2

d. y1 ≥ y2

9. The solution to y1 = y2 can be determined from the accompanying graph. Based on the information displayed, determine a number line graph of the solution to the following inequalities.

a. y1 < y2

b. y1 ≤ y2

c. y1 > y2

d. y1 ≥ y2

10. The solution to y1 = y2 can be determined from the accompanying graph. Based on the information displayed, determine a number line graph of the solution to the following inequalities.

a. y1 < y2

b. y1 ≤ y2

c. y1 > y2

d. y1 ≥ y2

PRACTICE *In Exercises 11–30, solve each inequality and express the solution in both interval notation and as a number line graph. Verify your solution by solving graphically.*

11. $2x - 3 \le 5$

12. $-3x - 5 < 4$

13. $-3x - 7 > -1$

14. $-5x + 7 \le 12$

15. $2x + 9 \le x + 8$

16. $3x + 7 \le 4x - 2$

17. $9x + 13 \ge 8x$

18. $7x - 16 < 6x$

19. $7 - x \le 3x - 1$

20. $2 - 3x \ge 6 + x$

21. $3(x - 8) < 5x + 6$

22. $9(x - 11) > 13 + 7x$

23. $8(5 - x) \le 10(8 - x)$

24. $17(3 - x) \ge 3 - 13x$

25. $\dfrac{5}{2}(7x - 15) + x \ge \dfrac{13}{2}x - \dfrac{3}{2}$

26. $\dfrac{5}{3}(x + 1) \le -x + \dfrac{2}{3}$

27. $\dfrac{3x - 3}{2} < 2x + 2$

28. $\dfrac{x + 7}{3} \ge x - 3$

29. $\dfrac{2(x + 5)}{3} \le 3x - 6$

30. $\dfrac{3(x - 1)}{4} > x + 1$

APPLICATIONS *In Exercises 31–39, express each solution as a statement of inequality.*

31. Calculating grades A student has test scores of 68%, 75%, and 79%. What must she score on the last exam to earn a B (80% or better)?

32. Calculating grades A student has test scores of 70%, 74%, and 84%. What does he need to maintain a C (70% or better)?

33. Fleet averages An auto manufacturer produces three sedan models in equal quantities. One model has an economy rating of 17 miles per gallon, and the second model is rated at 19 mpg. The manufacturer is required to have a fleet average of at least 21 mpg. What economy rating is required for the third model car?

34. Avoiding a service charge When the average daily balance of a customer's checking account falls below $500 in any week, the bank assesses a $5 service charge. Bill's account balances for the week were

Monday	$540.00
Tuesday	435.50
Wednesday	345.30
Thursday	310.00

What must Friday's balance be to avoid the service charge?

35. FDA guidelines The Food and Drug Administration recommends that an individual eat at least five servings of fruit each day. If for the first six days of the week, you have five, four, three, six, seven, and four servings, how many servings of fruit should you eat on the seventh day of the week in order to average at least five servings per day?

36. Car rental A rental car agency charges a daily rate of $37.50 per day plus $0.21 per mile. Sheila has budgeted $150 per day for car rental during her vacation. To the nearest tenth of a mile, how far can she travel per day to stay within her budget?

37. Rent-to-own The local Rent-to-Own store rents washing machines for a mere $0.67 a day. If a new washer costs $385, after how many days will it be cheaper to buy a machine instead of renting it?

38. Geometry The perimeter of an equilateral triangle is at most 57 feet. What could be the length of a side? (*Hint:* All three sides of an equilateral triangle are equal.)

39. Geometry The perimeter of a square is no less than 68 centimeters. How long can a side be?

REVIEW

40. Determine if each of the following represents a function:

a. $\{(-3, 1), (-2, 0), (0.4, 7)\ (-3, 0)\}$

b.
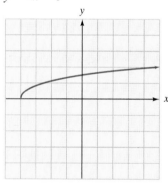

c. $y = 4x - 5$

d.

41. Identify the domain and range: $\{(1, 2), (-1, 3), (-2, 4)\}$.

42. Solve each of the equations both analytically (algebraically) and graphically. You may use either the Root/Zero or INTERSECT features. Copy your complete screen display to justify your work.

a. $2[3 - 2(3 - x)] = 2(3 + x)$

b. $|4x - 3| + 2 = 0$

c. $3x^2 + 19x = -28$

43. Verify that $-3/4$ is a solution of $4x^2 + 11x + 6 = 0$ using both the STOre and TABLE features of your calculator.

7.6 Compound Inequalities

■ AND STATEMENTS ■ OR STATEMENTS ■ GRAPHICAL SOLUTIONS

Compound inequalities are inequalities that are linked by the words *AND* or *OR*. When solving compound inequalities you will be required to determine either the **intersection** (common values) or the **union** (all represented values) of solutions to simple inequalities in order to determine the solution to the compound inequality statement.

■ AND STATEMENTS

The set of all values that are both greater than two *and* less than 5,

$$x > 2 \quad and \quad x < 5,$$

represents a compound inequality. The graph of each separate inequality is shown below.

$x > 2$

$x < 5$

The compound inequality of $x > 2$ *and* $x < 5$ is represented by the intersection of the two inequalities. The intersection of the two inequalities is the set of all values that satisfy (are common to) *both* inequality statements.

$x > 2$ *and* $x < 5$

This inequality of $x > 2$ *and* $x < 5$ can be written in the abbreviated form of $2 < x < 5$. It is still read as "x is greater than 2 and x is less than 5." This is often called a **continued inequality.**

When written in interval form, $(2, 5)$, we say that the interval is an **open interval** because neither endpoint is included in the solution.

EXAMPLE 1   Express each compound inequality as a number line graph and in interval notation.

STUDY TIP
Read directions carefully and follow them precisely—whether you are simply doing homework or are taking a test or exam.

a. $-1 < x < 3$ $(-1, 3)$

b. $-1 \leq x < 3$ $[-1, 3)$

c. $-1 < x \leq 3$ $(-1, 3]$

d. $-1 \leq x \leq 3$ $[-1, 3]$

Because there is a finite endpoint on each of these intervals, we say the intervals are **bounded.** The chart below summarizes bounded intervals.

	Summary of bounded intervals	
Inequality	**Number line graph**	**Interval notation**
$a < x < b$		(a, b) **open interval**
$a \leq x \leq b$		$[a, b]$ **closed interval**
$a < x \leq b$		$(a, b]$ **half-open interval**
$a \leq x < b$		$[a, b)$ **half-open interval**

EXAMPLE 2   Solve the compound inequality.

$$x - 4 < 5 \quad and \quad x + 2 \geq 5$$

Solution   Solve each inequality separately.

$$x - 4 < 5 \quad and \quad x + 2 \geq 5$$
$$x < 9 \quad and \quad x \geq 3$$

Graph each inequality. The compound inequality is true for all values of x that are less than 9 *and* greater than or equal to 3.

$x < 9$

$x \geq 3$

The intersection of these two graphs is

The solution to the inequality is $3 \leq x < 9$, which becomes $[3, 9)$ when written in interval notation.   ■

WARNING!   When writing compound inequalities be sure that the inequality makes sense when read from left to right. For example, $4 \leq x < -8$ does not make sense because $4 \nleq -8$.

EXAMPLE 3   Solve the compound inequality

$$4x - 5 > 3 \quad and \quad 4 - \frac{1}{2}x < 1$$

Solution   Solve each simple inequality.

$$4x - 5 > 3 \quad and \quad 4 - \frac{1}{2}x < 1$$
$$4x > 8 \quad and \quad -\frac{1}{2}x < -3$$
$$x > 2 \quad and \quad x > 6$$

The graphs are shown below.

$x > 2$

$x > 6$

The solution to the compound inequality is the set of all values that satisfy both inequalities: all real numbers greater than 2 and also greater than 6.

Expressed in interval notation this is $(6, \infty)$ ■

EXAMPLE 4 Solve the compound inequality

$$2(x - 5) \geq 2 \quad and \quad 4 - (3 - x) < 3$$

Solution Solve each simple inequality.

$$
\begin{array}{ccc}
2(x - 5) \geq 2 & and & 4 - (3 - x) < 3 \\
2x - 10 \geq 2 & and & 4 - 3 + x < 3 \\
2x \geq 12 & and & 1 + x < 3 \\
x \geq 6 & and & x < 2
\end{array}
$$

Express each inequality as a graph.

$x \geq 6$

$x < 2$

These two graphs have NO common points of intersection. There are no values for x that are both less than 2 and simultaneously greater than 6. The solution is the empty set, $\varnothing$.

$\varnothing$

EXAMPLE 5 Solve the compound inequality

$$2(x - 1) > -4 \quad and \quad 2(x - 1) \leq 4$$

Solution We can solve each inequality separately as in the previous examples, or we can combine the two inequalities into a continued inequality because both inequalities share a common side of $2(x - 1)$.

$-4 < 2(x - 1) \leq 4$	Be sure that the double inequality makes sense. Is $-4 \leq 4$?
$-4 < 2x - 2 \leq 4$	Remove grouping symbols.
$-4 + 2 < 2x - 2 + 2 \leq 4 + 2$	Add 2 to *all three parts of the inequality.*
$-2 < 2x \leq 6$	
$-\dfrac{2}{2} < \dfrac{2x}{2} \leq \dfrac{6}{2}$	Divide all three parts of the inequality by 2.
$-1 < x \leq 3$	

The solution expressed as a number line graph is

Expressed in interval notation, the solution is $(-1, 3]$.

■ OR STATEMENTS

Inequalities that involve the use of the word *OR* are also compound inequalities. When the inequality is an *or* statement, we are looking for values of the variable that will satisfy either inequality. These values are represented by the **union** of all possible solution values. For example, the solution to

$$x < 5 \quad or \quad x > 9$$

is graphed as

$x < 5$

$x > 9$

which is the combined set of values

The word *or* indicates that only one of the inequalities needs to be true to make the statement true. To express an *or* statement in interval notation, we will use an interval to describe each shaded section of the number line graph. The interval $(-\infty, 5)$ describes the portion of the number line graph represented by

and the interval $(9, \infty)$ describes

It is the union of these two intervals that produces the solution set. Thus we use the symbol $\cup$ (the symbol for union) to indicate that union.

$$(-\infty, 5) \cup (9, \infty)$$

EXAMPLE 6   Solve the compound inequality

$$5x + 3 > 4x \quad \textit{or} \quad 3(x + 2) - 1 < 2x$$

Solution   Solve each inequality separately.

$$
\begin{array}{lcl}
5x + 3 > 4x & \textit{or} & 3(x + 2) - 1 < 2x \\
3 > -x & \textit{or} & 3x + 6 - 1 < 2x \\
-3 < x & \textit{or} & 3x + 5 < 2x \\
& & 5 < -x \\
& & -5 > x
\end{array}
$$

Graph each inequality separately.

$x > -3$

$x < -5$

The original inequality is an *or* statement, so we want the union of these two sets of values for the solution.

The interval notation would be

$$(-\infty, -5) \cup (-3, \infty)$$

EXAMPLE 7   Solve the compound inequality

$$\frac{1}{2}(4x - 2) + 1 > 5 \quad or \quad 3\left(\frac{1}{3}x - 1\right) < 2$$

Solution   Solve each inequality.

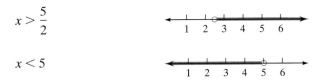

$$\frac{1}{2}(4x - 2) + 1 > 5 \quad or \quad 3\left(\frac{1}{3}x - 1\right) < 2$$

$$2x - 1 + 1 > 5 \quad or \quad x - 3 < 2$$

$$2x > 5 \quad or \quad x < 5$$

$$x > \frac{5}{2}$$

The graph of each inequality would be

$$x > \frac{5}{2}$$

$$x < 5$$

The union of these two inequalities would be every real number.

Expressed as an interval, this would be $(-\infty, \infty)$.

▪ GRAPHICAL SOLUTIONS

EXPLORING THE CONCEPT   To graphically solve the continued inequality $-4 \leq 2(x - 1) \leq 4$ in Example 5, graph y1 $= -4$, y2 $= 2(x - 1)$, and y3 $= 4$. We can now abbreviate our inequality statement to y1 $\leq$ y2 $\leq$ y3.

a. Sketch the graph in the standard viewing window.

b. Use the INTERSECT feature to find the intersection of y1 and y2.

c. Use the INTERSECT feature to find the intersection of y2 and y3.

d. On your paper, highlight the section of y2 that is **between** the graph of y1 and y3.

e. Determine the *x*-values for which y1 $\leq$ y2 $\leq$ y3.

f. Write the solution set of y1 $\leq$ y2 $\leq$ y3 in interval notation.

Solution   **a.**

b. The intersection of y1 and y2 is the coordinate $(-1, -4)$. The *x*-value of -1 is the critical point for the inequality.

c. The intersection of y2 and y3 is the coordinate $(3, 4)$. The *x*-value of 3 is the critical point for the inequality.

d.

e. When x is between -1 and 3, then y1 $\leq$ y2 $\leq$ y3.

f. The number line graph of the solution to the one-variable inequality would be

Converted to interval notation: $[-1, 3]$.

EXAMPLE 8 Graphically solve $-1 \leq \dfrac{4}{5}x + 2 < 6$.

Solution Graph y1 $= -1$:

$$y2 = \frac{4}{5}x + 2$$

$$y3 = 6$$

and use the INTERSECT feature to determine the critical points.

The critical points for the compound inequality are -3.75 and 5.

Finally, interpret the graphical display.

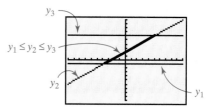

The solution consists of the critical points and all x-values *between* the critical points of -3.75 and 5.

    $[-3.75, 5]$

EXAMPLE 9 To hold the temperature of a room between 19° and 22° Celsius, what Fahrenheit temperatures must be maintained? (Recall that the Fahrenheit temperature, F, and Celsius temperature, C, are related by the formula $C = \frac{5}{9}(F - 32)$.)

Solution Write the compound inequality to model the problem:

$$19 \leq \frac{5}{9}(x - 32) \leq 22$$

Graph the inequality where y1 $= 19$, y2 $= \frac{5}{9}(x - 32)$, and y3 $= 22$, and use the INTERSECT feature to determine the critical points.

$[-10, 100]$ by $[15, 25]$

The critical points are the *x*-values of 71.6 and 66.2. Interpret the graph to determine y1 ≤ y2 ≤ y3.

The Fahrenheit temperature must be between 66.2° and 71.6°.

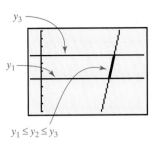

$y_1 \le y_2 \le y_3$

EXERCISE 7.6

VOCABULARY *In Exercises 1–4, fill in the blanks to make a true statement.*

1. compound inequalities are inequalities that are linked by the words *and* or *or*.

2. A compound inequality such as $2 < x < 5$ is also called a continued inequality.

3. *And* statements represent the _____ of the set of values that satisfy both inequalities.

4. *Or* statements represent the _____ of the set of values that satisfy both inequalities.

CONCEPTS *In Exercises 5–8, solve each compound inequality by interpreting the information displayed on the screen. Express the solution in interval notation.*

5.

[−10, 10] by [−10, 15]
a. y1 < y2 < y3
b. y2 < y1   or   y2 > y3

6.

[−10, 10] by [−10, 15]
a. y1 < y2 < y3
b. y2 < y1   or   y2 > y3

7.

[−15, 10] by [−5, 5]
a. y1 < y2 < y3
b. y2 < y1   or   y2 > y3

8.

[−10, 10] by [−10, 10]
a. y1 < y2 < y3
b. y2 < y1   or   y2 > y3

9. Match each compound inequality with the appropriate number line graph.

a. $-1 < x < 4$

i. (number line: open circles at −1 and 4, shaded between)

b. $x < -1$ or $x > 4$

ii. (number line: open circle at −1, closed circle at 4, shaded between)

c. $-1 < x \le 4$

iii. (number line: open circles at −1 and 4, shaded between)

d. $x \le -1$ or $x > 4$

iv. (number line: open circles at −1 and 4, shaded outside)

10. Match each number line graph with the appropriate interval notation.

a.   i. $(2, 3)$

b.   ii. $(-\infty, 2] \cup [3, \infty)$

c.   iii. $[2, 3]$

d.   iv. $(-\infty, 2) \cup (3, \infty)$

   v. $(-\infty, 2) \cup (3, \infty)$

PRACTICE   In Exercises 11–22, express each compound inequality as a number line graph and in interval notation.

11. $x \le 5$ and $x > -2$ **12.** $x \ge 3$ and $x < 11$ **13.** $x < 10$ and $x \le 4$

14. $x \ge -5$ and $x \ge -1$ **15.** $x \le 6$ and $x \ge 12$ **16.** $x < -3$ and $x > 1$

17. $x > 5$ or $x < -2$ **18.** $x \le 3$ or $x > 11$ **19.** $x > 10$ or $x \ge 4$

20. $x > 5$ or $x > 8$ **21.** $x < 6$ or $x \ge -2$ **22.** $x > 4$ or $x \le 5$

In Exercises 23–42, solve each compound inequality and express the solution both as a number line graph and in interval notation.

23. $5x - 3 > 12$ and $7x - 23 < 5$ **24.** $-3x - 1 \ge 5$ and $-2x + 6 \le 16$

25. $-9x \ge -x$ and $4 - 3x \le x$ **26.** $8 - 5x \ge -x$ and $4 - 2x \le 2x$

27. $7 - 4x \ge 8$ and $2 + 3x \le 8$ **28.** $8 - 9x \ge 4x$ and $4 - 3x \ge x$

29. $-3(x + 2) > 2(x + 1)$ and $-4(x - 1) < x + 8$ **30.** $\dfrac{1}{2}x + 2 \le \dfrac{1}{3}x - 4$ and $\dfrac{1}{4}x - \dfrac{1}{3} \le x + 2$

31. $-2 < -x + 3 < 5$ **32.** $4 < -x - 2 < 9$ **33.** $15 > 2x - 7 > 9$

34. $25 > 3x - 2 > 7$ **35.** $0 \le \dfrac{4 - x}{3} \le 2$ **36.** $-2 \le \dfrac{5 - 3x}{2} \le 2$

37. $3x + 2 < 8$ or $2x - 3 > 11$ **38.** $3x + 4 < -2$ or $3x + 4 > 10$

39. $-4(x + 2) \ge 12$ or $3x + 8 < 11$ **40.** $5(x - 2) \ge 0$ or $-3x < 9$

41. $4x < -24$ or $-3x > 18$ **42.** $1 + 2x \le 13$ or $x - 2 > 4$

In Exercises 43–50, solve each inequality graphically. Copy your screen display and in complete sentences explain how you interpreted the graphical display to determine the solution. Solutions should be expressed in interval notation.

43. $-2 < 3x + 4 \le 7$

44. $0 \le 5 - 2x \le 8$

45. $0 \le 1 - \dfrac{2}{3}x < 5$

46. $2 < x - 5 < 7$

47. $3x + 4 < -2$ or $3x + 4 \ge 7$

48. $5 - 2x \le 0$ or $5 - 2x \ge 8$

49. $1 - \dfrac{2}{3}x \le 0$ or $1 - \dfrac{2}{3}x > 5$

50. $x - 5 < 2$ or $x - 5 > 7$

APPLICATIONS *In Exercises 51–57, express each solution as an inequality. Solve algebraically and confirm graphically.*

51. Melting iron To melt iron, the temperature of a furnace must be at least 1540° C but no more than 1650° C. What range of Celsius temperatures must be maintained?

52. Phonograph records The radii of phonograph records must lie between 5.9 and 6.1 inches. What variation in circumference can occur? (*Hint:* The circumference of a circle is given by the formula $C = 2\pi r$, where r is the radius. Let $\pi = 3.14$)

53. Pythons A large snake, the African Rock Python, can grow to a length of 25 feet. To the nearest hundredth, what is the range of lengths in meters? (*Hint:* There are about 3.281 feet in one meter.)

54. Comparing weights The normal weight of a 6-foot-2-inch man is between 150 and 190 pounds. To

the nearest hundredth, what would such a person weigh in kilograms? (*Hint:* There are 2.2 pounds in one kilogram.)

55. Manufacturing stereos The time required to assemble a television set at the factory is 2 hours. A stereo receiver requires only 1 hour. The labor force at the factory can supply at least 640 and at most 810 hours of assembly time per week. When the factory is producing 3 times as many television sets as stereos, how many stereos could be manufactured in 1 week?

56. Geometry A rectangle's length is 3 feet less than twice its width, and its perimeter is between 24 and 48 feet. What might be its width?

57. Geometry A rectangle's width is 8 feet less than 3 times its length, and its perimeter is between 8 and 16 feet. What might be its length?

REVIEW

58. The solution to y1 = y2 can be determined from the accompanying graph. Based on the information displayed, determine a number line graph of the solution to the following inequalities.
    a. y1 < y2

    b. y1 ≤ y2

    c. y1 > y2

    d. y1 ≥ y2

59. Algebraically solve $\frac{2}{3}(x - 5) \le 6 - 12x$ and record the solution as a number line graph and in interval notation.

60. Construct a handrawn graph of $y = \frac{2}{3}(x - 5)$, designating the x- and y-intercepts.

61. Factor completely:
    a. $5x^4 - 125y^2$
    b. $6x^2 - 17x + 12$
    c. $4x^3 - 19x^2 - 5x$

7.7 Absolute Value Inequalities

■ INEQUALITIES OF THE FORM $|x| < k$ ■ INEQUALITIES OF THE FORM $|x| > k$

Recall the definition of absolute value.

> **Absolute Value**
>
> $|x| = x$ when $x \geq 0$
>
> $|x| = -x$ when $x < 0$

We can relate this to the number line by translating the expression $|x|$ to mean "find the number(s) that are a distance of x units from zero on the number line." Thus $|x| = 5$ translates to

We conclude that x must have a value of 5 or -5 in order for $|x| = 5$.

■ INEQUALITIES OF THE FORM $|x| < k$

Now consider $|x| < 5$. We can translate this expression to mean "find all numbers x whose distance is 5 or less units from zero. Thus, x is between -5 and 5, and

$|x| < 5$ is equivalent to $-5 < x < 5$

EXPLORING THE CONCEPT

Graphically solve the absolute value inequality $|x| < 5$.

a. Graph both sides of the inequality in the standard viewing window.

b. Determine the value of the critical points by using the INTERSECT feature.

c. Interpret the graphical display to determine the x-values that satisfy the inequality.

d. Express the solution as a number line graph and in interval notation.

Solution   **a.** Let y1 = $|x|$

    y2 = 5

b.

c.

Critical points: the x-values of -5 and 5.

$y_1 < y_2$

d.

$(-5, 5)$

In general, we have the following property:

> **Solving Absolute Value Inequalities**
> If k is a positive number then $|x| < k$ is equivalent to $-k < x < k$.

EXAMPLE 1 Solve: $|2x - 3| < 9$.

Solution *Analytical solution:*
We write the inequality as a continued inequality

$|2x - 3| < 9$ is equivalent to $-9 < 2x - 3 < 9$

and solve for x:

$-9 < 2x - 3 < 9$
$-6 < 2x < 12$ Add 3 to all three parts.
$-3 < x < 6$ Divide all parts by 2.

Any number between -3 and 6, not including either -3 or 6, is in the solution set. This is the interval $(-3, 6)$. The graph is shown in Figure 7-22.

Graphical solution:

$y_1 < y_2$

FIGURE 7-22

EXAMPLE 2 Solve: $|3x + 2| \le 5$.

Solution *Analytical solution:*
We write the expression as a continued inequality:

$|3x + 2| \le 5$ is equivalent to $-5 \le 3x + 2 \le 5$

and solve for x:

$-5 \le 3x + 2 \le 5$
$-7 \le 3x \le 3$ Subtract 2 from all three parts.
$-\dfrac{7}{3} \le x \le 1$ Divide all three parts by 3.

The solution set is the interval $\left[-\frac{7}{3}, 1\right]$, whose graph is shown in Figure 7-23.

Graphical solution:

$y_1 < y_2$

FIGURE 7-23

■ INEQUALITIES OF THE FORM $|x| > k$

The inequality $|x| > 5$ can be interpreted to mean that a point with coordinate x is greater than 5 units from the origin. (See Figure 7-24.)

Thus, $x < -5$ or $x > 5$.

FIGURE 7-24

EXPLORING THE CONCEPT   Graphically solve $|x| > 5$.

a. Graph both sides of the inequality in the standard viewing window.

b. Determine the value of the critical points through the use of the INTERSECT feature.

c. Interpret the graphical display to determine the x-values that satisfy the inequality.

d. Express the solution as a number line graph and in interval notation.

Solution   **a.** Let $y1 = |x|$

$y2 = 5$

■ STUDY TIP

When studying for a test focus on directions as well as problems. Words such as *evaluate, solve, simplify,* and *graph* should be the triggers that move you towards successful problem solving on tests.

b.

c.

Critical points: the x-values of -5 and 5.

d.

-5   5

$(-\infty, -5) \cup (5, \infty)$

Solving Absolute Value Inequalities

If k is a positive number then $|x| > k$ is equivalent to $x < -k$ or $x > k$.

-k   k

EXAMPLE 3   Solve the inequality $|5x - 10| > 20$.

Solution   *Analytical solution:*

We write the inequality as two separate inequalities

$|5x - 10| > 20$ is equivalent to $5x - 10 < -20$ or $5x - 10 > 20$

and solve each one for x:

$5x - 10 < -20$ or $5x - 10 > 20$

$5x < -10$ | $5x > 30$ Add 10 to both sides.

$x < -2$ | $x > 6$ Divide both sides by 5.

Thus, x is either less than -2 or greater than 6:

$x < -2$ or $x > 6$

This is the interval $(-\infty, -2) \cup (6, \infty)$. The graph appears in Figure 7-25.

Graphical solution:

-2   6

FIGURE 7-25

EXAMPLE 4   Solve the inequality $\left|\dfrac{3-x}{5}\right| \geq 6$.

Solution   *Analytical solution:*
We write the inequality as two separate inequalities

$\left|\dfrac{3-x}{5}\right| \geq 6$ is equivalent to $\dfrac{3-x}{5} \leq -6$ or $\dfrac{3-x}{5} \geq 6$

Graphical solution:

and solve each one for x:

$\dfrac{3-x}{5} \leq -6$ or $\dfrac{3-x}{5} \geq 6$

$3 - x \leq -30$	$3 - x \geq 30$	Multiply both sides by 5.
$-x \leq -33$	$-x \geq 27$	Subtract 3 from both sides.
$x \geq 33$	$x \leq -27$	Divide both sides by -1 and reverse the direction of the inequality symbol.

The solution set is the interval $(-\infty, -27] \cup [33, \infty)$, whose graph appears in Figure 7-26.

FIGURE 7-26

EXAMPLE 5   Solve the inequality $\left|\dfrac{2}{3}x - 2\right| - 3 > 6$.

Solution   *Analytical solution:*
We begin by adding 3 to both sides to isolate the absolute value on the left-hand side. We then proceed as follows:

$\left|\dfrac{2}{3}x - 2\right| - 3 > 6$

$\left|\dfrac{2}{3}x - 2\right| > 9$ Add 3 to both sides.

Graphical solution:

$\dfrac{2}{3}x - 2 < -9$ or $\dfrac{2}{3}x - 2 > 9$

$\dfrac{2}{3}x < -7$	$\dfrac{2}{3}x > 11$	Add 2 to both sides.
$2x < -21$	$2x > 33$	Multiply both sides by 3.
$x < -\dfrac{21}{2}$	$x > \dfrac{33}{2}$	Divide both sides by 2.

The solution set is $\left(-\infty, -\dfrac{21}{2}\right) \cup \left(\dfrac{33}{2}, \infty\right)$, whose graph appears in Figure 7-27.

FIGURE 7-27

WARNING! The expression involving the absolute value, $|ax + b|$, must be isolated before applying either of the properties for solving absolute value inequalities.

EXAMPLE 6   Solve the inequality $|3x - 5| \geq -2$.

Solution   *Analytical solution:*
Because the absolute value of any number is nonnegative and because any nonnegative number is larger than -2, the inequality is true for all x. The solution set is the interval $(-\infty, \infty)$, whose graph appears in Figure 7-28.

Graphical solution:

$y_1 > y_2$

FIGURE 7-28

Every portion of the graph of y1 is greater than (above) the graph of y2.

EXERCISE 7.7

VOCABULARY   *Fill in the blank to make a true statement.*

1. The _____ of a number can be defined as the distance between a number and zero on the number line.

CONCEPTS   *In Exercises 2–3, use the graphical display to determine the solution to the given inequality or equation.*

2. a. $|2x - 4| - 6 = 0$
   b. $|2x - 4| - 6 < 0$
   c. $|2x - 4| - 6 > 0$

3. a. $-|3x - 2| + 4 = -3$
   b. $-|3x - 2| + 4 < -3$
   c. $-|3x - 2| + 4 > -3$

In Exercises 4–9, use the graphical display to determine the solution to the given inequality or equation.

4. a. y1 = y2
   b. y1 < y2
   c. y1 > y2

5. a. y1 = y2
   b. y1 < y2
   c. y1 > y2

6. a. y1 = y2
   b. y1 < y2
   c. y1 > y2

7. a. y1 = y2
  b. y1 < y2
  c. y1 > y2

8. a. y1 = y2
  b. y1 < y2
  c. y1 > y2

9. a. y1 = y2
  b. y1 < y2
  c. y1 > y2

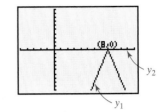

PRACTICE *In Exercises 10–50 solve each inequality and confirm graphically. Write the solution in interval notation and graph on a number line.*

10. $|2x| < 8$

11. $|3x| < 27$

12. $|x + 9| \leq 12$

13. $|x - 8| \leq 12$

14. $|3x + 2| \leq -3$

15. $|3x - 2| < 10$

16. $|4x - 1| \leq 7$

17. $|5x - 12| < -5$

18. $|3 - 2x| < 7$

19. $|4 - 3x| \leq 13$

20. $|5x| > 5$

21. $|7x| > 7$

22. $|x - 12| \geq 24$

23. $|x + 5| \geq 7$

24. $|3x + 2| > 14$

25. $|2x - 5| > 25$

26. $|4x + 3| > -5$

27. $|4x + 3| > 0$

28. $-|3x + 1| < -8$

29. $|8x - 3| > 0$

30. $|7x + 2| > -8$

31. $\left|\dfrac{x - 2}{3}\right| \leq 4$

32. $\left|\dfrac{x - 2}{3}\right| > 4$

33. $|3x + 1| + 2 < 6$

34. $|3x - 2| + 2 \geq 0$

35. $3|2x + 5| \geq 9$

36. $-2|3x - 4| < 16$

37. $|5x - 1| + 4 \leq 0$

38. $-|5x - 1| + 2 < 0$

39. $\left|\dfrac{1}{3}x + 7\right| - 5 > 6$

40. $\left|\dfrac{1}{2}x - 3\right| - 4 < 2$

41. $\left|\dfrac{1}{5}x - 5\right| + 4 > 4$

42. $\left|\dfrac{1}{6}x + 6\right| + 2 < 2$

43. $\left|\dfrac{3}{5}x + \dfrac{7}{3}\right| < 2$

44. $\left|\dfrac{7}{3}x - \dfrac{5}{3}\right| \geq 1$

45. $\left|3\left(\dfrac{x + 4}{4}\right)\right| > 0$

46. $3\left|\dfrac{1}{3}(x - 2)\right| + 2 \leq 3$

47. $\left|\dfrac{1}{7}x + 1\right| \leq 0$

48. $|2x + 1| + 2 \leq 2$

49. $\left|\dfrac{x - 5}{10}\right| < 0$

50. $\left|\dfrac{3}{5}x - 2\right| + 4 \leq 3$

In Exercises 51 and 52 experiment with various values of x and y with your calculator to determine a response to the questions. Be specific in explaining your conclusions.

51. Under what conditions is $|x| + |y| > |x + y|$?

52. Under what conditions is $|x| + |y| = |x + y|$?

53. If $k > 0$, explain the differences between the solution sets of $|x| < k$ and $|x| > k$.

54. If $k > 0$, explain why $|x| + k < 0$ would have $\varnothing$ as its solution.

55. If $k > 0$, explain why $|x| + k > 0$ would have the solution $(-\infty, \infty)$.

REVIEW

56. Solve $p = 2l + 2w$ for w.

57. Solve $|3x - 5| + 3 = 2$.

58. Determine the x-intercepts of the accompanying graph.

59. Is the graph in Exercise 58 a function? Explain why.

60. What is the range of the graph in Exercise 58? (Be careful! You will need to consider *all* ordered pairs, not just the labeled ones.)

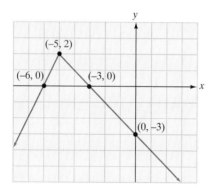

CHAPTER SUMMARY

CONCEPTS

REVIEW EXERCISES

Section 7.1

Negative Integral Exponents

If x is any nonzero number and n is a natural number, then

$$x^{-n} = \frac{1}{x^n}.$$

Properties of Exponents
If m and n are integers and there are no divisions by 0, then

$$x^m \cdot x^n = x^{m+n}$$

$$(x^m)^n = x^{mn}$$

$$\left(\frac{x}{y}\right)^n = \frac{x^n}{y^n}$$

$$x^0 = 1 \quad (x \neq 0)$$

$$x^{-n} = \frac{1}{x^n}$$

$$\frac{x^m}{x^n} = x^{m-n}$$

1. Evaluate each expression. Check your results with the calculator.
   a. $(-8)^2$ **b.** -8^2 **c.** $3^2 + 2^2$ **d.** $(3+2)^2$

2. Simplify each expression, leaving no negative exponents or parentheses.
   a. $(-z^2)(z^3 y^2)$ **b.** $(4x)x^2$ **c.** $(x^2 x^3)^3$ **d.** $(2x^2 y)^2$

3. Simplify each expression, leaving no negative exponents or parentheses.
   a. $\left(\dfrac{x^2 y}{xy^2}\right)^2$ **b.** $\left(\dfrac{3x}{6x^2}\right)^3$ **c.** $x^{-2} x^3$ **d.** $y^4 y^{-3}$

4. Simplify each expression, leaving no negative exponents or parentheses.
   a. $\dfrac{x^3}{x^{-7}}$ **b.** $(x^{-3} x^{-4})^{-2}$
   c. $\left(\dfrac{x^2}{x}\right)^{-5}$ **d.** $\left(\dfrac{15x^4}{5y^3}\right)^{-2}$

A number is written in scientific notation if it is written as the product of a number between 1 (including 1) and 10 and an integer power of 10.

5. Write each number in scientific notation.
   a. 728   **b.** 0.00942
   c. 0.018×10^{-2} **d.** 600×10^{-3}

6. Write each number in standard notation.
   a. 7.26×10^5 **b.** 3.91×10^{-4} **c.** $\dfrac{(4800)(20{,}000)}{600{,}000}$

7. The calculator displays

   Write this number in standard notation.

8. The calculator displays

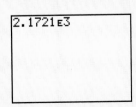

   Write this number in standard notation.

Section 7.2	Review of Factoring

POINTS TO REMEMBER

1. To factor means to write as a product. The expression $4(x + 5) + y(x + 5)$ is NOT expressed as a product; it is the sum of two terms $4(x + 5)$ and $y(x + 5)$.
2. Factoring out a negative 1 simply changes the sign of each term of the original polynomial.
3. The factors $(a + b)$ and $(b + a)$ are the same because the operation of addition is commutative.
4. The factors $(a - b)$ and $(b - a)$ are additive inverses (opposites). Thus $a - b = -(b - a)$.
5. The sum of squares, $x^2 + y^2$, does NOT factor over the set of integers. It is a prime polynomial.

Factoring Polynomials

1. Always check for and factor out the greatest common factor (GCF) first. If the leading coefficient is negative, factor out $a - 1$ with/as the GCF.
2. Count the number of terms in the polynomial and then follow the given procedures for that specific type of polynomial.

9. Factor each polynomial completely.
   a. $xy + 2y + 4x + 8$
   b. $ab + bc + 3a + 3c$
   c. $z^2 - 16$
   d. $2x^4 - 98$
   e. $y^2 + 21y + 20$
   f. $-y^2 + 5y + 24$
   g. $15x^2 - 57x - 12$
   h. $x^3 + 343$

Section 7.3	Equation Review

Points to Remember

1. The addition and subtraction properties of equality allow you to add or subtract the same quantity from both sides of an equation.
2. The multiplication and division properties of equality allow you to multiply or divide both sides of an equation by the same nonzero quantity.
3. Equations involving absolute value will have
   a. two solution parts if the absolute value is equal to a positive constant.
   b. one solution if the absolute value is equal to 0.
   c. no solution if the absolute value is equal to a negative number.

10. Solve each equation below algebraically and confirm results with either the TABLE or STOre feature of the graphing calculator.
   a. $2(x - 1) + 5 = \frac{1}{2}(2x - 3)$
   b. $\frac{x - 4}{5} + 1 = \frac{2x}{3}$
   c. $(x + 2)(x - 5) = (x - 4)(x - 1)$
   d. $2x^2 - x - 1 = 0$
   e. $6a^2 - 6 = 0$
   f. $x^3 + 5x^2 + 6x = 0$
   g. $|3x + 1| = 10$
   h. $\left|\frac{3}{2}x - 4\right| = -9$
   i. $\frac{4}{x} + \frac{5}{x} = 8$
   j. $\frac{x}{x + 2} + \frac{1}{x - 1} = 1$

4. When the equation involves a factorable polynomial, the zero product property allows you to set the factored polynomial equal to zero and then solve by setting each factor equal to zero.

5. When solving equations involving a rational expression, find the least common denominator for all rational expressions and multiply both sides of the equation by the LCD to clear the fractions.

11. Solve each equation for the indicated variable.
    a. $p = 2l + 2w$; w
    b. $4x + y = 2$; y
    c. $x + 2y = 6$; y

Section 7.4 — Graphing and Its Applications Revisited

Solutions to two variable equations are graphed on a rectangular coordinate system. For example, the ordered pairs $(3, 2)$, $(2, 3)$, and $(1, 4)$ are only three of an infinite number of solutions to the equation $x + y = 5$. These three ordered pairs are graphed on the coordinate system and connected to form a straight-line graph. ALL solutions to the equation $x + y = 5$ can be found on this straight line.

To determine the x-intercept for the equation of a line, substitute 0 for y and solve for x.

To determine the y-intercept for the equation of a line, substitute 0 for x and solve for y.

The standard viewing window for a graphing calculator is $[-10, 10]$ by $[-10, 10]$.

A function is a set of ordered pairs in which each x-value is assigned a unique y-value.

Both the INTERSECT and ROOT features can be used to graphically solve equations in one unknown.

12. Graph the linear equation using x- and y-intercepts.
    a. $2x + 5y = 10$

    b. $3x - 4y = 12$

    c. $3x - y = 6$

    d. $5x + y = 7$

13. Graphically solve each equation using either the INTERSECT or ZERO/ROOT feature of the calculator. Begin in the standard viewing window and adjust the window when necessary.
    a. $4(x - 1) = 8$
    b. $3(x + 7) = -6$
    c. $|3x + 1| = 10$
    d. $\left| \dfrac{3}{2}x - 4 \right| = 9$
    e. $x^2 - 11x + 30 = 0$
    f. $x^2 - 5x - 24 = 0$

14. Determine if the pictured graph is that of a function.
    a.

    b.

| Section 7.5 | Solving Simple Linear Inequalities |

The critical point of an inequality is the solution to the corresponding equation and divides the number line into regions that contain solution points versus the regions that contain no solution points.

Interval notation is a means of symbolically describing where the shading begins and ends on a number line graph.

ALWAYS reverse the direction of the inequality symbol when multiplying or dividing by a negative value.

The graphical display for solving an inequality graphically looks the same as the graphical display for solving the corresponding equation graphically. It is your *interpretation* that makes the difference.

15. Solve each linear inequality and express the solution in interval notation and as a number line graph. Verify results with the graphing calculator.
   a. $3x - 4 < 5$ **b.** $5 - 2x < 6$

   c. $2(x - 3) \leq 7$ **d.** $-3(x - 1) + 5 \geq 6$

In Review Exercises 16–17, solve the indicated equation or inequality based on the displayed graph.

16. a. $y_1 = y_2$
   b. $y_1 < y_2$
   c. $y_1 > y_2$

17. a. $y_1 = y_2$
   b. $y_1 < y_2$
   c. $y_2 > y_2$

| Section 7.6 | Compound Inequalities |

When solving compound inequalities linked by the word *and,* the intersection of the component simple inequalities comprises the solution.

When solving compound inequalities linked by the word *or,* the union of the component simple inequalities comprises the solution.

Compound inequalities can be solved graphically by graphing the *equations* corresponding to the simple inequalities and interpreting the results. Use of the INTERSECT feature of the calculator is helpful in locating critical points.

18. Solve each compound inequality and express the solution in interval notation and as a number line graph.
   a. $3 < 3x + 4 < 10$ **b.** $4 > 3x + 2 > -3$

   c. $6x + 1 < 5$ or $3(x - 9) > -2$ **d.** $5 - \frac{1}{2}x \geq -1$ or $6 - x < -1$

In Review Exercises 19–20, solve the compound inequality by interpreting the information displayed on the screen.

19. a. $y_1 < y_2 < y_3$
   b. $y_2 < y_1$ or $y_2 > y_3$

20. a. $y_1 < y_2 < y_3$
   b. $y_2 < y_1$ or $y_2 > y_3$

| Section 7.7 | **Absolute Value Inequalities** |

If k is a positive number, then $|x| < k$ is equivalent to the compound inequality $-k < x < k$.

If k is a positive number, then $|x| > k$ is equivalent to $x > k$ or $x < -k$.

21. Solve each absolute value inequality and confirm graphically. Write the solution in interval notation and graph on a number line.

  a. $|x + 3| \le 4$ **b.** $|2x + 3| < 11$

  c. $|2x - 4| > 22$ **d.** $|3x + 1| \ge 5$

  e. $|4x - 1| + 7 < 2$ **f.** $\left|\frac{1}{2}x + 3\right| - 8 > -11$

In Review Exercises 22–23, use the graphical display to determine the solution to the given inequality or equation.

22. a. $y_1 = y_2$ **23. a.** $y_1 = y_2$

  b. $y_1 < y_2$ **b.** $y_1 < y_2$

  c. $y_1 > y_2$ **c.** $y_1 > y_2$

In Review Exercises 24–27, write an equation or inequality to model the problem and solve either algebraically or graphically.

24. Cutting a board The carpenter in Illustration 1 saws a board into two pieces. He wants one piece to be 1 foot longer than twice the length of the shorter piece. Find the length of each piece.

ILLUSTRATION 1

25. Fencing a pasture   A farmer has 624 feet of fencing to enclose the pasture shown in Illustration 2. Because a river runs along one side, fencing will be needed on only three sides. Find the dimensions of the pasture if its length is double its width.

624 ft

ILLUSTRATION 2

26. Finding profit   The wholesale cost of a clock is $27. A store owner knows that in order to sell, the clock must be priced under $42. If p is the profit, express the possible profit as an inequality.

27. Buying compact discs   A student can afford to spend up to $330 on a stereo system and some compact discs. If the stereo costs $175 and the disks are $8.50 each, find the greatest number of discs the student can buy.

■ Chapter Test

In Problems 1 and 2, simplify, using the laws of exponents. Leave no parentheses or negative exponents in your final results.

1. $\dfrac{-12y^3(xz)^2}{18x^3(y^2z)^2}$

2. $\dfrac{x^2y^{-2}}{x^3z^2}$

In Problems 3–5, factor completely.

3. $a^3 - 125$

4. $2x^2 - 32$

5. $6x^2 + x - 2$

In Problems 6 and 7, solve each equation and verify your results with your calculator.

6. $9(x + 4) + 4 = 4(x - 5)$

7. $\dfrac{y - 1}{5} + 2 = \dfrac{2y - 3}{3}$

8. Solve $A = p + prt$ for t.

9. Solve $3x - 7y = 8$ for y.

In Problems 10 and 11, write an equation to model the problem and solve either algebraically or graphically.

10. A 20-foot pipe is to be cut into three pieces. One piece is to be twice as long as another, and the third piece is to be six times as long as the shortest. Find the length of the longest piece.

11. A rectangle with a perimeter of 26 centimeters is 5 centimeters longer than it is wide. Find its area.

In Problems 12 and 13, solve each equation and verify results with the graphing calculator.

12. $|2x + 3| = 11$

13. $\dfrac{3}{5} + \dfrac{7}{x + 2} = 2$

In Problems 14 and 15, graph the given equation.

14. $3x - y = 6$

15. $y = 5$

In Problems 16 and 17, use the graphical display to determine the solution to the equation or inequality.

16. a. $y_1 = y_2$
    b. $y_1 < y_2$
    c. $y_1 > y_2$

17. a. $y_1 = y_2$
    b. $y_1 < y_2$
    c. $y_1 > y_2$

18. Determine if the pictured graph is a function or not:

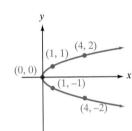

In Problems 19–25, solve each inequality and express the solution in interval notation and as a number line graph.

19. $\dfrac{1}{4}x - 2 \le -2$

20. $-2 \le 4 - x < 5$

21. $5 - 2x > 6$ or $x + 2 > 1$

22. $4x + 1 < x$ and $2x - 5 < 1$

23. $|2x - 5| < 4$

24. $\left| \dfrac{1}{3}x - 1 \right| \ge 5$

25. $\left| \dfrac{1}{2}x + 3 \right| + 4 \le 2$

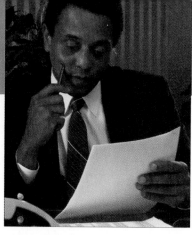

Statistician

Statisticians devise and carry out experiments and surveys and interpret the numerical results. They then apply their knowledge to subject areas such as economics, human behavior, natural science, and engineering.

Often, they are able to obtain accurate information about a group of people or things by surveying a small portion, called a **sample,** rather than the whole group. This technique requires statisticians to decide where and how to get the data, how to determine the type and size of the sample group, and how to develop the survey questionnaire or reporting form.

Qualifications

A bachelor's degree in statistics or mathematics is the minimum educational requirement for many beginning jobs in statistics. For other entry-level statistics jobs, however, it is preferable to have a bachelor's degree with a major in an applied field such as economics or natural science and a minor in statistics.

Job Outlook

Employment opportunities for persons who combine training in statistics with knowledge of a field of application such as manufacturing, engineering, scientific research, or business are expected to be favorable through the year 2005.

Example Application

A researcher wishes to estimate the mean (average) property tax paid by homeowners living in the city. To do so, he decides to select a **random sample** of homeowners and compute the mean tax paid by the homeowners in that sample. How large must the sample be for the researcher to be 95% certain that his computed sample mean will be within $35 of the population mean—that is, within $35 of the average tax paid by *all* homeowners in the city? Assume that the standard deviation, σ, of all tax bills is known to be $120.

Solution

From elementary statistics, the researcher has the formula

$$\frac{3.84\sigma^2}{N} < E^2$$

where σ^2, called the **population variance,** is the square of the standard deviation, E is the maximum acceptable error, and N is the sample size. He substitutes 120 for σ and 35 for E in the previous formula and solves for N.

$$\frac{(3.84)(120^2)}{N} < 35^2$$

$$\frac{55,296}{N} < 1225 \qquad \text{Simplify.}$$

$$55,296 < 1225N \qquad \text{Multiply both sides by } N.$$

$$\frac{55,296}{1225} < N \qquad \text{Divide both sides by 1225.}$$

$$45.1 < N \qquad \text{Simiplify.}$$

To be 95% certain that the sample mean will be within $35 of the population mean, the researcher must sample *more than* 45.1 homeowners: the sample must contain at least 46 homeowners.

1. What sample size must the researcher choose to keep the error less than $20?

2. What must the sample size be to keep the error less than $10?

3. If the researcher cuts the acceptable error in half, by what factor is the sample size increased?

4. If the researcher decides to double the acceptable error, by what factor is the sample size decreased?

(Answers: **1.** 139 **2.** 553 **3.** 4 **4.** $\frac{1}{4}$)

Functions

8

■ PERSPECTIVE

Graphs in Space

In an xy-coordinate system, graphs of equations containing the two variables x and y are lines or curves. Other equations have more than two variables, and graphing them often requires some ingenuity and perhaps the aid of a computer. Graphs of equations with the three variables x, y, and z are viewed in a three-dimensional coordinate system with three axes. The coordinates of points in a three-dimensional coordinate system are ordered triples (x, y, z). For example, the points $P(2, 3, 4)$ and $Q(-1, 2, 3)$ are plotted in Illustration 1.

Graphs of equations in three variables are not lines or curves, but flat planes or curved surfaces. Only the simplest of these equations can be conveniently graphed by hand; a computer provides the best images of others. The graph in Illustration 2 is called a paraboloid. Illustration 3 models a portion of the vibrating surface of a drum head.

ILLUSTRATION 1   ILLUSTRATION 2   ILLUSTRATION 3

Computer programs for producing three-dimensional graphs are readily available. Some of the more powerful are Maple, Mathematica, MathCad, and Derive. Perhaps your school has such a program

available for student use. With a brief introduction on the program's use, you can easily create several interesting graphs.

8.1 Relations and Functions

■ RELATIONS ■ FUNCTIONS ■ VERTICAL LINE TEST ■ FUNCTION NOTATION ■ LINEAR FUNCTIONS

■ RELATIONS

In a linear equation such as $y = 3x + 5$ we say that y is expressed in terms of x. The quantities x and y are related as an ordered pair (x, y) with x denoted as the independent variable and y as the dependent variable. A set of ordered pairs that establishes a correspondence between the numbers x and the values of y is called a **relation.**

Relation

A **relation** is a set of ordered pairs. The set of all values of x for which the relation is defined is called the **domain** of the relation. The set of all numbers, y, for which the relation is defined is called the **range** of the relation.

EXAMPLE 1   The following are examples of relations:

a. $A = \{(-2, 1), (0, 1), (1, 3), (4, 2)\}$

domain   range

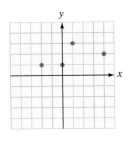

Therefore, the domain $= \{-2, 0, 1, 4\}$ and the range $= \{1, 2, 3\}$.

b. $B = \{(4, 1), (4, -2), (4, 3)\}$

domain   range

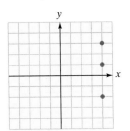

Therefore, the domain $= \{4\}$ and the range $= \{-2, 1, 3\}$.

c. $y = 3x + 5$

domain $= \{$all real numbers$\}$ (denoted symbolically as $\mathbb{R}$)

range $= \mathbb{R}$

Remember: When finding ordered pair solutions to $y = 3x + 5$ you may select *any* real number for x, and as a result produce a real number for y.

■ FUNCTIONS

■ STUDY TIP

Think of a function as a machine. The domain is the data that you put into the machine, and the range is the result that the machine puts back out.

The relations illustrated in Example 1 differ in one critical way. The x-values in relation B are all the same, while the y-values are all different. However, in both relation A and relation C, each x-value corresponds to a unique value of y. When each x-value is paired with a unique y-value, we call the relation a **function** and say that y is a function of x.

> **Function**
> A **function** is a relation that assigns to each number x exactly one value y.
>   The **domain** of the function is the set of all numbers x (the independent variable) for which the function is defined. The **range** is the set of all possible y-values (the dependent variable).

WARNING!   A function is always a relation because it consists of ordered pairs. However, a relation is not necessarily a function because a relation places no restriction on the relationship between the x- and y-values.

EXAMPLE 2   Determine which of the following relations is also a function.

a. $\{(-4, 5), (2, 3), (3, 5), (-2, 1)\}$

b. $\{(2, 8), (-1, 3), (4, 1), (2, -6)\}$

c. $y = 2x - 1$

d. $y^2 = x$

e. You are paid $5.25 per hour to sack groceries. If P represents your pay and n the number of hours worked, then $P = 5.75n$.

Solution **a.** Function; each x is assigned *exactly* one y-value.

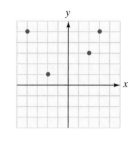

b. Not a function; when $x = 2$, $y = 8$ and $y = -6$.

c. Function; each value selected for x will produce exactly one y-value because arithmetic operations produce unique results.

d. Not a function; when $x = 4$, $y = 2$ or $y = -2$; when $x = 25$, $y = 5$ or $y = -5$.

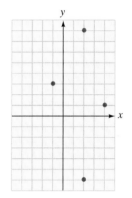

e. Function; for each hour, n, that you work, you will receive exactly one amount of pay.

■ VERTICAL LINE TEST

Examples 2b and 2d were not functions because each value of x corresponded to more than one value of y. When this happens, we can observe from the graph of the relation that the ordered pairs containing the same x-value lie in the path of a vertical line. Thus, the **vertical line test** can be used to determine whether the graph of an equation represents a function. If any vertical line intersects a graph more than once, the graph cannot represent a function, because to one number x there would correspond more than one value of y.

The graph in Figure 8-1(a) represents a function, because every vertical line that intersects the graph does so exactly once. However, the graph in Figure 8-1(b) does not represent a function, because some vertical lines intersect the graph more than once.

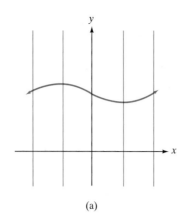

(a)   (b)

FIGURE 8-1

EXAMPLE 3   Use the vertical line test to determine which of the following relations is also a function.

a.   b.   c.

Solution   **a.** Function   **b.** Not a function

c. Function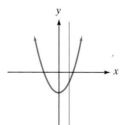

EXAMPLE 4   Determine the domain and range of each relation displayed.

a.   b.

Solution

a.

Domain: $\{x \mid -4 \leq x \leq 4\}$.

b.

Domain: $\mathbb{R}$.

Range: $\{y \mid -2 \leq y \leq 2\}$.

Range: $\{y \mid y \geq -2\}$.

■ FUNCTION NOTATION

When we say that y is a function of x, we are emphasizing that each x-value in the domain is assigned to a unique y-value in the range. The notation used to specify that y is a function of x is $y = f(x)$.

Function Notation
The notation $y = f(x)$ denotes that the variable y is a function of x.

The notation $f(x)$ is read as "f of x."

WARNING! The notation $f(x)$ does not mean "f times x."

We use function notation to simplify English statements that are used to give algebraic directions. For example, "find y when $x = 2$" can be expressed as "find $f(2)$." This notation also yields information about the ordered pair of the function. If $f(2) = 4$, then the ordered pair in the function is $(2, 4)$.

EXAMPLE 5 Let $f(x) = 4x + 3$. Find   **a.** $f(3)$,   **b.** $f(-1)$,   **c.** $f(0)$, and   **d.** $f(r)$.

Solution   **a.** We replace x with 3:

$$f(x) = 4x + 3$$
$$f(3) = 4(3) + 3$$
$$= 12 + 3$$
$$= 15$$

The ordered pair is $(3, 15)$.

b. We replace x with -1:

$$f(x) = 4x + 3$$
$$f(-1) = 4(-1) + 3$$
$$= -4 + 3$$
$$= -1$$

The ordered pair is $(-1, -1)$.

c. We replace x with 0:

$$f(x) = 4x + 3$$
$$f(0) = 4(0) + 3$$
$$= 0 + 3$$
$$= 3$$

The ordered pair is $(0, 3)$.

d. We replace x with r:

$$f(x) = 4x + 3$$
$$f(r) = 4(r) + 3$$
$$= 4r + 3$$

■

The letter f used in the notation $y = f(x)$ represents the word *function*. However, other letters can be used to represent functions. The notations $y = g(x)$ and $y = h(x)$ also denote functions involving the variable x.

EXAMPLE 6 If $g(x) = x^2 - 2x$, find **a.** $g\left(\dfrac{2}{5}\right)$ **b.** $g(s)$ **c.** $g(s - 1)$ **d.** $g(-t)$.

Solution **a.** We replace x with $\dfrac{2}{5}$.

$$g(x) = x^2 - 2x$$
$$g\left(\frac{2}{5}\right) = \left(\frac{2}{5}\right)^2 - 2\left(\frac{2}{5}\right)$$
$$= \frac{4}{25} - \frac{4}{5}$$
$$= -\frac{16}{25}$$

$g\left(\frac{2}{5}\right) = -\frac{16}{25}$ means the ordered pair $\left(\frac{2}{5}, -\frac{16}{25}\right)$ is a coordinate on the graph of $g(x) = x^2 - 2x$.

b. We replace x with s:

$$g(x) = x^2 - 2x$$
$$g(s) = s^2 - 2s$$

$g(s) = s^2 - 2s$ means the ordered pair $(s, s^2 - 2s)$ is a coordinate on the graph of $g(x) = x^2 - 2x$.

c. We replace x with $s - 1$:

$$g(x) = x^2 - 2x$$
$$g(s - 1) = (s - 1)^2 - 2(s - 1)$$
$$= (s^2 - 2s + 1) - 2s + 2$$
$$= s^2 - 4s + 3$$

$g(s - 1) = s^2 - 4s + 3$ means the ordered pair $(s - 1, s^2 - 4s + 3)$ is a coordinate on the graph of $g(x) = x^2 - 2x$.

d. We replace x with $-t$:

$$g(x) = x^2 - 2x$$
$$g(-t) = (-t)^2 - 2(-t)$$
$$= t^2 + 2t$$

$g(-t) = t^2 + 2t$ means the ordered pair $(t, t^2 - 2t)$ is a coordinate on the graph of $g(x) = x^2 - 2x$. ∎

EXAMPLE 7 Let $f(x) = 4x - 1$. Find **a.** $f(3) + f(2)$ and **b.** $f(a) - f(b)$.

Solution **a.** We find $f(3)$ and $f(2)$ separately.

$$f(x) = 4x - 1 \qquad\qquad f(x) = 4x - 1$$
$$f(3) = 4(3) - 1 \qquad\qquad f(2) = 4(2) - 1$$
$$= 12 - 1 \qquad\qquad\qquad = 8 - 1$$
$$= 11 \qquad\qquad\qquad\qquad = 7$$

```
Plot1 Plot2 Plot3
\Y1■4X-1
\Y2=
\Y3=
\Y4=
\Y5=
\Y6=
\Y7=
```

We then add the results to obtain

$$f(3) + f(2) = 11 + 7$$
$$= 18$$

```
Y1(3)+Y1(2)
              18
```

b. We find $f(a)$ and $f(b)$ separately.

$$f(x) = 4x - 1 \qquad\qquad f(x) = 4x - 1$$
$$f(a) = 4a - 1 \qquad\qquad f(b) = 4b - 1$$

We then subtract the results to obtain

$$f(a) - f(b) = (4a - 1) - (4b - 1)$$
$$= 4a - 1 - 4b + 1 \qquad \text{Remove parentheses.}$$
$$= 4a - 4b \qquad\qquad\qquad\qquad\qquad ∎$$

■ LINEAR FUNCTIONS

In Chapters 3 and 7 we determined ordered pair solutions for linear equations in x and y. These solutions were then graphed, and the graph of the equation was determined to be a straight line. All straight-line graphs, except for vertical lines, are linear functions.

> **Linear Functions**
> A **linear function** is a function defined by an equation that can be written in the form
> $$y = mx + b \qquad \text{or} \qquad f(x) = mx + b$$

EXAMPLE 8   Solve the equation $3x + 2y = 10$ for y to show that it defines a linear function. Then graph it to find its domain and range. (See Figure 8-2.)

Solution   We solve the equation for y as follows:

$$3x + 2y = 10$$

$$2y = -3x + 10 \qquad \text{Subtract } 3x \text{ from both sides.}$$

$$y = -\frac{3}{2}x + 5 \qquad \text{Divide both sides by 2.}$$

FIGURE 8-2

Domain: The set of all real numbers, $\mathbb{R}$.

Justification: If any x-value is selected on the x-axis, a corresponding y-value can be determined. Likewise, any x-value substituted into the function will yield exactly one y-value because the only two operations applied to the x-value are multiplication and addition.

Range: The set of all real numbers, $\mathbb{R}$.

Justification: If any y-value is selected on the y-axis, a corresponding x-value can be determined. ■

EXERCISE 8.1

VOCABULARY *In Exercises 1–7, fill in the blanks to make the given statements true.*

1. A set of ordered pairs that establishes a correspondence between numbers x and y is called a _____ .

2. The _____ is the set of all values of x for which a relation is defined.

3. The _____ is the set of all values of y for which a relation is defined.

4. A relation in which each x-value is paired with a *unique* y-value is called a _____ .

5. In a linear equation, solutions are expressed as _____ of numbers (x, y).

6. The notation $f(x)$ is read as "_____" and does not mean "f times x."

7. In the function defined as $f(x) = 3x - 5$, "$f(x)$" corresponds with the __-coordinate.

CONCEPTS

8.

9.

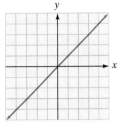

a. List the ordered pairs displayed.

b. State the domain of the relation.
c. State the range of the relation.
d. Is the pictured relation a function? Why or why not?

Use the graph to find each of the following:
a. $f(2)$
b. $f(-2)$
c. $f(0)$
d. $f(5)$

10. The graph of a function is pictured, with coordinates of a point displayed at the bottom of the graph screen. Find the value of each function specified below.

a.

$f(-14) =$ _____

b.

$f(22) =$ ___

c.

$f(-5) =$ _____

d.

$f(19) =$ ___

PRACTICE   In Exercises 11–14, determine which relations represent functions. Determine the domain and range.

11. $\{(2, 5), (3, -1), (-4, 7)\}$

12. $\{(5, -1), (7, -1), (3, -1)\}$

13. $\{(-1, 2), (-2, 3), (-1, -2)\}$

14. $\{(5, 7), (5, 8), (5, 9)\}$

In Exercises 15–22, tell whether each equation determines y to be a function of x.

15. $y = 2x + 3$ **16.** $y = 4x - 1$ **17.** $y = 3x^2$ **18.** $y = 3 + 7x^2$
19. $y^2 = x + 1$ **20.** $y^2 = 3 - 2x$ **21.** $y = |x|$ **22.** $x = |y|$

In Exercises 23–26, display the graph of each function in the Zinteger viewing window. Find the domain and the range.

23. $y = -x^2 + 5$ **24.** $y = (x - 10)^2$
25. $y = |x - 15| - 10$ **26.** $y = -|x + 20|$

In Exercises 27–34, determine if the relation displayed is also a function.

27.

28.

29.

30.

31.

32.

33.

34.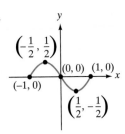

In Exercises 35–38, determine if the relation indicated by the table of values is also a function.

35.

x	y
4	1
3	2
−1	1
4	−1

36.

x	y
−1	5
2	−4
−2	−4
3	7

37.

x	y
−2	3
4	3
−1	3
0	3

38.

x	y
3	−2
3	4
3	−1
3	0

if x-repeats more than one

In Exercises 39–44, find a. f(2) b. f(−3) c. f(a) d. f(a + 2).

39. $f(x) = 3x + 2$

40. $f(x) = -4x + 1$

41. $f(x) = x^2 - 2x$

42. $f(x) = x^2 + x - 3$

43. $f(x) = \dfrac{3}{x - 4}$

44. $f(x) = \dfrac{x}{x^2 + 2}$

In Exercises 45–54, the function y = f(x) is displayed in either graphic or tabular form. Determine f(x).

45. $f(-4) = ?$

46. $f(0) = ?$

47. $f(-3) = ?$

48. $f(-2) = ?$

49. $f(3) = ?$

50. $f(8) = ?$

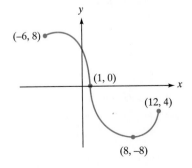

51. $f(-4) = ?$

X	Y₁
-8	14
-7	6
-6	0
-5	-4
-4	-6
-3	-6
-2	-4

X=-8

52. $f(5) = ?$

X	Y₁
1	14
2	24
3	36
4	50
5	66
6	84
7	104

X=1

53. $f(-3) = ?$

X	Y₁
-8	59
-7	44
-6	31
-5	20
-4	11
-3	4
-2	-1

X=-8

54. $f(-7) = ?$

X	Y₁
-8	59
-7	44
-6	31
-5	20
-4	11
-3	4
-2	-1

X=-8

55. Is it possible to determine $f(4)$ in Exercise 53? Explain why or why not.

56. Is it possible to determine $f(2)$, in Exercise 54? Explain why or why not.

In Exercises 57–64, $f(x) = 2x + 1$. Find each value.

57. $f(3) + f(2)$

58. $f(1) - f(-1)$

59. $f(b) - f(a)$

60. $f(b) + f(a)$

61. $f(b) - 1$

62. $f(b) - f(1)$

63. $f(0) - f\left(-\frac{1}{2}\right)$

64. $f(a) + f(2a)$

In Exercises 65–70, determine which equation defines a linear function. If the equation is a linear function, write it in $y = mx + b$ form and graph the equation.

65. $y = 3x^2 + 2$

66. $y = \dfrac{x - 3}{2}$

67. $x = 3y - 4$

68. $x = \dfrac{8}{y}$

69. $4x - 2y = 8$

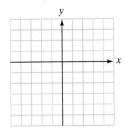

70. $6x - y = 2$

REVIEW

71. Is 5 part of the solution of the inequality $2x + 4 \geq 9$? Justify your response.

72. Simplify the expression $5(3x + 4y)$.

73. Evaluate the expression $3x + 2y$ when $x = -2$ and $y = 5$.

74. Is the ordered pair $(-2, 5)$ a solution of the equation $3x + 2y = 4$? Justify your response.

75. Is -7 a solution of the equation $x^2 - 49 = 0$? Justify your response.

8.2 Functions and Graphs

■ INCREASING AND DECREASING FUNCTIONS ■ INTERPRETING GRAPHS

■ INCREASING AND DECREASING FUNCTIONS

In Section 3.5 we used the graphing calculator to explore graphs of practical applications and models of linear functions. We will extend that concept now to describing and interpreting graphs.

When we describe a graph, we must agree to "read" the graph from left to right. For example, the graph of the linear function (Figure 8-3) is said to be an **increasing function** because as the x-value increases, the y-value also increases. Observe that the graph rises from left to right. Conversely, a **decreasing function** is one whose graph falls from left to right, and as the x-values increase, the y-values decrease. All functions are described in terms of increasing values of x because we are reading the graph from left to right.

FIGURE 8-3

EXAMPLE 1   Graph $y = 3\sqrt{x + 5}$ in the Zinteger viewing window. Is the function increasing or decreasing?

Solution   TRACE along the function from left to right. As x increases, so does y. We say the function is increasing on the interval $[-5, \infty)$.

■ TECHNOLOGY TIP
You may use any viewing window. Zinteger was chosen for its friendly values.

EXAMPLE 2   Determine the intervals where the function is increasing and where it is decreasing in Figure 8-4.

Solution   Remember to read from left to right! The function is increasing on the intervals $[-5, -2]$ and $[1, \infty)$, and decreasing on the interval $[-2, 1]$.

The points $(-2, 2)$ and $(1, 0)$ are called the **turning points** of the graph because the graph changes from increasing to decreasing. The turning point of the graph is also called the **relative maximum** or **relative minimum** and can be determined by using the maximum or minimum feature of your calculator.

■ TECHNOLOGY TIP
If using a TI-86, the maximum and minimum features are located by pressing [GRAPH] [MORE] [MATH] [MORE] and are denoted as FMIN and FMAX.

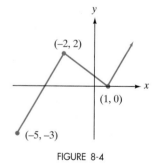

FIGURE 8-4

EXAMPLE 3   Find the turning point (relative minimum) for the graph of the function $f(x) = x^2 - 2x + 4$.

Solution   Graph $y1 = x^2 - 2x + 4$. Under your CALC menu, access the "minimum" feature and follow the prompts. Keep in mind that the prompts on your calculator are referencing x-values of the function. Thus, "lower bound" or "left bound" indicates an x-value that is lower, or to the left of, the desired x-coordinate.

We believe the minimum occurs when $x = 1$. We suspect that round-off error has occurred. To confirm the minimum of 1, use the VALUE (EVAL) feature to confirm that when $x = 1$, $y = 3$. (Other options that could have been used would include the TABLE, the VARS feature, or the STOre feature.) ∎

■ INTERPRETING GRAPHS

EXAMPLE 4   Suppose that the ball shown in Figure 8.5 is thrown straight up with a velocity of 128 feet per second. The function $f(x) = 128x - 16x^2$ gives the relation between the height, $f(x)$, of the ball and the time, x, as measured in seconds.

   a. Graph this function in an appropriate viewing window. Using the TRACE feature to assist you, describe how the time and height are related.

   b. What are the coordinates of the turning point of the graph?

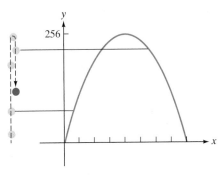

FIGURE 8-5

Solution   **a.** Tracing was used to determine an appropriate viewing window of $[-5, 10]$ by $[-10, 275]$. (Remember, you must be able to see x- and y-intercepts and critical features such as turning points.) Tracing from $x = 0$ to $x = 8$, we learn that as the seconds increase, so does the height until we reach the turning point. Then the function becomes decreasing because as the seconds *increase* the height *decreases.*

■ TECHNOLOGY TIP
You may want to review Section 3.5 for more information on setting appropriate viewing windows.

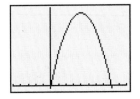

  b. The coordinate of the turning point is $(4, 256)$. The maximum feature under the CALC menu was used.

Again, verify through the TABLE or other feature of the calculator that the actual maximum is (4, 256). Interpreted within the context of the problem, this means that the ball will reach its maximum height of 256 feet in 4 seconds. ■

EXAMPLE 5 If x represents time and y represents location, the graph in Figure 8-6 can be used to illustrate a person walking through the mall. Describe what is happening from point A to B, B to C, C to D, and D to E.

FIGURE 8-6

Solution

A to B: The person is walking briskly because location has changed dramatically in a short amount of time.

B to C: Time is passing, but the location has not changed. The person could be talking with someone, looking in a window, or resting on a bench.

C to D: Location has changed. Moreover, the person has retraced previous steps to go back for something.

D to E: The person is again moving forward, but much more slowly. ■

EXERCISE 8.2

VOCABULARY *In Exercises 1–4, fill in the blanks to make the given statements true.*

1. A function is considered to be ___increasing___ if both the x- and y-values are increasing.

2. A function is considered to be ___decreasing___ if, as the x-values increase, the y-values decrease.

3. When looking at graphs, you should scan them from ___left___ to ___right___.

4. Points at which the graph of a function changes from increasing to decreasing are called ___turning___ points. They can also be referred to as relative ___maximum___ or relative ___minimum___.

CONCEPTS *Use the graph pictured in Illustration 1 to answer Exercises 5–8.*

Plane Deaths on the Wane, 1988–1993

SOURCE: National Transportation Safety Board.

ILLUSTRATION 1

5. Specify all yearly intervals (for example, 1980–1981) for which the number of plane deaths was increasing.

6. Specify all yearly intervals for which the number of plane deaths was decreasing.

7. What was the maximum number of plane deaths recorded for the entire period from 1980 to 1993?

8. Rounded to the nearest hundred, how many plane deaths occurred in 1985?

In Exercises 9–16, determine if the function is increasing or decreasing and over what interval it is increasing or decreasing. The viewing window is the standard viewing window.

9.

10.

11.

12.

13.

14.

15.

16.

PRACTICE   In Exercises 17 and 18, determine the interval(s) where the function is (a) increasing, (b) decreasing, and (c) neither (constant).

17.

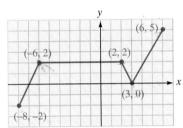

18.

19. In your own words, explain a "turning point" of a function.

20. What is the difference in being directed to (a) find a "turning point", (b) find the relative minimum, and (c) find the relative maximum. Explain carefully.

In Exercises 21–24, graph each function and determine the turning point using the max or min feature of your calculator. Verify with the TABLE to avoid round-off errors with displayed results, such as 4.99999989 that is actually 5.

21. $y = x^2 - 2x + 5$ **22.** $y = 5x^2 - 2x - 4$ **23.** $y = 4 - |x + 2|$ **24.** $y = 6 - |x - 5|$

In Exercises 25–30, use Illustration 2 to respond to the questions.

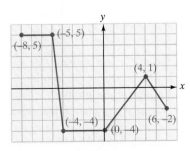

ILLUSTRATION 2

25. Is the graph a function: Why or why not?

26. For what intervals of x is the graph
  a. increasing?
  b. decreasing?
  c. constant?

27. On the interval [0, 6], what are the coordinates of the turning point? Is it a relative maximum or minimum?

28. What is the domain? What is the range?

29. Determine $f(-5)$.

30. Why can you *not* determine $f(7)$?

✍**31.** The function displayed in Illustration 3 describes the water level, y, in an aquarium over a period of time, x. In complete sentences, write a scenario that could explain the changes in water level.

✍**32.** The function in Illustration 4 describes the height, y, over a period of time, x, of a ball that a child has thrown into the air. In complete sentences, write a scenario that could explain what has happened to the ball.

ILLUSTRATION 3

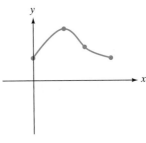

ILLUSTRATION 4

APPLICATIONS   *In Exercises 33–35, use your graphing calculator to respond to the questions.*

33. A traveling circus has a human cannonball act as its grand finale. The equation $y = -.01x^2 + .64x + 9.76$, where y = height in feet and x = horizontal distance traveled in feet, represents the flight path of the human cannonball. Display a graphical representation of this equation and use the TRACE feature to answer the questions. (Zinteger is the preferred viewing window.)

  a. What do the x-values represent?

  b. What do the y-values represent?

  c. The curve represents the flight path of the human cannonball. TRACE along the path to the right. How far, *approximately,* has the human cannonball traveled horizontally when he hits the ground?

  d. The human cannonball is traveling at speeds up to 65 mph. To land on the ground would mean certain death. If he uses a net for his landing, how far will he have traveled horizontally if the net is 11 feet above the ground?

  e. What is the maximum height that he reaches during the course of his flight?

  f. How far has he traveled horizontally when he reaches this maximum height?

  g. The human cannonball is shot out of the cannon head first, so all of the distances are measured from his head. TRACE along the curve to $x = 0$ and $y = 9.76$. Explain the meaning of these two values.

34. Bridges are often supported by arches in the shape of a parabola. The equation $y = \frac{10}{7}x - \frac{2}{49}x^2$, where y = height and x = distance from the base of the arch, provides a model for a specific parabolic arch that supports a bridge.

  a. What do the x-intercepts represent? (*Hint:* Your calculator has a ROOT or ZERO feature that should be used to determine the x-intercept.)

  b. What is the highest point under the arch?

  c. If an average vehicle is no more than 6 feet high, can it drive under the arch? Explain.

  d. If a two-lane road is 20 feet wide and has 2 shoulders of width 5 feet each, will it fit between the bases of the arch? Explain.

  e. Can an average-height vehicle drive in either lane and not scrape the paint off the roof of the car? That is to say, if this 20-foot wide road is centered under the arch, is the arch at least 6 feet above the ground at all points in its width?

35. The local community theater is considering increasing the price of its tickets to cover increases in costuming and stage effects. They must be careful because a ticket price that is too low will mean that expenses are not covered, and yet a ticket price that is too high will discourage people from attending. They estimate the total profit, y, by the formula $y = -x^2 + 35x - 150$, where x is the cost of the ticket.

  a. What is the maximum amount that can be charged for a ticket to maximize the profit?

  b. What is the maximum profit?

  c. If $13 is charged for each ticket, what will the profit be?

  d. If they predict a profit of $150.00 on a play, how much was charged per ticket? Scroll through the y-values in the TABLE to answer the question.)

REVIEW

36. Solve the inequality $3(x + 5) - 8 \leq 5x$ and specify your solution
   a. as a number line graph.

   b. using interval notation.
   c. using set-builder notation.

37. Describe how you would find the *x*- and *y*-intercepts of an equation in two variables both algebraically and graphically.

38. Solve the equation $3x^2 - 10x - 8 = 0$ both algebraically and graphically.

39. Graph the function
$$y = |x + 2| - 3$$

40. Solve the equation $|x + 2| - 3 = 5$ both algebraically and graphically.

8.3 Slope of a Nonvertical Line

■ INTRODUCTION/OVERVIEW ■ SLOPES OF HORIZONTAL AND VERTICAL LINES ■ SLOPES OF PARALLEL AND PERPENDICULAR LINES

■ INTRODUCTION/OVERVIEW

■ **STUDY TIP**
Use the graphing calculator to reinforce and check your algebraic work. DO NOT treat the algebra and the calculator as two separate entities.

Throughout this text you have been presented with equations of various forms and asked to interpret data that related to the equation. In order to successfully apply what you have learned to real-life situations, you must also be able to take specific data and construct the equation. This section and the next will help you develop the tools necessary for writing the most fundamental equation: the equation of a line.

Consider the following chart that relates the number of pounds of apples to the purchase price:

Pounds	1	2	3	4	5
Cost	$0.40	$0.80	$1.20	$1.60	$2.00

We can generalize this to: *x* pounds cost $0.40*x*. Thus the equation modeling this information is $y = 0.40x$, where *y* is the cost and *x* is the number of pounds of apples.

Graph this equation in the Zinteger viewing window and TRACE along the line. As *x* increases, so does *y*.

Thus the function is increasing. Note also that for each change of 1 unit in *x* (i.e., 1 pound of apples), there is a 0.4 change in *y* (an increase of $0.40 per pound). This comparison ratio of

$$\frac{\text{change in } y}{\text{change in } x}$$

determines the **rate** of increase in an increasing linear function and also the rate of decrease in a decreasing linear function. In linear functions, this rate is called the **slope** of the line. The "change in *y*-coordinates" is often called Δy (delta *y*) and the "change in *x*-coordinates" is often called Δx.

EXAMPLE 1 Graph $y = \frac{3}{4}x$ in the Zinteger viewing window. By tracing, determine Δy (change in y-coordinates) and Δx (change in x-coordinates) and compute the slope of the line.

Solution As we trace from $(0, 0)$ to $(4, 3)$ to $(8, 6)$ to $(12, 9)$, we observe that y is changing 3 units for every 4 units of change in x. (See Figure 8-7.) The ratio of $\frac{\Delta y}{\Delta x}$ is $\frac{3}{4}$.

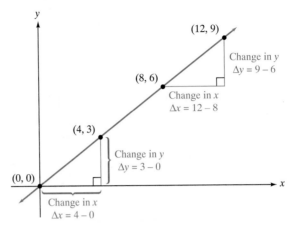

FIGURE 8-7

As we move from one point to another on a linear function, the ratio $\frac{\Delta y}{\Delta x}$ remains constant. Thus the slope of a nonvertical line can be determined by any two points on the line. Furthermore, if points $P(x_1, y_1)$ and $Q(x_2, y_2)$ are points on any nonvertical line we can say that

$$\frac{\Delta y}{\Delta x} = \frac{y_2 - y_1}{x_2 - x_1}$$

(See Figure 8-8.) Δy is called the **rise** of the line between P and Q and Δx is called the **run.** The slope, m, can also be defined as $m = \dfrac{\text{rise}}{\text{run}}$.

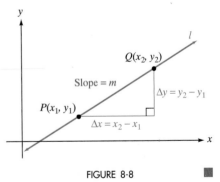

FIGURE 8-8

EXAMPLE 2 Use the graph in Figure 8-9(a) to determine the slope of the line passing thru P and Q.

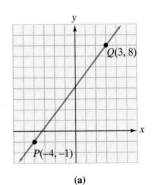

(a)

FIGURE 8-9

Solution   Use the graph in Figure 8-9(b) to determine the slope $\dfrac{\Delta y}{\Delta x}$.

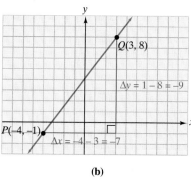

$$m = \frac{\Delta y}{\Delta x} = \frac{-9}{-7} = \frac{9}{7}$$

The slope is $\frac{9}{7}$.

(b)

FIGURE 8-9

EXAMPLE 3   If a line passes through the point $(0, 5)$ and has a slope of 4, determine one other point on the line and graph the line.

Solution   If the line has a slope of 4, then $\dfrac{\Delta y}{\Delta x} = \dfrac{4}{1}$. Plot the point $(0, 5)$ and count off $\dfrac{\Delta y}{\Delta x}$ $\left(\text{i.e., } m = \dfrac{\text{rise}}{\text{run}}\right)$. (See Figure 8-10.)

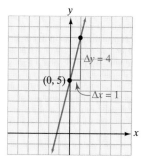

The second point would be $(1, 9)$.

FIGURE 8-10

EXAMPLE 4   In Figure 8-11, point P has coordinates $(-2, 5)$ and point Q has coordinates of $(6, -7)$. Find the slope of the line PQ.

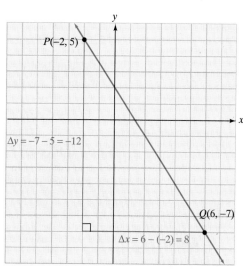

FIGURE 8-11

Solution   We can let $P(x_1, y_1) = P(-2, 5)$ and $Q(x_2, y_2) = Q(6, -7)$. Then

$$m = \frac{\Delta y}{\Delta x}$$

$$= \frac{y_2 - y_1}{x_2 - x_1}$$

$$= \frac{-7 - 5}{6 - (-2)}$$ Substitute -7 for y_2, 5 for y_1, 6 for x_2, and -2 for x_1.

$$= \frac{-12}{8}$$

$$= -\frac{3}{2}$$

The slope of the line is $-\frac{3}{2}$. We would obtain the same result if we had let $P(x_1, y_1) = P(6, -7)$ and $Q(x_2, y_2) = Q(-2, 5)$.   ■

WARNING!   When finding the slope of a line, always subtract the y-values and the x-values in the same order, or your answer will have the wrong sign.

EXPLORING THE CONCEPT   Graph the linear function $y = 4x + 5$ in the Zinteger viewing window.

   a. TRACE to determine the y-intercept.

   b. Using two other coordinates from the line, determine the slope.

   c. Draw conclusions about the slope and y-intercept as they relate to the equation $y = 4x + 5$.

Conclusion   **a.**

y-intercept: $(0, 5)$

   b.

$$m = \frac{21 - 9}{4 - 1} = \frac{12}{3} = 4$$

   c. In the equation $y = 4x + 5$, the coefficient of x is the slope of the line and the constant of 5 is the y-intercept.   ■

   A line is determined by a point and its slope.

Slope-Intercept Form
A linear equation in the form $y = mx + b$ is in **slope-intercept form.** The coefficient of x, denoted as m, represents the slope of the line, and the y-intercept is b.

EXAMPLE 5   Find the slope and the y-intercept of the line determined by the equation $3x - 4y = 12$.

Solution   Solve $3x - 4y = 12$ for y to write the equation in slope-intercept form.

$$3x - 4y = 12$$
$$-4y = -3x + 12$$
$$y = \frac{3}{4}x - 3$$

Therefore, the slope is $\frac{3}{4}$ and the coordinates of the y-intercept are $(0, -3)$.

An alternative approach would be to determine the y-intercept by letting $x = 0$ and solving for y.

$$3(0) - 4y = 12$$
$$-4y = 12$$
$$y = -3$$

The coordinates of the y-intercept are $(0, -3)$.

Compute the slope by determining one other ordered pair and using the slope formula.

If $y = 0$, then $x = 4$, and the point $(4, 0)$ is on the line (see Figure 8-12). We can then let $P(x_1, y_1) = P(0, -3)$, $Q(x_2, y_2) = Q(4, 0)$ and use the formula for slope to obtain

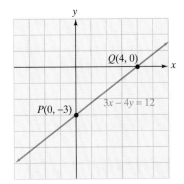

FIGURE 8-12

$$m = \frac{\Delta y}{\Delta x}$$

$$= \frac{y_2 - y_1}{x_2 - x_1}$$

$$= \frac{0 - (-3)}{4 - 0} \qquad \text{Substitute 0 for } y_2, -3 \text{ for } y_1, 4 \text{ for } x_2, \text{ and 0 for } x_1.$$

$$= \frac{3}{4}$$

The slope of the line is $\frac{3}{4}$.   ■

EXPLORING THE CONCEPT   Graph each of the following groups of lines in the Zinteger viewing window.

$$y = \frac{1}{2}x, y = x, y = 3x$$

$$y = -\frac{1}{2}x, y = -x, y = -3x$$

a. As the positive coefficient of x (slope) gets larger, what happens to the graph of the line?

b. As the negative coefficient of x gets smaller, what happens to the graph of the line?

c. What conclusions can you draw about the slope of a linear function and its classification as an increasing or decreasing function?

Conclusions   **a.** As the coefficient of x (slope) gets larger, the line becomes steeper. The line is also rising when viewed from left to right.

b. When the coefficient is negative, the line falls when viewed from left to right. The smaller the negative coefficient the faster the line falls.

c. Linear functions with positive slope are increasing functions, and linear functions with negative slopes are decreasing functions.   ■

■ SLOPES OF HORIZONTAL AND VERTICAL LINES

If $P(x_1, y_1)$ and $Q(x_2, y_2)$ are points on the horizontal line shown in Figure 8-13(a), then $y_1 = y_2$, and the numerator of the fraction

$$\frac{y_2 - y_1}{x_2 - x_1}$$ On a horizontal line, $x_2 \neq x_1$.

is 0. Thus, the value of the fraction is 0, and the slope of the horizontal line is 0.

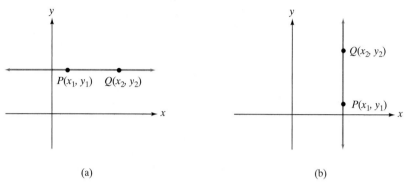

(a)   (b)

FIGURE 8-13

If $P(x_1, y_1)$ and $Q(x_2, y_2)$ are two points on the vertical line shown in Figure 8-13(b), then $x_1 = x_2$, and the denominator of the fraction

$$\frac{y_2 - y_1}{x_2 - x_1}$$ On a vertical line, $y_2 \neq y_1$.

is 0. Since the denominator of a fraction cannot be 0, a vertical line has no defined slope.

> **Slopes of Horizontal and Vertical Lines**
> Horizontal lines (lines with equations of the form $y = b$) have a slope of 0.
> Vertical lines (lines with equations of the form $x = a$) have no defined slope.

If a line rises as we follow it from left to right, as in Figure 8-14(a), its slope is positive. If a line drops as we follow it from left to right, as in Figure 8-14(b), its slope is negative. If a line is horizontal, as in Figure 8-14(c), its slope is 0. If a line is vertical, as in Figure 8-14(d), it has no defined slope.

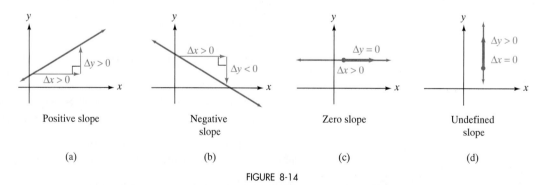

Positive slope   Negative slope   Zero slope   Undefined slope

(a)   (b)   (c)   (d)

FIGURE 8-14

■ SLOPES OF PARALLEL AND PERPENDICULAR LINES

We have observed that it only takes a point and a slope to determine a given line. We now want to see what occurs when two lines share the same slope value but pass through different points.

FIGURE 8-15

Graph the line l_1 that has a slope of $\frac{2}{3}$ and passes through $(0, 0)$ and the line l_2 that also has a slope of $\frac{2}{3}$ but passes through $(0, 5)$. (See Figure 8-15.)

Lines that have the same slope but pass through different points are parallel. It is also true that when two lines are parallel they have the same slope.

Slopes of Parallel Lines

Nonvertical parallel lines have the same slope, and lines having the same slope are parallel.

Since vertical lines are parallel, lines with no defined slope are parallel.

EXAMPLE 6   If the line passing through $P(3, -2)$ and $Q(-3, 4)$ is parallel to the line passing through $R(-2, 5)$ and $S(3, y)$, find y.

Solution   Since the lines PQ and RS are parallel, they have equal slopes (see Figure 8-16). To find y, we find the slope of each line, set them equal, and solve the resulting equation.

$$\overset{\textbf{Slope of } PQ}{} \qquad \overset{\textbf{Slope of } RS}{}$$

$$\frac{4 - (-2)}{-3 - 3} = \frac{y - 5}{3 - (-2)}$$

$$\frac{6}{-6} = \frac{y - 5}{5}$$

$$-1 = \frac{y - 5}{5}$$

$$-5 = y - 5$$

$$0 = y$$

Thus, $y = 0$. The line passing through $P(3, -2)$ and $Q(-3, 4)$ is parallel to the line passing through $R(-2, 5)$ and $S(3, 0)$.

FIGURE 8-16

EXPLORING THE CONCEPT

■ TECHNOLOGY TIP
Zinteger is a square viewing window. However, you may use any viewing window for this exercise, but it is critical that the window be "square." To ensure a square viewing window, apply Zsquare to the window you are using.

Graph each of the following pairs of lines in the Zinteger viewing window. Decide if the lines appear parallel, perpendicular, or neither. Then compare the slopes as indicated in the equation. What conclusions can you draw about the lines and the slopes of those lines that appear to be parallel or perpendicular?

a. $y = \frac{3}{7}x + 10$ and $y = \frac{3}{7}x - 8$

b. $y = \frac{3}{7}x + 10$ and $y = -\frac{3}{7}x - 8$

c. $y = \frac{3}{7}x + 10$ and $y = \frac{7}{3}x - 8$

d. $y = \frac{3}{7}x + 10$ and $y = -\frac{7}{3}x - 8$

Conclusions   **a.**

Lines appear parallel; both slopes are $\frac{3}{7}$; lines with the same slope are parallel.

b.

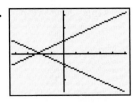

Lines are neither parallel nor perpendicular.

c.

Lines are neither parallel nor perpendicular.

■ **STUDY TIP**
Before beginning an exercise set, read over your notes *and* read the section in your textbook.

d.

Lines appear perpendicular; slopes are $\frac{3}{7}$ and $-\frac{7}{3}$, negative reciprocals of one another.

■

Slopes of Perpendicular Lines

If two nonvertical lines are perpendicular, their slopes are negative reciprocals.

If the slopes of two lines are negative reciprocals the lines are perpendicular.

Because a horizontal line is perpendicular to a vertical line, a line with a slope of 0 is perpendicular to a line with no definite slope.

$m_1 \cdot m_2 = -1$

EXERCISE 8.3

VOCABULARY *In Exercises 1–8, fill in the blanks to make the given statements true.*

1. The comparison ratio of change in __ over change in __ determines the slope of a line.

2. Slope is sometimes defined as ____ over run.

3. If, when viewed from left to right, a linear function is increasing, its slope is _____.

4. A linear equation in the form $y = mx + b$ is in _____ form. The coefficient of x, m, represents the _____ of the line, whereas b represents the _____.

5. The slope of a horizontal line is ____.

6. The slope of a vertical line is _____.

7. If two lines are parallel, their slopes are _____.

8. If two lines are perpendicular, their slopes are _____ of one another.

CONCEPTS   *In Exercises 9–12, find the slope of the line whose displayed TABLE is pictured.*

9.

10.

11.

12.

In Exercises 13–17, classify each of the lines as having positive, negative, 0, or no defined value for its slope.

13.

14.

15.

16.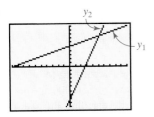

17.

PRACTICE   *In Exercises 18–23, when possible, find the slope of the line that passes through the given points.*

18. $(-1, 8), (6, 1)$

19. $(-5, -8), (3, 8)$

20. $(3, -1), (-6, 2)$

21. $(0, -8), (-5, 0)$

22. $(7, 5), (-9, 5)$

23. $(-7, -5), (-7, -2)$

In Exercises 24–27, determine whether the graph of y1 or y2 has the greater absolute value for slope.

24.

25.

26.

27.

28. Explain how, by merely looking at the displayed graphs in Exercises 13–17, you were able to determine if the slope was positive, negative, 0, or not defined.

In Exercises 29–34, find the slope and y-intercept of the line determined by the given equation.

29. $3x + 2y = 12$

30. $2x - y = 6$

31. $3x = 4y - 2$

32. $y = \dfrac{x - 4}{2}$

33. $x = \dfrac{3 - y}{4}$

34. $x = y$

In Exercises 35 and 36, find the slope of a line parallel to the given line.

35. $f(x) = 3x - 5$

36. $2x - y = 6$

In Exercises 37 and 38, find the slope of a line perpendicular to the given line.

37. $4x - y = 2$

38. $f(x) = \dfrac{1}{3}x + 1$

In Exercises 39–42, determine if the pair of lines determined by the given equation are parallel, perpendicular, or neither.

39. $f(x) = 3x - 4;\ f(x) = -\dfrac{1}{3}x + 2$

40. $f(x) = 5x + 2;\ f(x) = -5x + 2$

41. $2x - y = 7;\ 2x + y = 8$

42. $x - 3y = 5;\ 3x + y = 6$

43. Find the equation of the *x*-axis and its slope.

45. Explain why the graphing calculator will graph a horizontal line but not a vertical line.

44. Find the equation of the *y*-axis and its slope.

In Exercises 46–49, use the slope-intercept form of an equation ($y = mx + b$) to write the equation of a line with the given description.

46. slope $= 4$; *y*-intercept $= (0, -2)$

47. slope $= \dfrac{1}{5}$; *y*-intercept $= (0, 4)$

48. Perpendicular to a line whose slope is $\dfrac{2}{3}$ and has a *y*-intercept of $(0, 5)$.

49. Parallel to a line whose slope is 5 and has a *y*-intercept of $(0, -2)$.

APPLICATIONS

50. When a college started an aviation program, the administration agreed to predict enrollments using the graph of a straight line. If the enrollment during the first year was 12, and the enrollment during the fifth year was 26, find the rate of growth per year (the slope of the line). (See Illustration 1.)

ILLUSTRATION 1

51. A small business predicts sales according to a straight-line method. If sales were $50,000 in the first year and $110,000 in the third year, find the rate of growth in sales per year (the slope of the line).

52. The price of computer technology has been dropping steadily for the past 10 years. If a desktop PC cost $6700 10 years ago, and the same computing power cost $2200 3 years ago, find the rate of decrease per year. (Assume a straight-line model).

53. The first day of the fair, admission is $5 and it costs $0.75 to ride each ride. If *y* represents the total cost and *x* the number of rides, we can say that $y = 5 + 0.75x$ models the cost.
 a. What is the slope of this line and how does it relate to the context of the problem?
 b. If the initial cost was $7 and the cost per ride was $0.50, what would the equation be?

c. Would it be possible, in reality, for the equation that models the cost of attending the fair to ever have a negative value for the slope? Why or why not?

54. A local printer charges $200 plus $1.25 per invitation to print custom-designed wedding invitations.
a. Write an equation to model this application, where y represents the cost and x the number of invitations.
b. What does the slope of the linear equation represent?
c. Speculate what it would mean if your bill was only $200.

55. Assume that an antique purchased for $3000 increases in value in a straight-line fashion at a rate of $250 per year.
a. Write an equation to model this increase, where y represents the cost and x the subsequent years.

b. What does the slope of your equation represent?

c. In terms of the antique, what would it mean if the slope were negative instead of positive.

56. In June, the latest desktop computer technology could be purchased for $3000. Throughout the summer and into the fall, the cost of processor components began to decrease at a rate of $100 per month.
a. Assuming the decrease was a straight-line model, write an equation to model that cost, y, over x months.
b. What does the slope represent in your equation?

c. How would your equation change if prices were rising instead of falling?

8.4 Equations of Lines

■ REVIEW OF EQUATIONS IN SLOPE-INTERCEPT FORM ■ POINT-SLOPE FORM OF THE EQUATION OF A LINE ■ STRAIGHT-LINE DEPRECIATION

■ REVIEW OF EQUATIONS IN SLOPE-INTERCEPT FORM

In the previous section and in Section 3.5, we looked at linear applications whose basic information allowed us to write a model equation whose form was $y = mx + b$, slope-intercept form. This meant that we had to be given the y-intercept (the constant of the equation) and the rate at which the x variable was changing (the slope). From this basic information, we are able to determine other points that satisfy the model equation and subsequently graph the model.

EXAMPLE 1   The first day admission to the fair is $5 plus $0.75 per ride. Write the equation to model this situation and graph the line.

Solution   The constant cost is the $5 admission fee and the variable cost is $0.75 per ride.

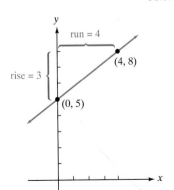

Let y = cost, x = number of rides

$$y = .75x + 5 \qquad \text{or} \qquad y = \frac{3}{4}x + 5$$

$$\text{slope} = \frac{3}{4} = \frac{\text{rise}}{\text{run}}$$

y-intercept: $(0, 5)$

■ POINT-SLOPE FORM OF THE EQUATION OF A LINE

In applications, we must be presented with enough information to construct an equation that models the situation. However, the information rarely comes in the straightforward form of slope and y-intercept. We know that it only requires two points to determine the graph of a line. It also only requires two points to determine the equation of a line.

The slope of a line is a constant. This means that when the coordinates of *any* two points are substituted into the slope formula, the result will be the same constant regardless of the coordinates used. Recall the slope formula,

$$m = \frac{y_2 - y_1}{x_2 - x_1}$$

If we have a line with slope m and the point (x_1, y_1) is on the line, then for the point (x, y) to be on the line it must satisfy the equation

$$m = \frac{y - y_1}{x - x_1}$$

If we clear the fraction by multiplying both sides of the equation by the quantity $x - x_1$, we have the equation

$$m(x - x_1) = y - y_1$$

or

$$y - y_1 = m(x - x_1) \qquad \text{Symmetric property of equality.}$$

We can see that if our line has slope m and a y-intercept of $(0, b)$, the equation above will generate the slope-intercept form:

$$y - y_1 = m(x - x_1)$$
$$y - b = m(x - 0)$$
$$y - b = mx$$
$$y = mx + b \qquad \text{Slope-intercept form}$$

We call $y - y_1 = m(x - x_1)$ the **point-slope form** of an equation of a line. It will be used to write equations of lines where the given information produces a point and a slope.

> **Point-Slope Form of the Equation of a Line**
> The equation of the line passing through $P(x_1, y_1)$ and with slope m is
> $$y - y_1 = m(x - x_1)$$

EXAMPLE 2 Use the point-slope form to write the equation of a line with a slope of $-\dfrac{2}{3}$ and passing through $P(-4, 5)$.

Solution We substitute $-\dfrac{2}{3}$ for m, -4 for x_1, and 5 for y_1 in the point-slope form and simplify.

$$y - y_1 = m(x - x_1)$$

$$y - 5 = -\frac{2}{3}[x - (-4)] \qquad \text{Substitute } -\tfrac{2}{3} \text{ for } m, -4 \text{ for } x_1, \text{ and } 5 \text{ for } y_1.$$

$$y - 5 = -\frac{2}{3}(x + 4) \qquad -(-4) = 4$$

$$y - 5 = -\frac{2}{3}x - \frac{8}{3} \qquad \text{Use the distributive property to remove parentheses.}$$

$$y = -\frac{2}{3}x + \frac{7}{3} \qquad \text{Add 5 to both sides.}$$

The equation of the line is $y = -\dfrac{2}{3}x + \dfrac{7}{3}$.

Verify your equation of $y = -\dfrac{2}{3}x + \dfrac{7}{3}$ with your calculator.

Because the slope of your equation is the same as the given slope, it is obvious that the information agrees. You must now use your calculator to determine if the given coordinate of $(-4, 5)$ is on the graph of $y = -\dfrac{2}{3}x + \dfrac{7}{3}$. You can either locate the coordinate $(-4, 5)$ in the TABLE or use the VALUE feature of your calculator.

■ **TECHNOLOGY TIP**
Remember that when using VALUE or EVAL, your x-value must be between the values defined for x-minimum and x-maximum on the window screen.

EXAMPLE 3   Use the point-slope form of the equation of a line to write the equation of the line passing through $P(-5, 4)$ and $Q(8, -5)$.

Solution   We must first find the slope of the line:

■ **STUDY TIP**
Place a piece of paper over the solution to an example in your text. Read the example and try to work the problem on your paper. If you get stuck, slide the paper down and jump-start yourself.

$$m = \frac{y_2 - y_1}{x_2 - x_1}$$

$$= \frac{-5 - 4}{8 - (-5)} \qquad \text{Substitute } -5 \text{ for } y_2, \ 4 \text{ for } y_1, \ 8 \text{ for } x_2, \text{ and } -5 \text{ for } x_1.$$

$$= -\frac{9}{13}$$

Because the line passes through both P and Q, we can choose either one and substitute its coordinates into the point-slope form. If we choose $P(-5, 4)$, we substitute -5 for x_1, 4 for y_1, and $-\frac{9}{13}$ for m and simplify.

$$y - y_1 = m(x - x_1)$$

$$y - 4 = -\frac{9}{13}[x - (-5)] \qquad \text{Substitution.}$$

$$y - 4 = -\frac{9}{13}(x + 5) \qquad -(-5) = 5$$

$$y - 4 = -\frac{9}{13}x - \frac{45}{13} \qquad \text{Use the distributive property to remove parentheses.}$$

$$y = -\frac{9}{13}x + \frac{7}{13} \qquad \text{Add 4 to both sides.}$$

■ **TECHNOLOGY TIP**
When you know which values of x you want to confirm in a TABLE, change the setting of the independent variable (x) from AUTO to ASK on the TABLE SETUP screen. This is shown below:

The equation of the line is $y = -\dfrac{9}{13}x + \dfrac{7}{13}$.

Now that we have the equation of the line, we can verify, using the TABLE feature, that the coordinates of the given points satisfy the equation of the line.

Because we have changed our setting for the independent variable from automatic to ASK, we must enter the x-values for which we desire corresponding y-values. We enter the x-value of -5 and press ENTER to see the corresponding y-value. We then enter the x-value of 8, press ENTER, and see the corresponding value for y.

The starting value for x and the table increment settings are immaterial.

The TABLE display confirms the fact that the points $(-5, 4)$ and $(8, -5)$, given initially, satisfy our equation of the line.   ■

EXAMPLE 4   Write the equation of a line passing through the point $P(-2, 5)$ and parallel to the line $y = 8x - 2$.

Solution   Since our line will be parallel to $y = 8x - 2$, we know that the slope of our line will be 8 ($m = 8$), the same as the slope of the given line. We now proceed as before, using $m = 8$ and the coordinates of the point $P(-2, 5)$; substitute **8** in for m, **−2** for x_1, and **5** for y_1.

$$y - y_1 = m(x - x_1)$$
$$y - 5 = 8[x - (-2)] \qquad \text{Substitution.}$$
$$y - 5 = 8(x + 2) \qquad -(-2) = 2$$
$$y - 5 = 8x + 16 \qquad \text{Use the distributive property to remove parentheses.}$$
$$y = 8x + 21 \qquad \text{Add 5 to both sides.}$$

Verify by graphing $y = 8x + 21$ and locating $(-2, 5)$ in the TABLE (choose AUTO from the TBLSET screen and use -2 as the initial, starting value) or use the VALUE feature from the graph screen to evaluate the function when $x = -2$.

■

EXAMPLE 5   Write the equation of the line passing through the point $P(-2, 5)$ and perpendicular to $y = 8x - 2$.

Solution   The slope of the given line is 8. Thus, the slope of the desired line must be $-\frac{1}{8}$, which is the negative reciprocal of 8.

We substitute -2 for x_1, 5 for y_1, and $-\frac{1}{8}$ for m into the point-slope form and simplify.

$$y - y_1 = m(x - x_1)$$
$$y - 5 = -\frac{1}{8}[x - (-2)] \qquad \text{Substitution.}$$
$$y - 5 = -\frac{1}{8}x - \frac{1}{4}$$
$$y = -\frac{1}{8}x + \frac{19}{4}$$

The displays below, illustrating the use of the TABLE feature and the VALUE feature, confirm that the point $(-2, 5)$ is on the line with equation $y = -\frac{1}{8}x + \frac{19}{4}$.

Graphing the lines with equations $y = -\frac{1}{8}x + \frac{19}{4}$ and $y = 8x - 2$ on a square screen shows that they appear to be perpendicular.

The final form for the equation of nonvertical lines can be slope-intercept ($y = mx + b$) or standard form ($Ax + By = C$). We consistently use slope-intercept form because it is the most convenient when using a graphing calculator.

We summarize the various forms for the equation of a line as follows:

> **Standard Form of a Linear Equation**
> $Ax + By = C$
> A and B cannot both be 0.
>
> **Slope-Intercept Form of a Linear Equation**
> $y = mx + b$
> The slope is m and the y-intercept is b.
> Remember, the *coordinates* of the y-intercept are $(0, b)$.
>
> **Point-Slope Form of a Linear Equation**
> $y - y_1 = m(x - x_1)$
> The slope is m and the line passes through (x_1, y_1).
>
> **A Horizontal Line**
> $y = b$
> The slope is 0, and the y-intercept is b.
>
> **A Vertical Line**
> $x = a$
> There is no defined slope, and the x-intercept is a.

■ STRAIGHT-LINE DEPRECIATION

For tax purposes, many businesses use the equation of a line to find the declining value of aging equipment. This method is called **straight-line depreciation.**

EXAMPLE 6   **Value of a lathe**   A machine shop buys a lathe for $1970 and expects it to last for 10 years. Then the lathe can be sold as scrap for an estimated **salvage value** of $270. If y represents the value of the lathe after x years of use and y and x are related by the equation of a line,

    a. Find the equation of the line.

    b. Find the value of the lathe after $2\frac{1}{2}$ years.

    c. Find the economic meaning of the y-intercept of the line.

    d. Find the economic meaning of the slope of the line.

Solution   **a.** To find the equation of the line, we first find two points on the line and then find the line's slope. We can then use either the point-slope form of a linear equation OR the slope-intercept form to find the equation of the line.

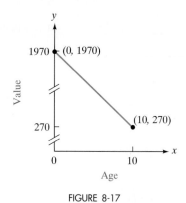

FIGURE 8-17

When the lathe is new, its age x is 0 and its value y is \$1970. When the lathe is 10 years old, $x = 10$ and its value is $y = \$270$. The line passes through the points (0, 1970) and (10, 270), as shown in Figure 8-17, so the slope of the line is

$$
\begin{aligned}
m &= \frac{y_2 - y_1}{x_2 - x_1} \\
&= \frac{270 - 1970}{10 - 0} \\
&= \frac{-1700}{10} \\
&= -170
\end{aligned}
$$

In this particular problem, we have two options to use when writing the equation of the line:

Option 1: We know $m = -170$ and can use the point-slope form to find the equation of the line. We can use either given point for (x_1, y_1), and choose (0, 1970). Substituting, we have

$$
\begin{aligned}
y - y_1 &= m(x - x_1) \\
y - \mathbf{1970} &= \mathbf{-170}(x - \mathbf{0}) \\
y &= -170x + 1970
\end{aligned}
$$

Option 2: We know $m = -170$ and that the y-intercept is (0, 1970). Thus, $b = 1970$. We substitute these values into the slope-intercept form of a linear equation:

$$
\begin{aligned}
y &= mx + b \\
y &= -170x + \mathbf{1970}
\end{aligned}
$$

b. To find the age of the lathe after $2\frac{1}{2}$ years, we substitute **2.5** for x into the equation of the line, $y = -170x + 1970$:

$$
\begin{aligned}
y &= -170(\mathbf{2.5}) + 1970 \\
y &= -425 + 1970 \\
y &= 1545
\end{aligned}
$$

Therefore, the lathe will be worth \$1545 when it is $2\frac{1}{2}$ years old.

We could have used the graphing calculator to answer the question. Enter the equation at the y1 prompt and use either the VALUE feature or the TABLE to find the value of y when x has a value of 2.5. This is illustrated in the screens displayed below:

Window values of
$[-10, 15]$ by $[-10, 2500]$

c. The y-intercept of the graph is (0, b), where b is the value of y when $x = 0$. Thus, b is the value of a 0-year old lathe, which is the lathe's original cost.

d. Each year, the value of the lathe decreases by \$170, because the slope of the line is -170. The slope of the depreciation line is the **annual depreciation rate.** ∎

EXERCISE 8.4

VOCABULARY *In Exercises 1–6, fill in the blanks to make each statement true.*

1. When a linear equation is written in the form $y = mx + b$, it is in _____ form. The coefficient of x is the _____ of the line, and b is the _____.

2. When a linear equation is written in the form $y - y_1 = m(x - x_1)$, it is in _____ form, where m is the _____ of the line and _____ are the coordinates of a point on the line.

3. When a linear equation is written in the form $Ax + By = C$, we say it is written in _____ form.

4. A horizontal line has the equation _____. Its slope is always _____.

5. A vertical line has the equation _____. Its slope is always _____.

6. The method of using an equation of a line to find the declining value of aging equipment is called _____.

CONCEPTS

7. A TABLE of values is given. Graph the values. Then use the graph and/or the TABLE to answer the questions below:
  a. Does the graph indicate a linear relationship?
  b. What is the slope?

  c. What is the y-intercept?

  d. The equation of the line is $y = \_\_ x + \_\_$

8. A house was purchased for $120,000. Its value in 5 years is predicted to be $150,000. Illustration 1 is a graph with points marked.
  a. State the coordinates of the two marked points.

  b. What is the average rate of appreciation per year? (*Hint:* Find the slope.)
  c. The y-intercept is _____.

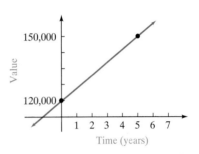

ILLUSTRATION 1

  d. The equation of the line is $y = \_\_\_\_ x +$

  _____

  e. The equation of the line was entered at the y1 prompt on the y= screen. Part of the TABLE is displayed.

  What will the value of the house be in 10 years? After how many years will the house have a value of $204,000?

9. Match each equation (stated in slope-intercept form) with its graphical display. (*Hint:* The sign of m, slope, and b, the y-intercept, will be of great help.)

$y = x + 3$ $y = x - 3$ $y = -x + 5$ $y = x + 5$ $y = 5x - 4$ $y = 5x + 4$

  a.

  b.

  c.

  d.

  e.

  f.
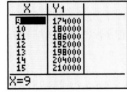

10. Match each equation with its graphical display.

$y = 3$ $y = -4$ $y = 8$ $x = 5$ $x = -6$ $x = -2$

a.

b.

c.

d.

e.

f.

PRACTICE *In Exercises 11–16, use the point-slope form to write the equation of the line with the given properties. Then transform the equation into standard form.*

11. $m = 5$, passing through $P(0, 7)$

12. $m = -8$, passing through $P(0, -2)$

13. $m = -3$, passing through $P(2, 0)$

14. $m = 4$, passing through $P(-5, 0)$

15. $m = \frac{3}{2}$, passing through $P(2, 5)$

16. $m = -\frac{2}{3}$, passing through $P(-3, 2)$

In Exercises 17–22, use the point-slope form to write the equation of the line passing through the two given points. Then transform the equation into slope-intercept form.

17. $P(0, 0)$, $Q(4, 4)$

18. $P(-5, -5)$, $Q(0, 0)$

19. $P(3, 4)$, $Q(0, -3)$

20. $P(4, 0)$, $Q(6, -8)$

21. $P(-2, 4)$, $Q(3, -5)$

22. $P(3, -5)$, $Q(-1, 12)$

In Exercises 23–28, write each equation in slope-intercept form to find the slope and the y-intercept. Then use the slope and the y-intercept to draw the line.

23. $y + 1 = x$

24. $x + y = 2$

25. $y = \frac{2}{3}x + \frac{9}{2}$

26. $5x - 4y = 10$

27. $3(y - 4) = -2(x - 3)$

28. $-4(2x + 3) = 3(3y + 8)$

In Exercises 29–34, write the equation of the line that passes through the given point and is parallel to the given line. Your equation should be written in slope-intercept form.

29. $P(0, 0)$, $y = 4x - 7$

30. $P(0, 0)$, $x = -3y - 12$

31. $P(2, 5)$, $4x - y = 7$

32. $P(-6, 3)$, $y + 3x = -12$

33. $P(4, -2)$, $x = \dfrac{5}{4}y - 2$

34. $P(1, -5)$, $x = -\dfrac{3}{4}y + 5$

In Exercises 35–40, write the equation of the line that passes through the given point and is perpendicular to the given line. Your equation should be written in slope-intercept form.

35. $P(0, 0)$, $y = 4x - 7$

36. $P(0, 0)$, $x = -3y - 12$

37. $P(2, 5)$, $4x - y = 7$

38. $P(-6, 3)$, $y + 3x = -12$

39. $P(4, -2)$, $x = \dfrac{5}{4}y - 2$

40. $P(1, -5)$, $x = -\dfrac{3}{4}y + 5$

41. Find the equation of the line perpendicular to the line $y = 3$ and passing through the midpoint of the segment joining $(2, 4)$ and $(-6, 10)$.

42. Find the equation of the line parallel to the line $y = -8$ and passing through the midpoint of the segment joining $(2, -4)$ and $(8, 12)$.

43. Find the equation of the line parallel to the line $x = 3$ and passing through the midpoint of the segment joining $(2, -4)$ and $(8, 12)$.

44. Find the equation of the line perpendicular to the line $x = 3$ and passing through the midpoint of the segment joining $(-2, 2)$ and $(4, -8)$.

45. Solve $Ax + By = C$ for y and thereby show that the slope of its graph is $-\frac{A}{B}$ and its y-intercept is $\left(0, \frac{C}{B}\right)$.

46. Show that the x-intercept of the graph of $Ax + By = C$ is $\left(\frac{C}{A}, 0\right)$.

APPLICATIONS *In Exercises 47–57, assume straight-line depreciation or straight-line appreciation.*

47. Finding a depreciation equation A taxicab was purchased for $24,300. Its salvage value at the end of its 7-year useful life is expected to be $1900 (see Illustration 2). Find the depreciation equation.

ILLUSTRATION 3

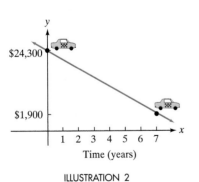

Time (years)

ILLUSTRATION 2

48. Finding a depreciation equation A computer system was purchased for $7900 and depreciated over its 4-year useful life. Its expected salvage value is $300. Find its depreciation equation.

49. Finding an appreciation equation An apartment building was purchased for $475,000. The owners expect the property to double in value in 10 years (see Illustration 3). Find the appreciation equation.

50. Finding an appreciation equation A house purchased for $112,000 is expected to double in value in 12 years. Find its appreciation equation.

51. Finding a depreciation equation Find the depreciation equation for the TV in the want ad in Illustration 4.

For Sale: 3-year-old 54-inch TV, $3,900 new. Asking $1,890. Call 875-5555. Ask for Mike.

ILLUSTRATION 4

52. **Depreciating a word processor** A word processor costs $555 when new and is expected to be worth $80 after 5 years. What will it be worth after 3 years?

53. **Finding salvage value** A copier cost $1050 when new and will be depreciated at the rate of $120 per year. If the useful life of the copier is 8 years, find its salvage value.

54. **Finding annual rate of depreciation** A truck that cost $27,600 when new will have no salvage value after 12 years. Find its annual rate of depreciation.

55. **Finding the value of antiques** An antique table is expected to appreciate $40 each year. If the table will be worth $450 in 2 years, what will it be worth in 13 years?

56. **Finding the value of antiques** An antique clock is expected to be worth $350 after 2 years and $530 after 5 years. What will the clock be worth after 7 years?

57. **Finding the purchase price of real estate** A cottage that was purchased 3 years ago is now appraised at $47,700. If the property has been appreciating $3500 per year, find its original purchase price.

58. **Charges for computer repair** A computer repair company charges a fixed amount, plus an hourly rate, for a service call. Use the information in Illustration 5 to find the hourly rate.

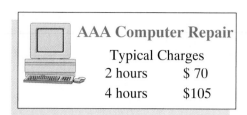

AAA Computer Repair	
Typical Charges	
2 hours	$ 70
4 hours	$105

ILLUSTRATION 5

59. **Charges for automobile repair** An auto repair shop charges an hourly rate, plus the cost of parts. If the cost of labor for a $1\frac{1}{2}$-hour radiator repair is $69, find the cost of labor for a 5-hour transmission overhaul.

60. **Finding printer charges** A printer charges a fixed setup cost, plus $1 for every 100 copies. If 700 copies cost $52, how much will it cost to print 1000 copies?

61. **Estimating the cost of a rain gutter** A neighbor says that an installer of rain gutters charges $60, plus a dollar amount per foot. If the neighbor paid $435 for the installation of 250 feet of gutter, how much will it cost you to have 300 feet installed?

62. **Predicting fires** A local fire department recognizes that city growth and the number of reported fires are related by a linear equation. City records show that 300 fires were reported in a year when the local population was 57,000 persons, and 325 fires were reported in a year when the population was 59,000 persons. How many fires can be expected when the population reaches 100,000 persons?

REVIEW In Exercises 63–66, solve each equation in two ways: (a) algebraically, showing all steps; and (b) graphically, copying the screen display and all displayed information from the ZERO/ROOT screen.

63. $2x^2 + x - 3 = 0$

64. $6x^2 - x - 2 = 0$

65. $4x^2 - 9 = 0$

66. $2x^2 - x = 0$

For each problem below, write an equation that models the words of the problem and solve it either algebraically (showing all steps in the solving process) or graphically (sketching the graph display, specifying window values, and copying all displayed information from the use of the INTERSECT or ZERO/ROOT feature). Regardless of the method of solution, write the answer to the question posed in a complete sentence.

67. **Mixing alloys** In 60 ounces of alloy for watch cases, there are 20 ounces of gold. How much copper must be added to the alloy so that a watch case weighing 4 ounces, made from the new alloy, will contain exactly 1 ounce of gold?

68. **Mixing coffee** To make a mixture of 80 pounds of coffee worth $272, a grocer mixes coffee worth $3.25 a pound with coffee worth $3.85 a pound. How many pounds of the cheaper coffee should the grocer use?

8.5 Variation

■ DIRECT VARIATION ■ INVERSE VARIATION ■ JOINT VARIATION ■ COMBINED VARIATION

■ DIRECT VARIATION

Direct variation is a common application of linear functions that vary directly or are directly proportional. Consider the formula, $C = \pi d$. Let $c = y_1$ and $d = x$ so that the formula may be examined graphically as $y_1 = \pi x$.

Looking at the TABLE, we see that when the diameter of the circle is 3 ($x = 3$), the circumference is approximately 9.4 ($y \approx 9.4$). As the diameter increases, the circumference also increases. When $x = 6$, $y \approx 18.8$, therefore, when the diameter doubles, so does the circumference. Thus, the function is increasing. Algebraically speaking, if we double the diameter we will create another circle with a circumference that is doubled. Therefore, if

$$C_1 = \pi(2d)$$
$$= 2\pi d$$
$$= 2C$$

Direct Variation

The words *y varies directly with x* mean that for some nonzero constant, k, $y = kx$. The constant, k, is called the **constant of variation.**

Observe that the graph of $y = kx$ is the graph of a straight line, where k is the slope:

$y = kx$; negative slope $y = kx$; positive slope

Instead of using the words *y varies directly as x,* you will sometimes find the words *y is directly proportional to x.* This results from dividing both sides of the equation, $y = kx$, by x:

$$y = kx$$

$$\frac{y}{x} = \frac{kx}{x}$$

$$\frac{y}{x} = k$$

which indicates a proportion exists between the quantities x and y and the constant k.

An example of direct variation is Hooke's law from physics. Hooke's law states that the distance a spring will stretch varies directly with the force that is applied to it. If d represents a distance and f represents a force, Hooke's law is expressed mathematically as

$$d = kf$$

where k is the constant of variation. If the spring stretches 10 inches when a weight of 6 pounds is attached, we can find k as follows:

$$d = kf$$

$$10 = k(6) \qquad \text{Substitute 10 for } d \text{ and 6 for } f.$$

$$\frac{5}{3} = k \qquad \text{Divide both sides by 6 and simplify.}$$

Graphical: Our equation now becomes $d = \dfrac{5}{3}f$ or $y = \dfrac{5}{3}x$. Graphed in Zinteger, we can TRACE to $y = 35$ to find the force (x) required to stretch a spring 3 inches. From the graph, we see that it would require a force of 21 pounds.

Analytical: To solve the problem analytically, we can solve the equation $d = kf$ for f, with $d = 35$ and $k = \frac{5}{3}$.

$$d = kf$$

$$35 = \frac{5}{3}f \qquad \text{Substitute 35 for } d \text{ and } \tfrac{5}{3} \text{ for } k.$$

$$105 = 5f \qquad \text{Multiply both sides by 3.}$$

$$21 = f \qquad \text{Divide both sides by 5.}$$

Thus, the force required to stretch the spring a distance of 35 inches is 21 pounds.

EXAMPLE 1 **Travel distance**   The distance traveled in a given time is directly proportional to speed. If a car travels 70 miles at 30 miles per hour, how far will it travel in the same time at 45 miles per hour?

Solution The words *distance is directly proportional to speed* can be expressed by the equation

$$d = ks$$

where d is distance, k is the constant of variation, and s is the speed. To find k, we substitute 70 for d, 30 for s, and solve for k.

$$d = ks$$

$$70 = k(30)$$

$$\frac{7}{3} = k \qquad \text{Divide both sides by 30 and simplify.}$$

Graphical: Our variation equation now becomes $d = \frac{7}{3}s$ or $y = \frac{7}{3}x$. This equation is graphed as shown below in the Zinteger viewing window. From the graph, we can determine the distance, y, for any integer speed, x, by tracing to that x-value.

For a speed of 30 miles per hour the distance traveled would be 70 miles.

For a speed of 45 miles per hour, the distance is 105 miles.

■ TECHNOLOGY TIP
Remember to adjust your window to include the desired *x*-value before using the VALUE (EVAL) feature.

For speeds that are not integers, use the VALUE (EVAL) option or the TABLE feature.

Analytical: To find the distance traveled at 45 miles per hour, we substitute $\frac{7}{3}$ for k and 45 for s in the equation and simplify.

$$d = ks$$

$$d = \frac{7}{3}(45)$$

$$d = 105$$

In the time it took to go 70 miles at 30 miles per hour, the car could travel 105 miles at 45 miles per hour.   ■

EXAMPLE 2 **Area of a circle** The area of a circle provides a familiar application of direct variation involving a power of x: $A = \pi r^2$. In the formula, our constant is π. Translated to a graphical format we have $y = \pi x^2$.

    a. Explore the graph of "y varies directly as the square of x" and determine which values of x are in the domain of the formula for the area of a circle.

    b. Graphically determine the area when the radius is 6.1 centimeters. Why is a graphical approach better than using the TABLE feature?

Solution **a.** Begin by graphing in a window that is twice the size of the Zdecimal window. Values of x less than zero are not applicable to the area formula because this would mean that the radius is a negative measurement. Thus, our domain is restricted to the interval $[0, \infty)$. For this domain, our function is increasing.

    b. In this particular window you cannot trace to $x = 6.1$. Therefore, you must use the VALUE (EVAL) feature. The area is approximately 116.89866 square centimeters when the radius is 6.1 centimeters.

    From the TABLE we can see that the accuracy level of our response has decreased because the TABLE rounds the area to 116.9.

X	Y1
6.1	116.9
7.1	158.37
8.1	206.12
9.1	260.16
10.1	320.47
11.1	387.08
12.1	459.96

X=6.1

■ INVERSE VARIATION

In the formula $w = \frac{12}{l}$, w gets smaller as l gets larger, and w gets larger as l gets smaller. Because these variables vary in opposite directions in a predictable way, we say that the variables **vary inversely,** or that they are **inversely proportional.** The constant 12 is the **constant of variation.**

    Inverse variation is an example of an application of rational functions. Variables will occur in the denominators of fractions, so you must be careful to watch for excluded values. Because of this, graphs should be viewed in both connected and in dot mode.

> **Inverse Variation**
> The words *y varies inversely with x* or *y is inversely proportional to x* mean that $y = \frac{k}{x}$ for some constant k. k is called the **constant of variation.**

EXAMPLE 3 **Illumination** The intensity I of light received from a light source varies inversely with the square of the distance from the light source. If the intensity from a light source 4 feet from an object is 8 candelas, find the intensity at a distance of 2 feet.

Solution The words *intensity varies inversely with the square of the distance d* can be expressed by the equation

$$I = \frac{k}{d^2}$$

To find k, we substitute 8 for I and 4 for d, and solve for k.

$$I = \frac{k}{d^2}$$

$$8 = \frac{k}{4^2}$$

$$128 = k \qquad \text{Multiply both sides by } 4^2.$$

Graphical: Our variation equation becomes $I = \dfrac{128}{d^2}$ or $y = \dfrac{128}{x^2}$, $(x \neq 0)$ for graphical purposes. The problem asks for the intensity at a distance of 2 feet, so we begin by graphing in the Zinteger viewing window. By tracing, we can determine that the intensity will be 32 candelas.

Note that for the application, the domain must be restricted to $[0, \infty)$; thus the portion of the graph in the second quadrant is not applicable to this particular application. For the domain $[0, \infty)$, the function is decreasing.

Analytical: To find the intensity when the object is 2 feet from the light source, we substitute 2 for d and 128 for k and simplify.

$$I = \frac{k}{d^2}$$

$$I = \frac{128}{2^2}$$

$$I = 32$$

The intensity at 2 feet is 32 candelas. ∎

■ **JOINT VARIATION**

There are times when one variable varies with the product of several variables. For example, the area of a triangle varies directly with the product of its base and height:

$$A = \frac{1}{2}bh$$

Such variation is called **joint variation.** Because more than two variables are involved, we will not examine these graphically.

> **Joint Variation**
> If one variable varies directly with the product of two or more variables, the relationship is called **joint variation.** If y varies jointly with x and z, then $y = kxz$. The constant is called the **constant of variation.**

EXAMPLE 4   The volume V of a cone varies jointly with its height h and the area of its base B. Express this relationship as an equation.

Solution   The words *V varies jointly with h and B* mean that *V varies directly as the product of h and B*. Thus,

$$V = khb$$

This relationship can also be read as *V is directly proportional to the product of h and B*. ∎

■ COMBINED VARIATION

Many problems involve a combination of direct and inverse variation. Such variation is called **combined variation.**

EXAMPLE 5   **Highway construction**   The time it takes to build a highway varies directly with the length of the road but inversely with the number of workers. If it takes 100 workers 4 weeks to build 2 miles of highway, how long will it take 80 workers to build 10 miles of highway?

Solution   We can let t represent the time in weeks, l represent the length in miles, and w represent the number of workers. The relationship among these variables can be expressed by the equation

$$t = \frac{kl}{w}$$

■ **STUDY TIP**
An effective way to see if you understand a problem is to restate it in your own words.

We substitute 4 for t, 100 for w, and 2 for l to find k:

$$4 = \frac{k(2)}{100}$$

$400 = 2k$ Multiply both sides by 100.

$200 = k$ Divide both sides by 2.

We now substitute 80 for w, 10 for l, and 200 for k in the equation $t = \frac{kl}{w}$, and simplify:

$$t = \frac{kl}{w}$$

$$t = \frac{200(10)}{80}$$

$$t = 25$$

It will take 25 weeks for 80 workers to build 10 miles of highway.   ■

Keep in mind that problems involving variation often consist of three parts:

1. Writing the variation equation from the words of the problem—this serves as a formula, or model.

2. Finding the value of k, the constant of variation (if it is not given).

3. Answering the question posed in the problem.

EXERCISE 8.5

VOCABULARY   *In Exercises 1–6, fill in the blanks to make the given statements true.*

1. The equation $y = kx$ is an example of _____ variation because as the value of x increases, the value of y _____.

2. This equation $y = kx$ is also a _____ function.

3. The equation $y = \frac{k}{x}$ is an example of _____ variation because as the value of x increases, the value of y _____.

4. The equation $y = \frac{k}{x}$ is also an example of a _____ function.

5. The equation $y = kxz$ is an example of _____ variation, because one variable varies directly with the value of two variables.

6. The equation $y = \frac{kx}{z}$ is an example of _____ variation. In this example, y varies _____ with x and _____ with z.

CONCEPTS *In Exercises 7–10, identify each graphical display as an example of direct variation, inverse variation, or neither.*

7.   **8.**   **9.**   **10.**

In Exercises 11–14, identify each TABLE display as an example of direct variation, inverse variation, or neither.

11.   **12.**   **13.**   **14.**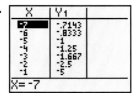

In Exercises 15–20, tell whether each equation is an example of direct variation, inverse variation, joint variation, or combined variation and identify the constant of variation.

15. $y = \dfrac{8}{x}$ **16.** $y = \dfrac{1}{2}xz$ **17.** $y = \dfrac{3x}{z}$

18. $y = \dfrac{1}{2}x$ **19.** $y = 3xz$ **20.** $y = \dfrac{3x}{z^2}$

PRACTICE *In Exercises 21–30, express each sentence as a formula.*

21. A varies directly with the square of p.

22. z varies inversely with the cube of t.

23. v varies inversely with the cube of r.

24. r varies directly with the square of s.

25. B varies jointly with m and n.

26. C varies jointly with x, y, and z.

27. P varies directly with the square of a, and inversely with the cube of j.

28. M varies inversely with the cube of n, and jointly with x and the square of z.

29. The force of attraction between two masses, m_1 and m_2, varies directly with the product of m_1 and m_2 and inversely with the square of the distance between them.

30. The force of wind on a vertical surface varies jointly with the area of the surface and the square of the velocity of the wind.

In Exercises 31–38, express each formula in words. In each formula, k is the constant of variation.

31. $L = kmn$ **32.** $P = \dfrac{km}{n}$ **33.** $E = kab^2$ **34.** $U = krs^2t$

35. $X = \dfrac{kx^2}{y^2}$ **36.** $Z = \dfrac{kw}{xy}$ **37.** $R = \dfrac{kL}{d^2}$ **38.** $e = \dfrac{kPL}{A}$

APPLICATIONS

39. Area of a circle The area of a circle varies directly with the square of its radius, and the constant of variation is π. Find the area of the circle with a radius of 6 inches.

40. Falling objects An object in free fall travels a distance s that is directly proportional to the square of the time t. If an object falls 1024 feet in 8 seconds, how far will it fall in 10 seconds?

41. **Finding distance**   The distance that a car can travel is directly proportional to the number of gallons of gasoline it consumes. If a car can go 288 miles on 12 gallons of gasoline, how far can it go on a full tank of 18 gallons?

42. **Farming**   A farmer's harvest in bushels varies directly with the number of acres planted. If 8 acres can produce 144 bushels, how many acres are required to produce 1152 bushels?

43. **Farming**   The length of time that a given number of bushels of corn will last when feeding cattle varies inversely with the number of animals. If x bushels will feed 25 cows for 10 days, how long will the feed last for 10 cows?

44. **Geometry**   For a fixed area, the length of a rectangle is inversely proportional to its width. A rectangle has a width of 18 feet and a length of 12 feet. If the length is increased to 16 feet and the area remains the same, find the width.

45. **Gas pressure**   Under a constant temperature, the volume occupied by a gas is inversely proportional to the pressure applied. If the gas occupies a volume of 20 cubic inches under a pressure of 6 pounds per square inch, find the volume when the gas is subjected to a pressure of 10 pounds per square inch.

46. **Value of a car**   The value of a car usually varies inversely with its age. If a car is worth $7000 when it is 3 years old, how much will it be worth when it is 7 years old?

47. **Organ pipes**   The frequency of vibration of air in an organ pipe is inversely proportional to the length of the pipe. (See Illustration 1.) If a pipe 2 feet long vibrates 256 times per second, how many times per second will a 6-foot pipe vibrate?

ILLUSTRATION 1

48. **Geometry**   The area of a rectangle varies jointly with its length and width. If both the length and width are tripled, by what factor is the area multiplied?

49. **Geometry**   The volume of a rectangular solid varies jointly with its length, width, and height. If the length is doubled, the width is tripled, and the height is doubled, by what factor is the volume multiplied?

50. **Costs of a trucking company**   The costs incurred by a trucking company vary jointly with the number of trucks in service and the number of hours they are used. When 4 trucks are used for 6 hours each, the costs are $1800. Find the costs of using 10 trucks, each for 12 hours.

51. **Storing oil**   The number of gallons of oil that can be stored in a cylindrical tank varies jointly with the height of the tank and the square of the radius of its base. The constant of proportionality is 23.5. Find the number of gallons that can be stored in the cylindrical tank in Illustration 2.

ILLUSTRATION 2

52. **Finding the constant of variation**   A quantity l varies jointly with x and y and inversely with z. If the value of l is 30 when $x = 15$, $y = 5$, and $z = 10$, find k.

53. **Electronics**   The voltage (in volts) measured across a resistor is directly proportional to the current (in amperes) flowing through the resistor. The constant of variation is the resistance (in ohms). If 6 volts is measured across a resistor carrying a current of 2 amperes, find the resistance.

54. **Electronics**   The power (in watts) lost in a resistor in the form of heat is directly proportional to the square of the current (in amperes) passing through it. The constant of proportionality is the resistance (in ohms). What power is lost in a 5-ohm resistor carrying a 3-ampere current?

55. **Building construction**   The deflection of a beam is inversely proportional to its width and the cube of its depth. If the deflection of a 4-inch-by-4-inch beam is 1.1 inches, find the deflection of a 2-inch-by-8-inch beam positioned as in Illustration 3.

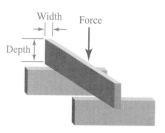

ILLUSTRATION 3

56. Building construction   Find the deflection of the beam in Exercise 55 when the beam is positioned as in Illustration 4.

ILLUSTRATION 4

57. Gas pressure   The pressure of a certain amount of gas is directly proportional to the temperature (measured in degrees Kelvin) and inversely proportional to the volume. A sample of gas at a pressure of 1 atmosphere occupies a volume of 1 cubic meter at a temperature of 273° Kelvin. When heated, the gas expands to twice its volume but the pressure remains constant. To what temperature is it heated?

58. Tension   A stone twirled at the end of a string is kept in its circular path by the tension of the string. The tension, T, is directly proportional to the square of the

speed, s, and inversely proportional to the radius, r, of the circle. In Illustration 5, the tension is 32 pounds when the speed is 8 feet/second and the radius is 6 feet. Find the tension when the speed is 4 feet/second and the radius is 3 feet.

ILLUSTRATION 5

✍ **59.** Discuss the similarities and differences between direct variation, inverse variation, joint variation, and combined variation.

✍ **60.** As temperature increases on the Fahrenheit scale, it also increases on the Celsius scale. Is this direct variation? Explain.

✍ **61.** As the cost of a purchase (less than $5) increases, the amount of change received from a five-dollar bill decreases. Is this inverse variation? Explain.

REVIEW   Specify the domain and the range of each function specified below (you may use either set notation or interval notation). Each function is represented in three ways—with a definition, a graphical display, and a TABLE. Use one or all three representations to aid you in determining domain and range.

62. $f(x) = x + 3$

63. $f(x) = x^2 + 3$

64. $f(x) = \sqrt{x + 3}$

65. $f(x) = |x + 3|$

CHAPTER SUMMARY

CONCEPTS

Section 8.1

A *relation* is a set of ordered pairs that establishes a correspondence between the numbers x and the values of y.

When each x-value is paired with a unique y-value we call the relation a *function*.

The *domain* is the set of all numbers x for which the function is defined.

The *range* is the set of all possible y-values.

When given the graph of an equation, the vertical line test can be used to determine if the graph is a function. If any vertical line intersects the graph at more than one point, the graph is *not* a function.

The notation $y = f(x)$ denotes that the variable y is a function of x.

All straight-line graphs, except for vertical lines, are functions. A *linear function* is defined by an equation that can be written in the form $y = mx + b$ or $f(x) = mx + b$.

REVIEW EXERCISES

Relations and Functions

1. State whether each equation defines y to be a function of x. Find the domain and range.

   a. $y = 4x - 1$ **b.** $x = 3y - 10$

   c. $y = 3x^2 + 1$ **d.** $y = \dfrac{4}{2 - x}$

   e. $x = \dfrac{y + 3}{2}$ **f.** $y^2 = 4x$

2. Use the vertical line test to determine whether each graph represents a function.

   a.   **b.**

   c.   **d.**

3. Assume that $f(x) = 3x + 2$ and $g(x) = x^2 - 4$. Find each value.

   a. $f(-3)$ **b.** $g(8)$ **c.** $g(-2)$ **d.** $f(5)$

4. Determine $f(x)$ from the displayed function $y = f(x)$.

   a.   **b.**

5. Tell which equations define linear functions.

   a. $y = 3x + 2$ **b.** $y = \dfrac{x + 5}{4}$

   c. $4x - 3y = 12$ **d.** $y = x^2 - 25$

| Section 8.2 | **Functions and Graphs** |

Graphs are always "read" from left to right. If, as the *x*-value increases the *y*-value also increases, the function is said to be *increasing*. The graph rises (when viewed from left to right).

If, as the *x*-values increase, the *y*-values decrease, the function is said to be decreasing. The graph falls (when viewed from left to right).

6. Determine the intervals over which the function is increasing, decreasing, and/or constant.

a.

b.

c.

d.

e.

f.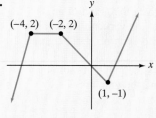

The *turning points* of a graph are the points where the graph changes from increasing to decreasing. It is also called a *relative maximum* or a *relative minimum* and can be determined with the graphing calculator.

7. Graph each function and determine the turning points (relative maximum/ relative minimum).

a. $y = -x^2 + 36$

b. $y = |x + 3| - 5$

c. $y = -2x^3 + 8x^2 - 6x$

d. $y = (x - 3)(2x + 5)(x - 1)$

8. Suppose that the ball shown in Illustration 1 is thrown straight up with a velocity of 128 feet per second. The function $y = 128x - 16x^2$ models the relationship between the number of feet above the ground, y, and the time, x, measured in seconds.

ILLUSTRATION 1

a. Graph in an appropriate viewing window (one in which all interesting features as well as x- and y-intercepts are visible).

b. Determine the relative maximum.

c. At the relative maximum, interpret the meanings of the x- and y-values of the coordinates within the context of the problem.

d. Explain why the two sets of coordinates (3.2, 245.76) and (4.8, 245.76) both appear on the graph.

e. Explain the meaning of the x-intercept.

9. A man wants to build the rectangular pen shown in Illustration 2 to house his dog. To save fencing, he intends to use one side of his garage and 80 feet of fencing.

ILLUSTRATION 2

a. Write an equation where y is the area and x is the width to model this problem.

b. Graph the equation.

c. Find the maximum area he can enclose.

d. What is the width when area is maximized?

Slope of a Nonvertical Line

The slope of a line is its rate of increase or decrease.

$$m = \frac{\Delta y}{\Delta x} = \frac{y_2 - y_1}{x_2 - x_1} = \frac{\text{rise}}{\text{run}}$$

10. Find the slope of the line passing through points P and Q.
  a. $P(2, 5)$ and $Q(5, 8)$.

  b. $P(-3, -2)$ and $Q(6, 12)$.

  c. $P(-3, 4)$ and $Q(-5, -6)$.

  d. $P(5, -4)$ and $Q(-6, -9)$.

A line is determined by a point and its slope. A linear equation in the form $y = mx + b$ is in *slope-intercept form*. The coefficient of x, m, represents the slope of the line, and b denotes the y-intercept.

11. Find the slope of the graph of each equation, if one exists.
  a. $2x - 3y = 18$ **b.** $2x + y = 8$

  c. $-2(x - 3) = 10$ **d.** $3x + 1 = 7$

12. Write the equation of the line with the given properties.
  a. Through $(0, -5)$ and with slope $\frac{2}{3}$.

  b. Through $(0, 2)$ and with slope $-\frac{1}{2}$.

Lines with positive slope are increasing functions, and the lines rise when viewed from left to right. Lines with negative slope are decreasing functions, and the lines fall when viewed from left to right.

13. Determine if the displayed graph has positive, negative, 0, or no defined slope.
  a.

  b.

Horizontal lines have equations of the form $y = b$ and have a slope of 0.

Vertical lines have equations of the form $x = a$ and have no defined slope.

  c.

  d.

If two lines are parallel, their slopes are equal. If two nonvertical lines are perpendicular, their slopes are negative reciprocals.

14. Tell whether the lines with the given slopes are parallel, perpendicular, or neither.
  a. $m_1 = 4$, $m_2 = -\frac{1}{4}$ **b.** $m_1 = 0.5$, $m_2 = \frac{1}{2}$

  c. $m_1 = 0.5$, $m_2 = -\frac{1}{2}$ **d.** $m_1 = 5$, $m_2 = -0.2$

Section 8.4	Equations of Lines

The equation of the line passing through $P(x_1, y_1)$ and with slope m is $y - y_1 = m(x - x_1)$. This is called the *point-slope form* of a linear equation. The *standard form* of a linear equation is $Ax + By = C$ (A and B cannot both be zero).

The equation of a line can also be used to find the declining value of an item (such as a car or aging equipment in a business). The method is called *straight-line depreciation*. The rate of depreciation (or appreciation, increase in value) is given by the slope of the line.

15. Write the equation of the line with the given properties. Write each equation in standard form.
  a. Slope of 3; passing through $P(-8, 5)$.
  b. Passing through $(-2, 4)$ and $(6, -9)$.
  c. Passing through $(-3, -5)$; parallel to the graph of $3x - 2y = 7$.

  d. Passing through $(-3, -5)$; perpendicular to the graph of $3x - 2y = 7$.

16. Write the equation of the pictured lines. Write each equation in standard form.
  a.   **b.**

17. A computer purchased in 1998 for $2500 has a value of $850 3 years later.
  a. What is the rate of depreciation?
  b. Write a linear equation to model this depreciation.
  c. Determine a good viewing window to display the graph of this equation.

18. In 1940 a cameo pin retailed for $15. The same cameo has increased in value in a linear pattern, and 58 years later is valued at $200. Find the rate of appreciation.

Section 8.5	Variation

The words *y varies directly as x* mean that for some nonzero constant k, $y = kx$.

The letter k is called the *constant of variation*. In direct variation, as one variable increases, so does a second variable.

The words *y varies inversely with x* mean that $y = \frac{k}{x}$ for some constant of variation, k. In inverse variation, as one variable increases, the second variable decreases. Problems involving direct and inverse variation can be examined graphically.

Joint variation occurs when more than two variables are involved; one variable varies with the product of several variables.

Problems that involve a combination of direct and inverse variation are called *combined variation* problems.

19. Assume that x varies directly as y. If $x = 12$ when $y = 2$, find the value of x when $y = 12$.

20. Assume that x varies inversely with y. If $x = 24$ when $y = 3$, find the value of y when $x = 12$.

21. Assume that x varies jointly with y and z. Find the constant of variation if $x = 24$ when $y = 3$ and $z = 4$.

22. Assume that x varies directly with t and inversely with y. Find the constant of variation if $x = 2$ when $t = 8$ and $y = 64$.

23. Physicians use the body-mass index (BMI), the accepted measure of a person's body composition, as a weight/height comparison. The BMI varies directly as an individual's weight in pounds and inversely as the square of their height in inches. If a person weighing 182 pounds at 69″ tall has a BMI of 26.95, find the BMI (to the nearest whole number) of a person weighing 132 pounds at 55″ height.

24. The volume of a cylinder varies jointly with its height and the area of its base. If the height is tripled and the area of the base is doubled, by what factor is the volume multiplied?

■ Chapter Test

1. Does the equation $y = |x + 1|$ determine y to be a function of x?

2. Determine the domain and range of the displayed function.

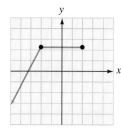

3. If $f(x) = x^2 + 4x$, find $f(-3)$.

4. Given the function at the right, determine the intervals over which the function is
   a. increasing
   b. decreasing
   c. constant

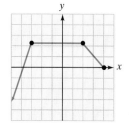

5. A child, standing on the deck of his house, throws a ball straight up with an initial velocity of 32 feet per second. The equation $y = -16x^2 + 32x + 10$ gives the height, y, of the ball x seconds after it is thrown.
   a. Graph in an appropriate viewing window.

   b. Find the maximum height reached by the ball.
   c. Find the time it takes the ball to hit the ground.

   d. What is the turning point of this function?

6. Find the slope of the graph of $x = \dfrac{3y - 8}{2}$.

7. Find the slope of the line passing through $P(-5, -6)$ and $Q(5, -2)$.

8. Write the equation of a line whose slope is $\frac{4}{5}$ and passes through $(0, 3)$.

9. Are the lines $y = \dfrac{2}{3}x + 4$ and $y = \dfrac{3}{2}x - 4$ parallel, perpendicular, or neither?

10. Write the equation of a line perpendicular to the y-axis passing through $(3, -1)$.

11. Write the equation of a line passing through $(4, 1)$ and $(-7, 6)$.

12. Does $y(x + 3) + 4 = x(y - 2)$ define a linear function? Explain your reasoning.

13. Express as a formula: y varies directly with w but inversely with the square of z.

14. The power (in watts) lost in a resistor in the form of heat is directly proportional to the square of the current (in amperes) passing through it. If k is the constant of variation, express this relationship with an equation.

15. In the previous equation, if the constant of proportionality is 5, answer the following questions.
   a. On the interval $[0, \infty)$ is the variation function increasing or decreasing?
   b. Locate the point $(3, 45)$ on the graph. To what do the values 3 and 45 correspond within the context of the problem?

16. x varies directly with y. If $x = 30$ when $y = 4$, find x when $y = 9$.

17. v varies inversely with t. If $v = 55$ when $t = 20$, find t when $v = 75$.

■ Cumulative Review: Chapters 7 and 8

VOCABULARY / CONCEPTS

1. A number written in _____ is the product of a number between 1 (including 1) and 10 and an integral power of 10.

2. Write each expression below with positive exponents.
   a. $\dfrac{1}{x^{-3}}$
   b. $\dfrac{x^8}{x^{-2}}$
   c. $x^{-5} \cdot x^0$
   d. $\dfrac{x^{-2}}{x^5}$

3. Write each number in scientific notation.
   a. 1,320,000
   b. 2,000,000
   c. 0.0013
   d. 0.00412

4. Write each number in standard notation.
   a. 1.3×10^{-2}
   b. 6.1×10^5
   c. 0.051×10^{-2}
   d. 3.82×10^6

5. The _____ of an equation is the value(s) of the variable(s) that make the equation a true statement.

6. A _____ is a set of ordered pairs in which each x-value in the domain is assigned a unique y-value in the range.

7. Each of the following represent solutions to an equation. Based on the given information, is the equation a function?

    a. $\{(1, 1), (1, -1), (0, 0), (4, 2), (4, -2)\}$

    b.

8. *AND* statements represent the _____ of the set of values that satisfy two simple inequalities.

9. *OR* statements represent the _____ of the set of values that satisfy two simple inequalities.

10. Write $|x + 2| \leq 4$ as a compound inequality.

11. State whether each of the following graphs is that of a function or a relation.

    a.   **b.**

    c.   **d.**

    e.   **f.**

12. If $f(x) = x^2 - 3x + 2$, find each of the following:

    a. $f(-1)$ **b.** $f(3)$

    c. $f(-3)$ **d.** $f(8)$

13. Write the equation $x = 3y - 4$ in $y = mx + b$ form.

14. The slope of a line is defined as the change in __ over the change in __.

15. A horizontal line is written in the form _____.

16. A vertical line is written in the form _____.

17. The slope of a horizontal line is __.

18. The slope of a vertical line is _____.

19. When a linear equation is written in the form $y = mx + b$, it is in _____ form. The coefficient of x is the _____ of the line, and b is the _____.

20. When a linear equation is written in the form $y - y_1 = m(x - x_1)$ it is in _____ form, where m is the _____ of the line and _____ are the coordinates of a point on the line.

PRACTICE

21. a. Simplify $\dfrac{(2x^{-4}y^2)^3}{4x^{-2}y^{-1}}$.

b. Evaluate both the original monomial and the simplified monomial when $x = \frac{1}{2}$ and $y = \frac{1}{5}$ using the STOre feature of your calculator.

In Exercises 22–26, solve each equation algebraically. Confirm your solutions with either the STOre or TABLE feature of the graphing calculator.

22. $\dfrac{8(x - 5)}{3} = 2(x - 4)$

23. $2x^2 - 7x = 4$

24. $2|x + 1| - 2 = 0$

25. $3 + |2x - 1| = 2$

26. $\dfrac{x}{x - 2} - \dfrac{5}{x - 2} = 3$

27. Solve the compound inequality $-2 \le \dfrac{5 - 3x}{2} \le 2$ and express the solution as a number line graph and in interval notation. Confirm your solution graphically.

28. Solve the inequality $3|2x + 5| \ge 9$. Express your solution in interval notation and as a number line graph. Confirm your solution graphically.

29. If $f(x) = 3x - 1$, find $f(2) - f(3)$.

30. For what intervals of x is the accompanying graph
a. increasing?

b. decreasing?

c. constant?

31. Graph the equation $2x + 3y = 6$ using x- and y-intercepts.

32. Determine the domain and range of the function specified below:

$$\{(1, 5), (2, 5), (-1, -1), (-3, -5)\}$$

33. Find the slope of the line whose displayed TABLE is pictured.

X	Y₁
0	-4
1	-3.5
2	-3
3	-2.5
4	-2
5	-1.5
6	-1

X=0

34. Find the slope and the y-intercept of the line with the equation $x = \dfrac{3}{4} - \dfrac{1}{4}y$.

35. Determine if the pair of lines determined by the given equations are parallel, perpendicular, or neither.

$$x - 2y = -18; \quad 2x + y = -6$$

In Exercises 36–40, write the equation of the line with the given properties. Then transform the equation into standard form.

36. The line with slope $\frac{2}{3}$ and passing through $(0, 5)$.

37. The line with slope 3 and passing through $(5, 1)$.

38. The line with slope 0, perpendicular to the line with equation $x = 5$ and passing through the point $(6, 2)$.

39. The line passing through the points with coordinates $(3, 2)$ and $(-1, 4)$.

40. The line through $(-6, 3)$ and perpendicular to the line with equation $y + 3x = -12$.

41. Solve the equation $-3x + 15 = \sqrt{x - 5}$ using the INTERSECT feature of your calculator. Check your solution in the TABLE.

42. Solve the equation $\sqrt{2x + 5} - 4 = 0$ using the ROOT/ZERO feature of your calculator. Check your solution in the TABLE.

43. Solve the equation $x^3 - x^2 - 12x = 0$ graphically using the INTERSECT or ROOT/ZERO feature of the graphing calculator. Check your solutions in the TABLE.

44. Graph the function $y = 3(x - 1)^2 + 6$ and determine the coordinates of the minimum.

45. Assume that an antique purchased for $300 increases at a rate of $15 per year.
  a. Write an equation to model this increase, where y represents the value and x the subsequent years.
  b. What does the slope of your equation represent?
  c. What is the predicted value of the antique in 20 years?

46. A computer service charges a fixed amount of $45 and $20 per hour for repair.
  a. If y represents total cost and x represents the number of hours, write an equation that models the cost of repairing computers.
  b. Graph the model in the window $[0, 30]$ by $[-10, 300]$.
  c. What is the cost to the owner if it takes 5 hours to repair his computer? How did you get your result?
  d. Set the TABLE to begin at $x = -2$. The value displayed for y is 5. Explain why negative x-values were not specified in the window given in part (b).
  e. If it took technicians 20 hours to repair a computer, what is the cost to the owner? How did you arrive at your result?

APPLICATIONS

47. Car rental It costs $45 per day to rent a car plus $0.24 per mile. If you had $100 to spend for one day of rental, how many miles could be driven?

48. Calculating grades A student has test scores of 78%, 80%, and 85%. What must she make on the fourth test to maintain an average of 70% or higher?

49. Area of a circle The area of a circle varies directly with the square of its radius, and the constant of variation is π. Find the area of a circle with a radius of 9 mm.

50. Value of a car The value of a car varies inversely with age. If a car is worth $8000 when it is 5 years old, how much will it be worth after 10 years?

CAREERS & MATHEMATICS

Economist   Economists study the way a society uses resources such as land, labor, raw materials, and machinery to provide goods and services. They analyze the results of their research to determine the costs and benefits of making, distributing, and using resources in a particular way. Some economists are theoreticians who use mathematical models to explain the causes of recession and inflation. Most economists, however, are concerned with practical applications of economic policy in a particular area.

Qualifications   Economists must thoroughly understand economic theory, mathematical methods of economic analysis, and basic statistical procedures. Training in computer science is highly recommended.

Job Outlook   Employment of economists is expected to grow faster than the average for all occupations during the next decade. Opportunities should be best for economists in business and industry, research organizations, and consulting firms.

Example Application   An electronics firm manufactures tape recorders, receiving $120 for each unit it makes. If x represents the number of recorders produced, then the income received is determined by the *revenue function,* given by the linear equation

$$R(x) = 120x$$

The manufacturer has determined that the *fixed costs* for advertising, insurance, utilities, and so on, are $12,000 per month, and the *variable cost* for materials is $57.50 for each machine produced. Thus, the *cost function* is given by the linear equation

$$C(x) = \textit{variable cost} + \textit{fixed costs}$$
$$= 57.50x + 12,000$$

The company's profit is the amount by which revenue exceeds costs. It is determined by the *profit function,* given by the equation

$$\textit{Profit} = \textit{revenue} - \textit{costs}$$
$$P(x) = R(x) - C(x)$$
$$= 120x - (57.50x + 12,000)$$
$$= 62.50x - 12,000$$

If $P(x) > 0$, the company is making money. If $P(x) < 0$, it is operating at a loss. How many recorders must the company manufacture to break even?

Solution   Graph the profit function $y = P(x) = 62.50x - 12,000$. The break-even point is that value of x that gives a profit of zero. It is the x-intercept of the graph of the profit function. (See Illustration 1.) To find it, set $P(x)$ equal to 0 and solve for x.

$$P(x) = 0$$
$$62.50x - 12,000 = 0$$
$$62.50x = 12,000 \qquad \text{Add 12,000 to both sides.}$$
$$x = 192 \qquad \text{Divide both sides by 62.50}$$

The company must manufacture and sell 192 tape recorders each month to break even.

ILLUSTRATION 1

EXERCISES

1. Find the revenue and the cost of manufacturing 192 units, and verify that the revenue and the cost are equal.

2. Determine the company's profit if it manufactures 150 units each month.

3. Determine the company's profit if it manufactures 400 units each month.

4. How many units must be manufactured each month to produce a total profit of $47,375?

(*Answers:* 1. $23,040 revenue and cost 2. $2625 loss 3. $13,000 profit 4. 90 units)

Rational Exponents and Radicals

■ PERSPECTIVE

Don't Buck the System

Pythagoras was a teacher. Although it was unusual for schools at that time, his classes were coeducational. According to some legends, Pythagoras married one of his students. He and his followers formed a secret society with two rules: membership was for life, and members could not reveal the secrets they knew.

Much of their teaching was good mathematics, but some ideas were strange. To them, numbers were sacred. Because beans were used as counters to represent numbers, Pythagoreans refused to eat beans. They also believed that the *only* numbers were the whole numbers. To them, fractions were not numbers; $\frac{2}{3}$ was just a way of comparing the whole numbers 2 and 3. They believed that whole numbers were the building blocks of the universe, just as we believe atoms are. The basic Pythagorean doctrine was "all things are number," and they meant *whole* number.

The Pythagorean theorem was an important dis-

covery of the Pythagorean school, yet it caused some division in the ranks. The right triangle in Illustration 1 has two legs of length 1. By the Pythagorean theorem, the length of the hypotenuse is $\sqrt{2}$. One of their own group, Hippasus of Metapontum, discovered that $\sqrt{2}$ is an irrational number: There are *no* whole numbers a and b that make the fraction $\frac{a}{b}$ exactly equal to $\sqrt{2}$. This discovery was not appreciated by the other Pythagoreans. How could everything in the universe be described with whole numbers, when the side of this simple triangle couldn't? The Pythagoreans had a choice. Either revise and expand their beliefs, or cling to the old. According to legend, the group was at sea at the time of the discovery. Rather than upset the system, they threw Hippasus overboard.

ILLUSTRATION 1

The ability to work with roots and radicals enables us to solve many more real-world problems. This chapter will examine the algebra of radicals and radical functions. At the conclusion of the chapter, we will extend our real-number system to include the complex number system.

9.1 Introduction to Roots and Radicals

■ SQUARE ROOTS AND RADICALS ■ CUBE ROOTS ■ *n*TH ROOTS ■ SQUARE ROOTS AND *n*TH ROOTS OF EXPRESSIONS WITH VARIABLES

■ SQUARE ROOTS AND RADICALS

The opposite of squaring a number is finding the **square root** of a number. For example, the square of 3, 3^2, is equivalent to the number 9. Conversely, the positive square root of 9 is 3. Some other examples of square roots are:

4 is a square root of 16 because $4^2 = 16$.

5 is a square root of 25 because $5^2 = 25$.

8 is a square root of 64 because $8^2 = 64$.

a^3 is the square root of a^6 because $(a^3)^2 = a^6$.

$7xy$ is a square root of $49x^2y^2$ because $(7xy)^2 = 49x^2y^2$.

All positive real numbers have two square roots, one positive and one negative. We symbolize the nonnegative or **principal** square root with a **radical sign,** $\sqrt{}$. An expression containing a radical sign is a **radical expression.** The expression under the radical sign is called the **radicand.**

radical sign $\rightarrow \sqrt{a} \leftarrow$ radicand

> ### Square Roots
>
> If $x > 0$, the *principal square root* of x is the positive square root of x and is denoted as $\sqrt{x}$. The negative or opposite of the principal square root is denoted as $-\sqrt{x}$.
>
> The principal square root of 0 is 0: $\sqrt{0} = 0$.

EXAMPLE 1   Simplify: **a.** $\sqrt{81}$ **b.** $-\sqrt{81}$ **c.** $\sqrt{225}$ **d.** $\sqrt{\dfrac{1}{4}}$

     e. $\sqrt{\dfrac{16}{121}}$ **f.** $\sqrt{0.04}$ **g.** $-\sqrt{0.0009}$ **h.** $\sqrt{-4}$

Solution   **a.** $\sqrt{81} = 9$ because $9^2 = 81$.

■ TECHNOLOGY TIP
The $\sqrt{}$ key can be used to find square roots. As long as the calculator is in **real** mode, an error message results when $\sqrt{-4}$ is entered.

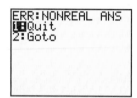

This confirms our definition that negative radicands are undefined in the real-number system.

b. $-\sqrt{81} = -9$. The negative in front of the radical indicates the negative, or opposite, of the square root of 81. The square root of 81 is 9, and its opposite is -9.

c. $\sqrt{225} = 15$ because $15^2 = 225$.

d. $\sqrt{\dfrac{1}{4}} = \dfrac{1}{2}$ because $\left(\dfrac{1}{2}\right)^2 = \dfrac{1}{4}$.

e. $\sqrt{\dfrac{16}{121}} = \dfrac{4}{11}$ because $\left(\dfrac{4}{11}\right)^2 = \dfrac{16}{121}$.

f. $\sqrt{0.04} = 0.2$ because $(0.2)^2 = 0.04$.

g. $-\sqrt{0.0009} = -0.03$ because $(0.03)^2 = 0.0009$ and its opposite is -0.03.

h. $\sqrt{-4}$ is undefined in the real-number system because there is no real number that can be multiplied *by itself* to obtain a result that is negative (a positive number times a positive number is positive and a negative number times a negative number is also positive). ■

The numbers 1, 4, 9, 16, and 25 are examples of **perfect squares,** so called because their square roots are rational numbers. Numbers like 3 are *not* perfect squares because $\sqrt{3}$ is not equivalent to a rational number. If using the calculator to evaluate $\sqrt{3}$, the number displayed is an *approximation*. It is an irrational number and is a nonrepeating, nonterminating decimal.

The decimal terminates *on the screen* because that is all the display that programming allows. It is imperative that when approximating roots you *clearly* indicate a result is an approximation. This can be done by using a symbol for approximation, either $\approx$ or $\doteq$. Therefore, $\sqrt{3} \doteq 1.732$.

■ TECHNOLOGY TIP
Press the MATH key and choose option number 4 to find the cube root of real numbers.

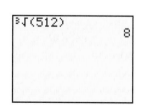

■ CUBE ROOTS

Finding roots can be extended from square roots to cube roots. Again, finding the cube root of a number is the opposite of cubing the number.

    4 is a cube root of 64 because $4^3 = 64$.

    $3x^2y$ is a cube root of $27x^6y^3$ because $(3x^2y)^3 = 27x^6y^3$.

    -2 is a cube root of -8 because $(-2)^3 = -8$.

To denote cube roots, we indicate the **index** or **order,** as shown below:

$$\text{index} \rightarrow \sqrt[3]{a} \leftarrow \text{radicand}$$

> **Cube Roots**
> The **cube root** of a is denoted as $\sqrt[3]{a}$ and is defined by $\sqrt[3]{a} = b$ if $b^3 = a$.

All real numbers have exactly one cube root. That root is either a rational number (a perfect cube) or an irrational number.

EXAMPLE 2 Simplify. If a root is irrational, approximate it to two decimal places.

a. $\sqrt[3]{125}$ b. $\sqrt[3]{\dfrac{1}{8}}$ c. $\sqrt[3]{-\dfrac{1}{8}}$

d. $\sqrt[3]{-27x^3}$ e. $\sqrt[3]{0.216x^3y^6}$ f. $\sqrt[3]{13}$

Solution

a. $\sqrt[3]{125} = 5$ because $5^3 = 5 \cdot 5 \cdot 5 = 125$.

b. $\sqrt[3]{\dfrac{1}{8}} = \dfrac{1}{2}$ because $\left(\dfrac{1}{2}\right)^3 = \dfrac{1}{2} \cdot \dfrac{1}{2} \cdot \dfrac{1}{2} = \dfrac{1}{8}$.

c. $\sqrt[3]{-\dfrac{1}{8}} = -\dfrac{1}{2}$ because $\left(-\dfrac{1}{2}\right)^3 = \left(-\dfrac{1}{2}\right)\left(-\dfrac{1}{2}\right)\left(-\dfrac{1}{2}\right) = -\dfrac{1}{8}$.

d. $\sqrt[3]{-27x^3} = -3x$ because $(-3x)^3 = (-3x)(-3x)(-3x) = -27x^3$.

e. $\sqrt[3]{0.216x^3y^6} = 0.6xy^2$ because $(0.6xy^2)^3 = (0.6xy^2)(0.6xy^2)(0.6xy^2) = 0.216x^3y^6$.

f. $\sqrt[3]{13} = 2.351334688\ldots$ The equality sign was used because the three dots at the end of the approximation indicate that the expression goes on indefinitely without repeating. Because the directions instructed us to state the irrational result correct to two decimal places, we may write $\sqrt[3]{13} \approx 2.35$. Notice that a symbol for approximation is used so that it is clear that 2.35 is an approximate value rather than an exact value.

■ nTH ROOTS

TECHNOLOGY TIP
Pressing the MATH key reveals an option for the xth root of a number, symbolized as $\sqrt[x]{}$. The index must be entered first. To simplify $\sqrt[4]{625}$, the index of 4 is entered before the $\sqrt[x]{}$ option is selected. The display given in Example 3 shows the evaluation of $\sqrt[4]{625}$. TI-85 and TI-86 users should customize $\sqrt[x]{}$ on their calculators. This is used for all roots other than the square root.

Just as there are square roots and cube roots, there are fourth, fifth, sixth roots, and so on. The notation $\sqrt[n]{a}$ indicates the **nth root** of a number, where n is a natural number greater than 1 ($n > 1$). The letter n is the index and specifies the order of the root. The index of 2 is omitted when writing square roots.

When the index is odd, the radical $\sqrt[n]{a}$ ($n > 1$) is often called an **odd root**. Every real number has exactly one real, nth root when n is odd. Moreover, when the radicand is positive, the root is positive. When the radicand is negative, the root is negative.

When the index is even, the radical $\sqrt[n]{a}$ ($n > 1$) is called an **even root**. The notation $\sqrt[n]{a}$ represents the positive, or principal root. When finding even roots we often use absolute value symbols to guarantee that the nth root is positive:

$$\sqrt[4]{(-3)^4} = |-3| = 3.$$

When the index is *even* the radicand must be positive for the radical to be defined within the real-number system. The basic definitions for the nth roots of real numbers are summarized as follows:

> **Definition of $\sqrt[n]{a^n}$**
> If n is a real number and $n > 1$, then
> if n is an odd natural number, $\sqrt[n]{a^n} = a$.
> if n is an even natural number, $\sqrt[n]{a^n} = |a|$.

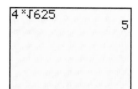

EXAMPLE 3   Simplify each of the following roots. Verify results with a calculator.

    a. $\sqrt[4]{625}$ **b.** $\sqrt[4]{-81}$ **c.** $\sqrt[3]{-27}$

    d. $-\sqrt[6]{64}$ **e.** $\sqrt[4]{27}$

Solution   **a.** $\sqrt[4]{625} = 5$ because $5^4 = 625$;
         $\sqrt[4]{625} = \sqrt[4]{5^4} = |5| = 5.$

b. $\sqrt[4]{-81}$ is undefined
   for real numbers.

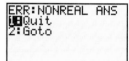

c. $\sqrt[3]{-27} = -3$ because $(-3)^3 = -27$;
   $\sqrt[3]{-27} = \sqrt[3]{(-3)^3} = -3.$

d. $-\sqrt[6]{64} = -2$ because $2^6 = 64$. The
   negative in front of the radical
   indicates we want the opposite of
   the indicated root.

e. $\sqrt[4]{27} \approx 2.2795$

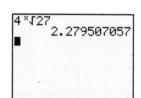

■ SQUARE ROOTS AND *n*TH ROOTS OF EXPRESSIONS WITH VARIABLES

If $x \neq 0$, the positive (principal) square root of x^2 is denoted as $\sqrt{x^2}$. *Incorrectly,* students often assume that $\sqrt{x^2} = x$. This assumption is a result of the idea that squaring a number and extracting a square root are inverse operations, with one operation "undoing" the other. The fallacy of this argument is demonstrated when taking the square root of a variable because we do not know if the variable represents a positive or a negative number. For example, if $x = 2$, then $\sqrt{2^2} = 2$, or $\sqrt{x^2} = x$. However, if $x = -2$, then $\sqrt{(-2)^2} \neq -2$; rather, $\sqrt{(-2)^2} = 2$. The use of absolute value guarantees extracting the principal root. Care must be taken when the index of a radical is even and variables are contained in the radicand.

EXAMPLE 4   Assuming x is a real number, simplify the following:

a. $\sqrt{16x^2}$ b. $\sqrt{(x+4)^2}$ c. $\sqrt{x^2 + 2x + 1}$

d. $\sqrt{x^4}$ e. $\sqrt[5]{x^5}$ f. $\sqrt[4]{16x^4}$

Solution   a. $16x^2 = \sqrt{(4x)^2} = |4x| = 4|x|$. Absolute value symbolism is necessary because x could represent a negative value. The use of absolute value ensures the principal, positive, root.

b. $\sqrt{(x+4)^2} = |x+4|$. Absolute value ensures the principal root.

c. $\sqrt{x^2 + 2x + 1} = \sqrt{(x+1)^2} = |x+1|$. Absolute value ensures the principal root.

d. $\sqrt{x^4} = \sqrt{(x^2)^2} = x^2$. Because x^2 is *always* positive (regardless of the value of x) absolute value symbols are unnecessary.

e. $\sqrt[5]{x^5} = \sqrt[5]{(x)^5} = x$. Because the index is odd, absolute value symbolism is unnecessary.

f. $\sqrt[4]{16x^4} = \sqrt[4]{(2x)^4} = |2x| = 2|x|$. Absolute value ensures the principal root.

The definitions concerning $\sqrt[n]{a}$ are summarized as follows:

Summary of Definitions of $\sqrt[n]{a}$

If n is a real number and $n > 1$, then

if $a > 0$, then $\sqrt[n]{a}$ is a positive number such that $\left(\sqrt[n]{a}\right)^n = a$.

if $a = 0$, then $\sqrt[n]{a} = 0$.

if $a < 0$ and n is odd, then $\sqrt[n]{a}$ is the real number such that $\left(\sqrt[n]{a}\right)^n = a$.

if $a < 0$ and n is even, then $\sqrt[n]{a}$ is not a real number.

WARNING!   Care must be taken when extracting even roots of radicands that contain variables. If the variable(s) is not restricted to positive numbers, absolute value symbolism *may* be needed.

EXERCISE 9.1

VOCABULARY   *In Exercises 1–4, fill in the blanks to make a true statement.*

1. When finding the square root of a number, we are always looking for the nonnegative or ___principal___ square root.

2. The expression under a radical sign is called the ___radicand___

3. If an indicated root is perfect, the result is an ___exact___ result. If an indicated root is *not* perfect, the calculator can be used to obtain an ___approximate___ result, which is an ___irrational___ number.

4. The order of a root is specified by the ___index___ of the radical.

CONCEPTS

5. Which of the indicated roots below are undefined for real numbers?
   a. $\sqrt[3]{-8}$ b. $\sqrt{-8}$ c. $\sqrt[5]{-36}$ d. $\sqrt[4]{-16}$

6. Classify each of the following as rational or irrational.
   a. $\sqrt{8}$ b. $\sqrt[3]{8}$ c. $\sqrt[5]{32}$ d. $\sqrt[4]{256}$

7. a. How many real cube roots does a positive number have?
   b. How many real cube roots does a negative number have?
   c. How many real square roots does a negative number have?

PRACTICE *In Exercises 8–49, simplify each radical, if possible. If the calculator is used, copy the screen display* ***exactly*** *as it appears.*

8. $\sqrt{121}$

9. $\sqrt{144}$ -12

10. $\sqrt[3]{1}$

11. $\sqrt[3]{-8}$
-2

12. $\sqrt[3]{8a^3}$

13. $\sqrt[3]{-27x^6}$
-3x²

14. $\sqrt{\dfrac{1}{9}}$

15. $-\sqrt{\dfrac{4}{25}}$ $\frac{2}{5}$

16. $-\sqrt[5]{243}$

17. $-\sqrt[4]{625}$
-5

18. $\sqrt{0.16}$

19. $\sqrt{-49}$
∅

20. $-\sqrt[5]{-\dfrac{1}{32}}$

21. $\sqrt[6]{-729}$
∅

22. $-\sqrt{64}$

23. $-\sqrt{1}$
~1

24. $\sqrt[3]{-125}$

25. $\sqrt[3]{512}$
8

26. $\sqrt[3]{125}$

27. $-\sqrt{-4}$
∅

28. $\sqrt[3]{-1000p^3q^3}$

29. $\sqrt[3]{343a^6b^3}$
7a²b

30. $\sqrt[4]{\dfrac{16}{625}}$

31. $\sqrt[5]{-\dfrac{243}{32}}$ $\frac{3}{2}$

32. $\sqrt{0.25}$

33. $\sqrt{-25}$
∅

34. $\sqrt[4]{81}$

35. $\sqrt[6]{64}$
2

36. $\sqrt{(-4)^2}$

37. $\sqrt{(-9)^2}$
9

38. $\sqrt[3]{-\dfrac{8}{27}}$

39. $\sqrt[3]{\dfrac{125}{216}}$ $\frac{5}{6}$

40. $\sqrt[3]{-\dfrac{1}{8}m^6n^3}$

41. $\sqrt[3]{\dfrac{27}{1000}a^6b^6}$ $\frac{3}{10a^2b^2}$ **42.** $\sqrt[5]{-32}$

43. $\sqrt[6]{729}$
3

44. $-\sqrt{\dfrac{25}{49}}$

45. $\sqrt{\dfrac{49}{81}}$ $\frac{7}{9}$

46. $\sqrt[3]{0.064}$

47. $\sqrt[3]{0.001}$
$\frac{1}{10}$ or .1

48. $\sqrt[4]{-256}$

49. $-\sqrt[4]{\dfrac{81}{256}}$ $-\frac{3}{4}$

In Exercises 50–67, simplify each radical. Assume that all variables are unrestricted and use absolute value symbols as necessary.

50. $\sqrt{4x^2}$

51. $\sqrt[4]{16y^4}$ 2|y|

52. $\sqrt[3]{8a^3}$

53. $\sqrt[6]{64x^6}$ 2|x|

54. $\sqrt[4]{x^{12}}$

55. $\sqrt[8]{x^{24}}$

56. $\sqrt{a^2 + 6a + 9}$

57. $\sqrt{x^2 + 10x + 25}$

58. $\sqrt[4]{\dfrac{1}{16}x^4}$ 2|x|

59. $\sqrt[4]{\dfrac{1}{81}x^8}$

60. $\sqrt[5]{-x^5}$

61. $\sqrt[3]{-x^6}$

62. $\sqrt{(t+5)^2}$

63. $\sqrt{(a+6)^2}$

64. $\sqrt[3]{-27a^6}$

65. $\sqrt[5]{-32x^5}$

66. $\sqrt{(-5b)^2}$

67. $\sqrt{(-8c)^2}$

In Exercises 68–75, a student's calculator screen is displayed.
(a) Write the problem that was being simplified as it would appear in a mathematics text.
(b) Identify the displayed result as a ***rational*** *or an* ***irrational*** *number.*

68.
```
5ˣ√32
              2
```

69.
```
6ˣ√4096
              4
```

70.

71.

72. `3√(25-8)`
`               12.36931688`

73. `5³√(27-9)`
`               13.10370697`

74. `5/(√(16)+1)`
`                          1`

75. `8/(√(49)+1)`
`                          1`

In Exercises 76–81, insert =, >, or < to describe the given relationship. Use your calculator as appropriate.

76. $\sqrt{25-16}$ ___ $\sqrt{25}-\sqrt{16}$ **77.** $\sqrt{25-9}$ ___ $\sqrt{25}-\sqrt{9}$ **78.** $\sqrt{25}+\sqrt{9}$ ___ $\sqrt{34}$

79. $\sqrt{36}+\sqrt{16}$ ___ $\sqrt{52}$ **80.** $\sqrt{25}-\sqrt{16}$ ___ $\sqrt{64}-\sqrt{15}$ **81.** $\sqrt{100}$ ___ $4\sqrt{25}$

APPLICATIONS

82. Medicine The approximate pulse rate, p (in beats per minute), of an adult who is t inches tall is given by the formula $p = \dfrac{590}{\sqrt{t}}$. The graph is displayed in the window $[0, 84]$ by $[-20, 500]$.

   a. From the general shape of the graph, as a person gets taller, what happens to the pulse rate?

   b. Interpret the screen display within the context of the problem:

83. Law enforcement The police sometimes use the formula $s = k\sqrt{L}$ to estimate the speed, s, in miles per hour of a car involved in an accident. In this formula, L, is the length of the skid in feet, and k is a constant related to the condition of the pavement. For wet pavement, $k \approx 3.24$. The formula can be written as $s = 3.24\sqrt{L}$. The graph is displayed in the window $[0, 600]$ by $[0, 100]$.

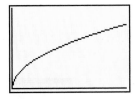

   a. From the general shape of the graph, what can you conclude about the relationship between the length of skid marks on the pavement and the speed of the vehicle?

   b. Interpret the accompanying graphical display within the context of the problem.

REVIEW *In Exercises 84–89, simplify each expression.*

84. $\left(\dfrac{3x}{y}\right)^2$ **85.** $(-5x^2y)^3$ **86.** $(3x^3y^2)^2$

87. $-(2a^2b^3c)^4$ **88.** $\left(\dfrac{2}{3}a\right)^3$ **89.** $(x+2y)^2$

9.2 Rational Exponents and Simplifying Radical Expressions

■ REVIEW OF EXPONENTS ■ RATIONAL EXPONENTS ■ PRODUCT AND QUOTIENT RULES FOR RADICALS ■ SIMPLIFYING RADICAL EXPRESSIONS

■ REVIEW OF EXPONENTS

Previously, we have defined integer exponents. Positive integer exponents indicate the number of times that a base is to be used as a factor in a product. For example,

x^5 means that the base, x, is to be used as a factor five times. Therefore, $x^5 = x \cdot x \cdot x \cdot x \cdot x$. A nonzero base raised to the zero power equals 1. For example, $8^0 = 1$ and $x^0 = 1$. Finally, a base raised to a negative exponent is defined as $x^{-n} = \dfrac{1}{x^n}$. The properties of exponents are summarized as follows:

Properties of Exponents
If there are no divisions by 0, then for all integers m and n,

1. $x^m x^n = x^{m+n}$ **2.** $(x^m)^n = x^{mn}$ **3.** $(xy)^n = x^n y^n$

4. $\left(\dfrac{x}{y}\right)^n = \dfrac{x^n}{y^n}$ **5.** $x^0 = 1 \ (x \neq 0)$ **6.** $x^{-n} = \dfrac{1}{x^n}$

7. $\dfrac{x^m}{x^n} = x^{m-n}$ **8.** $\left(\dfrac{x}{y}\right)^{-n} = \left(\dfrac{y}{x}\right)^n$

We now turn our attention to a definition for rational exponents.

■ RATIONAL EXPONENTS

EXPLORING THE CONCEPT

Simplify each expression using the graphing calculator. Include all rational exponents in parentheses to ensure correct results.

a. $3^{1/2} \cdot 3^{1/2}$ **b.** $\sqrt{3} \cdot \sqrt{3}$

c. $4^{1/3} \cdot 4^{1/3} \cdot 4^{1/3}$ **d.** $\sqrt[3]{4} \cdot \sqrt[3]{4} \cdot \sqrt[3]{4}$

Conclusion

a.
```
3^(1/2)*3^(1/2)
                3
```
b.
```
√(3)√(3)
                3
```
c.
```
4^(1/3)*4^(1/3)*
4^(1/3)
                4
```
d.
```
³√(4)³√(4)³√(4)
                4
```

This leads to the following definition:

Definition of $a^{1/n}$
If n is a positive integer greater than 1 and $\sqrt[n]{a}$ is a real number, then $\sqrt[n]{a} = a^{1/n}$.

We notice that the denominator of the rational number used as an exponent becomes the index of the radical.

EXAMPLE 1

Rewrite each expression as a radical and simplify, if possible.

a. $9^{1/2}$ **b.** $-\left(\dfrac{16}{9}\right)^{1/2}$ **c.** $(-64)^{1/3}$ **d.** $16^{1/4}$

e. $\left(\dfrac{1}{32}\right)^{1/5}$ **f.** $(-32x^5)^{1/5}$ **g.** $(3x^2 y)^{1/4}$

Solution **a.** $9^{1/2} = \sqrt{9} = 3$ **b.** $-\left(\dfrac{16}{9}\right)^{1/2} = -\sqrt{\dfrac{16}{9}} = -\dfrac{4}{3}$

c. $(-64)^{1/3} = \sqrt[3]{-64} = -4$ **d.** $16^{1/4} = \sqrt[4]{16} = 2$

e. $\left(\dfrac{1}{32}\right)^{1/5} = \sqrt[5]{\dfrac{1}{32}} = \dfrac{1}{2}$ **f.** $(-32x^5)^{1/5} = \sqrt[5]{-32x^5} = -2x$

g. $(3x^2y)^{1/4} = \sqrt[4]{3x^2y}$

The definition also allows us to move from radical form to exponential form.

EXAMPLE 2   Rewrite each expression with a rational exponent.

     a. $\sqrt{6}$ **b.** $\sqrt[4]{5x}$ **c.** $\sqrt[4]{5}\,x$

     d. $\sqrt[3]{18}$ **e.** $\sqrt[5]{\dfrac{1}{3}}$

Solution   **a.** $\sqrt{6} = 6^{1/2}$ **b.** $\sqrt[4]{5x} = (5x)^{1/4}$ **c.** $\sqrt[4]{5}\,x = 5^{1/4}x$

     d. $\sqrt[3]{18} = 18^{1/3}$ **e.** $\sqrt[5]{\dfrac{1}{3}} = \left(\dfrac{1}{3}\right)^{1/5}$

■ **TECHNOLOGY TIP**
When confirming analytical work with rational exponents with the calculator, enclose exponents in parentheses. The graphing calculator should only be used initially to confirm analytical processes.

     All of the rational exponents considered thus far have had numerators of 1. We can extend our definition of $a^{1/n}$ to include fractional exponents with numerators other than 1. This can easily be done by combining the definition of a rational exponent with previously discussed properties of exponents.
     Consider $4^{3/2}$. It can be rewritten as follows:

$$4^{3/2} = (4^{1/2})^3 = \left(\sqrt{4}\right)^3 = 2^3 = 8$$

This leads to the following definition:

> ### Definition of $a^{m/n}$
> If m and n are positive integers greater than 1, and m/n is in simplest form,
> $$a^{m/n} = \sqrt[n]{a^m} = \left(\sqrt[n]{a}\right)^m$$

This definition allows us to interpret $a^{m/n}$ in two ways:

1. $a^{m/n}$ means the **nth root** of the **mth power** of a.

2. $a^{m/n}$ means the **mth power** of the **nth root** of a.

The denominator of the rational exponent becomes the index of the radical. The numerator, m, becomes the power of either the radicand OR the entire radical. Therefore, $4^{3/2} = \left(\sqrt{4}\right)^3 = 2^3 = 8$ or $4^{3/2} = \sqrt{4^3} = \sqrt{64} = 8$. Either method leads to a correct result.

EXAMPLE 3   Simplify each of the following by first writing in radical form. Verify results with the calculator.

     a. $27^{2/3}$ **b.** $\left(\dfrac{1}{16}\right)^{3/4}$ **c.** $(-8x^3)^{4/3}$

Solution   **a.** $27^{2/3} = \left(\sqrt[3]{27}\right)^2$ or $27^{2/3} = \sqrt[3]{27^2}$

            $= 3^2$ $= \sqrt[3]{739}$

            $= 9$ $= 9$

```
27^(2/3)
              9
```

```
(1/16)^(3/4)►Fra
c
             1/8
```

b.
$$\left(\frac{1}{16}\right)^{3/4} = \left(\sqrt[4]{\frac{1}{16}}\right)^3 \quad \text{or} \quad \left(\frac{1}{16}\right)^{3/4} = \sqrt[4]{\left(\frac{1}{16}\right)^3}$$

$$= \left(\frac{1}{2}\right)^3 \qquad\qquad\qquad = \sqrt[4]{\frac{1}{4096}}$$

$$= \frac{1}{8} \qquad\qquad\qquad\qquad = \frac{1}{8}$$

c.
$$(-8x^3)^{4/3} = \left(\sqrt[3]{-8x^3}\right)^4 \quad \text{or} \quad (-8x^3)^{4/3} = \sqrt[3]{(-8x^3)^4}$$

$$= (-2x)^4 \qquad\qquad\qquad\qquad = \sqrt[3]{4096x^{12}}$$

$$= 16x^4 \qquad\qquad\qquad\qquad\quad = 16x^4 \qquad ■$$

To be consistent with the definition of negative integer exponents, we define $a^{-m/n}$ as follows:

Definition of $a^{-m/n}$

If m and n are positive integers, m/n is in simplest form, and $a^{1/n}$ is a real number, then

$$a^{-m/n} = \frac{1}{a^{m/n}} \quad \text{and} \quad \frac{1}{a^{-m/n}} = a^{m/n}, \quad a \neq 0$$

EXAMPLE 4 Simplify each expression and verify your results with the graphing calculator.

a. $64^{-1/2}$ **b.** $16^{-3/2}$ **c.** $(-27)^{-2/3}$ **d.** $(-16)^{-3/4}$

Solution **a.** $64^{-1/2} = \dfrac{1}{64^{1/2}} = \dfrac{1}{\sqrt{64}} = \dfrac{1}{8}$

```
64^(-1/2)►Frac
             1/8
```

b. $16^{-3/2} = \dfrac{1}{16^{3/2}} = \dfrac{1}{\left(\sqrt{16}\right)^3} = \dfrac{1}{4^3} = \dfrac{1}{64}$

```
16^(-3/2)►Frac
             1/64
```

c. $(-27)^{-2/3} = \dfrac{1}{(-27)^{2/3}} = \dfrac{1}{\left(\sqrt[3]{-27}\right)^2} = \dfrac{1}{(-3)^2} = \dfrac{1}{9}$

```
(-27)^(-2/3)►Fra
c
             1/9
```

d. $(-16)^{-3/4} = \dfrac{1}{(-16)^{3/4}} = \dfrac{1}{\left(\sqrt[4]{-16}\right)^3}$

Because the *index* is even and the *radicand* is negative, the expression is undefined in the real-number system. The calculator screen confirms our conclusion:

```
ERR:NONREAL ANS
1■Quit
2:Goto
```

■ PRODUCT AND QUOTIENT RULES FOR RADICALS

The radical expressions studied thus far have been either perfect roots, which simplify to **rational numbers,** or those that are **irrational numbers,** for which decimal approximations may be given.

We now turn our attention to simplifying radicals that do not equal rational numbers. To write radicals in simplest radical form, we introduce both a product rule and a quotient rule for radicals. These are developed next.

Because $a^{1/n} \cdot b^{1/n} = (ab)^{1/n}$, we may convert from exponential form to radical form:

$$\sqrt[n]{a} \cdot \sqrt[n]{b} = \sqrt[n]{ab}$$

For example,

$$\sqrt{5} \cdot \sqrt{5} = \sqrt{5^2} = 5 \qquad \text{and} \qquad \sqrt[3]{7x} \cdot \sqrt[3]{49x^2} = \sqrt[3]{7x \cdot 7^2 x^2} = \sqrt[3]{7^3 x^3} = 7x.$$

This leads to the following rule:

Product Rule for Radicals
If $\sqrt[n]{a}$ and $\sqrt[n]{b}$ are real numbers, then $\sqrt[n]{a} \cdot \sqrt[n]{b} = \sqrt[n]{ab}$. Because equality is symmetric, this could also be written $\sqrt[n]{ab} = \sqrt[n]{a} \cdot \sqrt[n]{b}$.

This rule states that as long as all radicals represent real numbers, *the nth root of the product of two numbers is equal to the product of the nth roots of the numbers.*

WARNING! The product rule applies to the *n*th root of the *product* of two numbers. **There is no such property for sums or differences.** For example,

$$\sqrt{9 + 4} \neq \sqrt{9} + \sqrt{4} \qquad \text{and} \qquad \sqrt{9 - 4} \neq \sqrt{9} - \sqrt{4}$$
$$\sqrt{13} \neq 3 + 2 \qquad\qquad\qquad \sqrt{5} \neq 3 - 2$$
$$\sqrt{13} \neq 5 \qquad\qquad\qquad\qquad \sqrt{5} \neq 1$$

Thus, $\sqrt{a + b} \neq \sqrt{a} + \sqrt{b}$ and $\sqrt{a - b} \neq \sqrt{a} - \sqrt{b}$.

Another property of radicals involves quotients. Because $\dfrac{a^{1/n}}{b^{1/n}} = \left(\dfrac{a}{b}\right)^{1/n}$ it

follows that $\dfrac{\sqrt[n]{a}}{\sqrt[n]{b}} = \sqrt[n]{\dfrac{a}{b}}$, $(b \neq 0)$. For example, $\dfrac{\sqrt{8x^3}}{\sqrt{2x}} = \sqrt{\dfrac{8x^3}{2x}} = \sqrt{4x^2} = 2x$,

$(x > 0)$, and $\dfrac{\sqrt[3]{54x^5}}{\sqrt[3]{2x^2}} = \sqrt[3]{\dfrac{54x^5}{2x^2}} = \sqrt[3]{27x^3} = 3x$. This leads to the following rule:

Quotient Rule for Radicals
If $\sqrt[n]{a}$ and $\sqrt[n]{b}$ are real numbers, then $\sqrt[n]{\dfrac{a}{b}} = \dfrac{\sqrt[n]{a}}{\sqrt[n]{b}}$, $b \neq 0$. Because equality is

symmetric, this could also be written $\dfrac{\sqrt[n]{a}}{\sqrt[n]{b}} = \sqrt[n]{\dfrac{a}{b}}$, $b \neq 0$.

As long as radicals represent real numbers, *the nth root of the quotient of two numbers is equal to the quotient of their nth roots.*

■ SIMPLIFYING RADICAL EXPRESSIONS

A radical is said to be in simplest radical form when each of the following statements is true:

Simplified Form of a Radical Expression
A radical expression is in simplest form when
1. no radical appears in the denominator of a fraction.
2. the radicand contains no fractions or negative numbers.
3. each factor in the radicand appears to a power that is less than the index of the radical.

The third statement in the preceding box implies that no factors of the radicand can contain perfect powers. When a number is written in exponential form, perfect roots often become more apparent. Factoring numbers into primes enables us to identify perfect roots more easily and frees us from memorizing long lists of perfect squares, cubes, fourths, etc. For example, to simplify $\sqrt[3]{343x^3y^6}$, we factor the constant, 343, into primes and rewrite the radicand as follows: $\sqrt[3]{343x^3y^6} = \sqrt[3]{7^3x^3y^6}$. Applying our definition of rational exponents allows us to continue thusly: $\sqrt[3]{7^3x^3y^6} = 7^{3/3}x^{3/3}y^{6/3} = 7xy^2$.

To write radical expressions in simplest radical form, use the following steps:

Simplifying a Radical Expression
1. Factor the radicand into primes. Use exponential notation rather than writing repeated factors.
2. Use the product rule and laws of exponents to rewrite the radical as a product of two radicals with radicands as specified below:
   a. The first radicand is comprised of factors that are perfect roots. The exponents of the factors are evenly divisible by the index of the radical.
   b. The second radicand is comprised of factors that have powers *smaller* than the index of the radical.
3. Simplify the radical containing perfect roots by rewriting the radicand with rational exponents and reducing those exponents.
4. The final result is the product of a rational number and a radical with a radicand that contains no factors that are perfect roots.

EXAMPLE 5 Simplify each of the following. Verify results with your calculator.

   a. $\sqrt{12}$ **b.** $\sqrt{98}$ **c.** $\sqrt[3]{54}$

Solution
a. $\sqrt{12} = \sqrt{2^2 \cdot 3^1}$ Factor the radicand into primes.

$= \sqrt{2^2} \cdot \sqrt{3^1}$ Use the product rule to write the radical as a product of two radicals—one of which contains perfect powers.

$= 2^{2/2} \cdot \sqrt{3}$ Use the definition of rational exponents to express the radical containing perfect powers in exponential form.

$= 2\sqrt{3}$ Simplify the exponent, thus expressing the radical as the product of a rational number and a radical that contains no factors that are perfect roots.

This result is read as "2 square roots of 3" and is the product of 2 and $\sqrt{3}$. Verification that the original radical and our result are equal is shown by the calculator. Both are equivalent to the same irrational number.

```
√(12)
        3.464101615
2√(3)
        3.464101615
```

b. $\sqrt{98} = \sqrt{7^2 \cdot 2}$

$= \sqrt{7^2} \cdot \sqrt{2}$

$= 7^{2/2}\sqrt{2}$

$= 7\sqrt{2}$

Verification:

```
√(98)
          9.89494937
7√(2)
          9.89494937
```

c. $\sqrt[3]{54} = \sqrt[3]{3^3 \cdot 2}$

$= \sqrt[3]{3^3} \cdot \sqrt[3]{2}$

$= 3^{3/3} \cdot \sqrt[3]{2}$

$= 3\sqrt[3]{2}$

Verification:

```
³√(54)
          3.77976315
3³√(2)
          3.77976315
```

■

Many students eliminate the step of rewriting the perfect root in exponential form. However, be careful about taking too many shortcuts.

EXAMPLE 6   Simplify: $\sqrt{\dfrac{15}{49x^2}}$, $(x > 0)$.

Solution   The quotient rule allows us to rewrite the radical as a quotient of square roots.

$$\sqrt{\frac{15}{49x^2}} = \frac{\sqrt{15}}{\sqrt{49x^2}}$$

$$= \frac{\sqrt{5^1 \cdot 3^1}}{\sqrt{7^2 \cdot x^2}} \qquad \text{Factor radicands.}$$

$$= \frac{\sqrt{15}}{7^{2/2}x^{2/2}} \qquad \begin{array}{l}\text{The exponents of 1 on the factors of 5 and 3 assure us}\\\text{that 15 has no factors that are perfect roots. Therefore,}\\\text{we simply write the radicand as 15.}\end{array}$$

$$= \frac{\sqrt{15}}{7x}$$

■

EXAMPLE 7   Simplify each expression. Assume that all variables represent positive numbers.

b, c

a. $\sqrt{128a^5}$ **b.** $\sqrt[3]{24x^5}$

c. $\dfrac{\sqrt{45xy^2}}{\sqrt{5x}}$ **d.** $\dfrac{\sqrt[3]{-432x^5}}{\sqrt[3]{8x}}$

Solution   **a.** $\sqrt{128a^5} = \sqrt{2^7 \cdot a^5}$ Factor the radicand into primes.

$= \sqrt{2^6 \cdot a^4}\sqrt{2^1 \cdot a^1}$ Use the product rule and laws of exponents to rewrite the radical as the product of two radicals, one of which contains perfect roots and the other containing factors with exponents smaller than the index of the radical.

$= 2^{6/2}a^{4/2}\sqrt{2a}$ Rewrite the radical containing perfect roots with rational exponents, thus extracting the root.

$= 2^3a^2\sqrt{2a}$ Simplify exponents.

$= 8a^2\sqrt{2a}$ Raise constant bases to specified powers.

b. $\sqrt[3]{24x^5} = \sqrt[3]{2^3 \cdot 3x^5}$ 　　　Factor the radicand.

$= \sqrt[3]{2^3 x^3}\sqrt[3]{3x^2}$ 　　Apply the product rule and laws of exponents to rewrite the radical as the product of two radicals.

$= 2^{3/3}x^{3/3}\sqrt[3]{3x^2}$ 　　Rewrite the radical containing perfect roots with rational exponents.

$= 2x\sqrt[3]{3x^2}$ 　　Simplify powers.

c. $\dfrac{\sqrt{45xy^2}}{\sqrt{5x}} = \sqrt{\dfrac{45xy^2}{5x}}$ 　　Use the quotient rule to rewrite the quotient of square roots as the square root of a quotient.

$= \sqrt{9y^2}$ 　　Simplify the fraction.

$= \sqrt{3^2 y^2}$ 　　Factor the radicand into primes.

$= 3^{2/2}y^{2/2}$ 　　Rewrite the radical using rational exponents.

$= 3y$ 　　Simplify exponents.

Note: As you become more comfortable with finding roots, $\sqrt{9y^2}$ may be replaced with the equivalent expression, $3y$, without the intermediate steps.

d. $\dfrac{\sqrt[3]{-432x^5}}{\sqrt[3]{8x}} = \sqrt[3]{\dfrac{-432x^5}{8x}}$ 　　Use the quotient rule to rewrite the quotient of cube roots as the cube root of a quotient.

$= \sqrt[3]{-54x^4}$ 　　Simplify the fraction in the radicand.

$= \sqrt[3]{(-3)^3 \cdot 2x^4}$ 　　Factor the radicand.

$= \sqrt[3]{(-3)^3 x^3}\sqrt[3]{2^1 x^1}$ 　　Use the product rule to rewrite the radical as the product of two radicals.

$= (-3)^{3/3}x^{3/3}\sqrt[3]{2x}$ 　　Rewrite the radical containing perfect roots in exponential form.

$= -3x\sqrt[3]{2x}$ 　　Simplify exponents.

EXERCISE 9.2

VOCABULARY *In Exercises 1–4, fill in the blanks to make the statement true.*

1. Positive integer exponents indicate the number of times a _____ is to be used as a _____.

2. Any nonzero base raised to the zero power equals ____.

3. Radical expressions that simplify to perfect roots are _____ numbers.

4. If a root is not perfect, it is a(n) _____ number, and its root may be approximated.

CONCEPTS *In Exercises 5–10, express each exponential expression as a radical.*

5. $7^{1/3}$ (verify with your calculator that your two expressions are equal).

6. $26^{1/2}$ (verify with your calculator that your two expressions are equal).

7. $3x^{1/4}$

8. $4ab^{1/6}$

9. $(3x)^{1/4}$

10. $(4ab)^{1/6}$

In Exercises 11–16, express each radical as an exponential expression.

11. $\sqrt[4]{3a}$ 　$= 3a^{\frac{1}{4}}$

12. $3\sqrt[4]{a}$

13. $\sqrt[6]{\dfrac{1}{7}abc}$

14. $\sqrt[7]{\dfrac{3}{8}p^2 q}$ 　$\left(\dfrac{3}{8}p^2 q\right)^{\frac{1}{7}}$

15. $\sqrt[3]{a^2 - b^2}$

16. $\sqrt{x^2 + y^2}$

PRACTICE *In Exercises 17–36, simplify each expression, if possible. Verify results using the calculator, copying the screen display exactly as it appears. Negative exponents should not appear in results.*

17. $4^{1/2}$

18. $25^{1/2}$

19. $8^{1/3}$

20. $125^{1/3}$

21. $\left(\dfrac{1}{8}\right)^{1/3}$

22. $\left(\dfrac{1}{16}\right)^{1/4}$

23. $-16^{1/4}$

24. $(-16)^{1/4}$

25. $0^{1/3}$ **26.** $(a^{1/3} + b^{1/3})^0$ **27.** $81^{3/4}$ **28.** $100^{3/2}$

29. $\left(\dfrac{1}{8}\right)^{2/3}$ **30.** $\left(\dfrac{4}{9}\right)^{3/2}$ **31.** $4^{-1/2}$ **32.** $8^{-1/3}$

33. $4^{-3/2}$ **34.** $25^{-5/2}$ **35.** $\left(\dfrac{27}{8}\right)^{-4/3}$ **36.** $\left(\dfrac{25}{49}\right)^{-3/2}$

In Exercises 37–48, simplify each expression, if possible. Assume that all variables represent positive numbers.

37. $(25x^4)^{3/2}$ **38.** $(27a^3b^3)^{2/3}$ **39.** $\left(\dfrac{8x^3}{27}\right)^{2/3}$ **40.** $\left(\dfrac{27}{64y^6}\right)^{2/3}$

41. $(16x^2)^{-3/2}$ **42.** $(81c^4)^{-3/2}$ **43.** $(-27y^3)^{-2/3}$ **44.** $(-8z^9)^{-2/3}$

45. $(-32p^5)^{-2/5}$ **46.** $(16q^6)^{-5/2}$ **47.** $\left(-\dfrac{8x^3}{27}\right)^{-1/3}$ **48.** $\left(\dfrac{16}{81y^4}\right)^{-3/4}$

In Exercises 49–78, express each radical in simplest radical form. Assume that all variables represent positive numbers.

49. $\sqrt{6} \cdot \sqrt{6}$ **50.** $\sqrt{11} \cdot \sqrt{11}$ **51.** $\sqrt[3]{5x^2} \cdot \sqrt[3]{25x}$ **52.** $\sqrt[4]{25a}\,\sqrt[4]{25a^3}$

53. $\dfrac{\sqrt{180ab^4}}{\sqrt{5ab^2}}$ **54.** $\dfrac{\sqrt{112ab^3}}{\sqrt{7ab}}$ **55.** $\dfrac{\sqrt[3]{189a^4}}{\sqrt[3]{7a}}$ **56.** $\dfrac{\sqrt[3]{243x^7}}{\sqrt[3]{9x}}$

57. $\sqrt{20}$ **58.** $\sqrt{8}$ **59.** $-\sqrt{200}$ **60.** $-\sqrt{250}$

61. $\sqrt[3]{80}$ **62.** $\sqrt[3]{270}$ **63.** $\sqrt[3]{-81}$ **64.** $\sqrt[3]{-72}$

65. $\sqrt[5]{96}$ **66.** $\sqrt[7]{256}$ **67.** $\sqrt{\dfrac{7}{9}}$ **68.** $\sqrt{\dfrac{3}{4}}$

69. $\sqrt[5]{\dfrac{3}{32}}$ **70.** $\sqrt[6]{\dfrac{5}{64}}$ **71.** $\sqrt{175a^2b^3}$ **72.** $\sqrt{128a^3b^5}$

73. $\sqrt[4]{32x^{12}y^4}$ **74.** $\sqrt[5]{64x^{10}y^5}$ **75.** $\sqrt{\dfrac{z^2}{16x^2}}$ **76.** $\sqrt{\dfrac{b^4}{64a^8}}$

77. $\sqrt[4]{\dfrac{5x}{16z^4}}$ **78.** $\sqrt[3]{\dfrac{11a^2}{125b^6}}$

REVIEW *Simplify*

79. $(3x^2y)^{-2}$ **80.** $\dfrac{3a^2b^{-4}c^3}{6a^{-2}b^{-3}c^3}$ **81.** $-\left(\dfrac{2xy}{4x^2y^3}\right)^{-3}$

Evaluate each of the following (a) algebraically, and (b) by using the STOre feature of the graphing calculator.

82. $6x^2y^{-3}$; $x = -2$, $y = -3$ **83.** $\left(\dfrac{3x}{6y}\right)^{-2}$; $x = -1$, $y = 4$

9.3   Arithmetic Operations with Radicals

■ ADDING AND SUBTRACTING RADICAL EXPRESSIONS ■ MULTIPLYING RADICAL EXPRESSIONS
■ RATIONALIZING DENOMINATORS

■ ADDING AND SUBTRACTING RADICAL EXPRESSIONS

■ **STUDY TIP**
In the few minutes before class begins, review your notes, have your list of questions ready, and make sure you have all the tools necessary for learning with you—notebook, text, calculator, and pencil(s).

Radical expressions with the same *index* and *radicand* can be added and subtracted using the distributive property. To simplify the expression $3\sqrt{2} + 2\sqrt{2}$, we use the distributive property to factor out the common factor of $\sqrt{2}$ and then simplify:

$$3\sqrt{2} + 2\sqrt{2} = (3 + 2)\sqrt{2}$$
$$= 5\sqrt{2}$$

Oftentimes, radicals with the same index but different radicands can be combined once each radical is written in simplest radical form. For example, to simplify the expression $\sqrt{27} - \sqrt{12}$, we simplify each radical and then combine, if possible:

$$\begin{aligned} \sqrt{27} - \sqrt{12} &= \sqrt{3^2 \cdot 3} - \sqrt{2^2 \cdot 3} \\ &= 3\sqrt{3} - 2\sqrt{3} \\ &= (3 - 2)\sqrt{3} \\ &= \sqrt{3} \end{aligned}$$

These examples suggest the following rules for adding and subtracting radicals.

> **Adding and Subtracting Radical Expressions**
> 1. Simplify each radical.
> 2. Use the distributive property to add or subtract radicals that have the same index and radicand.

EXAMPLE 1   Simplify: $2\sqrt{12} - 3\sqrt{48} + 3\sqrt{3}$.

Solution   We simplify each radical, and then combine those that have radicands and indices that are the same.

$$\begin{aligned} 2\sqrt{12} - 3\sqrt{48} + 3\sqrt{3} &= 2\sqrt{2^2 \cdot 3} - 3\sqrt{2^4 \cdot 3} + 3\sqrt{3} \\ &= 2(2)\sqrt{3} - 3(4)\sqrt{3} + 3\sqrt{3} \\ &= 4\sqrt{3} - 12\sqrt{3} + 3\sqrt{3} \\ &= (4 - 12 + 3)\sqrt{3} \\ &= -5\sqrt{3} \end{aligned}$$

Therefore, $2\sqrt{12} - 3\sqrt{48} + 3\sqrt{3} = -5\sqrt{3}$. Our result can be confirmed with the calculator:

```
2√(12)-3√(48)+3√
(3)
          -8.660254038
-5√(3)
          -8.660254038
```

EXAMPLE 2   Simplify: $\sqrt[3]{16} - \sqrt[3]{54} + \sqrt[3]{24}$.

Solution   We simplify each radical separately and then combine those that have radicands and indices that are the same.

$$\begin{aligned} \sqrt[3]{16} - \sqrt[3]{54} + \sqrt[3]{24} &= \sqrt[3]{2^3 \cdot 2} - \sqrt[3]{3^3 \cdot 2} + \sqrt[3]{2^3 \cdot 3} \\ &= 2\sqrt[3]{2} - 3\sqrt[3]{2} + 2\sqrt[3]{3} \\ &= (2 - 3)\sqrt[3]{2} + 2\sqrt[3]{3} \\ &= -\sqrt[3]{2} + 2\sqrt[3]{3} \end{aligned}$$

Therefore, $\sqrt[3]{16} - \sqrt[3]{54} + \sqrt[3]{24} = -\sqrt[3]{2} + 2\sqrt[3]{3}$.
Our result can be confirmed with the calculator:

```
3√(16)-3√(54)+3√
(24)
          1.624578091
-3√(2)+23√(3)
          1.624578091
```

EXAMPLE 3   Simplify: $\sqrt[3]{16x^4} + \sqrt[3]{54x^4} - \sqrt[3]{-128x^4}$.

Solution   We simplify each radical separately and then combine those that have radicands and indices that are the same.

$$\sqrt[3]{16x^4} + \sqrt[3]{54x^4} - \sqrt[3]{-128x^4} = \sqrt[3]{2^3 x^3 \cdot 2x} + \sqrt[3]{3^3 x^3 \cdot 2x} - \sqrt[3]{(-4)^3 x^3 \cdot 2x}$$
$$= 2x\sqrt[3]{2x} + 3x\sqrt[3]{2x} + 4x\sqrt[3]{2x}$$
$$= (2x + 3x + 4x)\sqrt[3]{2x}$$
$$= 9x\sqrt[3]{2x}$$

■ MULTIPLYING RADICAL EXPRESSIONS

The product rule ensures that radical expressions with the same index can be multiplied. Because radical expressions represent real numbers, properties of real numbers are applicable to radical expressions. The commutative and associative properties along with the product rule are used when multiplying radical expressions. It is suggested that *multiplication* be performed *before* simplification.

> **Multiplication of Radical Expressions**
> 1. The product rule allows the multiplication of radical expressions that have the same index.
> 2. Simplify radical expressions *after* the operation of multiplication has been performed.

EXAMPLE 4   Multiply: $3\sqrt{6} \cdot 2\sqrt{3}$.

Solution   We use the commutative and associative properties of multiplication coupled with the product rule to multiply.

$$3\sqrt{6} \cdot 2\sqrt{3} = (3 \cdot 2)\left(\sqrt{6} \cdot \sqrt{3}\right) \qquad \text{Commutative and associative properties.}$$
$$= 6\sqrt{18} \qquad \text{Simplify and apply the product rule.}$$
$$= 6\sqrt{3^2 \cdot 2} \qquad \text{Simplify the radical.}$$
$$= 6 \cdot 3\sqrt{2} \qquad \text{Extract the perfect root of 3.}$$
$$= 18\sqrt{2} \qquad \text{Multiply the integers 6 and 3.}$$

Therefore, $3\sqrt{6} \cdot 2\sqrt{3} = 18\sqrt{2}$. Our result can be confirmed with the calculator:

```
3√(6)*2√(3)
          25.45584412
18√(2)
          25.45584412
```

EXAMPLE 5   Multiply: $3\sqrt{3}\left(4\sqrt{8} - 5\sqrt{10}\right)$.

Solution $3\sqrt{3}\left(4\sqrt{8} - 5\sqrt{10}\right)$

$$= 3\sqrt{3} \cdot 4\sqrt{8} - 3\sqrt{3} \cdot 5\sqrt{10} \qquad \text{Use the distributive property.}$$
$$= 12\sqrt{24} - 15\sqrt{30} \qquad \text{Use the commutative and associative properties and the product rule to perform the indicated multiplication.}$$

$$= 12\sqrt{2^2 \cdot 2 \cdot 3} - 15\sqrt{2 \cdot 5 \cdot 3} \qquad \text{Factor radicands to simplify radicals.}$$

$$= 12 \cdot 2\sqrt{6} - 15\sqrt{30}$$

Notice that the factorization is used to identify perfect roots. Once they are extracted and we are assured the radicands have no other factors that are perfect squares, we multiply the factors in the radicand.

$$= 24\sqrt{6} - 15\sqrt{30}$$

Therefore, $3\sqrt{3}\left(4\sqrt{8} - 5\sqrt{10}\right) = 24\sqrt{6} - 15\sqrt{30}$. Our result can be confirmed with the calculator:

```
3√(3)(4√(8)-5√(1
0)
            -23.3706298
24√(6)-15√(30)
            -23.3706298
```

EXAMPLE 6   Multiply: $\left(\sqrt{7} + \sqrt{2}\right)\left(\sqrt{7} - 3\sqrt{2}\right)$.

Solution   Distribute the multiplication (the mnemonic FOIL may be used). We then compute products, simplify radicals, and add and subtract as applicable.

$$\left(\sqrt{7} + \sqrt{2}\right)\left(\sqrt{7} - 3\sqrt{2}\right) = \sqrt{7}\sqrt{7} - \sqrt{7} \cdot 3\sqrt{2} + \sqrt{2} \cdot \sqrt{7} - \sqrt{2} \cdot 3\sqrt{2}$$
$$= \sqrt{7^2} - 3\sqrt{14} + \sqrt{14} - 3\sqrt{2^2}$$
$$= 7 - 2\sqrt{14} - 3 \cdot 2$$
$$= 1 - 2\sqrt{14}$$

Therefore, $\left(\sqrt{7} + \sqrt{2}\right)\left(\sqrt{7} - 3\sqrt{2}\right) = 1 - 2\sqrt{14}$. Our result can be confirmed with the calculator:

```
(√(7)+√(2))(√(7)
-3√(2))
            -6.483314774
1-2√(14)
            -6.483314774
```

■ RATIONALIZING DENOMINATORS

For a radical expression to be in simplified form, there cannot be a fraction under the radical or a radical in the denominator of a fraction. The process of removing a radical from the denominator of a fraction is called **rationalizing the denominator.**

To rationalize a denominator, multiply both the numerator and the denominator by a radical that will make the denominator a perfect root. Factoring radicands in the denominator into primes is often helpful in identifying the required factors.

EXAMPLE 7   Rationalize each denominator.

a. $\sqrt{\dfrac{20}{7}}$ **b.** $\dfrac{4}{\sqrt[3]{2}}$ **c.** $\dfrac{\sqrt[3]{5x}}{\sqrt[3]{18x^2}}$

Solution   **a.** $\sqrt{\dfrac{20}{7}} = \dfrac{\sqrt{20}}{\sqrt{7}}$ Quotient rule.

Because 7 is prime and we are looking for a square root, multiply the numerator and the denominator by $\sqrt{7}$.

$$= \frac{\sqrt{20}}{\sqrt{7}} \cdot \frac{\sqrt{7}}{\sqrt{7}}$$

$$= \frac{\sqrt{140}}{\sqrt{7^2}}$$ Product rule.

$$= \frac{\sqrt{140}}{7}$$

The denominator is rationalized. We now simplify the numerator.

$$\frac{\sqrt{140}}{7} = \frac{\sqrt{2^2 \cdot 5 \cdot 7}}{7}$$

$$= \frac{2\sqrt{5 \cdot 7}}{7}$$

$$= \frac{2\sqrt{35}}{7}$$

Therefore, $\sqrt{\dfrac{20}{7}} = \dfrac{2\sqrt{35}}{7}$. Our result can be verified with the calculator.

b. Since the denominator is a cube root and is the prime number 2, we will multiply the numerator and the denominator by powers of 2 so that a perfect cube will be under the radical sign.

$$\frac{4}{\sqrt[3]{2}} \cdot \frac{\sqrt[3]{2^2}}{\sqrt[3]{2^2}} = \frac{4\sqrt[3]{2^2}}{\sqrt[3]{2^3}} \qquad \text{Product rule.}$$

$$= \frac{4\sqrt[3]{4}}{2} \qquad \text{Extract the perfect root of 2 from the denominator.}$$

$$= 2\sqrt[3]{4} \qquad \text{Reduce by dividing out the common factor of 2.}$$

Therefore, $\dfrac{4}{\sqrt[3]{2}} = 2\sqrt[3]{4}$. Our result can be verified by the calculator.

```
4/³√(2)
        3.174802104
2³√(4)
        3.174802104
```

c. $\dfrac{\sqrt[3]{5x}}{\sqrt[3]{18x^2}} = \dfrac{\sqrt[3]{5x}}{\sqrt[3]{3^2 \cdot 2x^2}}$ Factor the radicand in the denominator to help identify the factors necessary to make the denominator a perfect cube.

From the factorization, it is apparent that we need one more factor of 3, two more factors of 2, and one more factor of x. The necessary radical is $\sqrt[3]{3 \cdot 2^2 x}$. We now multiply the numerator and the denominator by this radical.

$$\frac{\sqrt[3]{5x}}{\sqrt[3]{3^2 \cdot 2x^2}} \cdot \frac{\sqrt[3]{3 \cdot 2^2 x}}{\sqrt[3]{3 \cdot 2^2 x}} = \frac{\sqrt[3]{5 \cdot 3 \cdot 2^2 x^2}}{\sqrt[3]{3^3 \cdot 2^3 x^3}} \qquad \text{Product rule.}$$

$$= \frac{\sqrt[3]{60x^2}}{3 \cdot 2x} \qquad \text{The prime factors in the numerator assure us there are no factors of 60 that are perfect cubes. We therefore multiply the factors to get a result of 60. We also extract the perfect cubes from the denominator.}$$

$$= \frac{\sqrt[3]{60x^2}}{6x} \qquad \text{Multiply } 3 \cdot 2 \text{ to get 6.} \qquad \blacksquare$$

To rationalize binomial denominators containing square roots, we multiply the numerator and denominator by the **conjugate** of the binomial in the denominator. Conjugate binomials are binomials with opposite signs between the two terms.

Conjugate Binomials

The conjugate of the binomial $a + b$ is $a - b$, and the conjugate of $a - b$ is $a + b$.

The use of conjugate binomials in rationalizing denominators is illustrated in the next two examples.

EXAMPLE 8   Rationalize the denominator of $\dfrac{1}{\sqrt{2} + 1}$.

Solution   We multiply the numerator and denominator of the fraction by $\sqrt{2} - 1$, which is the conjugate of the denominator.

$$\frac{1}{\sqrt{2} + 1} = \frac{1(\sqrt{2} - 1)}{(\sqrt{2} + 1)(\sqrt{2} - 1)}$$

Multiply the numerator and the denominator by the conjugate of the denominator.

$$= \frac{\sqrt{2} - 1}{(\sqrt{2})^2 - 1}$$

$(\sqrt{2} + 1)(\sqrt{2} - 1) =$
$(\sqrt{2})^2 - \sqrt{2} + \sqrt{2} - 1 = (\sqrt{2})^2 - 1$

$$= \frac{\sqrt{2} - 1}{2 - 1}$$

$(\sqrt{2})^2 = 2.$

$$= \sqrt{2} - 1$$

$\dfrac{\sqrt{2} - 1}{2 - 1} = \dfrac{\sqrt{2} - 1}{1} = \sqrt{2} - 1.$

Our result can be verified with the calculator:

```
1/(√(2)+1)
         .4142135624
√(2)-1
         .4142135624
```

EXAMPLE 9   Rationalize the denominator of $\dfrac{\sqrt{x} + \sqrt{2}}{\sqrt{x} - \sqrt{2}}$.

Solution   We multiply the numerator and denominator by $\sqrt{x} + \sqrt{2}$, which is the conjugate of $\sqrt{x} - \sqrt{2}$, and simplify.

■ **STUDY TIP**
As you complete a section of exercises, go back to the previous sections and review each section's title and objectives. Look for connections between sections and for the flow of the mathematics.

$$\frac{\sqrt{x} + \sqrt{2}}{\sqrt{x} - \sqrt{2}} = \frac{(\sqrt{x} + \sqrt{2})(\sqrt{x} + \sqrt{2})}{(\sqrt{x} - \sqrt{2})(\sqrt{x} + \sqrt{2})}$$

$$= \frac{x + \sqrt{2x} + \sqrt{2x} + 2}{x - 2}$$

Use the FOIL method.

$$= \frac{x + 2\sqrt{2x} + 2}{x - 2}$$

EXERCISE 9.3

VOCABULARY   *Fill in the blanks to make the statements true.*

1. Radicals with the same _____ and _____ can be added (or subtracted) using the _____ property.

2. Radicals that have the same _____ can be multiplied.

3. The process of removing a radical from the denominator of a fraction is called _____ the denominator.

4. Binomials with opposite signs between the terms are called _____ binomials.

CONCEPTS *In Exercises 5–10, classify each of the following statements as true or false.*

5. $\sqrt{5} + \sqrt{5} = 2\sqrt{5}$

6. $5\sqrt[3]{6} + 4\sqrt[3]{6} = 9\sqrt[3]{12}$

7. $\sqrt{3} - 3\sqrt{5} = -2\sqrt{5}$

8. The product of $\sqrt[3]{3} \cdot \sqrt[4]{5} = \sqrt[3]{15}$

9. To rationalize the denominator of $\dfrac{2}{\sqrt[3]{5}}$, the numerator and denominator must be multiplied by $\sqrt[3]{25}$.

10. To rationalize the denominator of $\dfrac{3}{1 + \sqrt{3}}$, multiply the numerator and the denominator by $1 + \sqrt{3}$.

PRACTICE *In Exercises 11–37, add or subtract as the signs indicate. When possible, check results with the calculator, recording screen displays exactly as they appear.*

11. $4\sqrt{2x} + 6\sqrt{2x}$

12. $6\sqrt[3]{5y} + 3\sqrt[3]{5y}$

13. $8\sqrt[5]{7a^2} - 7\sqrt[5]{7a^2}$

14. $10\sqrt[6]{12xyz} - \sqrt[6]{12xyz}$

15. $\sqrt{2} - \sqrt{8}$

16. $\sqrt{20} - \sqrt{125}$

17. $3\sqrt{24} + \sqrt{54}$

18. $\sqrt{18} + 2\sqrt{50}$

19. $\sqrt[3]{32} - \sqrt[3]{108}$

20. $\sqrt[3]{80} - \sqrt[3]{10,000}$

21. $\sqrt[4]{32} + \sqrt[4]{162}$

22. $14\sqrt[4]{32} - 15\sqrt[4]{162}$

23. $23\sqrt[4]{768} + \sqrt[4]{48}$

24. $\sqrt{18} + \sqrt{300} - \sqrt{243}$

25. $\sqrt{80} - \sqrt{128} + \sqrt{288}$

26. $\sqrt{25y^2z} - \sqrt{16y^2z}$

27. $\sqrt{25yz^2} + \sqrt{9yz^2}$

28. $\sqrt{36xy^2} + \sqrt{49xy^2}$

29. $3\sqrt{2x} - \sqrt{8x}$

30. $2\sqrt[3]{64a} + 2\sqrt[3]{8a}$

31. $3\sqrt[4]{x^4y} - 2\sqrt[4]{x^4y}$

32. $\sqrt{y^5} - \sqrt{9y^5} - \sqrt{25y^5}$

33. $\sqrt{8y^7} + \sqrt{32y^7} - \sqrt{2y^7}$

34. $\sqrt[5]{x^6y^2} + \sqrt[5]{32x^6y^2} + \sqrt[5]{x^6y^2}$

35. $\sqrt[3]{xy^4} + \sqrt[3]{8xy^4} - \sqrt[3]{27xy^4}$

36. $\sqrt{x^2 + 2x + 1} + \sqrt{x^2 + 2x + 1}$

37. $\sqrt{4x^2 + 12x + 9} + \sqrt{9x^2 + 6x + 1}$

In Exercises 38–67, multiply and simplify. When possible, check results with the calculator, recording screen displays exactly as they appear.

38. $\sqrt{2}\sqrt{8}$

39. $\sqrt{3}\sqrt{27}$

40. $(3\sqrt[3]{9})(2\sqrt[3]{3})$

41. $(2\sqrt[3]{16})(-\sqrt[3]{4})$

42. $\sqrt{ab^3}\sqrt{ab}$

43. $\sqrt{8x}\sqrt{2x^3y}$

44. $\sqrt{5ab}\sqrt{5a}$

45. $\sqrt{15rs^2}\sqrt{10r}$

46. $\sqrt[3]{5r^2s}\sqrt[3]{2r}$

47. $\sqrt[3]{3xy^2}\sqrt[3]{9x^3}$

48. $\sqrt{x(x+3)}\sqrt{x^3(x+3)}$

49. $\sqrt{y^2(x+y)}\sqrt{(x+y)^3}$

50. $3\sqrt{5}(4 - \sqrt{5})$

51. $2\sqrt{7}(3\sqrt{7} - 1)$

52. $3\sqrt{2}(4\sqrt{3} + 2\sqrt{7})$

53. $-\sqrt{3}(\sqrt{7} - \sqrt{5})$

54. $-2\sqrt{5x}(4\sqrt{2x} - 3\sqrt{3})$

55. $3\sqrt{7t}(2\sqrt{7t} + 3\sqrt{3t^2})$

56. $(\sqrt{2} + 1)(\sqrt{2} - 3)$

57. $(2\sqrt{3} + 1)(\sqrt{3} - 1)$

58. $(4\sqrt{x} + 3)(2\sqrt{x} - 5)$

59. $(7\sqrt{y} + 2)(3\sqrt{y} - 5)$

60. $(\sqrt{3x} - \sqrt{2y})(\sqrt{3x} + \sqrt{2y})$

61. $(\sqrt{3m} + \sqrt{2n})(\sqrt{3m} + \sqrt{2n})$

62. $(2\sqrt{3a} - \sqrt{b})(\sqrt{3a} + 3\sqrt{b})$

63. $(5\sqrt{p} - \sqrt{3q})(\sqrt{p} + 2\sqrt{3q})$

64. $(3\sqrt{2r} - 2)^2$

65. $(2\sqrt{3t} + 5)^2$

66. $-2(\sqrt{3x} + \sqrt{3})^2$

67. $3(\sqrt{5x} - \sqrt{3})^2$

In Exercises 68–93, rationalize each denominator. Verify results with the calculator when applicable.

68. $\sqrt{\dfrac{1}{7}}$

69. $\sqrt{\dfrac{5}{3}}$

70. $\dfrac{\sqrt{5}}{\sqrt{8}}$

71. $\dfrac{\sqrt{3}}{\sqrt{50}}$

72. $\dfrac{1}{\sqrt[3]{2}}$ **73.** $\dfrac{2}{\sqrt[3]{6}}$ **74.** $\dfrac{\sqrt[3]{2}}{\sqrt[3]{9}}$ **75.** $\dfrac{\sqrt[3]{9}}{\sqrt[3]{54}}$

76. $\dfrac{\sqrt{8x^2y}}{\sqrt{xy}}$ **77.** $\dfrac{\sqrt{9xy}}{\sqrt{3x^2y}}$ **78.** $\dfrac{\sqrt{10xy^2}}{\sqrt{2xy^3}}$ **79.** $\dfrac{\sqrt{5ab^2c}}{\sqrt{10abc}}$

80. $\dfrac{1}{\sqrt[4]{4}}$ **81.** $\dfrac{1}{\sqrt[5]{2}}$ **82.** $\dfrac{1}{\sqrt[5]{16}}$ **83.** $\dfrac{4}{\sqrt[4]{32}}$

84. $\dfrac{1}{\sqrt{2}-1}$ **85.** $\dfrac{3}{\sqrt{3}-1}$ **86.** $\dfrac{\sqrt{3}+1}{\sqrt{3}-1}$ **87.** $\dfrac{\sqrt{2}-1}{\sqrt{2}+1}$

88. $\dfrac{\sqrt{7}-\sqrt{2}}{\sqrt{2}+\sqrt{7}}$ **89.** $\dfrac{\sqrt{3}+\sqrt{2}}{\sqrt{3}-\sqrt{2}}$ **90.** $\dfrac{2}{\sqrt{x}+1}$ **91.** $\dfrac{3}{\sqrt{x}-2}$

92. $\dfrac{\sqrt{x}-\sqrt{y}}{\sqrt{x}+\sqrt{y}}$ **93.** $\dfrac{\sqrt{x}+\sqrt{y}}{\sqrt{x}-\sqrt{y}}$

REVIEW   *In Exercises 94–96, solve each equation (a) algebraically and (b) graphically (copy the screen display and specify WINDOW values that are different from Zstandard).*

94. $3(x + 4) - 2x = 5x - 6$ **95.** $x^2 + 5x = 6$ **96.** $|x + 4| = 6$

In Exercises 97–99, solve each inequality. Specify the solution (a) as a number line graph and (b) using interval notation.

97. $2x - 1 > 9$ **98.** $-8 \le -4x \le 20$ **99.** $x + 1 < 0$ or $2x - 6 > 0$

9.4   Radical Equations and Their Applications

■ THE POWER RULE ■ EQUATIONS CONTAINING RADICALS ■ THE PYTHAGOREAN THEOREM
■ THE DISTANCE FORMULA

■ THE POWER RULE

To solve equations that contain radicals we will use the **power rule.**

> **Power Rule**
> If x, y, and n are real numbers, and $x = y$, then $x^n = y^n$.

If we raise both sides of an equation to the same power, the resulting equation is true. However, the resulting equation *may* or *may not* be equivalent to the original equation. This is illustrated below.

1. Consider the equation $x = 3$, which has a solution set of $\{3\}$. If we square both sides of the equation, we obtain

2. $x^2 = 9$, which has a solution set of $\{3, -3\}$.

The equations in 1 and 2 above are not equivalent because they have different solution sets. All we are guaranteed when raising both sides of an equation to a power is that all solutions of the original equation are solutions of the new equation. However, the new equation may have extra, extraneous roots. For this reason, roots *must* be checked in the original equation.

■ EQUATIONS CONTAINING RADICALS

A **radical equation** is an equation containing a radical with a variable in the radicand. Example 1 illustrates the solution of a radical equation.

EXAMPLE 1   Solve the equation $\sqrt{x + 3} = 4$. Check the analytic solution by solving graphically.

Solution   To eliminate the radical, apply the power rule by squaring both *sides* of the equation.

$$\left(\sqrt{x + 3}\right)^2 = 4^2 \qquad \text{Square both sides of the equation (apply the power rule).}$$

$$x + 3 = 16 \qquad \text{Subtract 3 from both sides.}$$

$$x = 13$$

Check the analytic solution by using the INTERSECT feature of the calculator. ■

■ TECHNOLOGY TIP
Recall that to use the INTERSECT feature for solving equations, enter the left side of the equation at the y1 prompt and the right side at the y2 prompt. Access the INTERSECT feature from the calc menu and respond to the prompts.

To algebraically solve a radical equation, follow these steps:

> **Solving Radical Equations**
> 1. Isolate one radical on one side of the equation.
> 2. Raise both *sides* of the equation to the same power as the index of the radical.
> 3. Simplify both sides of the equation.
> 4. **a.** If there are no radicals left in the equation, solve.
>   **b.** If the equation still contains radicals, repeat steps 1–3.
> 5. Check all proposed solutions and reject any extraneous roots.

EXAMPLE 2   Solve the equation $\sqrt{3x + 1} + 1 = x$.

Solution

$$\sqrt{3x + 1} + 1 = x$$

$$\sqrt{3x + 1} = x - 1 \qquad \text{Subtract 1 from both sides of the equation to isolate the radical.}$$

$$\left(\sqrt{3x + 1}\right)^2 = (x - 1)^2 \qquad \text{Raise both sides of the equation to the same power as the index of the radical to eliminate the radical.}$$

$$3x + 1 = x^2 - 2x + 1 \qquad \text{Recall that } (x - 1)^2 = (x - 1)(x - 1) \text{ and the distributive property must be applied.}$$

$$0 = x^2 - 5x \qquad \text{Subtract } 3x \text{ and 1 from both sides because the equation is quadratic and must be solved by factoring.}$$

$$0 = x(x - 5) \qquad \text{Factor the polynomial } x^2 - 5x.$$

$$x = 0 \quad \text{or} \quad x - 5 = 0 \qquad \text{Set each factor equal to zero and solve for the variable.}$$

$$x = 5$$

The equation *appears* to have the two solutions of 0 and 5. A graphical check will confirm one, both, or neither of the solutions.

The graphical display confirms the root of 5 and also shows that 0 *is not* a root. Therefore, the solution is simply 5. We suggest that you confirm the solution of 5 analytically *and* confirm that 0 is *not* a solution by the same method (substitution).

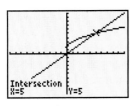

■

EXAMPLE 3   Solve the equation $\sqrt[3]{x^3 + 7} = x + 1$. Check by solving the equation graphically.

Solution   To eliminate the radical, we cube both *sides* of the equation.

$$\sqrt[3]{x^3 + 7} = x + 1$$

$$\left(\sqrt[3]{x^3 + 7}\right)^3 = (x + 1)^3 \qquad \text{Cube both sides to eliminate the radical.}$$

$$x^3 + 7 = x^3 + 3x^2 + 3x + 1 \qquad \text{Remember, } (x + 1)^3 = (x + 1)(x + 1)(x + 1).$$

$$0 = 3x^2 + 3x - 6 \qquad \text{Subtract } x^3 \text{ and 7 from both sides of the equation.}$$

$$0 = x^2 + x - 2 \qquad \text{Divide both sides by 3 to make factoring easier.}$$

$$0 = (x + 2)(x - 1) \qquad \text{Factor the trinomial.}$$

$$x + 2 = 0 \qquad \text{or} \qquad x - 1 = 0 \qquad \text{Set each factor equal to zero and solve.}$$

$$x = -2 \qquad\qquad\qquad x = 1$$

The solutions appear to be -2 and 1. A graphical solution will confirm one, both, or neither of the solutions.

The INTERSECT feature confirms both solutions of -2 and 1. Remember, move the cursor close to the desired point of intersection. The graphs displayed are in the window $[-4, 4]$ by $[-4, 4]$. ∎

     When equations contain two or more radicals, the algebra involved can become rather lengthy. Be extra careful when solving these equations algebraically.

EXAMPLE 4   Solve the equation $\sqrt{x} + \sqrt{x + 2} = 2$.

Solution   To remove the radicals, we must square both sides of the equation. This is easier to do if one radical is on each side of the equation. We begin by subtracting $\sqrt{x}$ from both sides to isolate a radical.

$$\sqrt{x} + \sqrt{x + 2} = 2$$

$$\sqrt{x + 2} = 2 - \sqrt{x} \qquad \text{Subtract } \sqrt{x} \text{ from both sides.}$$

$$\left(\sqrt{x + 2}\right)^2 = \left(2 - \sqrt{x}\right)^2 \qquad \text{Square both sides to eliminate one square root.}$$

$$x + 2 = 4 - 4\sqrt{x} + x \qquad \left(2 - \sqrt{x}\right)\left(2 - \sqrt{x}\right) = 4 - 2\sqrt{x} - 2\sqrt{x} + x = 4 - 4\sqrt{x} + x.$$

$$2 = 4 - 4\sqrt{x} \qquad \text{Subtract } x \text{ from both sides.}$$

$$-2 = -4\sqrt{x} \qquad \text{Subtract 4 from both sides.}$$

$$\frac{1}{2} = \sqrt{x} \qquad \text{Divide both sides by } -4.$$

$$\frac{1}{4} = x \qquad \text{Square both sides.}$$

The graphical display and use of the INTERSECT feature of the calculator confirm the analytic solution of $\frac{1}{4}$. ∎

EXAMPLE 5   Solve the equation $\sqrt{x + 2} + \sqrt{2x} = \sqrt{18 - x}$.

Solution   Because the right side of the equation contains a single radical, begin by squaring both sides. We then proceed as follows:

■ **TECHNOLOGY TIP**
The STOre feature, VARS capability, and TABLE features of the calculator can all be used to confirm solutions to equations. Experiment with your calculator and decide which method(s) you prefer.

$$\sqrt{x + 2} + \sqrt{2x} = \sqrt{18 - x}$$
$$\left(\sqrt{x + 2} + \sqrt{2x}\right)^2 = \left(\sqrt{18 - x}\right)^2 \qquad \text{Square both sides to eliminate one square root.}$$

$$x + 2 + 2\sqrt{x + 2}\sqrt{2x} + 2x = 18 - x$$
$$2\sqrt{x + 2}\sqrt{2x} = 16 - 4x \qquad \text{Subtract } 3x \text{ and 2 from both sides.}$$

$$\sqrt{x + 2}\sqrt{2x} = 8 - 2x \qquad \text{Divide both sides by 2.}$$
$$\left(\sqrt{x + 2}\sqrt{2x}\right)^2 = (8 - 2x)^2 \qquad \text{Square both sides to eliminate the other square roots.}$$

$$(x + 2)2x = 64 - 32x + 4x^2$$
$$2x^2 + 4x = 64 - 32x + 4x^2$$
$$0 = 2x^2 - 36x + 64 \qquad \text{Write the equation in standard quadratic form.}$$

$$0 = x^2 - 18x + 32 \qquad \text{Divide both sides by 2.}$$
$$0 = (x - 16)(x - 2) \qquad \text{Factor the trinomial.}$$
$$x - 16 = 0 \qquad \text{or} \qquad x - 2 = 0 \qquad \text{Set each factor equal to 0.}$$
$$x = 16 \qquad | \qquad x = 2$$

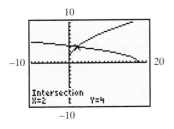

The INTERSECT feature of the calculator confirms the solution of 2.   ■

■ **THE PYTHAGOREAN THEOREM**

A **right triangle** is a triangle that contains a right (90°) angle. The two sides that form the right angle are called **legs** of the triangle. The side opposite the right angle is called the **hypotenuse** of the triangle.

Pythagorean Theorem
In any right triangle, the square of the length of the hypotenuse is equal to the sum of the squares of the lengths of the legs.

Symbolically, if a and b represent the legs of a right triangle and c is the hypotenuse, $a^2 + b^2 = c^2$.

 WARNING!   Many students simply remember "$a^2 + b^2 = c^2$," disregarding the critical fact that a and b represent the lengths of the LEGS of the triangle and c is the length of the hypotenuse. Remember the *words* of the Pythagorean theorem: $(\text{leg})^2 + (\text{leg})^2 = (\text{hypotenuse})^2$.

EXAMPLE 6   **Fighting fires**   To fight a forest fire, the forestry department plans to clear a rectangular fire break around the fire, as shown in Figure 9-1. Crews are equipped with mobile communications with a 3000-yard range. Can crews at points A and B remain in radio contact?

EXERCISE 9.4

VOCABULARY *Fill in each blank to make a true statement.*

1. A radical equation is an equation that contains a
_____ with a variable in the _____.

2. A _____ triangle is one that contains a _____ angle.

3. The legs of a right triangle are the sides that form the
_____ angle.

4. The _____ is the longest side of a right triangle
and is opposite the _____ angle.

5. The Pythagorean theorem states that, in a right triangle,
the sum of the squares of the lengths of the ____ is
equal to the square of the length of the _____.

6. The distance formula can be used to find the distance
between two points in the _____ coordinate
system.

CONCEPTS

7. To isolate the variable x, match each equation with the
necessary process.
   a. $x^2 = 4$
   b. $\sqrt{x} = 4$
   c. $x + 5 = 4$
   d. $x - 5 = 4$
   e. $5x = 4$
   f. $\dfrac{x}{5} = 4$

   i. Divide each side of the equation by 5.
   ii. Add 5 to both sides of the equation.
   iii. Extract the square root of both sides of the equation.
   iv. Square both sides of the equation.
   v. Subtract 5 from both sides of the equation.
   vi. Multiply both sides of the equation by 5.

8. When solving radical equations, we raise both sides of
the equation to the same power as the index of the
radical. Consider the equation $\sqrt{2x - 3} = x - 9$.
Which of the following is the square of the right *side*
of the equation?
   a. $x^2 - 18$
   b. $x^2 - 81$
   c. $x^2 - 9x + 81$
   d. $x^2 - 18x + 81$

PRACTICE *In Exercises 9–34, solve each equation algebraically. Check results by solving graphically.*

9. $\sqrt{6x + 1} + 2 = 7$

10. $\sqrt{6x + 13} - 2 = 5$

11. $\sqrt[3]{7n - 1} = 3$

12. $\sqrt[3]{12m + 4} = 4$

13. $\sqrt[4]{10p + 1} = \sqrt[4]{11p - 7}$

14. $\sqrt[4]{10x + 2} = 2\sqrt[4]{2}$

15. $r - 9 = (2r - 3)^{1/2}$

16. $-x - 3 = 2(5 - x)^{1/2}$

17. $(-5x + 24)^{1/2} = 6 - x$

18. $(-x + 2)^{1/2} = x - 2$

19. $\sqrt{x}\sqrt{x + 16} = 15$

20. $\sqrt{x}\sqrt{x + 6} = 4$

21. $(x^3 - 7)^{1/3} = x - 1$

22. $(x^3 + 56)^{1/3} - 2 = x$

23. $\sqrt{2y + 1} = 1 - 2\sqrt{y}$

24. $\sqrt{u} + 3 = \sqrt{u - 3}$

25. $\sqrt{y + 7} + 3 = \sqrt{y + 4}$

26. $1 + \sqrt{z} = \sqrt{z + 3}$

27. $\sqrt{6x + 1} - 3\sqrt{x} = -1$

28. $\sqrt{4x + 1} - \sqrt{6x} = -1$

29. $\sqrt{x - 1} + \sqrt{x + 2} = 3$

30. $\sqrt{16x + 1} + \sqrt{8x + 1} = 12$

31. $\sqrt{\sqrt{x} + \sqrt{x + 8}} = 2$

32. $\sqrt{\sqrt{2y} - \sqrt{y - 1}} = 1$

33. $\dfrac{6}{\sqrt{x + 5}} = \sqrt{x}$

34. $\dfrac{\sqrt{2x}}{\sqrt{x + 2}} = \sqrt{x - 1}$

Solve the equations in Exercises 35–38 graphically. Copy your graphical display to justify your result.

35. $\sqrt{x} + \sqrt{3x + 2} + 4 = \sqrt{2x - 5} - \sqrt{5x - 1}$

36. $\sqrt{x + 4} + \sqrt{x + 2} + 4 = \sqrt{2x} - \sqrt{5x - 1} + 2$

37. $\sqrt{x} + \sqrt{4x - 4} = \sqrt{6x - 8} - \sqrt{2x - 4}$

38. $\sqrt{2x - 9} + 3 - \sqrt{x} = \sqrt{5x + 4} - \sqrt{3x - 11}$

39. Consider the two equations, $\sqrt{2x} = 5$ and $\sqrt[3]{2x} = 5$. Explain the similarities and differences in the solving processes for these two equations.

In Exercises 40–43, the lengths of two sides of the right triangle ABC shown in Illustration 1 are given. Find the length of the missing side.

40. $a = 6$ ft and $b = 8$ ft

41. $a = 10$ cm and $c = 26$ cm

42. $b = 18$ m and $c = 82$ m

43. $a = 14$ in. and $c = 50$ in.

ILLUSTRATION 1

In Exercises 44–49, find the distance between P and Q.

44. $P(-2, -8)$, $Q(3, 4)$

45. $P(-5, -2)$, $Q(7, 3)$

46. $P(6, 8)$, $Q(12, 16)$

47. $P(10, 4)$, $Q(2, -2)$

48. $Q(-3, 5)$, $P(-5, -5)$

49. $Q(2, -3)$, $P(4, -8)$

50. Geometry Show that a triangle with vertices at $(-2, 4)$, $(2, 8)$, and $(6, 4)$ is isosceles.

51. Geometry The square in Illustration 2 has an area of 18 square units, and its diagonals lie on the x- and y-axes. Find the coordinates of each corner of the square.

52. Geometry Show that a triangle with vertices at $(2, 3)$, $(-3, 4)$, and $(1, -2)$ is a right triangle. (*Hint:* If the Pythagorean relation holds, then the triangle is a right triangle.)

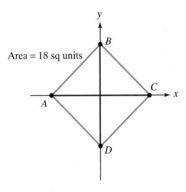

ILLUSTRATION 2

APPLICATIONS

53. Sailing Refer to the sailboat in Illustration 3. How long must a rope be to fasten the top of the mast to the bow?

54. Carpentry The gable end of the roof shown in Illustration 4 is divided in half by a vertical brace. Find the distance from eaves to peak.

ILLUSTRATION 4

ILLUSTRATION 3

In Exercises 55–58, the baseball diamond shown in Illustration 5 is a square, 90 feet on a side.

55. Baseball   How far must a catcher throw the ball to throw out a runner stealing second base?

56. Baseball   In baseball, the pitcher's mound is 60 feet, 6 inches from home plate. How far from the mound is second base?

57. Baseball   If the third baseman fields a ground ball 10 feet directly behind third base, how far must he throw the ball to throw the batter out at first base?

58. Baseball   A shortstop fields a grounder at a point one-third of the way from second base to third base. How far will he have to throw the ball to make an out at first base?

59. Packing a tennis racquet   The diagonal d of a rectangular box with dimensions $a \times b \times c$ is given by $d = \sqrt{a^2 + b^2 + c^2}$. Will the tennis racquet pictured in Illustration 6 fit into the box?

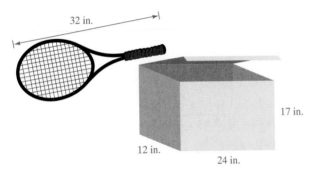

ILLUSTRATION 6

60. Shipping packages   A delivery service won't accept a package for shipping if any dimension exceeds 21 inches. An archeologist wants to ship a 36-inch femur. Will it fit in a 3-inch-tall box that has a 21-inch-square base?

61. Shipping packages   Can the archeologist in Exercise 60 ship the femur in a cubical box 21 inches on an edge?

62. Reach of a ladder   The base of the 37-foot ladder in Illustration 7 is 9 feet from the wall. Will the top reach a window ledge that is 35 feet above the ground?

ILLUSTRATION 7

ILLUSTRATION 5

63. Telephone service   The telephone cable in Illustration 8 runs from A to B to C to D. How much cable is required to run from A to D directly?

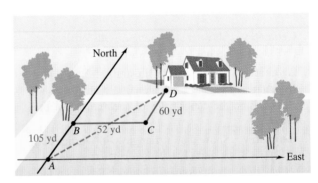

ILLUSTRATION 8

64. Electric service   The power company routes its lines as in Illustration 9. How much wire could be saved by going directly from A to E?

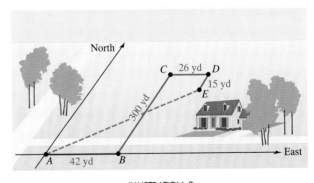

ILLUSTRATION 9

65. Supporting a weight   A weight placed on the tight wire in Illustration 10 pulls the center down 1 foot. By how much is the wire stretched? Round the answer to the nearest hundredth of a foot.

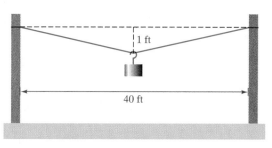

ILLUSTRATION 10

66. Geometry The side, *s*, of a square with area *A* square feet is given by the formula $s = \sqrt{A}$. Find the perimeter of a square with an area of 49 square feet.

67. Volume of a cube The total surface area, *A*, of a cube is related to its volume, *V*, by the formula $A = 6\sqrt[3]{V^2}$. Find the volume of a cube with a surface area of 24 square centimeters.

68. Area of many cubes A grain of table salt is a cube with a volume of approximately 6×10^{-6} cubic inches, and there are about 1.5 million grains of salt in one cup. Find the total surface area of the salt in one cup. (see Exercise 67).

REVIEW In Exercises 69–72, determine if each of the graphs is a function or strictly a relation.

69.   **70.**   **71.**   **72.**

9.5   Radical Functions

■ FUNCTION NOTATION ■ DOMAIN AND RANGE ■ INTERPRETING GRAPHS OF RADICAL FUNCTIONS

■ FUNCTION NOTATION

Recall that a **relation** is a correspondence that assigns to each number *x* one or more values, *y*. A **function** is a relation that assigns to each value of *x* EXACTLY one value, *y*.

When we state that "*y* is a function of *x*," we are asserting that each *x*-value in the domain is assigned to a UNIQUE *y*-value in the range. The notation used to specify that "*y* is a function of *x*" is $f(x)$. This notation is useful for evaluation. For example, if $f(x) = \sqrt{x+2}$, then $f(2) = \sqrt{2+2} = \sqrt{4} = 2$.

EXAMPLE 1   If $f(x) = \sqrt{x+4}$, find each of the following: **a.** $f(0)$ **b.** $f(5)$ **c.** $f(7)$

Solution   **a.** Because $f(x) = \sqrt{x+4}$, to find the value of the function when $x = 0$, $f(0)$, we substitute 0 in for *x* and simplify.

$$f(x) = \sqrt{x+4}$$
$$f(0) = \sqrt{0+4}$$
$$f(0) = \sqrt{4}$$
$$f(0) = 2$$

This can be confirmed graphically using the value feature of the calculator.

b. Because $f(x) = \sqrt{x + 4}$, to find the value of the function when $x = 5$, $f(5)$, we substitute 5 in for x and simplify.

$$f(x) = \sqrt{x + 4}$$
$$f(5) = \sqrt{5 + 4}$$
$$f(5) = \sqrt{9}$$
$$f(5) = 3$$

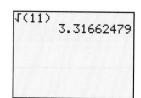

This can be confirmed graphically using the value feature of the calculator.

c. Because $f(x) = \sqrt{x + 4}$, to find the value of the function when $x = 7$, $f(7)$, we substitute 7 in for x and simplify.

$$f(x) = \sqrt{x + 4}$$
$$f(7) = \sqrt{7 + 4}$$
$$f(7) = \sqrt{11}$$

The radical expression $\sqrt{11}$ is an exact result. However, its value can be approximated. This can be done from the home screen, as illustrated below.

Use of the value option and the graphical display also allows us an approximation.

The notation $y = f(x)$ gives two different notations for the same quantity. Therefore, $y = \sqrt[3]{x + 3} - 2$ is the same statement as $f(x) = \sqrt[3]{x + 3} - 2$. Either notation specifies a pairing of two values using a definition, or rule.

EXAMPLE 2 If $f(x) = \sqrt[3]{x + 3} - 2$, find $f(-4)$, $f(-3)$, $f(-2)$, $f(5)$, and $f(7)$.

Solution Graphing the function in the window Zdecimal $\times$ 2, $[-9.4, 9.4]$ by $[-6.2, 6.2]$, and tracing, we note the following:

■ **TECHNOLOGY TIP**
On the Zdecimal window, multiply each value (except scale) by any constant factor, n, to increase the size of the screen. The cursor moves will be in units of $0.1n$.

$f(-4) = y = -3$

$f(-3) = y = -2$

$f(-2) = y = -1$

$f(5) = y = 0$

$f(7) = y \approx 0.154$

■ DOMAIN AND RANGE

The domain of a function is the set of all possible values for the independent variable (x). The range is the set of all possible values for the dependent variable (y). Three tools that help determine the domain and range are the definition of the function, its graphical display, and the table of values. Experiment with all three to ensure success in determining the domain and the range of a function.

EXAMPLE 3   Determine the domain and the range of the function $y = \sqrt{x} + 2$.

Solution

The graph appears in the first quadrant. From the display, we first make a guess (an *educated* guess) that the domain is $x \geq 0$ and $y \geq 0$ because both x and y are positive in the first quadrant.

The TABLE display supports this guess, in a limited way; however, it should be noted that both the graph and the TABLE information are limited because of the display capabilities of the graphing calculator.

Examining the function, $f(x) = \sqrt{x} + 2$, we note that x *must* be 0 or positive to be defined in the real-number system. Therefore, our domain, $x \geq 0$, appears to be correct.

Scrolling down the TABLE confirms the fact that each x-value in the domain is paired with a single y-value in the range. These y-values increase as x increases. However, scrolling *up* the TABLE (looking at x-values smaller than 0), we see that an ERROR message is consistently displayed for these negative x-values. Again, this serves as confirmation of our domain. However, scrolling in this direction also reveals that the *smallest* y-value displayed is 2. We therefore modify our original conjecture of a range ($y \geq 0$).

We can confirm this algebraically since the smallest possible value for x is 0:

$$y = \sqrt{x} + 2$$
$$y = \sqrt{0} + 2$$
$$y = 0 + 2$$
$$y = 2$$

Therefore, our domain is $\{x \mid x \geq 0\}$ and our range is $\{y \mid y \geq 2\}$. In interval notation this would be expressed as

domain: $[0, \infty)$ and range: $[2, \infty)$ ■

Suggestions for Using the Graphing Calculator to Determine Domain and Range

1. Examine the graphical display in a window that shows the x-intercept(s), y-intercept, and any peaks or valleys of the graph. A good starting place is to note the quadrants that contain graphical display.

2. Examine the values displayed in the TABLE for the function. It is suggested that values be examined for $x = 0$, for positive x-values, and for x-values that are negative.

3. Examine the definition (or rule) of the function. THINK about any values for x for which the function would not be defined. Think about any maximum/minimum values for either x or y (often the peaks and valleys of the graphical display are helpful).

EXAMPLE 4   State the domain and the range of the function $y = \sqrt{x + 2}$.

Solution   We begin, as before, by examining the graphical display, the TABLE, and the function itself.

The graphical display reveals the graph in quadrants I and II, which suggests negative x-values are defined for the function. It is also apparent that the domain is *not* the set of all real numbers.

Close examination of the definition of the function, $y = \sqrt{x + 2}$, reveals that the radicand of $x + 2$ must be 0 or positive. Therefore, $x + 2 \geq 0$ if and only if $x \geq -2$. We conclude that our domain must be $\{x | x \geq -2\}$. This is confirmed when we scroll up the TABLE to negative values for x.

The range appears to be all positive numbers and 0. Again, examination of the graphical display, function definition, and the TABLE confirm our conjecture.

Therefore, we conclude that the domain is $\{x | x \geq -2\}$ and the range is $\{y | y \geq 0\}$. In interval notation this would be

domain: $[-2, \infty)$ and range: $[0, \infty)$. ■

■ INTERPRETING GRAPHS OF RADICAL FUNCTIONS

EXPLORING THE
CONCEPT
a, b, c

The distance x (in feet) that an object will fall in y seconds is given by the formula $y = \sqrt{\frac{x}{16}}$. Graph the equation in the window $[-75.2, 75.2]$ by $[-6.2, 6.2]$. The scale on the x-axis is 10, whereas the scale on the y-axis is 1.

Examining the general shape of the graph and being cognizant that x represents height and y represents time, we see that as x increases, so does y. Interpreted within the context of our problem, this means that as the object falls a greater distance, the time increases.

TRACE along the graph to answer the questions that appear below AND answer each question using analytic methods.

a. If a stone is dropped from a height of 16 feet, how long will it take to hit the ground?

b. Approximately how long will it take a stone to reach the ground if it is dropped from a height of 32 feet?

c. After 2 seconds has passed, how far has the stone fallen?

To avoid confusion between $\sqrt{3}i$ and $\sqrt{3i}$, $i\sqrt{3}$ is an acceptable way to write the result. When the calculator is used, the following is displayed:

```
√(-3)
      1.732050808i
```

The result displayed is an approximation, whereas $i\sqrt{3}$ is an exact result. ■

WARNING! If a and b are both negative, then $\sqrt{a} \cdot \sqrt{b} \neq \sqrt{ab}$. For example, $\sqrt{-16} \cdot \sqrt{-4} \neq \sqrt{64}$, rather $\sqrt{-16} \cdot \sqrt{-4} = 4i \cdot 2i = 8i^2 = -8$.

■ POWERS OF i

The powers of i produce an interesting pattern.

EXPLORING THE CONCEPT

a. Use the definition of i^2 and the laws of exponents to evaluate $i^1, i^2, i^3, \ldots i^8$.

b. State the pattern revealed from part (a).

c. Suggest a method for simplifying powers of i.

Conclusion

a. $i^1 = i$

$i^2 = -1$

$i^3 = i^2 \cdot i = -1 \cdot i = -1$

$i^4 = i^2 \cdot i^2 = -1 \cdot -1 = 1$

$i^5 = i^3 \cdot i^2 = -i \cdot -1 = i$

$i^6 = i^2 \cdot i^4 = -1 \cdot 1 = -1$

$i^7 = i^3 \cdot i^4 = -i \cdot 1 = -i$

$i^8 = i^4 \cdot i^4 = 1 \cdot 1 = 1$

b. The powers of i repeat in a pattern of four results: $i, -1, -i, 1$.

c.

Powers of i

If n is a natural number that leaves a remainder of r when divided by 4, then $i^n = i^r$.

■

EXAMPLE 2 Simplify each of the powers of i: **a.** i^9 **b.** i^{24} **c.** i^{55}

Solution **a.** Dividing 9 by 4 leaves a remainder of 1. Therefore, $i^9 = i^1 = i$.

b. Dividing 24 by 4 leaves no remainder. Therefore, $i^{24} = i^0 = 1$. Remember, any base (except 0) raised to the zero power is 1.

c. Dividing 55 by 4 leaves a remainder of 3. Therefore, $i^{55} = i^3 = -1$. ■

■ COMPLEX NUMBERS

Defining $\sqrt{-1}$ equal to i allows the expansion of the real-number system to the **complex-number** system.

Complex Numbers

A complex number is *any* number that can be written in the form $a + bi$, where a and b are real numbers and $i = \sqrt{-1}$. If $b = 0$, then $a + bi$ is a real number. If $b \neq 0$ and $a = 0$, the complex number $a + bi = 0 + bi = bi$ and is an imaginary number.

In the complex number $a + bi$, a is the *real part* and b is the *imaginary part*.

The relationship between the real numbers, the imaginary numbers, and the complex numbers is shown in Figure 9-4 below.

COMPLEX NUMBERS

Real numbers $a + 0i$	Imaginary numbers $0 + bi$ $(b \neq 0)$
$3, \dfrac{7}{3}, \pi, 125.345$	$4i, -12i, \sqrt{-4}$

$$4 + 7i,\ 5 - 16i,\ \frac{1}{32 - 12i},\ 15 + \sqrt{-25}$$

FIGURE 9-4

Two complex numbers are equal if and only if their real parts are equal and their imaginary parts are equal.

Equality of Complex Numbers

The complex numbers $a + bi$ and $c + di$ are equal if and only if $a = c$ and $b = d$.

EXAMPLE 3   Classify each statement as true or false. Verify your response both analytically and with the graphing calculator.

a. $2 + 3i = \sqrt{4} + \sqrt{-9}$ **b.** $6 + 4i = \dfrac{12}{2} + \sqrt{-8}$

Solution

a. $2 + 3i = \sqrt{4} + \sqrt{-9}$
$= 2 + \sqrt{-1} \cdot \sqrt{9}$
$= 2 + i \cdot 3$
$= 2 + 3i$

■ **TECHNOLOGY TIP**
The equal sign is located under the TEST menu. After entering the equation and pressing ENTER, either a 1 or a zero will appear. A 1 indicates the statement is true, whereas a 0 indicates a false statement.

The original statement is true.

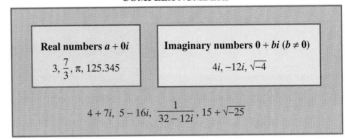

b. $6 + 4i = \dfrac{12}{2} + \sqrt{-8}$
$= 6 + \sqrt{-1} \cdot \sqrt{2^2 \cdot 2}$
$= 6 + i \cdot 2\sqrt{2}$
$= 6 + 2i\sqrt{2}$

■ **STUDY TIP**
Remember to extract the i value before performing arithmetic operations with imaginary numbers.

The original statement is false.

■ ARITHMETIC OF COMPLEX NUMBERS

Complex numbers can be added or subtracted by adding or subtracting their real parts and complex parts.

> **Addition and Subtraction of Complex Numbers**
> If $a + bi$ and $c + di$ are complex numbers, their sum is
> $$(a + bi) + (c + di) = (a + c) + (b + d)i$$
> The difference is defined as
> $$(a + bi) - (c + di) = a + bi - c - di = (a - c) + (b - d)i$$

EXAMPLE 4   Add or subtract the given complex numbers as the signs indicate. Verify results with your calculator.

   a. $(8 + 4i) + (12 + 8i)$ **b.** $(7 - 4i) + (9 + 2i)$

   c. $(-6 + i) - (3 - 4i)$ **d.** $(2 - 4i) - (-4 + 3i)$

Solution   **a.**
$$\begin{aligned}(8 + 4i) + (12 + 8i) &= 8 + 4i + 12 + 8i \\ &= (8 + 12) + (4i + 8i) \\ &= (8 + 12) + (4 + 8)i \\ &= 20 + 12i\end{aligned}$$

```
(8+4i)+(12+8i)
          20+12i
```

   b.
$$\begin{aligned}(7 - 4i) + (9 + 2i) &= 7 - 4i + 9 + 2i \\ &= (7 + 9) + (-4i + 2i) \\ &= (7 + 9) + (-4 + 2)i \\ &= 16 + (-2)i \\ &= 16 - 2i\end{aligned}$$

```
(7-4i)+(9+2i)
          16-2i
```

   c.
$$\begin{aligned}(-6 + i) - (3 - 4i) &= -6 + i - 3 + 4i \\ &= (-6 - 3) + (i + 4i) \\ &= (-6 - 3) + (1 + 4)i \\ &= -9 + 5i\end{aligned}$$

```
(-6+i)-(3-4i)
          -9+5i
```

   d.
$$\begin{aligned}(2 - 4i) - (-4 + 3i) &= 2 - 4i + 4 - 3i \\ &= (2 + 4) + (-4i - 3i) \\ &= (2 + 4) + (-4 - 3)i \\ &= 6 + (-7)i \\ &= 6 - 7i\end{aligned}$$

```
(2-4i)-(-4+3i)
          6-7i
```

To multiply complex numbers (say, by another complex number, an imaginary number, or a real number), use the distributive property. Remember, $i^2 = -1$ and powers of i are *not* retained in results.

Multiplying Complex Numbers

Complex numbers are multiplied as if they were binomials, with $i^2 = -1$.

$$(a + bi)(c + di) = ac + a \cdot di + bi \cdot c + bi \cdot di$$
$$= ac + adi + bci + bdi^2$$
$$= ac + (ad + bc)i + bd(-1)$$
$$= (ac - bd) + (ad + bc)i$$

EXAMPLE 5   Multiply and verify results with your calculator.

   a. $6(3 + 4i)$ **b.** $6i(5 - 2i)$ **c.** $(2 + 3i)(3 - 2i)$

   d. $(2 - 3i)^2$ **e.** $(-4 + 2i)(2 + i)$

Solution   **a.** $6(3 + 4i) = 6 \cdot 3 + 6 \cdot 4i$
$$= 18 + 24i$$

```
6(3+4i)
              18+24i
```

b. $6i(5 - 2i) = 6i \cdot 5 + 6i \cdot -2i$
$$= 30i - 12i^2$$
$$= 30i - 12(-1)$$
$$= 30i + 12$$
$$= 12 + 30i$$

```
6i(5-2i)
              12+30i
```

c. $(2 + 3i)(3 - 2i) = 2 \cdot 3 + 2 \cdot -2i + 3i \cdot 3 + 3i \cdot -2i$
$$= 6 - 4i + 9i - 6i^2$$
$$= 6 + 5i - 6(-1)$$
$$= 6 + 5i + 6$$
$$= 12 + 5i$$

```
(2+3i)(3-2i)
              12+5i
```

d. $(2 - 3i)^2 = (2 - 3i)(2 - 3i)$
$$= 2 \cdot 2 + 2 \cdot -3i - 3i \cdot 2 - 3i \cdot -3i$$
$$= 4 - 6i - 6i + 9i^2$$
$$= 4 - 12i + 9(-1)$$
$$= -5 - 12i$$

```
(2-3i)²
              -5-12i
```

e. $(-4 + 2i)(2 + i) = -4 \cdot 2 - 4 \cdot i + 2i \cdot 2 + 2i \cdot i$
$$= -8 - 4i + 4i + 2i^2$$
$$= -8 + 2(-1)$$
$$= -8 - 2$$
$$= -10$$

```
(-4+2i)(2+i)
              -10
```

Notice that the product of two complex numbers can be simply a real number. ■

To divide a complex number by a nonzero real number, both the real part and the imaginary part are divided by that real number.

Dividing a Complex Number by a Real Number

If $a + bi$ is a complex number and c is a nonzero real number, then

$$\frac{a + bi}{c} = \frac{a}{c} + \frac{b}{c}i$$

EXAMPLE 6   Divide and verify results with your calculator:

$$\textbf{a. } \frac{6 + 9i}{3} \qquad \textbf{b. } \frac{-8 - 3i}{-2}$$

Solution   **a.** $\dfrac{6 + 9i}{3} = \dfrac{6}{3} + \dfrac{9}{3}i$

$$= 2 + 3i$$

```
(6+9i)/3►Frac
              2+3i
```

b. $\dfrac{-8 - 3i}{-2} = \dfrac{-8}{-2} - \dfrac{3}{-2}i$

$$= 4 + \frac{3}{2}i$$

```
(-8-3i)/-2►Frac
             4+3/2i
```

■ RATIONALIZING DENOMINATORS

Because the definition of i is that $i = \sqrt{-1}$, i is not left in the denominator of a fractional result. Multiplying the numerator and the denominator by i, when the denominator is a monomial containing an *odd* power of i, removes i from the denominator. Recall that even powers of i may be replaced by either 1 or -1, as appropriate. If the denominator is a binomial, the numerator and the denominator must be multiplied by its **complex conjugate** to remove i.

Complex Conjugates

The complex numbers $a + bi$ and $a - bi$ are called **complex conjugates** of each other. Their product is always a real number, as illustrated below ($i^2 = -1$):

$$(a + bi)(a - bi) = a \cdot a + a \cdot -bi + bi \cdot a + bi \cdot -bi$$
$$= a^2 - abi + abi - b^2i^2$$
$$= a^2 - b^2(-1)$$
$$= a^2 + b^2$$

EXAMPLE 7   Rationalize the denominators and express results in $a + bi$ form. Verify results with your calculator.

$$\textbf{a. } \frac{6i}{i^5} \qquad \textbf{b. } \frac{2}{3 + i} \qquad \textbf{c. } \frac{3 - i}{2 + i} \qquad \textbf{d. } \frac{4 + \sqrt{-16}}{2 + \sqrt{-4}}$$

Solution   **a.** $\dfrac{6i}{i^5} = \dfrac{6i}{i^5} \cdot \dfrac{i}{i}$ Multiply the numerator and the denominator by i so that the denominator will have an even power of i.

$$= \frac{6i^2}{i^6} \qquad \text{Substitution: } i^6 = i^2 = -1.$$

$$= \frac{6(-1)}{-1}$$

$$= 6$$

```
(6i)/(i^5)
                    6
```

An alternative method is as follows:

$$\frac{6i}{i^5} = \frac{6}{i^4} = \frac{6}{1} = 6$$

Either method yields a correct result, as verified above.

b. $\dfrac{2}{3 + i} = \dfrac{2}{3 + i} \cdot \dfrac{3 - i}{3 - i}$

$$= \frac{2(3 - i)}{(3 + i)(3 - i)}$$

$$= \frac{6 - 2i}{9 - 3i + 3i - i^2}$$

$$= \frac{6 - 2i}{9 - (-1)}$$

$$= \frac{6 - 2i}{10}$$

$$= \frac{6}{10} - \frac{2}{10}i$$

$$= \frac{3}{5} - \frac{1}{5}i$$

```
2/(3+i)▶Frac
              3/5-1/5i
```

c. $\dfrac{3 - i}{2 + i} = \dfrac{3 - i}{2 + i} \cdot \dfrac{2 - i}{2 - i}$

$$= \frac{(3 - i)(2 - i)}{(2 + i)(2 - i)}$$

$$= \frac{6 - 3i - 2i + i^2}{4 - 2i + 2i - i^2}$$

$$= \frac{6 - 5i - 1}{4 - (-1)}$$

$$= \frac{5 - 5i}{5}$$

$$= \frac{5}{5} - \frac{5}{5}i$$

$$= 1 - i$$

```
(3-i)/(2+i)▶Frac
              1-i
```

d. $\dfrac{4 + \sqrt{-16}}{2 + \sqrt{-4}} = \dfrac{4 + 4i}{2 + 2i}$

$= \dfrac{4 + 4i}{2 + 2i} \cdot \dfrac{2 - 2i}{2 - 2i}$

$= \dfrac{(4 + 4i)(2 - 2i)}{(2 + 2i)(2 - 2i)}$

$= \dfrac{8 - 8i + 8i - 8i^2}{4 - 4i + 4i - 4i^2}$

$= \dfrac{8 - 8(-1)}{4 - 4(-1)}$

$= \dfrac{8 + 8}{4 + 4}$

$= \dfrac{16}{8}$

$= 2$

 WARNING! To avoid mistakes, always put complex numbers in $a + bi$ form before performing any complex arithmetic.

EXERCISE 9.6

VOCABULARY *Fill in the blanks to make a true statement.*

1. The imaginary number $\sqrt{-1}$ is defined as the imaginary unit __.

2. Defining $\sqrt{-1}$ allows us to expand the real number system to the _____ number system.

3. Two complex numbers are equal if and only if their ____ parts are equal and their _____ parts are equal.

4. When rationalizing a denominator that contains a complex number, multiply the numerator and the denominator by the _____ to remove i.

CONCEPTS

5. Powers of i repeat in a cycle of 4 values:
$i^1 =$ __ $i^2 =$ ___ $i^3 =$ ___ $i^4 =$ __

In Exercises 6–17, simplify each expression.

6. $\sqrt{-25}$ **7.** $\sqrt{-36}$ **8.** $-\sqrt{-75}$ **9.** $-\sqrt{-200}$

10. $\sqrt{-98}$ **11.** $\sqrt{-18}$ **12.** $\sqrt{-27}$ **13.** $-\sqrt{-32}$

14. i^{21} **15.** i^{19} **16.** i^{27} **17.** i^{22}

In Exercises 18–23, classify each equation as true or false. Justify your results analytically AND by using the TEST option of your calculator.

18. $3 + 7i = \sqrt{9} + (5 + 2)i$ **19.** $\sqrt{4} + \sqrt{25}i = 2 - (-5)i$ **20.** $8 + 5i = 2^3 + \sqrt{25}i^3$

21. $4 - 7i = -4i^2 + 7i^3$ **22.** $\sqrt{4} + \sqrt{-4} = 2 - 2i$ **23.** $\sqrt{-9} - i = 4i$

PRACTICE *In Exercises 24–65, perform the indicated operations. All results should be written in $a + bi$ form. Verify results with your calculator.*

24. $(3 + 4i) + (5 - 6i)$ **25.** $(5 + 3i) - (6 - 9i)$ **26.** $(7 - 3i) - (4 + 2i)$

27. $(8 + 3i) + (-7 - 2i)$ **28.** $(1 + i) - 2i + (5 - 7i)$ **29.** $(-9 + i) - 5i + (2 + 7i)$

30. $(5 + 3i) - (3 - 5i) + \sqrt{-1}$

31. $(8 + 7i) - \left(-7 - \sqrt{-64}\right) + (3 - i)$

32. $\left(-8 - \sqrt{3}i\right) - \left(7 - 3\sqrt{3}i\right)$

33. $\left(2 + 2\sqrt{2}i\right) + \left(-3 - \sqrt{2}i\right)$

34. $3i(2 - i)$

35. $-4i(3 + 4i)$

36. $(2 - 4i)(3 + 2i)$

37. $(3 - 2i)(4 - 3i)$

38. $\left(2 + \sqrt{2}i\right)\left(3 - \sqrt{2}i\right)$

39. $\left(5 + \sqrt{3}i\right)\left(2 - \sqrt{3}i\right)$

40. $\left(8 - \sqrt{-1}\right)\left(-2 - \sqrt{-16}\right)$

41. $\left(-1 + \sqrt{-4}\right)\left(2 + \sqrt{-9}\right)$

42. $(2 + i)^2$

43. $(3 - 2i)^2$

44. $i(5 + i)(3 - 2i)$

45. $i(-3 - 2i)(1 - 2i)$

46. $(2 + i)(2 - i)(1 + i)$

47. $(3 + 2i)(3 - 2i)(i + 1)$

48. $\dfrac{4}{5i^3}$

49. $\dfrac{3}{2i}$

50. $\dfrac{3i}{8\sqrt{-9}}$

51. $\dfrac{5i^3}{2\sqrt{-4}}$

52. $\dfrac{-3}{5i^5}$

53. $\dfrac{-4}{6i^7}$

54. $\dfrac{5}{2 - i}$

55. $\dfrac{26}{3 - 2i}$

56. $\dfrac{13i}{5 + i}$

57. $\dfrac{2i}{5 + 3i}$

58. $\dfrac{-12}{7 - \sqrt{-1}}$

59. $\dfrac{4}{3 + \sqrt{-1}}$

60. $\dfrac{5i}{6 + 2i}$

61. $\dfrac{-4i}{2 - 6i}$

62. $\dfrac{3 - 2i}{3 + 2i}$

63. $\dfrac{2 + 3i}{2 - 3i}$

64. $\dfrac{\sqrt{5} - \sqrt{3}i}{\sqrt{5} + \sqrt{3}i}$

65. $\dfrac{\sqrt{3} + \sqrt{2}i}{\sqrt{3} - \sqrt{2}i}$

66. a. Show, using analytical methods, that $1 - 5i$ and its conjugate are solutions of $x^2 - 2x = -26$.
b. Use the STOre feature of your calculator to verify your analytical results.

67. a. Show, using analytical methods, that $3 - 2i$ and its conjugate are solutions of $x^2 = 6x - 13$.
b. Use the STOre feature of your calculator to verify your analytical results.

68. a. Show, using analytical methods, that i and its conjugate are solutions of $x^4 = 3x^2 + 4$.
b. Use the STOre feature of your calculator to verify your analytical results.

69. a. Show, using analytical methods, that $2 + i$ and its conjugate are *not* solutions of $x^2 + x + 1 = 0$.
b. Use the STOre feature of your calculator to verify your analytical results.

REVIEW *In Exercises 70–75, simplify each of the polynomials.*

70. $(3x + 2) + (5x - 8)$

71. $(3x^2 + 5x - 2) - (8x^2 - 2x + 4)$

72. $3x^2(x - 4)$

73. $(3x + 2)(5x - 8)$

74. $\dfrac{3x^2 - 6x + 9}{3}$

75. $\dfrac{x^2 - 1}{x - 1}$

CHAPTER SUMMARY

CONCEPTS	REVIEW EXERCISES

Section 9.1

The expression $\sqrt[n]{a}$ specifies the nth root of a number, when $n > 1$. The letter n is the *index*, or order of the root.

The expression under the radical sign is the *radicand*. The index of 2 is omitted when writing square roots.

$$\sqrt[n]{a^n} = a \text{ if } n \text{ is odd}$$
$$\sqrt[n]{a^n} = |a| \text{ if } n \text{ is even}$$

When n is odd, the radicand may be positive, negative, or 0. When n is even, the radicand must be 0 or positive to be defined in the real-number system.

Introduction to Roots and Radicals

1. Simplify each radical.

   a. $\sqrt{49}$ **b.** $-\sqrt{121}$

   c. $-\sqrt{36}$ **d.** $\sqrt{225}$

   e. $\sqrt[3]{-27}$ **f.** $-\sqrt[3]{216}$

   g. $\sqrt[4]{625}$ **h.** $\sqrt[5]{-32}$

2. Simplify each radical. Assume that all variables represent positive numbers.

   a. $\sqrt{25x^2}$ **b.** $\sqrt{x^2 + 4x + 4}$

   c. $\sqrt[3]{27a^6b^2}$ **d.** $\sqrt[4]{256a^8y^4}$

Section 9.2

The link between rational exponents and radicals is stated below:

$$a^{m/n} = \sqrt[n]{a^m} = \left(\sqrt[n]{a}\right)^m$$

A radical expression is simplified when no radical appears in the denominator of a fraction, the radicand contains no fractions or negative numbers, and each factor in the radicand appears to a power that is less than the index of the radical.

The product rule states

$$\sqrt[n]{ab} = \sqrt[n]{a} \cdot \sqrt[n]{b}$$

whereas the quotient rule specifies

$$\sqrt[n]{\frac{a}{b}} = \frac{\sqrt[n]{a}}{\sqrt[n]{b}}$$

Rational Exponents and Simplifying Radical Expressions

3. Simplify each expression. Assume that all variables represent positive numbers.

   a. $(-8)^{1/3}$ **b.** $-8^{2/3}$ **c.** $\left(\frac{4}{9}\right)^{3/2}$

   d. $\frac{1}{25^{5/2}}$ **e.** $\left(\frac{4}{9}\right)^{-3/2}$ **f.** $(27x^3y)^{1/3}$

   g. $(81x^4y^2)^{1/4}$ **h.** $(25x^3y^4)^{3/2}$ **i.** $(8u^2v^3)^{-2/3}$

4. Simplify each expression. Assume that all variables represent positive numbers.

   a. $\sqrt{240}$ **b.** $\sqrt[3]{54}$ **c.** $\sqrt[4]{32}$

   d. $\sqrt[5]{96}$ **e.** $\sqrt[3]{16x^5y^4}$ **f.** $\sqrt[3]{54x^7y^3}$

   g. $\frac{\sqrt{32x^3}}{\sqrt{2x}}$ **h.** $\frac{\sqrt[3]{16x^5}}{\sqrt[3]{2x^2}}$ **i.** $\sqrt[3]{\frac{2a^2b}{27x^3}}$

   j. $\sqrt{\frac{17xy}{64a^4}}$

Section 9.3

To add or subtract radical expressions, simplify each radical and use the distributive property to add or subtract radicals that have the same index and radicand.

Arithmetic Operations with Radicals

5. Simplify and combine radicals as indicated. Assume that all variables represent positive numbers.

   a. $2\sqrt[3]{3} - \sqrt[3]{24}$ **b.** $\sqrt[4]{32} + 2\sqrt[4]{162}$

   c. $2x\sqrt{8} + 2\sqrt{200x^2} + \sqrt{50x^2}$

   d. $\sqrt[3]{54} - 3\sqrt[3]{16} + 4\sqrt[3]{128}$

The product rule ensures that radical expressions with the same index can be multiplied. It is suggested that multiplication be done before simplification.

For a radical expression to be in simplified form, there cannot be a fraction under the radical or a radical in the denominator of a fraction.

To rationalize the denominator, multiply both the numerator and the denominator by a radical that will make the denominator a perfect root.

To rationalize binomial denominators, multiply the numerator and the denominator by the conjugate of the binomial in the denominator.

6. Simplify each expression. Assume that all variables represent positive numbers.

a. $\sqrt[3]{3} \cdot \sqrt[3]{9}$

b. $-\sqrt[3]{2x^2} \cdot \sqrt[3]{4x}$

c. $\sqrt{2}(\sqrt{8} - 3)$

d. $\sqrt{5}(\sqrt{2} - 1)$

e. $(\sqrt{2} + 1)(\sqrt{2} - 1)$

f. $(2\sqrt{u} + 3)(3\sqrt{u} - 4)$

7. Rationalize each denominator.

a. $\dfrac{1}{\sqrt{3}}$

b. $\dfrac{\sqrt[3]{uv}}{\sqrt[3]{u^5 v^7}}$

c. $\dfrac{2}{\sqrt{2} - 1}$

d. $\dfrac{2x - 32}{\sqrt{x} + 4}$

Section 9.4 — Radical Equations and Their Applications

A radical equation is an equation that contains one or more radicals, with a variable in the radicand. The power rule allows algebraic solutions of radical equations: If x, y, and n are real numbers and $x = y$, then $x^n = y^n$.

Raising both sides of an equation to the same power can introduce extraneous roots. Therefore, all supposed solutions *must* be checked.

The *Pythagorean theorem* states that the sum of the squares of the legs of a right triangle is equal to the square of the hypotenuse.

$$(\text{leg})^2 + (\text{leg})^2 = (\text{hypotenuse})^2$$

The distance formula allows us to find the distance between any two points in the rectangular coordinate system:

$$d = \sqrt{(x_2 - x_1)^2 + (y_2 - y_1)^2},$$

where (x_1, y_1) and (x_2, y_2) are the coordinates of two points.

8. Solve each equation both algebraically and graphically.

a. $\sqrt{x + 3} = \sqrt{2x - 19}$

b. $x = \sqrt{12x - 27}$

c. $\sqrt{x + 1} + \sqrt{x} = 2$

d. $\sqrt[3]{x^3 + 8} = x + 2$

9. Sailing   A technique called *tacking* allows a sailboat to make progress into the wind. A sailboat follows the course in Illustration 1. Find d, the distance the boat advances into the wind.

ILLUSTRATION 1

10. Find the distance between the points $P(-2, -7)$ and $Q(4, 1)$.

Section 9.5	Radical Functions

A function is a relation that assigns to each value of x exactly one value, y.

The domain of a function is the set of all possible values for the independent variable, x, whereas the range is the set of all possible values for the dependent variable, y.

The definition of the function, the TABLE, and the graph of the function are often useful in determining domains and ranges.

11. If $f(x) = \sqrt{2x + 4} - 1$ and $g(x) = \sqrt[3]{2x}$, find each of the following. Round irrational results to the nearest hundredth.
   a. $f(6)$ **b.** $g(4)$ **c.** $g(32)$
   d. $f(3)$ **e.** $g(8)$

12. State the domain and the range of each of the following.
   a. $f(x) = \sqrt{x - 5} - 4$ **b.** $f(x) = \sqrt{2x + 4}$

   c. $f(x) = \sqrt[3]{x + 2}$ **d.** $f(x) = \sqrt[3]{x + 4} - 3$

The horizontal distance d (measured in miles) is related to the height h (measured in feet) of the observer by the formula $d = 1.4\sqrt{h}$.

13. View from a submarine   A submarine periscope extends 4.7 feet above the surface. How far is the horizon?

14. View from a submarine   How far out of the water must a submarine periscope extend to provide a 4-mile horizon?

Section 9.6	$\sqrt{-1}$ and Complex Numbers

The letter i is used to denote the imaginary number defined below:
$$\sqrt{-1} = i \quad \text{and} \quad i^2 = -1$$
The powers of i repeat in the pattern, i, -1, $-i$, 1. This allows us to simplify powers of i. If n is a natural number that leaves a remainder of r when divided by 4, then $i^n = i^r$.

Defining i allows the expansion of the real-number system to the complex-number system. A complex number is any number that can be written in the form $a + bi$.

Two complex numbers $a + bi$ and $c + di$ are equal if and only if $a = c$ and $b = d$.

Complex numbers can be added or subtracted by adding or subtracting their real parts and their imaginary parts.

To multiply complex numbers, use the distributive property.

15. Simplify and express results in $a + bi$ form.
   a. $(5 + 4i) + (7 - 12i)$ **b.** $(-6 - 40i) - (-8 + 28i)$
   c. $\left(-32 + \sqrt{-144}\right) - \left(64 + \sqrt{-81}\right)$
   d. $\left(-8 + \sqrt{-8}\right) + \left(6 - \sqrt{-32}\right)$

16. Simplify and express results in $a + bi$ form.
   a. $(2 - 7i)(-3 + 4i)$ **b.** $(-5 + 6i)(2 + i)$
   c. $\left(5 - \sqrt{-27}\right)\left(-6 + \sqrt{-12}\right)$
   d. $\left(2 + \sqrt{-128}\right)\left(3 - \sqrt{-98}\right)$

The imaginary unit i is never left in the denominator of a fraction. To rationalize the denominator, multiply the numerator and the denominator by i when the denominator is a monomial containing an odd power of i. If the denominator is a binomial, the numerator and the denominator must be multiplied by its complex conjugate to remove i.

17. Rationalize the denominators.

a. $\dfrac{3}{4i}$

b. $\dfrac{-2}{5i^3}$

c. $\dfrac{6}{2+i}$

d. $\dfrac{7}{3-i}$

e. $\dfrac{3}{5+\sqrt{-4}}$

f. $\dfrac{2}{3-\sqrt{-9}}$

■ Chapter Test

In Problems 1–5, simplify each expression. Assume that all variables represent positive numbers.

1. $\sqrt{250x^3y^5}$

2. $\dfrac{\sqrt[3]{24x^{15}y^4}}{\sqrt[3]{y}}$

3. $\sqrt{\dfrac{3a^5}{48a^7}}$

4. $\sqrt[3]{27x^3}$

5. $\sqrt{18x^4y^9}$

6. Simplify each expression. Write results without using negative exponents.

a. $16^{1/4}$

b. $36^{-3/2}$

c. $\left(-\dfrac{8}{27}\right)^{-2/3}$

In Problems 7–10, simplify each expression containing radicals. Assume that all variables represent positive numbers.

7. $\sqrt{12} - \sqrt{27}$

8. $2\sqrt[3]{40} - \sqrt[3]{5000} + 4\sqrt[3]{625}$

9. $2\sqrt{48y^5} - 3y\sqrt{12y^3}$

10. $2\sqrt{xy}\left(3\sqrt{x} + \sqrt{xy^3}\right)$

11. Rationalize each denominator.

a. $\dfrac{1}{\sqrt{5}}$

b. $\dfrac{6}{\sqrt[3]{9}}$

c. $\dfrac{-4\sqrt{2}}{\sqrt{5}+3}$

d. $\dfrac{3x-1}{\sqrt{3x}-1}$

In Problems 12 and 13, solve each equation both algebraically and graphically.

12. $\sqrt[3]{6x+4} - 4 = 0$

13. $1 - \sqrt{x} = \sqrt{x-3}$

14. Find the distance between P and Q if $P(-2, 5)$ and $Q(22, 12)$.

15. Shipping crates The diagonal brace on the shipping crate shown in Illustration 1 below is 53 inches. Find the height, h, of the crate.

h in.   53 in.

45 in.

ILLUSTRATION 1

16. If $f(x) = 2\sqrt{x} + 1$, find each of the following. Round any irrational results to the nearest thousandth.

a. $f(3)$

b. $f(9)$

c. $f(-20)$

d. $f(13)$

17. Find the domain and the range of each function specified below.

a. $f(x) = \sqrt{x-6} + 2$

b. $f(x) = -\sqrt{x}$

18. Add or subtract the complex numbers as the signs indicate.

a. $(2 + 4i) + (-3 + 7i)$

b. $\left(3 - \sqrt{-9}\right) - (-1 + \sqrt{-16})$

19. Multiply the complex numbers below.

a. $2i(3 - 4i)$

b. $(3 + 2i)(-4 - i)$

20. Rationalize the denominators of each of the expressions below.

a. $\dfrac{1}{i\sqrt{2}}$

b. $\dfrac{2+i}{3-i}$

Electrical/Electronic Engineer

Electrical engineers design, develop, test, and supervise the manufacture of electrical and electronic equipment. Electrical engineers who work with electronic equipment are often called **electronic engineers.**

Electrical engineers generally specialize in a major area, such as power-distributing equipment, integrated circuits, computers, electrical equipment manufacturing, or communications. Besides manufacturing and research, development, and design, many are employed in administration and management, technical sales, or teaching.

Qualifications

A bachelor's degree in engineering is generally acceptable for beginning engineering jobs. College graduates with a degree in natural science or mathematics also may qualify for some jobs.

Engineers should be able to work as part of a team and should have creativity, an analytical mind, and a capacity for detail. In addition, engineers should be able to express themselves well—both orally and in writing.

Job Outlook

Employment of electrical engineers is expected to increase faster than the average for all occupations through the year 2005. Although increased demand for computers, communications equipment, and electronics is expected to be the major contributor to this growth, demand for electrical and electronic consumer goods, along with increased research and development in new types of power generation, should create additional jobs.

Example Application

In a radio, an inductor and a capacitor are used in a resonant circuit to select a desired station at frequency f and reject all others. The inductance L and the capacitance C determine the inductive reactance X_L and the capacitive reactance X_C of that circuit, where

$$X_L = 2\pi fL \qquad \text{and} \qquad X_C = \frac{1}{2\pi fC}$$

The radio station selected will be at the frequency f for which the inductive reactance and capacitive reactance are equal. To find that frequency, solve for f in terms of L and C.

Solution

Since X_L and X_C are to be equal, we can solve the system of equations

$$\begin{cases} X_L = 2\pi fL \\ X_C = \dfrac{1}{2\pi fC} \\ X_L = X_C \end{cases}$$

by substitution to express f in terms of L and C.

$$X_L = X_C$$

$$2\pi fL = \frac{1}{2\pi fC}$$

$(2\pi fL)(2\pi fC) = 1$ Multiply both sides by $2\pi fC$.

$4\pi^2 f^2 LC = 1$ Simplify.

$f^2 = \dfrac{1}{4\pi^2 LC}$ Divide both sides by $4\pi^2 LC$.

$$f = \sqrt{\frac{1}{4\pi^2 LC}}$$ Take the positive square root of both sides.

$$f = \frac{1}{2\pi\sqrt{LC}}$$ Simplify.

EXERCISES

1. At what frequency will a 0.0001 farad capacitor and a 0.005 henry inductor resonate: (*Hint:* $C = 0.0001$ and $L = 0.005$.)

2. If the inductor and the capacitor in Exercise 1 were both doubled, what would the resonant frequency become?

3. At what frequency will a 0.0008 farad capacitor and a 0.002 henry inductor resonate?

4. If the inductor of Exercise 3 were doubled and the capacitance reduced by one-half, what would the resonant frequency be?

(*Answers:* **1.** 225 hertz **2.** 113 hertz **3.** 126 hertz **4.** 126 hertz)

Quadratics

From Ridicule to Respect

The Pythagoreans (ca. 500 B.C.) understood the universe as a harmony of whole numbers. They did not classify fractions as numbers and were upset that $\sqrt{2}$ was not the ratio of whole numbers. For 2000 years, little progress was made in the understanding of the various kinds of numbers.

The father of algebra, François Vieta (1540–1603), understood the whole numbers, fractions, and certain irrational numbers. But he was unable to accept negative numbers and certainly not imaginary numbers.

Girolamo Cardano (1501–1576) was more daring, even though negative and imaginary numbers puzzled him. He attacked the problem *Divide 10 into two parts whose product is 40.* "Of course, that is impossible," said Cardano. "Nevertheless, we will operate." (Cardano had some medical training!) He found solutions: the two numbers $5 + i\sqrt{15}$ and $5 - i\sqrt{15}$.

(Can you show that their sum is 10 and their product is 40?) Cardano was impressed. "These numbers," he said, "are truly sophisticated." Unfortunately, he decided that this kind of arithmetic would be useless. Cardano was ahead of his time, but he didn't appreciate the value of his discovery.

René Descartes (1596–1650) thought these numbers to be nothing more than figments of his imagination, so he called them *imaginary numbers*. Leonhard Euler (1707–1783) used the letter i for $\sqrt{-1}$; Augustin Cauchy (1789–1857) used the term *conjugate*; and Karl Gauss (1777–1855) first used the word *complex*.

Today, we accept complex numbers without question, and use them freely in science, economics, medicine, and industry. But it took many centuries and the work of many mathematicians to make them respectable.

This chapter studies methods appropriate for finding solutions of quadratic (second-degree) equations that are not factorable. Graphs and tables are used to model problem solutions of quadratic equations. The discussion then extends to families of graphs and further to quadratic and other nonlinear inequalities.

10.1 Square Root Property and Completing the Square

■ REVIEW OF SOLUTIONS OF QUADRATICS BY FACTORING ■ SQUARE ROOT PROPERTY
■ COMPLETING THE SQUARE

■ REVIEW OF SOLUTIONS OF QUADRATICS BY FACTORING

Earlier in this text quadratic equations were solved by factoring. Recall that a **quadratic equation (second-degree equation)** is one that *can be* written in the form $ax^2 + bx + c = 0$, where a, b, and c are real numbers and $a \neq 0$.

To solve a quadratic equation by factoring, the equation was first written in standard form and the zero factor property was applied. Consider the solution of the quadratic equation $x^2 = 16$.

$$x^2 - 16 = 0$$ Subtract 16 from both sides so that the equation is in standard form.

$$(x - 4)(x + 4) = 0$$ Factor the polynomial side.

$$x - 4 = 0 \quad \text{or} \quad x + 4 = 0$$ Apply the zero factor property and set each side equal to zero.

$$x = 4 \quad \text{or} \quad x = -4$$ Solve.

The solution set is $\{4, -4\}$. Use of the STOre feature confirms both solutions.

```
4→X:X²
              16
-4→X:X²
              16
```

Many quadratic equations do not factor easily. For example, it would be difficult to factor $x^2 - 3 = 0$ because it cannot be factored using only integers. Therefore, we will develop methods for solving all quadratic equations.

■ SQUARE ROOT PROPERTY

■ TECHNOLOGY TIP
Change the MODE setting of your calculator from real to $a + bi$. Because the set of real numbers is a subset of the set of complex numbers, the calculator will display all roots (both real and complex).

Reconsider the quadratic equation $x^2 = 16$. We can take the square root of both sides to solve; however, we must consider both the positive AND the negative square roots of 16. This is necessary to find *all* values of the variable that make the statement $x^2 = 16$ true. Therefore,

$$x^2 = 16$$
$$x = \pm\sqrt{16}$$
$$x = \pm 4$$

This illustrates the **square root property,**

> **Square Root Property**
> If c is a real number, the equation $x^2 = c$ has two solutions, $x = \sqrt{c}$ and $x = -\sqrt{c}$. This is often expressed as $x = \pm\sqrt{c}$.

EXAMPLE 1   Solve the equation $x^2 - 12 = 0$.

Solution   We can write the equation as $x^2 = 12$ and use the square root property.

■ STUDY TIP
Try to schedule a free hour immediately after your algebra class to go over your notes and begin the homework assignment. It is important to reinforce the concepts learned in class as soon as possible after class.

$$x^2 - 12 = 0$$
$$x^2 = 12 \qquad \text{Add 12 to both sides.}$$
$$x = \pm\sqrt{12} \qquad \text{Apply the square root property.}$$
$$x = \pm 2\sqrt{3} \qquad \text{Simplify the radical.}$$

The solutions of the equation are $2\sqrt{3}$ and $-2\sqrt{3}$.

The STOre feature of the calculator can be used to check the solutions:

The solutions to the equation $x^2 - 12 = 0$ could also have been found using the ROOT/ZERO feature of the graphing calculator. These are approximations rather than exact results:

   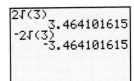

EXAMPLE 2   Solve the equation $(x - 3)^2 - 1 = 15$.

Solution   We begin by adding 1 to both sides so that the equation will be in the form $x^2 = c$ (quantity squared = constant).

$$(x - 3)^2 = 16$$ Add 1 to both sides.

$$x - 3 = \pm\sqrt{16}$$ Apply the square root property.

$$x - 3 = \pm 4$$ Simplify the radical.

$$x = 3 \pm 4$$ Solve for x (add 3 to both sides).

$$x = 3 + 4 \quad \text{or} \quad x = 3 - 4$$ Consider both cases indicated by the $\pm$ sign.

$$x = 7 \qquad\qquad x = -1$$

The solution set is $\{7, -1\}$.

The STOre feature can be used to check solutions:

The solutions could also be found using the INTERSECT feature of the calculator:

EXAMPLE 3   Solve $(3x + 2)^2 = -16$.

Solution
$$(3x + 2)^2 = -16$$

$$3x + 2 = \pm\sqrt{-16}$$ Square root property.

$$3x + 2 = \pm 4i$$ Simplify the radical.

$$3x = -2 \pm 4i$$ Subtract 2 from both sides.

$$x = \frac{-2 \pm 4i}{3}$$ Divide both sides by 3.

Stated in $a + bi$ form, the solutions are $-\frac{2}{3} + \frac{4i}{3}$ and $-\frac{2}{3} - \frac{4i}{3}$.

The STOre feature can be used to check the solutions (remember, the calculator was set earlier to $a + bi$ MODE):

When we try to solve the equation graphically, we obtain the following display in the window $[-10, 10]$ by $[-20, 10]$:

The graphs of the two sides of the equation *do not* intersect. Previously, this meant that the equation was a contradiction and had no solution. However, the check above contradicts that fact. Examination of the roots shows that they are complex. The rectangular coordinate system includes real numbers *only*. It will only show real-number solutions. The graphs shown above can be interpreted to mean that the roots (solutions) of the equation are not real numbers. ■

■ COMPLETING THE SQUARE

The square root property can be used to solve any quadratic equation that is in the form $x^2 = c$ (or quantity squared = constant). The process of **completing the square** allows us to solve any quadratic equation and is based on these special products:

$$x^2 + 2ax + a^2 = (x + a)^2 \qquad \text{and} \qquad x^2 - 2ax + a^2 = (x - a)^2$$

EXPLORING THE CONCEPT   **a.** Rewrite each perfect square trinomial as a binomial quantity squared.

$$x^2 + 10x + 25 = (x + 5)^2$$
$$x^2 - 6x + 9 = (x - 3)^2$$
$$x^2 + 8x + 16 = (x + 4)^2$$
$$x^2 - 12x + 36 = (x - 6)^2$$

b. Compare the *terms* of the binomial quantity with the terms of the original trinomials.

c. Compare the middle term of the trinomial with the binomial terms.

Conclusions   **a.** $(x + 5)^2$
$(x - 3)^2$
$(x + 4)^2$
$(x - 6)^2$

b. The terms of the binomial quantities are the square roots of the first and third terms of the trinomial.

$x^2 + 10x + 25$	$x^2 - 6x + 9$
$(x)^2 + 10x + (5)^2$	$(x)^2 - 6x + (3)^2$
$(x + 5)^2$	$(x - 3)^2$
$x^2 + 8x + 16$	$x^2 - 12x + 36$
$(x)^2 + 8x + (4)^2$	$(x)^2 - 12x + (6)^2$
$(x + 4)^2$	$(x - 6)^2$

c. The middle term of the *trinomial* is twice the product of the terms in the *binomial*.

$$x^2 + 10x + 25 = (x + 5)^2 \qquad x^2 - 6x + 9 = (x - 3)^2$$
$$10x = 2(5 \cdot x) \qquad\qquad -6x = 2(-3 \cdot x)$$
$$x^2 + 8x + 16 = (x + 4)^2 \qquad x^2 - 12x + 36 = (x - 6)^2$$
$$8x = 2(4 \cdot x) \qquad\qquad -12x = 2(-6 \cdot x)$$

In the process of completing the square, we will algebraically manipulate a trinomial to that we can write it in the form of a perfect square trinomial.

EXAMPLE 4   Use the process of completing the square to solve $x^2 + 2x = 5$.

Solution   To make $x^2 + 2x$ a perfect square trinomial, we will add the square of $\frac{1}{2}$ the coefficient of x to both sides of the equation.

$$\tfrac{1}{2}(2) = 1 \qquad\qquad \text{Find } \tfrac{1}{2} \text{ the coefficient of } x.$$
$$x^2 + 2x + (1)^2 = 5 + (1)^2 \qquad \text{Add the square of } \tfrac{1}{2} \text{ the coefficient of } x \text{ to both sides of the equation.}$$
$$(x + 1)^2 = 6 \qquad\qquad \text{Factor the perfect square trinomial on the left and simplify the right side of the equation.}$$
$$x + 1 = \pm\sqrt{6} \qquad\qquad \text{Apply the square root property.}$$
$$x = -1 \pm \sqrt{6} \qquad\qquad \text{Solve for } x.$$

Solution set: $\left\{-1 + \sqrt{6}, -1 - \sqrt{6}\right\}$.
Check solutions using the STOre feature:

The solutions can also be approximated graphically:

EXAMPLE 5   Solve $2x^2 + 4x + 1 = 0$ by completing the square.

Solution   Notice that the trinomials in the Exploring the Concepts exercise and in Example 4 had leading coefficients of 1. To make the coefficient of x^2 equal to 1, we begin by dividing through the equation by 2.

$$2x^2 + 4x + 1 = 0$$
$$\frac{2x^2}{2} + \frac{4x}{2} + \frac{1}{2} = \frac{0}{2} \qquad \text{Divide through the equation by 2.}$$

$$x^2 + 2x + \frac{1}{2} = 0 \qquad \text{Simplify.}$$

$$x^2 + 2x = -\frac{1}{2} \qquad \text{Subtract } \tfrac{1}{2} \text{ from both sides.}$$

$$x^2 + 2x + (1)^2 = -\frac{1}{2} + (1)^2 \qquad \text{Get } \tfrac{1}{2} \text{ the coefficient of } x \text{ and add its square to both sides of the equation.}$$

$$(x + 1)^2 = \frac{1}{2} \qquad \text{Factor the polynomial side of the equation and simplify constants on the right.}$$

$$x + 1 = \pm\sqrt{\frac{1}{2}} \qquad \text{Square root property.}$$

$$x + 1 = \pm\frac{\sqrt{2}}{2} \qquad \text{Simplify the radical (make sure the denominator is rationalized).}$$

$$x = -1 \pm \frac{\sqrt{2}}{2} \qquad \text{Solve for } x.$$

The solution set is $\left\{ -1 + \dfrac{\sqrt{2}}{2}, -1 - \dfrac{\sqrt{2}}{2} \right\}$.

The STOre feature can be used to check the solutions:

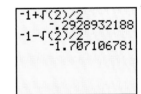

The solutions can also be approximated using the ROOT/ZERO feature of the calculator:

The methods used in Example 5 suggest steps that could be used when completing the square to solve quadratic equations of the form $ax^2 + bx + c = 0$.

Completing the Square

1. Make sure the coefficient of x^2 is 1. If it is not, divide both *sides* of the equation by the coefficient of x^2 to make it 1.

2. The constant term should be on the right side of the equal sign (terms containing variables will be located on the left).

3. Complete the square by
   a. Identifying the coefficient of the linear term (x).
   b. Find one-half of this coefficient.
   c. Add the square of the coefficient to *both* sides of the equation.

4. Factor the polynomial (on the left) as the square of a binomial.

5. Use the square root property to solve the resulting equation.

EXAMPLE 6   Solve $3x^2 + 4x + 2 = 0$ by completing the square.

Solution

$$\frac{3x^2}{3} + \frac{4x}{3} + \frac{2}{3} = \frac{0}{3}$$ Divide through the equation by the coefficient of x^2.

$$x^2 + \frac{4}{3}x + \frac{2}{3} = 0$$ Simplify.

$$x^2 + \frac{4}{3}x = -\frac{2}{3}$$ The constant term should be on the right side of the equation.

$$x^2 + \frac{4}{3}x + \left(\frac{2}{3}\right)^2 = -\frac{2}{3} + \left(\frac{2}{3}\right)^2$$ Take one-half of the coefficient of x and add its square to both sides of the equation.

$$\left(x + \frac{2}{3}\right)^2 = -\frac{2}{9}$$ Write the polynomial side as a binomial quantity squared and simplify the right side.

$$x + \frac{2}{3} = \pm\sqrt{-\frac{2}{9}}$$ Apply the square root property.

$$x + \frac{2}{3} = \pm\frac{\sqrt{2}}{3}i$$ Simplify the radical.

$$x = -\frac{2}{3} \pm \frac{\sqrt{2}}{3}i$$ Solve for x.

The solution set is $\left\{ -\frac{2}{3} + \frac{\sqrt{2}}{3}i, -\frac{2}{3} - \frac{\sqrt{2}}{3}i \right\}$.

The STOre feature can be used to check the solutions:

The fact that the solutions are complex and not real is confirmed by the graphical display:

EXERCISE 10.1

VOCABULARY   *Fill in the blanks to make the given statements true.*

1. A quadratic equation is one that can be written in the form _____. The constant, represented by __, cannot equal 0. Otherwise, the equation would be linear rather than quadratic.

2. The _____ property allows us to find the roots of equations that can be put into the form $x^2 = c$.

3. The _____ of an equation is/are the value(s) of the variable that make the equation a true statement.

4. The process of _____ allows us to solve any quadratic equation.

5. The _____ feature of the graphing calculator can be used to check both real and complex roots of quadratic equations.

6. Quadratic equations that have _____ roots cannot be confirmed using the graphical display of the graphing calculator.

CONCEPTS

7. The graphs of several quadratic functions are displayed below. Specify the roots as *real* or *imaginary*.

a.

b.

c.

d.

8. a. Graph the quadratic equation $y = x^2 - 9$ by completing the table of values, graphing the points whose coordinates are represented in the table, and connecting the points with a smooth curve.

x	y
0	
2	
-2	
3	
-3	
4	
-4	

b. Graphically, the solutions of the equation can be determined by looking at the x-intercepts. What are the x-intercepts, the solutions, of the equation $x^2 - 9 = 0$?

In Exercises 9 and 10, consider the specified equations:

a. $x^2 = 49$ **b.** $x^2 = -16$ **c.** $x^2 = 6$ **d.** $x^2 = 8$ **e.** $x^2 = -5$

9. Which of the given equations have solutions that are real numbers? Explain how you arrived at your conclusion.

10. Which of the given equations have solutions that are imaginary numbers? Explain how you arrived at your conclusion.

In Exercises 11–14, use the square root property to solve each equation. Check roots using the STOre feature of the graphing calculator.

11. $x^2 = 25$ **12.** $x^2 = 12$ **13.** $x^2 = 100$ **14.** $x^2 = -36$

In Exercises 15–20, (a) add the proper constant to make each binomial a perfect square trinomial and (b) factor the resulting trinomial.

15. $x^2 + 10x$ **16.** $x^2 + 8x$ **17.** $x^2 - 6x$
18. $x^2 - 12x$ **19.** $x^2 - 5x$ **20.** $x^2 + 3x$

PRACTICE *In Exercises 21–30, use the square root property to solve each quadratic equation. Check solutions using the STOre feature of the graphing calculator.*

21. $3x^2 - 16 = 0$ **22.** $5x^2 - 49 = 0$ **23.** $(x + 1)^2 = 27$ **24.** $(x + 1)^2 = 8$

25. $(x + 3)^2 = -18$ **26.** $(x + 4)^2 = -25$ **27.** $(2x + 3)^2 = 40$
28. $(3x + 5)^2 = 28$ **29.** $(3x + 1)^2 = -13$ **30.** $(2x - 4)^2 = -16$

In Exercises 31–40, solve each equation by completing the square. Check solutions by using the STOre feature or graphically.

31. $x^2 + 2x - 8 = 0$ **32.** $x^2 + 6x + 5 = 0$ **33.** $2x^2 - x = -1$ **34.** $2x^2 - 5x = 2$

35. $3x^2 - 11x = -10$ **36.** $2x^2 = x - 6$ **37.** $5x^2 = -3x - 2$

38. $x^2 = 2x - 6$ **39.** $5x^2 + 2x = 6$ **40.** $3x^2 + 4x = 2$

✎**41.** When solving equations graphically, either the INTERSECT or ROOT/ZERO feature of the calculator may be used.
   a. Discuss which feature you prefer and WHY you prefer it.
   b. Discuss the advantages and disadvantages of each feature.

✎**42.** Solve the equation $x^2 - 12 = 0$ both analytically (by hand) and graphically. Which process gives an exact result? Explain. (*Hint:* Use the STOre feature to confirm the exact roots and to show that the irrational approximations are truly just approximations.)

43. Solve the equation $x^2 - x - 12 = 0$
   a. by factoring
   b. by completing the square
   c. graphically

44. Find a quadratic equation with roots of 3 and 5.

45. Find a quadratic equation with solutions of −4 and 6.

46. Find a third-degree equation with solutions of 2, 3, and −4.

47. Find a fourth-degree equation with solutions of 3, −3, 4, and −4.

APPLICATIONS

48. Building fences Mr. Maier is constructing a dog run for his miniature dachshund (who *hates* the outdoors). He'd like the length to be 3 times the width, and the area is to be 147 square feet.
   a. What are the dimensions of the dog run?

   b. How many feet of fencing must he buy?

49. Building sheds Mr. MacLean is building a shed for his yard tools. He wants it to be in the shape of a rectangle, with the length being 4 feet more than the width. The area needs to be 96 square feet. What are the dimensions of the shed?

50. Pricing concert tickets Tickets to a rock concert cost $4, and the projected attendance is 300 people. It is further projected that for every $0.10 increase in ticket price, the average attendance will decrease by 5.
   a. The chart below models the data from the problem. Fill in the last row of the chart, letting x represent the number of price increases.

Ticket price	Attendance	Revenue
$4	300	4 · 300
$4 + 1(.10)	300 − 1(5)	[4 + 1(.10)][300 − 1(5)]
$4 + 2(.10)	300 − 2(5)	[4 + 2(.10)][300 − 2(5)]
$4 + 3(.10)	300 − 3(5)	[4 + 3(.10)][300 − 3(5)]
.	.	.
.	.	.
.	.	.

   b. How many price increases will yield nightly receipts of $1248?

   c. What would the price of a ticket be?

51. Setting bus fares A bus company has 3000 passengers daily, paying a $0.25 fare. For each nickel increase in fare, the company estimates it will lose 80 passengers.
   a. The chart below models the data from the problem. Fill in the last row of the chart, letting x represent the number of price increases.

Fare	Number of passengers	Daily revenue
.25	3000	.25(3000)
.25 + 1(.05)	3000 − 1(80)	[.25 + 1(.05)][3000 − 1(80)]
.25 + 2(.05)	3000 − 2(80)	[.25 + 2(.05)][3000 − 2(80)]
.25 + 3(.05)	3000 − 3(80)	[.25 + 3(.05)][3000 − 3(80)]
.	.	.
.	.	.
.	.	.

   b. How many price increases will yield a daily revenue of $994?
   c. What will the fare price be?

52. Computing revenue The *Gazette*'s profit is $20 per year for each of its 3000 subscribers. Management estimates that the profit per subscriber will increase $0.01 for each additional subscriber over the current 3000. How many subscribers will bring a total profit of $120,000? (*Hint:* Use the methods developed in Exercises 50 and 51 above to work this problem.)

53. Finding interest rates A woman invests $1000 in a mutual fund for which interest is compounded annually at a rate r. After 1 year she deposits an additional $2000. After 2 years the balance in the account is $1000(1 + r)^2 + $2000(1 + r)$. If this amount is $3368.10, find r.

54. Framing a picture The frame surrounding the picture in Illustration 1 has a constant width, and the area of the frame equals the area of the picture. How wide is the frame?

ILLUSTRATION 1

REVIEW

55. The pictured table is that of a linear function.

  a. State the x-intercept of the function.

  b. State the y-intercept of the function.

  c. Write an equation of the function in slope-intercept form.

In Exercises 56–58, classify each pictured display as that of a linear function, an absolute value function, or a radical function.

56.

57.

58.

10.2   The Quadratic Formula and the Discriminant

■ THE QUADRATIC FORMULA   ■ THE DISCRIMINANT

■ THE QUADRATIC FORMULA

Although all quadratic equations can be solved by completing the square, the method is sometimes tedious. The **quadratic formula** will give the solutions of a quadratic equation with much less effort. To develop this formula, we will complete the square.

A quadratic equation is an equation that can be written in the form $ax^2 + bx + c = 0$, with $a \neq 0$. If we solve this standard form by completing the square, we are solving every possible quadratic equation at one time. The solution of the standard quadratic equation results in the quadratic formula.

To solve the equation $ax^2 + bx + c = 0$, with $a \neq 0$, we proceed as follows:

$$ax^2 + bx + c = 0$$

$$\frac{ax^2}{a} + \frac{bx}{a} + \frac{c}{a} = \frac{0}{a} \qquad \text{Since } a \neq 0, \text{ we can divide both sides by } a.$$

$$x^2 + \frac{bx}{a} = -\frac{c}{a} \qquad \text{Simplify and subtract } \tfrac{c}{a} \text{ from both sides.}$$

c. $3x^2 + 2x + 2 = 0$

$$a = 3, b = 2, c = 2 \qquad \text{Identify the values of } a, b, \text{ and } c.$$

$$x = \frac{-2 \pm \sqrt{(2)^2 - 4(3)(2)}}{2(3)} \qquad \text{Substitute.}$$

$$x = \frac{-2 \pm \sqrt{4 - 24}}{6} \qquad \text{Simplify.}$$

$$x = \frac{-2 \pm \sqrt{-20}}{6} \qquad \text{Simplify.}$$

$$x = \frac{-2 \pm 2i\sqrt{5}}{6} \qquad \text{Simplify the radical.}$$

$$x = \frac{2(-1 \pm i\sqrt{5})}{6} \qquad \text{Factor the numerator so that the fraction can be simplified by dividing out common factors.}$$

$$x = \frac{-1 \pm i\sqrt{5}}{3}$$

The solutions are $x = \dfrac{-1 + i\sqrt{5}}{3}$ and $x = \dfrac{-1 - i\sqrt{5}}{3}$. These could also be expressed in $a + bi$ form as $-\dfrac{1}{3} + \dfrac{\sqrt{5}}{3}i$ and $-\dfrac{1}{3} - \dfrac{\sqrt{5}}{3}i$.

Check: Using the STOre feature:

d. Begin by putting the equation $9x^2 + 12x = -4$ in standard form and identifying the values of a, b, and c.

$$9x^2 + 12x + 4 = 0$$

$$a = 9, b = 12, c = 4$$

$$x = \frac{-12 \pm \sqrt{(12)^2 - 4(9)(4)}}{2(9)} \qquad \text{Substitute.}$$

$$x = \frac{-12 \pm \sqrt{144 - 144}}{18} \qquad \text{Simplify.}$$

$$x = \frac{-12 \pm \sqrt{0}}{18} \qquad \text{Simplify.}$$

$$x = \frac{-12 \pm 0}{18} \qquad \text{Simplify the radical.}$$

Because either adding or subtracting the number 0 to -12 does not change the value of the numerator, this equation has only one root, $x = \dfrac{-12}{18} = -\dfrac{2}{3}$, which is often called a *double root*.

Check: Using the STOre feature: Using the graphical capabilities:

Notice that round-off error has occurred in the graphical solution: $-.6666664$ *is not equal* to $-\frac{2}{3}$.

> **Remember:**
> 1. The quadratic equation must be in standard form before a, b, and c are identified.
> 2. To ensure that the quantity represented by b is squared, enclose b in parentheses *before* squaring.
> 3. The numerator of the quadratic formula is the *entire* expression $-b \pm \sqrt{b^2 - 4ac}$, and the radicand is the *entire* expression $b^2 - 4ac$.
> 4. When simplifying results, factor the numerator and divide out common factors.

■ THE DISCRIMINANT

Sometimes it is helpful to know the type of solutions a quadratic equation will have, whether the roots are real (rational or irrational), imaginary, or equal. Consider the solutions of the quadratic equations in Example 1.

Equation	Roots	Number/Type of roots
$x^2 - 4x - 3 = 0$	$2 + \sqrt{7}, 2 - \sqrt{7}$	two distinct real, irrational roots
$x^2 - 8x + 7 = 0$	$7, 1$	two distinct real, rational roots
$3x^2 + 2x + 2 = 0$	$\dfrac{1 + i\sqrt{5}}{3}, \dfrac{-1 - i\sqrt{5}}{3}$	two distinct imaginary roots
$x^2 + 12x + 4 = 0$	$-\dfrac{2}{3}$	one real, rational root, often called a *double root*

What caused the differences in the types of roots? Looking back at the solutions, it is apparent that the discriminator was the quantity underneath the radical sign. From the quadratic formula, we see that this expression is $b^2 - 4ac$. This quantity is called the **discriminant.**

> **Discriminant**
> IF $b^2 - 4ac > 0$ AND is a perfect square,
>   then the quadratic equation will have two distinct real roots (Example 1b). Moreover, if a, b, and c are rational, the roots will be rational; if a, b, and c are irrational, the roots will be irrational.
> IF $b^2 - 4ac > 0$ AND is *not* a perfect square,
>   then the quadratic equation will have two distinct real, irrational roots (Example 1a).
> IF $b^2 - 4ac = 0$,
>   then the quadratic equation will have one real root, often called a *double root* (Example 1d). Again, if a, b, and c are rational, the roots will be rational; if a, b, and c are irrational, the roots will be irrational.
> IF $b^2 - 4ac < 0$,
>   then the quadratic equation will have two distinct imaginary roots. These roots are complex conjugates (Example 1c).

EXAMPLE 2 Use the discriminant to determine the number and type of roots of each quadratic equation.

  a. $15x^2 = 14x + 8$ **b.** $25x^2 + 10x = -1$

  c. $x^2 - 2x + 5 = 0$ **d.** $x^2 - 2x = 5$

Solution **a.** Put the equation in standard form and identify a, b, and c.

$$15x^2 - 14x - 8 = 0$$
$$a = 15, \, b = -14, \, c = -8$$

Evaluate the discriminant, $b^2 - 4ac$, for the given values of a, b, and c.

$$b^2 - 4ac = (-14)^2 - 4(15)(-8)$$
$$= 196 + 480$$
$$= 676$$

The discriminant is positive, which means the equation will have two distinct, real roots. Next, we must determine if the roots are rational or irrational. This is done by checking to see if 676 is a perfect square. If it is, the roots are rational because the equation's coefficients are rational; if it is not, the roots are irrational.

$$\sqrt{676} = 26$$

therefore, the roots are rational. This equation has two distinct real, rational roots.

b. Put the equation in standard form and identify a, b, and c.

$$25x^2 + 10x + 1 = 0$$
$$a = 25, \, b = 10, \, c = 1$$

Evaluate the discriminant, $b^2 - 4ac$, for the given values of a, b, and c.

$$b^2 - 4ac = (10)^2 - 4(25)(1)$$
$$= 100 - 100$$
$$= 0$$

The value of the discriminant is 0, which means the equation will have one real, rational root. This root is often called a *double root*. This root is rational because the coefficients are rational.

c. Because the equation

$$x^2 - 2x + 5 = 0$$

is in standard form, we simply identify a, b, and c.

$$a = 1, \, b = -2, \, c = 5$$

Evaluate the discriminant, $b^2 - 4ac$, for the given values of a, b, and c.

$$b^2 - 4ac = (-2)^2 - 4(1)(5)$$
$$= 4 - 20$$
$$= -16$$

The discriminant is negative, which means the equation will have two distinct, imaginary roots. These roots are complex conjugates.

d. Put the equation in standard form and identify a, b, and c.

$$x^2 - 2x - 5 = 0$$
$$a = 1, b = -2, c = -5$$

Evaluate the discriminant, $b^2 - 4ac$, for the given values of a, b, and c.

$$b^2 - 4ac = (-2)^2 - 4(1)(-5)$$
$$= 4 + 20$$
$$= 24$$

The discriminant is positive, which means the equation will have two distinct, real roots. Next, we must determine if the roots are rational or irrational. This is done by checking to see if 24 is a perfect square. If it is, the roots are rational; if it is not, the roots are irrational.

$$\sqrt{24} \approx 4.898979486$$

therefore, the roots are irrational. This equation has two distinct real, irrational roots. ■

■ **TECHNOLOGY TIP**
It is often convenient to store values for a, b, and c; evaluate the discriminant; and then recall the store command to change values for a, b, and c.

Solving Quadratic Equations
1. Using the quadratic formula or completing the square will provide solutions to any quadratic equation.
2. If $b = 0$ (there is no x, or linear term), use the square root property.
3. If the polynomial $ax^2 + bx + c$ is factorable, factoring may be used.
4. If required solutions are to be exact, algebraic methods are preferred.
5. If approximations are acceptable, graphical methods may be more expedient.
6. *All* exact solutions may be checked, or confirmed, using the STOre feature of the calculator.

EXERCISE 10.2

VOCABULARY   *Fill in the blanks to make a true statement.*

1. When a quadratic equation is in standard form, $ax^2 + bx + c = 0$, the quadratic formula states

$x = $ _____.

2. When solving a quadratic equation, the roots are $1 + \sqrt{5}$ and $1 - \sqrt{5}$. This is an _____ result. (exact/approximate)

3. When solving the same quadratic equation as in Exercise 2, a student used the ROOT/ZERO feature of the graphing calculator and stated that the roots are 3.236068 and −1.236068. These results are _____. (exact/approximate)

4. The formula for the discriminant is _____.

5. If the value of the discriminant is positive and a perfect square and the coefficients are rational, the equation will have two distinct real, _____ roots.

6. If the value of the discriminant is positive and is not a perfect square, the equation has two distinct real, _____ roots.

7. If the value of the discriminant is 0 and the coefficients are rational, the equation has one real _____ root, often called a _____ root.

8. If the value of the discriminant is negative, the equation has two distinct _____ roots. These are _____.

CONCEPTS *Consider the equation* $y = x^2 - x - 12$ *whose graph is pictured below. A table of values is also provided.*

x	y
-4	8
-3	0
-2	-6
-1	-10
0	-12
1	-12
2	-10
3	-6
4	0
5	8

9. State the x-coordinates of the x-intercepts.

10. Complete the solution of $x^2 - x - 12 = 0$, where $a = 1$, $b = -1$, $c = -12$.

$$x = \frac{-(-1) \pm \sqrt{\boxed{}^2 - 4\,\boxed{}\,\boxed{}}}{2\,\boxed{}}$$

$$x = \frac{\boxed{} \pm \sqrt{1 + 48}}{\boxed{}}$$

$$x = \frac{1 \pm \sqrt{49}}{2}$$

$$x = \frac{1 \pm \boxed{}}{2}$$

$$x = \boxed{} \quad \text{or} \quad x = \boxed{}$$

11. The solutions, or roots, of the equation are the same as _____.

PRACTICE *In Exercises 12–35, use the quadratic formula to solve each quadratic equation. Check your solutions, or roots, by using the STOre feature of the calculator.*

12. $x^2 + 3x + 2 = 0$

13. $x^2 - 3x + 2 = 0$

14. $x^2 + 9 = 0$

15. $x^2 + 16 = 0$

16. $x^2 + 12x = -36$

17. $y^2 - 18y = -81$

18. $3x^2 = -16$

19. $2x^2 = -25$

20. $x^2 + 2x + 2 = 0$

21. $x^2 + 3x + 3 = 0$

22. $5x^2 + 5x + 1 = 0$

23. $4w^2 + 6w + 1 = 0$

24. $2x^2 + x + 1 = 0$

25. $3x^2 + 2x + 1 = 0$

26. $16y^2 + 8y - 3 = 0$

27. $16x^2 + 16x + 3 = 0$

28. $3x^2 - 4x = -2$

29. $2x^2 + 3x = -3$

30. $\dfrac{x^2}{2} + \dfrac{5}{2}x = -1$

31. $-3x = \dfrac{x^2}{2} + 2$

32. $3x^2 - 2x = -3$

33. $5x^2 = 2x - 1$

34. $8u = -4u^2 - 3$

35. $4t + 3 = 4t^2$

In Exercises 36–43, use the discriminant to determine what type of solutions exist for each quadratic equation. ***DO NOT SOLVE THE EQUATION.***

36. $4x^2 - 4x + 1 = 0$

37. $5x^2 + x + 2 = 0$

38. $2x^2 = 4x - 1$

39. $x(2x - 3) = 20$

40. $6x^2 - 5x - 6 = 0$

41. $3x^2 + 10x - 2 = 0$

42. $9x^2 = 12x - 4$

43. $x(x - 3) = -10$

In Exercises 44–51, find the value(s) of k that will make the solutions of each given quadratic equation equal.

44. $x^2 + kx + 9 = 0$

45. $9x^2 + 4 = -kx$

46. $(k-1)x^2 + (k-1)x + 1 = 0$

47. $(k+4)x^2 + 2kx + 9 = 0$

48. $kx^2 - 12x + 4 = 0$

49. $9x^2 - kx + 25 = 0$

50. $(k+3)x^2 + 2kx + 4 = 0$

51. $(k+15)x^2 + (k-30)x + 4 = 0$

52. Use the discriminant to determine whether the solutions of $1492x^2 + 1776x - 1984 = 0$ are real numbers.

53. Use the discriminant to determine whether the solutions of $1776x^2 - 1492x + 1984 = 0$ are real numbers.

54. Determine k such that the solutions of $3x^2 + 4x = k$ are imaginary numbers.

55. Determine k such that the solutions of $kx^2 - 4x = 7$ are imaginary numbers.

In Exercises 56–63, solve each equation graphically. Round to the nearest hundredth where necessary. Specify window values used as well as calculator feature (INTERSECT or ROOT/ZERO).

56. $2.3x^2 + 16.2x = 21.5$

57. $3.4x^2 + 10.2x = 16.5$

58. $\dfrac{2x}{x+1} - \dfrac{3x}{x+2} = -3$

59. $\dfrac{3x}{x-1} - \dfrac{x}{x+2} = 5$

60. $(x+5)(2x+3) = 2x + 5$

61. $(x+1)(x+3) = 2x^2 + 3x$

62. $0.2x + \dfrac{5}{x} = 8$

63. $0.4x + \dfrac{6}{x} = 6$

64. Discuss ways to use the graphing calculator to check solutions of quadratic equations. Your discussion should address the use of the VARS, STOre, and TABLE features as well as the use of the graphical display. Be sure to address roots that are rational, irrational, and imaginary.

65. Does the quadratic formula hold for equations with complex coefficients? Solve $x^2 + 2ix - 5 = 0$ using the quadratic formula. Verify the solutions.

66. Does the quadratic formula hold for equations with complex coefficients? Solve $2x^2 - 4ix + 6 = 0$ using the quadratic formula. Verify the solutions.

In Exercises 67–70, solve each quadratic equation and verify that the sum of the solutions is $-\frac{b}{a}$ and that the product of the solutions is $\frac{c}{a}$.

67. $12x^2 - 5x - 2 = 0$

68. $8x^2 - 2x - 3 = 0$

69. $2x^2 + 5x + 1 = 0$

70. $3x^2 + 9x + 1 = 0$

APPLICATIONS

71. Discriminant
   a. Solve the quadratic equation $x^2 + 2\sqrt{3}x + 2 = 0$ by completing the square.
   b. Identify a, b, and c and find the discriminant.
   c. What type of root(s) would you expect, given the discriminant value in part (b)?
   d. Explain why your answers in parts (a) and (c) are not a contradiction.

72. Discriminant
   a. Solve the quadratic equation $x^2 - 2\sqrt{2}x + 2 = 0$ by completing the square.
   b. Identify a, b, and c and find the discriminant.
   c. What type of root(s) would you expect, given the discriminant value in b?
   d. Explain why your answers in parts (a) and (c) are not a contradiction.

REVIEW

73. a. Graph the function $f(x) = x^2 + 5x - 6$.
   b. Find the *x*-intercepts of the function.
   c. Solve the equation $x^2 + 5x - 6 = 0$ by factoring.

   d. Compare the *x*-intercepts of the function (found in part (b)) and the solutions of the equation (found in part (c)).

74. a. Graph the function $f(x) = 3x + 6$.
   b. Find the *x*-intercept of the function.
   c. Solve the equation $3x + 6 = 0$.
   ✐**d.** Compare the *x*-intercept of the function and the solution of the equation.

75. a. Graph the function $f(x) = |3x + 6| - 2$.
   b. Find the *x*-intercepts of the function.
   c. Solve the equation $|3x + 6| - 2 = 0$.
   ✐**d.** Compare the *x*-intercepts of the function and the solutions of the equation.

76. a. Graph the function $f(x) = \sqrt{5x - 10} - 4$.
   b. Find the *x*-intercept of the function.
   c. Solve the equation $\sqrt{5x - 10} - 4 = 0$.
   ✐**d.** Compare the *x*-intercept of the function and the solution of the equation.

10.3 Applications of Quadratic Functions and Solutions of Equations Using Quadratic Methods

■ APPLICATIONS OF QUADRATIC FUNCTIONS ■ SOLUTIONS OF EQUATIONS USING QUADRATIC METHODS

■ APPLICATIONS OF QUADRATIC FUNCTIONS

In the previous sections we examined solutions of quadratic equations. We found that the roots of the equations are the *x*-intercepts of the graph of the function associated with the equation. Examination of quadratic functions reveals information about the general behavior of quadratic models. We will relate the algebraic processes previously developed to the applications presented in this section.

The graph of a quadratic function is a curve called a **parabola.** All graphs of quadratic functions are parabolic in shape, opening either up or down, as shown in the following graphs:

The turning point of the curve is also the highest or the lowest point (depending on the opening of the parabola). This point is called the **vertex** of the parabola.

EXAMPLE 1   Suppose the ball shown in Figure 10-1 is thrown straight up with a velocity of 128 feet per second. The quadratic function $y = 128x - 16x^2$ gives the relationship between *x* and *y*. The variable *x* represents the time measured in seconds, whereas *y* represents the number of feet the ball is above the ground.

   a. Graph the function in an appropriate window. Make sure the *x*-intercepts, the *y*-intercept, and the vertex are clearly visible.

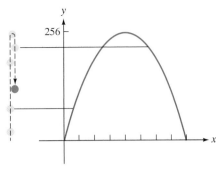

FIGURE 10-1

b. How long does it take for the ball to hit the ground?

c. What is the maximum height of the ball?

d. How high is the ball after 2 seconds?

e. At what other time does the ball reach the same height?

Solution

a. Graphing the quadratic in the standard viewing window reveals little information. The methods previously developed (tracing to find maximum or minimum points and x-intercepts) are laborious and time-consuming. To obtain a good graph, think about what x and y represent. The variable x is time, measured in seconds, whereas y (the value of the function at x, or $f(x)$) is the height. Set the TABLE to begin at 0 and to be incremented by 1. Examining the table, we see that 256 is a maximum (for the given increment).

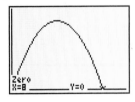

Using this information, reset the window to an x-minimum of 0 and an x-maximum of 10 (we are guessing, using our table values to formulate an educated guess). Our y-minimum is 0, and we will use 300 as a maximum. Both scales have been set equal to 0 to eliminate distortion of the axes.

b. Recall that our function is $y = 128x - 16x^2$. The ball will hit the ground when $y = 0$. We are looking for the x-value corresponding to the value of 0 for y. These points are the x-intercepts of the graph. The ROOT/ZERO feature enables us to find the x-intercept. Interpreting the display within the context of the problem, we can conclude that the ball will hit the ground after 8 seconds.

The equation $0 = 128x - 16x^2$ could have been solved algebraically by any of the methods presented in Sections 10.1 and 10.2. Of course, whether solving algebraically or graphically, we only find the x-intercept that "makes sense" within the context of the problem and the graphical display. The other x-intercept is 0, which is, of course, the initial placement of the ball; it has traveled 0 feet in 0 seconds.

■ TECHNOLOGY TIP
The max option is usually located under the same menu as the ROOT/ZERO and INTERSECT options. On the TI-83, this is under the CALC menu.

c. To find the maximum height of the ball, the maximum option of the calculator can be used. Bounds are set to the left and the right of the maximum point.

There may be round-off error (as in the third screen), so be careful. Because the variable y represents the height of the ball, the maximum height is 256 feet.

Currently, we have no algebraic process in place to find a maximum.

d. To find the height after 2 seconds, we are merely finding the value of the function $y = 128x - 16x^2$ when $x = 2$. The value feature is the perfect option to use. Interpretation of the display reveals that the height of the ball is 192 feet after 2 seconds.

Algebraically, we would simply substitute the value of 2 in for x in the definition of the function and solve for y.

e. To find the other x-value when $y = 192$, we are asked to find the other solution of the equation $192 = 128x - 16x^2$. Entering the value of 192 at the y2 prompt shows *both* points of intersection with the curve. Using the INTERSECT feature and interpreting the screen display, we see that the ball reaches a height of 192 feet again after 6 seconds have elapsed.

Algebraically, we would simply solve the equation $192 = 128x - 16x^2$ using methods introduced previously.

EXAMPLE 2   A man wants to build the rectangular pen shown in Figure 10-2 to house his dog. To save fencing, he intends to use one side of his garage. Find the maximum area that he can enclose with 80 feet of fencing.

FIGURE 10-2

Solution   We can let the width of the desired area be represented by w. Then, the length is represented by $80 - 2w$. The area of a rectangular region is the length times the width. Therefore, $A = w(80 - w)$.

We enter the function $y = x(80 - x)$ at the y= screen, remembering that the variable x represents the width of the desired region and y represents the area.

To find a good window, we again scroll through TABLE values while observing x- and y-values. We set the window to [0, 40] by [0, 1000] to obtain the pictured graph.

Use of the maximum feature of the calculator and interpretation of the graphical display allows us to conclude that the pen should be 20 feet wide and 40 feet long for a maximum area of 800 square feet.

In the previous examples, the maximum feature of the graphing calculator was used to find the vertex of graphs of parabolas that opened downward. No algebraic methods have been developed, thus far, to find these vertices. The following Exploring the Concept feature will help you develop a method for finding the vertices of a parabola.

EXPLORING THE CONCEPT

a. Graph $f(x) = -x^2 - 2x + 24$ in the window $[-10, 10]$ by $[-10, 30]$ and use the maximum feature of the graphing calculator to find the vertex.

b. Find the x-intercepts using either algebraic methods or the graphing calculator.

c. Find the average of the x-intercepts.

d. Find the value of the function when x equals the average of the x-intercepts.

e. Compare the results of part (d) to part (a).

Conclusions

a.

The vertex has coordinates $(-1, 25)$. (Round-off error has occurred in the calculator graph. Verify the ordered pair using the STOre feature.)

b.

The x-intercepts are -6 and 4.
   Algebraically:

$$0 = -x^2 - 2x + 24$$
$$x^2 + 2x - 24 = 0$$
$$(x + 6)(x - 4) = 0$$
$$x + 6 = 0 \quad \text{and} \quad x - 4 = 0$$
$$x = -6 \quad \text{and} \quad x = 4$$

c. The average of the x-intercepts is $\dfrac{-6 + 4}{2} = \dfrac{-2}{2} = -1$.

d. Since $f(x) = -x^2 - 2x + 24$, then

$$f(-1) = -(-1)^2 - 2(-1) + 24$$
$$= -1 + 2 + 24$$
$$= 25$$

Therefore, $f(-1) = 25$.

e. The vertex of the parabola can be found by averaging the x-intercepts and then finding the value of the function when x equals the average of the x-intercepts.

The average of the x-intercepts is the x-coordinate of the vertex, whereas the value of the function at this average is the y-coordinate of the vertex. ■

■ SOLUTIONS OF EQUATIONS USING QUADRATIC METHODS

Many nonquadratic equations can be put into quadratic form and then solved with the techniques used for solving quadratic equations. For example, to solve $x^4 - 5x^2 + 4 = 0$, we can proceed as follows:

$$x^4 - 5x^2 + 4 = 0$$
$$(x^2)^2 - 5(x^2) + 4 = 0$$
$$u^2 - 5u + 4 = 0 \qquad \text{Let } u = x^2$$
$$(u - 4)(u - 1) = 0$$
$$u - 4 = 0 \qquad \text{or} \qquad u - 1 = 0$$
$$u = 4 \qquad \text{or} \qquad u = 1$$

Since $u = x^2$, it follows that $x^2 = 4$ or $x^2 = 1$. Thus, we can use the square root property to find the roots. Because

$$x^2 = 4 \qquad \text{or} \qquad x^2 = 1$$

it follows that

$$x = 2 \qquad \text{or} \qquad x = -2 \qquad x = 1 \qquad \text{or} \qquad x = -1.$$

This equation has four solutions: 1, −1, 2, and −2.

The equation could also be solved graphically. The ROOT/ZERO feature would simply have to be used four times to find all of the roots. It should be remembered that the x-intercepts of a function are the real solutions of the equation.

EXAMPLE 3   Solve the equation $x - 7x^{1/2} + 12 = 0$.

Solution   The equation is not quadratic. However, if u^2 is substituted for x and u is substituted for $x^{1/2}$, the equation $x - 7x^{1/2} + 12 = 0$ becomes a quadratic equation that can be solved by factoring:

$$u^2 - 7u + 12 = 0 \qquad \text{Substitute } u^2 \text{ for } x \text{ and } u \text{ for } x^{1/2}.$$
$$(u - 3)(u - 4) = 0 \qquad \text{Factor.}$$
$$u - 3 = 0 \qquad \text{or} \qquad u - 4 = 0 \qquad \text{Set each factor equal to zero.}$$
$$u = 3 \qquad \text{or} \qquad u = 4 \qquad \text{Solve.}$$

Because $x = u^2$, it follows that

$$x = 3^2 \quad \text{or} \quad x = 4^2$$
$$x = 9 \quad \text{or} \quad x = 16$$

The equation has two solutions.

The equation could also be solved graphically. The ROOT/ZERO feature would simply have to be used twice to find both of the roots. It should be remembered that the x-intercepts of a function are the real solutions of the equation.

EXAMPLE 4   Solve the equation $x^{2/3} - 5x^{1/3} + 6 = 0$.

Solution   Because $x^{2/3} = (x^{1/3})^2$, we will let $t = x^{1/3}$, which means that $t^2 = x^{2/3}$. Substituting these values in makes our equation $t^2 - 5t + 6 = 0$, which can be solved by factoring:

$$(t - 3)(t - 2) = 0$$
$$t - 3 = 0 \quad \text{or} \quad t - 2 = 0$$
$$t = 3 \quad \text{or} \quad t = 2$$

Because $t = x^{1/3}$, we have

$$x^{1/3} = 3 \quad \text{or} \quad x^{1/3} = 2$$

Cubing both sides of these equations yields the equations $x = 27$ or $x = 8$. These are the solutions of the equation.

The equation could also be solved graphically. The ROOT/ZERO feature would simply have to be used twice to find both of the roots. It should be remembered that the x-intercepts of a function are the real solutions of the equation.

(The window values used in the calculator graphs shown are $[-2, 40]$ by $[-2, 2]$.)

EXERCISE 10.3

VOCABULARY   *Fill in the blanks to make a true statement.*

1. The roots of a quadratic equation are the _____ of the graph of the associated quadratic function.

2. The graph of a quadratic function is a curve called a _____. This curve will open either ___ or _____.

3. The turning point of the graph of a quadratic function is called the _____. It is the _____ or _____ point, depending on how the curve opens.

CONCEPTS

A ball is dropped from the Sears Tower in Chicago, a height of approximately 1454 feet. The pull of gravity is approximately 32 ft/sec². If we disregard air resistance, the approximate height of the ball from the ground is $h = -16t^2 + 1454$, where h is height and t is time in seconds. A graph modeling the relationship between the height of the ball and elapsed time is shown in Illustration 1.

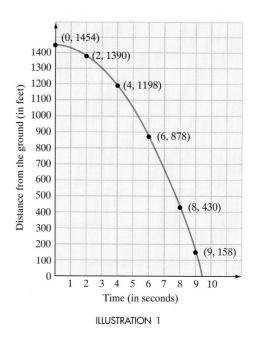

ILLUSTRATION 1

Match each phrase in Exercises 4–7 with one of these positions on the parabola: x-intercept, y-intercept, point with coordinates (4, 1198).

4. The position of the ball when it first reaches the ground.

5. The position of the ball 4 seconds after release.

6. The position of the ball before it is released.

7. The ball is 1198 feet from the ground after 4 seconds.

In Exercises 8 and 9, use Illustration 1 to answer the questions.

8. After how many seconds is the ball 430 feet from the ground?

9. After 6 seconds, how far is the ball from the ground?

PRACTICE   *In Exercises 10 and 11, use the given TABLES to answer the questions.*

10. The function $f(x) = x^2 - 2x - 15$ was entered at the y1= prompt.

X	Y₁			X	Y₁	
-4	9			3	-12	
-3	0			4	-7	
-2	-7			5	0	
-1	-12			6	9	
0	-15			7	20	
1	-16			8	33	
2	-15			9	48	
X=-4				X=3		

   a. State the coordinates of the *x*-intercepts.

   b. State the coordinates of the *y*-intercept.
   c. State the coordinates of the vertex (turning point).

   d. Graph the equation on graph paper.
   e. $f(2) = ?$
   f. If $f(x) = -12$, then $x = ?$ or $x = ?$

11. The function $f(x) = -x^2 + 8x - 16$ was entered at the y1= prompt.

X	Y₁			X	Y₁	
-1	-25			6	-4	
0	-16			7	-9	
1	-9			8	-16	
2	-4			9	-25	
3	-1			10	-36	
4	0			11	-49	
5	-1			12	-64	
X=-1				X=6		

   a. State the coordinates of the *x*-intercept.
   b. State the coordinates of the *y*-intercept.
   c. State the coordinates of the vertex (turning point).

   d. Graph the equation on graph paper.
   e. $f(2) = ?$
   f. If $f(x) = -1$, then $x = ?$ or $x = ?$

In Exercises 12–15, use the given graph of the area of a circle ($A = \pi r^2$), where y represents the area and x represents the radius.

12. As the radius of the circle increases, what happens to the area?

13. If the area is approximately 78.5 square units, what is the radius of the circle? Express this information using an ordered pair and using function notation.

14. If the radius of the circle is 7 units, what is the area? Express this information using an ordered pair and using function notation.

✍**15.** Explain why the authors chose to display only part of the parabolic curve.

In Exercises 16–20, use the given graph relating the height (in feet) of a model rocket to its time to answer the given questions. The equation $y = -16x^2 + 750x$ is graphed in a window [0, 60] by [0, 10000].

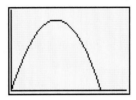

✍**16.** Two x-intercepts are apparent, one with coordinates (0, 0) and the other with coordinates (46.875, 0). Interpret each of these values within the context of the problem.

✍**18.** Interpret the given graphical displays within the context of the problem.

✍**17.** The pictured display gives approximations of the coordinates of the vertex of the parabola. Interpret these values (rounded to the nearest unit) within the context of the problem.

✍**19.** After 20 seconds, how high is the rocket? Explain in detail how you obtained your result.

✍**20.** After how many seconds (approximately) is the rocket 8000 feet high? Explain in detail how you obtained your result.

APPLICATIONS *Solve each of the following either algebraically or graphically. Make sure that anyone reading your solution could logically follow your reasoning.*

21. If a ball is thrown straight up with an initial velocity of 48 feet per second, its height, y, after x seconds is given by the equation $y = 48x - 16x^2$.
   a. Graph the equation in a good viewing window.

   b. What is the maximum height attained by the ball?

   c. To the nearest tenth of a second, how long does it take to reach its maximum height?
   d. How long does it take the ball to reach the ground?

22. From the top of the building shown in Illustration 2, a ball is thrown straight up with an initial velocity of 32 feet per second. The equation $s = -16t^2 + 32t + 48$ gives the height, s, of the ball t seconds after it was thrown.

23. A farmer wants to fence in three sides of a rectangular field (as shown in Illustration 3) with 1000 feet of fencing. The other sides of the rectangle will be a river. If the enclosed area is to be a maximum, determine the size of the field.

ILLUSTRATION 2

ILLUSTRATION 3

a. Find the maximum height of the ball.
b. Find the time it takes for the ball to hit the ground.

PRACTICE   *In Exercises 24–51, solve each equation either graphically or by using algebraic methods.*

24. $x^4 - 17x^2 + 16 = 0$

25. $x^4 - 10x^2 + 9 = 0$

26. $x^4 - 3x^2 = -2$

27. $x^4 - 29x^2 = -100$

28. $x^4 = 6x^2 - 5$

29. $x^4 = 8x^2 - 7$

30. $x^4 = 6x^2 - 5$

31. $2x^4 + 24 = 26x^2$

32. $2x + x^{1/2} - 3 = 0$

33. $2x - x^{1/2} - 1 = 0$

34. $3x + 5x^{1/2} + 2 = 0$

35. $3x - 4x^{1/2} + 1 = 0$

36. $x^{2/3} + 5x^{1/3} + 6 = 0$

37. $x^{2/3} - 7x^{1/3} + 12 = 0$

38. $x^{2/3} - 2x^{1/3} - 3 = 0$

39. $x^{2/3} + 4x^{1/3} - 5 = 0$

40. $x + 5 + \dfrac{4}{x} = 0$

41. $x - 4 + \dfrac{3}{x} = 0$

42. $x + 1 = \dfrac{20}{x}$

43. $x + \dfrac{15}{x} = 8$

44. $\dfrac{1}{x-1} + \dfrac{3}{x+1} = 2$

45. $\dfrac{6}{x-2} - \dfrac{12}{x-1} = -1$

46. $\dfrac{1}{x+2} + \dfrac{24}{x+3} = 13$

47. $\dfrac{3}{x} + \dfrac{4}{x+1} = 2$

48. $x^{-4} - 2x^{-2} + 1 = 0$

49. $4x^{-4} + 1 = 5x^{-2}$

50. $x + \dfrac{2}{x-2} = 0$

51. $x + \dfrac{x+5}{x-3} = 0$

REVIEW   *In Exercises 52–59, graph each of the following functions on graph paper. You may use your graphing calculator to help you obtain points; however, make sure your hand-drawn graph is complete.*

52. Graph the function $y = \dfrac{1}{2}x + 4$

53. Graph the function $y = \dfrac{1}{2}x - 3$.

54. Graph the function $y = (x - 3)^2 + 4$.

55. Graph the function $y = (x - 3)^2 - 3$.

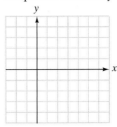

56. Graph the function $y = \sqrt{x} + 4$.

57. Graph the function $y = \sqrt{x} - 3$.

58. Graph the function $y = |x| + 4$.

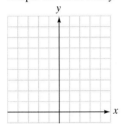

59. Graph the function $y = |x| - 3$.

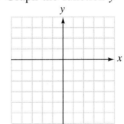

10.4 Graphs of Quadratic Functions

■ GRAPHS OF $f(x) = ax^2$ ■ GRAPHS OF $f(x) = ax^2 + k$ ■ GRAPHS OF $f(x) = a(x - h)^2$
■ SUMMARY OF GRAPHS OF $f(x) = a(x - h)^2 + k$ ■ GRAPHS OF $f(x) = ax^2 + bx + c$

■ GRAPHS OF $f(x) = ax^2$

A **quadratic function** is a second-degree polynomial function of the form $y = f(x) = ax^2 + bx + c$ $(a \neq 0)$, where a, b, and c are real numbers. The graphs of quadratic functions are called **parabolas.**

EXPLORING THE CONCEPT

a. Examine the graphs of the quadratic functions

$$f(x) = x^2 \qquad g(x) = 3x^2 \qquad h(x) = \frac{1}{3}x^2 \qquad t(x) = -3x^2$$

b. Complete a table of values for each of the functions for x-values of -2, -1, 0, 1, 2.

c. State the vertex of each of the parabolas.

d. Does the graph of each parabola open up or down?

e. What effect does changing the value of a (the coefficient of the squared term) have on the graphs?

Conclusions   **a.**

$f(x) = x^2$

$g(x) = 3x^2$

$h(x) = \frac{1}{3}x^2$

$t(x) = -3x^2$

b.

$f(x) = x^2$

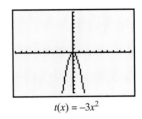

$g(x) = 3x^2$

$h(x) = \frac{1}{3}x^2$

$t(x) = -3x^2$

  c. The vertex of $f(x) = (0, 0)$.
     The vertex of $g(x) = (0, 0)$.
     The vertex of $h(x) = (0, 0)$.
     The vertex of $t(x) = (0, 0)$.

  d. $f(x)$, $g(x)$, and $h(x)$ open up; $t(x)$ opens down.

  e. When $a > 0$ (positive), the parabola opens up; when $a < 0$ (negative), the parabola opens down. The graph of $h(x) = \frac{1}{3}x^2$ is wider than the graph of $f(x) = x^2$. The graph of $g(x) = 3x^2$ is narrower than the graph of $f(x) = x^2$. The graph of $t(x) = -3x^2$ is the same width as $g(x) = 3x^2$; however, the vertex is a maximum rather than a minimum.

    The results are summarized below:

> ## Horizontal Stretches of Graphs of Quadratic Functions
> If f is a function and a a real number, then
> - the graph of $f(x) = ax^2$ opens up when a is positive and down when a is negative.
> - if $|a| > 1$, then the graph of $f(x) = ax^2$ is narrower than the graph of $f(x) = x^2$; it is shrunk by a factor of $|a|$.
> - if $0 < |a| < 1$, then the graph of $f(x) = ax^2$ is wider than the graph of $f(x) = x^2$; it is increased by a factor of $|a|$.

■ GRAPHS OF $f(x) = ax^2 + k$

The next Exploring the Concept section examines the effect of adding or subtracting a constant to the squared term of a quadratic function.

EXPLORING THE CONCEPT

a. Examine the graphs of the quadratic functions

$$f(x) = x^2 \qquad g(x) = x^2 + 4 \qquad h(x) = x^2 - 5$$

b. Complete a table of values for each of the functions for x-values of $-2, -1, 0, 1, 2$.

c. State the vertex of each of the parabolas.

d. What effect does adding or subtracting a constant have on the function?

Conclusions

a.

$f(x) = x^2$

$g(x) = x^2 + 4$

$h(x) = x^2 - 5$

b.

$f(x) = x^2$ $g(x) = x^2 + 4$ $h(x) = x^2 - 5$

c. The vertex of $f(x) = (0, 0)$.
The vertex of $g(x) = (0, 4)$.
The vertex of $h(x) = (0, -5)$.

d. The constant simply translates (moves) the parabola up or down the y-axis. The translation is up when a constant is added; down when a constant is subtracted. ■

The results are summarized below:

> **Vertical Translations of Graphs of Quadratic Functions**
> If f is a function and k is a positive number, then
> - the graph of $f(x) + k$ is identical to the graph of $y = f(x)$ except that it is translated k units upward.
> - the graph of $f(x) - k$ is identical to the graph of $y = f(x)$, except that it is translated k units downward.

EXAMPLE 1

Write the equation of each quadratic function described:

a. a quadratic function that is shifted 5 units vertically and is stretched so that it is 5 times wider than the graph of $f(x) = x^2$

b. a quadratic function that is shrunk so as to be $\frac{1}{8}$ as wide as $f(x) = x^2$ and is translated 4 units down

Solution   **a.** To shift the graph of $f(x) = x^2$ five units vertically, we add 5. The function is now $f(x) = x^2 + 5$. To be stretched so that the graph is five times wider, $a = \frac{1}{5}$. Therefore, the function is $f(x) = \frac{1}{5}x^2 + 5$.

The graphing calculator confirms our translations and stretches are correct:

Notice the vertex of $f(x) = x^2$ is $(0, 0)$, whereas the vertex of $f(x) = \frac{1}{5}x^2 + 5$ is $(0, 5)$.

b. To shift the graph of $f(x) = x^2$ down 4 units, subtract 4. The function is now $f(x) = x^2 - 4$. To be $\frac{1}{8}$ as wide, multiply x^2 by 8. Therefore, the function is $f(x) = 8x^2 - 4$.

The graphing calculator confirms our conjectures.

Notice the vertex of $f(x) = x^2$ is $(0, 0)$, whereas the vertex of $f(x) = 8x^2 - 4$ is $(0, -4)$.   ■

■ GRAPHS OF $f(x) = a(x - h)^2$

What happens to the graph of $f(x) = x^2$ when a constant is added to x *before* squaring?

EXPLORING THE CONCEPT   **a.** Examine the graphs of the quadratic functions

$$f(x) = x^2 \qquad g(x) = (x + 3)^2 \qquad h(x) = (x - 4)^2$$

b. Complete a table of values for each of the functions for x-values of -2, -1, 0, 1, 2.

c. State the vertex of each of the parabolas.

d. What effect does adding or subtracting a constant to x before squaring have on the function?

Conclusions   **a.**

  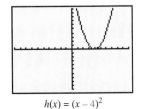

$f(x) = x^2$ $g(x) = (x + 3)^2$ $h(x) = (x - 4)^2$

b.

X	Y₁			X	Y₁			X	Y₁	
-2	4			-2	1			-2	36	
-1	1			-1	4			-1	25	
0	0			0	9			0	16	
1	1			1	16			1	9	
2	4			2	25			2	4	
3	9			3	36			3	1	
4	16			4	49			4	0	
X= -2				X= -2				X= -2		

$f(x) = x^2$ $\qquad\qquad$ $g(x) = (x + 3)^2$ $\qquad\qquad$ $h(x) = (x - 4)^2$

c. The vertex of $f(x) = (0, 0)$.
The vertex of $g(x) = (-3, 0)$.
The vertex of $h(x) = (4, 0)$.

d. The constant simply translates (moves) the graph of $f(x) = x^2$ to the left when a constant is added to x before squaring, and to the right when a constant is subtracted before squaring. ∎

The results are summarized below:

> ### Horizontal Translations of Graphs of Quadratic Functions
> If f is a function, and h is a constant,
> - the graph of $f(x - h)$ is identical to the graph of $y = f(x)$ except that it is translated h units to the right when h is positive.
> - the graph of $f(x - h)$ is identical to the graph of $y = f(x)$, except that it is translated h units to the left when h is negative.

EXAMPLE 2 Write the equation of each quadratic function described:

a. a quadratic function that is translated 3 units to the right and 6 units down from the graph of $f(x) = x^2$

b. a quadratic function that is translated 8 units left and 5 units up from the graph of $f(x) = x^2$

Solution **a.** To shift the graph of $y = x^2$ three units horizontally to the right, we subtract 3 from x before squaring. The function is now $f(x) = (x - 3)^2$. To shift the function down six units, we subtract 6. The desired function is $f(x) = (x - 3)^2 - 6$. The graphing calculator confirms our translations are correct.

X	Y₁	Y₂
-2	4	19
-1	1	10
0	0	3
1	1	-2
2	4	-5
3	9	-6
4	16	-5
X= -2		

The vertex of $f(x) = (x - 3)^2 - 6$ is $(3, -6)$. The coordinates specified at the bottom of the pictured screen are those of the y-intercept.

b. To shift the graph of $f(x) = x^2$ eight units horizontally to the left, we add 8 to the value of x before squaring. The function is now $f(x) = (x + 8)^2$. To shift the graph up five units, we add 5. The desired function is $f(x) = (x + 8)^2 + 5$.

The graphing calculator confirms our translations are correct.

The vertex of $f(x) = (x + 8)^2 + 5$ is $(-8, 5)$.

■ SUMMARY OF GRAPHS OF $f(x) = a(x - h)^2 + k$

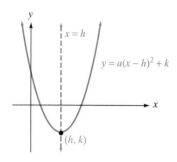

FIGURE 10-3

When a quadratic function is written in the form $f(x) = a(x - h)^2 + k$, its vertex is easily identified as (h, k). Graphs of quadratic functions are symmetric about an imaginary vertical line through the vertex. This line is called the **axis of symmetry** and has *equation* $x = h$ (see Figure 10-3). It is *not* a part of the graph of the quadratic function, but is often helpful when graphing.

> **Vertex and Axis of Symmetry of a Parabola**
> The graph of the equation $y = a(x - h)^2 + k$, where $a \neq 0$, is a parabola with vertex at (h, k).
>   The parabola opens upward when $a > 0$ and downward when $a < 0$. The axis of symmetry is the line $x = h$.

EXAMPLE 3   Graph the quadratic function $f(x) = 3(x + 5)^2 - 2$. State the coordinates of the vertex and the equation of the axis of symmetry.

Solution   The vertex is $(-5, -2)$, and the axis of symmetry is $x = -5$. The parabola has been shrunk by a factor of $\frac{1}{3}$ and is narrower than the graph of $f(x) = x^2$. However, specific coordinates need to be found, graphed, and connected with a smooth curve.
    It is suggested that two x-values larger than -5 and two x-values smaller than -5 be used (see Figure 10-4).

x	y
-7	10
-6	1
-5	-2
-4	1
-3	10

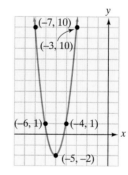

FIGURE 10-4

Remember, the vertex must be on the graph of the parabola and coordinates of enough points must be found on either side of the vertex to form a smooth curve.

■ GRAPHS OF $f(x) = ax^2 + bx + c$

When the equation of a quadratic function is in the form $f(x) = y = a(x - h)^2 + k$, the vertex and axis of symmetry are easily identifiable. For this reason, when graphing quadratic functions, we often put them in this form by completing the square on x.

EXAMPLE 4   Graph: $f(x) = 2x^2 - 4x - 1$.

Solution   We complete the square on x to write the function in the form $f(x) = a(x - h)^2 + k$.

$$f(x) = 2x^2 - 4x - 1$$

$$f(x) = 2(x^2 - 2x) - 1$$ Factor 2 from $2x^2 - 4x$ so that the coefficient of x^2 is 1.

$$f(x) = 2(x^2 - 2x + 1) - 1 - 2$$ Complete the square on x. Since this adds 2 to the right-hand side, we also subtract 2 from that side to maintain equality.

$$f(x) = 2(x - 1)^2 - 3$$ Factor $x^2 - 2x + 1$ and combine terms.

To graph, we identify the vertex $(1, -3)$, and complete a table of values by choosing two x-values smaller than the x-value of 1, and two x-values larger than 1. We then plot the points and connect them with a smooth curve (see Figure 10-5).

x	$f(x)$
-1	5
0	-1
1	-3
2	-1
3	5

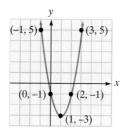

FIGURE 10-5

The graphing calculator, coupled with our knowledge of completing the square, can be used to write the function in the desired form, $f(x) = a(x - h)^2 + k$.

Graph the function with the calculator. Make sure your display shows a complete graph. Because the parabola opens up, use the minimum feature to find the vertex.

Be aware that there can be round-off error. Set the TABLE to begin at $x = 1$ (what we believe the x-coordinate of the vertex to be) and increment by 1. Scroll up until the vertex is centered on the screen.

Notice the symmetry of y-values about $y = -3$.

Therefore, to write the equation $y = 2x^2 - 4x - 1$ in the form $f(x) = a(x - h)^2 + k$, we replace h and k with the x- and y-values of the vertex:

$$f(x) = a(x - 1)^2 - 3$$

The only value we need is a. Recall that when completing the square we began by factoring out the coefficient of x^2. This is *always* the value of a. Therefore, the desired equation is $f(x) = 2(x - 1)^2 - 3$.

Experiment with the different forms of quadratic functions, and how you can algebraically and graphically move from one form to the other.

EXERCISE 10.4

VOCABULARY *Fill in each blank to make a true statement.*

1. A quadratic function that is a second-degree polynomial function can be written in the form _____, where __ ≠ 0.

2. The turning point, the highest or lowest point of a parabola, is called the _____.

3. The axis of symmetry is an imaginary line passing through the vertex (h, k) and has equation _____.

4. When $a > 0$, the graph of $y = f(x) = a(x - h)^2 + k$ opens ___.

5. When $a < 0$, the graph of $y = f(x) = a(x - h)^2 + k$ opens _____.

CONCEPTS *In Exercises 6–11, match each graph below with the phrase that best describes it.*

a.

b.

c.

d.

e.

f.
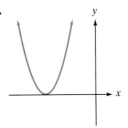

6. The graph of the function $f(x) = x^2$.

7. The graph of a function of the form $f(x) = (x - h)^2$.

8. The graph of a function of the form $f(x) = x^2 + k$.

9. The graph of a function of the form $f(x) = -x^2$.

10. The graph of a function of the form $f(x) = (x + h)^2$.

11. The graph of a function of the form $f(x) = x^2 - k$.

PRACTICE *In Exercises 12–23, graph each function on graph paper. Be sure to provide a table of values.*

12. $y = x^2$

13. $y = -x^2$

14. $y = x^2 + 2$

15. $y = x^2 - 3$

16. $y = -(x - 2)^2$

17. $y = (x + 2)^2$

18. $y = -3x^2 + 2x$

19. $y = 5x + x^2$

20. $y = x^2 + x - 6$

21. $y = x^2 - x - 6$

22. $y = 12x^2 + 6x - 6$

23. $y = -2x^2 + 4x + 3$

*In Exercises 24–33, tell if the parabola opens up or down, state the coordinates of the vertex, the axis of symmetry, and write the equation in the form $y = a(x - h)^2 + k$. **Do not graph the equation.***

24. $y = x^2 - 2x + 3$

25. $y = 2x^2 - 8x + 7$

26. $y = 2x^2 + 12x + 14$

27. $y = -3x^2 - 6x$

28. $y = -3x^2$

29. $y = 3x^2 - 3$

30. $y = 2x^2 - 4x$

31. $y = 3x^2 + 6x$

32. $y = -4x^2 + 16x + 5$

33. $y = 5x^2 + 20x + 25$

In Exercises 34–37, use your calculator to solve each equation. Solutions should be approximations to the nearest hundredth.

34. $2x^2 + x - 4 = 0$

35. $2x^2 - 5x - 3 = 0$

36. $0.5x^2 - 0.7x - 3 = 0$

37. $2x^2 - 0.5x - 2 = 0$

38. If $a > 0$, do changes in h and k in the equation $f(x) = a(x - h)^2 + k$ affect the domain, the range, or both? In what way? Be specific in your discussion.

39. If $a < 0$, do changes in h and k in the equation $f(x) = a(x - h)^2 + k$ affect the domain, the range, or both? In what way? Be specific in your discussion.

40. The graph of $x = y^2 - 2y$ is a parabola.
   a. Graph the equation.
   b. Explain why the graph is not a function

41. Consider the linear function $y = ax + b$, with $y = x$ as the parent function.
   a. Using different values for a (both positive and negative), describe the effect these differing values have on the function when it is compared to $y = x$.

   b. Using different values for b (both positive and negative), describe the effect these differing values have on the function when compared to $y = x$.

42. Consider the absolute value function $y = a|x - h| + k$, with $y = |x|$ as the parent function.
  a. Using different values for a (both positive and negative), describe the effect these differing values have on the function when it is compared to $y = |x|$.

  b. Using different values for k (both positive and negative), describe the effect these differing values have on the function when compared to $y = |x|$.

  c. Using different values for h (both positive and negative), describe the effect these differing values have on the function when compared to $y = |x|$.

43. Consider the square root function $y = a\sqrt{x - h} + k$, with $y = \sqrt{x}$ as the parent function.
  a. Using different values for a (both positive and negative), describe the effect these differing values have on the function when it is compared to $y = \sqrt{x}$.

  b. Using different values for k (both positive and negative), describe the effect these differing values have on the function when compared to $y = \sqrt{x}$.

  c. Using different values for h (both positive and negative), describe the effect these differing values have on the function when compared to $y = \sqrt{x}$.

APPLICATIONS

44. Maximizing revenue The revenue received for selling x radios is given by the formula
$$R = -\frac{x^2}{728} + 9x.$$
  a. Graph the function on the calculator and determine a good window.
  b. Find the maximum of the function.
  c. How many radios must be sold to obtain the maximum revenue?
  d. What is the maximum revenue?

45. Maximizing revenue The revenue received for selling x stereos is given by the formula
$$R = -\frac{x^2}{5} + 80x - 1000.$$
  a. Graph the function on the calculator and determine a good window.
  b. Find the maximum of the function.
  c. How many stereos must be sold to obtain the maximum revenue?
  d. What is the maximum revenue?

REVIEW Solve each inequality below. Specify solutions as number line graphs AND by using interval notation.

46. $-3x - 1 \le 8$.

47. $3(x - 2) > 2(x + 7)$

48. $-2 - x + 4 < 5$

49. $3x + 2 < 8$ or $2x - 3 > 11$

50. $5x + 12 < -2x < 3x$

10.5 Quadratic and Other Inequalities

■ SOLVING QUADRATIC INEQUALITIES ■ SOLVING OTHER NONLINEAR INEQUALITIES ■ GRAPHS OF INEQUALITIES IN TWO VARIABLES

■ SOLVING QUADRATIC INEQUALITIES

To solve a quadratic inequality such as $x^2 + x - 6 < 0$, we want to find all the values of x that will make the inequality a true statement.

If the inequality were an equality (an equation), its solutions could be found easily by factoring. We will find the values of x that make the *equation* $x^2 + x - 6 = 0$ true.

$$x^2 + x - 6 = 0$$
$$(x + 3)(x - 2) = 0$$
$$x + 3 = 0 \quad \text{or} \quad x - 2 = 0$$
$$x = -3 \quad \text{or} \quad x = 2$$

Placing the solutions to the equation associated with the inequality serves to partition the number line. We will call these points (solutions of the associated equations) **critical points.** Because they are not a part of the solution to the *inequality* $x^2 + x - 6 < 0$, we will enclose them with open circles.

The number line is now partitioned into three regions that we will call A, B, and C,

If the value of the polynomial $x^2 + x - 6$ is less than zero (or negative) for *one* point in a region, it will be negative for *all* points in the region. If the value of $x^2 + x - 6$ is positive for one value in the region, it will be positive for all values in the region.

We will now test a point from each region in the factored form of the polynomial (simply because that is easiest). Because we are looking for the polynomial to be negative when it is evaluated, we simply need to remember that if the product contains an *even* number of negative signs, the result is positive. If the product contains an *odd* number of negative signs, the result is negative.

Region	Test point	$(x + 3)(x - 2) < 0$	True/False
region A	test -5	$(-5 + 3)(-5 - 2) < 0?$ $(-2)(-7) < 0$ ✘	false
region B	test 0	$(0 + 3)(0 - 2) < 0?$ $(3)(-2) < 0$ ✔	true
region C	test 6	$(6 + 3)(6 - 2) < 0?$ $(9)(4) < 0$ ✘	false

Because our point tested true in region B, we shade that region on the number line graph.

The solution can be expressed in interval notation as $(-3, 2)$.

Interpreting the graphical representation of $x^2 + x - 6$ can also provide the solution to the inequality. The *x*-intercepts are the solutions of the *equation* $x^2 + x - 6 = 0$. Use of the ROOT/ZERO feature provides their values (-3 and 2).

The *x*-axis has equation $y = 0$. Therefore, the part of the graph that depicts values of the function that are positive is located **above** the *x*-axis in quadrants I and II, where y has positive values. The part of the graph **below** the *x*-axis (the part of the parabola containing the vertex, which is heavily shaded) depicts values of the function that are negative (because they are located **below** the *x*-axis in quadrants III and IV where y has negative values). The **x-values** that produce this part of the graph are those located *between* the critical points of -3 and 2. Therefore the solution of the inequality $x^2 + x - 6 < 0$ is $\{x \mid -3 < x < 2\}$ or

The TABLE also provides a way to actually see how the critical points truly do partition the number line. To examine values of the function $y = x^2 + x - 6$ for x-values *smaller* than -3, we set the TABLE to begin at -7. All x-values less than -3 yield y-values (functional values) that are positive.

To examine values of the function between the two critical points, scroll down the TABLE until $x = -3$ is at the top. We can now see that when x has a value between -3 and 2, the y-values are negative. Therefore, the x-values between -3 and 2 are solutions of the inequality $x^2 + x - 6 < 0$.

Continuing to scroll down the TABLE, we see that for x-values larger than 2 (the critical point on the right), y-values (the value of the function) are positive. Therefore, these x-values are *not* part of the solution.

Methods for solving quadratic inequalities algebraically are summarized as follows:

Solving Quadratic Inequalities Algebraically
1. Ensure one side of the inequality is set equal to 0.
2. Change the inequality sign to an equal sign. Solve this equation.
3. The solutions of the equation associated with the inequality are critical points and should be located on a number line.
4. If the original inequality is $>$ or $<$, enclose the critical points with open circles. If the original inequality is $\geq$ or $\leq$, enclose the critical points with closed circles.
5. The critical points have partitioned the number line into regions. Test a point from each region in the original inequality.
6. Shade regions for which the test points test true. These are solutions of the inequality.

EXAMPLE 1   Solve the inequality $x^2 + 2x \geq 3$.

Solution   Rewrite the inequality so that one side is equal to 0:

$$x^2 + 2x - 3 \geq 0$$
$$x^2 + 2x - 3 = 0$$

Write the equation associated with the inequality and solve it.

$$(x + 3)(x - 1) = 0$$
$$x + 3 = 0 \quad \text{or} \quad x - 1 = 0$$
$$x = -3 \quad \text{or} \quad x = 1$$

The critical points are -3 and 1. They are located on a number line and enclosed with closed circles because the inequality includes equality ($\geq$).

We will now test a point from each region:

Region	Test point	$(x + 3)(x - 1) \geq 0$	True/False
region A	test -4	$(-4 + 3)(-4 - 1) \geq 0?$ $(-1)(-5) \geq 0$ ✔	true
region B	test 0	$(0 + 3)(0 - 1) \geq 0?$ $(3)(-1) \geq 0$ ✘	false
region C	test 8	$(8 + 3)(8 - 1) \geq 0?$ $(11)(7) \geq 0$ ✔	true

Shade the regions that tested true. Express the solution using interval notation $(-\infty, -3] \cup [1, \infty)$.

We can confirm our solutions graphically. After finding the x-intercepts (using the ROOT/ZERO feature of the calculator), it can easily be seen that x-values *smaller* than -3 and *larger* than 1 yield the heavily shaded portions of the graphical display. Therefore, the solution found algebraically is correct.

■ SOLVING OTHER NONLINEAR INEQUALITIES

EXAMPLE 2   Use a graphical approach to solve $(x + 3)(x - 2)(x - 6) < 0$.

Solution   The x-intercepts are -3, 2, and 6 and can be found by using the ROOT/ZERO feature of the calculator or by solving the associated equation $(x + 3)(x - 2)(x - 6) = 0$. (The window used is $[-10, 10]$ by $[-30, 50]$.) These points (-3, 2, and 6) are the critical points of the graph, and they are located on a number line and enclosed with open circles because the original example is strictly an inequality.

Looking at the graphical display, we see that the graph of $(x + 3)(x - 2)(x - 6)$ is below the x-axis (less than 0) to the left of the critical point -3 and between the critical points 2 and 6. These parts of the graph are heavily shaded. Therefore, the solution of $(x + 3)(x - 2)(x - 6) < 0$ is

This would be written in interval notation as $(-\infty, -3) \cup (2, 6)$. We suggest that you check your solution by looking at values for y1 in the TABLE, as was done in the introductory problem of this section.

When working with inequalities that contain rational expressions, care must be taken to *not* include any values in the solution that would make a fraction undefined.

Therefore, we suggest that you FIRST find these excluded values and locate them on the number line. They, too, serve as critical points and **are always enclosed with an open circle because they are *never* part of the solution.** Otherwise, these inequalities are solved using the same methods described previously for solving quadratic inequalities.

EXAMPLE 3   Solve: $\dfrac{x^2 - 3x + 2}{x - 3} \geq 0$.

Solution   When the denominator, $x - 3$, equals 0, the rational expression is undefined. Therefore, x cannot have a value of 3; 3 is an excluded value and is a critical point on our number line graph. Critical points that are excluded values are **always** enclosed by open circles.

Write the associated equation and solve for x.

$$\dfrac{x^2 - 3x + 2}{x - 3} = 0$$

$$x^2 - 3x + 2 = 0$$

$$(x - 2)(x - 1) = 0$$

$$x - 2 = 0 \quad \text{or} \quad x - 1 = 0$$

$$x = 2 \quad \text{or} \quad x = 1$$

Multiply both sides by the quantity $(x - 3)$ to clear the fraction. Also, realize that the *only* way the rational expression can equal 0 is for the numerator to be 0.

Solve the equation.

The x-values of 2 and 1 are solutions of the equation. They are also critical points, and they are located on the number line and are enclosed with closed circles because the inequality is $\geq$.

Test a point from each region in the original inequality.

Region	Test point	$\dfrac{(x - 2)(x - 1)}{x - 3} \geq 0$	True/False
region A	test 0	$\dfrac{(-2)(-1)}{-3} \geq 0$ ✗	false
region B	test 1.5	$\dfrac{(-0.5)(0.5)}{-1.5} \geq 0$ ✔	true
region C	test 2.5	$\dfrac{(0.5)(1.5)}{-0.5} \geq 0$ ✗	false
region D	test 5	$\dfrac{(3)(4)}{2} \geq 0$ ✔	true

Shade the regions that tested true.

Write the solution using interval notation: $[1, 2] \cup (3, \infty)$. ■

Interpreting the graphical displays of rational functions can be difficult and takes a lot of practice. Consider the graphical display of $\dfrac{x^2 - 3x + 2}{x - 3} \geq 0$.

■ **T E C H N O L O G Y   T I P**

The graph style icon can be changed by placing the cursor on it and pressing ENTER until the dotted segment appears.

The vertical "line" that appears at $x = 3$ is simply where the calculator connected two pieces of the graph over the excluded value of $x = 3$. It is *not* a part of the graphical display. This becomes evident if you try to TRACE along it. A better display results if the calculator is changed from connected to dot mode.

However, you must remember that 3 is an excluded value and is not part of the solution. It does, however, help partition the number line because it is a critical point. Because the x-intercepts are located so close together, care must be taken when setting bounds using the ROOT/ZERO option.

The only parts of the graph above the x-axis are the small piece between the two roots (shown here in the window $[0, 3]$ by $[-1, 1]$) and the piece of graph "hanging" in the first quadrant.

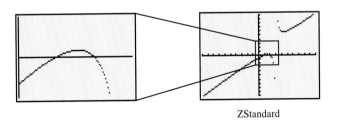

ZStandard

Therefore, the solutions pictured on the number line are confirmed.

To examine values of $y1 = \dfrac{x^2 - 3x + 2}{x - 3}$ in the TABLE, increment by $\frac{1}{2}$ and scroll to values smaller than 1, between 1 and 2, and larger than 3. Examination of these values confirms our number line graph of the solutions.

■ GRAPHS OF INEQUALITIES IN TWO VARIABLES

The graph of an equation in two variables consists of the points (ordered pairs) that satisfy the equation. The graph of an inequality in two variables consists of an area of the plane bounded by the equation. For example, the inequality $y \geq x - 5$ consists of the area of the plane containing all ordered pairs such that $y > x - 5$ and all the ordered pairs on the line $y = x - 5$.

EXAMPLE 4   Graph the inequality $y \geq x - 5$.

Solution   Since $y \geq x - 5$ means $y = x - 5$ or $y > x - 5$, we begin by graphing the line $y = x - 5$. The graph appears in Figure 10-6.

$y = x - 5$

x	y
0	-5
5	0

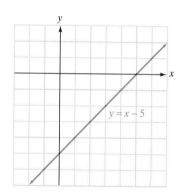

FIGURE 10-6

Since $y \geq x - 5$ also indicates y can be greater than $x - 5$, there are coordinates of points other than those shown in Figure 10-6 that satisfy the inequality. For example, the coordinates of the origin satisfy the inequality. We can verify this by letting x and y be 0 in the given inequality:

$y \geq x - 5$

$0 \geq 0 - 5$ Substitute 0 for x and 0 for y.

$0 \geq -5$

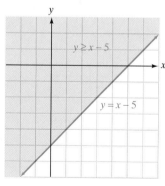

FIGURE 10-7

Because $0 \geq -5$, the coordinates of the origin satisfy the original inequality. In fact, the coordinates of every point on the same side of the line as the origin satisfy the inequality. The graph of $y \geq x - 5$ is the **half-plane,** the area of the plane bounded by a line, that is shaded in Figure 10-7. Since the boundary line $y = x - 5$ is included, we draw it with a solid line.   ■

If the inequality to be graphed had been a strict inequality, strictly greater than or strictly less than, we would have to indicate on the graph that the equation was *not* part of the solution set. To do this, the boundary line (or boundary curve as in the next example) would be designated as a dashed line to show that the points on it are *not* part of the solution.

EXAMPLE 5   Graph the inequality $y < -x^2 + 4$.

Solution   Using our knowledge of quadratics, we know that the graph of $y = -x^2 + 4$ is a parabola with vertex at $(0, 4)$. Because the coefficient of $x^2(a)$ is negative, the parabola opens downward. We construct a table of values and graph the parabola (see Figure 10-8).

x	y
-2	0
-1	3
0	4
1	3
2	0

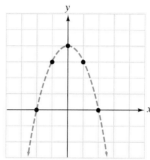

FIGURE 10-8

Because we are graphing an inequality ($<$), we draw the curve with a dashed line rather than a solid line. This is because points on the curve *do not* satisfy the inequality.

To determine whether to shade inside the parabolic curve, or outside the curve, choose a point to test in the original inequality. Test the origin, $(0, 0)$:

$$0 < -0^2 + 4?$$

$$0 < 4 \ \checkmark$$

The point tests true; therefore, shade inside the curve (Figure 10-9):

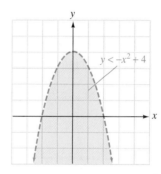

$y < -x^2 + 4$

FIGURE 10-9

The graph style icon on the calculator can be set to shade "below" because the inequality is $<$. The calculator-generated graph confirms our hand-drawn solution.

EXAMPLE 6   Graph the inequality $x \le |y|$.

Solution   Using our knowledge of graphs of absolute value equations, we carefully choose values so as to form the "V" shape that is typical of the graphs. We will choose values for y and substitute them into the inequality to find corresponding x-values. We use a solid line because equality is included (see Figure 10-10).

x	y
0	0
1	1
1	-1
2	2
2	-2

FIGURE 10-10

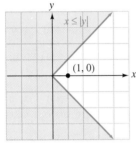

FIGURE 10-11

To determine whether to shade inside the figure or outside, we choose a test point $(1, 0)$, and substitute into the original inequality:

$$x \leq |y|$$
$$1 \leq |0|$$
$$1 \leq 0 \; \textbf{✗}$$

We see that the point tests false. Therefore, we *do not* shade the area inside the figure, but, rather shade the area outside the figure (see Figure 10-11). ▪

EXERCISE 10.5

VOCABULARY *Fill in each blank to make a true statement.*

1. When solving nonlinear inequalities, solve the associated equation to find the _____ points.

2. The solutions of the equation associated with a nonlinear inequality _____ the number line.

3. If the inequality symbol is $>$ or $<$, solutions to the associated equation are enclosed with _____ circles. If the symbol is $\geq$ or $\leq$, solutions to the associated equation are enclosed with _____ circles.

4. When a nonlinear inequality contains one or more rational expressions, values that make the denominator equal zero must be _____. They also serve to partition the number line and are always enclosed with an _____ circle.

CONCEPTS *In Exercises 5–8, state the excluded values for each rational inequality.*

5. $\dfrac{x - 3}{x} \geq 0$

6. $\dfrac{x^2 - 5x + 6}{x - 5} < 0$

7. $\dfrac{x^2 + x - 12}{(x - 2)(x + 3)} < 0$

8. $\dfrac{x - 5}{x^2 + 2} \geq 0$

In Exercises 9–12, interpret the graphical display and state the solution of the quadratic inequality as either all real numbers ($\mathbb{R}$) or $\varnothing$.

9.

$3x^2 - 30x + 78 \geq 0$

10.
$-x^2 + 4x - 8 \geq 0$

11.
$\dfrac{1}{5}x^2 - \dfrac{2}{5}x + \dfrac{6}{5} < 0$

12.

$-2x^2 - 16x - 35 \leq 0$

In Exercises 13–16, match each graphical display with the appropriate inequality.

a.

b.

c.

d.

13. $y \geq x^2 - 2x - 8$

14. $y \geq -x^2 + 2x + 24$

15. $y \leq |x - 2| - 3$

16. $y \leq -|x + 2| + 4$

PRACTICE *In Exercises 17–50, solve each inequality. Give each result in interval notation and graph the solution set.*

17. $x^2 - 5x + 4 < 0$

18. $x^2 - 3x - 4 > 0$

19. $x^2 - 8x + 15 > 0$

20. $x^2 + 2x - 8 < 0$

21. $x^2 + x - 12 \leq 0$

22. $x^2 + 7x + 12 \geq 0$

23. $x^2 + 2x \geq 15$

24. $x^2 - 8x \leq -15$

25. $-x^2 + 8x > 18$

26. $x^2 + 6x \geq -9$

27. $x^2 \geq 9$

28. $x^2 \geq 16$

29. $2x^2 - 50 < 0$

30. $3x^2 - 243 < 0$

31. $\dfrac{1}{x} < 2$

32. $\dfrac{1}{x} > 3$

33. $\dfrac{4}{x} \geq 2$

34. $-\dfrac{6}{x} < 12$

35. $\dfrac{x^2 - x - 12}{x - 1} < 0$

36. $\dfrac{x^2 + x - 6}{x - 4} \geq 0$

37. $\dfrac{x^2 + x - 20}{x + 2} \geq 0$

38. $\dfrac{x^2 - 10x + 25}{x + 5} < 0$

39. $\dfrac{6x^2 - 5x + 1}{2x + 1} > 0$

40. $\dfrac{6x^2 + 11x + 3}{3x - 1} < 0$

41. $\dfrac{3}{x - 2} < \dfrac{4}{x}$

42. $\dfrac{-6}{x + 1} \geq \dfrac{1}{x}$

43. $\dfrac{-5}{x + 2} \geq \dfrac{4}{2 - x}$

44. $\dfrac{-6}{x - 3} < \dfrac{5}{3 - x}$

45. $\dfrac{x}{x + 4} \leq \dfrac{1}{x + 1}$

46. $\dfrac{x}{x + 9} \geq \dfrac{1}{x + 1}$

47. $\dfrac{x}{x + 16} > \dfrac{1}{x + 1}$

48. $\dfrac{x}{x + 25} < \dfrac{1}{x + 1}$

49. $(x + 2)^2 > 0$

50. $(x - 3)^2 < 0$

In Exercises 51–70, graph each inequality on graph paper.

51. $2x - y > 2$

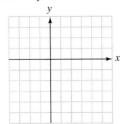

52. $3x - 2y > 6$

53. $4x + 3y \leq 12$

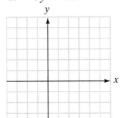

54. $5x + 4y \geq 20$

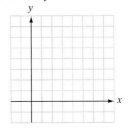

55. $3(x + y) + x < 6$

56. $2(x - y) - y \geq 4$

57. $4x - 4(x + 2y) \geq -6y - 8$

58. $3y + 3(x - y) < 9$

59. $y < x^2 + 1$

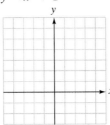

60. $y > x^2 - 3$

61. $y \leq x^2 + 5x + 6$

62. $y \geq x^2 + 5x + 4$

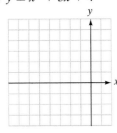

63. $x \geq y^2 - 3$

64. $x \leq y^2 + 1$

65. $-x^2 - y + 6 > -x$

66. $y > (x + 3)(x - 2)$

67. $y < |x + 4|$

68. $y \geq |x - 3|$

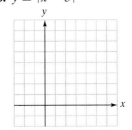

69. $y \leq -|x| + 2$

70. $y > |x| - 2$

REVIEW *Solve each of the following variation problems.*

71. The area of a circle varies directly with the square of its radius, and the constant of variation is π. Find the area of a circle with a radius of 8 inches.

72. An object in free fall travels a distance, s, that is directly proportional to the square of the time, t. If an object falls 375 feet in 5 seconds, how far does it fall in 7 seconds?

73. The value of a car usually varies inversely with its age. If a car is worth $9000 when it is 3 years old, how much will it be worth when it is 9 years old?

74. The distance that a car can travel is directly proportional to the number of gallons of gasoline it consumes. If a car can go 360 miles on 15 gallons of gasoline, how far can it go on a full tank of 18 gallons?

C H A P T E R   S U M M A R Y

CONCEPTS

REVIEW EXERCISES

Section 10.1

Square Root Property and Completing the Square

A quadratic equation is one that can be written in the form $ax^2 + bx + c = 0$, $a \neq 0$.

These equations were previously solved by factoring.

The square root property allows solutions of equations that may or may not be factorable. It states: If c is a real number, the equation $x^2 = c$ has two solutions, $x = \sqrt{c}$ and $x = -\sqrt{c}$. This is often expressed as $x = \pm\sqrt{c}$.

The process of completing the square may be used to solve *any* quadratic equation. In this process the coefficient of x^2 must be 1 and the constant term is moved to the right side of the equation. The "square" is completed by adding $\frac{1}{2}$ the coefficient of the linear term, squared, to both sides of the equation. The polynomial is then factored as a binomial quantity squared, and the square root property is used to determine the roots.

1. Solve each equation by factoring or by using the square root property.

   a. $12x^2 + x - 6 = 0$ **b.** $15x^2 + 2x - 8 = 0$

   c. $6x^2 + 17x + 5 = 0$ **d.** $(x + 2)^2 = 36$

2. Solve each equation by completing the square.

   a. $x^2 + 6x + 8 = 0$ **b.** $2x^2 - 9x + 7 = 0$

Section 10.2	The Quadratic Formula and the Discriminant

If $a \neq 0$, the solutions of $ax^2 + bc + c = 0$ are given by the formula:

$$x = \frac{-b \pm \sqrt{b^2 - 4ac}}{2a}.$$

The value of the discriminant, $b^2 - 4ac$, can be used to determine the number and type of roots of a quadratic equation.

If $b^2 - 4ac > 0$ and is a perfect square, the equation will have two distinct real, rational roots if the coefficients are rational.

If $b^2 - 4ac > 0$ and is not a perfect square, the equation will have two distinct real, irrational roots.

If $b^2 - 4ac = 0$, the equation will have one real rational root, often called a *double root*.

If $b^2 - 4ac < 0$, the equation will have two distinct imaginary roots. They are complex conjugates.

3. Solve each equation using the quadratic formula.
  a. $x^2 - 8x - 9 = 0$
  b. $x^2 - 10x = 0$
  c. $2x^2 + 13x - 7 = 0$
  d. $3x^2 - 7 = 20x$

4. Use the discriminant to determine what type of solutions exist for each equation.
  a. $3x^2 + 4x - 3 = 0$
  b. $4x^2 - 5x + 7 = 0$

5. Find the values of k that will make the solutions of $(k - 8)x^2 + (k + 16)x = -49$ equal.

6. Find the values of k such that the solutions of $3x^2 + 4x = k + 1$ are real numbers.

7. Dimensions of a rectangle A rectangle is 2 centimeters longer than it is wide. If both the length and width are doubled, its area is increased by 72 square centimeters. Find the dimensions of the original rectangle.

Section 10.3	Applications of Quadratic Functions and Solutions of Equations Using Quadratic Methods

Solutions of quadratic equations are the x-intercept(s) of the graph of the equation.

Many nonquadratic equations can be put into quadratic form and then solved using methods previously presented for solving quadratic equations. A careful substitution of variables can be used to put the equation in quadratic form.

8. If a rocket is launched straight up into the air with an initial velocity of 112 feet per second, its height after t seconds is given by the formula $h = 112t - 16t^2$, where h represents the height of the rocket in feet.
  a. Graph the function on your graphing calculator in an appropriate viewing window.

  b. After how many seconds does the rocket touch the ground?

  c. After 2 seconds have passed, how high is the rocket?

  d. At what other time is the rocket the same height as it is after 2 seconds?

  e. What is the maximum height of the rocket?

9. Solve each equation.
  a. $x - 13x^{1/2} + 12 = 0$
  b. $a^{2/3} + a^{1/3} - 6 = 0$
  c. $\dfrac{1}{x + 1} - \dfrac{1}{x} = -\dfrac{1}{x + 1}$
  d. $\dfrac{6}{x + 2} - \dfrac{6}{x + 1} = 5$

| Section 10.4 | Graphs of Quadratic Functions |

The graphs of quadratic functions are parabolas. The parabolas will either open up or down. The turning point (the highest or lowest point) is called the *vertex* of the parabola.

If $a \neq 0$, the graph of $y = a(x - h)^2 + k$ is a parabola with vertex at (h, k). It opens upward when $a > 0$ and downward when $a < 0$.

The axis of symmetry is $x = h$.

To graph a parabola, graph the vertex, two points with x-coordinates smaller than that of the vertex, and two points with x-coordinates greater than the x-coordinate of the vertex. Connect all points with a smooth curve.

To put a quadratic equation into the form $y = a(x - h)^2 + k$, either complete the square in x or use the capabilities of the graphing calculator in combination with the algebra.

10. Graph each equation. State the coordinates of the vertex, the axis of symmetry, and write the equation in the form $y = a(x - h)^2 + k$.

a. $y = 2x^2 - 3$ **b.** $y = -2x^2 - 1$

c. $y = -4x^2 + 16x - 15$ **d.** $y = 5x^2 + 10x - 1$

11. a. Find the dimensions of the rectangle of maximum area that can be constructed with 100 feet of fencing.

b. Find the maximum area.

| Section 10.5 | Quadratic and Other Inequalities |

To solve quadratic inequalities, begin by ensuring that one side of the inequality equals 0. Locate solutions to the associated equation on the number line. These are critical points that partition the number line into regions. Test a point from each region in the original inequality and shade regions on the number line that test true. These are the solutions of the inequality.

When solving other nonlinear inequalities, find any excluded values and place them on the number line as critical points. These points are always enclosed with an open circle. The remainder of the solution is completed as when solving quadratic inequalities.

12. Solve each quadratic inequality below. Express solutions both as a number line graph and using interval notation.

a. $x^2 + 2x - 35 > 0$ **b.** $x^2 + 7x - 18 < 0$

13. Solve each inequality below. Express solutions both as a number line graph and using interval notation.

a. $(x - 2)(x + 1)(x + 4) < 0$

b. $\dfrac{3}{x} \leq 5$

c. $\dfrac{2x^2 - x - 28}{x - 1} > 0$

To graph an inequality in two variables, begin by graphing the associated equation. Use a dashed line if the inequality is $>$ or $<$. The graph of the equation separates the plane into two regions. Test a point in one of the regions. If it tests true, shade the region containing the point. If it tests false, shade the other region.

14. Graph each inequality.

a. $3x + y \geq 3$

b. $y < 4$

c. $y < \frac{1}{2}x^2 - 1$

d. $y \geq -|x|$

■ Chapter Test

1. Solve by factoring.
  a. $x^2 + 3x + 2 = 0$

  b. $12x^2 + 5x = 2$

2. Solve by using the square root property.
  a. $(x - 3)^2 = -2$

  b. $3x^2 = 27$

3. Solve by completing the square or using the quadratic formula.
  a. $x^2 + 3x - 18 = 0$ **b.** $x(6x + 19) = -15$ **c.** $x^2 + 4x + 1 = 0$ **d.** $x^2 - 5x - 3 = 0$

$-3/2 \quad -5/3$ $\quad -2+\sqrt{3} \quad -2-\sqrt{3}$ $\quad \dfrac{5\pm\sqrt{37}}{2}$

4. Determine whether the solutions of $3x^2 + 5x + 17 = 0$ are real or imaginary. ***Do not solve the equation.***

5. For what value(s) of k are the solutions of $4x^2 - 2kx + k - 1 = 0$ equal?

6. One leg of a right triangle is 14 inches longer than the other, and the hypotenuse is 26 inches. Find the length of the shorter leg.

7. Solve the equation $2x - 3x^{1/2} + 1 = 0$.

8. Consider the equation $y = \dfrac{1}{2}x^2 + 5$.

  a. Rewrite the equation in the form $y = a(x - h)^2 + k$.

  b. State the vertex.
  c. State the axis of symmetry.
  d. Graph the figure.

9. Graph the inequality $y \leq -x^2 + 3$.

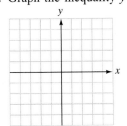

In Problems 10 and 11, solve each inequality and express the solution as both a number line graph and in interval notation.

10. $x^2 - 2x - 8 > 0$

11. $\dfrac{x - 2}{x + 3} \le 0$

■ Cumulative Review: Chapters 9 and 10

VOCABULARY / CONCEPTS

1. The expression written under the radical sign is called the _____.

2. If an indicated root of a number is perfect, the result is always a _____ and is an exact result. If an indicated root is not perfect, the calculator can be used to obtain an approximation, which is an _____ number.

3. The Pythagorean theorem states that the sum of the squares of the _____ of a right triangle equals the square of the _____.

4. Powers of i repeat in a cycle of 4 values:__, ____, ___, __.

5. The _____ of an equation is/are the value(s) of the variable that make the equation a true statement.

6. Graphically, the __-intercepts are the solutions of an equation when one side is equal to 0.

7. The graph of a quadratic function is a curve called a _____. The turning point of this figure is called its _____.

8. Classify each of the following as *rational* or *irrational*.
   a. $\sqrt{64}$ **b.** $\sqrt{32}$
   c. $\sqrt[5]{32}$ **d.** $\sqrt[4]{18}$

9. Express each radical as an exponential expression.
   a. $3\sqrt{ab}$ **b.** $\sqrt{3ab}$

   c. $\sqrt[3]{x^5}$ **d.** $\sqrt[6]{\dfrac{1}{3}p^5 q}$

10. Classify each of the following as *true* or *false*.
   a. $\sqrt{3} + \sqrt{3} = \sqrt{9} = 3$
   b. $5\sqrt[3]{2} - \sqrt[3]{2} = 4\sqrt[3]{2}$
   c. $\sqrt{25 - 9} = \sqrt{25} - \sqrt{9}$

11. Match each graphical display with its equation.
   a. $y = \sqrt{x} + 5$ **b.** $y = \sqrt{x + 5}$
   c. $y = -\sqrt{x} + 5$ **d.** $y = \sqrt{x - 5}$

(i)

(ii)

(iii)

(iv)

12. Use the given display to fill in the blanks below. A radical function was entered at the y1 prompt.

X	Y1
-1	ERROR
0	1
1	2
2	2.4142
3	2.7321
4	3
5	3.2361

X= -1

   a. $f(1) =$ __
   b. An approximation to the nearest thousandth of $f(3) =$ _____
   c. The ERROR message displayed for an x-value of -1 implies that -1 is not a part of the _____ of the function.

13. Simplify each of the following:
   a. $\sqrt{-4}$
   b. i^{15}
   c. $6 + \sqrt{-25}$
   d. $(8 + 3i) - 2(5 - 3i)$

14. The graphs of four quadratic functions are displayed below. Specify the roots as real or imaginary.

   a.   **b.**

   c.   **d.**

15. State the vertex of each parabola whose equation is given below.
   a. $y = 3(x - 2)^2 + 4$
   b. $y = -2(x + 4)^2 - 5$

16. Write the equation of the parabola whose graph is shifted 3 units to the right and 4 units up from the graph of $y = x^2$.

17. Write the equation of the parabola that is shifted 5 units left and 6 units down from the graph of $y = x^2$.

18. Write the equation of the parabola that has a vertical stretching by a factor of $\dfrac{1}{6}$ when compared to the parabola whose graph is $y = x^2$.

19. Which of the graphs below is of the inequality $y < (x - 1)^2 + 3$?

   a.   **b.**

   c.   **d.**

20. Which inequality graphed below is the solution of $\dfrac{x + 2}{x - 1} \le 0$?

   a.
           -2 1

   b.
           -2 1

   c.
           -2 1

   d.
           -2 1

PRACTICE *In Exercises 21–27, simplify each expression (being sure to rationalize denominators as applicable).*

21. $\sqrt{175a^4b^3c^5}$

22. $\sqrt[4]{\dfrac{5x}{16z^5}}$

23. $\sqrt{18} + \sqrt{400} - \sqrt{243}$

24. $\sqrt[3]{6x^2y^4} \cdot \sqrt[3]{9xy^2}$

25. $\dfrac{\sqrt[3]{2}}{\sqrt[3]{3}}$

26. $\dfrac{\sqrt{3} + 1}{\sqrt{3} - 1}$

27. $\dfrac{-4i}{2 - 3i}$

28. If $f(x) = \sqrt{2x + 2}$ and $g(x) = \sqrt{3x + 3} - 1$, find each of the following (express your result in simplest radical form).
   a. $f(1)$ **b.** $g(2)$ **c.** $f(5)$

29. Solve the equation $\sqrt{x} + 6 = -x + 8$ algebraically. Verify your result(s) using the STOre feature of the calculator. Copy your screen display exactly as it appears to justify your work.

30. Graph $y = \sqrt{x} + 3$ on graph paper and specify the domain and the range.

31. Solve the quadratic equation $2x^2 - 5x - 3 = 0$
   a. by factoring
   b. by completing the square
   c. by using the quadratic formula

32. Find a quadratic equation with roots of -3 and $\dfrac{2}{3}$ (express the equation with integral coefficients).

33. The function
$f(x) = x^2 - 5x + 6$
was entered at the y1 prompt on the calculator. Its TABLE display is pictured. Use the display to answer the following questions.

X	Y1
0	6
1	2
2	0
3	0
4	2
5	6
6	12

X=0

   a. State the coordinates of the x-intercept(s).

   b. State the coordinates of the y-intercept(s).
   c. $f(5) = $ ___.
   d. If $f(x) = 6$, then $x = $ ___ or ___.

34. Solve the equation $x^4 = 6x^2 - 5$.

35. Solve the equation $x^{2/3} - 2x^{1/3} - 3 = 0$.

36. Solve the equation $\dfrac{1}{x + 2} + \dfrac{24}{x + 3} = 13$.

37. Solve the equation $3(x - 4)^2 + 2 = 5$
   a. algebraically
   b. graphically

38. Graph the function $y = (x + 3)^2 - 4$ on graph paper. Provide a table of values for your graph.

39. Graph the function $y = -2x^2 - 1$ on graph paper. Provide a table of values for your graph.

40. Consider the equation $y = 2x^2 - 8x + 13$.
   a. Does the graph of the parabolic curve open up or down?
   b. State the coordinates of the vertex.
   c. State the equation of the axis of symmetry.
   d. Write the equation in the form
   $y = a(x - h)^2 + k$.

41. Consider the equation $(x - 8)^2 + 3 = 8$.
   a. Solve the equation graphically, rounding solutions to the nearest hundredth, if necessary.
   b. Solve the equation algebraically, expressing results in simplest radical form.

42. Solve the equation below graphically. Copy your screen display to justify your results. Include the WINDOW settings if they are different from Zstandard.

$$\sqrt{x} - \sqrt{5x + 4} + 6 = \sqrt{18x - 48} + \sqrt{16x}$$

43. Solve the equation below graphically. Copy your screen display to justify your results. Include the WINDOW settings if they are different from Zstandard (ensure the graphical display is clear in the defined window). Confirm your solution(s) with the STOre feature of the calculator.

$$\sqrt[4]{x} = \sqrt{\frac{x}{4}}$$

44. Solve the given equation graphically. Round results to the nearest hundredth.

$$\sqrt{1.3x} + 4 = 1.2x + 3.2$$

45. Consider the rounded solution from Exercise 44. If you used the STOre feature to check your solution, what would be the result? Explain fully.

46. Solve the equation $2.3x^2 - 5.2x - 7 = 0$ graphically. Round results to the nearest hundredth.

47. If a ball is thrown straight up in the air with an initial velocity of 50 ft/sec, its height, y, after x seconds is given by the equation $y = 50x - 16x^2$.
   a. Graph the equation in a good viewing window.
   b. What is the maximum height attained by the ball (rounded to the nearest foot)?
   c. To the nearest tenth of a second, how long does it take the ball to reach this height?
   d. To the nearest tenth of a second, how long does it take the ball to reach the ground?

APPLICATIONS

48. Anchoring flagpoles A cable is to be used to support a flagpole that is 17 feet tall. The cable is 18 feet long. How far from the base of the pole should the anchor stake be placed? Round to the nearest tenth.

49. Bowling The velocity, v, of an object after it has fallen d feet is given by the equation $v^2 = 64d$. An inexperienced bowler lifts the ball 4 feet. With what velocity does it strike the alley?

50. Investments A woman invests $2000 in a mutual fund for which interest is compounded annually at a rate r. After 1 year she deposits an additional $3000. After 2 years, the balance in the account is $2000(1 + r)^2 + 3000(1 + r)$. If this amount is $5572.80, find r.

51. Maximizing revenue The revenue received for selling x stereos is given by the formula

$$R = -\frac{x^2}{1000} + 10x.$$

   a. How many stereos must be sold to obtain the maximum revenue?
   b. Find the maximum revenue.

CAREERS & MATHEMATICS

Photographer

Photographers use cameras and film to portray people, objects, places, and events. Some specialize in scientific, medical, or engineering photography and provide illustrations and documentation for publications and research reports. Others specialize in portrait, fashion, or industrial photography and provide the pictures for catalogs and other publications. Photojournalists capture newsworthy events, people, and places, and their work is seen in newspapers and magazines as well as on television.

Qualifications

Employers want applicants with a broad technical understanding of photography, as well as imagination, creativity, and a good sense of timing. Some knowledge of mathematics, physics, and chemistry is helpful for understanding lenses, films, lighting, and development processes.

Job Outlook

Job opportunities will be excellent through the year 2005. Business and industry will need photographers to provide visual aids for meetings and reports, sales campaigns, and public relations work. Law enforcement agencies and scientific and medical research organizations will also require photographers with appropriate technical skills.

Example Application

In each photographic lens, there is an adjustable circular opening called the **aperture,** which controls the amount of light that passes through the lens. Various lenses—wide-angle, close-up, and telephoto—are distinguished by their **focal length.** The diameter, d, of the aperture and the focal length, f, of the lens determine the **f-number** of the lens by the formula

$$f\text{-number} = \frac{f}{d}$$

Thus, a lens with a focal length of 12 centimeters and an aperture with a diameter of 6 centimeters has an f-number of $\frac{12}{6}$, or 2. It would be an $f/2$ lens.

Find the f-number of this lens if the *area* of its aperture is cut in half so as to admit only half as much light.

Solution   First determine the area of a circle with diameter d. Then substitute 6 for d to find the area of the aperture.

$$A = \pi r^2 \qquad \text{The formula for the area of a circle.}$$

$$A = \pi \left(\frac{d}{2}\right)^2 \qquad \text{Substitute } \tfrac{d}{2} \text{ for } r.$$

$$A = \pi \left(\frac{6}{2}\right)^2 \qquad \text{Substitute 6 for } d.$$

$$A = 9\pi$$

To find the diameter of a circle with area equal to one-half of 9π, substitute $\frac{9\pi}{2}$ for A in the formula for the area of a circle and solve for d.

$$A = \pi r^2 \qquad \text{The formula for the area of a circle.}$$

$$\frac{9\pi}{2} = \pi \left(\frac{d}{2}\right)^2 \qquad \text{Substitute } \tfrac{9\pi}{2} \text{ for } A \text{ and } \tfrac{d}{2} \text{ for } r.$$

$$\frac{9\pi}{2} = \frac{\pi d^2}{4}$$

$$18 = d^2 \qquad \text{Multiply both sides by 4.}$$

$$3\sqrt{2} = d \qquad \sqrt{18} = \sqrt{9}\sqrt{2} = 3\sqrt{2}.$$

559

If the diameter of the aperture were reduced from 6 centimeters to $3\sqrt{2}$ centimeters, the area (and the light admitted) would be cut in half. The f-number of the lens is now found by letting $f = 12$ (as before) and $d = 3\sqrt{2}$ in the formula for an f-number.

$$f\text{-number} = \frac{f}{d} = \frac{12}{3\sqrt{2}} = 2\sqrt{2} \approx 2.8$$

An $f/2.8$ lens admits one-half the light admitted by an $f/2$ lens.

If the light were cut in half again, the f-number would be $2.8\sqrt{2}$, or 4. The next f-number, representing another halving of the light admitted, would be $4\sqrt{2}$, or 5.6. These numbers,

$$f/2, \ f/2.8, \ f/4, \ f/5.6, \ f/8, \ f/11, \ f/16$$

are called **f-stops.** They are well known to all professional photographers.

EXERCISES

1. What would be the f-number of a lens with focal length of 20 cm and aperture with diameter of 5 cm?

2. If the focal length of the lens of Exercise 1 were doubled and the aperture held constant, what would be the f-number?

3. What diameter would give a lens with focal length 55 mm an f-number of $f/3.5$?

4. The **speed** of a lens is the square of the reciprocal of the f-number. How many times faster is an $f/2$ than an $f/4.5$ lens?

(Answers: 1. $f/4$ 2. $f/8$ 3. 15.7 mm 4. 5 times faster)

560

Functions and Their Inverses

11

■ PERSPECTIVE

Another Graph Paper

Because an exponential function increases rapidly, it is difficult to fit the graph into a reasonable space. For example, between $x = 0$ and $x = 2$, the exponential function $y = 10^x$ changes from a low of 1 to a high of 100.

On the **semilogarithmic graph paper** in Illustration 1, the vertical scale is marked so that the distance between 1 and 2 is the logarithm of 2, the distance between 1 and 3 is log 3, and so on. On this logarithmic scale, the distance between 1 and 2 represents a doubling of y. That same distance, between 2 and 4, also represents a doubling, as does that between 4 and 8,

and between 8 and 16, and so on. On a logarithmic y-axis, equal vertical distances represent increasing y by the same *factor*.

Illustration 2 shows the graphs of three exponential functions in the first quadrant of a Cartesian coordinate system. All of them quickly run off of the grid. Illustration 3 shows the graphs of the same exponential functions on a semilog graph. Note how easily the coordinate system accommodates large values of y. Note also that the graph of an exponential function on a semilog coordinate system is a straight line.

ILLUSTRATION 1

ILLUSTRATION 2

ILLUSTRATION 3

In this chapter we will examine functions and their inverses. The functions that will be examined in detail are exponential and logarithmic functions. Exponential functions are functions that are used to compute compound interest, model population growth, and find radioactive decay. Logarithmic functions are used to measure the acidity of solutions, drug dosages, the gain of an amplifier, the intensity of an earthquake, and safe noise levels in factories.

11.1 Algebra of Functions

■ SUM, DIFFERENCE, PRODUCT, AND QUOTIENTS OF FUNCTIONS ■ COMPOSITION OF FUNCTIONS
■ IDENTITY FUNCTION

■ SUM, DIFFERENCE, PRODUCT, AND QUOTIENTS OF FUNCTIONS

A function is a rule that can be described by using an algebraic formula. Functions can be added, subtracted, multiplied, and divided to form new functions.

Operations on Functions

If the domains and ranges of functions f and g are subsets of the real numbers, then

the **sum** of f and g, denoted as $f + g$, is defined by

$$(f + g)(x) = f(x) + g(x)$$

the **difference** of f and g, denoted as $f - g$, is defined by

$$(f - g)(x) = f(x) - g(x)$$

the **product** of f and g, denoted as $f \cdot g$, is defined by

$$(f \cdot g)(x) = f(x)g(x)$$

the **quotient** of f and g, denoted as $\frac{f}{g}$, is defined by

$$\left(\frac{f}{g}\right)(x) = \frac{f(x)}{g(x)} \quad (g(x) \neq 0)$$

The domain of each of these functions is the set of real numbers x that are in the domain of both f and g. In the case of the quotient, there is the further restriction that $g(x) \neq 0$.

EXAMPLE 1   Let $f(x) = 2x^2 + 1$ and $g(x) = 5x - 3$. Find each of the following functions and its domain: **a.** $f + g$ and   **b.** $f - g$.

Solution   **a.** $(f + g)(x) = f(x) + g(x)$

$$= (2x^2 + 1) + (5x - 3)$$
$$= 2x^2 + 5x - 2$$

The domain of $f + g$ is the set of real numbers that are in the domain of both f and g. Because the domain of both f and g is the set of real numbers, the domain of $f + g$ is also the set of real numbers.

b. $(f - g)(x) = f(x) - g(x)$

$$= (2x^2 + 1) - (5x - 3)$$
$$= 2x^2 + 1 - 5x + 3 \qquad \text{Remove parentheses.}$$
$$= 2x^2 - 5x + 4 \qquad \text{Combine like terms.}$$

Because the domain of both f and g is the set of real numbers, the domain of $f - g$ is also the set of real numbers.   ■

EXPLORING THE CONCEPT   Enter $f(x) = \frac{1}{10}x - 15$ and $g(x) = \frac{1}{100}x^2 + 5$ on the y= screen of your calculator.

At the Y3= prompt, enter Y1 + Y2 to indicate the addition $f(x) + g(x)$. Display the graphs of the two functions and their sum in the ZInteger viewing window.

a. TRACE to each indicated x-value and record the corresponding y-value. (*Hint:* The left and right cursor keys move you *along* the graph while the up and down cursor keys move you from one graph to another. The number displayed at the top of the screen indicates which function you are tracing on.)

TECHNOLOGY TIP
Press ZOOM, ZINTEGER, and move the cursor until the display reads $x = 0$ and $y = 0$ and press ENTER.

Y1:	$(0, \underline{\quad})$
Y2:	$(0, \underline{\quad})$
Y1 + Y2 = Y3:	$(0, \underline{\quad})$
Y1:	$(8, \underline{\quad})$
Y2:	$(8, \underline{\quad})$
Y1 + Y2 = Y3:	$(8, \underline{\quad})$

Y1:	$(-5, \underline{\quad})$
Y2:	$(-5, \underline{\quad})$
Y1 + Y2 = Y3:	$(-5, \underline{\quad})$

b. What pattern do you observe?

c. The same functions are displayed in the TABLE at the right. Does the same pattern occur for all values of x displayed in the TABLE?

X	Y1	Y2	Y3
-5	-15.5	5.25	-10.25
-4	-15.4	5.16	-10.24
-3	-15.3	5.09	-10.21
-2	-15.2	5.04	-10.16
-1	-15.1	5.01	-10.09
0	-15	5	-10
1	-14.9	5.01	-9.89

X=-5

d. What do you predict will happen to the pattern when $f(x) - g(x)$ is considered?

e. What do you predict will happen to the pattern when $f(x) \cdot g(x)$ is considered?

f. What do you predict will happen to the pattern when $f(x)/g(x)$ is considered?

Conclusions

a.
Y1: $(0, -15)$	Y1: $(-5, -15.5)$	Y1: $(8, -14.2)$
Y2: $(0, 5)$	Y2: $(-5, 5.25)$	Y2: $(8, 5.64)$
Y3: $(0, -10)$	Y3: $(-5, -10.25)$	Y3: $(8, -8.56)$

b. The sum of the values of Y1 and Y2 is equal to the value of Y3.

c. The same pattern of sums occurs for all x-values displayed in the TABLE.

d. If the functions are subtracted, $f(x) - g(x)$, then the difference of the values of Y1 and Y2 should be equal to the value of Y3.

e. If the functions are multiplied, $f(x) \cdot g(x)$, then the product of the values of Y1 and Y2 should be equal to the value of Y3.

f. If the functions are divided, $f(x)/g(x)$, then the quotient of the values of Y1 and Y2 should be equal to the value of Y3. ∎

EXAMPLE 2 Let $f(x) = 2x^2 + 1$ and $g(x) = 5x - 3$. Find each function and its domain:
a. $f \cdot g$ and **b.** $\frac{f}{g}$.

Solution **a.** $(f \cdot g)(x) = f(x)g(x)$
$$= (2x^2 + 1)(5x - 3)$$
$$= 10x^3 - 6x^2 + 5x - 3$$

The domain $f \cdot g$ is the set of real numbers that are in the domain of both f and g. Because the domain of both f and g is the set of real numbers, the domain of $f \cdot g$ is also the set of real numbers.

b. $\left(\dfrac{f}{g}\right)(x) = \dfrac{f(x)}{g(x)}$
$$= \dfrac{2x^2 + 1}{5x - 3}$$

Because the denominator of the fraction cannot be 0, $x \neq \frac{3}{5}$. Thus, the domain of $\frac{f}{g}$ is the set of real numbers, except $\frac{3}{5}$.

The fraction $\frac{3}{5}$ is said to be the **excluded value** of the quotient and is determined by setting the denominator equal to zero and solving for x.

$$5x - 3 = 0$$
$$5x = 3$$
$$x = \frac{3}{5}$$

The value of $x = \frac{3}{5}$ will thus make the denominator equal to zero. ◼

◼ COMPOSITION OF FUNCTIONS

Composition of functions is the "building" of a new function from two previously defined functions. Consider the following two tables that can be used to convert yards to feet and feet to inches.

Domain of f	x = yards	1	2	3	x
Range of f	$f(x)$ = feet	3	6	9	$3x$

The function rule for this conversion is $f(x) = 3x$.

Domain of g	f = feet	3	6	9	f
Range of g	$g(f)$ = inches	36	72	108	$12f$

The function rule for this conversion is $g(f) = 12f$.

Study the two charts. Mathematically speaking, the range of the function f is the domain of the function g. Remember that in the function notation $f(x)$, x represents the values in the domain and $f(x)$ the values in the range. Thus, in symbolic notation we have

$$g(f(x))$$

for our composition function.

If we create a chart that combines the two functions and converts yards directly into inches we have:

Domain of f	x = yards	1	2	3	x
Range of g	$g(f(x))$ = inches	36	72	108	$12(3x)$

The function rule for this conversion is $g(f(x)) = g(3x) = 12(3x) = 36x$. The notation $g(f(x))$ can also be written as $(g \circ f)(x)$ and expressed as "g composition f."

EXAMPLE 3 Let $f(x) = 2x^2 + 1$ and $g(x) = x - 3$. Find each of the following composition functions and its domain: **a.** $(f \circ g)(x)$ and **b.** $(g \circ f)(x)$.

Solution **a.** $(f \circ g)(x) = f(g(x))$
$$= f(x - 3)$$
$$= 2(x - 3)^2 + 1$$
$$= 2(x^2 - 6x + 9) + 1$$
$$= 2x^2 - 12x + 18 + 1$$
$$= 2x^2 - 12x + 19$$

Note that the graph of $(f \circ g)(x)$ is the function $y = 2x^2 + 1$ shifted/translated 3 units to the right.

$f(x) = 2x^2 + 1$

Domain: $\mathbb{R}$

$g(x) = x - 3$

Domain: $\mathbb{R}$

$f(g(x)) = Y1(Y2)$

Domain: $\mathbb{R}$

b. $(g \circ f)(x) = g(f(x))$

$\qquad\qquad = g(2x^2 + 1)$

$\qquad\qquad = (2x^2 + 1) - 3$

$\qquad\qquad = 2x^2 - 2$

$g(f(x)) = Y2(Y1)$

Domain: $\mathbb{R}$

From parts (a) and (b) we can draw the conclusion that, unlike addition and multiplication of functions, in general, composition of functions is not a commutative operation. ∎

EXAMPLE 4   A boy throws a stone into a pond causing a circular ripple to move out from the point of impact. If the radius of the circle is increasing at a rate of 3 inches per second, write a function that defines the area of the circle in terms of time and determine the area of the circle 5 seconds after impact.

Solution   To find the area of the circle we use the formula $A = A(r) = \pi r^2$. The notation $A(r)$ indicates the area is a function of the radius. Remember, we need the area to be a function of time, and it is the operation of composition that will allow us to do this.

Because the radius is increasing at a rate of 3 inches per second, the radius is a function of time. Thus, $r = r(t) = 3t$. The area of the circle in terms of time (not radius) can be defined as the following composition function:

$$A(r) \circ r(t) = A(r(t)) = A(3t)$$
$$= \pi(3t)^2$$
$$= \pi \cdot 9t^2$$
$$= 9\pi t^2$$

Thus, the area of the circle in terms of time is: $A(t) = 9\pi t^2$.

To find the area after 5 seconds we have

$$A(r(5)) = 9\pi(5)^2 = 225\pi \qquad 225\pi \text{ is approximately 707 square inches}$$

$Y1 = 9\pi x^2$

$[0,10]$ by $[0,1000]$

∎

■ IDENTITY FUNCTION

A special function, called the **identity function,** is defined by the equation $I(x) = x$. Under this function, the value that corresponds to any real number x is x itself. If f is any function, the composition of f with the identity function is just the function f:

$$(f \circ I)(x) = (I \circ f)(x) = f(x)$$

EXAMPLE 5   Let f be any function and I be the identity function, $I(x) = x$. Show that
a. $(f \circ I)(x) = f(x)$ and   **b.** $(I \circ f)(x) = f(x)$.

Solution   **a.** $(f \circ I)(x)$ means $f(I(x))$. Because $I(x) = x$, we have

$$(f \circ I)(x) = f(I(x)) = f(x)$$

b. $(I \circ f)(x)$ means $I(f(x))$. Because I passes any number unchanged, we have

$$I(f(x)) = f(x) \qquad \text{and} \qquad (I \circ f)(x) = I(f(x)) = f(x) \qquad ■$$

EXERCISE 11.1

VOCABULARY   *Fill in the blanks to make a true statement.*

1. The _____ value of the quotient of two functions is the value for which the quotient is undefined.

2. _____ of functions builds a new function from two previously defined functions.

3. The _____ function is defined by the equation $I(x) = x$. For any function f, $f \circ I = f$.

CONCEPTS   *In Exercises 4 and 5, use your results from Exploring the Concept to determine $f + g$, $f - g$, $f \cdot g$, and $\frac{f}{g}$.*

4. $f = \{(4, 5), (8, -2), (-3, 1), (-6, -2)\}$
$g = \{(4, 8), (8, 3), (-3, 5), (-6, 4)\}$

5. $f = \{(-2, 1), (5, 3), (7, -1), (-5, -2), (4, -3)\}$
$g = \{(-5, 6), (4, 1), (-2, 8), (7, 6), (5, 3)\}$

In Exercises 6–10, $f(x) = 4x^2 - 8x$ and $g(x) = 4x$.

6. What are the domains and ranges of each of the functions $f(x)$ and $g(x)$?

7. Graph the indicated function in the standard viewing window of your calculator. How do the domains and ranges compare to the domains and ranges of $f(x)$ and $g(x)$?
a. $f(x) + g(x)$
b. $f(x) - g(x)$
c. $f(x) \cdot g(x)$

8. State the domain of $f(x)/g(x)$ and sketch the graph in the standard viewing window of your calculator as well as on graph paper using the x- and y-intercepts.

9. A table of values for $f(x)/g(x)$ is displayed. Why does $x = 0$ yield an error message?

✎**10.** What should be done to your hand-drawn graph to reflect the fact that $x = 0$ is not on the graph?

11. A function that can be traced without lifting your pencil is called a **continuous function.** The point at which a break in the graph occurs is the **point of discontinuity.** What is the point of discontinuity for $f(x)/g(x)$?

✎**12.** How is the point of discontinuity related to the information given when you stated the domain of the function $f(x)/g(x)$?

■ **TECHNOLOGY TIP**

Set your calculator viewing window to ZInteger. From the FORMAT screen turn off the axes, then view the graph and its point of discontinuity.

PRACTICE *In Exercises 13–16, $f(x) = 3x - 2$ and $g(x) = 2x^2 + 1$. Find each function and its domain.*

13. $f - g$

14. $f + g$

15. $\frac{f}{g}$

16. $f \cdot g$

In Exercises 17–20, $f(x) = x^2 - 1$ and $g(x) = x^2 - 4$. Find each function and its domain.

17. $f - g$

18. $f + g$

19. $\frac{f}{g}$

20. $f \cdot g$

In Exercises 21–24, find $\frac{f(x)}{g(x)}$ and its domain.

21. $f(x) = 9x^6 - 12x^4$
   $g(x) = 3x^2$

22. $f(x) = 4x^2 - 10x^4 + 6x^5$
   $g(x) = 2x^2$

23. $f(x) = 2x^2 + 25x + 63$
   $g(x) = 2x + 7$

24. $f(x) = x^3 + 5x^2 + 4x - 10$
   $g(x) = x - 1$

In Exercises 25–36, $f(x) = 2x + 1$ and $g(x) = x^2 - 1$. Find each value.

25. $(f \circ g)(2)$

26. $(g \circ f)(2)$

27. $(g \circ f)(-3)$

28. $(f \circ g)(-3)$

29. $(f \circ g)(0)$

30. $(g \circ f)(0)$

31. $(g \circ f)\left(\frac{1}{2}\right)$

32. $(f \circ g)\left(\frac{1}{2}\right)$

33. $(f \circ g)(x)$

34. $(g \circ f)(x)$

35. $(g \circ f)(2x)$

36. $(f \circ g)(2x)$

37. Is composition of functions associative? Choose functions f, g, and h and determine if $[f \circ (g \circ h)](x) = [(f \circ g) \circ h](x)$.

38. If $f(x) = x + 1$ and $g(x) = 2x - 5$, show that $(f \circ g)(x) \neq (g \circ f)(x)$.

39. If $f(x) = x^2 + 2x - 3$, find $f(a)$, $f(h)$, and $f(a + h)$. Then show that $f(a + h) \neq f(a) + f(h)$.

40. If $f(x) = x^2 + 2$, find $\frac{f(x + h) - f(x)}{h}$.

41. If $f(x) = x^3 - 1$, find $\frac{f(x + h) - f(x)}{h}$.

APPLICATIONS

42. Volume Sand is dropped off a conveyor belt to form a conical pile whose height is always twice its base radius. The volume of the conical pile, expressed as a function of the radius, r, is $V(r) = \frac{2}{3}\pi r^2$ cubic feet. If the base radius of the pile is increasing at a rate of 1.5 feet per minute, express the volume as a function of time. What is the volume after 4 minutes?

✎**43. Volume** If the volume of a right circular cone is $V = \frac{1}{3}Bh$ (where B is the area of the base and h is the height), explain how to obtain the formula $V(r) = \frac{2}{3}\pi r^3$ used in Exercise 42.

44. Volume A party store uses helium to fill balloons. If a spherical balloon is filled so that the radius of the sphere is increasing at the rate of 1.25 inches per second, express the volume of the sphere as a function of time. What is the volume of the balloon to the nearest hundredth after 5 seconds? $\left(\textit{Hint: } V(r) = \frac{4}{3}\pi r^3\right)$

45. **Area** A radiator is leaking antifreeze onto the garage floor, creating a circular pool. Suppose that the radius is growing at a constant rate of 2 inches per minute. Express the area of the circular pool as a function of time, t, where t is the number of minutes elapsed since the leak began. Find the area of the pool after 7 minutes.

46. **World record** On April 18, 1999, students at Clarksville High School in Clarksville, Tennessee, broke the record for the longest paper-clip chain created within a 24-hour period. The students added $1\frac{1}{4}$-inch-long paper clips at the rate of 38,480 per hour; therefore, the length of the chain can be expressed as a function of time, t (in hours). The previous world record was 16.88 miles in a 24-hour period. How long was the chain made by the Clarksville students? (*Hint:* Length is a function of the number of paper clips and rate is a function of hours.)

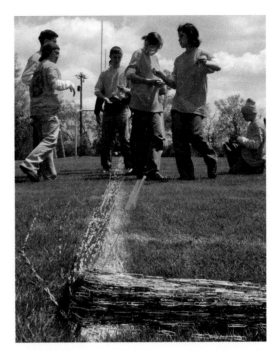

Record-breaking paper-clip chain (Courtesy of *The Leaf-Chronicle*)

REVIEW

47. Determine the slope of a line passing through the points $(3, 5)$ and $(-2, 1)$.

48. Write the equation of the line determined by the two points in Exercise 47.

49. Determine the domain and range of each of the following functions:
    a. $f(x) = 4x + 2$
    b. $f(x) = \sqrt{x - 5} + 2$

    c. $f(x) = (x - 5)^2 + 2$

50. What is the difference between finding the roots of the equation $-2(x + 3)^2 + 2 = 0$ and finding the x-intercepts of the function $f(x) = -2(x + 3)^2 + 2$?

11.2 Inverse Functions

■ INVERSE RELATIONS ■ ONE-TO-ONE FUNCTIONS ■ INVERSE FUNCTIONS

■ INVERSE RELATIONS

In the previous section, we constructed a composite function by comparing tables of measurement conversions between yards to feet and feet to inches. The composite function was formed by combining the two tables to create a new function that converted yards to inches. The composition of two functions *always* produces another function.

Consider what happens when we take the table of measurements used to convert yards to feet.

Domain	a = yards	1	2	3	a
Range	b = feet	3	6	9	$3a$

If we want to convert feet to yards, the table becomes

Domain	b = feet	3	6	9	$3a$
Range	a = yards	1	2	3	a

By reversing the a and b elements we have created an inverse relationship where inverse basically means "undoing." Although these two relations that are inverses of one another are also functions, that will not always be the case. In general, if a relation contains the point (a, b), then the **inverse relation** will contain the point (b, a).

All relations have inverse relations because the definition of a relation is that it is a set of ordered pairs.

> **Inverse Relations**
> If R is any relation and R^{-1} is the relation obtained from R by interchanging the components of each ordered pair of R, then R^{-1} is called the **inverse relation of R.**
>
> The domain of R^{-1} is the range of R, and the range of R^{-1} is the domain of R.

EXAMPLE 1   If the relation defined by converting x minutes to y seconds produces the ordered pairs $\left\{\left(\frac{1}{2}, 30\right), (1, 60), \left(\frac{3}{2}, 90\right), (2, 120)\right\}$, find the inverse relation that is defined by the conversion of seconds to minutes.

Solution   Inverse relations are determined by reversing the domain and range elements. Thus, the inverse relation would be

$$\left\{\left(30, \tfrac{1}{2}\right), (60, 1), \left(90, \tfrac{3}{2}\right), (120, 2)\right\}$$

This means that 30 seconds is equal to $\frac{1}{2}$ a minute, 60 seconds is equal to 1 minute, and so on. ∎

EXAMPLE 2   Show that $y = |x|$ and $x = |y|$ are inverse relations and then graph each relation.

Solution   Look at a table of values for $y = |x|$.

x	−3	−2	−1	0	1	2	3
y	3	2	1	0	1	2	3

Now reverse the domain and range values:

x	3	2	1	0	1	2	3
y	−3	−2	−1	0	1	2	3

Each of these domain and range pairs satisfies $x = |y|$. Thus, if (a, b) is an ordered pair in the relation $y = |x|$, we have $b = |a|$, and (b, a) is an ordered pair in the relation $x = |y|$.

■ TECHNOLOGY TIP
To have the calculator **draw** the inverse of a function, enter the function at the y= editor. From the home screen, go to the DRAW menu and select DrawInv. With DrawInv now displayed on the home screen, enter y1 from the VARS menu (press VARS, cursor over to choose Y-VARS, press ENTER for Function and ENTER again for the function entered at y1). Both the function entered at y1 and its inverse will be graphed unless you "turn off" the graph of y1.

The graph of $y = |x|$ in the standard viewing window is displayed. To *draw* (not *graph*) the inverse, $x = |y|$, on your calculator, use the DrawInv feature.

■ ONE-TO-ONE FUNCTIONS

In order for a function, $f(x)$, to have an inverse that is also a function, the function $f(x)$ must be a one-to-one function. Recall that a function is a relation (set of ordered pairs) in which each value x in the domain is paired with a unique value y in the range. In order for a function to be **one-to-one,** we must extend the definition so that each y-value is paired with a unique x-value.

> **One-to-One Function**
> A function is called **one-to-one** if and only if each value of y in the range corresponds to only one number x in the domain.

We use a **vertical line test** to determine if each x in the domain has a unique y in the range. A similar test, called the **horizontal line test,** can be used to determine if each y in the range has a unique x in the domain.

> **Horizontal Line Test**
> If any horizontal line that intersects the graph of a function does so only once, the function is one-to-one.

EXAMPLE 3   Determine if the graphs of the following relations are functions and then determine if they are one-to-one functions.

a.

b.

c.

d.

e.

Solution **a.** The graph is that of a function because it passes the vertical line test. It is one-to-one because it also passes the horizontal line test.

b. The graph is that of a function because it passes the vertical line test. It is *not* one-to-one because it does not pass the horizontal line test.

c. The graph is simply a relation.

d. The graph is that of a function because it passes the vertical line test. It is one-to-one because it also passes the horizontal line test.

e. The graph is that of a function because it passes the vertical line test. It is one-to-one because it also passes the horizontal line test. ■

EXPLORING THE CONCEPT Use your calculator to *draw* the inverse of each of the following functions: **a.** $y = 3x + 5$ **b.** $y = x^2$ **d.** $y = 3\sqrt{x}$ **e.** $y = 2^x$. (The functions were labeled in this manner because they correspond to the graphs in Example 3.)

Compare your inverses to the graphs of a, b, d, and e in Example 3.

Which of the inverses are function?

Which of the graphs are one-to-one functions?

What conclusion can you draw?

Conclusion **a.**   **b.**

d.   **e.**

The graphs in Example 3 that are one-to-one functions are a, d, and e. Similarly, the inverses that are functions are also a, d, and e.

Conclusion: Functions that are one-to-one have inverses that are also functions. ■

■ INVERSE FUNCTIONS

Inverse functions are determined in the same way that inverse relations are determined—by switching the domain and range values. We will also use the notation $f^{-1}(x)$, read f *inverse* of x, to denote the inverse of the function f.

> **Inverse Function**
> The inverse of the one-to-one function $f(x)$ is denoted as $f^{-1}(x)$. It consists of all ordered pairs (b, a) where (a, b) belongs to $f(x)$.

EXAMPLE 4   Find the inverse function for $f(x) = 4x + 2$.

Solution   We know that $f^{-1}(x)$ exists because the graph of $y = 4x + 2$ passes both the Vertical and Horizontal Line tests. To find $f^{-1}(x)$, we interchange the x and y variables to obtain

$$x = 4y + 2$$

Solve this for y so that the $f^{-1}(x)$ notation may be used:

$$x = 4y + 2$$
$$x - 2 = 4y \qquad \text{Subtract 2 from both sides of the equation.}$$
$$\frac{x - 2}{4} = y \qquad \text{Divide both sides by 4.}$$
$$\frac{1}{4}x - \frac{1}{2} = y$$

Thus, the function $f(x) = 4x + 2$ has $f^{-1}(x) = \dfrac{1}{4}x - \dfrac{1}{2}$ as its inverse.   ■

Steps for Finding the Inverse of a Function

1. If the function is given as a set of ordered pairs, interchange the x- and y-values to obtain the inverse function.
2. If the function is given as an equation:
   a. Interchange the variables x and y.
   b. Solve the resulting equation for y, if possible.
   c. This equation is $y = f^{-1}(x)$, which defines the inverse function.

To reinforce the concept stated in the definition that if the ordered pair (a, b) belongs to $f(x)$ then the ordered pair (b, a) belongs to $f^{-1}(x)$, do the following, using the function given in Example 4, $f(x) = 4x + 2$:

Find $y = f(3)$.

$$y = 4(3) + 2$$
$$y = 14$$

Thus, $(3, 14)$ belongs to $f(x)$.
Now find $y = f^{-1}(14)$.

$$y = \frac{1}{4}(14) - \frac{1}{2}$$
$$y = \frac{7}{2} - \frac{1}{2}$$
$$y = 3$$

Thus, $(14, 3)$ belongs to $f^{-1}(x)$.

EXPLORING THE CONCEPT   As we have seen, we know that inverse functions "undo" each other. We now want to explore what occurs when we take the composition of a function and its inverse.

$$\text{Let } f(x) = 4x + 2 \qquad \text{and} \qquad f^{-1}(x) = \frac{1}{4}x - \frac{1}{2}$$

a. Find $(f \circ f^{-1})(3)$ b. Find $(f^{-1} \circ f)(3)$

c. Find $(f \circ f^{-1})(x)$ d. Find $(f^{-1} \circ f)(x)$

e. What conclusions can you draw about $f \circ f^{-1}$ and $f^{-1} \circ f$?

f. Given $g(x)$ and $g^{-1}(x)$, what would be the value of $g[g^{-1}(4)]$?

Conclusions   **a.** $(f \circ f^{-1})(3) = f[\, f^{-1}(3)]$

$$= f\left[\frac{1}{4} \cdot 3 - \frac{1}{2}\right]$$

$$= f\left[\frac{1}{4}\right]$$

$$= 4\left(\frac{1}{4}\right) + 2$$

$$= 1 + 2$$

$$= 3$$

$$(f \circ f^{-1})(3) = 3$$

b. $(f^{-1} \circ f)(3) = f^{-1}[\, f(3)]$

$$= f^{-1}[4 \cdot 3 + 2]$$

$$= f^{-1}[14]$$

$$= \frac{1}{4}(14) - \frac{1}{2}$$

$$= 3$$

$$(f^{-1} \circ f)(3) = 3$$

c. $(f \circ f^{-1})(x) = f[\, f^{-1}(x)]$

$$= f\left[\frac{1}{4} \cdot x - \frac{1}{2}\right]$$

$$= 4\left(\frac{1}{4}x - \frac{1}{2}\right) + 2$$

$$= x - 2 + 2$$

$$= x$$

$$(f \circ f^{-1})(x) = x$$

d. $(f^{-1} \circ f)(x) = f^{-1}[\, f(x)]$

$$= f^{-1}[4x + 2]$$

$$= \frac{1}{4}(4x + 2) - \frac{1}{2}$$

$$= x + \frac{1}{2} - \frac{1}{2}$$

$$= x$$

$$(f^{-1} \circ f)(x) = x$$

e. *Conclusions:* Composition of a function and its inverse will always yield x. Thus, composition of a function and its inverse is commutative. (This is the **only** condition under which composition is commutative!) To show that two functions are inverses of one another, you must show that $f[f^{-1}(x)] = f^{-1}[f(x)] = x$.

f. $g[g^{-1}(4)] = 4$ since $g[g^{-1}(x)] = x$. ■

EXAMPLE 5   If $f(x) = \dfrac{1}{3}x + 8$, find $f^{-1}(x)$ and graph both functions and $y = x$ in the ZInteger viewing window.

Solution

$$f(x) = \frac{1}{3}x + 8$$

$$y = \frac{1}{3}x + 8$$

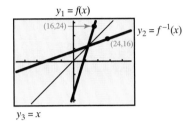

$$x = \frac{1}{3}y + 8 \qquad \text{Interchange } x \text{ and } y.$$

$$x - 8 = \frac{1}{3}y \qquad \text{Subtract 8 from both sides.}$$

$$3x - 24 = y \qquad \text{Multiply both sides by 3.}$$

$$f^{-1}(x) = 3x - 24$$

The graph of any function, $f(x)$, and its inverse, $f^{-1}(x)$, will be symmetric about the line $y = x$. This is always true because we defined inverse functions as one-to-one functions such that if (a, b) belongs to the function then (b, a) belongs to the inverse, and the points (a, b) and (b, a) are symmetric with respect to the line $y = x$. ■

EXERCISE 11.2

VOCABULARY   *Fill in each blank to make a true statement.*

1. If a relation contains a point (a, b), then the _____ will contain the point (b, a).

2. A function that is _____ is a function in which each y-value in the range is paired with a unique x-value in the domain.

3. The _____ test is used to determine if each y-value in the range is paired with a unique x-value in the domain.

4. The notation _____ is used to denote the inverse of the function $f(x)$.

CONCEPTS

5. True or false: Every relation has an inverse. Justify your response.

6. True or false: Every function has an inverse function. Justify your response.

In Exercises 7–10, determine the inverse relation of each set of ordered pairs (x, y).

7. $\{(4, 5), (2, 8), (-3, 7), (6, -9)\}$

8. $\{(1, 2), (2, 3), (1, 3), (1, 5)\}$

9. $\{(-4, 2), (-3, 2), (0, 2), (1, 2)\}$

10. $\{(1, 4), (1, 5), (1, 6), (1, 7)\}$

11. The following relation defines the conversion of kilograms to pounds. Find the inverse relation that defines the conversion of pounds to kilograms.

Kilograms	1	2	3	3.5	4
Pounds	2.2	4.4	6.6	7.7	8.8

12. The following relation defines the conversion of degrees in Fahrenheit to degrees in Celsius. Find the inverse relation that defines the conversion of degrees in Celsius to degrees in Fahrenheit.

Fahrenheit	23	32	41	50	59	68	77
Celsius	-5	0	5	10	15	20	25

PRACTICE   *In Exercises 13–24, determine which graphs represent one-to-one functions.*

13.

14.

15.

16.

17.

18.

19.

20.

21.

x	4	5	6	1
y	2	−1	3	−1

22.

x	2	3	4	5
y	4	4	4	4

23.

x	1	3	5	7
y	2	4	6	8

24.

x	−2	1	3	−4
y	1	−2	−4	5

In Exercises 25–32, find the inverse of the relation determined by the given equation and tell whether that inverse relation is a function. If the inverse relation is a function, express it in the form $y = f^{-1}(x)$.

25. $y = 3x + 1$ **26.** $y + 1 = 5x$ **27.** $x + 4 = 5y$ **28.** $x = 3y + 1$

29. $y = \dfrac{x - 4}{5}$ **30.** $y = \dfrac{2x + 6}{3}$ **31.** $4x - 5y = 20$ **32.** $3x + 5y = 15$

In Exercises 33–40, find the inverse of each linear function. Graph both the function and its inverse on graph paper on the same set of axes, as well as the axis of symmetry, $y = x$.

33. $y = 4x + 3$ **34.** $x = 3y - 1$ **35.** $x = \dfrac{y - 2}{3}$ **36.** $y = \dfrac{x + 3}{4}$

37. $3x - y = 5$

38. $2x + 3y = 9$

39. $3(x + y) = 2x + 4$

40. $-4(y - 1) + x = 2$

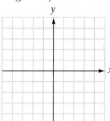

In Exercises 41–48, graph each function, f(x) and its inverse $f^{-1}(x)$ in the ZInteger viewing window. Then graph $f(f^{-1}(x))$. (Hint: Graph the composition y1(y2) where y1 = f(x) and y2 = $f^{-1}(x)$). Compare your calculator display to your hand-drawn graphs in Exercises 33–40.

41. $y = 4x + 3$

42. $x = 3y - 1$

43. $x = \dfrac{y - 2}{3}$

44. $y = \dfrac{x + 3}{4}$

45. $3x - y = 5$

46. $2x + 3y = 9$

47. $3(x + y) = 2x + 4$

48. $-4(y - 1) + x = 2$

In Exercises 49–52, show that each of the pairs of equations are inverse functions. For each pair, $f(f^{-1}(x)) = x$.

49. $y = 2x + 6$

$\quad y = \dfrac{1}{2}x - 3$

50. $y = \dfrac{2}{3}x + 1$

$\quad y = \dfrac{3x - 3}{2}$

51. $y = \dfrac{x + 5}{3}$

$\quad y = 3x - 5$

52. $y = x - 7$

$\quad y = x + 7$

In Exercises 53–62, find the inverse of the relation determined by each equation. Tell whether the inverse relation is a function. Verify your inverse by both graphing the inverse on graph paper and drawing the inverse (via the DrawInv feature) on your calculator.

53. $y = x^2 + 4$

54. $x = y^2 - 2$

55. $x = y^2 - 4$

56. $y = x^2 + 5$

57. $y = x^3$

58. $xy = 4$

59. $y = \pm\sqrt{x}$

60. $y = \sqrt[3]{x}$

61. $y = \sqrt{x}$

62. $4y^2 = x - 3$

REVIEW

63. Evaluate:
  a. 4^3
  b. -4^4
  c. 4^{-4}
  d. $4^{1/2}$
  e. $-4^{1/2}$
  f. $(-4)^{-1/2}$

64. Express in radical form:
  a. $a^{2/5}$
  b. $4^{-2/5}$
  c. $8a^{1/3}$
  d. $(8a)^{1/3}$

65. Graph each of the following and determine their domains and ranges.
  a. $f(x) = x^2$

  b. $f(x) = x^3$

Domain: $\mathbb{R}$, range: $\mathbb{R}$

  c. $f(x) = \sqrt{x}$

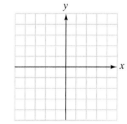

✐**66.** Write the equation of a function that has the same size/shape as $f(x) = x^2$ but is reflected in the x-axis, translated 3 units left and 4 units up.

11.3 Exponential Functions

■ BASIC CONCEPTS OF EXPONENTS ■ GRAPHS OF EXPONENTIAL FUNCTIONS ■ APPLICATIONS OF EXPONENTIAL FUNCTIONS ■ BASE-*e* EXPONENTIAL FUNCTIONS ■ SOLVING EXPONENTIAL EQUATIONS

■ BASIC CONCEPTS OF EXPONENTS

Natural-number exponents were defined in Section 1.2 and rational exponents were defined in Section 9.2. From these definitions we note that

$$x^5 = x \cdot x \cdot x \cdot x \cdot x$$

and

$$x^{2/5} = \sqrt[5]{x^2}$$

We now want to extend our definition of exponents to include all real-number values. It should be noted that all exponent rules previously addressed for natural- and rational-number exponents are also valid for all real-number exponents.

Consider $3^{\sqrt{2}}$. Because $\sqrt{2}$ is approximately 1.414213562, we can approximate $3^{\sqrt{2}}$ by expressing it as $3^{1.4}$, $3^{1.41}$, $3^{1.414}$, etc., depending on the desired degree of accuracy.

   vs.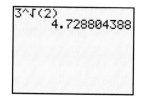

■ TECHNOLOGY TIP
The VALUE feature under the CALC menu was used. Let $x = \sqrt{2}$ at the prompt. This graph is displayed in the standard viewing window.

This can also be supported using a graph and letting $y = 3^x$:

EXAMPLE 1   Evaluate the following powers in one of two ways: using the laws of exponents when possible or using the exponentiation power of the calculator. Then, verify your results graphically.

a. $4^{1/2}$ **b.** $8^{2/3}$ **c.** $5^{\sqrt{2}}$ **d.** $0.5^{\sqrt{2}}$

Solution   **a.** $4^{1/2} = \sqrt{4} = 2$

Definition of $a^{1/n} = \sqrt[n]{a}$ (Section 9.2)
When $y = 4^x$ and $x = \frac{1}{2}$, then $y = 2$.

[−5,5] by [−5,5]

b. $8^{2/3} = \left(\sqrt[3]{8}\right)^2 = (2)^2 = 4$ Definition of $a^{m/n} = \left(\sqrt[n]{a}\right)^m$ (Section 9.2)

When $y = 8^x$ and $x = \frac{2}{3}$, then $y = 4$.

[−5,5] by [−5,5]

c.

[−5,5] by [−5,10]

When $y = 5^x$ and $x = \sqrt{2}$, then $y \approx 9.738517$.

d. .5^-√(2)
2.665144143

When $y = 0.5^x$ and $x = -\sqrt{2}$, then $y \approx 2.665144$.

[−5,5] by [−5,5]

GRAPHS OF EXPONENTIAL FUNCTIONS

Functions with variables as exponents are called **exponential functions.**

> **Exponential Function**
> A function of the form $f(x) = b^x$, where $b > 0$ and $b \neq 1$, is called an **exponential function** with base b.

You should note that the base, b, of an exponential function cannot be 1 because $f(x) = 1^x = 1$ for all values of x. Thus, if $b = 1$, we have a constant function, not an exponential function. Be careful also not to confuse functions with real number powers with exponential functions.

Exponential Function	*NOT an Exponential Function*
$f(x) = 2^x$	$f(x) = x^2$
$f(x) = 3^x$	$f(x) = x^3$
$f(x) = \left(\frac{1}{2}\right)^x$	$f(x) = x^{1/2}$

EXPLORING THE CONCEPT — Set the viewing window of your calculator to $[-4.7, 4.7]$ by $[-10, 10]$. Examine each of the exponential functions where the base b is greater than 1.

$$f(x) = 2^x \qquad f(x) = 3^x \qquad f(x) = 8^x$$

a. Begin with your cursor at $x = -3$ and TRACE along the graph of each function. As x increases, what happens to y? Verify your conclusion by scrolling through the TABLE and observing the behavior of the y-values.

b. TRACE (or scroll through the TABLE) to determine if the y-value can ever be equal to 0. Why or why not?

c. What is the domain and the range of each function?

d. Compare the three graphs. How are they similar? How are they different? (*Hint:* Place your cursor at $(0, 0)$ and Zoom In once.)

Conclusions **a.** As x increases, so does y; however, y increases very rapidly as compared to x.

b. The y-value will never equal zero because $y = b^x$ $(b \neq 0)$, and a nonzero base raised to any power is a nonzero number.

c. Domain $= (-\infty, \infty)$ and range $= (0, \infty)$.

d. All graphs represent increasing functions. All graphs have the same basic shape and they all pass through $(0, 1)$. All graphs $y = b^x$ contain the point $(1, b)$. For $x > 0$, the graph rises faster for larger bases.

The window should be $[-1.175, 1.175]$ by $[-2.5, 2.5]$ after Zooming In once.

Because the graphs approach the x-axis but never cross it (i.e., $y \neq 0$), we say that the x-axis is the horizontal asymptote of the function. Thus, the equation of the asymptote is $y = 0$.

We can also observe from these graphs that the y-intercept is $(0, 1)$. This is true for all exponential functions $f(x) = b^x$, since

$f(x) = b^0$ Let $x = 0$ and find the y-intercept.

$f(x) = 1$ $b^0 = 1$ since $b \neq 0$.   ■

Exponential functions of the form $f(x) = b^x$ where $b > 1$ are called **exponential growth** functions because y increases rapidly as x increases.

EXAMPLE 2   Graph $y = 4^x$ on graph paper and verify with your calculator.

Solution   Construct a table of values based on the y-intercept of $(0, 1)$.

x	y
-1	$\frac{1}{4}$
$-\frac{1}{2}$	$\frac{1}{2}$
0	1
$\frac{1}{2}$	2
1	4

Note: Values of $\pm\frac{1}{2}$ were selected because we can compute exact values of $4^{1/2}$ and $4^{-1/2}$. Integer values could have been selected instead, but the y-axis would need to be increased vertically.

     ■

EXPLORING THE CONCEPT   Set your viewing window to $[-4.7, 4.7]$ by $[-10, 10]$. Examine each of the exponential functions where the base b is between 0 and 1.

$$f(x) = \left(\frac{1}{10}\right)^x \qquad f(x) = \left(\frac{1}{2}\right)^x \qquad f(x) = \left(\frac{4}{5}\right)^x$$

a. Begin with your cursor at the leftmost visible point on the graph and TRACE along each function. As x increases, what happens to y? Verify your conclusions by scrolling through the TABLE and observing the behavior of the y-values.

b. TRACE (or scroll through the TABLE) to determine if the y-value will ever be 0. Is the line $y = 0$ the horizontal asymptote?

c. What is the domain and range of each function?

d. Compare the three graphs. How are they similar? How are they different? (*Hint:* Place your cursor at $(0, 0)$ and Zoom In once).

Conclusions   a. As x increases, y decreases.

b. Because y never equals 0 (a nonzero base raised to a nonzero power will never be zero), the line $y = 0$ is the horizontal asymptote.

c. Domain $= (-\infty, \infty)$ and range $= (0, \infty)$.

d. All graphs represent decreasing functions. They all have the same basic shape and pass through $(0, 1)$. All graphs $y = b^x$ contain $(1, b)$. The closer the b value is to zero, the faster the graph decreases. The window should be $[-1.175, 1.175]$ by $[-2.5, 2.5]$ after Zooming In once.

EXAMPLE 3   Graph $y = \left(\frac{1}{4}\right)^x$ on graph paper and verify with your calculator.

Solution   Construct a table of values based on a y-intercept of $(0, 1)$.

x	y
-1	4
$-\frac{1}{2}$	2
0	1
$\frac{1}{2}$	$\frac{1}{2}$
1	$\frac{1}{4}$
2	$\frac{1}{4}$

Exponential functions of the form $f(x) = b^x$, where $0 < b < 1$, are called **exponential decay** functions because y decreases as x increases.

■ APPLICATIONS OF EXPONENTIAL FUNCTIONS

A mathematical description of an observed event is called a **model** of that event. Many events that change with time are modeled by functions defined by equations of the form

$$y = f(t) = ab^{kt} \qquad \text{Remember that } ab^{kt} \text{ means } a(b^{kt})$$

where a, b, and k are constants and t represents time. If the function is an increasing function, then y is said to **grow exponentially.** If the function is a decreasing function, then y is said to **decay exponentially.**

One example of exponential growth is **compound interest.** When interest is allowed to accumulate in a savings account, the balance will grow exponentially according to the following model.

Formula for Compound Interest

If the present value P is deposited in an account, and interest is paid k times a year at an annual interest rate r, the amount of money, A, in the account after t years is given by $A = P\left(1 + \dfrac{r}{k}\right)^{kt}$.

EXAMPLE 4 **Saving for college** To save for college, parents invest $12,000 for their newborn child in a mutual fund that should average 10% annual interest. If the quarterly dividends are reinvested, how much will be available in 18 years?

Solution We substitute 12,000 for P, 0.10 for r, and 18 for t into the formula for compound interest and find A. Since interest is paid quarterly, $k = 4$.

$$A = P\left(1 + \frac{r}{k}\right)^{kt}$$
$$A = 12{,}000\left(1 + \frac{0.10}{4}\right)^{4(18)}$$
$$= 12{,}000(1 + 0.025)^{72}$$
$$= 12{,}000(1.025)^{72}$$
$$= 71{,}006.74$$

```
12000(1+0.10/4)^
(4*18)
        71006.73675
```

In 18 years, the account will contain more than $71,000. ■

In business applications, the initial amount of money deposited is called the **present value** *(PV)*. The amount to which the money will grow is called the **future value** *(FV)*. The interest rate used for each compounding period is the **period interest rate** *(i)*, and the number of times interest is compounded is the **number of compounding periods** *(n)*. Using these definitions, an alternative formula for the compound interest formula is as follows:

Formula for Compound Interest

$$FV = PV(1 + i)^n$$

This alternative formula appears on business calculators. To use this formula to solve Example 3, we proceed as follows:

$$FV = PV(1 + i)^n$$
$$FV = 12{,}000(1 + 0.025)^{72} \qquad i = \tfrac{0.10}{4} = 0.025 \text{ and } n = 4(18) = 72.$$
$$= 71{,}006.74$$

EXAMPLE 5   **Watching money grow**   If $1 is deposited in an account earning 6% annual interest, compounded monthly, graphically determine how much money will be in the account after 100 years.

Solution   Using the formula $A = P\left(1 + \dfrac{r}{k}\right)^{kt}$, we can substitute 1 for P, 0.06 for r, and 12 for k (the interest is compounded monthly). Thus,

$$A = 1\left(1 + \frac{.06}{12}\right)^{12t}$$

To evaluate graphically, we let $A = y$ and $k = x$.

$$y1 = 1\left(1 + \frac{.06}{12}\right)^{12x}$$

and use the VALUE feature to determine the value of y after $x = 100$ years.

After 100 years, the initial deposit of $1 will have grown to $397.44.   ∎

■ BASE-e EXPONENTIAL FUNCTIONS

When banks compound interest quarterly, the account balance increases four times a year. In other situations, quantities grow continuously. For example, it is possible for a bank to compound money continuously, and the population growth of a town increases steadily (not in abrupt jumps at every quarter).

We use the irrational number e, named after the Swiss mathematician Leonhard Euler, to predict the future value of a quantity that increases or decreases continuously. The value of e can be best approximated by the expression $\left(1 + \dfrac{1}{x}\right)^{x}$. As x gets increasingly larger, $\left(1 + \dfrac{1}{x}\right)^{x}$ gets closer and closer to the value of e, which is approximately 2.71828182846.

If we compound interest continuously using the formula $A = P\left(1 + \dfrac{r}{k}\right)^{kt}$, it can be shown that $\left(1 + \dfrac{r}{k}\right)^{kt}$ approaches the value of e (mathematical justification of this will not be done until a later course in algebra). Thus, $A = P\left(1 + \dfrac{r}{k}\right)^{kt}$, which represents a finite number of compounding periods, can be rewritten as

$$A = Pe^{rt}$$

where compounding occurs continuously.

> ### Formula for Continuous Compounding
> $A = Pe^{rt}$, where P is the present value of the quantity, A is the future value, r is the rate of interest, and t is the time in years.

Previously in Example 4, we determined that $71,006.74 would be in a college fund if $12,000 were compounded quarterly for 18 years at a 10% interest rate. We will now see how the future value of the $12,000 is affected if we compound continuously.

EXAMPLE 6   **Saving for college**   Compute the future value of $12,000 invested for 18 years at 10% compounded continuously.

Solution:

$A = Pe^{rt}$

$A = 12{,}000e^{.10(18)}$

$A = \$72{,}595.77$

After 18 years, the account will contain $72,595.77. This is $1589.03 more than the result in Example 4, which was compounded quarterly.   ■

In the **Malthusian model for population growth,** the future population of a colony is related to the present population by the formula for exponential growth. This is the same formula as was used for continuous compounding. However, P represents the current population and A is the future population.

EXAMPLE 7   **City planning**   The population of a city is currently 15,000, but changing economic conditions are causing the population to decrease 2% each year. If this trend continues by the Malthusian model, find the population in 30 years.

Solution   Since the population is decreasing 2% each year, the annual growth rate is −2%, or −0.02. We can substitute −0.02 for r, 30 for t, and 15,000 for P in the formula for exponential growth and find A.

$A = Pe^{rt}$

$A = 15{,}000e^{(-.02)(30)}$

$A = 15{,}000e^{(-0.6)}$

$A = 8232.17$

In 30 years, city planners expect a population of approximately 8232 persons.   ■

■ SOLVING EXPONENTIAL EQUATIONS

We can make several observations based on our examinations of the exponential growth function $f(x) = b^x$, where $b > 1$, and the exponential decay function $f(x) = b^x$, where $0 < b < 1$.

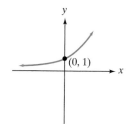

a. Both functions have the set of all real numbers as the domain.

b. Both functions have as their range all real numbers greater than 0.

c. The exponential growth function $f(x) = b^x$, where $b > 1$, is an increasing function.

d. The exponential decay function $f(x) = b^x$, where $0 < b < 1$, is a decreasing function.

e. Both functions have a y-intercept of $(0, 1)$ and no x-intercept because the x-axis is the horizontal asymptote.

f. Both functions are one-to-one (each x has a unique y-value, and each y-value has a unique x-value), and thus exponential functions have an inverse function.

It is the last observation—exponential functions are one-to-one—that leads us to the property we use to solve exponential equations.

Exponential Property of Equality
If $b > 0$ and $b \neq 1$, then $b^r = b^s$ is equivalent to $r = s$.
If $r \neq s$, then $b^r \neq b^s$.

EXAMPLE 8   Solve each of the following:

$$\textbf{a. } 2^x = 4 \quad \textbf{b. } 3^x = 3^{4x+5} \quad \textbf{c. } \left(\frac{1}{16}\right)^x = 4^{3x-1}$$

Solution   **a.** $2^x = 4$

$\quad\quad 2^x = 2^2$ Rewrite the equation so that both sides have the same base.

$\quad\quad\; x = 2$ Based on the property, if $b^r = b^s$, then $r = s$.

b. $3^x = 3^{4x+5}$

$\quad\quad x = 4x + 5$ If $b^r = b^s$, then $r = s$.

$\quad\quad x = -\dfrac{5}{3}$

Verify graphically:

[–3,3] by [–3,3]

c. $\left(\dfrac{1}{16}\right)^x = 4^{3x-1}$

$\quad \left(\dfrac{1}{2^4}\right)^x = (2^2)^{3x-1}$ Write each base as a power of two.

$\quad (2^{-4})^x = 2^{2(3x-1)}$ $\frac{1}{a^m} = a^{-m}$ for $m > 0$.

$\quad\quad 2^{-4x} = 2^{6x-2}$ $(a^m)^n = a^{mn}$

$\quad\quad -4x = 6x - 2$ If $b^r = b^s$, then $r = s$.

$\quad\quad -10x = -2$

$\quad\quad\quad\; x = \dfrac{1}{5}$

Verify graphically:

[–3,3] by [–3,3]

EXAMPLE 9   Use a graphing calculator to solve the equation $x^2 + 8x + 11 = 0.8^x$. Approximate the solution to the nearest thousandth.

Solution   Use the INTERSECT feature to solve graphically:

$\{-7.152, -1.472\}$

EXERCISE 11.3

VOCABULARY *Fill in each blank to make a true statement.*

1. Functions with variables as exponents are called _____ functions.

2. An exponential _____ function is in the form $f(x) = b^x$, where $b > 1$.

3. An exponential _____ function is in the form $f(x) = b^x$, where $0 < b < 1$.

4. The *x*-axis forms the horizontal _____ for all exponential functions because exponential functions never have a *y*-value of 0.

CONCEPTS

5. Evaluate each of the following with the laws of exponents when possible or by the exponentiation power of the calculator.
   a. $25^{1/2}$
   b. $16^{3/4}$
   c. $25^{-\sqrt{2}}$
   d. $\left(\dfrac{1}{3}\right)^{\sqrt{3}}$

6. Several exponential expressions have been simplified using the calculator and then verified graphically. Match the correct graphical display to its corresponding simplification.

   a.

   b.

   c.

   i.

   ii.
   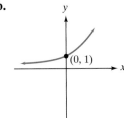

   iii.

7. Classify each function as "exponential function" or "not an exponential function."
   a. $f(x) = x^4$
   b. $f(x) = 4^x$
   c. $f(x) = \left(\dfrac{1}{2}\right)^x$
   d. $f(x) = \dfrac{1}{2}x^2$

8. Label each graph as an increasing or decreasing function and as exponential growth or exponential decay.
   a.
   b.

9. Under what conditions will an exponential function represent exponential decay?

10. Under what conditions will an exponential function represent exponential growth?

11. Explain why all exponential functions will have a *y*-intercept of $(0, 1)$.

12. The irrational number *e* has an approximate value of 2.71828182846, it can be best approximated by the expression $\left(1 + \dfrac{1}{x}\right)^x$. Fill in the table below and watch the expression $\left(1 + \dfrac{1}{x}\right)^x$ grow increasingly closer to the value of *e* as *x* grows larger.

x	$\left(1 + \dfrac{1}{x}\right)^{x}$
1	
10	
100	
1000	
10,000	
100,000	
1,000,000	

13. Solve each of the following equations by rewriting each side of the equation as an exponential expression with a common base.
  a. $2^x = 16$ **b.** $3^x = 243$

14. Solve each of the following equations by rewriting each side of the equation as an exponential expression with a common base.
  a. $4^x = 1024$ **b.** $27^x = 81$

PRACTICE *In Exercises 15–22, graph each exponential function on graph paper. Verify your graphs with your graphing calculator.*

15. $f(x) = 3^x$

16. $f(x) = 4^x$

17. $f(x) = 5^x$

18. $f(x) = 6^x$

19. $f(x) = \left(\dfrac{1}{3}\right)^x$

20. $f(x) = \left(\dfrac{1}{4}\right)^x$

21. $f(x) = \left(\dfrac{1}{5}\right)^x$

22. $f(x) = \left(\dfrac{1}{6}\right)^x$

In Exercises 23–30, refer to the following graphs of the functions and graph each of the given functions on graph paper. The pictured functions are graphed in the window $[-5, 10]$ by $[-2, 10]$.

$f(x) = 2^x$ $f(x) = 3^x$

23. Graph each function below on graph paper and verify with your calculator.
  a. $f(x) = 2^x + 2$ **b.** $f(x) = 3^x - 2$

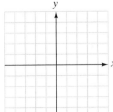

24. In each part of Exercise 23, a constant was either added to or subtracted from the basic exponential function of $f(x) = b^x$. What effect did the constant have on the graph of the function?

25. Graph each function below on graph paper and verify with your calculator.

    a. $f(x) = 3(2^x)$ **b.** $f(x) = 2(3^x)$

✍**26.** In each part of Exercise 25 the basic function was multiplied by an integer value. What effect did this have on the graph of the function?

27. Graph each function on graph paper and verify with your calculator.

    a. $f(x) = 2^{-x}$ **b.** $f(x) = 3^{-x}$

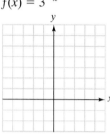

✍**28.** In Exercise 27, what effect does the negative exponent have on the graph of the function? Also explain why the function changes from one of exponential growth, in Exercise 25, to exponential decay in Exercise 27.

29. Graph each of the functions and verify with your calculator.

    a. $f(x) = 2^{x+5}$ **b.** $y = 3^{x-4}$

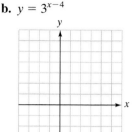

✍**30.** In Exercise 29, how is the graph affected by adding/subtracting a constant to the exponent?

In Exercises 31–40, solve each equation algebraically and verify with the graphing calculator.

31. $2^3 = 2^{5x+7}$ **32.** $3^5 = 3^{4-2x}$ **33.** $9 = 3^{1/2x+5}$

34. $16 = 4^{3x-2}$ **35.** $32 = 4^{6x-5}$ **36.** $27 = 9^{2x+4}$

37. $4^{5x-1} = 8^{3x+2}$ **38.** $125^{x-3} = 25^{2x+7}$ **39.** $125^{6-2x} = 625^{3x+1}$

40. $81^{4x-1} = 27^{6x-2}$

APPLICATIONS   *In Exercises 41–46, assume there are no deposits or withdrawals.*

41. Compound interest   An initial deposit of $10,000 earns 8% interest, compounded quarterly. How much will be in the account after 10 years?

42. Compound interest   An initial deposit of $10,000 earns 8% interest, compounded monthly. How much will be in the account after 10 years?

43. Comparing savings plans   How much more interest could $1000 earn in 5 years, compounded quarterly, if the annual interest rate were $5\frac{1}{2}\%$ instead of 5%?

44. Comparing savings plans   Which institution in Illustration 1 provides the better investment?

45. Compound interest   If $1 had been invested on July 4, 1776 at 5% interest, compounded annually, what would it be worth on July 4, 2076?

46. Frequency of compounding $10,000 is invested in each of two accounts, both paying 6% annual interest. In the first account, interest compounds quarterly; and in the second account, interest compounds daily. Find the difference between the accounts after 20 years.

> ### *Fidelity Savings & Loan*
> earn 5.25%
> compounded monthly

> ### Union Trust
> Money Market account
> paying 5.35%
> compounded annually

ILLUSTRATION 1

47. Fish population   A population of fish is growing according to the Malthusian model. How many fish will there be in 10 years if the annual growth rate is 3% and the initial population is 2700 fish?

48. Town population The population of a town is 1350. The town is expected to grow according to the Malthusian model, with an annual growth rate of 6%. Find the population of the town in 20 years.

49. World population The population of the world is approximately 5 billion. If the population is growing according to the Malthusian model, with an annual growth rate of 1.8%, what will be the population of the world in 30 years?

50. World population The population of the world is approximately 5 billion. If the population is growing according to the Malthusian model, with an annual growth rate of 1.8%, what will be the population of the world in 60 years?

51. Bacteria cultures A colony of 6 million bacteria is growing in a culture medium (see Illustration 2). The population P after t hours is given by the formula $P = (6 \times 10^6)(2.3)^t$. Find the population after 4 hours.

52. Discharging a battery The charge remaining in a battery decreases as the battery discharges. The charge C (coulombs) after t days is given by the formula $C = (3 \times 10^{-4})(0.7)^t$. Find the charge after 5 days.

| 12:00 PM | 4:00 PM |

ILLUSTRATION 2

53. Radioactive decay A radioactive material decays according to the formula $A = A_0\left(\dfrac{2}{3}\right)^t$, where A_0 is the initial amount present and t is measured in years. Find the amount present in 5 years.

54. Town population The population of North Rivers is decreasing exponentially according to the formula $P = 3745(0.93)^t$, where t is measured in years from the present date. Find the population in 6 years, 9 months.

55. Depreciation A camping trailer originally purchased for \$4570 is continuously losing value at the rate of 6% per year. Find its value when it is $6\frac{1}{2}$ years old.

56. Depreciation A boat purchased for \$7500 has been decreasing in value at the rate of 2% each year. It is now 8 years, 3 months old. Find its value.

REVIEW

57. Evaluate:
  a. 7^0 **b.** 25^2 **c.** 25^{-2}
  d. $25^{1/2}$ **e.** $25^{-1/2}$

58. Simplify:
  a. $\sqrt{240x^5}$ **b.** $\left(\sqrt{3x+1}\right)^2$

59. Solve: $\sqrt{3x+1} + 1 = x$.

60. Solve: $\dfrac{24}{x} + \dfrac{12}{x+1} = 11$.

61. Graphically determine the maximum height of a ball thrown straight up from the top of a building with an initial velocity of 32 feet per second if the equation $h = -16t^2 + 32t + 48$ models the situation (t = time, h = height). How long does it take the ball to reach the maximum height?

11.4 Logarithmic Functions

■ BASIC CONCEPTS OF EXPONENTS AND LOGARITHMS ■ CHANGE-OF-BASE FORMULA ■ GRAPHS OF LOGARITHMIC FUNCTIONS ■ APPLICATIONS OF LOGARITHMS

■ BASIC CONCEPTS OF EXPONENTS AND LOGARITHMS

When we studied the laws of exponents, we referred to expressions such as 4^5 as exponential expressions. Five is the exponent to which the base of 4 must be raised to get 1024:

$$4^5 = 1024$$

Five is also called the **logarithm** to the base 4 of 1024 and is written as

$$\log_4 1024 = 5$$

> **Logarithm**
> If $b > 0$ and $b \neq 1$, then $b^y = x$ is equivalent to $\log_b x = y$.

Remember, a logarithm is just an exponent! The following comparisons can be made between logarithms and exponents:

Logarithmic Form	*Exponential Form*
$\log_7 1 = 0$	$7^0 = 1$
$\log_5 25 = 2$	$5^2 = 25$
$\log_5 \frac{1}{25} = -2$	$5^{-2} = \frac{1}{25}$
$\log_{16} 4 = \frac{1}{2}$	$16^{1/2} = 4$
$\log_{10} 1000 = 3$	$10^3 = 1000$
$\log_{10} \frac{1}{100} = -2$	$10^{-2} = \frac{1}{100}$
$\log_e 148.4131591 = 5$	$e^5 = 148.4131591$

> **WARNING!**   It is impossible to find the logarithm of a negative number or zero. Note that because there is no real number x such that $2^x = -8$, there is no meaning to the expression $\log_2 (-8) = x$.

Logarithms of base 10, such as $\log_{10} 1000 = 3$, are called **common logs.** These logarithms often appear in mathematics and science. The base 10 is understood and generally not written. Thus, $\log 1000 = \log_{10} 1000 = 3$.

Because the number e appears in mathematical models of natural events, base-e logarithms are called **natural logarithms.** Natural logarithms are usually denoted by the symbol $\ln x$ instead of $\log_e x$. Thus, $\log_e 148.4131591 = 5$ becomes $\ln 148.4131591 = 5$.

EXAMPLE 1   Find y in each equation.

a. $\log_3 81 = y$ **b.** $\log_5 \dfrac{1}{5} = y$

c. $\log 10{,}000 = y$ **d.** $\ln 8 = y$

Solution   **a.** Change the equation $\log_3 81 = y$ into the equivalent equation $3^y = 81$ and solve for y.

$$3^y = 81$$
$$3^y = 3^4$$
$$y = 4$$

b. $\log_5 \dfrac{1}{5} = y$ is equivalent to $5^y = \dfrac{1}{5}$.

$$5^y = \frac{1}{5}$$
$$5^y = 5^{-1}$$
$$y = -1$$

c. $\log 10{,}000 = y$ is a common logarithm of base 10 and is equivalent to $10^y = 10000$.

$$10^y = 10000$$
$$10^y = 10^4$$
$$y = 4$$

d. Use your calculator to compute ln 5. Be sure to use the natural log key, LN, and not the common log key, LOG.

```
ln(5)
          1.609437912
```

EXAMPLE 2   Find the value of x in each equation.

a. $\log_x 81 = 4$ **b.** $\log_4 x = 3$

c. $\log x = -2$ **d.** $\log_{1/3} \frac{1}{27} = x$

Solution   **a.** $\log_x 81 = 4$ is equivalent to $x^4 = 81$.

$$x^4 = 81$$
$$x^4 = 3^4$$
$$x = 3$$

b. $\log_4 x = 3$ is equivalent to $4^3 = x$. Thus, $x = 64$.

c. $\log x = -2$ is a base 10 common logarithm and is equivalent to $10^{-2} = x$. Thus, $x = \frac{1}{100}$.

d. $\log_{1/3} \frac{1}{27} = x$ is equivalent to $\left(\frac{1}{3}\right)^x = \frac{1}{27}$.

$$\left(\frac{1}{3}\right)^x = \frac{1}{27}$$
$$(3^{-1})^x = 3^{-3}$$
$$3^{-x} = 3^{-3}$$
$$-x = -3$$
$$x = 3$$

■ CHANGE-OF-BASE FORMULA

Both the LOG and LN keys on your calculator allow for easy calculator approximation of both common and natural logarithms. When the logarithm does not have a base of 10 (or e), we will need the **change-of-base formula** that allows the conversion of a logarithm of any base to a base 10 logarithm for calculator computation.

> **Change-of-Base Formula**
>
> If a, b, and x are positive real numbers, then $\log_b x = \dfrac{\log_a x}{\log_a b}$ $(a \neq 1, b \neq 1)$.

EXAMPLE 3   Use the change-of-base formula to perform the following computations on your calculator. When possible, verify analytically.

a. $\log_7 1 = y$ **b.** $\log_5 25 = y$

c. $\log_{16} 4 = y$ **d.** $\log_3 5 = y$

Solution **a.** $\log_7 1 = \dfrac{\log_{10} 1}{\log_{10} 7} = \dfrac{\log 1}{\log 7}$

Thus, $\log_7 1 = 0$.

Analytic verification: $7^0 = 1$.

b. $\log_5 25 = \dfrac{\log 25}{\log 5}$

Thus, $\log_5 25 = 2$.

Analytic verification: $5^2 = 25$.

c. $\log_{16} 4 = \dfrac{\log 4}{\log 16}$

Thus, $\log_{16} 4 = 0.5$.

Analytic verification: $16^{1/2} = 4$.

d. $\log_3 5 = \dfrac{\log 5}{\log 3}$

Thus, $\log_3 5 \approx 1.464973521$.

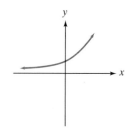

■ GRAPHS OF LOGARITHMIC FUNCTIONS

All of the exponential functions defined by $y = b^x$ in the previous section had the same basic shape for their graphs, as illustrated at the right. Because the graphs of exponential functions also pass the horizontal line test, we can conclude that exponential functions are one-to-one and thus have an inverse function. The inverse function is $x = b^y$, which was defined at the beginning of this section as being equivalent to $\log_b x = y$.

Before writing this inverse in $y = f^{-1}(x)$ form, we will construct a graph of $y = 2^x$ and its inverse, $x = 2^y$ (see Figure 11-1).

$y = 2^x$

x	y
-2	$\frac{1}{4}$
-1	$\frac{1}{2}$
0	1
1	2
2	4

Domain: $\mathbb{R}$
Range: $\{y \mid y > 0\}$

$x = 2^y$

x	y
$\frac{1}{4}$	-2
$\frac{1}{2}$	-1
1	0
2	1
4	2

Domain: $\{x \mid x > 0\}$
Range: $\mathbb{R}$

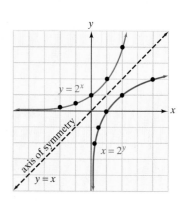

FIGURE 11-1

Based on the definition of logarithm, $2^y = x$ is equivalent to $\log_2 x = y$. Thus, the inverse of $f(x) = 2^x$ is $f^{-1}(x) = \log_2 x$.

To graph $y = \log_b x$, we find several points (x, y) that satisfy the equation $x = b^y$ and join them with a smooth curve.

EXAMPLE 4   Graph the functions   **a.** $y = \log_3 x$ and   **b.** $y = \log_{1/3} x$.

Solution   **a.** The equation $y = \log_3 x$ is equivalent to $3^y = x$. Select points (x, y) that satisfy this equation and plot a smooth curve. (Remember to use the change-of-base formula to verify with the calculator.)

x	y
$\frac{1}{9}$	-2
$\frac{1}{3}$	-1
1	0
3	1
9	2

[–1,10] by [–3,3]

b. The equation $y = \log_{1/3} x$ is equivalent to $\left(\frac{1}{3}\right)^y = x$. Select points (x, y) that satisfy this equation and plot a smooth curve. Verify with your calculator.

x	y
9	-2
3	-1
1	0
$\frac{1}{3}$	1
$\frac{1}{9}$	2

[-1,10] by [-3,3]

We can draw the following conclusions about graphs of logarithmic functions based on these two graphs.

1. If $b > 1$, then $f(x) = \log_b x$ has the shape

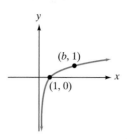

and passes through the points $(0, 1)$ and $(b, 1)$.

2. If $0 < b < 1$ then $f(x) = \log_b x$ has the shape

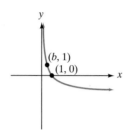

and passes through the points $(1, 0)$ and $(b, 1)$.

▣ APPLICATIONS OF LOGARITHMS

Chemistry: In chemistry, common logarithms are used to express the acidity of solutions. The more acidic a solution, the greater the concentration of hydrogen ions. This concentration is indicated indirectly by the pH scale, or hydrogen ion index. The pH of a solution is defined by

$$pH = -\log [H^+],$$

where $[H^+]$ is the hydrogen ion concentration in gram-ions per liter.

EXAMPLE 5   Use your calculator to find the pH of pure water, which has a hydrogen ion concentration of 10^{-7} grams-ions per liter.

Solution   The hydrogen ion concentration of pure water is 10^{-7} grams-ions per liter, so its pH is

$$pH = -\log (10^{-7})$$

Thus, the pH is 7.

EXAMPLE 6   Find the hydrogen ion concentration of seawater if its pH is 8.5. Note that substances with a pH value greater than 7 are alkaline, and those less than 7 are acidic.

Solution   To find the hydrogen ion concentration, solve for $[H^+]$.

$$8.5 = -\log [H^+]$$

$$-8.5 = \log [H^+] \qquad \text{Multiply both sides by } -1.$$

$$[H^+] = 10^{-8.5} \qquad \text{Rewrite the equation in the equivalent exponential form.}$$

$$[H^+] \approx 3.16227766 \times 10^{-9} \qquad \text{Evaluate } 10^{-8.5} \text{ with the calculator.}$$

If a quantity grows exponentially at a certain annual rate, the time required for the quantity to double is called the **doubling time** and is computed using the formula

$$t = \frac{\ln 2}{r}$$

where r is the annual rate and t is the time required for the quantity to double.

EXAMPLE 7   The population of the earth is growing at the rate of approximately 2% per year. If this rate continues, how long will it take the population to double?

Solution   Because the population is growing at the rate of 2% per year, we substitute 0.02 for r into the formula for doubling time and simplify.

$$t = \frac{\ln 2}{r}$$

$$t = \frac{\ln 2}{0.02}$$

$$\approx 34.65735903$$

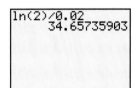

At the current rate of growth, the population of the earth will double in about 35 years.   ■

EXERCISE 11.4

VOCABULARY *Fill in each blank to make the statement true.*

1. A _____ is an exponent.

2. If $b > 0$ and $b \neq 1$, then $\log_b x = y$ is equivalent to _____.

3. A logarithm of base ___ is called a common logarithm.

4. A _____ logarithm is a log of base __ and uses the notation _____ instead of $\log_e x$.

5. The _____ formula allows conversion of a logarithm of any base to a base 10 logarithm for calculator computation.

CONCEPTS

6. Change from logarithmic form to exponential form.

   a. $\log_4 256 = 4$ **b.** $\log_{121} 11 = \dfrac{1}{2}$

   c. $\log 10 = 1$ **d.** $\log_{1/4} \dfrac{1}{16} = 2$

7. Write each exponential expression in logarithmic form.

   a. $8^2 = 64$ **b.** $4^{-2} = \dfrac{1}{16}$

   c. $\left(\dfrac{1}{2}\right)^{-5} = 32$ **d.** $10^{-4} = \dfrac{1}{10,000}$

8. Use the change-of-base formula to correctly match each logarithmic expression to its calculator evaluation.
  i. $\log_7 3$ ii. $\log 8$
  iii. $\log_3 243$ iv. $\log_4 64$
  a.   **b.**

  c.   **d.**

PRACTICE *In Exercises 9–48, find each value of x.*

9. $\log_2 8 = x$

10. $\log_3 9 = x$

11. $\log_4 64 = x$

12. $\log_6 216 = x$

13. $\log_{1/2} \dfrac{1}{8} = x$

14. $\log_{1/3} \dfrac{1}{81} = x$

15. $\log_9 3 = x$

16. $\log_{125} 5 = x$

17. $\log_{1/2} 8 = x$

18. $\log_{1/2} 16 = x$

19. $\log_8 x = 2$

20. $\log_7 x = 0$

21. $\log_7 x = 1$

22. $\log_2 x = 3$

23. $\log_{25} x = \dfrac{1}{2}$

24. $\log_4 x = \dfrac{1}{2}$

25. $\log_5 x = -2$

26. $\log_3 x = -4$

27. $\log_{36} x = -\dfrac{1}{2}$

28. $\log_{27} x = -\dfrac{1}{3}$

29. $\log_{100} \dfrac{1}{1000} = x$

30. $\log_{5/2} \dfrac{4}{25} = x$

31. $\log_{27} 9 = x$

32. $\log_{12} x = 0$

33. $\log_x 5^3 = 3$

34. $\log_x 5 = 1$

35. $\log_x \dfrac{9}{4} = 2$

36. $\log_x \dfrac{\sqrt{3}}{3} = \dfrac{1}{2}$

37. $\log_x \dfrac{1}{64} = -3$

38. $\log_x \dfrac{1}{100} = -2$

39. $\log_{2\sqrt{2}} x = 2$

40. $\log_4 8 = x$

41. $2^{\log_2 5} = x$

42. $3^{\log_3 4} = x$

43. $x^{\log_4 6} = 6$

44. $x^{\log_3 8} = 8$

45. $\log 10^3 = x$

46. $\log 10^{-2} = x$

47. $10^{\log x} = 100$

48. $10^{\log x} = \dfrac{1}{10}$

In Exercises 49–60, use a calculator to find each value, if possible. Give answers to four decimal places.

49. $\log 3.25$

50. $\log 0.57$

51. $\log 0.00467$

52. $\log 375.876$

53. $\ln 0.93$

54. $\ln 7.39$

55. $\ln 37.896$

56. $\ln 0.00465$

57. $\log (\ln 1.7)$

58. $\ln (\log 9.8)$

59. $\ln (\log 0.1)$

60. $\log (\ln 0.01)$

In Exercises 61–68, use a calculator to find each value of y, if possible. Give answers to four decimal places.

61. $\log y = 1.4023$

62. $\ln y = 2.6490$

63. $\ln y = 4.24$

64. $\log y = 0.926$

65. $\log y = -3.71$

66. $\ln y = -0.28$

67. $\log y = \ln 8$

68. $\ln y = \log 7$

In Exercises 69–72, graph the function defined by each equation on graph paper. Check your graph with your graphing calculator.

69. $y = \log_3 x$

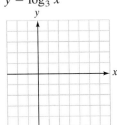

70. $y = \log_{1/3} x$

71. $y = \log_{1/2} x$

72. $y = \log_4 x$

In Exercises 73–76, graph each pair of inverse functions on graph paper. Check your graph with your graphing calculator.

73. $y = 2^x$, $y = \log_2 x$

74. $y = \left(\dfrac{1}{2}\right)^x$, $y = \log_{1/2} x$

75. $y = \left(\dfrac{1}{4}\right)^x$, $y = \log_{1/4} x$

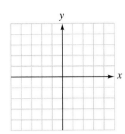

76. $y = 4^x$, $y = \log_4 x$

In Exercises 77–80, use a graphing calculator to do each experiment. Describe what you find.

77. Graph $y = k + \log x$ for many values of k.

78. Graph $y = \log (x + k)$ for many values of k.

79. Graph $y = k \log x$ for many values of k.

80. Graph $y = \log kx$ for many values of k.

In Exercises 81–88, use the change-of-base formula to find each logarithm to four decimal places.

81. $\log_3 7$

82. $\log_7 3$

83. $\log_{1/3} 3$

84. $\log_{1/2} 6$

85. $\log_3 8$

86. $\log_5 10$

87. $\log_{\sqrt{2}} \sqrt{5}$

88. $\log_\pi e$

APPLICATIONS *In Exercises 89–96, solve each problem.*

89. pH of a solution Find the pH of a solution with a hydrogen ion concentration of 1.7×10^{-5} gram-ions.

90. Hydrogen ion concentration of calcium hydroxide Find the hydrogen ion concentration of a saturated solution of calcium hydroxide whose pH is 13.2.

91. Aquariums To test for safe pH levels in a fresh-water aquarium, a test strip is compared with the scale shown in Illustration 1. Find the corresponding safe range in the hydrogen ion concentration.

AquaTest pH Kit

Safe range

6.4   6.8   7.2   7.6   8.0

ILLUSTRATION 1

92. pH of sour pickles The hydrogen concentration of sour pickles is 6.31×10^{-4}. Find the pH.

REVIEW

97. Given the function $y = \sqrt{2x + 5}$
   a. Is the function increasing or decreasing?
   b. What is the domain of the function?

93. Population growth How long will it take the population of River City (Illustration 2) to double?

River City

A growing community

• 6 parks   • 12% annual growth
• 10 churches   • Low crime rate

ILLUSTRATION 2

94. Doubling money How long will it take $1000 to double if it is invested at an annual rate of 5% compounded continuously?

95. Population growth A population growing at an annual rate r will triple in a time t given by the formula $t = \dfrac{\ln 3}{r}$. How long will it take the population of a town growing at the rate of 12% per year to triple?

96. Tripling money Find the length of time for $25,000 to triple if invested at 6% annual interest, compounded continuously.

98. Given the radical equation $5 + \sqrt{2x + 5} = x$
   a. Solve analytically.
   b. Justify your analytical solution with a graphical solution.

✍**99.** Why is $x = 2$ not a solution to the equation $5 + \sqrt{2x + 5} = x$ but *is* a solution to $x^2 - 12x + 20 = 0$?

100. Solve $x^2 - 12x + 20 = 0$ by
  a. completing the square
  b. quadratic formula
  c. zero-factor property
  d. graphing

11.5 Properties of Logarithms

■ BASIC PROPERTIES OF LOGARITHMS ■ ADDITIONAL PROPERTIES OF LOGARITHMS ■ APPLICATIONS

■ BASIC PROPERTIES OF LOGARITHMS

In Section 11.4 we examined some basic properties of logarithms, which we will now summarize. Because logarithms are exponents, it is not surprising that the properties of exponents have counterparts in logarithmic notation.

Properties of Logarithms
If b is a positive number and $b \neq 1$, then
1. $\log_b 1 = 0$ **2.** $\log_b b = 1$
3. $\log_b b^x = x$ **4.** $b^{\log_b x} = x$ $(x > 0)$

These properties follow directly from the definition of logarithms:

1. $\log_b 1 = 0$ because $b^0 = 1$. Thus, $\log_5 1 = 0$ because $5^0 = 1$.
2. $\log_b b = 1$ because $b^1 = b$. Thus, $\log_3 3 = 1$ because $3^1 = 3$.
3. $\log_b b^x = x$ because $b^x = b^x$. Thus, $\log_7 7^3 = 3$ because $7^3 = 7^3$.
4. $b^{\log_b x} = x$ because $\log_b x$ is the exponent to which b is raised to get x. Thus, $b^{\log_b 7} = 7$ because $\log_b 7$ is the power to which b is raised to get 7.

■ ADDITIONAL PROPERTIES OF LOGARITHMS

Before the widespread use of calculators, logarithms provided the only reasonable way of performing certain long and involved calculations. The advent of hand-held calculators has virtually made these computations obsolete. However, the properties of logarithms that are addressed in this section will provide a foundation for future mathematics courses.

The first property deals with the logarithm of a product and is called the **product property of logarithms.**

Product Property of Logarithms
If M, N, and b are positive numbers and $b \neq 1$, then
$$\log_b MN = \log_b M + \log_b N$$

To prove this, we let $x = \log_b M$ and $y = \log_b N$. Because of the definition of logarithms, these equations can be written in the form

$M = b^x$ and $N = b^y$

Then $MN = b^x b^y$ and the properties of exponents give

$MN = b^{x+y}$

Using the definition of logarithms gives

$\log_b MN = x + y$

Substituting the values of x and y completes the proof:

$$\log_b MN = \log_b M + \log_b N$$

WARNING!   The product property of logarithms asserts that the logarithm of the product of two numbers is equal to the sum of their logarithms. The logarithm of a sum or a difference usually does not simplify. In general,

$$\log_b (M + N) \neq \log_b M + \log_b N \quad \text{and} \quad \log_b (M - N) \neq \log_b M - \log_b N$$

EXAMPLE 1   Use the product property to write each logarithmic expression as a sum of logarithms. Assume that variables represent positive numbers.

    a. $\log_b xyz$ **b.** $\log 6$

Solution   **a.** $\log_b xyz = \log_b (xy)z$ Associative property of multiplication.

                 $= \log_b xy + \log_b z$ Product property of logarithms.

                 $= \log_b x + \log_b y + \log_b z$ Product property of logarithms.

    b. $\log 6 = \log (2 \cdot 3)$

             $= \log 2 + \log 3$

Calculator verification:
```
log(6)
          .7781512504
log(2)+log(3)
          .7781512504
```

EXAMPLE 2   Use the product property of logarithms to write each sum as a single logarithm.

    a. $\log_b x^3 + \log_b y^{1/2}$ **b.** $\log_b (x + 3) + \log_b (x - 3)$

Solution   **a.** $\log_b x^3 + \log_b y^{1/2} = \log_b (x^3 y^{1/2})$ Product property of logarithms.

                         $= \log_b \left(x^3 \sqrt{y}\right)$ Product property of logarithms.

    b. $\log_b (x + 3) + \log_b (x - 3) = \log_b [(x + 3)(x - 3)]$

                                  $= \log_b (x^2 - 9)$

Quotient Property of Logarithms
If M, N, and b are positive numbers and $b \neq 1$, then

$$\log_b \frac{M}{N} = \log_b M - \log_b N$$

To prove this, we again let $x = \log_b M$ and $y = \log_b N$. These equations can be written as

$$M = b^x \quad \text{and} \quad N = b^y$$

Then $\dfrac{M}{N} = \dfrac{b^x}{b^y}$ and the properties of exponents give

$$\frac{M}{N} = b^{x-y}$$

Using the definition of logarithms gives

$$\log_b \frac{M}{N} = x - y$$

Substituting the values of x and y completes the proof:

$$\log_b \frac{M}{N} = \log_b M - \log_b N$$

WARNING! The quotient property of logarithms asserts that the logarithm of the quotient of two numbers is equal to the difference of their logarithms. The logarithm of a quotient is not the quotient of the logarithms:

$$\log_b \frac{M}{N} \neq \frac{\log_b M}{\log_b N}$$

EXAMPLE 3   Use the quotient property of logarithms to write each difference as a single logarithm. Assume all variables represent positive numbers.

a. $\log_b (yx) - \log_b z$ **b.** $\log 20 - \log 2$ **c.** $\log_2 8 + \log_2 5 - \log_2 6$

Solution   **a.** $\log_b (yx) - \log_b z = \log \dfrac{yx}{z}$ Quotient property of logarithms.

b. $\log 20 - \log 2 = \log \dfrac{20}{2}$ Quotient property of logarithms.

$= \log 10$ Simplify.

$= 1$ $\log_{10} 10 = y$ implies $10^y = 10$.

c. $\log_2 8 + \log_2 5 - \log_2 6 = \log_2 (8 \cdot 5) - \log_2 6$ Product property of logarithms.

$= \log_2 \dfrac{40}{6}$ Quotient property of logarithms.

$= \log_2 \dfrac{20}{3}$ Simplify.

■ TECHNOLOGY TIP
Remember to apply the change-of-base formula because the logarithms are base 2 instead of base 10.

Calculator verification:

```
log(8)/log(2)+lo
g(5)/log(2)-log(
6)/log(2)
          2.736965594
log(20/3)/log(2)

          2.736965594
```

Power Property of Logarithms
If M, N, and b are positive numbers and $b \neq 1$, then

$$\log_b M^p = p \log_b M$$

To prove this, we let $x = \log_b M$, write the expression in exponential form, and raise both sides to the pth power:

$$M = b^x$$
$$(M)^p = (b^x)^p$$
$$M^p = b^{px}$$

Using the definition of logarithms gives

$$\log_b M^p = px$$

Substituting the value for x completes the proof:

$$\log_b M^p = p\log_b M$$

EXAMPLE 4   Use the power property to write each difference as a single logarithm. Assume all variables represent positive numbers.

a. $\log_b \left(x^2\sqrt[3]{y}\right)$ **b.** $\log_b \dfrac{z^3\sqrt{x-2}}{y}$

Solution   **a.** $\log_b \left(x^2\sqrt[3]{y}\right) = \log_b x^2 + \log_b \sqrt[3]{y}$ Product property.

$\qquad = 2\log_b x + \log_b y^{1/3}$ Power property.

$\qquad = 2\log_b x + \dfrac{1}{3}\log_b y$ Power property.

b. $\log_b \dfrac{z^3\sqrt{x-2}}{y} = \log\left(\dfrac{z^3}{y}\cdot\sqrt{x-2}\right)$ Arithmetic of fractions.

$\qquad = \log\dfrac{z^3}{y} + \log\sqrt{x-2}$ Product property.

$\qquad = \log z^3 - \log y + \log\sqrt{x-2}$ Quotient property.

$\qquad = \log z^3 - \log y + \log(x-2)^{1/2}$ $\sqrt[m]{x^n}=x^{n/m}$

$\qquad = 3\log z - \log y + \dfrac{1}{2}\log(x-2)$ Power property.

EXAMPLE 5   Rewrite each logarithmic expression using the product, quotient, and power properties. Justify the equality of your expression with the calculator.

a. $\log 5 + \log 3$ **b.** $2\log 3$

c. $\log 10 - \log 2$ **d.** $\frac{1}{2}\log 5$

Solution   **a.** $\log 5 + \log 3 = \log(5\cdot 3)$ Product property.

$\qquad = \log 15$

```
log(5)+log(3)
         1.176091259
log(15)
         1.176091259
```

b. $2\log 3 = \log 3^2$ Power property.

$\qquad = \log 9$

```
2log(3)
         .9542425094
log(9)
         .9542425094
```

c. $\log 10 - \log 2 = \log\dfrac{10}{2}$ Quotient property.

$\qquad = \log 5$

```
log(10)-log(2)
         .6989700043
log(5)
         .6989700043
```

d. $\dfrac{1}{2} \log 5 = \log 5^{1/2}$ Power rule.

$= \log \sqrt{5}$

■ APPLICATIONS

Electrical engineering   Common logarithms are used in electrical engineering to express the voltage gain (or loss) of an electronic device such as an amplifier. The unit of gain (or loss), called the **decibel,** is defined by the following logarithmic relation.

> **Decibel Voltage Gain**
> If E_O is the output voltage of a device and E_I is the input voltage, the decibel voltage gain is given by
>
> $$\text{db gain} = 20 \log \frac{E_O}{E_I}$$

EXAMPLE 6   If the input to an amplifier is 0.5 volt and the output is 40 volts, find the decibel voltage gain of the amplifier.

Solution   We can find the decibel voltage gain by substituting **0.5** for E_I and **40** for E_O into the formula for db gain:

$$\text{db gain} = 20 \log \frac{E_O}{E_I}$$

$$\text{db gain} = 20 \log \frac{40}{0.5}$$

$$= 20 \log 80$$

$$\approx 38 \qquad \text{Use a calculator.}$$

The amplifier provides a 38-decibel voltage gain.   ■

Geology   In seismology, common logarithms are used to measure the intensity of earthquakes on the **Richter scale.** The intensity of an earthquake is given by the following logarithmic function.

> **Richter Scale**
> If A is the amplitude of an earthquake (the earth movement, measured in micrometers), and P is the period (the time of one oscillation of the earth's surface measured in seconds), then the intensity R on the Richter scale is
>
> $$R = \log \frac{A}{P}$$

EXAMPLE 7   Find the measure on the Richter scale of an earthquake with an amplitude of 10,000 micrometers (1 centimeter) and a period of 0.1 second.

Solution   We substitute **10,000** for A and **0.1** for P in the Richter scale formula and simplify:

$$R = \log \frac{A}{P}$$

$$R = \log \frac{10,000}{0.1}$$

$$= \log 100,000$$

$$= \log 10^5$$

$$= 5 \log 10 \qquad \text{Use the power property of logarithms.}$$

$$= 5 \qquad \text{Use property 2 of logarithms.}$$

The earthquake measures 5 on the Richter scale.   ■

Physiology   In physiology, experiments suggest that the relationship between the loudness and the intensity of sound is a logarithmic one, known as the **Weber-Fechner law.**

> **Weber-Fechner Law**
> If L is the apparent loudness of a sound, I is the actual intensity, and k is a constant, then
> $$L = k \ln I$$

EXAMPLE 8   Find the increase in actual intensity that will cause the apparent loudness of a sound to double.

Solution   If the original loudness L_0 is caused by an actual intensity I_0, then

$$L_0 = k \ln I_0$$

To double the apparent loudness, we multiply both sides of the equation by 2 and use the power property of logarithms:

$$2L_0 = 2k \ln I_0$$

$$= k \ln (I_0)^2$$

Thus, to double the apparent loudness of a sound, the intensity must be squared.

                                                                              ■

EXERCISE 11.5

VOCABULARY   *Fill in the blanks to make a true statement.*

1. The _____ property guarantees that $\log_b MN = \log_b M + \log_b N$.

2. If $\log_b \frac{M}{N} = \log_b M - \log_b N$, then the _____ property has been applied.

3. The power property of logarithms ensures that $\log_b M^P = $ _____.

CONCEPTS   *In Exercises 4–23, classify each of the following statements as true or false. If false, correct the statement to a true statement. All variables are positive and $b \neq 1$.*

4. $\log_b 0 = 1$

5. $\log_b (x + y) \neq \log_b x + \log_b y$

6. $\log_b xy = (\log_b x)(\log_b y)$

7. $\log_b ab = \log_b a + 1$

8. $\log_7 7^7 = 7$

9. $\log^{\log_7 7} = 7$

10. $\dfrac{\log_b A}{\log_b B} = \log_b A - \log_b B$

11. $\log_b (A - B) = \dfrac{\log_b A}{\log_b B}$

12. $3 \log_b \sqrt[3]{a} = \log_b a$

13. $\dfrac{1}{3} \log_b a^3 = \log_b a$

14. $\log_b \dfrac{1}{a} = -\log_b a$

15. $\log_b 2 = \log_2 b$

16. If $\log_a b = c$, then $\log_b a = c$

17. If $\log_a b = c$, then $\log_b a = \dfrac{1}{c}$

18. $\log_b (-x) = -\log_b x$

19. If $\log_b a = c$, then $\log_b a^p = pc$

20. $\log_b \dfrac{1}{5} = -\log_b 5$

21. $\log_{4/3} y = -\log_{3/4} y$

22. $\log_b y + \log_{1/b} y = 0$

23. $\log_{10} 10^3 = 3(10^{\log_{10} 3})$

✍24. Explain why ln (log 0.9) is undefined.

✍25. Explain why $\log_b$ (ln 1) is undefined.

PRACTICE   *In Exercises 26–31, use your calculator to verify that the left side of the equation is equal to the right side of the equation. Identify the property that was applied.*

26. log [(2.5)(3.7)] = log 2.5 + log 3.7

27. $\ln \dfrac{11.3}{6.1} = \ln 11.3 - \ln 6.1$

28. $\ln (2.25)^4 = 4 \ln 2.25$

29. $\log 45.37 = \dfrac{\ln 45.37}{\ln 10}$

30. $\log \sqrt{24.3} = \dfrac{1}{2} \log 24.3$

31. $\ln 8.75 = \dfrac{\log 8.75}{\log e}$

In Exercises 32–43, assume that x, y, and z are positive numbers. Use the properties of logarithms to write each expression in terms of the logarithms of x, y, and z.

32. $\log_b xyz$

33. $\log_b xz$

34. $\log_b \dfrac{2x}{y}$

35. $\log_b \dfrac{x}{yz}$

36. $\log_b x^3 y^2$

37. $\log_b xy^2 z^3$

38. $\log_b (xy)^{1/2}$

39. $\log_b x^3 y^{1/2}$

40. $\log_b x\sqrt{z}$

41. $\log_b \sqrt{xy}$

42. $\log_b \dfrac{\sqrt[3]{x}}{\sqrt[4]{yz}}$

43. $\log_b \sqrt[4]{\dfrac{x^3 y^2}{z^4}}$

In Exercises 44–51, assume that x, y, and z are positive numbers. Use the properties of logarithms to write each expression as a log of a single quantity.

44. $\log_b (x + 1) - \log_b x$

45. $\log_b x + \log_b (x + 2) - \log_b 8$

46. $2 \log_b x + \dfrac{1}{2} \log_b y$

47. $-2 \log_b x - 3 \log_b y + \log_b z$

48. $-3 \log_b x - 2 \log_b y + \dfrac{1}{2} \log_b z$

49. $3 \log_b (x + 1) - 2 \log_b (x + 2) + \log_b x$

50. $\log_b \left(\dfrac{x}{z} + x \right) - \log_b \left(\dfrac{y}{z} + y \right)$

51. $\log_b (xy + y^2) - \log_b (xz + yz) + \log_b z$

APPLICATIONS

52. Finding output voltage   The db gain of an amplifier is 35. Find the output voltage when the input voltage is 0.05 volt.

53. db gain of an amplifier   Find the db gain of the amplifier shown in Illustration 1.

Input   Output

0.1   30

Volts   Volts

ILLUSTRATION 1

54. db gain of an amplifier   An amplifier produces an output of 80 volts when driven by an input of 0.12 volts. Find the amplifier's db gain.

55. Earthquakes   An earthquake has amplitude of 5000 micrometers and a period of 0.2 seconds. Find its measure on the Richter scale.

56. Earthquakes   Find the period of an earthquake with amplitude of 80,000 micrometers that measures 6 on the Richter scale.

57. Earthquakes   An earthquake with a period of $\frac{1}{4}$ second measures 4 on the Richter scale. Find its amplitude.

58. Earthquakes   By what factor must the amplitude of an earthquake change to increase its severity by 1 point on the Richter scale? Assume that the period remains constant.

59. Change in loudness   If the intensity of a sound is doubled, find the apparent change in loudness.

60. Change in intensity   If the intensity of a sound is tripled, find the apparent change in loudness.

61. Change in loudness   What change in intensity of sound will cause an apparent tripling of the loudness?

62. Change in intensity   What increase in the intensity of a sound will cause the apparent loudness to be multiplied by 4?

63. Charging a battery   The time (in hours) required to charge a certain battery to a level C (expressed as a fraction of the battery's full charge) is $t = -3.7 \ln(1 - C)$. How long would it take to bring the battery to 80% of its full charge?

64. Power output   If P_O is the output of an amplifier (in watts) and P_1 is the power input, then the decibel voltage gain is db voltage gain $= 10 \log \frac{P_O}{P_1}$. One brand of amplifier produces a 70-watt output with a 0.035-watt input. Find the db voltage gain.

65. Depreciation   In business, equipment is often depreciated using the double declining-balance method. In this method, a piece of equipment with a life expectancy of N years, costing $\$C$, will depreciate to a value of $\$V$ in n years, where n is given by the formula

$$n = \frac{\log V - \log C}{\log\left(1 - \dfrac{2}{N}\right)}$$

A computer that cost $\$37,000$ has a life expectancy of 5 years. If it has depreciated to a value of $\$8000$, how old is it?

66. Depreciation   A typewriter worth $\$470$ when new had a life expectancy of 12 years. If it is now worth $\$189$, how old is it?

✎**67.** Explain why an earthquake measuring 7 on the Richter scale is much worse than an earthquake measuring 6.

✎**68.** Graph the logistic function $y = \dfrac{1}{1 + e^{-2x}}$ and discuss its graph.

REVIEW

69. Graphically solve each inequality.

   a. $\frac{3}{4}(x + 5) - 3 \le 0$ **b.** $\frac{3}{4}|x + 5| - 3 \le 0$

   c. $\frac{3}{4}(x + 5)^2 - 3 \le 0$

70. Graph each inequality in two variables on graph paper.

   a. $\frac{3}{4}(x + 5) - 3 \le y$ **b.** $\frac{3}{4}|x + 5| - 3 \le y$

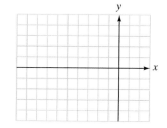

c. $\dfrac{3}{4}(x + 5)^2 - 3 \leq y$

✎**71.** Explain the difference in graphically solving an inequality in one variable and graphing an inequality in two variables.

72. In September of 1998, a leading computer firm sold a 450-MHz computer for \$3200. From September through January of 1999, the computer decreased in value 13% each month. The value, $V(x)$, of the computer can be represented by the model $V(x) = 3200(.87^x)$, where $x = 0$ is the value in September and $x = 4$ is the value in January.
  a. Graph the model in an appropriate viewing window.
  b. Determine the value of the computer in each of the 4 months.

11.6 Exponential and Logarithmic Equations

■ EXPONENTIAL EQUATIONS ■ LOGARITHMIC EQUATIONS ■ APPLICATIONS OF EXPONENTIAL AND LOGARITHMIC EQUATIONS

■ EXPONENTIAL EQUATIONS

In Section 11.3 we solved exponential equations $b^r = b^s$ using the exponential property of equality: "for $b > 0$ and $b \neq 1$, $b^r = b^s$ is equivalent to $r = s$." Thus, problems such as $3^x = 81$ can be rewritten as

$$3^x = 3^4$$

and it follows

$$x = 4$$

If, however, we have an equation in which one expression cannot easily be written as a power of the other expression, such as $3^x = 7$, we will use the following property of equality:

> **Logarithmic Property of Equality**
> If $M > 0$ and $N > 0$, $\log_b M = \log_b N$, if and only if $M = N$

Both the exponential and logarithmic properties of equality follow from the fact that exponential and logarithmic functions are one-to-one.

EXAMPLE 1   Solve the exponential equation $3^x = 5$.

Solution   Logarithms of equal numbers are equal, so we can take the common logarithm of each side of the equation. The power property of logarithms then provides a way of moving the variable x from its position as an exponent to a position as a coefficient.

Analytic solution:

$$3^x = 5$$

$$\log 3^x = \log 5 \qquad \text{Take the common logarithm of each side.}$$

$$x \log 3 = \log 5 \qquad \text{Power property of logarithms.}$$

$$x = \frac{\log 5}{\log 3} \qquad \text{Divide both sides by log 3.}$$

$$\approx 1.464973521 \qquad \text{Use a calculator.}$$

To four decimal places, $x = 1.4650$.

Graphical solution: Enter the left side of the equation at the y1 prompt and the right side at the y2 prompt. Use the INTERSECT feature to find the approximate value of *x*.

The ROOT/ZERO feature could also be used. However, first rewrite the equation as $3^x - 5 = 0$ and enter the left side of the equation at the y1 prompt.

 WARNING! A careless reading of the equation $x = \frac{\log 5}{\log 3}$ leads to a common error. The right-hand side of the equation calls for a division, *not a subtraction*. $\frac{\log 5}{\log 3}$ means (log 5) ÷ (log 3). It is the expression $\log \frac{5}{3}$ that means $\log 5 - \log 3$.

EXAMPLE 2   Solve the exponential equation $6^{x-3} = 2^x$.

Solution   *Analytic solution:*

$$6^{x-3} = 2^x$$

$$\log 6^{x-3} = \log 2^x \qquad \text{Take the common logarithm of each side.}$$

$$(x - 3) \log 6 = x \log 2 \qquad \text{Power property of logarithms.}$$

$$x \log 6 - 3 \log 6 = x \log 2 \qquad \text{Use the distributive property.}$$

$$x \log 6 - x \log 2 = 3 \log 6 \qquad \text{Add 3 log 6 and subtract } x \log 2 \text{ from both sides.}$$

$$x(\log 6 - \log 2) = 3 \log 6 \qquad \text{Factor out } x \text{ on the left-hand side.}$$

$$x = \frac{3 \log 6}{\log 6 - \log 2} \qquad \text{Divide both sides by log 6} - \log 2.$$

$$x \approx 4.892789261 \qquad \text{Use a calculator.}$$

Graphical solution: Enter the left side of the equation at the y1 prompt and the right side at the y2 prompt. Use the INTERSECT feature to find the approximate value of *x*.

The ROOT/ZERO feature could also be used. However, first rewrite the equation as $6^{x-3} - 2^x = 0$ and enter the left side of the equation at the y1 prompt.

EXAMPLE 3 Solve the exponential equation $2^{x^2+2x} = \dfrac{1}{2}$.

Solution *Analytic solution:* Because $\frac{1}{2} = 2^{-1}$, we can write the equation in the form $2^{x^2+2x} = 2^{-1}$. Equal quantities with equal bases have equal exponents, so we have

$$x^2 + 2x = -1$$

$$x^2 + 2x + 1 = 0 \qquad\qquad \text{Add 1 to both sides.}$$

$$(x + 1)(x + 1) = 0 \qquad\qquad \text{Factor the trinomial.}$$

$$x + 1 = 0 \qquad \text{or} \qquad x + 1 = 0 \qquad \text{Set each factor equal to 0.}$$

$$x = -1 \qquad\qquad\qquad x = -1$$

Therefore, the solution is -1.

Graphical solution: Enter the left side of the equation at the y1 prompt and the right side at the y2 prompt. Use the INTERSECT feature to find the approximate value of x.

The ROOT/ZERO feature could also be used. However, first rewrite the equation as $2^{x^2+2x} - \frac{1}{2} = 0$ and enter the left side of the equation at the y1 prompt.

■ LOGARITHMIC EQUATIONS

A wide variety of logarithmic equations can be solved by applying the logarithmic properties. It will be important to observe the domain of the variable in the original equation so that extraneous values are not included in the solution.

EXAMPLE 4 Solve the logarithmic equation $\log x + \log (x - 3) = 1$.

Solution *Analytic solution:*

$$\log x + \log (x - 3) = 1$$

$$\log x (x - 3) = 1 \qquad\qquad \text{Product property of logarithms.}$$

$$x(x - 3) = 10^1 \qquad\qquad \begin{array}{l}\text{Use the definition of logarithms to} \\ \text{change the equation to exponential form.}\end{array}$$

$$x^2 - 3x - 10 = 0 \qquad\qquad \begin{array}{l}\text{Remove parentheses and subtract 10} \\ \text{from both sides.}\end{array}$$

$$(x + 2)(x - 5) = 0 \qquad\qquad \text{Factor the trinomial.}$$

$$x + 2 = 0 \qquad \text{or} \qquad x - 5 = 0 \qquad \text{Set each factor equal to 0.}$$

$$x = -2 \qquad\qquad\qquad x = 5$$

Check: Use the STOre feature to check. The value -2 is extraneous.

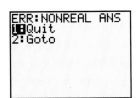

We can conclude that 5 is the only solution.

Graphical solution: Enter the left side of the equation at the y1 prompt and the right side at the y2 prompt. Use the INTERSECT feature to find the approximate value of x. Note that the graphs do not intersect at $x = -2$, verifying that -2 is an extraneous solution.

EXAMPLE 5 Solve the logarithmic equation $\log_b (3x + 2) - \log_b (2x - 3) = 0$.

Solution *Analytic solution:*

$$\log_b (3x + 2) - \log_b (2x - 3) = 0$$
$$\log_b (3x + 2) = \log_b (2x - 3) \qquad \text{Add } \log_b (2x - 3) \text{ to both sides.}$$
$$3x + 2 = 2x - 3 \qquad \text{If } \log_b r = \log_b s, \text{ then } r = s.$$
$$x = -5 \qquad \text{Subtract } 2x \text{ and } 2 \text{ from both sides.}$$

Check: We could check using the STOre feature of the calculator, or analytically, as follows:

$$\log_b (3x + 2) - \log_b (2x - 3) = 0$$
$$\log_b [3(-5) + 2] - \log_b [2(-5) - 3] = ? \ 0$$
$$\log_b (-13) - \log_b (-13) = ? \ 0$$

Because the logarithm of a negative number does not exist, -5 is an extraneous solution and must be discarded. This equation has no roots.

EXAMPLE 6 Solve the logarithmic equation $\dfrac{\log (5x - 6)}{\log x} = 2$.

Solution We can multiply both sides of the equation by $\log x$ to get $\log (5x - 6) = 2 \log x$ and apply the power property of logarithms to get $\log (5x - 6) = \log x^2$. Because $(5x - 6)$ and x^2 have equal logarithms, we can set these expressions equal and solve for x:

$$5x - 6 = x^2$$
$$0 = x^2 - 5x + 6$$
$$0 = (x - 3)(x - 2)$$
$$x - 3 = 0 \qquad \text{or} \qquad x - 2 = 0$$
$$x = 3 \qquad\qquad\qquad x = 2$$

Check: We use the STOre feature of the calculator to check these supposed roots. This verifies the fact that both 3 and 2 are actual roots. The roots could also be verified graphically.

APPLICATIONS OF EXPONENTIAL AND LOGARITHMIC EQUATIONS

Carbon-14 dating Experiments have determined the time it takes for half of a sample of a given radioactive material to decompose. This time, called the material's **half-life,** is a constant.

When a living organism dies, the oxygen/carbon dioxide cycle common to all living things ceases, and carbon-14, a radioactive isotope with a half-life of 5700 years, is no longer absorbed. By measuring the amount of carbon-14 present in an ancient object, archeologists can estimate the object's age by using the radioactive decay formula.

Radioactive Decay Formula

If A is the amount of radioactive material present at time t, A_0 was the amount present at $t = 0$, and h is the material's half-life, then

$$A = A_0 2^{-t/h}$$

EXAMPLE 7 How old is a wooden statue that contains only one-third of its original carbon-14 content?

Solution To find the time t when $A = \frac{1}{3}A_0$, we substitute $\dfrac{A_0}{3}$ for A and **5700** for h into the radioactive decay formula and solve for t:

$$A = A_0 2^{-t/h}$$

$$\frac{A_0}{3} = A_0 2^{-t/5700}$$

$$A_0 = 3A_0 2^{-t/5700} \qquad \text{Multiply both sides by 3.}$$

$$1 = 3(2^{-t/5700}) \qquad \text{Divide both sides by } A_0.$$

$$\log 1 = \log 3(2^{-t/5700}) \qquad \text{Take the common logarithm of each side.}$$

$$0 = \log 3 + \log 2^{-t/5700} \qquad \log_b 1 = 0 \text{ and product property of logarithms.}$$

$$-\log 3 = -\frac{t}{5700}\log 2 \qquad \text{Subtract log 3 from both sides and use the power property of logarithms.}$$

$$5700\left(\frac{\log 3}{\log 2}\right) = t \qquad \text{Multiply both sides by } -\frac{5700}{\log 2}.$$

$$t \approx 9034.286254 \qquad \text{Use a calculator.}$$

The wooden statue is approximately 9000 years old.

Population growth When there is sufficient food and space, populations of living organisms tend to increase exponentially according to the Malthusian growth model.

Malthusian Growth Model

If P is the population at some time t, P_0 is the initial population at $t = 0$, and k depends on the rate of growth, then

$$P = P_0 e^{kt}$$

EXAMPLE 8 The bacteria in a laboratory culture increased from an initial population of 500 to 1500 in 3 hours. How long will it take for the population to reach 10,000?

Solution   We substitute **500** for P_0, **1500** for P, and **3** for t and simplify to find k:

$$P = P_0 e^{kt}$$
$$1500 = 500(e^{k3}) \qquad \text{Substitute 1500 for } P, \text{ 500 for } P_0, \text{ and 3 for } t.$$
$$3 = e^{3k} \qquad \text{Divide both sides by 500.}$$
$$\ln 3 = \ln e^{3k} \qquad \text{Take the natural logarithm of both sides.}$$
$$\ln 3 = 3k \qquad \ln e^{3k} = 3k \ln e = 3k.$$
$$k = \frac{\ln 3}{3} \qquad \text{Divide both sides by 3.}$$

To find when the population will reach 10,000, we substitute **10,000** for P, **500** for P_0, and $\frac{\ln 3}{3}$ for k in the equation $P = P_0 e^{kt}$ and solve for t:

$$P = P_0 e^{kt}$$
$$10{,}000 = 500 e^{[(\ln 3)/3]t}$$
$$20 = e^{[(\ln 3)/3]t} \qquad \text{Divide both sides by 500.}$$
$$\left(\frac{\ln 3}{3}\right)t = \ln 20 \qquad \text{Change the equation to logarithmic form.}$$
$$t = \frac{3 \ln 20}{\ln 3} \qquad \text{Multiply both sides by } \tfrac{3}{\ln 3}.$$
$$\approx 8.180499084 \qquad \text{Use a calculator.}$$

The culture will reach 10,000 bacteria in a little more than 8 hours.

EXERCISE 11.6

CONCEPTS

 1. Explain why $x = -4$ could not be a solution to $\log(2x + 1) - \log(3x - 2) = 0$.

2. Substitute in several values for b in $\log x = b$ and then explain why $x \neq 0$.

In Exercises 3–6, an exponential or logarithmic function, f, has been graphed at the y1 prompt. Use the graphical display to find the solution y1 = 0. Round solutions to the nearest thousandth.

3.

4.

5.

6.

PRACTICE   *In Exercises 7–22, solve each exponential equation. Give all answers to four decimal places. Check all solutions.*

7. $4^x = 5$

8. $7^x = 12$

9. $13^{x-1} = 2$

10. $5^{x+1} = 3$

11. $2^{x+1} = 3^x$

12. $5^{x-3} = 3^{2x}$

13. $2^x = 3^x$

14. $3^{2x} = 4^x$

15. $7^{x^2} = 10$

16. $8^{x^2} = 11$

17. $8^{x^2} = 9^x$

18. $5^{x^2} = 2^{5x}$

19. $2^{x^2-2x} = 8$

20. $3^{x^2-3x} = 81$

21. $3^{x^2-4x} = \dfrac{1}{81}$

22. $7^{x^2+3x} = \dfrac{1}{49}$

In Exercises 23–40, solve each logarithmic equation. Check each solution with the calculator.

23. $\log 2x = \log 4$

24. $\log 3x = \log 9$

25. $\log (3x + 1) = \log (x + 7)$

26. $\log (x^2 + 4x) = \log (x^2 + 16)$

27. $\log (2x - 3) - \log (x + 4) = 0$

28. $\log (3x + 5) - \log (2x + 6) = 0$

29. $\log \dfrac{4x + 1}{2x + 9} = 0$

30. $\log \dfrac{5x + 2}{2(x + 7)} = 0$

31. $\log x + \log (x - 48) = 2$

32. $\log x + \log (x + 9) = 1$

33. $\log (x + 90) = 3 - \log x$

34. $\log (x - 90) = 3 - \log x$

35. $\log x^2 = (\log x)^2$

36. $\log (\log x) = 1$

37. $\dfrac{\log (3x - 4)}{\log x} = 2$

38. $\dfrac{\log (8x - 7)}{\log x} = 2$

39. $\dfrac{\log (5x + 6)}{2} = \log x$

40. $\dfrac{1}{2} \log (4x + 5) = \log x$

APPLICATIONS *In Exercises 41–58, solve each problem.*

41. Tritium decay The half-life of tritium is 12.4 years. How long will it take for 25% of a sample of tritium to decompose?

42. Radioactive decay In two years, 20% of a radioactive element decays. Find its half-life.

43. Thorium decay An isotope of thorium, ^{227}Th, has a half-life of 18.4 days. How long will it take 80% of the sample to decompose?

44. Lead decay An isotope of lead, ^{201}Pb, has a half-life of 8.4 hours. How many hours ago was there 30% more of the original amount of the substance?

45. Carbon-14 dating The bone fragment shown in the photograph contains 60% of the carbon-14 that it is assumed to have had initially. How old is the bone?

Archaeological Consulting/Gary Breschini and Trudy Haversat

46. Carbon-14 dating Only 10% of the carbon-14 in a small wooden bowl remains. How old is the bowl?

47. Compound interest If $500 is deposited in an account paying 8.5% annual interest compounded semiannually, how long will it take for the account to increase to $800?

48. Continuous compound interest In Exercise 47, how long will it take if the interest is compounded continuously?

49. Compound interest If $1300 is deposited in a savings account paying 9% interest compounded quarterly, how long will it take the account to increase to $2100?

50. Compound interest A sum of $5000 deposited in an account grows to $7000 in 5 years. Assuming continuous compounding, what interest rate is being paid?

51. Comparing investments Which of these investment opportunities is best: 4.9% compounded annually, 4.8% compounded quarterly, or 4.7% compounded monthly?

52. Bacterial growth A bacteria culture grows according to the formula $P = P_0 a^t$. If it takes 5 days for the culture to triple in size, how long will it take to double in size?

53. Rodent control The rodent population in a city is currently estimated at 30,000. If it is expected to double every 5 years, when will the population reach 1 million?

54. Population growth The population of a city is expected to triple every 15 years. When can the city planners expect the population to be double the present census of 140 persons.

55. Bacteria culture A bacteria culture doubles in size every 24 hours. By how much will it have increased in 36 hours?

56. Oceanography The intensity I of a light a distance x meters beneath the surface of a lake decreases exponentially. From Illustration 1, find the depth at which the intensity will be 20%.

57. Newton's law of cooling Water initially at 100° C is left to cool in a room of temperature 60° C. After 3 minutes, the water temperature is 90°. If the water temperature, T, is a function of time t given by $T = 60 + 40e^{kt}$, find k.

58. Newton's law of cooling Refer to Exercise 57 and find the time for the water temperature to reach 70° C.

ILLUSTRATION 1

In Exercises 59–62, use a graphing calculator to solve each equation.

59. $\log x + \log (x - 15) = 2$

60. $\log x + \log (x + 3) = 1$

61. $2^{x+1} = 7$

62. $\ln (2x + 5) - \ln 3 = \ln (x - 1)$

63. Explain how to solve the equation $2^x = 7$.

64. Explain how to solve the equation $x^2 = 7$.

REVIEW

65. A ball is thrown into the air with an initial velocity of 80 feet per second. The equation that models the relation between height h after t seconds is $h = -16t^2 + 80t + 3$.

a. Graph the model in an appropriate viewing window.

b. Based on the graph, how high is the ball 3 seconds after it is released?

c. At what other point in time is the ball the same height as it is after 3 seconds?

d. How long does it take the ball to hit the ground?

66. What is the slope of the line that passes through the y-intercept and the vertex of the parabola in Exercise 65?

67. What is the equation of the line defined in Exercise 66?

CHAPTER SUMMARY

CONCEPTS

REVIEW EXERCISES

Section 11.1

Algebra of Functions

$(f + g)(x) = f(x) + g(x)$
$(f - g)(x) = f(x) - g(x)$
$(f \cdot g)(x) = f(x) \cdot g(x)$
$\left(\dfrac{f}{g}\right)(x) = \dfrac{f(x)}{g(x)},$
provided $g(x) \neq 0$.

The functional operations of addition and multiplication are commutative.

1. Let $f(x) = 2x$ and $g(x) = x^2 - 1$. Find each of the following:

a. $f + g$

b. $f - g$

c. $f \cdot g$

d. $\frac{f}{g}$

The composition of two functions f and g, denoted $(f \circ g)(x)$ or $f(g(x))$, takes the range of the function g and makes it the domain of the function f.

Composition is *not* a commutative operation.

2. Let $f(x) = 2x$ and $g(x) = x^2 + 1$. Find each of the following:

 a. $(f \circ g)(x)$ **b.** $(g \circ f)(x)$

 c. $(f \circ g)(2)$ **d.** $(g \circ f)(-1)$

3. Suppose an oil rig is leaking oil into the Gulf of Mexico with the leak forming a circular layer of oil on the surface of the water. If the radius of the circle is increasing at a rate of 5 meters per minute, then the function $f(t) = 5t$ defines the radius as a function of time. Express the area of the circle as a function of time and determine the approximate area (to the nearest foot) 40 minutes after the leak begins.

Section 11.2   Inverse Functions

If a relation contains the point (a, b), then the inverse will contain the point (b, a).

Every relation has an inverse relation.

4. If the table defines the relation that converts feet to miles, construct the inverse relation that converts miles to feet.

Feet	1320	2640	3960	5280	6600
Miles	$\frac{1}{4}$	$\frac{1}{2}$	$\frac{3}{4}$	1	$1\frac{1}{4}$

In order for a function to be one-to-one, each y-value in the range must be paired with a unique x-value in the domain. The horizontal line test will verify this relationship on a graph.

5. Graph each function and determine if it is one-to-one.

 a. $y = 2(x - 3)$ **b.** $y = x^2 - 4$

6. Determine if the relation specified is a one-to-one function.

 a. $\{(4, 1), (2, -3), (3, 2), (1, 4)\}$ **b.** $\{(2, 3), (3, 2), (4, 2), (1, 5)\}$

In order for a function, f, to have an inverse function, f^{-1}, f must be one-to-one. To algebraically determine the inverse function, reverse the x and y variables in the original function.

7. Algebraically, determine $f^{-1}(x)$ and then graph the function $f(x)$, its inverse $f^{-1}(x)$, and the line of symmetry $f(x) = x$. Verify with your calculator.

 a. $y = 6x - 3$ **b.** $y = \frac{1}{2}x + 3$

| **Section 11.3** | **Exponential Functions** |

An exponential function is a function of the form $f(x) = b^x$, where $b > 0$ and $b \neq 1$. The domain is the set of reals and the range is the set of positive numbers.

8. Graph each exponential function and verify with the calculator.

   a. $f(x) = \left(\dfrac{6}{5}\right)^x$ **b.** $f(x) = \left(\dfrac{1}{5}\right)^x$

   c. $f(x) = e^x$

Exponential property of equality: If $b > 0$ and $b \neq 1$, then $b^r = b^s$ is equivalent to $r = s$. If $r \neq s$, then $b^r \neq b^s$.

9. Solve for x. Verify solutions graphically.

   a. $2^{x+5} = 2^{x^2-4x}$ **b.** $3^{x^2} = 9^{2x}$

   c. $2^{x^2+4x} = \dfrac{1}{8}$ **d.** $\left(\dfrac{1}{5}\right)^x = 5^{x-3}$

10. Solve graphically. Approximate the solution to the nearest thousandth:
$3^x = -x^x + 5$

| **Section 11.4** | **Logarithmic Functions** |

A logarithm is an exponent.

11. **Compound interest** An initial deposit of \$1000 earns 6% interest, compounded twice a year. How much will be in the account in 1 year?

12. **Interest compounded continuously** An initial deposit of \$1000 earns 6% interest, compounded continuously. How much will be in the account in one year?

13. Find each value.

   a. $\log_3 9$ **b.** $\log_9 \dfrac{1}{3}$

   c. $\log_5 .04$ **d.** $\ln e^9$

14. Find x.

   a. $\log_2 x = 3$ **b.** $\log_3 x = -2$

   c. $\log_x 9 = 2$ **d.** $\log_x .125 = -3$

   e. $\log_3 \sqrt{3} = x$ **f.** $\log_2 \sqrt{2} = x$

The change-of-base formula is necessary for computation or graphing involving logarithms other than base 10 or base e. It states: If a, b, and x are positive real numbers, then

$$\log_b x = \frac{\log_a x}{\log_a b}, \ (a \neq 1, b \neq 1)$$

15. Evaluate each of the following on your calculator using the change-of-base formula:

   a. $\log_3 9$ **b.** $\log_5 7$ (to the nearest hundredth)

The exponential functions $y = 2^x$ and $x = 2^y$ are inverses of one another. Because $x = 2^y$ can be expressed as $\log_2 x = y$ using the definition of logarithms, then $y = 2^x$ and $\log_2 x = y$ are inverses of one another.

16. Graph each function and its inverse. Verify with your calculator.

a. $f(x) = 3^x$ and $f^{-1}(x) = \log_3 x$ **b.** $f(x) = \left(\dfrac{1}{2}\right)^x$ and $f^{-1}(x) = \log_{1/2} x$

c. $f(x) = \left(\dfrac{2}{5}\right)^x$ and $f^{-1}(x) = \log_{2/5} x$

17. pH of grapefruit The pH of grapefruit juice is approximately 3.1. Find its hydrogen ion concentration.

18. Formula for pH Some chemistry textbooks define the pH of a solution with the formula

$$\text{pH} = \log_{10} \frac{1}{[\text{H}^+]}$$

Show that this definition is equivalent to the one given in this text.

Section 11.5	**Properties of Logarithms**

Properties of logarithms:
If $b > 0$ and $b \neq 1$, then

1. $\log_b 1 = 0$

2. $\log_b b = 1$

3. $\log_b b^x = x$

4. $b^{\log_b x} = x, \quad x > 0$

M, N are both positive:

5. $\log_b MN = \log_b M + \log_b N$

6. $\log_b \dfrac{M}{N} = \log_b M - \log_b N$

7. $\log_b M^p = p \log_b M$

19. Write each expression in terms of logarithms of x, y, and z.

a. $\log_b \dfrac{x^2 y^3}{z^4}$ **b.** $\log_b \sqrt{\dfrac{x}{yz^2}}$

20. Write each expression as the logarithm of a single quantity.

a. $3 \log_b x - 5 \log_b y + 7 \log_b x$

b. $\dfrac{1}{2} \log_b x + 3 \log_b y - 7 \log_b z$

| Section 11.6 | Exponential and Logarithmic Equations |

Logarithmic property of equality: If $M > 0$ and $N > 0$, $\log_b M = \log_b N$ if and only if $M = N$.

21. Solve for x and verify solutions graphically.

   a. $3^x = 7$ **b.** $5^{x+2} = 625$

   c. $2^x = 3^{x-1}$ **d.** $\log x + \log (29 - x) = 2$

   e. $\dfrac{\log (7x - 12)}{\log x} = 2$ **f.** $\log 3 - \log (x - 1) = -1$

   g. $\ln x = \ln (x - 1)$ **h.** $\ln x = \ln (x - 1) + 1$

   i. $\ln x = \log_{10} x$ *(Hint:* Remember to use the change-of-base formula.)

22. Carbon-14 dating A wooden statue excavated in Egypt has a carbon-14 content that is two-thirds of that found in living wood. If the half-life of carbon-14 is 5700 years, how old is the statue?

23. Radioactive decay One-third of a radioactive material decays in 20 years. Find its half-life.

■ Chapter Test

In Problems 1–4, $f(x) = 4x$ and $g(x) = x - 1$. Find each function.

1. $(g \circ f)(1)$ **2.** $(f \circ g)(0)$ **3.** $(f \circ g)(-1)$ **4.** $(g \circ f)(-2)$

In Problems 5 and 6, $f(x) = 4x$ and $g(x) = x - 1$. Find each function.

5. $(f \circ g)(x)$ **6.** $(g \circ f)(x)$

In Problems 7 and 8, find the inverse of each function.

7. $3x + 2y = 12$ **8.** $y = 3x^2 + 4, \quad (x \leq 0)$

9. Graph the function $y = \frac{1}{4}x^2 - 3$ on graph paper and use the horizontal line test to decide whether the function is one-to-one.

In Problems 10 and 11, graph each function on graph paper.

10. $y = 2^x + 1$ **11.** $y = 2^{-x}$

                                                      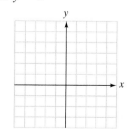

In Problems 12 and 13, solve each equation.

12. A radioactive material decays according to the formula $A = A_0 (2)^{-t}$. How much of a 3-gram sample will be left in 6 years?

13. An account contains $2000 and has been earning 8% interest, compounded continuously. How much will be in the account in 10 years?

14. Graph the function $y = e^x$ on graph paper.

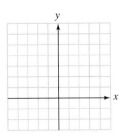

In Problems 15–17, find x.

15. $\log_4 16 = x$

16. $\log_x 81 = 4$

17. $\log_3 x = -3$

In Problems 18 and 19, graph each function on graph paper.

18. $y = -\log_3 x$

19. $y = \ln x$

In Problems 20 and 21, write each expression in terms of the logarithms of a, b, and c.

20. $\log a^2bc^3$

21. $\ln \sqrt{\dfrac{a}{b^2c}}$

In Problems 22 and 23, write each expression as a logarithm of a single quantity.

22. $\dfrac{1}{2} \log (a + 2) + \log b - 2 \log c$

23. $\dfrac{1}{3} (\log a - 2 \log b) - \log c$

In Problems 24 and 25, use the change-of-base formula to find each logarithm. Do not attempt to simplify the result.

24. $\log_5 24$

25. $\log_\pi e$

In Problems 26–29, tell whether each statement is true.

26. $\log_a ab = 1 + \log_a b$

27. $\dfrac{\log a}{\log b} = \log a - \log b$

28. $\log a^{-3} = \dfrac{1}{3 \log a}$

29. $\ln (-x) = -\ln x$

30. Find the pH of a solution with a hydrogen ion concentration of 3.7×10^{-7}. (*Hint:* pH $= -\log$ [H$^+$].)

31. Find the db gain of an amplifier when $E_0 = 60$ volts and $E_I = 0.3$ volt. (*Hint:* db gain $= 20 \log E_0 /E_I$.)

In Problems 32 and 33, solve each equation. Do not simplify the logarithms.

32. $5^x = 3$

33. $3^{x-1} = 100^x$

In Problems 34 and 35, solve each equation.

34. $\log (5x + 2) = \log (2x + 5)$

35. $\log x + \log (x - 9) = 1$

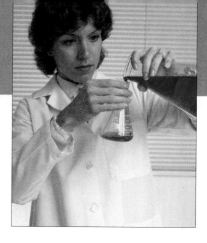

Chemist Chemists search for new knowledge about substances and put it to practical use. Over half of all chemists work in research and development. In basic research, chemists investigate the properties, composition, and structure of matter and the laws that govern the combination of elements and the reactions of substances. Their research has resulted in the development of a tremendous variety of synthetic materials, of ingredients that have improved other substances, and of processes that help save energy and reduce pollution. In applied research and development, they create new products or improve existing ones, often using knowledge gained from basic research.

Qualifications A bachelor's degree in chemistry or a related discipline is sufficient for many beginning jobs as a chemist. However, graduate training is required for most research jobs, and most college teaching jobs require a Ph.D. degree.

Job Outlook The employment of chemists is expected to grow about as fast as the average for all occupations through the year 2005.

The majority of job openings are expected to be in private industry, primarily in the development of new products. In addition, industrial companies and government agencies will need chemists to help solve problems related to energy shortages, pollution control, and health care.

Example Application A certain weak acid (0.1 M concentration) will break down into free cations (the hydrogen ion, H^+) and anions (A^-). When this acid dissociates, the following equilibrium equation is established.

$$\frac{[H^+][A^-]}{[HA]} = 4 \times 10^{-4}$$

where $[H^+]$, the hydrogen ion concentration, is equal to $[A^-]$, the anion concentration. $[HA]$ is the concentration of the undissociated acid itself. Find $[H^+]$ at equilibrium.

Solution Let x be the concentration of H^+. Then x is also $[A^-]$. From chemistry, it turns out that the concentration $[HA]$ of the undissociated acid is $0.1 - x$. Substituting these concentrations into the equation gives the equation

$$\frac{x^2}{0.1 - x} = 4 \times 10^{-4}$$

which can be solved as follows:

$$\frac{x^2}{0.1 - x} = 0.0004 \qquad \text{Because } 4 \times 10^{-4} = 0.0004.$$

$$x^2 = 0.0004(0.1 - x) \qquad \text{Multiply both sides by } 0.1 - x.$$

$$x^2 = 0.00004 - 0.0004x \qquad \text{Remove parentheses.}$$

$$x^2 + 0.0004x - 0.00004 = 0 \qquad \text{Add } 0.0004x - 0.00004 \text{ to both sides.}$$

Substitute into the quadratic formula with $a = 1$, $b = 0.0004$, and $c = -0.00004$. We only need to consider the positive root.

$$x = \frac{-b + \sqrt{b^2 - 4ac}}{2a}$$

$$x = \frac{-0.0004 + \sqrt{(0.0004)^2 - 4(1)(-0.00004)}}{2(1)}$$

$$x = \frac{-0.0004 + 0.01266}{2}$$

$$x = 0.00613$$

At equilibrium, $[H^+]$ is approximately 6.13×10^{-3} M.

EXERCISES   *Use your calculator to solve these problems.*

1. A saturated solution of hydrogen sulfide (concentration 0.1 M) dissociates into cation H^+ and anion HS^-. When this solution dissociates, the following equilibrium equation is established.

$$\frac{[H^+][HS^-]}{[HHS]} = 1.0 \times 10^{-7}$$

Find $[H^+]$.

2. An HS^- anion dissociates into cation H^+ and anion S^{--} with equilibrium equation

$$\frac{[H^+][S^{--}]}{[HS^-]} = 1.3 \times 10^{-13}$$

Assume the concentration of HS^- to be 1×10^{-4} M. Find $[H^+]$.

3. Suppose the concentration of the acid of the example were 0.2 M. Find the hydrogen ion concentration, $[H^+]$, at equilibrium.

4. Show that the equation of the example has only one meaningful solution by showing that its other solution is negative.

(*Answers:*   **1.** 9.995×10^{-5} **2.** 3.605×10^{-9} **3.** 8.75×10^{-3})

Systems of Equations and Inequalities

12

■ PERSPECTIVE

Stay in Bed!

As a child, René Descartes, the inventor of analytic geometry, was frail and often sick. To improve his health, eight-year-old René was sent to a Jesuit school. The headmaster encouraged him to sleep in the morning as long as he wished, often past noon. As a young man, Descartes spent several years as a soldier and world traveler, but his interests included mathematics and philosophy, as well as science, literature, and writing, and taking it easy: The habit of sleeping late continued throughout his life. Descartes claimed that his times of most productive thinking were spent lying in bed, long past breakfast. According to one story, Descartes

first thought of analytic geometry as he watched the path of a fly walking on his bedroom ceiling.

Descartes might have lived longer if he had stayed in bed. In 1649, Sweden's Queen Christina decided that she needed a tutor in philosophy, and she requested the service of Descartes. Tutoring would not have been difficult, except that the stubborn queen scheduled her lessons before dawn in her library, with the windows open. The cold Stockholm weather and the early hour was too much for a man who was used to sleeping until noon. Within a few months, Descartes developed a fever and died, probably of pneumonia.

This chapter will consider **systems** of both equations and inequalities and methods for finding solutions to these systems. A system is a group of equations or inequalities. Solutions of systems are those values of the variable(s) that make all the equations (or inequalities) in the system true.

12.1 Solving Systems of Two Equations in Two Variables

■ SOLVING SYSTEMS GRAPHICALLY ■ THE SUBSTITUTION METHOD ■ THE ADDITION METHOD

Consider the two linear equations in two variables $x + y = 3$ and $3x - y = 1$. Although there are an infinite number of ordered pairs that satisfy each of these equations, only one ordered pair, (1, 2), satisfies both equations.

The pair of equations,

$$\begin{cases} x + y = 3 \\ 3x - y = 1 \end{cases}$$

is called a **system of equations.** Because the ordered pair (1, 2) satisfies both equations, it is the **solution of the system.** Notice that because the original equations each contained the variables x and y the solution is an ordered pair, (x, y).

■ SOLVING SYSTEMS GRAPHICALLY

One way to solve a system is by graphing each equation in the same rectangular coordinate system. The coordinates of the point (or points) where the graphs intersect are the solutions of the system. We suggest that you always check solutions in the original equations.

EXAMPLE 1   Solve the system $\begin{cases} x + 2y = 4 \\ 2x - y = 3 \end{cases}$ graphically.

Solution   We graph both equations on a single set of coordinates axes, as shown in Figure 12-1. Although an infinite number of ordered pairs (x, y), satisfy the equation $x + 2y = 4$, and an infinite number of ordered pairs, (x, y) satisfy the equation $2x - y = 3$, only one ordered pair, (2, 1), satisfies both equations in the system.

$x + 2y = 4$		
x	y	(x, y)
4	0	$(4, 0)$
0	2	$(0, 2)$
-2	3	$(-2, 3)$

$2x - y = 3$		
x	y	(x, y)
1	-1	$(1, -1)$
0	-3	$(0, -3)$
-1	-5	$(-1, -5)$

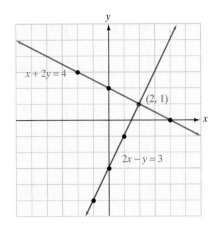

FIGURE 12-1

We can confirm our solution by substituting 2 for x and 1 for y in each of the equations. The resulting statements are true.

Check:

$$
\begin{array}{ll}
x + 2y = 4 & 2x - y = 3 \\
2 + 2(1) = 4 & 2(2) - 1 = 3 \qquad \text{Substitute 2 for } x \text{ and 1 for } y \\
2 + 2 = 4 & 4 - 1 = 3 \\
4 = 4\checkmark & 3 = 3\checkmark
\end{array}
$$

We can use the INTERSECT feature of the calculator to solve the system. However, we must first solve each equation in the system for the variable y.

$$
\begin{array}{ll}
x + 2y = 4 & \\
2y = -x + 4 & \text{Subtract } x \text{ from both sides.} \\
y = -\dfrac{1}{2}x + 2 & \text{Divide through the equation by 2.} \\
2x - y = 3 & \\
-y = -2x + 3 & \text{Subtract } 2x \text{ from both sides.} \\
y = 2x - 3 & \text{Multiply through the equation by } -1 \text{ to make } y \text{ positive.}
\end{array}
$$

Enter $y = -\dfrac{1}{2}x + 2$ at the y1 prompt and $y = 2x - 3$ at the y2 prompt. Use the INTERSECT feature to find the *coordinates* of the ordered pair solution of the system, $(2, 1)$, for this example.

Our solution can be confirmed with the TABLE. Set the TABLE to begin at the x-value of 2 (the increment is immaterial). The expressions entered at the y1 and y2 prompts both evaluate to 1 when x has a value of 2.

When a system of equations (as in Example 1) has a solution, the system is a **consistent system of equations.** A system with no solution is an **inconsistent system** and is illustrated in the next example.

EXAMPLE 2   Solve the system $\begin{cases} 2x + 3y = 6 \\ 4x + 6y = 24 \end{cases}$

Solution   We graph both equations on the same set of coordinate axes, as shown in Figure 12-2. The lines do not intersect, so the system does not have a solution. It is an inconsistent system.

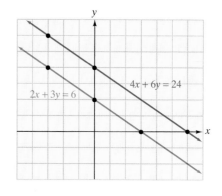

$2x + 3y = 6$		
x	y	(x, y)
3	0	$(3, 0)$
0	2	$(0, 2)$
-3	4	$(-3, 4)$

$4x + 6y = 24$		
x	y	(x, y)
6	0	$(6, 0)$
0	4	$(0, 4)$
-3	6	$(-3, 6)$

FIGURE 12-2

To look at the graphical representations on the calculator, begin by solving each equation for the variable y.

$$2x + 3y = 6$$
$$3y = -2x + 6 \qquad \text{Subtract } 2x \text{ from both sides.}$$
$$y = -\frac{2}{3}x + 2 \qquad \text{Divide through the equation by 3.}$$
$$4x + 6y = 24$$
$$6y = -4x + 24 \qquad \text{Subtract } 4x \text{ from both sides.}$$
$$y = -\frac{2}{3}x + 4 \qquad \text{Divide through the equation by 6 and reduce fractions.}$$

Because the slopes are the same, the lines are parallel and have no points of intersection. *Do not conclude that lines are parallel simply from the graphical display. Support the decision of parallelism analytically and confirm graphically.*

When the equations of a system have different graphs (as in Examples 1 and 2), the equations are **independent equations.** Two equations with the same graph are **dependent equations.**

EXAMPLE 3   Solve the system $\begin{cases} 2y - x = 4 \\ 2x + 8 = 4y \end{cases}$

Solution   We graph each equation on the same set of coordinate axes, as shown in Figure 12-3.

$2y - x = 4$		
x	y	(x, y)
-4	0	$(-4, 0)$
0	2	$(0, 2)$

$2x + 8 = 4y$		
x	y	(x, y)
-4	0	$(-4, 0)$
0	2	$(0, 2)$

FIGURE 12-3

Because the graphs coincide, the system has infinitely many solutions. Any ordered pair, (x, y) that satisfies one equation satisfies the other equation as well. Because the two equations have the same graph, they are dependent equations. ■

When two linear equations in two variables are graphed, the following possibilities can occur:

If the lines are different and intersect, the equations are independent and the system is consistent. *One solution exists.*

If the lines are different and parallel, the equations are independent and the system is inconsistent. *No solution exists.*

If the lines coincide, the equations are dependent and the system is consistent. *Infinitely many simultaneous solutions exist.*

■ THE SUBSTITUTION METHOD

A system of two equations in two variables can be solved algebraically by either the substitution or the elimination (often called *addition*) method. We will first consider the substitution method.

Substitution Method for Solving Systems of Equations
1. If necessary, solve one equation for one of its variables.
2. Substitute the resulting expression for the variable obtained in step 1 into the *other* equation and solve that equation.
3. Find the value of the other variable by substituting the value of the variable found in step 2 in any equation containing both variables.
4. Check the solution in both of the original equations.

EXAMPLE 4 Use the substitution method to solve the system $\begin{cases} \dfrac{4}{3}x + \dfrac{1}{2}y = -\dfrac{2}{3} \\ \dfrac{1}{2}x + \dfrac{2}{3}y = \dfrac{5}{3} \end{cases}$

Solution We first find an equivalent system without fractions by multiplying each side of each equation by 6 to obtain the system

1. $\begin{cases} 8x + 3y = -4 \\ 3x + 4y = 10 \end{cases}$
2.

Because no variable in either equation has a coefficient of 1, it is impossible to avoid fractions when solving for a variable. If we choose Equation 2 and solve it for x, we get

$$3x + 4y = 10$$

$$3x = -4y + 10 \qquad \text{Subtract } 4y \text{ from both sides.}$$

3. $\qquad x = -\dfrac{4}{3}y + \dfrac{10}{3} \qquad \text{Divide both sides by 3.}$

We can then substitute $-\frac{4}{3}y + \frac{10}{3}$ for x in Equation 1 and solve for y.

$$8x + 3y = -4$$

$$8\left(-\dfrac{4}{3}y + \dfrac{10}{3}\right) + 3y = -4 \qquad \text{Substitute } -\frac{4}{3}y + \frac{10}{3} \text{ for } x.$$

$$8(10 - 4y) + 9y = -12 \qquad \text{Multiply both sides by 3.}$$

$$80 - 32y + 9y = -12 \qquad \text{Use the distributive property to remove parentheses.}$$

$$-23y = -92 \qquad \text{Combine like terms and subtract 80 from both sides.}$$

$$y = 4 \qquad \text{Divide both sides by } -23.$$

We can find x by substituting 4 for y in Equation 3 and simplifying:

$$x = -\dfrac{4}{3}y + \dfrac{10}{3}$$

$$= -\dfrac{4}{3}(4) + \dfrac{10}{3} \qquad \text{Substitute 4 for } y.$$

$$= -\dfrac{6}{3} \qquad\qquad -\frac{16}{3} + \frac{10}{3} = -\frac{6}{3}.$$

$$= -2$$

The solution is the pair $(-2, 4)$. Verify that this solution satisfies each equation in the original system. We can verify the solutions in both equations by using the STOre feature of the graphing calculator.

■ THE ADDITION METHOD

The addition (elimination) method is another way to algebraically solve a system of equations. In this method, we combine the equations of the system in a way that will eliminate terms involving one of the variables.

Addition (Elimination) Method for Solving Systems of Equations
1. Write both equations of the system in general form.
2. Multiply the terms of one or both of the equations by constants chosen to make the coefficients of x (or y) differ only in sign.
3. Add the equations and solve the equation that results, if possible.
4. Substitute the value obtained in step 3 into either of the original equations and solve for the remaining variable.
5. The results obtained in steps 3 and 4 are the solution of the system.
6. Check the solution in both of the original equations.

EXAMPLE 5   Use the addition method to solve the system $\begin{cases} \dfrac{4}{3}x + \dfrac{1}{2}y = -\dfrac{2}{3} \\ \dfrac{1}{2}x + \dfrac{2}{3}y = \dfrac{5}{3} \end{cases}$

Solution   This is the system discussed in Example 4. To solve it by addition, we find an equivalent system with no fractions by multiplying both sides of each equation by 6 to obtain

4. $\begin{cases} 8x + 3y = -4 \\ 3x + 4y = 10 \end{cases}$
5.

To make the y terms drop out when we add the equations, we multiply both sides of Equation 4 by 4 and both sides of Equation 5 by -3 to get

$$\begin{cases} 32x + 12y = -16 \\ -9x - 12y = -30 \end{cases}$$

When these equations are added, the y terms drop out, and we get

$$23x = -46$$
$$x = -2 \qquad \text{Divide both sides by 23.}$$

To find y, we substitute -2 for x in either Equation 4 or Equation 5. If we substitute -2 for x in Equation 5, we get

$$3x + 4y = 10$$
$$3(-2) + 4y = 10 \qquad \text{Substitute } -2 \text{ for } x.$$
$$-6 + 4y = 10 \qquad \text{Simplify.}$$
$$4y = 16 \qquad \text{Add 6 to both sides.}$$
$$y = 4 \qquad \text{Divide both sides by 4.}$$

The solution is $(-2, 4)$.   ∎

EXAMPLE 6   Solve the system $\begin{cases} y = 2x + 4 \\ 8x - 4y = 7 \end{cases}$

Solution   Because the first equation is already solved for y, we use the substitution method.

$$8x - 4y = 7$$
$$8x - 4(2x + 4) = 7 \qquad \text{Substitute } 2x + 4 \text{ for } y.$$

We then solve this equation for x:

$$8x - 8x - 16 = 7 \qquad \text{Use the distributive property to remove parentheses.}$$
$$-16 \neq 7 \qquad \text{Combine like terms.}$$

This impossible result shows that the equations in the system are independent and that the system is inconsistent. Because the system has no solution, the graphs of the equations in the system would be parallel.   ■

EXAMPLE 7   Solve the system $\begin{cases} 4x + 6y = 12 \\ -2x - 3y = -6 \end{cases}$

Solution   Because the equations are written in general form, we use the addition method and multiply both sides of the second equation by 2 to get

$$\begin{cases} 4x + 6y = 12 \\ -4x - 6y = -12 \end{cases}$$

After adding the left-hand sides and the right-hand sides, we get

$$0x + 0y = 0$$
$$0 = 0$$

Here, both the x and y terms drop out. The true statement $0 = 0$ shows that the equations in this system are dependent and that the system is consistent. Note that the equations of the system are equivalent because, when the second equation is multiplied by -2, it becomes the first equation. The line graphs of these equations would coincide. Any ordered pair that satisfies one of the equations also satisfies the other.   ■

EXERCISE 12.1

VOCABULARY   *Fill in the blanks to make a true statement.*

1. If two or more equations are considered at the same time, they are called a _____ of equations.

2. When a system of equations has one or more solutions, it is called a _____ system.

3. If a system has no solutions, it is called an _____ system.

4. If two equations have different graphs, they are called _____ equations.

5. Two equations with the same graph are called _____ equations.

6. If the two equations in a system are equivalent, the systems are called _____ systems.

CONCEPTS   *In Exercises 7–12, state whether the ordered pair is a solution of the system. Justify your response either analytically or by using the STOre feature of the calculator.*

7. $(4, 5)$; $\begin{cases} 2x - 3y = -7 \\ 4x - 5y = 25 \end{cases}$

8. $(2, 3)$; $\begin{cases} 3x - 2y = 0 \\ 5x - 3y = -1 \end{cases}$

9. $\left(\dfrac{1}{2}, 3\right)$; $\begin{cases} 2x + y = 4 \\ 4x - 3y = 11 \end{cases}$

10. $\left(2, \dfrac{1}{3}\right)$; $\begin{cases} x - 3y = 1 \\ -2x + 6y = -6 \end{cases}$

11. $\left(-\dfrac{2}{5}, \dfrac{1}{4}\right)$; $\begin{cases} 5x - 4y = -6 \\ 8y = 10x + 12 \end{cases}$

12. $\left(-\dfrac{1}{3}, \dfrac{3}{4}\right)$; $\begin{cases} 3x + 4y = 2 \\ 12y = 3(2 - 3x) \end{cases}$

13. Equations were entered at the y1 and y2 prompts of the calculator. State the solution of the system by interpreting the given screen display.

14. Equations were entered at the y1 and y2 prompts of the calculator. State the solution of the system by interpreting the given TABLE display.

15. Equations were entered at the y1 and y2 prompts of the calculator.
 a. State the solution of the system by interpreting the TABLE results.
 b. Are the equations *independent* or *dependent?*
 c. If you were looking at the graphical representation of the equations, would the lines be *parallel* or *coincident?*

16. Equations were entered at the y1 and y2 prompts of the calculator.

a. State the solution of the system by interpreting the TABLE results.

X	Y₁	Y₂
-2	-1	-1
-1	1	1
0	3	3
1	5	5
2	7	7
3	9	9
4	11	11

X=-2

b. Are the equations *independent* or *dependent*?

c. If you were looking at the graphical representation of the equations, would the lines be *parallel* or *coincident*?

PRACTICE *In Exercises 17–28, solve each system by graphing, if possible. Use hand-drawn graphs. Check your graphs with your calculator.*

17. $\begin{cases} x - y = 4 \\ 2x + y = 5 \end{cases}$

18. $\begin{cases} 2x + y = 1 \\ x - 2y = -7 \end{cases}$

19. $\begin{cases} x = 13 - 4y \\ 3x = 4 + 2y \end{cases}$

20. $\begin{cases} 3x = 7 - 2y \\ 2x = 2 + 4y \end{cases}$

21. $\begin{cases} x = 3 - 2y \\ 2x + 4y = 6 \end{cases}$

22. $\begin{cases} 3x = 5 - 2y \\ 3x + 2y = 7 \end{cases}$

23. $\begin{cases} y = 3 \\ x = 2 \end{cases}$

24. $\begin{cases} 2x + 3y = -15 \\ 2x + y = -9 \end{cases}$

25. $\begin{cases} x = \dfrac{11 - 2y}{3} \\ y = \dfrac{11 - 6x}{4} \end{cases}$

26. $\begin{cases} x = \dfrac{1 - 3y}{4} \\ y = \dfrac{12 + 3x}{2} \end{cases}$

27. $\begin{cases} \dfrac{5}{2}x + y = \dfrac{1}{2} \\ 2x - \dfrac{3}{2}y = 5 \end{cases}$

28. $\begin{cases} \dfrac{5}{2}x + 3y = 6 \\ y = \dfrac{24 - 10x}{12} \end{cases}$

In Exercises 29–32, use a graphing calculator to solve each system. Give answers to the nearest hundredth.

29. $\begin{cases} y = 3.2x - 1.5 \\ y = -2.7x - 3.7 \end{cases}$

30. $\begin{cases} y = -0.45x + 5 \\ y = 5.55x - 13.7 \end{cases}$

31. $\begin{cases} 1.7x + 2.3y = 3.2 \\ y = 0.25x + 8.95 \end{cases}$

32. $\begin{cases} 2.75x = 12.9y - 3.79 \\ 7.1x - y = 35.76 \end{cases}$

In Exercises 33–44, solve each system by substitution, if possible. Check results using the STOre feature of the calculator.

33. $\begin{cases} y = x \\ x + y = 4 \end{cases}$

34. $\begin{cases} y = x + 2 \\ x + 2y = 16 \end{cases}$

35. $\begin{cases} x - y = 2 \\ 2x + y = 13 \end{cases}$

36. $\begin{cases} x - y = -4 \\ 3x - 2y = -5 \end{cases}$

37. $\begin{cases} x + 2y = 6 \\ 3x - y = -10 \end{cases}$

38. $\begin{cases} 2x - y = -21 \\ 4x + 5y = 7 \end{cases}$

39. $\begin{cases} 3x = 2y - 4 \\ 6x - 4y = -4 \end{cases}$

40. $\begin{cases} 8x = 4y + 10 \\ 4x - 2y = 5 \end{cases}$

41. $\begin{cases} 3x - 4y = 9 \\ x + 2y = 8 \end{cases}$

42. $\begin{cases} 3x - 2y = -10 \\ 6x + 5y = 25 \end{cases}$

43. $\begin{cases} 2x + 2y = -1 \\ 3x + 4y = 0 \end{cases}$

44. $\begin{cases} 5x + 3y = -7 \\ 3x - 3y = 7 \end{cases}$

In Exercises 45–56, solve each system by addition (elimination), if possible. Check results using the STOre feature of the calculator.

45. $\begin{cases} x - y = 3 \\ x + y = 7 \end{cases}$

46. $\begin{cases} x + y = 1 \\ x - y = 7 \end{cases}$

47. $\begin{cases} 2x + y = -10 \\ 2x - y = -6 \end{cases}$

48. $\begin{cases} x + 2y = -9 \\ x - 2y = -1 \end{cases}$

49. $\begin{cases} 8x - 4y = 16 \\ 2x - 4 = y \end{cases}$

50. $\begin{cases} 2y - 3x = -13 \\ 3x - 17 = 4y \end{cases}$

51. $\begin{cases} x = \dfrac{3}{2}y + 5 \\ 2x - 3y = 8 \end{cases}$

52. $\begin{cases} x = \dfrac{2}{3}y \\ y = 4x + 5 \end{cases}$

53. $\begin{cases} \dfrac{x}{2} + \dfrac{y}{2} = 6 \\ \dfrac{x}{2} - \dfrac{y}{2} = -2 \end{cases}$

54. $\begin{cases} \dfrac{x}{2} - \dfrac{y}{3} = -4 \\ \dfrac{x}{2} + \dfrac{y}{9} = 0 \end{cases}$

55. $\begin{cases} \dfrac{3}{4}x + \dfrac{2}{3}y = 7 \\ \dfrac{3}{5}x - \dfrac{1}{2}y = 18 \end{cases}$

56. $\begin{cases} \dfrac{2}{3}x - \dfrac{1}{4}y = -8 \\ \dfrac{1}{2}x - \dfrac{3}{8}y = -9 \end{cases}$

✍**57.** Tell which method you would use to solve the following system. Why?

$$\begin{cases} y = 3x + 1 \\ 3x + 2y = 12 \end{cases}$$

58. Use a graphing calculator to solve the system

$$\begin{cases} 11x - 20y = 21 \\ -4x + 7y = 21 \end{cases}$$

✍**59.** Two students are solving the equation $2(x + 5) - 6 = 8$ graphically. They both use the INTERSECT feature of the graphing calculator to determine the solution. Their screen is pictured at the right. Blaine states the solution is $x = 2$, $y = 8$, or

(2, 8). David states the solution is 2. Who is correct and *why* is he correct? Explain fully.

60. Solve the system $\begin{cases} y = 3(x - 1)^2 + 2 \\ y = 2x + 1 \end{cases}$
  a. algebraically
  b. graphically

REVIEW

61. Solve the equation $2(4x + 5) - 3 = 23$
  a. algebraically
  b. using the INTERSECT feature of the graphing calculator. Specify the viewing window used if it is different from ZStandard.

62. Solve the equation $x^2 + 6x + 8 = 0$
  a. algebraically
  b. using the ROOT/ZERO or INTERSECT feature of the calculator. Specify the viewing window used if it is different from ZStandard.

63. Solve the equation $3\sqrt{x} + 4 = 22$
  a. algebraically
  b. using the INTERSECT feature of the calculator. Specify the viewing window used if it is different from ZStandard.

12.2 Applications of Systems of Two Equations in Two Variables

The use of a system of equations often makes application problems easier to set up. However, it must be remembered that if two variables are used, then two equations must also be specified. The following steps are helpful when setting up applications that involve systems:

■ **TECHNOLOGY TIP**
If the calculator is to be used to find solutions, we suggest that the variables x and y be used to avoid confusion. Common sense is the biggest help in setting a good window. THINK about what each variable represents and set the viewing window accordingly. Remember, if finding the point of intersection with the calculator becomes too frustrating, algebraic methods may be used, with the calculator providing a check.

Solving Application Problems Using a System of Equations
1. Read the problem several times and analyze the facts. Occasionally, a sketch, chart, or diagram will help in visualizing the problem.
2. Pick different variables to represent the unknown quantities and clearly state what each variable represents.
3. Write two equations involving each of the two variables. This will produce a system of two equations in two variables.
4. Solve the system using the most convenient method.
5. Verify solutions in the words of the problem.

EXAMPLE 1 **Farming** A farmer raises wheat and soybeans on 215 acres. If he wants to plant 31 more acres in wheat than in soybeans, how many acres of each should he plant?

Solution   If w represents the number of acres of wheat and s represents the number of acres of soybeans to be planted, we have

The number of acres planted in wheat	$+$	the number of acres planted in soybeans	$=$	215 acres
w	$+$	s	$=$	215

Since the farmer wants to plant 31 more acres in wheat than in soybeans, we have

The number of acres planted in wheat	$-$	the number of acres planted in soybeans	$=$	31 acres
w	$-$	s	$=$	31

We can now solve the system

1. $\quad \{ w + s = 215$
2. $\quad \{ w - s = 31$

by using the addition method.

$$
\begin{aligned}
w + s &= 215 \\
\underline{w - s} &= \underline{31} \\
2w &= 246 \\
w &= 123 \qquad \text{Divide both sides by 2.}
\end{aligned}
$$

To find s, we substitute 123 for w in Equation 1.

$$
\begin{aligned}
w + s &= 215 \\
\mathbf{123} + s &= 215 \qquad \text{Substitute 123 for } w. \\
s &= 92 \qquad \text{Add } -123 \text{ to both sides.}
\end{aligned}
$$

The farmer should plant 123 acres of wheat and 92 acres of soybeans.

Check: The total acreage planted is $123 + 92$, or 215 acres. The area planted in wheat is 31 acres greater than that planted in soybeans, because $123 - 92 = 31$. ■

EXAMPLE 2   **Lawn care**   An installer of underground irrigation systems wants to cut a 20-foot length of plastic tubing into two pieces. The longer piece is to be 2 feet longer than twice the shorter piece. Find the length of each piece.

Solution   We can let s represent the length of the shorter piece and l represent the length of the longer piece (see Figure 12-4).

FIGURE 12-4

The length of tubing is 20 feet, so we have

The length of the shorter piece	$+$	the length of the longer piece	$=$	20 feet
s	$+$	l	$=$	20

Because the longer piece is 2 feet longer than twice the shorter piece, we have

The length of the longer piece	=	2	$\cdot$	the length of the shorter piece	+	2 feet
l	=	2	$\cdot$	s	+	2

We can use the substitution method to solve the system.

1.
2. $\begin{cases} s + l = 20 \\ l = 2s + 2 \end{cases}$

$$s + 2s + 2 = 20 \qquad \text{Substitute } 2s + 2 \text{ for } l.$$
$$3s + 2 = 20 \qquad \text{Combine like terms.}$$
$$3s = 18 \qquad \text{Subtract 2 from both sides.}$$
$$s = 6 \qquad \text{Divide both sides by 3.}$$

The shorter piece should be 6 feet long.

To find the length of the longer piece, we substitute 6 for s in Equation 1 and solve for l.

$$s + l = 20$$
$$6 + l = 20$$
$$l = 14 \qquad \text{Subtract 6 from both sides.}$$

The longer piece should be 14 feet long.

Check: The sum of 6 and 14 is 20. 14 is 2 more than twice 6.   ■

EXAMPLE 3   **Gardening**   Tom has 150 feet of fencing to enclose a rectangular garden. If the length is to be 5 feet less than 3 times the width, find the area of the garden.

Solution   We can let l represent the length of the garden and w represent the width (see Figure 12-5). The perimeter of a rectangle is two lengths plus two widths, so we have

FIGURE 12-5

2	$\cdot$	the length of the garden	+	2	$\cdot$	the width of the garden	=	150 feet
2	$\cdot$	l	+	2	$\cdot$	w	=	150

Because the length is 5 feet less than 3 times the width,

The length of the garden	=	3	$\cdot$	the width of the garden	−	5 feet
l	=	3	$\cdot$	w	−	5

We can use the substitution method to solve this system.

1.
2. $\begin{cases} 2l + 2w = 150 \\ l = 3w - 5 \end{cases}$

$$2(3w - 5) + 2w = 150 \qquad \text{Substitute } 3w - 5 \text{ for } l \text{ in Equation 1.}$$
$$6w - 10 + 2w = 150 \qquad \text{Remove parentheses.}$$
$$8w - 10 = 150 \qquad \text{Combine like terms.}$$
$$8w = 160 \qquad \text{Add 10 to both sides.}$$
$$w = 20 \qquad \text{Divide both sides by 8.}$$

The width is 20 feet.

To find the length, we substitute 20 for w in Equation 2 and simplify.

$$l = 3w - 5$$
$$= 3(20) - 5$$
$$= 60 - 5$$
$$= 55$$

The dimensions of the rectangle are 55 feet by 20 feet, and the area of a rectangle is given by the formula

$$A = l \cdot w \qquad \text{Area = length times width.}$$

so we have

$$A = 55 \cdot 20$$
$$= 1100$$

The garden covers an area of 1100 square feet.

Check: Because the dimensions of the garden are 55 feet by 20 feet, the perimeter is

$$P = 2l + 2w$$
$$= 2(55) + 2(20)$$
$$= 110 + 40$$
$$= 150$$

It is also true that 55 feet is 5 feet less than 3 times 20 feet. ■

EXAMPLE 4 **Manufacturing**   The setup cost of a machine that mills brass plates is $750. After setup, it costs $0.25 to mill each plate. Management is considering the purchase of a larger machine that can produce the same plate at a cost of $0.20 per plate. If the setup cost of the larger machine is $1200, how many plates would the company have to produce to make the purchase worthwhile?

Solution   We begin by finding the number of plates (called the **break point**) that will cost equal amounts to produce on either machine. We can let c represent the cost of milling p plates. Then we have

The cost of making p plates on machine 1	=	the setup cost of machine 1	+	the cost per plate on machine 1	·	the number of plates to be made
c	=	750	+	0.25	·	p

the cost of making p plates on machine 2	=	the setup cost of machine 2	+	the cost per plate on machine 2	·	the number of plates to be made
c	=	1200	+	0.20	·	p

Because the costs are equal, we can use the substitution method to solve the system

$$\begin{cases} c = 750 + 0.25p \\ c = 1200 + 0.20p \end{cases}$$

$$750 + 0.25p = 1200 + 0.20p \qquad \text{Substitute } 750 + 0.25p \text{ for } c \text{ in the second equation.}$$

$$0.25p = 450 + 0.20p \qquad \text{Subtract 750 from both sides.}$$

$$0.05p = 450 \qquad \text{Subtract } 0.20p \text{ from both sides.}$$

$$p = 9000 \qquad \text{Divide both sides by 0.05.}$$

Current machine		New, larger machine	
$c = 750 + 0.25p$		$c = 1200 + 0.20p$	
p	c	p	c
0	750	0	1,200
1,000	1,000	4,000	2,000
5,000	2,000	12,000	3,600

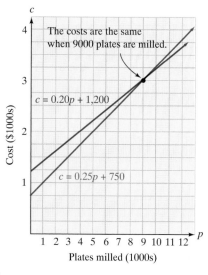

FIGURE 12-6

If 9000 plates are milled, the cost will be the same on either machine. If more than 9000 plates are milled, the cost will be cheaper on the newer machine, and it should be purchased. ∎

EXAMPLE 5   **Investments**   Terri and Juan earned $1150 from a 1-year investment of $15,000 to purchase a large-screen television set. If Terri invested some of the money at 8% interest and Juan invested the rest at 7%, how much did each invest?

Solution   We can let x represent the amount of money invested by Terri and y represent the amount of money invested by Juan. Because the total investment is $15,000, we have

The amount invested by Terri	+	the amount invested by Juan	=	$15,000
x	+	y	=	15,000

The income on x dollars invested at 8% is $0.08x$, the income on y dollars invested at 7% is $0.07y$, and the combined income is $1150, so we have

The income on the 8% investment	+	the income on the 7% investment	=	$1150
$0.08x$	+	$0.07y$	=	1150

Thus, we have the system

1. $\begin{cases} x + y = 15{,}000 \\ 0.08x + 0.07y = 1150 \end{cases}$
2.

To solve the system, we use the addition method.

$$
\begin{array}{rl}
-8x - 8y = -120{,}000 & \quad \text{Multiply both sides of Equation 1 by } -8. \\
\underline{8x + 7y = 115{,}000} & \quad \text{Multiply both sides of Equation 2 by 100.} \\
-y = -5000 & \\
y = 5000 & \quad \text{Multiply both sides by } -1.
\end{array}
$$

To find x, we substitute 5000 for y in Equation 1 and simplify.

$$x + y = 15{,}000$$
$$x + \mathbf{5000} = 15{,}000$$
$$x = 10{,}000 \qquad \text{Subtract 5000 from both sides.}$$

Terri invested \$10,000 and Juan invested \$5000.

Check: \$10,000 + \$5000 = \$15,000
$$0.08(\$10{,}000) = \$800$$
$$0.07(\$5000) = \$350$$

The combined interest is \$1150. ■

EXAMPLE 6   **Boating**   A boat traveled 30 kilometers downstream in 3 hours and made the return trip in 5 hours. Find the speed of the boat in still water.

Solution   We can let s represent the speed of the boat in still water and let c represent the speed of the current. Then the rate of speed of the boat while going downstream is $s + c$. The rate of the boat while going upstream is $s - c$. We can organize the information of the problem as in Table 12-1.

	Distance	= Rate	· Time
Downstream	30	$s + c$	3
Upstream	30	$s - c$	5

TABLE 12-1

Because $d = r \cdot t$, the information in the table gives two equations in two variables.

$$\begin{cases} 30 = 3(s + c) \\ 30 = 5(s - c) \end{cases}$$

After removing parentheses and rearranging terms, we have

1. $\begin{cases} 3s + 3c = 30 \\ 5s - 5c = 30 \end{cases}$
2.

To solve this system by addition, we multiply Equation 1 by 5 and Equation 2 by 3, add the equations, and solve for s.

$$15s + 15c = 150$$
$$\underline{15s - 15c = 90}$$
$$30s = 240$$
$$s = 8 \qquad \text{Divide both sides by 30.}$$

The speed of the boat in still water is 8 kilometers per hour. Check the result. ■

EXAMPLE 7   **Medical technology**   A laboratory technician has one batch of antiseptic that is 40% alcohol and a second batch that is 60% alcohol. She would like to make 8 liters of solution that is 55% alcohol. How many liters of each batch should she use?

Solution   We can let x represent the number of liters to be used from batch 1, let y represent the number of liters to be used from batch 2, and organize the information of the problem as in Table 12-2.

	Fractional part that is alcohol ·	Number of liters of solution =	Number of liters of alcohol
Batch 1	0.40	x	$0.40x$
Batch 2	0.60	y	$0.60y$
Mixture	0.55	8	$0.55(8)$

TABLE 12-2

The information in Table 12-2 provides two equations.

1. $x + y = 8$ — The number of liters of batch 1 plus the number of liters of batch 2 equals the total number of liters in the mixture.

2. $0.40x + 0.60y = 0.55(8)$ — The amount of alcohol in batch 1 plus the amount of alcohol in batch 2 equals the amount of alcohol in the mixture.

We can use addition to solve this system.

$$-40x - 40y = -320 \quad \text{Multiply both sides of Equation 1 by } -40.$$
$$\underline{40x + 60y = 440} \quad \text{Multiply both sides of Equation 2 by 100.}$$
$$20y = 120$$
$$y = 6 \quad \text{Divide both sides by 20.}$$

To find x, we substitute 6 for y in Equation 1 and simplify.

$$x + y = 8$$
$$x + 6 = 8$$
$$x = 2 \quad \text{Subtract 6 from both sides.}$$

The technician should use 2 liters of the 40% solution and 6 liters of the 60% solution. Check the result. ■

EXERCISE 12.2

Use two equations in two variables to solve each problem.

1. Integer problem One integer is twice another, and their sum is 96. Find the integers.

2. Integer problem The sum of two integers is 38, and their difference is 12. Find the integers.

3. Integer problem Three times one integer plus another integer is 29. If the first integer plus twice the second is 18, find the integers.

4. Integer problem Twice one integer plus another integer is 21. If the first integer plus 3 times the second is 33, find the integers.

5. Raising livestock A rancher raises five times as many cows as horses. If he has 168 animals, how many head of cattle does he have?

6. Grass seed mixture A landscaper used 100 pounds of grass seed containing twice as much bluegrass as rye. She adds 15 more pounds of bluegrass to the mixture before seeding a lawn. How many pounds of bluegrass did she use?

7. Buying painting supplies Two partial receipts for paint supplies appear in Illustration 1. How much does each gallon of paint and each brush cost?

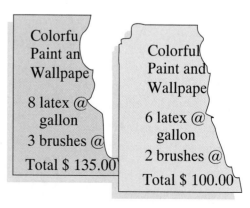

ILLUSTRATION 1

8. **Buying baseball equipment**   One catcher's mitt and 10 outfielder's gloves cost $239.50. How much does each cost if one catcher's mitt and 5 outfielder's gloves cost $134.50?

9. **Buying contact lens cleaner**   Two bottles of contact lens cleaner and three bottles of soaking solution cost $29.40, and three bottles of cleaner and two bottles of soaking solution cost $28.60. Find the cost of each.

10. **Buying clothing**   Two pairs of shoes and four pairs of socks cost $109, and three pairs of shoes and five pairs of socks cost $160. Find the cost of a pair of socks.

11. **Cutting pipe**   A plumber wants to cut the pipe shown in Illustration 2 into two pieces so that one piece is 5 feet longer than the other. How long should each piece be?

25 ft

ILLUSTRATION 2

12. **Cutting lumber**   A carpenter wants to cut a 20-foot board into two pieces so that one piece is four times as long as the other. How long should each piece be?

13. **Splitting the lottery**   Maria and Susan pool their resources to buy several lottery tickets. They agree that Susan should get $50,000 more than Maria because she gave most of the money. If they win $250,000, how much will Susan get?

14. **Figuring inheritances**   According to a will, a man left his older son $10,000 more than twice as much as he left his younger son. If the estate is worth $497,500, how much will the younger son get?

15. **Geometry**   The perimeter of the rectangle shown in Illustration 3 is 110 feet. Find its dimensions.

$l = w + 5$

ILLUSTRATION 3

16. **Geometry**   A rectangle is three times as long as it is wide, and its perimeter is 80 centimeters. Find its dimensions.

17. **Geometry**   A rectangle has a length that is 2 feet more than twice its width. If its perimeter is 34 feet, find its area.

18. **Geometry**   A 50-meter path surrounds the rectangular garden shown in Illustration 4. The width of the garden is two-thirds its length. Find its area.

ILLUSTRATION 4

19. **Choosing a furnace**   A high-efficiency 90+ furnace costs $2250 and costs an average of $412 per year to operate in Rockford, IL. An 80+ furnace costs only $1715 but costs $466 per year to operate. Find the break point.

20. **Making tires**   A company has two molds to form tires. One mold has a setup cost of $600 and the other a setup cost of $1100. The cost to make each tire on the first machine is $15, and the cost to make each tire on the second machine is $13. Find the break point.

21. **Choosing a furnace**   If you intended to live in a house for 7 years, which furnace would you choose in Exercise 19?

22. **Making tires**   If you planned a production run of 500 tires, which mold would you use in Exercise 20?

23. **Investing money**   Bill invested some money at 5% annual interest, and Janette invested some at 7%. If their combined annual interest was $310 on a total investment of $5000, how much did Bill invest?

24. **Investing money**   Peter invested some money at 6% annual interest, and Martha invested some at 12%. If their combined investment was $6000 and their combined annual interest was $540, how much money did Martha invest?

25. **Buying tickets**   Students can buy tickets to a basketball game for $1. However, the admission for nonstudents is $2. If 350 tickets are sold and the total receipts are $450, how many student tickets were sold?

26. **Buying tickets**   If receipts for the movie advertised in Illustration 5 were $720 for an audience of 190 people, how many senior citizens attended?

Admissions: $4
Seniors: $3
Showtimes: 7, 9, 11

ILLUSTRATION 5

27. **Boating** A boat can travel 24 miles downstream in 2 hours and can make the return trip in 3 hours. Find the speed of the boat in still water.

28. **Aviation** With the wind, a plane can fly 3000 miles in 5 hours. Against the same wind, the trip takes 6 hours. Find the airspeed of the plane (the speed in still air).

29. **Aviation** An airplane can fly downwind a distance of 600 miles in 2 hours. However, the return trip against the same wind takes 3 hours. Find the speed of the wind.

30. **Finding the speed of a current** It takes a motorboat 4 hours to travel 56 miles down a river, and takes 3 hours longer to make the return trip. Find the speed of the current.

31. **Mixing chemicals** A chemist has one solution that is 40% alcohol and another that is 55% alcohol. How much of each must she use to make 15 liters of a solution that is 50% alcohol?

32. **Mixing pharmaceuticals** A nurse has a solution that is 25% alcohol and another that is 50% alcohol. How much of each must he use to make 20 liters of a solution that is 40% alcohol?

33. **Mixing nuts** A merchant wants to mix the peanuts with the cashews shown in Illustration 6 to get 48 pounds of mixed nuts to sell at $4 per pound. How many pounds of each should the merchant use?

ILLUSTRATION 6

34. **Mixing peanuts and candy** A merchant wants to mix peanuts worth $3 per pound with jelly beans worth $1.50 per pound to make 30 pounds of a mixture worth $2.10 per pound. How many pounds of each should he use?

35. **Markdown** A set of golf clubs has been marked down 40% to a sale price of $384. Let r represent the retail price and d represent the discount. Then use the following equations to find the original retail price.

36. **Markup** A stereo system retailing at $565.50 has been marked up 45% from wholesale. Let w represent the wholesale cost and m represent the markup. Then use the following equations to find the wholesale cost.

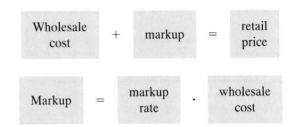

37. **Selling radios** An electronics store put two types of car radios on sale. One model sold for $87, and the other sold for $119. During the sale, the receipts for the 25 radios sold were $2495. How many cheap radios were sold?

38. **Selling ice cream** At a store, ice cream cones cost $0.90 and sundaes cost $1.65. One day the receipts for a total of 148 cones and sundaes were $180.45. How many cones were sold?

39. **Investing money** An investment of $950 at one rate of interest and $1200 at a higher rate together generate an annual income of $205.50. If the investment rates differ by 1%, find the lower rate. (*Hint:* Treat 1% as 0.01).

40. **Motion problem** A man drives for a while at 45 miles per hour. Realizing that he is running late, he increases his speed to 60 miles per hour and completes his 405-mile trip in 8 hours. How long did he drive at 45 miles per hour?

✐41. Which of the problems in Exercises 1–40 could you have done more easily using a single equation with one variable? What makes a solution method "easier"?

REVIEW *In Exercises 42–45, solve each inequality and confirm graphically.*

42. $|3x + 6| < 5$

43. $|3x + 6| \geq 5$

44. $|2x + 5| - 2 > -6$

45. $|2x + 5| - 2 < -6$

12.3 Solving Systems of Three Equations in Three Variables

■ SOLVING SYSTEMS OF THREE EQUATIONS IN THREE VARIABLES ■ PROBLEM SOLVING

■ SOLVING SYSTEMS OF THREE EQUATIONS IN THREE VARIABLES

In this section, we extend the definition of a linear equation to include any equation of the form $ax + by + cz = d$. The solution of a system of three linear equations with three variables is an ordered triple of numbers. For example, the solution of the system

$$\begin{cases} 2x + 3y + 4z = 20 \\ 3x + 4y + 2z = 17 \\ 3x + 2y + 3z = 16 \end{cases}$$

is the ordered triple $(1, 2, 3)$. Verify that the numbers in the triple satisfy each of the three equations in the system.

The graph of an equation of the form $ax + by + cz = d$ is a flat surface called a **plane.** A system of three linear equations in three variables is consistent or inconsistent, depending on how the three planes corresponding to the three equations intersect. Figure 12-7 illustrates some of the possibilities.

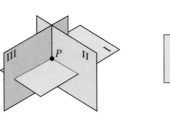

The three planes intersect at a single point P: One solution

(a)

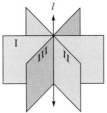

The three planes have a line l in common: An infinite number of solutions

(b)

The three planes have no point in common: No solutions

(c)

FIGURE 12-7

To solve a system of three linear equations in three variables, we follow these steps.

Steps to Follow When Solving Three Equations in Three Variables

1. Pick any two equations and eliminate a variable.
2. Pick a different pair of equations and eliminate the same variable.
3. Solve the resulting pair of two equations in two variables.
4. To find the value of the third variable, substitute the values of the two variables found in step 3 into any equation containing all three variables and solve the equation.
5. Check the solution in all three of the original equations.

Example 1 discusses a consistent system of three equations in three variables. Example 2 discusses an inconsistent system.

EXAMPLE 1 Solve the system $\begin{cases} 2x + y + 4z = 12 \\ x + 2y + 2z = 9 \\ 3x - 3y - 2z = 1 \end{cases}$

Solution We are given the system

1. $\begin{cases} 2x + y + 4z = 12 \\ \end{cases}$
2. $\begin{cases} x + 2y + 2z = 9 \\ \end{cases}$
3. $\begin{cases} 3x - 3y - 2z = 1 \end{cases}$

If we pick Equations 2 and 3 and add them, the variable z is eliminated:

2. $x + 2y + 2z = 9$
3. $\underline{3x - 3y - 2z = 1}$
4. $4x - y \qquad = 10$

We now pick a different pair of equations (Equations 1 and 3) and eliminate z again. If each side of Equation 3 is multiplied by 2 and the resulting equation is added to Equation 1, z is again eliminated:

1. $2x + y + 4z = 12$
$\underline{6x - 6y - 4z = 2}$
5. $8x - 5y \qquad = 14$

Equations 4 and 5 form a system of two equations in two variables:

4. $\begin{cases} 4x - y = 10 \\ \end{cases}$
5. $\begin{cases} 8x - 5y = 14 \end{cases}$

To solve this system, we multiply Equation 4 by -5 and add the resulting equation to Equation 5 to eliminate y:

$-20x + 5y = -50$
5. $\underline{\quad 8x - 5y = \quad 14}$
$-12x \qquad = -36$
$\qquad\qquad x = 3 \qquad$ Divide both sides by -12.

To find y, we substitute 3 for x in any equation containing x and y (such as Equation 5) and solve for y:

5. $\quad 8x - 5y = 14$
$\quad 8(3) - 5y = 14 \qquad$ Substitute 3 for x.
$\quad 24 - 5y = 14 \qquad$ Simplify.
$\qquad -5y = -10 \qquad$ Subtract 24 from both sides.
$\qquad\quad y = 2 \qquad$ Divide both sides by -5.

To find z, we substitute 3 for x and 2 for y in an equation containing x, y, and z (such as Equation 1) and solve for z:

1. $\quad 2x + y + 4z = 12$
$\quad 2(3) + 2 + 4z = 12 \qquad$ Substitute 3 for x and 2 for y.
$\qquad 8 + 4z = 12 \qquad$ Simplify.
$\qquad\qquad 4z = 4 \qquad$ Subtract 8 from both sides.
$\qquad\qquad z = 1 \qquad$ Divide both sides by 4.

The solution of the system is $(x, y, z) = (3, 2, 1)$. Verify that these values satisfy each equation in the original system. ∎

EXAMPLE 2   Solve the system $\begin{cases} 2x + y - 3z = -3 \\ 3x - 2y + 4z = 2 \\ 4x + 2y - 6z = -7 \end{cases}$

Solution   We are given the system of equations

1. $\begin{cases} 2x + y - 3z = -3 \\ 2. \quad 3x - 2y + 4z = 2 \\ 3. \quad 4x + 2y - 6z = -7 \end{cases}$

We can multiply Equation 1 by 2 and add the resulting equation to Equation 2 to eliminate y:

$$4x + 2y - 6z = -6$$
2. $\underline{3x - 2y + 4z = \quad 2}$
4. $7x \qquad - 2z = -4$

We now add Equations 2 and 3 to eliminate y again:

2. $3x - 2y + 4z = \quad 2$
3. $\underline{4x + 2y - 6z = -7}$
5. $7x \qquad - 2z = -5$

Equations 4 and 5 form the system

4. $\begin{cases} 7x - 2z = -4 \\ 5. \quad 7x - 2z = -5 \end{cases}$

Because $7x - 2z$ cannot equal both -4 and -5, this system is inconsistent. Thus, it has no solution.   ■

■ PROBLEM SOLVING

EXAMPLE 3   **Integer problem**   The sum of three integers is 2. The third integer is 2 greater than the second and 17 greater than the first. Find the integers.

Solution   Let a, b, and c represent the integers. Because their sum is 2, we know that

$$a + b + c = 2$$

Because the third integer is 2 greater than the second and 17 greater than the first, we know that

$$c - b = 2$$
$$c - a = 17$$

Put these three equations together to form a system of three equations in three variables:

1. $\begin{cases} a + b + c = 2 \\ 2. \qquad -b + c = 2 \\ 3. \quad -a \qquad + c = 17 \end{cases}$

We add Equations 1 and 2 to get Equation 4:

4. $a + 2c = 4$

Because Equations 3 and 4 have only the variables a and c, Equations 3 and 4 form a system of two equations in two variables.

3. $\begin{cases} -a + c = 17 \\ 4. \quad a + 2c = 4 \end{cases}$

We add Equations 3 and 4 to get the equation

$$3c = 21$$
$$c = 7$$

We substitute 7 for c in Equation 4 to find a:

4. $a + 2c = 4$
$\qquad a + 2(7) = 4$
$\qquad\quad a + 14 = 4$ Simplify.
$\qquad\qquad\quad a = -10$ Add -14 to both sides.

We substitute 7 for c in Equation 2 to find b:

2. $-b + c = 2$
$\qquad -b + 7 = 2$
$\qquad\quad -b = -5$ Subtract 7 from both sides.
$\qquad\qquad b = 5$ Divide both sides by -1.

Thus, the integers are -10, 5, and 7. These integers have a sum of 2; 7 is 2 greater than 5; and 7 is 17 greater than -10.   ■

EXAMPLE 4   **Manufacturing**   A company manufactures three types of hammers—good, better, and best. The cost of manufacturing each type is $4, $6, and $7, respectively, and the hammers sell for $6, $9, and $12. Each day, the cost for manufacturing 100 hammers is $520, and the daily revenue from their sales is $810. How many of each type are manufactured?

Solution   We can let x represent the number of good hammers manufactured, y represent the number of better hammers manufactured, and z represent the number of best hammers manufactured. Then

The number of good hammers	+	the number of better hammers	+	the number of best hammers	=	the total number of hammers
x	+	y	+	z	=	100

The cost of good hammers	+	the cost of better hammers	+	the cost of best hammers	=	the total cost
$4x$	+	$6y$	+	$7z$	=	520

The revenue from good hammers	+	the revenue from better hammers	+	the revenue from best hammers	=	the total revenue
$6x$	+	$9y$	+	$12z$	=	810

These three equations give the system

1. $\begin{cases} x + \ y + \quad z = 100 \\ 4x + 6y + \ \ 7z = 520 \\ 6x + 9y + 12z = 810 \end{cases}$
2.
3.

If we multiply Equation 1 by -7 and add the result to Equation 2, we get

$$-7x - 7y - 7z = -700$$
$$\underline{4x + 6y + 7z = 520}$$
4. $-3x - y = -180$

We can eliminate z again if we multiply Equation 1 by -12 and add the result to Equation 3.

$$-12x - 12y - 12z = -1200$$
$$\underline{6x + 9y + 12z = 810}$$
5. $-6x - 3y = -390$

If we multiply Equation 4 by -3 and add it to Equation 5, we get

$$9x + 3y = 540$$
$$\underline{-6x - 3y = -390}$$
$$3x = 150$$
$$x = 50 \qquad \text{Divide both sides by 3.}$$

To find y, we substitute 50 for x in Equation 4:

4. $-3x - y = -180$

$-3(\mathbf{50}) - y = -180 \qquad$ Substitute 50 for x.

$-150 - y = -180$

$-y = -30 \qquad$ Add 150 to both sides.

$y = 30 \qquad$ Divide both sides by -1.

To find z, we substitute 50 for x and 30 for y in Equation 1:

1. $x + y + z = 100$

$\mathbf{50} + \mathbf{30} + z = 100$

$z = 20 \qquad$ Subtract 80 from both sides.

The company manufactures 50 good hammers, 30 better hammers, and 20 best hammers each day. ∎

EXERCISE 12.3

VOCABULARY Fill in each blank to make a true statement.

1. The graph of an equation in the form $ax + by + cz = d$ is a flat surface called a _____.

2. When three planes intersect at a single point, the system has ____ solution(s).

3. If three planes intersect in a line, the system has a(n) _____ number of solutions.

4. When three planes are parallel, the system has ___ solutions.

PRACTICE In Exercises 5–16, solve each system.

5. $\begin{cases} x + y + z = 4 \\ 2x + y - z = 1 \\ 2x - 3y + z = 1 \end{cases}$

6. $\begin{cases} x + y + z = 4 \\ x - y + z = 2 \\ x - y - z = 0 \end{cases}$

7. $\begin{cases} 2x + 2y + 3z = 10 \\ 3x + y - z = 0 \\ x + y + 2z = 6 \end{cases}$

8. $\begin{cases} x - y + z = 4 \\ x + 2y - z = -1 \\ x + y - 3z = -2 \end{cases}$

9. $\begin{cases} x + y + 2z = 7 \\ x + 2y + z = 8 \\ 2x + y + z = 9 \end{cases}$

10. $\begin{cases} x + 2y + 2z = 10 \\ 2x + y + 2z = 9 \\ 2x + 2y + z = 11 \end{cases}$

11. $\begin{cases} 2x + y - z = 1 \\ x + 2y + 2z = 2 \\ 4x + 5y + 3z = 3 \end{cases}$

12. $\begin{cases} 4x + 3z = 4 \\ 2y - 6z = -1 \\ 8x + 4y + 3z = 9 \end{cases}$

13. $\begin{cases} 2x + 3y + 4z - 6 = 0 \\ 2x - 3y - 4z + 4 = 0 \\ 4x + 6y + 8z - 12 = 0 \end{cases}$

14. $\begin{cases} x - 3y + 4z - 2 = 0 \\ 2x + y + 2z - 3 = 0 \\ 4x - 5y + 10z - 7 = 0 \end{cases}$

15. $\begin{cases} x + \dfrac{1}{3}y + z = 13 \\[2mm] \dfrac{1}{2}x - y + \dfrac{1}{3}z = -2 \\[2mm] x + \dfrac{1}{2}y - \dfrac{1}{3}z = 2 \end{cases}$

16. $\begin{cases} x - \dfrac{1}{5}y - z = 9 \\[2mm] \dfrac{1}{4}x + \dfrac{1}{5}y - \dfrac{1}{2}z = 5 \\[2mm] 2x + y + \dfrac{1}{6}z = 12 \end{cases}$

APPLICATIONS   In Exercises, 17–28, solve each problem.

17. Integer problem   The sum of three integers is 18. The third integer is four times the second, and the second integer is 6 more than the first. Find the integers.

18. Integer problem   The sum of three integers is 48. If the first integer is doubled, the sum is 60. If the second integer is doubled, the sum is 63. Find the integers.

19. Geometry problem   The sum of the angles in any triangle is 180°. In triangle ABC, angle A is 100° less than the sum of angles B and C, and angle C is 40° less than twice angle B. Find each angle.

20. Geometry problem   The sum of the angles of any four-sided figure is 360°. In the quadrilateral shown in Illustration 1, angle A = angle B, angle C is 20° greater than angle A, and angle D = 40°. Find the angles.

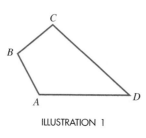

ILLUSTRATION 1

21. Nutritional planning   According to Table 1, a unit of food A contains 1 gram of fat, 1 gram of carbohydrate, and 2 grams of protein. Food B contains 2 grams of fat, 1 gram of carbohydrate, and 1 gram of protein. Food C contains 2 grams of fat, 1 gram of carbohydrate, and 2 grams of protein. How many units of each food must be used to provide exactly 11 grams of fat, 6 grams of carbohydrate, and 10 grams of protein?

Food	Fat	Carbohydrates	Protein
A	1	1	2
B	2	1	1
C	2	1	2

TABLE 1

22. Nutritional planning   As shown in Table 2, a unit of food A contains 2 grams of fat, 1 gram of carbohydrate, and 2 grams of protein. Food B contains 3 grams of fat, 2 grams of carbohydrate, and 1 gram of protein. Food C contains 1 gram of fat, 1 gram of carbohydrate, and 2 grams of protein. How many units of each food must be used to provide exactly 14 grams of fat, 9 grams of carbohydrate, and 9 grams of protein?

Food	Fat	Carbohydrates	Protein
A	2	1	2
B	3	2	1
C	1	1	2

TABLE 2

23. Making statues   An artist makes three types of ceramic statues at a monthly cost of $650 for 180 statues. The manufacturing costs for the three types are $5, $4, and $3. If the statues sell for $20, $12, and $9, respectively, how many of each type should be made to produce $2100 in monthly revenue?

24. Manufacturing footballs   A factory manufactures three types of footballs at a monthly cost of $2425 for 1125 footballs. The manufacturing costs for the three types of footballs are $4, $3, and $2. These footballs sell for $16, $12, and $10, respectively. How many of each type are manufactured if the monthly profit is $9275? (*Hint:* Profit = income − cost.)

25. Concert tickets   Tickets for a concert cost $5, $3, and $2. Twice as many $5 tickets were sold as $2 tickets. The receipts for 750 tickets were $2625. How many of each price ticket were sold?

26. Mixing nuts   The owner of a candy store wants to mix some peanuts worth $3 per pound, some cashews worth $9 per pound, and some Brazil nuts worth $9 per pound to get 50 pounds of a mixture that will sell for $6 per pound. She used 15 fewer pounds of cashews than peanuts. How many pounds of each did she use?

27. Chainsaw sculpting   A northwoods sculptor carves three types of statues with a chainsaw. The time required for carving, sanding, and painting a totem pole, a bear, and a deer are shown in Table 3. How many of each should be produced to use all available labor hours?

	Totem pole	Bear	Deer	Time available
Carving	2 hours	2 hours	1 hour	14 hours
Sanding	1 hour	2 hours	2 hours	15 hours
Painting	3 hours	2 hours	2 hours	21 hours

TABLE 3

28. Making clothing A clothing manufacturer makes coats, shirts, and slacks. The time required for cutting, sewing, and packaging each item are shown in Table 4.

How many of each should be made to use all available labor hours?

	Coats	Shirts	Slacks	Time available
Cutting	20 min	15 min	10 min	115 hr
Sewing	60 min	30 min	24 min	280 hr
Packaging	5 min	12 min	6 min	65 hr

TABLE 4

REVIEW

29. Does the equation $y = |x + 2|$ determine y to be a function of x? Why or why not? If you answered yes, state the domain and the range.

30. Find the slope of the graph of $x = \dfrac{2y - 3}{2}$.

31. Write the equation of the line that has a slope of $\dfrac{3}{5}$ and passes through the point $(1, 2)$.

32. Express as a formula: y varies directly as the square of x and inversely with z.

33. If $f(x) = -x^2 - 5x + 2$, find $f(-3)$.

12.4 Using Matrices to Solve Systems of Equations

■ BASIC DEFINITIONS ■ GAUSSIAN ELIMINATION ■ SYSTEMS WITH MORE EQUATIONS THAN VARIABLES ■ SYSTEMS WITH MORE VARIABLES THAN EQUATIONS

■ BASIC DEFINITIONS

Another method for solving systems of equations depends on the coefficients of the variables and the constants. We will form a rectangular array of numbers using these coefficients and constants to solve systems of equations. This rectangular array is called a **matrix.**

Matrix
A matrix is any rectangular array of numbers.

Some examples of matrices are

$$A = \begin{bmatrix} 1 & 2 & 3 \\ 4 & 5 & 6 \end{bmatrix} \quad B = \begin{bmatrix} 1 & 2 \\ 3 & 4 \\ 5 & 6 \end{bmatrix} \quad C = \begin{bmatrix} 2 & 4 & 6 \\ 8 & 10 & 12 \\ 14 & 16 & 18 \end{bmatrix}$$

The numbers in each matrix are called its **elements.** Because matrix A has two rows and three columns, it is called a 2×3 matrix (read "2 by 3" matrix). Matrix B is a 3×2 matrix, because the matrix has three rows and two columns. Matrix C is a 3×3 matrix (three rows and three columns). Any matrix with the same number of rows and columns is called a **square matrix.** Matrix C is an example of a square matrix.

To show how to use matrices to solve systems of linear equations, we consider the following system

$$\begin{cases} x - 2y - z = 6 \\ 2x + 2y - z = 1 \\ -x - y + 2z = 1 \end{cases}$$

which can be represented by the following matrix, called an **augmented matrix:**

$$\begin{bmatrix} 1 & -2 & -1 & \vdots & 6 \\ 2 & 2 & -1 & \vdots & 1 \\ -1 & -1 & 2 & \vdots & 1 \end{bmatrix}$$

The 3×3 matrix to the left of the dashed line, called the **coefficient matrix,** is determined by the coefficients of x, y, and z in the equations of the system. The 3×1 matrix to the right of the dashed line is determined by the constants in the equations. Each row of the augmented matrix represents exactly one equation of the system:

$$\begin{bmatrix} 1 & -2 & -1 & \vdots & 6 \\ 2 & 2 & -1 & \vdots & 1 \\ -1 & -1 & 2 & \vdots & 1 \end{bmatrix} \leftrightarrow \begin{cases} x - 2y - z = 6 \\ 2x + 2y - z = 1 \\ -x - y + 2z = 1 \end{cases}$$

■ GAUSSIAN ELIMINATION

To solve a system by **Gaussian elimination,** we transform the augmented matrix into the following matrix that has all 0s below its main diagonal, formed by elements a, e, and h.

$$\begin{bmatrix} a & b & c & \vdots & d \\ 0 & e & f & \vdots & g \\ 0 & 0 & h & \vdots & i \end{bmatrix} \qquad (a, b, c, \dots, i \text{ are real numbers.})$$

We can often write a matrix in this form, called **triangular form,** by using three operations called **elementary row operations.**

Elementary Row Operations
1. Any two rows of a matrix can be interchanged.
2. Any row of a matrix can be multiplied by a nonzero constant.
3. Any row of a matrix can be changed by adding a constant multiple of another row to it.

- A type 1 row operation corresponds to interchanging two equations of the system.
- A type 2 row operation corresponds to multiplying both sides of an equation by a nonzero constant.
- A type 3 row operation corresponds to adding a multiple of one equation to another.

None of these operations will change the solution of the given system of equations.

After we have written the matrix in triangular form, we can solve the corresponding system of equations by a substitution process, as shown in Example 1.

EXAMPLE 1   Solve the system $\begin{cases} x - 2y - z = 6 \\ 2x + 2y - z = 1 \\ -x - y + 2z = 1 \end{cases}$

Solution   We can represent the system with the following augmented matrix:

$$\begin{bmatrix} 1 & -2 & -1 & \vdots & 6 \\ 2 & 2 & -1 & \vdots & 1 \\ -1 & -1 & 2 & \vdots & 1 \end{bmatrix}$$

To get 0s under the 1 in the first column, we use a type 3 row operation twice:

Multiply row 1 by -2 and add to row 2. Multiply row 1 by 1 and add to row 3.

$$\begin{bmatrix} 1 & -2 & -1 & | & 6 \\ 2 & 2 & -1 & | & 1 \\ -1 & -1 & 2 & | & 1 \end{bmatrix} \approx \begin{bmatrix} 1 & -2 & -1 & | & 6 \\ 0 & 6 & 1 & | & -11 \\ -1 & -1 & 2 & | & 1 \end{bmatrix} \approx \begin{bmatrix} 1 & -2 & -1 & | & 6 \\ 0 & 6 & 1 & | & -11 \\ 0 & -3 & 1 & | & 7 \end{bmatrix}$$

The symbol "$\approx$" is read as "is row equivalent to." Each of the matrices above represents a system of equations, and they are all equivalent.

To get a 0 under the 6 in the second column of the last matrix, we use another type 3 row operation:

Multiply row 2 by $\frac{1}{2}$ and add to row 3.

$$\begin{bmatrix} 1 & -2 & -1 & | & 6 \\ 0 & 6 & 1 & | & -11 \\ 0 & -3 & 1 & | & 7 \end{bmatrix} \approx \begin{bmatrix} 1 & -2 & -1 & | & 6 \\ 0 & 6 & 1 & | & -11 \\ 0 & 0 & \frac{3}{2} & | & \frac{3}{2} \end{bmatrix}$$

Finally, we use a type 2 row operation:

Multiply row 3 by $\frac{2}{3}$.

$$\begin{bmatrix} 1 & -2 & -1 & | & 6 \\ 0 & 6 & 1 & | & -11 \\ 0 & 0 & \frac{3}{2} & | & \frac{3}{2} \end{bmatrix} \approx \begin{bmatrix} 1 & -2 & -1 & | & 6 \\ 0 & 6 & 1 & | & -11 \\ 0 & 0 & 1 & | & 1 \end{bmatrix}$$

The final matrix represents the system of equations

1. $\begin{cases} x - 2y - z = 6 \\ \textbf{2.} \quad 0x + 6y + z = -11 \\ \textbf{3.} \quad 0x + 0y + z = 1 \end{cases}$

From equation 3, we can read that $z = 1$. To find y, we substitute 1 for z in equation 2 and solve for y:

2. $6y + z = -11$
$6y + \mathbf{1} = -11$ Substitute 1 for z.
$6y = -12$ Subtract 1 from both sides.
$y = -2$ Divide both sides by 6.

■ TECHNOLOGY TIP
The solution $(3, -2, 1)$ can be verified using the STOre feature of the calculator.

```
3→X: -2→Y:1→Z:X-2
Y-Z
              6
2X+2Y-Z
              1
-X-Y+2Z
              1
```

Thus, $y = -2$. To find x, we substitute 1 for z and -2 for y in equation 1 and solve for x:

1. $x - 2y - z = 6$
$x - 2(-2) - \mathbf{1} = 6$ Substitute 1 for z and -2 for y.
$x + 4 - 1 = 6$ $(-2)(-2) = +4$.
$x + 3 = 6$ Simplify.
$x = 3$ Subtract 3 from both sides.

Thus, $x = 3$. The solution to the given system is $(3, -2, 1)$. Verify that this triple satisfies each equation of the original system. ■

EXAMPLE 2 Solve the system $\begin{cases} x - 2y - z = 6 \\ 2x + 2y - z = 1 \\ -x - y + 2z = 1 \end{cases}$ using the calculator to perform matrix operations.

Solution   This is the same system solved in Example 1. The augmented matrix associated with the system is

$$\begin{bmatrix} 1 & -2 & -1 & \vdots & 6 \\ 2 & 2 & -1 & \vdots & 1 \\ -1 & -1 & 2 & \vdots & 1 \end{bmatrix}$$

Enter this matrix in your calculator as matrix *A* using the matrix edit feature of your calculator located under the MATRIX menu. Pressing the ENTER key after each entry will automatically move the cursor to the next entry location, or you may use the arrow keys to move to the desired location on the screen.

row 2, column 1 entry

Return to the home screen (2nd, QUIT) and display your matrix *A* (MATRIX, 1:[A], ENTER).

We will now follow the steps in Example 1, replacing the analytic arithmetic computations with parallel calculator computations. The indicated matrix operations are located under the MATH submenu of the MATRIX menu. Both the analytical directions and the corresponding calculator command will be given.

To get 0s under the 1 in the first column, we do a type 3 row operation twice.

Analytical operation: $-2 \cdot$ row 1 + row 2 *replaces* row 2

Calculator operation: *row + (−2, [A], 1, 2) STO▶ [B]

multiplier
row being multiplied
added row (and the row being replaced)
new matrix location
STOre feature

 WARNING!   Row operations do not change the matrix stored in memory! The new matrix must be stored each time row operations are performed. The new matrix was stored in [B] to preserve the original matrix.

Analytical operation: $1 \cdot$ row 1 + row 3 *replaces* row 3

Calculator operation: *row + (1, [B], 1, 3) STO▶ [B]

To get a 0 under the 6 in the second column of the last matrix, we use another type 3 row operation.

Analytical operation: $\frac{1}{2} \cdot$ row 2 + row 3 *replaces* row 3

Calculator operation: *row $+ \left(\frac{1}{2}, [B], 2, 3\right)$ STO➤ [B]

Finally, we use a type 2 row operation.

Analytical operation: $1 \cdot$ row 1 + row 3 *replaces* row 3

Calculator operation: *row $+ (1, [B], 1, 3)$ STO➤ [B]

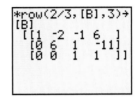

The final matrix represents the system of equations:

1. $x - 2y - z = 6$

2. $6y + z = -11$

3. $z = 1$

From $z = 1$ in equation 3, we back substitute into equation 2 to find $y = -2$. Using both $z = 1$ and $y = -2$, we back substitute into equation 1 to find $x = 3$. Thus, the solution to the system is the ordered triple $(3, -2, 1)$. ■

Note: Not all of the available row operations were used. The following row operations indicate the order in which information is to be entered.

rowSwap (matrix, row 1, row 2)	Swaps row 1 and row 2.
row+ (matrix, row 1, row 2)	Adds row 1 and row 2 and stores result in row 2.
***row** (value, matrix, row)	Multiplies a row by the indicated value.
***row+** (value, matrix, row 1, row 2)	Multiplies the matrix row 1 by the indicated value, adds this product to row 2, and stores the result in row 2.

■ SYSTEMS WITH MORE EQUATIONS THAN VARIABLES

We can use matrices to solve systems with more equations than variables.

EXAMPLE 3   Solve the system $\begin{cases} x + y = -1 \\ 2x - y = 7 \\ -x + 2y = -8 \end{cases}$

Solution This system, with three equations and two variables, can be represented by a 3×3 augmented matrix:

$$\left[\begin{array}{rr|r} 1 & 1 & -1 \\ 2 & -1 & 7 \\ -1 & 2 & -8 \end{array} \right]$$

To get 0s under the 1 in the first column, we do a type 3 row operation twice:

$$\left[\begin{array}{rr|r} 1 & 1 & -1 \\ 2 & -1 & 7 \\ -1 & 2 & -8 \end{array} \right] \overset{\text{Multiply row 1 by } -2}{\underset{\text{and add to row 2.}}{\approx}} \left[\begin{array}{rr|r} 1 & 1 & -1 \\ 0 & -3 & 9 \\ -1 & 2 & -8 \end{array} \right] \overset{\text{Multiply row 1 by 1}}{\underset{\text{and add to row 3.}}{\approx}} \left[\begin{array}{rr|r} 1 & 1 & -1 \\ 0 & -3 & 9 \\ 0 & 3 & -9 \end{array} \right]$$

We can do other row operations to get

$$\left[\begin{array}{rr|r} 1 & 1 & -1 \\ 0 & -3 & 9 \\ 0 & 3 & -9 \end{array} \right] \overset{\text{Add row 2 to}}{\underset{\text{to row 3.}}{\approx}} \left[\begin{array}{rr|r} 1 & 1 & -1 \\ 0 & -3 & 9 \\ 0 & 0 & 0 \end{array} \right] \overset{\text{Multiply row 2}}{\underset{\text{by } -\frac{1}{3}.}{\approx}} \left[\begin{array}{rr|r} 1 & 1 & -1 \\ 0 & 1 & -3 \\ 0 & 0 & 0 \end{array} \right]$$

The final matrix represents the system

$$\begin{cases} x + y = -1 \\ 0x + y = -3 \\ 0x + 0y = 0 \end{cases}$$

The third equation can be discarded, because $0x + 0y = 0$ for all x and y. From the second equation, we can read that $y = -3$. To find x, we substitute -3 for y in the first equation and solve for x:

$$x + y = -1$$
$$x - 3 = -1 \qquad \text{Substitute } -3 \text{ for } y.$$
$$x = 2 \qquad \text{Add 3 to both sides.}$$

The solution is $(2, -3)$. Verify that this solution satisfies all three equations of the original system. ■

If the last row of the final matrix in Example 3 had been of the form $0x + 0y = k$, where $k \neq 0$, the system would have no solution. No values of x and y could make the expression $0x + 0y$ equal to a nonzero constant k.

■ SYSTEMS WITH MORE VARIABLES THAN EQUATIONS

We can also solve many systems with more variables than equations.

EXAMPLE 4   Solve the system $\begin{cases} x + y - 2z = -1 \\ 2x - y + z = -3 \end{cases}$

Solution   This system has two equations and three variables. We can start to solve it by doing a type 3 row operation to get a 0 under the 1 in the first column.

$$\begin{bmatrix} 1 & 1 & -2 & \vdots & -1 \\ 2 & -1 & 1 & \vdots & -3 \end{bmatrix} \approx \begin{bmatrix} 1 & 1 & -2 & \vdots & -1 \\ 0 & -3 & 5 & \vdots & -1 \end{bmatrix}$$

Multiply row 1 by -2
and add to row 2.

We then do a type 2 row operation:

$$\begin{bmatrix} 1 & 1 & -2 & \vdots & -1 \\ 0 & -3 & 5 & \vdots & -1 \end{bmatrix} \approx \begin{bmatrix} 1 & 1 & -2 & \vdots & -1 \\ 0 & 1 & -\frac{5}{3} & \vdots & \frac{1}{3} \end{bmatrix}$$

Multiply row 2 by $-\frac{1}{3}$.

The final matrix represents the system

$$\begin{cases} x + y - 2z = -1 \\ \quad\quad y - \dfrac{5}{3}z = \dfrac{1}{3} \end{cases}$$

We add $\frac{5}{3}z$ to both sides of the second equation to obtain

$$y = \frac{1}{3} + \frac{5}{3}z$$

We substitute $\frac{1}{3} + \frac{5}{3}z$ for y in the first equation and simplify to get

$$x + y - 2z = -1$$

$$x + \frac{1}{3} + \frac{5}{3}z - 2z = -1 \qquad \text{Substitute } \tfrac{1}{3} + \tfrac{5}{3}z \text{ for } y.$$

$$x + \frac{1}{3} - \frac{1}{3}z = -1 \qquad \tfrac{5}{3}z - 2z = \tfrac{5}{3}z - \tfrac{6}{3}z = -\tfrac{1}{3}z.$$

$$x - \frac{1}{3}z = -\frac{4}{3} \qquad \text{Subtract } \tfrac{1}{3} \text{ from both sides.}$$

$$x = -\frac{4}{3} + \frac{1}{3}z \qquad \text{Add } \tfrac{1}{3}z \text{ to both sides.}$$

A solution to this system must have the form

$$\left(-\frac{4}{3} + \frac{1}{3}z, \frac{1}{3} + \frac{5}{3}z, z \right)$$

for all values of z. This system has an infinite number of solutions, a different one for each value of z. For example,

- If $z = 0$, the corresponding solution is $\left(-\frac{4}{3}, \frac{1}{3}, 0\right)$.
- If $z = 1$, the corresponding solution is $(-1, 2, 1)$.

Verify that both of these solutions satisfy each equation of the given system. ■

EXERCISE 12.4

VOCABULARY *In Exercises 1–5, fill in the blanks to make a true statement.*

1. A rectangular array of numbers is called a _____.

2. The numbers in the rectangular array are called its _____.

3. A _____ matrix has the same number of rows and columns.

4. An _____ matrix includes the constant values of the system, whereas a _____ matrix only includes the coefficients.

5. _____ is used to transform an augmented matrix into a matrix that has all 0s below its main diagonal.

CONCEPTS

6. Write the coefficient matrix for the system:

$$2x + y + 2z = 1$$
$$x - 2y + 3z = 4$$
$$2x - 3y + z = 0$$

7. Write the augmented matrix for the system in Exercise 6.

✍ 8. Explain why any two rows of a matrix can be interchanged.

✍ 9. Explain why you can multiply any row of a matrix by a nonzero constant. Support your reason with a specific property that was introduced earlier in the text.

✍ 10. Explain why you **cannot** multiply any column of a matrix by a nonzero constant.

✍ 11. Explain how to check the solution of a system of equations.

✍ 12. Explain how to perform a type 3 row operation.

PRACTICE *In Exercises 13–24, use matrices to solve each system of equations. Row operations may be done either by hand or with the calculator. Verify each solution with the calculator.*

13. $x + y = 2$
    $x - y = 0$

14. $x + y = 3$
    $x - y = -1$

15. $x + 2y = -4$
    $2x + y = 1$

16. $2x - 3y = 16$
    $-4x + y = -22$

17. $3x + 4y = -12$
    $9x - 2y = 6$

18. $5x - 4y = 10$
    $x - 7y = 2$

19. $x + y + z = 6$
    $x + 2y + z = 8$
    $x + y + 2z = 9$

20. $x - y + z = 2$
    $x + 2y - z = 6$
    $2x - y - z = 3$

21. $2x + y + 3z = 3$
$-2x - y + z = 5$
$4x - 2y + 2z = 2$

22. $3x + 2y + z = 8$
$6x - y + 2z = 16$
$-9x + y - z = -20$

23. $3x - 2y + 4z = 4$
$x + y + z = 3$
$6x - 2y - 3z = 10$

24. $2x + 3y - z = -8$
$x - y - z = -2$
$-4x + 3y + z = 6$

In Exercises 25–32, use matrices to solve each system of equations. Row operations may be done either by hand or with the calculator. Verify each solution with the calculator. If a system has no solution, so indicate.

25. $x + y = 3$
$3x - y = 1$
$2x + y = 4$

26. $x - y = -5$
$2x + 3y = 5$
$x + y = 1$

27. $2x - y = 4$
$x + 3y = 2$
$-x - 4y = -2$

28. $3x - 2y = 5$
$x + 2y = 7$
$-3x - y = -11$

29. $2x + y = 7$
$x - y = 2$
$-x + 3y = -2$

30. $3x - y = 2$
$-6x + 3y = 0$
$-x + 2y = -4$

31. $x + 3y = 7$
$x + y = 3$
$3x + y = 5$

32. $x + y = 3$
$x - 2y = -3$
$x - y = 1$

In Exercises 33–36, use matrices to solve each system of equations.

33. $x + 2y + 3z = -2$
$-x - y - 2z = 4$

34. $2x - 4y + 3z = 6$
$-4x + 6y + 4z = -6$

35. $x - y = 1$
$y + z = 1$
$x + z = 2$

36. $x + z = 1$
$x + y = 2$
$2x + y + z = 3$

APPLICATIONS

37. Chemistry   The atomic number of lead is four more than three times the atomic number of iron. If the atomic number of lead is decreased by twice the atomic number of iron, the result is the atomic number of zinc, which is 30. Find the atomic numbers of lead and iron.

38. Geometry   In triangle ABC, angle B is 20 more than the difference of angles C and A. Angle C is 20 less than the sum of A and B. If the sum of the angles in the triangle is 180 degrees, find the number of degrees in each angle.

REVIEW

39. If $f(x) = 3x^2 - 2x$ and $g(x) = 4x - 2$, find:
   a. $f \circ g$
   b. fg
   c. $f - g$

40. Given $f(x) = 3x^2 - 2x$,
   a. determine if the function is one-to-one and if it is, find its inverse.
   b. determine the coordinates of the vertex
   c. determine the coordinates of the x-intercepts
   d. determine the coordinate of the y-intercept
   e. graph the function on graph paper
   f. is the function increasing or decreasing on the interval $[1, \infty)$?

41. Explain the difference between finding the x-intercepts of $f(x)$ in Exercise 40 and solving the equation $0 = 3x^2 - 2x$.

42. Explain the relationship between the turning point of the function $f(x) = 3x^2 - 2x$ and its vertex.

12.5 Using Determinants to Solve Systems of Equations

■ DETERMINANTS ■ CRAMER'S RULE

■ DETERMINANTS

A determinant is a number that is associated with a square matrix. For any square matrix A, the symbol $|A|$ represents the determinant of A.

Value of a 2 × 2 Determinant

If a, b, c, and d are numbers, the determinant of $A = \begin{bmatrix} a & b \\ c & d \end{bmatrix}$ is

$$\begin{vmatrix} a & b \\ c & d \end{vmatrix} = ad - bc.$$

The determinant of a 2×2 matrix is the number that is equal to the product of the numbers on the major diagonal

$$\begin{vmatrix} a & b \\ c & d \end{vmatrix}$$

minus the product of the numbers on the other diagonal

$$\begin{vmatrix} a & b \\ c & d \end{vmatrix}$$

EXAMPLE 1 Evaluate the determinants **a.** $\begin{vmatrix} 3 & 2 \\ 6 & 9 \end{vmatrix}$ and **b.** $\begin{vmatrix} -5 & \frac{1}{2} \\ -1 & 0 \end{vmatrix}$

Solution **a.**
$$\begin{vmatrix} 3 & 2 \\ 6 & 9 \end{vmatrix} = 3(9) - 2(6)$$
$$= 27 - 12$$
$$= 15$$

b.
$$\begin{vmatrix} -5 & \frac{1}{2} \\ -1 & 0 \end{vmatrix} = -5(0) - \frac{1}{2}(-1)$$
$$= 0 + \frac{1}{2}$$
$$= \frac{1}{2}$$

A 3×3 determinant is evaluated by expanding by **minors.** To find the minor of a_1, we cross out the elements of the determinant that are in the same row and column as a_1:

$$\begin{vmatrix} a_1 & b_1 & c_1 \\ a_2 & b_2 & c_2 \\ a_3 & b_3 & c_3 \end{vmatrix}$$

The minor of a_1 is $\begin{vmatrix} b_2 & c_2 \\ b_3 & c_3 \end{vmatrix}$

To find the minor of b_1, we cross out the elements of the determinant that are in the same row and column as b_1:

$$\begin{vmatrix} a_1 & b_1 & c_1 \\ a_2 & b_2 & c_2 \\ a_3 & b_3 & c_3 \end{vmatrix}$$

The minor of b_1 is $\begin{vmatrix} a_2 & c_2 \\ a_3 & c_3 \end{vmatrix}$

To find the minor of c_1, we cross out the elements of the determinant that are in the same row and column as c_1:

$$\begin{vmatrix} a_1 & b_1 & c_1 \\ a_2 & b_2 & c_2 \\ a_3 & b_3 & c_3 \end{vmatrix}$$

The minor of c_1 is $\begin{vmatrix} a_2 & b_2 \\ a_3 & b_3 \end{vmatrix}$

Value of a 3 × 3 Determinant

$$\begin{vmatrix} a_1 & b_1 & c_1 \\ a_2 & b_2 & c_2 \\ a_3 & b_3 & c_3 \end{vmatrix} = a_1 \underset{\substack{\uparrow \\ \text{Minor} \\ \text{of } a_1}}{\begin{vmatrix} b_2 & c_2 \\ b_3 & c_3 \end{vmatrix}} - b_1 \underset{\substack{\uparrow \\ \text{Minor} \\ \text{of } b_1}}{\begin{vmatrix} a_2 & c_2 \\ a_3 & c_3 \end{vmatrix}} + c_1 \underset{\substack{\uparrow \\ \text{Minor} \\ \text{of } c_1}}{\begin{vmatrix} a_2 & b_2 \\ a_3 & b_3 \end{vmatrix}}$$

EXAMPLE 2   Evaluate the determinant $\begin{vmatrix} 1 & 3 & -2 \\ 2 & 1 & 3 \\ 1 & 2 & 3 \end{vmatrix}$

Solution

$$\begin{vmatrix} 1 & 3 & -2 \\ 2 & 1 & 3 \\ 1 & 2 & 3 \end{vmatrix} = 1 \underset{\substack{\uparrow \\ \text{Minor} \\ \text{of } 1}}{\begin{vmatrix} 1 & 3 \\ 2 & 3 \end{vmatrix}} - 3 \underset{\substack{\uparrow \\ \text{Minor} \\ \text{of } 3}}{\begin{vmatrix} 2 & 3 \\ 1 & 3 \end{vmatrix}} + (-2) \underset{\substack{\uparrow \\ \text{Minor} \\ \text{of } -2}}{\begin{vmatrix} 2 & 1 \\ 1 & 2 \end{vmatrix}}$$

$$= 1(3 - 6) - 3(6 - 3) - 2(4 - 1)$$
$$= -3 - 9 - 6$$
$$= -18$$

■ **TECHNOLOGY TIP**
The determinant of a matrix can be computed with the calculator. The determinant command, det(, is located under the MATH submenu of the MATRIX menu.

```
[A]
      [[1  3  -2]
       [2  1  3 ]
       [1  2  3 ]]
det([A])
                -18
```

We can evaluate a 3 × 3 determinant by expanding it along any row or column. To determine the signs between the terms of the expansion of a 3 × 3 determinant, we use the following array of signs.

Array of Signs for a 3 × 3 Determinant

$$\begin{array}{ccc} + & - & + \\ - & + & - \\ + & - & + \end{array}$$

EXAMPLE 3   Evaluate the determinant $\begin{vmatrix} 1 & 3 & -2 \\ 2 & 1 & 3 \\ 1 & 2 & 3 \end{vmatrix}$ by expanding on the middle column.

Solution   This is the determinant of Example 2. To expand it along the middle column, we use the signs of the middle column of the array of signs:

$$\begin{vmatrix} 1 & 3 & -2 \\ 2 & 1 & 3 \\ 1 & 2 & 3 \end{vmatrix} = -3 \overset{\text{Minor of 3}}{\begin{vmatrix} 2 & 3 \\ 1 & 3 \end{vmatrix}} + 1 \overset{\text{Minor of 1}}{\begin{vmatrix} 1 & -2 \\ 1 & 3 \end{vmatrix}} - 2 \overset{\text{Minor of 2}}{\begin{vmatrix} 1 & -2 \\ 2 & 3 \end{vmatrix}}$$

$$= -3(6 - 3) + 1[3 - (-2)] - 2[3 - (-4)]$$
$$= -3(3) + 1(5) - 2(7)$$
$$= -9 + 5 - 14$$
$$= -18$$

As expected, we get the same value as in Example 2. ■

■ CRAMER'S RULE

Gabriel Cramer (1704–1752) Although other mathematicians had worked with determinants, it was the work of Cramer that popularized them.

The method of using determinants to solve systems of equations is called **Cramer's rule,** named after the 18th-century mathematician Gabriel Cramer. To develop Cramer's rule, we consider the system

$$\begin{cases} ax + by = e \\ cx + dy = f \end{cases}$$

where x and y are variables and a, b, c, d, e, and f are constants.

If we multiply both sides of the first equation by d and multiply both sides of the second equation by $-b$, we can add the equations and eliminate y:

$$\begin{array}{r} adx + bdy = ed \\ -bcx - bdy = -bf \\ \hline adx - bcx = ed - bf \end{array}$$

To solve for x, we factor out x on the left-hand side and divide each side by $ad - bc$:

$$(ad - bc)x = ed - bf$$
$$x = \frac{ed - bf}{ad - bc} \qquad (ad - bc \neq 0)$$

We can find y in a similar manner. After eliminating the variable x, we get

$$y = \frac{af - ec}{ad - bc} \qquad (ad - bc \neq 0)$$

Determinants provide an easy way of remembering these formulas. Note that the denominator for both x and y is

$$\begin{vmatrix} a & b \\ c & d \end{vmatrix} = ad - bc$$

The numerators can be expressed as determinants also:

$$x = \frac{ed - bf}{ad - bc} = \frac{\begin{vmatrix} e & b \\ f & d \end{vmatrix}}{\begin{vmatrix} a & b \\ c & d \end{vmatrix}} \qquad \text{and} \qquad y = \frac{af - ec}{ad - bc} = \frac{\begin{vmatrix} a & e \\ c & f \end{vmatrix}}{\begin{vmatrix} a & b \\ c & d \end{vmatrix}}$$

If we compare these formulas with the original system

$$\begin{cases} ax + by = e \\ cx + dy = f \end{cases}$$

we note that in the expressions for x and y above, the denominator determinant is formed by using the coefficients a, b, c, and d of the variables in the equations. The numerator determinants are the same as the denominator determinant, except that the column of coefficients of the variable for which we are solving is replaced with the column of constants e and f.

Cramer's Rule for Two Equations in Two Variables

The solution of the system $\begin{cases} ax + by = e \\ cx + dy = f \end{cases}$ is given by

$$x = \frac{D_x}{D} = \frac{\begin{vmatrix} e & b \\ f & d \end{vmatrix}}{\begin{vmatrix} a & b \\ c & d \end{vmatrix}} \quad \text{and} \quad y = \frac{D_y}{D} = \frac{\begin{vmatrix} a & e \\ c & f \end{vmatrix}}{\begin{vmatrix} a & b \\ c & d \end{vmatrix}}$$

If every determinant is 0, the system is consistent but the equations are dependent.

If $D = 0$ and D_x or D_y is nonzero, the system is inconsistent.

EXAMPLE 4   Use Cramer's rule to solve the system $\begin{cases} 4x - 3y = 6 \\ -2x + 5y = 4 \end{cases}$

Solution   The value of x is the quotient of two determinants. The denominator determinant is made up of the coefficients of x and y:

$$D = \begin{vmatrix} 4 & -3 \\ -2 & 5 \end{vmatrix}$$

■ **TECHNOLOGY TIP**

To solve for x, the numerator determinant will be computed from matrix B and the denominator determinant from matrix A:

```
[A]
      [[4   -3]
       [-2  5 ]]
[B]
      [[6   -3]
       [4   5 ]]
```

To solve for y, the numerator determinant will be computed from matrix C and the denominator determinant from matrix A:

```
[A]
      [[4   -3]
       [-2  5 ]]
[C]
      [[4   6]
       [-2  4]]
```

To solve for x, we form the numerator determinant from the denominator determinant by replacing its first column (the coefficients of x) with the column of constants (6 and 4).

To solve for y, we form the numerator determinant from the denominator determinant by replacing the second column (the coefficients of y) with the column of constants (6 and 4).

To find the values of x and y, we evaluate each determinant:

$$x = \frac{\begin{vmatrix} 6 & -3 \\ 4 & 5 \end{vmatrix}}{\begin{vmatrix} 4 & -3 \\ -2 & 5 \end{vmatrix}} = \frac{6(5) - (-3)(4)}{4(5) - (-3)(-2)} = \frac{30 + 12}{20 - 6} = \frac{42}{14} = 3$$

```
det([B])/det([A]
)
                 3
det([C])/det([A]
)
                 2
```

$$y = \frac{\begin{vmatrix} 4 & 6 \\ -2 & 4 \end{vmatrix}}{\begin{vmatrix} 4 & -3 \\ -2 & 5 \end{vmatrix}} = \frac{4(4) - 6(-2)}{4(5) - (-3)(-2)} = \frac{16 + 12}{20 - 6} = \frac{28}{14} = 2$$

The solution to this system is (3, 2). Verify that $x = 3$ and $y = 2$ satisfy each equation in the given system.   ■

EXAMPLE 5 Use Cramer's rule to solve the system $\begin{cases} 7x = 8 - 4y \\ 2y = 3 - \frac{7}{2}x \end{cases}$

Solution We multiply both sides of the second equation by 2 to eliminate the fraction and write the system in the form

$$\begin{cases} 7x + 4y = 8 \\ 7x + 4y = 6 \end{cases}$$

When we attempt to use Cramer's rule to solve this system, we find that the determinant in the denominator is 0:

$$\begin{vmatrix} 7 & 4 \\ 7 & 4 \end{vmatrix} = 7(4) - 4(7) = 0$$

Because the two equations are different (independent) and because the determinant in the denominator is 0, the system is inconsistent. It has no solutions. ■

Cramer's Rule for Three Equations in Three Variables

The solution of the system $\begin{cases} ax + by + cz = j \\ dx + ey + fz = k \\ gx + hy + iz = l \end{cases}$ is given by

$$x = \frac{D_x}{D}, \qquad y = \frac{D_y}{D}, \qquad \text{and} \qquad z = \frac{D_z}{D}$$

where

$$D = \begin{vmatrix} a & b & c \\ d & e & f \\ g & h & i \end{vmatrix} \qquad D_x = \begin{vmatrix} j & b & c \\ k & e & f \\ l & h & i \end{vmatrix}$$

$$D_y = \begin{vmatrix} a & j & c \\ d & k & f \\ g & l & i \end{vmatrix} \qquad D_z = \begin{vmatrix} a & b & j \\ d & e & k \\ g & h & l \end{vmatrix}$$

If every determinant is 0, the system is consistent but the equations are dependent.

If $D = 0$ and D_x or D_y or D_z is nonzero, the system is inconsistent.

EXAMPLE 6 Use Cramer's rule to solve the system $\begin{cases} 2x + y + 4z = 12 \\ x + 2y + 2z = 9 \\ 3x - 3y - 2z = 1 \end{cases}$

Solution The denominator determinant is the determinant formed by the coefficients of the variables. The numerator determinants are formed by replacing the coefficients of the variable being solved for by the column of constants. We form the quotients for x, y, and z and evaluate the determinants:

$$x = \frac{\begin{vmatrix} 12 & 1 & 4 \\ 9 & 2 & 2 \\ 1 & -3 & -2 \end{vmatrix}}{\begin{vmatrix} 2 & 1 & 4 \\ 1 & 2 & 2 \\ 3 & -3 & -2 \end{vmatrix}} = \frac{12\begin{vmatrix} 2 & 2 \\ -3 & -2 \end{vmatrix} - 1\begin{vmatrix} 9 & 2 \\ 1 & -2 \end{vmatrix} + 4\begin{vmatrix} 9 & 2 \\ 1 & -3 \end{vmatrix}}{2\begin{vmatrix} 2 & 2 \\ -3 & -2 \end{vmatrix} - 1\begin{vmatrix} 1 & 2 \\ 3 & -2 \end{vmatrix} + 4\begin{vmatrix} 1 & 2 \\ 3 & -3 \end{vmatrix}} = \frac{12(2) - (-20) + 4(-29)}{2(2) - (-8) + 4(-9)} = \frac{-72}{-24} = 3$$

$$y = \frac{\begin{vmatrix} 2 & 12 & 4 \\ 1 & 9 & 2 \\ 3 & 1 & -2 \end{vmatrix}}{\begin{vmatrix} 2 & 1 & 4 \\ 1 & 2 & 2 \\ 3 & -3 & -2 \end{vmatrix}} = \frac{2\begin{vmatrix} 9 & 2 \\ 1 & -2 \end{vmatrix} - 12\begin{vmatrix} 1 & 2 \\ 3 & -2 \end{vmatrix} + 4\begin{vmatrix} 1 & 9 \\ 3 & 1 \end{vmatrix}}{-24} = \frac{2(-20) - 12(-8) + 4(-26)}{-24} = \frac{-48}{-24} = 2$$

$$z = \frac{\begin{vmatrix} 2 & 1 & 12 \\ 1 & 2 & 9 \\ 3 & -3 & 1 \end{vmatrix}}{\begin{vmatrix} 2 & 1 & 4 \\ 1 & 2 & 2 \\ 3 & -3 & -2 \end{vmatrix}} = \frac{2\begin{vmatrix} 2 & 9 \\ -3 & 1 \end{vmatrix} - 1\begin{vmatrix} 1 & 9 \\ 3 & 1 \end{vmatrix} + 12\begin{vmatrix} 1 & 2 \\ 3 & -3 \end{vmatrix}}{-24} = \frac{2(29) - (-26) + 12(-9)}{-24} = \frac{-24}{-24} = 1$$

The solution to this system is (3, 2, 1). ■

EXERCISE 12.5

VOCABULARY *In Exercises 1 and 2, fill in the blanks to make a true statement.*

1. A _____ is a number that is associated with a square matrix.

2. _____ rule uses determinants to solve systems of equations.

PRACTICE *In Exercises 3–20, evaluate each determinant. Verify with your calculator.*

3. $\begin{vmatrix} 2 & 3 \\ -2 & 1 \end{vmatrix}$

4. $\begin{vmatrix} 3 & -2 \\ -2 & 4 \end{vmatrix}$

5. $\begin{vmatrix} -1 & 2 \\ 3 & -4 \end{vmatrix}$

6. $\begin{vmatrix} -1 & -2 \\ -3 & -4 \end{vmatrix}$

7. $\begin{vmatrix} x & y \\ y & x \end{vmatrix}$

8. $\begin{vmatrix} x+y & x-y \\ x-y & x+y \end{vmatrix}$

9. $\begin{vmatrix} 1 & 0 & 1 \\ 0 & 1 & 0 \\ 1 & 1 & 1 \end{vmatrix}$

10. $\begin{vmatrix} 1 & 2 & 0 \\ 0 & 1 & 2 \\ 0 & 0 & 1 \end{vmatrix}$

11. $\begin{vmatrix} -1 & 2 & 1 \\ 2 & 1 & -3 \\ 1 & 1 & 1 \end{vmatrix}$

12. $\begin{vmatrix} 1 & 2 & 3 \\ 1 & 2 & 3 \\ 1 & 2 & 3 \end{vmatrix}$

13. $\begin{vmatrix} 1 & -2 & 3 \\ -2 & 1 & 1 \\ -3 & -2 & 1 \end{vmatrix}$

14. $\begin{vmatrix} 1 & 1 & 2 \\ 2 & 1 & -2 \\ 3 & 1 & 3 \end{vmatrix}$

15. $\begin{vmatrix} 1 & 2 & 3 \\ 4 & 5 & 6 \\ 7 & 8 & 9 \end{vmatrix}$

16. $\begin{vmatrix} 1 & 4 & 7 \\ 2 & 5 & 8 \\ 3 & 6 & 9 \end{vmatrix}$

17. $\begin{vmatrix} a & 2a & -a \\ 2 & -1 & 3 \\ 1 & 2 & -3 \end{vmatrix}$

18. $\begin{vmatrix} 1 & 2b & -3 \\ 2 & -b & 2 \\ 1 & 3b & 1 \end{vmatrix}$

19. $\begin{vmatrix} 1 & a & b \\ 1 & 2a & 2b \\ 1 & 3a & 3b \end{vmatrix}$

20. $\begin{vmatrix} a & b & c \\ 0 & b & c \\ 0 & 0 & c \end{vmatrix}$

In Exercises 21–46, use Cramer's rule to solve each system of equations, if possible.

21. $\begin{cases} x + y = 6 \\ x - y = 2 \end{cases}$

22. $\begin{cases} x - y = 4 \\ 2x + y = 5 \end{cases}$

23. $\begin{cases} 2x + y = 1 \\ x - 2y = -7 \end{cases}$

24. $\begin{cases} 2x + 3y = 0 \\ 4x - 6y = -4 \end{cases}$

25. $\begin{cases} 4x - 3y = -1 \\ 8x + 3y = 4 \end{cases}$

26. $\begin{cases} 3x - y = -3 \\ 2x + y = -7 \end{cases}$

27. $\begin{cases} y = \dfrac{11 - 3x}{2} \\ x = \dfrac{11 - 4y}{6} \end{cases}$

28. $\begin{cases} x = \dfrac{12 - 6y}{5} \\ y = \dfrac{24 - 10x}{12} \end{cases}$

29. $\begin{cases} y = \dfrac{-2x + 1}{3} \\ 3x - 2y = 8 \end{cases}$

30. $\begin{cases} 2x + 3y = -1 \\ x = \dfrac{y - 9}{4} \end{cases}$

31. $\begin{cases} x = \dfrac{5y - 4}{2} \\ y = \dfrac{3x - 1}{5} \end{cases}$

32. $\begin{cases} y = \dfrac{1 - 5x}{2} \\ x = \dfrac{3y + 10}{4} \end{cases}$

33. $\begin{cases} x + y + z = 4 \\ x + y - z = 0 \\ x - y + z = 2 \end{cases}$

34. $\begin{cases} x + y + z = 4 \\ x - y + z = 2 \\ x - y - z = 0 \end{cases}$

35. $\begin{cases} x + y + 2z = 7 \\ x + 2y + z = 8 \\ 2x + y + z = 9 \end{cases}$

36. $\begin{cases} x + 2y + 2z = 10 \\ 2x + y + 2z = 9 \\ 2x + 2y + z = 1 \end{cases}$

37. $\begin{cases} 2x + y - z = 1 \\ x + 2y + 2z = 2 \\ 4x + 5y + 3z = 3 \end{cases}$

38. $\begin{cases} 2x + 3y + 4z = 6 \\ 2x - 3y - 4z = -4 \\ 4x + 6y + 8z = 12 \end{cases}$

39. $\begin{cases} 2x + y + z = 5 \\ x - 2y + 3z = 10 \\ x + y - 4z = -3 \end{cases}$

40. $\begin{cases} 3x + 2y - z = -8 \\ 2x - y + 7z = 10 \\ 2x + 2y - 3z = -10 \end{cases}$

41. $\begin{cases} 4x + 3z = 4 \\ 2y - 6z = -1 \\ 8x + 4y + 3z = 9 \end{cases}$

42. $\begin{cases} x - 3y + 4z - 2 = 0 \\ 2x + y + 2z - 3 = 0 \\ 4x - 5y + 10z - 7 = 0 \end{cases}$

43. $\begin{cases} x + y = 1 \\ \dfrac{1}{2}y + z = \dfrac{5}{2} \\ x - z = -3 \end{cases}$

44. $\begin{cases} 3x + 4y + 14z = 7 \\ -\dfrac{1}{2}x - y + 2z = \dfrac{3}{2} \\ x + \dfrac{3}{2}y + \dfrac{5}{2}z = 1 \end{cases}$

45. $\begin{cases} 2x - y + 4z + 2 = 0 \\ 5x + 8y + 7z = -8 \\ x + 3y + z + 3 = 0 \end{cases}$

46. $\begin{cases} x + 2y + 2z + 3 = 0 \\ x + \dfrac{1}{2}y + z - \dfrac{1}{2} = 0 \\ 2x + 2y + z + 1 = 0 \end{cases}$

In Exercises 47–50, evaluate each determinant and solve the resulting equation.

47. $\begin{vmatrix} x & 1 \\ 3 & 2 \end{vmatrix} = 1$

48. $\begin{vmatrix} x & -x \\ 2 & -3 \end{vmatrix} = -5$

49. $\begin{vmatrix} x & -2 \\ 3 & 1 \end{vmatrix} = \begin{vmatrix} 4 & 2 \\ x & 3 \end{vmatrix}$

50. $\begin{vmatrix} x & 3 \\ x & 2 \end{vmatrix} = \begin{vmatrix} 3 & 2 \\ 1 & 1 \end{vmatrix}$

REVIEW

51. Solve the system $\begin{cases} 2x + y = 1 \\ x - 2y = 7 \end{cases}$ using the indicated method:
  a. elimination method
  b. substitution method
  c. graphing, using your calculator.

52. Algebraically determine the inverse function of $f(x) = 6x - 5$.

53. Graph each function on graph paper and verify with your calculator.

a. $f(x) = \left(\dfrac{4}{5}\right)^x$

b. $g(x) = \log_{4/5} x$

54. Explain why $f(x)$ and $g(x)$ in Exercise 53 are inverses of one another.

12.6 Systems of Inequalities

■ SOLVING SYSTEMS OF LINEAR INEQUALITIES   ■ SOLVING SYSTEMS OF QUADRATIC INEQUALITIES

■ SOLVING SYSTEMS OF LINEAR INEQUALITIES

To solve **systems of inequalities** containing two variables, we want to find all ordered pairs that are solutions of all the inequalities in the system. We will, therefore, graph the equations associated with each inequality in the same coordinate plane (using dotted lines for strict inequalities, $<$ or $>$, and solid lines when equality is included, $\leq$ or $\geq$). We then use a test point to decide which region for each inequality should be shaded. The region that is common to all graphs of the system will be the most heavily shaded and contains all the points that are solutions to the system. The steps are specified below.

> **Solving Systems of Inequalities**
> **1.** Graph the boundary line for *each* region by graphing the equation associated with the inequality. If the inequality allows the possibility of equality, draw the boundary line as a solid line. If equality is not allowed, draw the boundary line as a dotted line.
>
> **2.** After graphing the boundary line for a particular inequality, pick a point that is on one side of the boundary line. Replace x and y with the coordinates of that point. If the inequality is true, shade the side that contains that point. If the inequality is false, shade the region that does NOT contain the test point.
>
> **3.** Follow steps 1 and 2 for each inequality in the system.
>
> **4.** The region common to every graph is the solution of the system. It should be the most heavily shaded area.
>
> **5.** Pick a test point and verify it in each of the original inequalities in the system.

EXAMPLE 1   Graph the solution set of the system $\begin{cases} 2x + y < 4 \\ -2x + y > 2 \end{cases}$

Solution   We graph each inequality on the same set of coordinate axes, as in Figure 12-8.

$2x + y = 4$			$-2x + y = 2$		
x	y	(x, y)	x	y	(x, y)
0	4	$(0, 4)$	-1	0	$(-1, 0)$
1	2	$(1, 2)$	0	2	$(0, 2)$
2	0	$(2, 0)$	2	6	$(2, 6)$

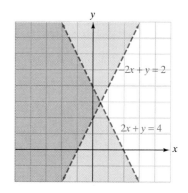

FIGURE 12-8

We then find the graph of each inequality.

- The graph of $2x + y < 4$ includes all points below the line $2x + y = 4$. Because the boundary is not included, we draw it as a dotted line.
- The graph of $-2x + y > 2$ includes all points above the line $-2x + y = 2$. Because the boundary is not included, we draw it as a dotted line.

The area that is shaded twice represents the set of simultaneous solutions of the given system of inequalities. Any point in the doubly shaded region has coordinates that will satisfy both inequalities of the system.

Pick a point in the doubly shaded region and show that it satisfies both inequalities. ∎

EXAMPLE 2   Graph the solution set of the system $\begin{cases} x \leq 2 \\ y > 3 \end{cases}$

Solution   We graph each inequality on the same set of coordinate axes, as in Figure 12-9.

$x = 2$			$y = 3$		
x	y	(x, y)	x	y	(x, y)
2	0	$(2, 0)$	0	3	$(0, 3)$
2	2	$(2, 2)$	1	3	$(1, 3)$
2	4	$(2, 4)$	4	3	$(4, 3)$

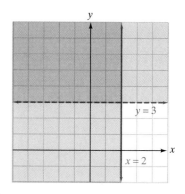

FIGURE 12-9

We then find the graph of each inequality.

- The graph of $x \leq 2$ includes all points on the line $x = 2$ and all points to the left of the line. Because the boundary line is included, we draw it as a solid line.
- The graph $y > 3$ includes all points above the line $y = 3$. Because the boundary is not included, we draw it as a dotted line.

The area that is shaded twice represents the set of simultaneous solutions of the given system of inequalities. Any point in the doubly shaded region has coordinates that will satisfy both inequalities of the system. Pick a point in the doubly shaded region and show that this is true. ■

EXAMPLE 3   Graph the solution set of the system $\begin{cases} y < 3x - 1 \\ y \geq 3x + 1 \end{cases}$

Solution   We graph each inequality, as in Figure 12-10.

- The graph of $y < 3x - 1$ includes all of the points below the dotted line $y = 3x - 1$.
- The graph of $y \geq 3x + 1$ includes all of the points on and above the solid line $y = 3x + 1$.

Because the graphs of these inequalities do not intersect, the solution set is empty. There are no solutions.

FIGURE 12-10 ■

EXAMPLE 4   Graph the solution set of the system $\begin{cases} x \geq 0 \\ y \geq 0 \\ x + 2y \leq 6 \end{cases}$

Solution   We graph each inequality, as in Figure 12-11.

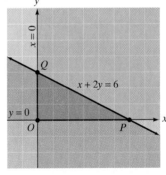

- The graph of $x \geq 0$ includes all of the points on the y-axis and to the right.
- The graph of $y \geq 0$ includes all of the points on the x-axis and above.
- The graph of $x + 2y \leq 6$ includes all of the points on the line $x + 2y = 6$ and below.

The solution is the region that is shaded three times. This includes triangle OPQ and the triangular region it encloses.

FIGURE 12-11 ■

■ SOLVING SYSTEMS OF QUADRATIC INEQUALITIES

The process for solving systems of quadratic inequalities is exactly the same as that presented previously.

EXAMPLE 5   Graph the solution set of the system $\begin{cases} y < x^2 \\ y > \dfrac{x^2}{4} - 2 \end{cases}$

Solution   The graph of $y = x^2$ is the parabola shown in Figure 12-12, which opens upward and has its vertex at the origin. The points with coordinates that satisfy the inequality $y < x^2$ are those points below the parabola.

$$y = x^2$$

x	y	(x, y)
0	0	$(0, 0)$
1	1	$(1, 1)$
-1	1	$(-1, 1)$
2	4	$(2, 4)$
-2	4	$(-2, 4)$

$$y = \frac{x^2}{4} - 2$$

x	y	(x, y)
0	-2	$(0, -2)$
2	-1	$(2, -1)$
-2	-1	$(-2, -1)$
4	2	$(4, 2)$
-4	2	$(-4, 2)$

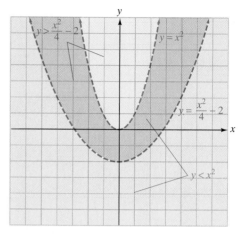

FIGURE 12-12

The graph of $y = \dfrac{x^2}{4} - 2$ is a parabola opening upward, with vertex at $(0, -2)$.

However, this time the points with coordinates that satisfy the inequality are those points above the parabola. The graph of the solution set of the system is the area between the parabolas. ∎

EXERCISE 12.6

In Exercises 1–26, find the solution set of each system of inequalities, when possible.

1. $\begin{cases} x + 2y \le 3 \\ 2x - y \ge 1 \end{cases}$

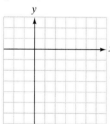

2. $\begin{cases} 2x + y \ge 3 \\ x - 2y \le -1 \end{cases}$

3. $\begin{cases} x + y < -1 \\ x - y > -1 \end{cases}$

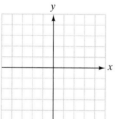

4. $\begin{cases} x + y > 2 \\ x - y < -2 \end{cases}$

5. $\begin{cases} 2x - y < 4 \\ x + y \ge -1 \end{cases}$

6. $\begin{cases} x - y \ge 5 \\ x + 2y < -4 \end{cases}$

7. $\begin{cases} x > 2 \\ y \le 3 \end{cases}$

8. $\begin{cases} x \ge -1 \\ y > -2 \end{cases}$

9. $\begin{cases} 2x - 3y < 0 \\ y > x - 1 \end{cases}$

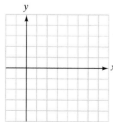

10. $\begin{cases} 3x - y \ge -1 \\ y \ge 3x + 1 \end{cases}$

11. $\begin{cases} x + y < 1 \\ x + y > 3 \end{cases}$

12. $\begin{cases} x + y > 2 \\ x + y < 4 \end{cases}$

13. $\begin{cases} x > 0 \\ y > 0 \end{cases}$

14. $\begin{cases} x \le 0 \\ y < 0 \end{cases}$

15. $\begin{cases} 3x + 4y > -7 \\ 2x - 3y \ge 1 \end{cases}$

16. $\begin{cases} 3x + y \le 1 \\ 4x - y > -8 \end{cases}$

17. $\begin{cases} x < 3y - 1 \\ y \ge 2x - 3 \end{cases}$

18. $\begin{cases} y \ge x + 2 \\ x \le y - 2 \end{cases}$

19. $\begin{cases} 2x + y < 7 \\ y > 2(1 - x) \end{cases}$

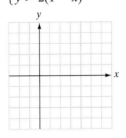

20. $\begin{cases} 2x + y \ge 6 \\ y \le 2(2x - 3) \end{cases}$

21. $\begin{cases} 2x - 4y > -6 \\ 3x + y \ge 5 \end{cases}$

22. $\begin{cases} 2x - 3y < 0 \\ 2x + 3y \ge 12 \end{cases}$

23. $\begin{cases} 3x - y \le -4 \\ 3y > -2(x + 5) \end{cases}$

24. $\begin{cases} 3x + y < -2 \\ y > 3(1 - x) \end{cases}$

25. $\begin{cases} \dfrac{x}{2} + \dfrac{y}{3} \ge 2 \\[2mm] \dfrac{x}{2} - \dfrac{y}{2} < -1 \end{cases}$

26. $\begin{cases} \dfrac{x}{3} - \dfrac{y}{2} < -3 \\[2mm] \dfrac{x}{3} + \dfrac{y}{2} > -1 \end{cases}$

In Exercises 27–30, use the graphing method to find the region that satisfies all of the inequalities of the system.

27. $\begin{cases} x \ge 0 \\ y \ge 0 \\ x + y \le 3 \end{cases}$

28. $\begin{cases} x - y \le 6 \\ x + 2y \le 6 \\ x \ge 0 \end{cases}$

29. $\begin{cases} x \ge 0 \\ y \ge 0 \\ x \le 5 \\ y \le x \end{cases}$

30. $\begin{cases} x \ge 0 \\ y \ge 0 \\ y \le 2 + x \\ y \ge 4x - 2 \end{cases}$

In Exercises 31–33, solve each system of inequalities by graphing.

31. $\begin{cases} x^2 - 6x - y \le 5 \\ x^2 - 6x - y \ge -5 \end{cases}$

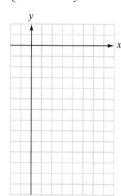

32. $\begin{cases} x^2 - x - y \ge 2 \\ 4x - 3y < 0 \end{cases}$

33. $\begin{cases} x^2 - 6x - y < -5 \\ x^2 - 6x + y < -5 \end{cases}$

APPLICATIONS *In Exercises 34–37, graph each system of inequalities and give two possible solutions to each problem.*

34. Buying compact disks Melodic Music has compact disks on sale for either $10 or $15. If a customer wants to spend at least $30 but no more than $60 on CDs, find a system of inequalities whose graph will show the possible combinations of $10 CDs ($x$) and $15 CDs ($y$) that the customer can buy.

35. Buying boats Dry Boatworks wholesales aluminum boats for $800 and fiberglass boats for $600. Northland Marina wants to order at least $2400 but no more than $4800 worth of boats. Find a system of inequalities whose graph will show the possible combination of aluminum boats (x) and fiberglass boats (y) that can be ordered.

36. Buying furniture A distributor wholesales desk chairs for $150 and side chairs for $100. Best Furniture wants to order no more than $900 worth of chairs and wants to order more side chairs than desk chairs. Find a system of inequalities whose graph will show the possible combinations of desk chairs (x) and side chairs (y) that can be ordered.

37. Ordering furnace equipment J. Bolden Heating Company wants to order no more than $2000 worth of electronic air cleaners and humidifiers from a wholesaler that charges $500 for air cleaners and $200 for humidifiers. If Bolden wants more humidifiers than air cleaners, find a system of inequalities whose graph will show the possible combinations of air cleaners (x) and humidifiers (y) that can be ordered.

✍**38.** Explain how the graphing calculator can be used to solve a system of inequalities.

✍**39.** Explain when a system of inequalities will have no solution.

REVIEW

40. Solve the equation $\sqrt[3]{6x + 4} - 4 = 0$ both algebraically and graphically.

41. Solve the equation $1 - \sqrt{x} = \sqrt{x - 3}$ both algebraically and graphically.

42. Find the distance between P and Q if $P(-3, 6)$ and $Q(20, 12)$. Round the result to the nearest tenth.

43. Find the domain and the range of the function $f(x) = \sqrt{x + 2} - 4$.

44. Using the function specified in Exercise 43, find $f(2)$.

C H A P T E R   S U M M A R Y

CONCEPTS

REVIEW EXERCISES

Section 12.1

The solution(s) to a system of equations are the value(s) of the variable(s) that make all the equations in the system true.

To solve a system of equations graphically, carefully graph each equation of the system on the same set of axes. If the lines intersect, the coordinates of the point of intersection give the solution to the system.

To solve a system of equations by substitution, solve one of the equations of the system for one of its variables, substitute the resulting expression into the other equation, and solve for the other variable. Then substitute the value obtained back into one of the original equations to solve for the other variable.

To solve a system of equations by addition (elimination), first multiply one or both of the equations by suitable constants, if necessary, to eliminate one of the variables when the equations are added. The equation that results can be solved for its single variable. Then substitute the value obtained back into one of the original equations and solve for the other variable.

Solving Systems of Two Equations in Two Variables

1. Tell whether the ordered pair is a solution of the given system.

   a. $(1, 5);$ $\begin{cases} 3x - y = -2 \\ 2x + 3y = 17 \end{cases}$

   b. $\left(14, \frac{1}{2}\right);$ $\begin{cases} 2x + 4y = 30 \\ \dfrac{x}{4} - y = 3 \end{cases}$

2. Solve each system below graphically.

   a. $\begin{cases} x + y = 7 \\ 2x - y = 5 \end{cases}$

   b. $\begin{cases} \dfrac{x}{3} + \dfrac{y}{5} = -1 \\ x - 3y = -3 \end{cases}$

3. Use the substitution method to solve each system.

   a. $\begin{cases} x = 3y + 5 \\ 5x - 4y = 3 \end{cases}$

   b. $\begin{cases} 6(x + 3) = y - 1 \\ 5(y - 1) = x + 2 \end{cases}$

4. Use the addition (elimination) method to solve each system.

   a. $\begin{cases} 2x + y = 1 \\ 5x - y = 20 \end{cases}$

   b. $\begin{cases} 11x + 3y = 27 \\ 8x + 4y = 36 \end{cases}$

Section 12.2

Many application problems can be solved by specifying a system of equations. If two variables are used to represent unknown quantities, two equations must be written.

Applications of Systems of Two Equations in Two Variables

5. **Integer problem**   One number is 5 times another, and their sum is 18. Find the numbers.

6. **Geometry**   The length of a rectangle is 3 times its width, and its perimeter is 24 feet. Find its dimensions.

7. **Buying grapefruit**   A grapefruit costs 15 cents more than an orange. Together, they cost 85 cents. Find the cost of the grapefruit.

8. **Utility bills**   A man's electric bill for January was $23 less than his gas bill. The two utilities charged him a total of $109. Find the amount of his gas bill.

9. **Investing money** Carlos invested part of $3000 in a 10% certificate account and the rest in a 6% passbook account. The total annual interest from both accounts is $270. How much did he invest at 6%?

Section 12.3

Solving Systems of Three Equations in Three Variables

A system of three equations in three variables can be solved by using the addition (elimination) method.

10. Solve the system $\begin{cases} x + y + z = 6 \\ x - y - z = -4 \\ -x + y - z = -2 \end{cases}$

Section 12.4

Using Matrices to Solve Systems of Equations

A matrix is any rectangular array of numbers. Elementary row operations that can be performed on the matrix are:
1. Any two rows can be interchanged.
2. Any row can be multiplied by a nonzero constant.
3. Any row can be changed by adding a constant multiple of another row to it.

11. Solve each system of equations by using matrices.

a. $\begin{cases} x + 2y = 4 \\ 2x - y = 3 \end{cases}$

b. $\begin{cases} x + y + z = 6 \\ 2x - y + z = 1 \\ 4x + y - z = 5 \end{cases}$

Section 12.5

Using Determinants to Solve Systems of Equations

A determinant is any number that is associated with a square matrix. Cramer's rule can be used to solve systems of linear equations.

12. Evaluate each determinant.

a. $\begin{vmatrix} 2 & 3 \\ -4 & 3 \end{vmatrix}$

b. $\begin{vmatrix} -1 & 2 & -1 \\ 2 & -1 & 3 \\ 1 & -2 & 2 \end{vmatrix}$

13. Use Cramer's rule to solve each system of equations.

a. $\begin{cases} 3x + 4y = 10 \\ 2x - 3y = 1 \end{cases}$

b. $\begin{cases} x + 2y + z = 0 \\ 2x + y + z = 3 \\ x + y + 2z = 5 \end{cases}$

Section 12.6

Systems of Inequalities

Systems of inequalities can be solved graphically. Carefully graph each inequality on the same set of axes and shade appropriately. The most heavily shaded area is the solution of the system.

14. Graph the solution set of each system of inequalities.

a. $y \geq x + 1$
$\quad 3x + 2y < 6$

b. $y \geq x^2 - 4$
$\quad y < x + 3$

■ Chapter Test

1. Solve the system $\begin{cases} 3x + y = 7 \\ x - 2y = 0 \end{cases}$ graphically.

2. Solve the system $\begin{cases} y = x - 1 \\ 2x + y = -7 \end{cases}$ by substitution.

3. Solve the system $\begin{cases} 3x - y = 2 \\ 2x + y = 8 \end{cases}$ by addition (elimination).

In Problems 4 and 5, solve each system using any method.

4. $\begin{cases} \dfrac{x}{6} + \dfrac{y}{10} = 3 \\ \dfrac{5x}{16} - \dfrac{3y}{16} = \dfrac{15}{8} \end{cases}$

5. $\begin{cases} 4x + 3 = -3y \\ -\dfrac{x}{7} + \dfrac{4y}{21} = 1 \end{cases}$

6. Solve the system $\begin{cases} 2x + 3y + z = -5 \\ -x + 2y - z = -6 \\ 3x + y + 2z = 4 \end{cases}$ by addition.

In Problems 7–10, evaluate each determinant.

7. $\begin{vmatrix} 2 & -3 \\ 4 & 5 \end{vmatrix}$

8. $\begin{vmatrix} -3 & -4 \\ -2 & 3 \end{vmatrix}$

9. $\begin{vmatrix} 1 & 2 & 0 \\ 2 & 0 & 3 \\ 1 & -2 & 2 \end{vmatrix}$

10. $\begin{vmatrix} 2 & -1 & 1 \\ 3 & 1 & 0 \\ 0 & 1 & 2 \end{vmatrix}$

In Problems 11–14, consider the system $\begin{cases} x - y = -6 \\ 3x + y = -6 \end{cases}$ *which is to be solved with Cramer's rule.*

11. When solving for x, what is the numerator determinant? (**Do not evaluate it.**)

12. When solving for y, what is the denominator determinant? (**Do not evaluate it.**)

13. Solve the system for x.

14. Solve the system for y.

In Problems 15–18, consider the system $\begin{cases} x + y + z = 4 \\ x + y - z = 6 \\ 2x - 3y + z = -1 \end{cases}$

15. Solve for x.

16. Solve for y.

17. Solve for z.

18. Write the augmented matrix that represents the system.

In Problems 19 and 20, use graphing to solve each system.

19. $\begin{cases} 2x - 3y \geq 6 \\ y \leq -x + 1 \end{cases}$

20. $\begin{cases} y \geq x^2 \\ y < x + 3 \end{cases}$

■ Cumulative Review: Chapters 11 and 12

VOCABULARY / CONCEPTS

1. If a function contains the point (a, b), then the inverse of the function contains the point _____.

2. The notation _____ is used to denote the inverse of the function $f(x)$.

3. Determine the inverse relation of the set of ordered pairs $\{(1, 3), (2, 5), (-3, -2), (-4, -6)\}$.

In Exercises 4–6, determine which of the following represent one-to-one functions.

4.

5.

6.

7. Functions with variables as exponents are called _____ functions.

8. Classify each of the following as "an exponential function" or "not an exponential function."
  a. $f(x) = x^3$
  b. $f(x) = 3^x$
  c. $f(x) = \left(\frac{1}{2}\right)^x$

9. Find x: $\log_3 9 = x$.

10. Use the properties of logarithms to write the given expression as a single logarithm:
$-5 \log_b x - 3 \log_b y + \log_b z$

11. If two or more equations are considered at the same time, they are called a _____ of equations.

12. When a system of equations has one or more solutions, it is called a _____ system.

13. If a system has no solutions, it is called an _____ system.

14. Equations were entered at the y1 and y2 prompts of the calculator. State the solution of the system by interpreting the given graphical display.

15. A rectangular array of numbers is called a _____.

16. A _____ matrix has the same number of rows and columns.

17. An _____ matrix includes the constant values of the system, whereas a _____ matrix only includes the coefficients.

18. Write the coefficient matrix for the system
$$\begin{cases} x + 2y + 2z = 10 \\ 2x + y + 2z = 9 \\ 2x + 2y + z = 11 \end{cases}$$

19. Write the augmented matrix for Exercise 18.

20. _____ rule uses determinants to solve systems of equations.

PRACTICE   *In Exercises 21–24, $f(x) = 3x - 2$ and $g(x) = x^2 + 1$. Find each value.*

21. $(f \circ g)(3)$

22. $(g \circ f)(3)$

23. $(f \circ g)\left(\dfrac{1}{4}\right)$

24. $(g \circ f)(h)$

25. If $f(x) = x^2 + 5x + 6$ and $g(x) = x + 3$, find each of the following:
   a. $f + g$
   b. $f - g$
   c. $f \cdot g$
   d. $\dfrac{f}{g}$

26. If $f(x) = x^2 - 4$ and $g(x) = x + 2$, find each function and its domain.
   a. $f + g$
   b. $f - g$
   c. $f \cdot g$
   d. $\dfrac{f}{g}$

27. Given the function $x = 2y - 1$,
   a. find the equation of the inverse function.

   b. graph the function and its inverse on the same coordinate axes.
   c. graph the axis of symmetry, $y = x$.

28. Solve the equation $3^x = 9^3$ by rewriting each side of the equation as an exponential expression with a common base.

29. Graph the function $f(x) = \log_{1/2} x$ on graph paper. Verify your graph with the graphing calculator.

30. Solve the equation $125^{x-3} = 25^{2x+7}$ algebraically. Verify your solution with the graphing calculator.

31. Solve the system below by *substitution*.
$$\begin{cases} 2x - y = -21 \\ 4x + 5y = 7 \end{cases}$$

32. Solve the system below by *addition (elimination)*.
$$\begin{cases} 2y - 3x = -13 \\ 3x - 17 = 4y \end{cases}$$

33. Solve the system below algebraically.
$$\begin{cases} x + 2y + 2z = 10 \\ 2x + y + 2z = 9 \\ 2x + 2y + z = 11 \end{cases}$$

In Exercises 34 and 35, use matrices to solve each system of equations. Row operations may be done either by hand or with the calculator. Verify solutions with the calculator.

34. $\begin{cases} 2x - 3y = 16 \\ -4x + y = -12 \end{cases}$

35. $\begin{cases} 3x - 2y + 4z = 4 \\ x + y + z = 3 \\ 6x - 2y - 3z = 10 \end{cases}$

In Exercises 36 and 37, use Cramer's rule to solve each system of equations.

36. $\begin{cases} 3x - y = -3 \\ 2x + y = -7 \end{cases}$

37. $\begin{cases} 2x + 3y + 4z = 6 \\ 2x - 3y - 4z = -4 \\ 4x + 6y + 8z = 12 \end{cases}$

In Exercises 38–40, graph the solution of each system of inequalities on graph paper when possible.

38. $\begin{cases} 3x + y \leq 1 \\ 4x - y > -8 \end{cases}$

39. $\begin{cases} x \geq 0 \\ y < 0 \end{cases}$

40. $\begin{cases} y \geq x + 2 \\ x \leq y - 2 \end{cases}$

41. Use a calculator to solve the equation $\log x + \log (x + 3) = 1$.

43. Solve the equation $2^{x^2 - 2x} = 8$
   a. algebraically   **b.** graphically

42. Solve the equation $\log 3x = \log 9$
   a. algebraically   **b.** graphically

44. Solve each system below graphically.
   a. $\begin{cases} x^2 - 6x - y = 5 \\ x^2 - 6x + y = -5 \end{cases}$
   b. $\begin{cases} x^2 - x - y = 2 \\ 4x - 3y = 0 \end{cases}$

APPLICATIONS

45. Ants   The number of ants in an anthill increases according to the function $y = 200e^{.5x}$, where y is the number of ants and x is time in days.
   a. How many days will it take the population of ants to increase to 1000?
   b. After 10 days, approximately how many ants are in the anthill?

46. Radioactive decay   A radioactive material decays according to the formula $A = A_0\left(\frac{2}{3}\right)^t$, where A_0 is the initial amount present and t is time, in years. If the initial sample contains 5 grams of a radioactive material, how many grams are present after 10 years?

47. Buying clothing   Two skirts and three sweaters cost $192, and three skirts and one sweater cost $184. Find the cost of a single skirt.

48. Mixing candy   A merchant wants to mix two types of candy to produce 30 pounds of mixture costing $2 per pound. If the more expensive candy in the mixture costs $2.50 per pound and the less expensive candy costs $1.25 per pound, how much of each type should he mix?

CAREERS & MATHEMATICS

Traffic Engineer

Traffic engineers supervise the design and construction of roads and gather and analyze data on traffic patterns. They determine efficient routes, designate the location of turn lanes and traffic control signals, and specify the timing of those signals to allow smooth traffic flow.

Qualifications

The minimal requirement for entry-level positions is a bachelor's degree with emphasis on mathematics and the physical sciences. Additional coursework in civil engineering is preferred. Because traffic engineers do a wide variety of tasks, many skills are learned on the job.

Job Outlook

For prepared individuals, employment opportunities in transportation are expected to be plentiful for many years to come.

Example Application

Before recommending traffic control signals for the intersection of the one-way streets shown in Illustration 1, a traffic engineer places sensors across the road to record traffic flow. The illustration shows the number of vehicles passing each of the four counters during 1 hour. From those data, find the number of vehicles passing straight through the intersection and also the number that turn from one road to the other.

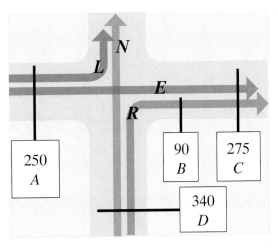

ILLUSTRATION 1

Solution

As shown in the illustration, we let

L represent the number of vehicles turning left,

R represent the number of vehicles turning right,

N represent the number of vehicles headed north, and

E represent the number of vehicles headed east.

Because the counter at A registers the total number of vehicles headed east and turning left, we have the equation

$L + E = 250$

Counter B records only vehicles turning right, so we have the equation

$R = 90$

673

Similarly, the two remaining counters provide two more equations, and we have the system

1. $L + E = 250$ Counter *A* records the total of vehicles turning left and those heading east.

2. $R = 90$ Counter *B* records the number turning right.

3. $R + E = 275$ Counter *C* records the total of vehicles turning right and those heading east.

4. $R + N = 340$ Counter *D* records the total of vehicles turning right and those heading north.

From equation 2, we know that $R = 90$. To find *E*, we substitute 90 for *R* in equation 3 and solve for *E*.

3. $R + E = 275$

    $90 + E = 275$ Substitute 90 for *R*.

       $E = 275 - 90$ To undo the addition of 90, subtract 90 from both sides.

       $E = 185$

To find *N*, we substitute 90 for *R* in equation 4 and solve for *N*.

4. $R + N = 340$

    $90 + N = 340$ Substitute 90 for *R*.

       $N = 250$ Subtract 90 from both sides.

Finally, to find *L*, we substitute 185 for *E* in equation 1 and solve for *L*.

1. $L + E = 250$

    $L + 185 = 250$

       $L = 65$ Subtract 185 from both sides.

The solutions are $L = 65$, $R = 90$, $N = 250$, and $E = 185$. The traffic engineer knows that 65 vehicles turned left, 90 turned right, 250 passed through heading north, and 185 headed east.

EXERCISES   The intersection of Marsh Street and one-way Fleet Avenue has three counters to record the traffic, as shown in Illustration 2.

1. Find *E*, the number of vehicles passing through the intersection headed east.
Find *S*, the number of vehicles turning south.
Find *N*, the number of northbound vehicles turning east.

2. The traffic engineer suspects that the counters are in error. Why?

ILLUSTRATION 2

(*Answers:* **1.** $E = 230$, $S = 120$, $N = -10$ **2.** The traffic count cannot be negative.)

674

Appendix A Conic Sections

■ THE CIRCLE ■ THE PARABOLA ■ ELLIPSES CENTERED AT THE ORIGIN ■ THE HYPERBOLA

The graphs of second-degree equations in *x* and *y* fall into one of several categories: a pair of lines, a point, a circle, a parabola, an ellipse, a hyperbola, or no graph at all (see Figure A-1). Because all of these graphs can be formed by the intersection of a plane and a right-circular cone, they are called **conic sections.**

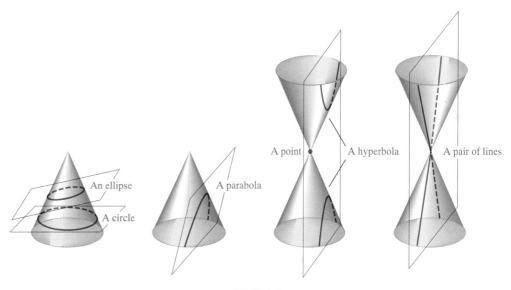

FIGURE A-1

Conic sections have many applications. For example, a parabola can be rotated to generate a dish-shaped surface called a **paraboloid.** Any light or sound placed at a certain point, called the *focus* of the paraboloid, is reflected outward in parallel paths. This property makes parabolic surfaces ideal for flashlight and headlight reflectors.

Using the same property in reverse makes parabolic surfaces good antennas, because signals captured by such an antenna are concentrated at the focus. A parabolic mirror is capable of concentrating the rays of the sun at a single point and thereby generating tremendous heat. This fact is used in the design of certain solar furnaces. Any object that is thrown upward and outward travels in a parabolic path. In architecture, many arches are parabolic in shape because of their strength, and the cable that supports a suspension bridge hangs in the form of a parabola.

Ellipses have optical and acoustical properties that are useful in architecture and engineering. For example, many arches are portions of an ellipse because the shape is pleasing to the eye. Gears are often cut into elliptical shapes to provide nonuniform motion. The planets and some comets have elliptical orbits.

Hyperbolas serve as the basis of a navigational system known as LORAN (LOng RAnge Navigation). They are also used to find the source of a distress signal, are the basis for the design of hypoid gears, and describe the orbits of some comets.

■ THE CIRCLE

The Circle
A **circle** is the set of all points in a plane that are a fixed distance from a point called its **center.**
The fixed distance is called the **radius** of the circle.

To develop the general equation of a circle, we must write the equation of a circle with a radius of r and with a center at some point $C(h, k)$, as in Figure A-2. This task is equivalent to finding all points $P(x, y)$ such that the length of line segment CP is r. We can use the distance formula to find r.

$$r = \sqrt{(x - h)^2 + (y - k)^2}$$

We then square both sides to obtain

$$r^2 = (x - h)^2 + (y - k)^2$$

This equation is called the **standard form of the equation of a circle** with a radius of r and center at the point with coordinates (h, k).

FIGURE A-2

Standard Equation of a Circle with Center at (h, k)

Any equation that can be written in the form

$$(x - h)^2 + (y - k)^2 = r^2$$

has a graph that is a circle with radius r and center at point (h, k).

If $r = 0$, the graph reduces to a single point called a **point circle.** If $r < 0$, then a circle does not exist. If both h and k are 0, then the center of the circle is the origin.

Standard Equation of a Circle with Center at (0, 0)

Any equation that can be written in the form

$$x^2 + y^2 = r^2$$

has a graph that is a circle with radius r and center at the origin.

EXAMPLE 1   Graph the circle $x^2 + y^2 - 4x + 2y = 20$.

Solution   Because the equation is not in standard form, the coordinates of the center and the length of the radius are not obvious. To put the equation in standard form, we complete the square on both x and y as follows:

$$x^2 + y^2 - 4x + 2y = 20$$
$$x^2 - 4x + y^2 + 2y = 20$$
$$x^2 - 4x + 4 + y^2 + 2y + 1 = 20 + 4 + 1 \qquad \text{Add 4 and 1 to both sides to complete the squares.}$$
$$(x - 2)^2 + (y + 1)^2 = 25 \qquad \text{Factor } x^2 - 4x + 4 \text{ and } y^2 + 2y + 1.$$
$$(x - 2)^2 + [y - (-1)]^2 = 5^2$$

The radius of the circle is 5, and the coordinates of its center are $h = 2$ and $k = -1$. We plot the center of the circle and draw a circle with a radius of 5 units, as shown in Figure A-3.

■ TECHNOLOGY TIP

To graph an equation that is *not* a function, you will need to solve the equation for y and graph each of the two resulting equations. Be sure to use a square viewing window such as ZDecimal × 2 or ZInteger so that the graph looks like a circle and not an ellipse.

$$y1 = -1 + \sqrt{25 - (x - 2)^2}$$
$$y2 = -1 - \sqrt{25 - (x - 2)^2}$$

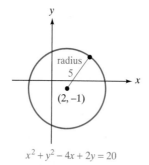

$x^2 + y^2 - 4x + 2y = 20$

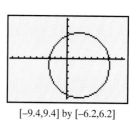

$[-9.4, 9.4]$ by $[-6.2, 6.2]$

FIGURE A-3

EXAMPLE 2   **Radio translators**   The effective broadcast area of a television station is bounded by the circle $x^2 + y^2 = 3600$, where x and y are measured in miles. A translator station picks up the signal and retransmits it from the center of a circular area bounded by $(x + 30)^2 + (y - 40)^2 = 1600$. Find the location of the translator and the greatest distance from the main transmitter that the signal can be received.

Solution   The coverage of the television station is bounded by $x^2 + y^2 = 60^2$, a circle centered at the origin with a radius of 60 miles, as shown in Figure A-4. Because the translator is at the center of the circle $(x + 30)^2 + (y - 40)^2 = 1600$, it is located at $(-30, 40)$, a point 30 miles west and 40 miles north of the television station. The radius of the translator's coverage is $\sqrt{1600}$, or 40 miles.

As shown in Figure A-4, the greatest distance of reception is the sum of A, the distance of the translator from the television station, and 40 miles, the radius of the translator's coverage.

To find A, we use the distance formula to find the distance between $(x_1, y_1) = (-30, 40)$ and the origin, $(0, 0)$.

$$A = \sqrt{(x_2 - x_1)^2 + (y_2 - y_1)^2}$$
$$A = \sqrt{(-30 - 0)^2 + (40 - 0)^2}$$
$$= \sqrt{(-30)^2 + 40^2}$$
$$= \sqrt{2500}$$
$$= 50$$

Thus, the translator is located 50 miles from the television station, and it broadcasts the signal an additional 40 miles. The greatest reception distance is $50 + 40$, or 90 miles.

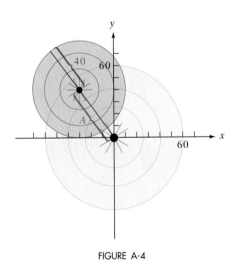

FIGURE A-4

■ THE PARABOLA

We have seen that equations of the form $y = a(x - h)^2 + k$, with $a \neq 0$, represent parabolas with vertex at the point (h, k). They open upward when $a > 0$ and downward when $a < 0$.

Equations of the form $x = a(y - k)^2 + h$ also represent parabolas with vertex at the point (h, k). However, they open to the right when $a > 0$ and to the left when $a < 0$. Parabolas that open to the right or left do not represent functions, because their graphs do not pass the vertical line test.

Several types of parabolas are summarized in the following chart. (In all cases, $a > 0$.)

Equations of Parabolas ($a > 0$)

Parabola Opening	Vertex at Origin	Vertex at (h, k)
Up	$y = ax^2$	$y = a(x - h)^2 + k$
Down	$y = -ax^2$	$y = -a(x - h)^2 + k$
Right	$x = ay^2$	$x = a(y - k)^2 + h$
Left	$x = -ay^2$	$x = -a(y - k)^2 + h$

EXAMPLE 3   Graph the equation $x = -2(y - 2)^2 + 3$.

Solution   We make a table of ordered pairs, plot each pair, and draw the parabola as in Figure A-5. Because the equation is of the form $x = -a(y - k)^2 + h$, the parabola opens to the left and has its vertex at the point with coordinates (3, 2).

■ **TECHNOLOGY TIP**

The equation of a parabola that opens left or right is a relation and not a function. Solve the equation for y and graph both equations to produce the complete graph of the relation.

$y1 = 2 + \sqrt{((x - 3)/-2)}$

$y2 = 2 - \sqrt{((x - 3)/-2)}$

$x = -2(y - 2)^2 + 3$

x	y	(x, y)
-5	0	$(-5, 0)$
1	1	$(1, 1)$
3	2	$(3, 2)$
1	3	$(1, 3)$
-5	4	$(-5, 4)$

FIGURE A-5

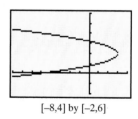

$[-8,4]$ by $[-2,6]$

■ **ELLIPSES CENTERED AT THE ORIGIN**

The Ellipse

An **ellipse** is the set of all points P in the plane the sum of whose distances from two fixed points is a constant. See Figure A-6, in which $d_1 + d_2$ is a constant.

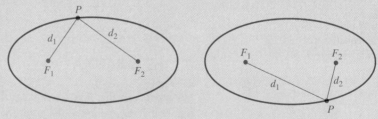

FIGURE A-6

Any equation that can be written in the form

$$\frac{x^2}{a^2} + \frac{y^2}{b^2} = 1 \qquad (a > b)$$

has a graph that is an **ellipse** centered at the origin, as in Figure A-7(a). The x-intercepts are the **vertices** $V(a, 0)$ and $V'(-a, 0)$. (Read V' as "V prime.") The y-intercepts are the points $(0, b)$ and $(0, -b)$.

Any equation that can be written in the form

$$\frac{x^2}{b^2} + \frac{y^2}{a^2} = 1 \qquad (a > b)$$

has a graph that is also an ellipse centered at the origin, as in Figure A-7(b). The y-intercepts are the vertices $V(0, a)$ and $V'(0, -a)$. The x-intercepts are the points $(b, 0)$ and $(-b, 0)$.

The point midway between the vertices is the **center** of the ellipse.

(a)   (b)

FIGURE A-7

Each of the two points is called a **focus.** Midway between the foci is the **center** of the ellipse.

The **major axis of a horizontal ellipse** is the line segment between the coordinates $(-a, 0)$ and $(a, 0)$. The **major axis of a vertical ellipse** is the line segment between the coordinates $(0, a)$ and $(0, -a)$.

EXAMPLE 4   Graph the equation $\dfrac{x^2}{36} + \dfrac{y^2}{9} = 1$.

Solution   The graph of the equation will be a horizontal ellipse with x-intercepts of $(6, 0)$ and $(-6, 0)$ and y-intercepts of $(0, 3)$ and $(0, -3)$. (See Figure A-8.) Plot the intercepts and connect with a smooth curve.

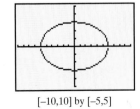

[−10,10] by [−5,5]

FIGURE A-8

EXAMPLE 5   **Landscape design**   A landscape architect is designing an elliptical pool that will fit in the center of a 20-by-30-foot rectangular garden, leaving at least 5 feet of space on all sides. Find the equation of the ellipse.

Solution   We place the rectangular garden in a coordinate system, as in Figure A-9. To maintain 5 feet of clearance at the ends of the ellipse, the vertices must be the points $V(10, 0)$ and $V'(-10, 0)$. Similarly, the y-intercepts are the points $(0, 5)$ and $(0, -5)$.

The equation of the ellipse has the form

$$\frac{x^2}{a^2} + \frac{y^2}{b^2} = 1$$

with $a = 10$ and $b = 5$. Thus, the equation of the boundary of the pool is

$$\frac{x^2}{100} + \frac{y^2}{25} = 1$$

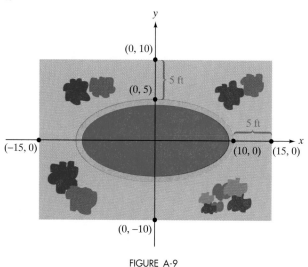

FIGURE A-9

■ THE HYPERBOLA

The Hyperbola

A **hyperbola** is the set of all points P in the plane for which the difference of the distances of each point on the hyperbola from two fixed points is a constant. See Figure A-10, in which $d_1 - d_2$ is a constant. Each of the two points is called a **focus**. Midway between the foci is the **center** of the hyperbola.

FIGURE A-10

Any equation that can be written in the form

$$\frac{x^2}{a^2} - \frac{y^2}{b^2} = 1$$

has a graph that is a **hyperbola** centered at the origin, as in Figure A-11. The x-intercepts are the **vertices** $V(a, 0)$ and $V'(-a, 0)$. There are no y-intercepts.

Although it is possible to draw any hyperbola by plotting many points and connecting them in a smooth curve, there is an easier way. Plot the points $(a, 0)$ and $(-a, 0)$, the x-intercepts, and $(0, b)$ and $(0, -b)$. Construct a rectangle whose sides pass horizontally through $(0, b)$ and $(0, -b)$ and vertically through $(a, 0)$ and $(-a, 0)$. This rectangle is called the **fundamental rectangle of the hyperbola.** The extended diagonals of the rectangle are the **asymptotes** of the hyperbola. As the points on the branches of the hyperbola move further away from the origin, they get closer to, but never cross, these extended diagonals. See Figure A-11.

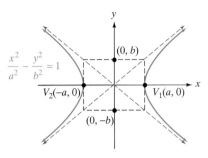

FIGURE A-11

EXAMPLE 6   Graph the hyperbola whose equation is $\dfrac{x^2}{25} - \dfrac{y^2}{9} = 1$.

Solution   We can see from the equation that $a = \pm 25$ and $b = \pm 3$. Plot the points $(5, 0)$, $(-5, 0)$, $(0, 3)$ and $(0, -3)$. Construct the fundamental rectangle and its extended diagonals. Sketch the hyperbola (see Figure A-12).

■ **TECHNOLOGY TIP**

Solve $\dfrac{x^2}{25} - \dfrac{y^2}{9} = 1$ for y and graph the pair of functions that determine the relation.

$y1 = \sqrt{(9x^2 - 225)}/5$

$y2 = -\sqrt{(9x^2 - 225)}/5$

$[-9.4, 9.4]$ by $[-6.2, 6.2]$

FIGURE A-12

Equation of a Hyperbola Centered at the Origin

Any equation that can be written in the form

$$\frac{y^2}{a^2} - \frac{x^2}{b^2} = 1$$

has a graph that is a **hyperbola** centered at the origin, as in Figure A-13. The y-intercepts are the **vertices** $V(0, a)$ and $V'(0, -a)$. There are no x-intercepts. The **asymptotes** of the hyperbola are the extended diagonals of the rectangle in the figure.

FIGURE A-13

EXAMPLE 7   Graph the equation $9y^2 - 4x^2 = 36$.

Solution   To write the equation in standard form, we divide both sides by 36 to obtain

$$\frac{9y^2}{36} - \frac{4x^2}{36} = 1$$

$$\frac{y^2}{4} - \frac{x^2}{9} = 1 \qquad \text{Simplify each fraction.}$$

Note that $a = \pm 2$ and $b = \pm 3$. Plot the points $(3, 0)$, $(-3, 0)$, $(0, 2)$ and $(0, -2)$. Sketch the fundamental rectangle and its extended diagonals. Sketch the hyperbola. (See Figure A-14.)

■ **TECHNOLOGY TIP**
Solve $9y^2 - 4x^2 = 36$ for y and graph the pair of functions that determine the relation.

$$y1 = \sqrt{(4x^2 + 36)}/3$$
$$y2 = -\sqrt{(4x^2 + 36)}/3$$

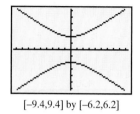

[−9.4,9.4] by [−6.2,6.2]

FIGURE A-14

There is a special type of hyperbola (also centered at the origin) that does not intersect either the x- or the y-axis. These hyperbolas have equations of the form $xy = k$, where $k \neq 0$.

EXAMPLE 8   Graph the equation $xy = -8$.

Solution   We make a table of ordered pairs, plot each pair, and join the points with a smooth curve to obtain the hyperbola. (See Figure A-15.)

$$xy = -8$$

x	y	(x, y)
1	−8	$(1, -8)$
2	−4	$(2, -4)$
4	−2	$(4, -2)$
8	−1	$(8, -1)$
−1	8	$(-1, 8)$
−2	4	$(-2, 4)$
−4	2	$(-4, 2)$
−8	1	$(-8, 1)$

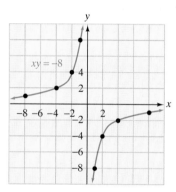

FIGURE A-15

The result in Example 8 suggests the following theorem.

Theorem
Any equation of the form $xy = k$, where $k \neq 0$, has a graph that is a **hyperbola** that does not intersect either the x- or the y-axis.

EXAMPLE 9   **Atomic structure**   In an experiment that led to the discovery of the atomic structure of matter, Lord Rutherford (1872–1937) shot high-energy alpha particles toward a thin sheet of gold. Because many were reflected, Rutherford showed the existence of a gold atom. The alpha particle in Figure A-16 is repelled by the nucleus at the origin and travels along the hyperbolic path given by $4x^2 - y^2 = 16$. How close does the particle come to the nucleus?

Solution   To find the distance from the nucleus at the origin, we must find the coordinates of the vertex V. To do so, we write the equation of the particle's path in standard form:

$$4x^2 - y^2 = 16$$

$$\frac{4x^2}{16} - \frac{y^2}{16} = \frac{16}{16} \qquad \text{Divide both sides by 16.}$$

$$\frac{x^2}{4} - \frac{y^2}{16} = 1 \qquad \text{Simplify.}$$

$$\frac{x^2}{2^2} - \frac{y^2}{4^2} = 1 \qquad \text{Write 4 as } 2^2 \text{ and 16 as } 4^2.$$

This equation is in the form $\dfrac{x^2}{a^2} - \dfrac{y^2}{b^2} = 1$, with $a = 2$. Thus, the vertex of the path is $(2, 0)$. The particle is never closer than 2 units from the nucleus. ■

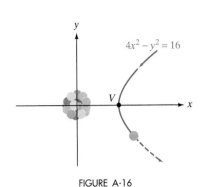

FIGURE A-16

APPENDIX A EXERCISE

VOCABULARY   In Exercises 1–7, fill in the blanks to make a true statement.

1. Graphs formed by the intersection of a plane and a right-circular cone are called _____.

2. The set of all points in a plane that are a fixed distance from a point called the center form a _____.

3. The standard form for a circle whose center is at (h, k) is _____.

4. The equation $x^2 + y^2 = r^2$ is the standard form of the equation of a circle with center at _____.

5. An ellipse whose major axis is along the y-axis will be an ellipse centered at the _____ whose orientation is _____.

6. A hyperbola of the form $\dfrac{x^2}{a^2} - \dfrac{y^2}{b^2} = 1$ is centered at the _____ and has no __-intercepts.

7. A hyperbola of the form $\dfrac{y^2}{a^2} - \dfrac{x^2}{b^2} = 1$ is centered at the _____ and has no __-intercepts.

CONCEPTS   In Exercises 8–14, match each equation to its description without graphing.

8. $\dfrac{x^2}{9} + \dfrac{y^2}{16} = 1$

9. $x = (y - 2)^2 + 5$

10. $x^2 + y^2 = 1$

11. $\dfrac{x^2}{9} - \dfrac{y^2}{16} = 1$

12. $(x - 1)^2 + (y - 2)^2 = 4$

13. $\dfrac{y^2}{16} - \dfrac{x^2}{9} = 1$

14. $\dfrac{x^2}{16} + \dfrac{y^2}{9} = 1$

a. a vertical ellipse centered at $(0, 0)$

b. a circle of radius 1 with center $(0, 0)$

c. a hyperbola opening left/right

d. a horizontal ellipse centered at $(0, 0)$

e. a parabola opening right with vertex at $(5, 2)$

f. a circle of radius 2 with center at $(1, 2)$

g. a hyperbola opening up/down

PRACTICE   In Exercises 15–19, find the center and radius of each circle and then graph it. Verify with your calculator.

15. $x^2 + y^2 = 9$

16. $x^2 + y^2 = 16$

17. $(x - 2)^2 + (y - 4)^2 = 4$

18. $(x - 3)^2 + (y - 2)^2 = 4$

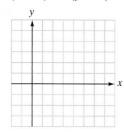

19. $x^2 + y^2 - 2x + 4y = -1$

In Exercises 20–25, find the vertex of each parabola and then graph it. Verify with your calculator.

20. $x = y^2$

21. $x = 4y^2$

22. $x = -3(y + 2)^2 - 2$

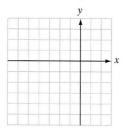

23. $x = 2(y - 3)^2 - 4$

24. $x = -y^2 - 2y + 3$

25. $x = y^2 + 4y + 5$

In Exercises 26–29, determine the x- and y-intercepts and graph each ellipse. Verify with your calculator.

26. $\dfrac{x^2}{4} + \dfrac{y^2}{9} = 1$

27. $x^2 + \dfrac{y^2}{9} = 1$

28. $25x^2 + 9y^2 = 225$

29. $4x^2 + 16y^2 = 64$

In Exercises 30–37, graph the hyperbola by sketching the fundamental rectangle and asymptotes, where applicable.

30. $\dfrac{x^2}{9} - \dfrac{y^2}{4} = 1$

31. $\dfrac{y^2}{4} - \dfrac{x^2}{9} = 1$

32. $\dfrac{y^2}{25} - \dfrac{x^2}{4} = 1$

33. $\dfrac{x^2}{4} - \dfrac{y^2}{25} = 1$

34. $25x^2 - y^2 = 25$

35. $9x^2 - 4y^2 = 36$

36. $xy = 8$

37. $xy = -10$

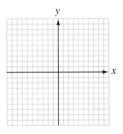

APPLICATIONS

38. Meshing gears   For design purposes, the large gear in Illustration 1 is the circle $x^2 + y^2 = 16$. The smaller gear is a circle centered at $(7, 0)$ and tangent to the larger circle. Find the equation of the smaller gear.

39. Width of a walkway   The walkway in Illustration 2 is bounded by the two circles $x^2 + y^2 = 2500$ and $(x - 10)^2 + y^2 = 900$, measured in feet. Find the largest and the smallest width of the walkway.

ILLUSTRATION 1

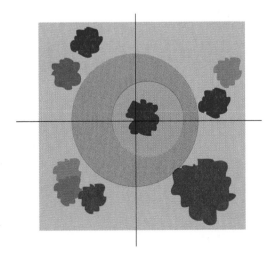

ILLUSTRATION 2

40. Broadcast ranges   Radio stations applying for licensing may not use the same frequency if their broadcast areas overlap. One station's coverage is bounded by $x^2 + y^2 - 8x - 20y + 16 = 0$, and the other's by $x^2 + y^2 + 2x + 4y - 11 = 0$. May they be licensed for the same frequency?

41. Highway curves   Highway design engineers want to join two sections of highway with a curve that is one-quarter of a circle, as in Illustration 3. The equation of the circle is $x^2 + y^2 - 16x - 20y + 155 = 0$, where distances are measured in kilometers. Find the locations (relative to the center of town at the origin) of the intersections of the highway with State and Main.

ILLUSTRATION 3

42. Orbit of a comet   If the orbit of the comet shown in Illustration 4 is given by the equation $2y^2 - 9x = 18$, how far is it from the sun at the vertex of the orbit? Distances are in astronomical units (AU).

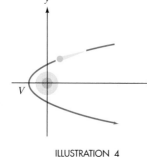

ILLUSTRATION 4

43. Satellite antenna   The cross section of the satellite antenna in Illustration 5 is a parabola given by the equation $x = \frac{1}{16}y^2$, with distances measured in feet. If the dish is 8 feet wide, how deep is it?

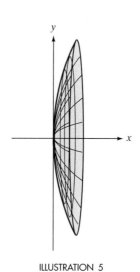

ILLUSTRATION 5

44. Designing an underpass   The arch of the underpass in Illustration 6 is part of an ellipse. Find the equation of the arch.

ILLUSTRATION 6

45. Calculating clearance   Find the height of the elliptical arch in Exercise 44 at a point 10 feet from the center of the roadway.

46. Area of an ellipse   The area A of the ellipse $\dfrac{x^2}{a^2} + \dfrac{y^2}{b^2} = 1$ is given by $A = \pi ab$. Find the area of the ellipse $9x^2 + 16y^2 = 144$.

47. Area of a track   The elliptical track in Illustration 7 is bounded by the ellipses $4x^2 + 9y^2 = 576$ and $9x^2 + 25y^2 = 900$. Find the area of the track. (See Exercise 46)

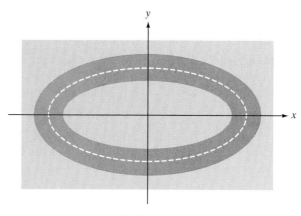

ILLUSTRATION 7

48. Alpha particles   The particle in Illustration 8 approaches the nucleus at the origin along the path $9y^2 - x^2 = 81$. How close does the particle come to the nucleus?

ILLUSTRATION 8

49. LORAN   By determining the difference of the distances between the ship in Illustration 9 and the two radio transmitters, the LORAN system places the ship on the hyperbola $x^2 - 4y^2 = 576$. If the ship is also 5 miles out to sea, find its coordinates.

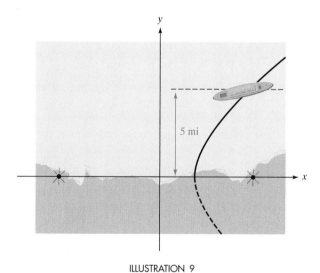

5 mi

ILLUSTRATION 9

50. Sonic boom   The position of the sonic boom caused by the faster-than-sound aircraft in Illustration 10 is the hyperbola $y^2 - x^2 = 25$ in the coordinate system shown. How wide is the hyperbola 5 miles from its vertex?

5 mi

ILLUSTRATION 10

51. Electrostatic repulsion   Two similarly charged particles are shot together for an almost head-on collision, as in Illustration 11. They repel each other and travel the two branches of the hyperbola given by $x^2 - 4y^2 = 4$. How close do they get?

ILLUSTRATION 11

Appendix B Binomial Theorem

■ PASCAL'S TRIANGLE ■ FACTORIAL NOTATION ■ THE BINOMIAL THEOREM ■ THE nTH TERM OF A BINOMIAL EXPANSION

We have discussed how to raise binomials to positive integral powers. For example, we know that

$$(a + b)^2 = a^2 + 2ab + b^2$$

and that

$$
\begin{aligned}
(a + b)^3 &= (a + b)(a + b)^2 \\
&= (a + b)(a^2 + 2ab + b^2) \\
&= a^3 + 2a^2b + ab^2 + a^2b + 2ab^2 + b^3 \\
&= a^3 + 3a^2b + 3ab^2 + b^3
\end{aligned}
$$

To show how to raise binomials to positive integral powers without doing the actual multiplications, we consider the following binomial expansions:

$$(a + b)^0 = 1$$
$$(a + b)^1 = a + b$$
$$(a + b)^2 = a^2 + 2ab + b^2$$
$$(a + b)^3 = a^3 + 3a^2b + 3ab^2 + b^3$$
$$(a + b)^4 = a^4 + 4a^3b + 6a^2b^2 + 4ab^3 + b^4$$
$$(a + b)^5 = a^5 + 5a^4b + 10a^3b^2 + 10a^2b^3 + 5ab^4 + b^5$$
$$(a + b)^6 = a^6 + 6a^5b + 15a^4b^2 + 20a^3b^3 + 15a^2b^4 + 6ab^5 + b^6$$

Several patterns appear in these expansions:

1. Each expansion has one more term than the power of the binomial.

2. The degree of each term in each expansion is equal to the exponent of the binomial that is being expanded.

3. The first term in each expansion is a, raised to the power of the binomial.

4. The exponents of a decrease by 1 in each successive term. The exponents of b, beginning with $b^0 = 1$ in the first term, increase by 1 in each successive term. Thus, the variables have the pattern

$$a^n, \ a^{n-1}b, \ a^{n-2}b^2, \ \ldots , \ ab^{n-1}, \ b^n$$

■ PASCAL'S TRIANGLE

To see another pattern, we write the coefficients of each binomial expansion in the following triangular array:

```
                 1
              1     1
           1     2     1
        1     3     3     1
     1     4     6     4     1
   1     5    10    10     5     1
 1     6    15    20    15     6     1
```

In this array, called **Pascal's triangle,** each entry between the 1s is the sum of the closest pair of numbers in the line immediately above it. For example, the first 15 in the bottom row is the sum of the 5 and 10 immediately above it. Pascal's triangle continues with the same pattern forever. The next two lines are

$$1 \quad 7 \quad 21 \quad 35 \quad 35 \quad 21 \quad 7 \quad 1$$
$$1 \quad 8 \quad 28 \quad 56 \quad 70 \quad 56 \quad 28 \quad 8 \quad 1$$

EXAMPLE 1   Expand $(u - v)^4$.

Solution   We note that the expression $(u - v)^4$ can be written in the form $[u + (-v)]^4$. The variables in this expansion are

$$u^4, \quad u^3(-v), \quad u^2(-v)^2, \quad u(-v)^3, \quad (-v)^4$$

and the coefficients are given in Pascal's triangle in the row whose second entry is 4:

$$1 \quad 4 \quad 6 \quad 4 \quad 1$$

Hence, the required expansion is

$$
\begin{aligned}
(u - v)^4 &= u^4 + 4u^3(-v) + 6u^2(-v)^2 + 4u(-v)^3 + (-v)^4 \\
&= u^4 - 4u^3v + 6u^2v^2 - 4uv^3 + v^4
\end{aligned}
$$
∎

■ FACTORIAL NOTATION

Although Pascal's triangle gives the coefficients of the terms in a binomial expansion, it is not the best way to expand a binomial. To develop another way to expand a binomial, we introduce **factorial notation.**

> **Factorial Notation**
> If n is a natural number, the symbol $n!$ (read as **n factorial** or as **factorial n**) is defined as
> $$n! = n(n - 1)(n - 2)(n - 3) \cdots (3)(2)(1)$$

EXAMPLE 2   Find: **a.** $2!$ **b.** $5!$ **c.** $-9!$ and   **d.** $(n - 2)!$

Solution   **a.** $2! = 2 \cdot 1 = 2$

b. $5! = 5 \cdot 4 \cdot 3 \cdot 2 \cdot 1 = 120$

■ TECHNOLOGY TIP
The factorial symbol "!" is located under the submenu PRB of the MATH menu.

c. $-9! = -9 \cdot 8 \cdot 7 \cdot 6 \cdot 5 \cdot 4 \cdot 3 \cdot 2 \cdot 1 = -362{,}880$

d. $(n - 2)! = (n - 2)(n - 3)(n - 4) \cdot \cdots \cdot 3 \cdot 2 \cdot 1$

∎

WARNING!   According to the previous definition, part d is meaningful only if $n - 2$ is a natural number.

We define zero factorial as follows.

> **Zero Factorial**
> $0! = 1$

We note that

$$5 \cdot 4! = 5 \cdot 4 \cdot 3 \cdot 2 \cdot 1 = 5!$$
$$7 \cdot 6! = 7 \cdot 6 \cdot 5 \cdot 4 \cdot 3 \cdot 2 \cdot 1 = 7!$$
$$10 \cdot 9! = 10 \cdot 9 \cdot 8 \cdot 7 \cdot 6 \cdot 5 \cdot 4 \cdot 3 \cdot 2 \cdot 1 = 10!$$

These examples suggest the following theorem.

> **Theorem**
> If n is a positive integer, then $n(n-1)! = n!$.

THE BINOMIAL THEOREM

We now state the binomial theorem.

> **The Binomial Theorem**
> If n is any positive integer, then
>
> $$(a+b)^n = a^n + \frac{n!}{1!(n-1)!}a^{n-1}b + \frac{n!}{2!(n-2)!}a^{n-2}b^2$$
> $$+ \frac{n!}{3!(n-3)!}a^{n-3}b^3 + \cdots + \frac{n!}{r!(n-r)!}a^{n-r}b^r$$
> $$+ \cdots + b^n$$

In the binomial theorem, the exponents of the variables follow the familiar pattern:

- The sum of the exponents of a and b in each term is n,
- the exponents of a decrease, and
- the exponents of b increase.

Only the method of finding the coefficients is different. Except for the first and last terms, the numerator of each coefficient is $n!$. If the exponent of b in a particular term is r, the denominator of the coefficient of that term is $r!(n-r)!$.

EXAMPLE 3   Use the binomial theorem to expand $(x-y)^4$.

Solution   We can write $(x-y)^4$ in the form $[x + (-y)]^4$, substitute directly into the binomial theorem, and simplify:

$$(x-y)^4 = [x + (-y)]^4$$
$$= x^4 + \frac{4!}{1!(4-1)!}x^3(-y) + \frac{4!}{2!(4-2)!}x^2(-y)^2 + \frac{4!}{3!(4-3)!}x(-y)^3 + (-y)^4$$
$$= x^4 - \frac{4 \cdot 3!}{1!3!}x^3y + \frac{4 \cdot 3 \cdot 2!}{2!2!}x^2y^2 - \frac{4 \cdot 3!}{3!1!}xy^3 + y^4$$
$$= x^4 - 4x^3y + 6x^2y^2 - 4xy^3 + y^4$$

THE nTH TERM OF A BINOMIAL EXPANSION

To find the fourth term of the expansion of $(a+b)^9$, we could raise the binomial $a + b$ to the ninth power and look at the fourth term. However, this task would be very tedious. By using the binomial theorem, we can construct the fourth term without finding the complete expansion of $(a+b)^9$.

EXAMPLE 4   Find the sixth term in the expansion of $(x - y)^7$.

Solution   We find the sixth term of $[x + (-y)]^7$. In the sixth term, the exponent of $(-y)$ is 5. Thus, the variables in the sixth term are

$$x^2(-y)^5 \qquad \text{The sum of the exponents must be 7.}$$

The coefficient of these variables is

$$\frac{n!}{r!(n - r)!} = \frac{7!}{5!(7 - 5)!}$$

The complete sixth term is

$$\frac{7!}{5!(7 - 5)!}x^2(-y)^5 = -\frac{7 \cdot 6 \cdot 5!}{5! \cdot 2 \cdot 1}x^2 y^5$$
$$= -21x^2 y^5$$

APPENDIX B EXERCISE

VOCABULARY   *In Exercises 1–3, fill in the blanks to make a true statement.*

1. A triangular array of numbers that determines the coefficients of a binomial expansion is called _____.

2. If n is a natural number, then $n!$ is called _____ notation, which is defined as _____.

3. Zero factorial, $0!$, is equal to __.

CONCEPTS

✏ **4.** Explain how to construct Pascal's triangle.

5. Without looking at your text, write the first ten rows of Pascal's triangle.

6. Find the sum of the numbers in each of the first ten rows of Pascal's triangle. What is the pattern?

7. Find the sum of the numbers in the designated diagonal rows of Pascal's triangle shown in Illustration 1. What is the pattern?

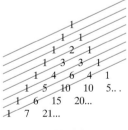

ILLUSTRATION 1

✏ **8.** Explain how to find the variables in the expansion of $(r + s)^4$.

PRACTICE   *In Exercises 9–14, evaluate each expression and verify with your calculator.*

9. $3!$

10. $-5!$

11. $11!(5)$

12. $8(7!)$

13. $\dfrac{5!}{3!(5 - 3)!}$

14. $\dfrac{6!}{4!(6 - 4)!}$

In Exercises 15–24, use the binomial theorem to expand each expression.

15. $(x + y)^3$

16. $(x - y)^3$

17. $(2x + y)^4$

18. $(x + 2y)^4$

19. $(x - 2y)^5$

20. $(2x - y)^5$

21. $(2x + 3y)^4$

22. $(3x - 2y)^4$

23. $\left(\dfrac{x}{2} - \dfrac{y}{2}\right)^4$

24. $\left(\dfrac{x}{2} + \dfrac{y}{2}\right)^4$

In Exercises 25–30, use the binomial theorem to find the required term of each expansion.

25. $(a + b)^4$; second term

26. $(a + b)^4$; third term

27. $(x - 3y)^8$; third term

28. $(x - 3y)^8$; fifth term

29. $(2x + 3)^6$; sixth term

30. $(2x + 3)^6$; fourth term

Appendix C Arithmetic and Geometric Sequences

■ ARITHMETIC SEQUENCES ■ ARITHMETIC MEANS ■ THE SUM OF THE FIRST *n* TERMS OF AN ARITHMETIC SEQUENCE ■ SUMMATION NOTATION ■ GEOMETRIC SEQUENCES ■ GEOMETRIC MEANS ■ THE SUM OF THE FIRST *n* TERMS OF A GEOMETRIC SEQUENCE

A **sequence** is a function whose domain is the set of natural numbers. For example, the function $f(n) = 3n + 2$, where n is a natural number, is a sequence. Because a sequence is a function whose domain is the set of natural numbers, it is easy to write its values as a list. If the natural numbers are substituted for n, the function $f(n) = 3n + 2$ generates the list

5, 8, 11, 14, 17, . . .

It is common to call the list, as well as the function, a sequence. Each number in the list is called a **term** of the sequence. Other examples of sequences are

a. $1^3, 2^3, 3^3, 4^3, \ldots$ b. 4, 8, 12, 16, . . .
c. 2, 3, 5, 7, 11, . . . d. 1, 1, 2, 3, 5, 8, 13, 21, . . .

The sequence in part (a) is the ordered list of the cubes of the natural numbers. The sequence in part (b) is the ordered list of the positive multiples of 4. The sequence in part (c) is the ordered list of the prime numbers. The sequence in part (d) is called the **Fibonacci sequence,** after the 12th-century mathematician Leonardo of Pisa—also known as Fibonacci. Beginning with the 2, each term of the Fibonacci sequence is the sum of the two preceding terms.

■ ARITHMETIC SEQUENCES

One important type of sequence is the **arithmetic sequence.**

> **Arithmetic Sequence**
> An **arithmetic sequence** is a sequence of the form
> $$a, a + d, a + 2d, a + 3d, \ldots, a + (n - 1)d, \ldots$$
> where a is the **first term,** $a + (n - 1)d$ is the **nth term,** and d is the **common difference.**

We note that the second term of an arithmetic sequence has an addend of $1d$, the third term has an addend of $2d$, the fourth term has an addend of $3d$, and the nth term has an addend of $(n - 1)d$. We also note that the difference between any two consecutive terms in an arithmetic sequence is d.

EXAMPLE 1 The first three terms of an arithmetic sequence are 3, 8, and 13.
Find **a.** the 67th term and **b.** the 100th term.

Solution We first find d, the common difference. It is the difference between successive terms:

$$d = 8 - 3 = 13 - 8 = 5$$

To generate a table of terms in the sequence, change the calculator to **sequ**ence MODE. At the y= screen, enter $u(n) = 3 + (n - 1)5$ to designate a sequence whose common difference is 5 and initial term is 3 and $u(n \text{ min}) = 3$ to designate that the sequence begins at the value of three. Now view the TABLE.

a. We substitute 3 for a, 67 for n, and 5 for d in the formula for the nth term and simplify:

$$n\text{th term} = a + (n - 1)d$$
$$67\text{th term} = 3 + (67 - 1)5$$
$$= 3 + 66(5)$$
$$= 333$$

b. We substitute 3 for a, 100 for n, and 5 for d in the formula for the nth term and simplify:

$$n\text{th term} = a + (n - 1)d$$
$$100\text{th term} = 3 + (100 - 1)5$$
$$= 3 + 99(5)$$
$$= 498$$

■

■ ARITHMETIC MEANS

If numbers are inserted between two numbers a and b to form an arithmetic sequence, the inserted numbers are called **arithmetic means** between a and b. If a single number is inserted between the numbers a and b, that number is called **the arithmetic mean** between a and b.

EXAMPLE 2   Insert two arithmetic means between 6 and 27.

Solution   In this example, the first term is $a = 6$, and the fourth term (or the last term) is $l = 27$. We must find the common difference such that the terms

$$6, 6 + d, 6 + 2d, 27$$

form an arithmetic sequence. To find d, we can substitute 6 for a and 4 for n into the formula for the nth term:

$$n\text{th term} = a + (n - 1)d$$
$$4\text{th term} = 6 + (4 - 1)d$$
$$27 = 6 + 3d \qquad \text{Simplify.}$$
$$21 = 3d \qquad \text{Subtract 6 from both sides.}$$
$$7 = d \qquad \text{Divide both sides by 3.}$$

The two arithmetic means between 6 and 27 are

$$6 + d = 6 + 7 \qquad \text{or} \qquad 6 + 2d = 6 + 2(7)$$
$$= 13 \qquad\qquad\qquad\qquad = 6 + 14$$
$$\qquad\qquad\qquad\qquad\qquad = 20$$

The numbers 6, 13, 20, and 27 are the first four terms of an arithmetic sequence.

■

■ THE SUM OF THE FIRST n TERMS OF AN ARITHMETIC SEQUENCE

There is a formula that gives the sum of the first n terms of an arithmetic sequence. To develop this formula, we let S_n represent the sum of the first n terms of an arithmetic sequence:

$$S_n = \quad a \quad + \quad [a + d] \quad + \quad [a + 2d] \quad + \cdots + \quad [a + (n - 1)d]$$

We write the same sum again, but in reverse order:

$$S_n = [a + (n - 1)d] + [a + (n - 2)d] + [a + (n - 3)d] + \cdots + \quad a$$

We add these two equations together, term by term, to get

$$2S_n = [2a + (n - 1)d] + [2a + (n - 1)d] + [2a + (n - 1)d] + \cdots + [2a + (n - 1)d]$$

Because there are n equal terms on the right-hand side of the preceding equation, we can write

$$2S_n = n[2a + (n - 1)d]$$
$$2S_n = n[a + a + (n - 1)d] \qquad 2a = a + a.$$
$$2S_n = n[a + l] \qquad \text{Substitute } l \text{ for } a + (n - 1)d, \text{ because}$$
$$\qquad\qquad\qquad\qquad a + (n - 1)d \text{ is the last term of the sequence.}$$

$$S_n = \frac{n(a + l)}{2} \qquad \text{Divide both sides by 2.}$$

This reasoning establishes the following theorem.

Sum of the First n Terms of an Arithmetic Sequence
The sum of the first n terms of an arithmetic sequence is given by the formula

$$S_n = \frac{n(a + l)}{2} \qquad \text{with } l = a + (n - 1)d$$

where a is the first term, l is the last (or nth) term, and n is the number of terms in the sequence.

EXAMPLE 3   Find the sum of the first 40 terms of the arithmetic sequence 4, 10, 16, ...

Solution   In this example, we let $a = 4$, $n = 40$, $d = 6$, and $l = 4 + (40 - 1)6 = 238$ and substitute these values into the formula for S_n:

$$S_n = \frac{n(a + l)}{2}$$

$$S_{40} = \frac{40(4 + 238)}{2}$$

$$= 20(242)$$

$$= 4840$$

The sum of the first 40 terms is 4840.   ■

■ SUMMATION NOTATION

There is a shorthand notation for indicating the sum of a finite (ending) number of consecutive terms in a sequence. This notation, called **summation notation,** involves the Greek letter Σ (sigma). The expression

$$\sum_{k=2}^{5} 3k \qquad \text{Read as "the summation of } 3k \text{ as } k \text{ runs from 2 to 5."}$$

designates the sum of all terms obtained if we successively substitute the numbers 2, 3, 4, and 5 for k, called the **index of the summation.** Thus, we have

$$\sum_{k=2}^{5} 3k = 3(2) + 3(3) + 3(4) + 3(5)$$

$$= 6 + 9 + 12 + 15$$

$$= 42$$

EXAMPLE 4   Find each sum: **a.** $\displaystyle\sum_{k=3}^{5}(2k + 1)$ **b.** $\displaystyle\sum_{k=2}^{5}k^2$

Solution   **a.** $\displaystyle\sum_{k=3}^{5}(2k + 1) = [2(3) + 1] + [2(4) + 1] + [2(5) + 1]$

$$= 7 + 9 + 11$$

$$= 27$$

■ TECHNOLOGY TIP
With the calculator in sequence MODE, you can compute the sum of the sequence. The **sum(** command is located under the MATH submenu of the LIST menu and the **seq(** command is found under the OPS submenu of the LIST menu.

b. $\displaystyle\sum_{k=2}^{5}k^2 = 2^2 + 3^2 + 4^2 + 5^2$

$$= 4 + 9 + 16 + 25$$

$$= 54$$

■ GEOMETRIC SEQUENCES

Another important sequence is the geometric sequence.

> **Geometric Sequence**
> A **geometric sequence** is a sequence of the form
> $$a, ar, ar^2, ar^3, \ldots, ar^{n-1}, \ldots$$
> where a is the **first term,** ar^{n-1} is the **nth term,** and r is the **common ratio.**

We note that the second term of a geometric sequence has a factor of r^1, the third term has a factor of r^2, the fourth term has a factor of r^3, and the nth term has a factor of r^{n-1}. We also note that the quotient obtained when any term is divided by the previous term is r.

EXAMPLE 5   The first three terms of a geometric sequence are 16, 4, and 1. Find the seventh term of the sequence.

Solution We substitute 16 for a, $\frac{1}{4}$ for r, and 7 for n in the formula for the nth term and simplify:

$$n\text{th term} = ar^{n-1}$$

$$7\text{th term} = 16\left(\frac{1}{4}\right)^{7-1} \qquad r = \frac{4}{16} = \frac{1}{4}.$$

$$= 16\left(\frac{1}{4}\right)^6$$

$$= 16\left(\frac{1}{4096}\right)$$

$$= \frac{1}{256}$$

```
Plot1 Plot2 Plot3
nMin=1
·u(n)◘16(1/4)^(n
-1)
u(nMin)◘(16)
\v(n)=
v(nMin)=
\w(n)=
```

```
 n  | u(n)
 1  | 16
 2  | 4
 3  | 1
 4  | .25
 5  | .0625
 6  | .01563
    | .00391
n=7
```

GEOMETRIC MEANS

If numbers are inserted between two numbers a and b to form a geometric sequence, the inserted numbers are called **geometric means** between a and b. If a single number is inserted between the numbers a and b, that number is called a **geometric mean** between a and b.

EXAMPLE 6 Insert two geometric means between 7 and 1512.

Solution In this example, the first term is $a = 7$ and the fourth term (or last term) is $l = 1512$. To find the common ratio r such that the terms

$$7, \quad 7r, \quad 7r^2, \quad 1512$$

form a geometric sequence, we substitute 4 for n and 7 for a into the formula for the nth term of a geometric sequence and solve for r.

$$n\text{th term} = ar^{n-1}$$

$$4\text{th term} = 7r^{4-1}$$

$$1512 = 7r^3$$

$$216 = r^3 \qquad \text{Divide both sides by 7.}$$

$$6 = r \qquad \text{Take the cube root of both sides.}$$

The two geometric means between 7 and 1512 are

$$7r = 7(6) = 42$$

and

$$7r^2 = 7(6)^2 = 7(36) = 252$$

The numbers 7, 42, 252, and 1512 are the first four terms of a geometric sequence.

THE SUM OF THE FIRST n TERMS OF A GEOMETRIC SEQUENCE

There is a formula that gives the sum of the first n terms of a geometric sequence. To develop this formula, we let S_n represent the sum of the first n terms of a geometric sequence.

1. $S_n = a + ar + ar^2 + ar^3 + \cdots + ar^{n-1}$

We multiply both sides of Equation 1 by r to get

2. $S_n r = ar + ar^2 + ar^3 + \cdots + ar^{n-1} + ar^n$

We now subtract Equation 2 from Equation 1 and solve for S_n:

$$S_n - S_n r = a - ar^n$$

$$S_n(1 - r) = a - ar^n \qquad \text{Factor out } S_n \text{ from the left side.}$$

$$S_n = \frac{a - ar^n}{1 - r} \qquad \text{Divide both sides by } 1 - r.$$

This reasoning establishes the following theorem.

Sum of the First *n* Terms of a Geometric Sequence

The sum of the first n terms of a geometric sequence is given by the formula

$$S_n = \frac{a - ar^n}{1 - r} \qquad (r \neq 1)$$

where S_n is the sum, a is the first term, r is the common ratio, and n is the number of terms.

EXAMPLE 7   Find the sum of the first six terms of the geometric sequence $250, 50, 10, \ldots$.

Solution   In this sequence, $a = 250$, $r = \frac{1}{5}$, and $n = 6$. We substitute these values into the formula for the sum of the first n terms of a geometric sequence and simplify.

$$S_n = \frac{a - ar^n}{1 - r} = \frac{250 - 250\left(\dfrac{1}{5}\right)^6}{1 - \dfrac{1}{5}}$$

$$= \frac{250 - 250\left(\dfrac{1}{15,625}\right)}{\dfrac{4}{5}}$$

$$= \frac{5}{4}\left(250 - \frac{250}{15,625}\right)$$

$$= \frac{5}{4}\left(\frac{3,906,000}{15,625}\right)$$

$$= 312.48$$

The sum of the first six terms is 312.48.   ■

EXAMPLE 8   The mayor of Eagle River (population 1500) predicts a growth rate of 4% each year for the next ten years. Find the population of Eagle River ten years from now.

Solution   Let P_0 be the initial population of Eagle River. After 1 year, there will be a different population P_1. The initial population (P_0) plus the growth (the product of P_0 and the rate of growth, r) will equal this new population, P_1:

$$P_1 = P_0 + P_0 r = P_0(1 + r)$$

The population after 2 years will be P_2, and

$$P_2 = P_1 + P_1 r$$
$$= P_1(1 + r) \qquad \text{Factor out } P_1.$$
$$= P_0(1 + r)(1 + r) \qquad \text{Remember that } P_1 = P_0(1 + r).$$
$$= P_0(1 + r)^2$$

The population after 3 years will be P_3, and

$$P_3 = P_2 + P_2 r$$
$$= P_2(1 + r) \qquad \text{Factor out } P_2.$$
$$= P_0(1 + r)^2(1 + r) \qquad \text{Remember that } P_2 = P_0(1 + r)^2.$$
$$= P_0(1 + r)^3$$

The yearly population figures

$$P_0, \quad P_1, \quad P_2, \quad P_3, \ldots$$

or

$$P_0, \quad P_0(1 + r), \quad P_0(1 + r)^2, \quad P_0(1 + r)^3, \ldots$$

form a geometric sequence with a first term of P_0 and a common ratio of $1 + r$. The population of Eagle River after 10 years is P_{10}, which is the 11th term of this sequence:

$$n\text{th term} = ar^{n-1}$$
$$P_{10} = 11\text{th term} = P_0(1 + r)^{10}$$
$$= 1500(1 + 0.04)^{10}$$
$$= 1500(1.04)^{10}$$
$$\approx 2220 \qquad \text{Use a calculator.}$$

The estimated population ten years from now is 2220 people. ■

APPENDIX C EXERCISE

VOCABULARY *In Exercises 1–5, fill in the blanks to make a true statement.*

1. A _____ is a function whose domain is the set of natural numbers.

2. An _____ sequence is a sequence in which each term after the first is formed by adding a fixed number.

3. The arithmetic _____ is the number inserted between two numbers a and b to form an arithmetic sequence.

4. The Greek letter sigma, __, is used in summation notation.

5. A _____ sequence is a sequence in which each term after the first is obtained by multiplying the preceding term by a constant nonzero number called the

_____ .

CONCEPTS

6. Determine the common difference in each arithmetic sequence.
   a. 1, 4, 7, . . .   **b.** 2, 6, 10, . . .

7. If a is the first term and d is the common difference, write the first five terms of the arithmetic sequence.
   a. $a = 3, d = 2$ **b.** $a = -5, d = -3$

8. Determine the common ratio in each geometric sequence.
   a. 3, 6, 12, . . .   **b.** 3, −6, 12, . . .

9. If a is the first term and r is the common ratio, write the first five terms of the geometric sequence.
   a. $a = 3, r = 2$ **b.** $a = -5, r = \frac{1}{5}$

✍**10.** Explain why 8, 3, -2, and -7 are the first four terms of an arithmetic sequence when we appear to be subtracting a constant instead of adding a constant to produce the next term.

✍**11.** Explain why 12, 6, and 2 are the first three terms of a geometric sequence when we appear to be dividing instead of multiplying to produce the next term.

PRACTICE

12. Find the 30th term of the arithmetic sequence with $a = 7$ and $d = 12$.

13. Find the 55th term of the arithmetic sequence with $a = -5$ and $d = 4$.

14. Find the first term of the arithmetic sequence with a common difference of 11 and whose 27th term is 263.

15. Find the common difference of the arithmetic sequence with a first term of -164 if its 36th term is -24.

16. Insert three arithmetic means between 2 and 11.

17. Insert four arithmetic means between 5 and 25.

18. Find the arithmetic mean between 10 and 19.

19. Find the arithmetic mean between 5 and 23.

In Exercises 20–23, find the sum of the first n terms of each arithmetic sequence.

20. 1, 4, 7, . . . , $n = 30$

21. 2, 6, 10, . . . , $n = 28$

22. $f(n) = 2n + 1$, nth term is 31, n is a natural number

23. $f(n) = 4n + 3$, nth term is 23, n is a natural number

24. Find the sum of the first 50 natural numbers.

25. Find the sum of the first 100 natural numbers.

26. Find the tenth term of the geometric sequence with $a = 7$ and $r = 2$.

27. Find the 12th term of the geometric sequence with $a = 64$ and $r = \frac{1}{2}$.

28. Find the common ratio of the geometric sequence with a first term of -8 and a sixth term of -1944.

29. Find the common ratio of the geometric sequence with a first term of 12 and a sixth term of $\frac{3}{8}$.

30. Insert three positive geometric means between 2 and 162.

31. Insert four geometric means between 3 and 96.

32. Find the negative geometric mean between 2 and 128.

33. Find the positive geometric mean between 3 and 243.

In Exercises 34–37, find the sum of the first n terms of each geometric sequence.

34. 2, 6, 18, . . .; $n = 6$

35. 2, -6, 18, . . .; $n = 6$

36. The third term is -2 and the fourth term is 1, $n = 6$

37. The third term is -3 and the fourth term is 1, $n = 5$

In Exercises 38–41, find each sum.

38. $\sum_{k=1}^{4} 6k$

39. $\sum_{k=2}^{5} 3k$

40. $\sum_{k=3}^{4} (k^2 + 3)$

41. $\sum_{k=2}^{6} (k^2 + 1)$

APPLICATIONS

42. Saving money Fred puts $60 into a safety deposit box. After each succeeding month, he puts $50 more in the safety deposit box. Write the first six terms of an arithmetic sequence that gives the monthly amounts in his savings, and find his savings after 10 years.

43. Installment loan Freda borrowed $10,000 interest-free from her mother. Freda agreed to pay back the loan in monthly installments of $275. Write

the first six terms of an arithmetic sequence that shows the balance due after each month, and find the balance due after 17 months.

44. Designing a patio Each row of bricks in the triangular patio in Illustration 1 is to have one more brick than the previous row, ending with the longest row of 150 bricks. How many bricks will be needed?

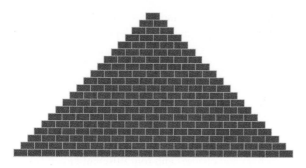

ILLUSTRATION 1

45. Falling object The equation $s = 16t^2$ represents the distance s in feet that an object will fall in t seconds. After 1 second, the object has fallen 16 feet. After 2 seconds, the object has fallen 64 feet, and so on. Find the distance that the object will fall during the second and third seconds.

46. Falling object Refer to Exercise 45. How far will the object fall during the 12th second?

47. Interior angles The sums of the angles of several polygons are given in Table 1. Assuming that the pattern continues, find the sum of the interior angles of an octagon (8 sides) and dodecagon (12 sides).

Figure	Number of sides	Sum of angles
Triangle	3	180°
Quadrilateral	4	360°
Pentagon	5	540°
Hexagon	6	720°
Octagon	8	
Dodecagon	12	

TABLE 1

48. Population growth The population of Union is predicted to increase by 6% each year. What will be the population of Union 5 years from now if its current population is 500?

49. Population growth The population of Hicksville is decreasing by 10% each year. If its current population is 98, what will be the population 8 years from now?

50. Declining savings John has $10,000 in a safety deposit box. Each year he spends 12% of what is left in the box. How much will be in the box after 15 years?

51. Savings growth Sally has $5000 in an investment account earning 12% annual interest. How much will be in her account 10 years from now? (Assume that Sally makes no deposits or withdrawals.)

52. House appreciation A house appreciates by 6% each year. If the house is worth $70,000 today, how much will it be worth 12 years from now?

53. Motorboat depreciation A motorboat that cost $5000 when new depreciates at a rate of 9% per year. How much will the boat be worth in 5 years?

54. Inscribed squares Each inscribed square in Illustration 2 joins the midpoints of the sides of the next larger square. The area of the first square, the largest, is 1. Find the area of the 12th square.

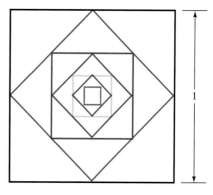

ILLUSTRATION 2

55. Genealogy The family tree in Illustration 3 spans three generations and lists seven people. How many names would be listed in a family tree that spans ten generations?

ILLUSTRATION 3

Appendix D Permutations and Combinations

■ THE MULTIPLICATION PRINCIPLE FOR EVENTS ■ PERMUTATIONS ■ COMBINATIONS
■ ALTERNATIVE FORM OF THE BINOMIAL THEOREM

■ THE MULTIPLICATION PRINCIPLE FOR EVENTS

A student goes to the cafeteria for lunch. He has a choice of three different sandwiches (hamburger, hot dog, or ham and cheese) and four different beverages (cola, root beer, orange, or milk). How many different lunches can he choose?

The student has three choices of sandwich, and for any one of these choices, he has four choices of drink. The different options are shown in the *tree diagram* in Figure D-1. The tree diagram shows that there is a total of 12 different lunches to choose from. One possibility is a hamburger with a cola, and another is a hot dog with milk.

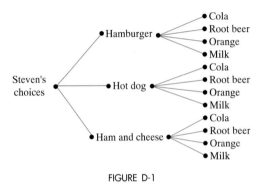

FIGURE D-1

A situation that can have several different outcomes—such as choosing a sandwich—is called an **event.** Choosing a sandwich and choosing a beverage can be thought of as two events. The preceding example illustrates the **multiplication principle for events.**

> **Multiplication Principle for Events**
> Let E_1 and E_2 be two events. If E_1 can be done in a_1 ways, and if—after E_1 has occurred—E_2 can be done in a_2 ways, then the event "E_1 followed by E_2" can be done in $a_1 \cdot a_2$ ways.

EXAMPLE 1   In how many ways can we arrange five books on a shelf?

Solution   We can fill the first space with any of the 5 books, the second space with any of the remaining 4 books, the third space with any of the remaining 3 books, the fourth space with any of the remaining 2 books, and the fifth space with the remaining 1 (or last) book, According to the multiplication principle for events, the number of ways that the books can be arranged is

$$5 \cdot 4 \cdot 3 \cdot 2 \cdot 1 = 120$$

■ PERMUTATIONS

When computing the number of possible arrangements of objects such as books on a shelf or flags on a pole, we are finding the number of **permutations** of those objects. In Example 1, we found that the number of permutations of five books, using all five of them, is 120.

The symbol $P(n, r)$, read as "the number of permutations of n things r at a time," is often used to express permutation problems. In the case of the number of arrangements of five books on the shelf, we find that $P(5, 5) = 120$ because we are taking 5 objects 5 at a time.

EXAMPLE 2   If Sarah has seven flags, each of a different color, to hang on a flagpole, how many different signals can she send by using three flags.

Solution   Sarah must find $P(7, 3)$ (the number of permutations of 7 things 3 at a time). In the top position, Sarah can hang any of the 7 flags; in the middle position, any one of the remaining 6 flags; and in the bottom position, any one of the remaining 5 flags. According to the multiplication principle for events,

$$P(7, 3) = 7 \cdot 6 \cdot 5$$
$$= 210$$

Sarah can send 210 signals using only three of the available seven flags. ■

Although it is correct to write $P(7, 3) = 7 \cdot 6 \cdot 5$, there is an advantage in changing the form of this answer to obtain a formula for computing $P(7, 3)$:

$$P(7, 3) = 7 \cdot 6 \cdot 5$$
$$= \frac{7 \cdot 6 \cdot 5 \cdot 4 \cdot 3 \cdot 2 \cdot 1}{4 \cdot 3 \cdot 2 \cdot 1} \qquad \text{Multiply both the numerator and denominator by } 4 \cdot 3 \cdot 2 \cdot 1.$$
$$= \frac{7!}{4!}$$
$$= \frac{7!}{(7 - 3)!}$$

The generalization of this idea gives the following formula:

Formula for Finding P(n, r)
The number of permutations of n things r at a time is given by the formula

$$P(n, r) = \frac{n!}{(n - r)!}$$

EXAMPLE 3   Compute: **a.** $P(8, 2)$ **b.** $P(n, n)$ **c.** $P(n, 0)$

Solution   **a.** $P(8, 2) = \dfrac{8!}{(8 - 2)!}$

$$= \frac{8 \cdot 7 \cdot 6!}{6!}$$
$$= 8 \cdot 7$$
$$= 56$$

```
8 nPr 2
              56
```

■ **TECHNOLOGY TIP**
The permutation of n things taken r at a time, $P(n, r)$, can also be written as $_nP_r$. To compute $P(8, 2) = {_8}P_2$, enter 8 then retrieve the $_nP_r$ command from the probability (PRB) submenu of the MATH feature and follow with the 2, as indicated on the screen accompanying the problem.

b. $P(n, n) = \dfrac{n!}{(n - n)!}$

$$= \frac{n!}{0!}$$
$$= \frac{n!}{1}$$
$$= n!$$

c. $P(n, 0) = \dfrac{n!}{(n - 0)!}$

$$= \frac{n!}{n!}$$
$$= 1$$

■

Parts (c) and (d) of Example 3 establish the following formulas:

> **Formulas for Finding P(n, n) and P(n, 0)**
> The number of permutations of n things n at a time and n things 0 at a time are given by the formula
>
> $$P(n, n) = n! \quad \text{and} \quad P(n, 0) = 1$$

EXAMPLE 4   **a.** In how many ways can a television executive arrange the Saturday night lineup of 6 programs if there are 15 programs to choose from?

b. If there are only 6 programs to choose from?

Solution   **a.** To find the number of permutations of 15 programs 6 at a time, we use the formula $P(n, r) = \frac{n!}{(n-r)!}$ with $n = 15$ and $r = 6$.

$$P(15, 6) = \frac{15!}{(15-6)!} = \frac{15 \cdot 14 \cdot 13 \cdot 12 \cdot 11 \cdot 10 \cdot 9!}{9!}$$
$$= 15 \cdot 14 \cdot 13 \cdot 12 \cdot 11 \cdot 10$$
$$= 3,603,600$$

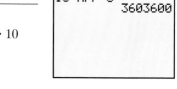

b. To find the number of permutations of 6 programs 6 at a time, we use the formula $P(n, n) = n!$ with $n = 6$.

$$P(6, 6) = 6! = 720$$

■ COMBINATIONS

Suppose that Raul must read 4 books from a reading list of 10 books. The order in which he reads them is not important. For the moment, however, let's assume that order is important and find the number of permutations of 10 things 4 at a time:

$$P(10, 4) = \frac{10!}{(10-4)!}$$
$$= \frac{10 \cdot 9 \cdot 8 \cdot 7 \cdot 6!}{6!}$$
$$= 10 \cdot 9 \cdot 8 \cdot 7$$
$$= 5040$$

> 10 nPr 4
>   5040

If order is important, there are 5040 ways of choosing 4 books when there are 10 books to choose from.

However, because the order in which Raul reads the books does not matter, the previous result of 5040 is too big. Because there are 4! ways of ordering the 4 books that are chosen, the result of 5040 is exactly 4! (or 24) times too big. Thus, the number of choices that Raul has is the number of permutations of 10 things 4 at a time, divided by 24:

$$\frac{P(10, 4)}{24} = \frac{5040}{24}$$
$$= 210$$

Raul has 210 ways of choosing 4 books to read from the list of 10 books.

In situations where order is *not* important, we are interested in **combinations,** not permutations. The symbols $C(n, r)$ and $\binom{n}{r}$ both mean the number of combinations of n things r at a time.

If a selection of r books is chosen from a total of n books, the number of possible selections is $C(n, r)$, and there are $r!$ arrangements of the r books in each selection. If we consider the selected books as an ordered grouping, the number of orderings is $P(n, r)$. Thus, we have

$$r! \cdot C(n, r) = P(n, r)$$

We can divide both sides of this equation by $r!$ to get the formula for finding $C(n, r)$:

$$C(n, r) = \binom{n}{r} = \frac{P(n, r)}{r!} = \frac{n!}{r!(n - r)!}$$

Formula for Finding C(n, r)

The number of combinations of n things r at a time is given by

$$C(n, r) = \frac{n!}{r!(n - r)!}$$

EXAMPLE 5 Compute: **a.** $C(8, 5)$ **b.** $C(n, n)$ **c.** $C(n, 0)$

Solution **a.** $C(8, 5) = \dfrac{8!}{5!(8 - 5)!}$

$$= \frac{8 \cdot 7 \cdot 6 \cdot 5!}{5!3!}$$

$$= 8 \cdot 7$$

$$= 56$$

■ **TECHNOLOGY TIP**
The notation $C(n, r)$ and $\binom{n}{r}$ are also equivalent to $_nC_r$. $_nC_r$ is located under the probability submenu of the MATH menu.

```
8 nCr 5
             56
```

b. $C(n, n) = \dfrac{n!}{n!(n - n)!}$ **c.** $C(n, 0) = \dfrac{n!}{0!(n - 0)!}$

$$= \frac{1}{0!} \qquad\qquad = \frac{n!}{0!n!}$$

$$= \frac{1}{1} \qquad\qquad = \frac{1}{0!}$$

$$= 1 \qquad\qquad\quad = \frac{1}{1}$$

$$= 1$$

The notation $C(n, 0)$ indicates that we choose 0 things from the available n things. ■

Parts (b) and (c) of Example 5 establish the following formulas:

Formulas for Finding C(n, n) and C(n, 0)

The number of combinations of n things n at a time is 1. The number of combinations of n things 0 at a time is 1.

$$C(n, n) = 1 \quad\text{and}\quad C(n, 0) = 1$$

EXAMPLE 6 If 15 students want to pick a committee of 4 students to plan a party, how many different committees are possible?

Solution Because the ordering of people on each possible committee is unimportant, we find the number of combinations of 15 people 4 at a time:

$$C(15, 4) = \frac{15!}{4!(15-4)!}$$

$$= \frac{15 \cdot 14 \cdot 13 \cdot 12 \cdot 11!}{4 \cdot 3 \cdot 2 \cdot 1 \cdot 11!}$$

$$= \frac{15 \cdot 14 \cdot 13 \cdot 12}{4 \cdot 3 \cdot 2 \cdot 1}$$

$$= 1365$$

```
15 nCr 4
              1365
```

There are 1365 possible committees. ∎

EXAMPLE 7   A committee in Congress consists of 10 Democrats and 8 Republicans. In how many ways can a subcommittee be chosen if it is to contain 5 Democrats and 4 Republicans?

Solution There are $C(10, 5)$ ways of choosing the 5 Democrats and $C(8, 4)$ ways of choosing the 4 Republicans. By the multiplication principle for events, there are $C(10, 5) \cdot C(8, 4)$ ways of choosing the subcommittee:

$$C(10, 5) \cdot C(8, 4) = \frac{10!}{5!(10-5)!} \cdot \frac{8!}{4!(8-4)!}$$

$$= \frac{10 \cdot 9 \cdot 8 \cdot 7 \cdot 6 \cdot 5!}{120 \cdot 5!} \cdot \frac{8 \cdot 7 \cdot 6 \cdot 5 \cdot 4!}{24 \cdot 4!}$$

$$= \frac{10 \cdot 9 \cdot 8 \cdot 7 \cdot 6}{120} \cdot \frac{8 \cdot 7 \cdot 6 \cdot 5}{24}$$

$$= 17,640$$

```
10 nCr 5*8 nCr 4
              17640
```

There are 17,640 possible subcommittees. ∎

■ ALTERNATIVE FORM OF THE BINOMIAL THEOREM

We have seen that the expansion of $(x + y)^3$ is

$$(x + y)^3 = 1x^3 + 3x^2y + 3xy^2 + 1y^3$$

and that

$$C(3, 0) = \binom{3}{0} = 1, \quad C(3, 1) = \binom{3}{1} = 3, \quad C(3, 2) = \binom{3}{2} = 3, \quad \text{and} \quad C(3, 3) = \binom{3}{3} = 1$$

■ TECHNOLOGY TIP
To easily generate all the coefficients for the binomial theorem, enter $y1 = 6_nC_r x$ and generate the TABLE beginning at $x = 0$.

Putting these facts together gives the following way of writing the expansion of $(x + y)^3$:

$$(x + y)^3 = \binom{3}{0}x^3 + \binom{3}{1}x^2y + \binom{3}{2}xy^2 + \binom{3}{3}y^3$$

Likewise, we have

$$(x + y)^4 = \binom{4}{0}x^4 + \binom{4}{1}x^3y + \binom{4}{2}x^2y^2 + \binom{4}{3}xy^3 + \binom{4}{4}y^4$$

```
 X    | Y1
 0    | 1
 1    | 6
 2    | 15
 3    | 20
 4    | 15
 5    | 6
 6    | 1
X=0
```

The generalization of this idea allows us to state the binomial theorem in an alternative form, using combinatorial notation.

Alternative Form of the Binomial Theorem

If n is any positive integer, then

$$(a + b)^n = \binom{n}{0}a^n + \binom{n}{1}a^{n-1}b + \binom{n}{2}a^{n-2}b^2 + \cdots +$$

$$\binom{n}{r}a^{n-r}b^r + \cdots + \binom{n}{n}b^n$$

EXAMPLE 8   Use the alternative form of the binomial theorem to expand $(x + y)^6$.

Solution $(x + y)^6 = \binom{6}{0}x^6 + \binom{6}{1}x^5y + \binom{6}{2}x^4y^2 + \binom{6}{3}x^3y^3 + \binom{6}{4}x^2y^4 + \binom{6}{5}xy^5 + \binom{6}{6}y^6$

$= x^6 + 6x^5y + 15x^4y^2 + 20x^3y^3 + 15x^2y^4 + 6xy^5 + y^6$ ■

EXAMPLE 9   Use the alternative form of the binomial theorem to expand $(2x - y)^3$.

Solution $(2x - y)^3 = [2x + (-y)]^3$

$= \binom{3}{0}(2x)^3 + \binom{3}{1}(2x)^2(-y) + \binom{3}{2}(2x)(-y)^2 + \binom{3}{3}(-y)^3$

$= 1(2x)^3 + 3(4x)^2(-y) + 3(2x)(-y)^2 + (-y)^3$

$= 8x^3 - 12x^2y + 6xy^2 - y^3$ ■

APPENDIX D EXERCISE

VOCABULARY   *In Exercises 1–4, fill in the blanks to make a true statement.*

1. A situation that can have several different outcomes is called an _____.

2. The _____ principle for events states that if event 1 can be done in a ways and event 2 can be done in b ways, then event 1 followed by event 2 can be done in ___ ways.

3. A _____ computes the number of possible arrangements of objects; order is important.

4. A _____ computes the number of possible groups of objects; order is not important.

CONCEPT

5. Label each of the following sets as a group of permutations or combinations:
  a. $\{1, 2, 3\}, \{1\}, \{2\}, \{1, 2\}, \{1, 3\}, \{2, 3\}$

  b. $\{1, 2, 3\}, \{2, 1, 3\}, \{2, 3, 1\}, \{3, 1, 2\}, \{3, 2, 1\}$

6. Based on the two calculator screens, determine
  a. $C(8, 4)$ **b.** $C(8, 5)$ **c.** $C(8, 8)$

7. Based on the two calculator screens, determine
  a. $P(8, 4)$ **b.** $P(8, 5)$ **c.** $P(8, 8)$

✍**8.** Explain why $C(8, 8)$ is less than $P(8, 8)$.

PRACTICE   *In Exercises 9–18, evaluate each expression. Verify with your calculator.*

9. $P(5, 5)$ **10.** $P(4, 4)$ **11.** $P(5, 3)$ **12.** $P(3, 2)$

13. $P(2, 2) \cdot P(3, 3)$ **14.** $P(3, 2) \cdot P(3, 3)$ **15.** $\dfrac{P(5, 3)}{P(4, 2)}$ **16.** $\dfrac{P(6, 2)}{P(5, 4)}$

17. $\dfrac{P(5, 3) \cdot P(7, 3)}{P(5, 1)}$ **18.** $\dfrac{P(8, 3)}{P(5, 3) \cdot P(4, 3)}$

In Exercises 19–30, evaluate each expression. Verify with your calculator.

19. $C(5, 3)$ **20.** $C(5, 4)$ **21.** $\dbinom{6}{3}$ **22.** $\dbinom{6}{4}$

23. $\dbinom{5}{4}\dbinom{5}{3}$ **24.** $\dbinom{6}{5}\dbinom{6}{4}$ **25.** $\dfrac{C(38, 37)}{C(19, 18)}$ **26.** $\dfrac{C(25, 23)}{C(40, 39)}$

27. $C(12, 0)C(12, 12)$ **28.** $\dfrac{C(8, 0)}{C(8, 1)}$ **29.** $C(n, 2)$ **30.** $C(n, 3)$

In Exercises 31–36, use the alternative form of the binomial theorem to expand each expression. Use the TABLE feature of the calculator to verify your coefficients.

31. $(x + y)^4$ **32.** $(x - y)^2$

33. $(2x + y)^3$ **34.** $(2x + 1)^4$

35. $(3x - 2)^4$ **36.** $(3 - x^2)^3$

In Exercises 37–40, find the indicated term of the binomial expansion. The TABLE feature of the calculator can again be used to verify your coefficient.

37. $(x - 5y)^5$; fourth term   **38.** $(2x - y)^5$; third term

39. $(x^2 - y^3)^4$; second term   **40.** $(x^3 - y^2)^4$; fourth term

APPLICATIONS

41. Arranging an evening Kristy intends to go out to dinner and then see a movie. In how many ways can she arrange her evening if she has a choice of five movies and seven restaurants?

42. Travel choices Paula has five ways to travel from New York to Chicago, three ways to travel from Chicago to Denver, and four ways to travel from Denver to Los Angeles. How many choices are available to Paula if she travels from New York to Los Angeles?

43. Making license plates How many six-digit license plates can be manufactured? (Note that there are ten choices—0, 1, 2, 3, 4, 5, 6, 7, 8, 9—for each digit.)

44. Making license plates How many six-digit license plates can be manufactured if no digit can be repeated?

45. Making license plates How many six-digit license plates can be manufactured if no license number can begin with 0 and if no digit can be repeated?

46. Making license plates How many license plates can be manufactured with two letters followed by four digits?

47. Phone numbers How many seven-digit phone numbers are available in area code 815 if no phone number can begin with 0 or 1?

48. Phone numbers How many ten-digit phone numbers are available if area codes 000 and 911 cannot be used and if no local number can begin with 0 or 1?

49. Lining up In how many ways can six girls be placed in a line?

50. Arranging books In how many ways can seven books be placed on a shelf?

51. Arranging books In how many ways can four novels and five biographies be arranged on a shelf if the novels are placed first?

52. Making a ballot In how many ways can six candidates for mayor and four candidates for the county board be arranged on a ballot if all of the candidates for mayor must be placed first?

53. Combination locks How many permutations does a combination lock have if each combination has 3 numbers, no two numbers of any combination are equal, and the lock has 25 numbers?

54. Combination locks   How many permutations does a combination lock have if each combination has 3 numbers, no two numbers of any combination are equal, and the lock has 50 numbers?

55. Arranging appointments   The receptionist at a dental office has only three appointment times available before next Tuesday, and there are ten patients with a toothache. In how many ways can the receptionist fill those appointments?

56. Computers   In many computers, a *word* consists of 32 *bits*, a string of 32 1s and 0s. How many different words are possible?

57. Palindromes   A palindrome is any word, such as *madam* or *radar*, that reads the same backward and forward. How many five-digit numerical palindromes (like 12321) are there? (*Hint:* A leading zero would be dropped.)

58. Call letters   The call letters of a U.S. commercial radio station have either three or four letters, and the first is either a *W* or a *K*. How many radios stations could this system support?

59. Planning a picnic   A class of 14 students wants to pick a committee of 3 students to plan a picnic. How many committees are possible?

60. Choosing books   Jeffrey must read 3 books from a reading list of 15 books. How many choices does he have?

61. Forming a committee   The number of three-person committees that can be formed from a group of persons is ten. How many persons are in the group?

62. Forming a committee   The number of three-person committees that can be formed from a group of persons is 20. How many persons are in the group?

63. Winning a lottery   In one state lottery, anyone who picks the correct six numbers (in any order) wins. With the numbers 0 through 99 available, how many choices are possible?

64. Taking a test   The instructions on a test read, "Answer any 10 of the following 15 questions. Then choose 1 of the remaining questions for homework, and turn in its solution tomorrow." In how many ways can the questions be chosen?

65. Forming a committee   In how many ways can we select a committee of two boys and two girls from a group containing three boys and four girls?

66. Forming a committee   In how many ways can we select a committee of three boys and two girls from a group containing five boys and three girls?

67. Choosing clothes   In how many ways can we select 2 shirts and 3 neckties from a group of 12 shirts and 10 neckties?

68. Choosing clothes   In how many ways can we select five dresses and two coats from a wardrobe containing nine dresses and three coats?

Appendix E Answers to Selected Exercises

Exercise 1.1 (page 11)

1. natural **3.** rational **5.** identity, identity **7.** a, c **9.** b, c **11.** Factor each denominator into prime factors. Multiply together the primes that occur, using each one the greatest number of times it occurs as a factor in any one factorization. **13.** Reversing the order produces different answers: $\frac{8}{4} = 2$ while $\frac{4}{8} = \frac{1}{2}$. **15.** Enter $4 + 17/19\blacktriangleright$Frac. **17.** $\frac{4}{3}$ **19.** $\frac{9}{8}$ **21.** $\frac{3}{2}$ **23.** 10 **25.** $\frac{13}{3}$ **27.** $\frac{13}{48}$ **29.** 28 **31.** $\frac{1}{5}$ **33.** $\frac{6}{5}$ **35.** $\frac{5}{24}$ **37.** $\frac{1}{6}$ **39.** $\frac{29}{3}$ **41.** $\frac{26}{5}$ **43.** $\frac{5}{4}$ **45.** The parentheses override the order of operations. **47.** $\frac{1545}{8667} = \frac{515}{2889}$ **49.** approximately 2427 **51. a.** $\frac{1}{5} \cdot 750 = 150$ **b.** $\frac{4}{5} \cdot 750 = 600$ **53.** $\frac{7}{10}$ **55.** $\frac{39}{50}$

Exercise 1.2 (page 24)

1. exponent **3.** base, exponent **5. a.** $x \cdot x$ **b.** $3 \cdot z \cdot z \cdot z \cdot z$ **c.** $5 \cdot x \cdot 5 \cdot x$ **d.** $5 \cdot 2 \cdot x \cdot 2 \cdot x \cdot 2 \cdot x$ **7.** "x multiplied by itself n times" means $x \cdot (n \text{ factors of } x) = x^{n+1}$ **9.** Smaller numbers would be produced. **11.** Evaluate means to give a value or number as the answer. Simplify means to write in a simpler alternative form. **13.** 25 **15.** 343 **17.** 36 **19.** 1000 **21.** 18 **23.** 18 **25.** 29 **27.** 44 **29.** 12 **31.** 27 **33.** 20 **35.** 2 **37.** 32 **39.** 14 **41.** 100 **43.** 10 **45.** 19 **47.** 9 **49.** 2 **51.** $\frac{2}{21}$ **53.** 10 **55.** 16 **57.** 8 **59.** 43 **61.** 15 **63.** 169 square units **65.** 216 cubic units **67.** 31,208 students **69.** $8.00 **71.** $\frac{35}{24}$ **73.** $\frac{61}{24}$

Exercise 1.3 (page 36)

1. absolute value **3.** associative **5. a.** 4 **b.** -6 **c.** -5 **7. a.** 4 **b.** -12 **c.** 12 **d.** -4 **9.** Change the sign of the number to be subtracted and add. **11.** 2 **13.** -12 **15.** 0.5 **17.** $\frac{12}{35}$ **19.** $-\frac{1}{12}$ **21.** 7 **23.** -1 **25.** -7 **27.** -8 **29.** -18 **31.** 1.3 **33.** $\frac{1}{2}$ **35.** $-\frac{28}{5}$ **37.** 4.2 **39.** 10 **41.** 0 **43.** 8 **45.** 64 **47.** 3 **49.** 4 **51.** 3 **53.** -3 **55.** 13 **57.** 1 **59.** $\frac{1}{9}$ **61.** -15.8 **63.** -7.1 **65.** -4 degrees **67.** 2000 **69.** 0 yards gained or lost **71.** 18 degrees difference **73.** $83,425.57 **75.** $3\frac{13}{20}$ tons **77.** $\frac{7}{6}, \frac{27}{40}$ **79.** Various examples are possible involving two decreases for the negative numbers and two increases for the positive numbers. **81.** Answers agree when the numbers have the same sign. **83.** 47 **85.** 46 **87.** $(90 + 89 + 82 + 76 + 71)/5 = 81.6$

Exercise 1.4 (page 43)

1. product **3.** Divisors and factors represent the same values. **5.** The product of two numbers with common signs will be a positive number. The product of two numbers with different signs will be a negative number. **7.** The quotient of two numbers with common signs will be a positive number. The quotient of two numbers with different signs will be a negative number. **9.** 0 **11.** 16 **13.** 2 **15.** 72 **17.** -24 **19.** 2 **21.** 1 **23.** 1 **25.** -3 **27.** -1 **29.** 4 **31.** 9 **33.** -54 **35.** -27 **37.** $\frac{3}{2}$ **39.** 6 **41.** $-\frac{4}{3}$ **43.** -13 **45.** $-\frac{1}{6}$ **47.** $-\frac{11}{12}$ **49.** $-\frac{7}{36}$ **51.** $-\frac{11}{48}$ **53.** 1, 3, or 5 **55.** $30 **57.** 15 weeks **59.** a yearly loss of $\frac{8}{10}$ of a million **61.** $2.65 **63.** $\frac{2}{15}$ **65. a.** $-\frac{5}{2}$ **b.** $\frac{3}{2}$

Exercise 1.5 (page 51)

1. irrational **3.** area **5.** volume **7. a.** 3 **b.** 0, 3 **c.** -3 **d.** $-3, -\frac{1}{2}, 0, 3, 5.\overline{6}$ **e.** $\sqrt{2}$ **9.** False. Two-thirds is equal to 0.6666 . . . where the 6 repeats infinitely. **11. a.** 0.05 **b.** 0.045 **c.** 0.0875 **13.** 26 cm **15.** 16 in. **17.** 16 cm^2 **19.** 228 cm^2 **21.** $\approx$65.97344573 cm **23.** $\approx$153.938 m^2 **25.** 36 ft^3 **27.** 527.788 in^3 **29.** 96 in^3 **31.** 2704 in^2 **33.** 8856 in^3 **35.** 31.81 in^2 **37.** No; the netting will cover 100 ft^2 but 264 ft^2 is actually needed. **39.** $247.50 **41.** $2144.88 **43. a.** 2° **b.** 50° **c.** The table does not display 72° C. **45.** $43.13 **47.** 12,480 **49.** Approximately 52 students do not have snacks. **51.** $\approx$$780.68 **53.** the Guernsey **55.** The second contractor has the lower bid. **57. a.** -9 **b.** 9 **c.** 9 **d.** 36 **59.** $\frac{11}{9}$ or $1\frac{2}{9}$ gallons

Chapter 1 Summary (page 56)

1. a. $\frac{10}{21}$ **b.** $\frac{11}{21}$ **c.** $\frac{1}{3}$ **d.** 1 **e.** $\frac{49}{6}$ or $8\frac{1}{6}$ **f.** $\frac{67}{15}$ or $4\frac{7}{15}$ **g.** $\frac{91}{6}$ or $15\frac{1}{6}$ **h.** $\frac{30}{13}$ or $2\frac{4}{13}$ **2. a.** $\frac{2}{15}$ **b.** 1st day: 120; 2nd day: 400; 3rd day: 80 **3. a.** 81 **b.** $\frac{4}{9}$ **c.** -25 **d.** 25 **4. a.** 6 **b.** 3 **c.** 38 **d.** 3 **e.** $\frac{23}{12}$ **f.** 15 **5. a.** $\left(-\frac{1}{4}\right)^3$ **b.** $\frac{4-8}{-12+1}$ **c.** $4 - \frac{8}{-12} + 1$ **6.** 6.85 hours **7. a.** -7 **b.** 1 **c.** $-\frac{1}{7}$ **d.** 1 **e.** $\frac{1}{4}$ **f.** $-\frac{2}{3}$ **8. a.** 6 **b.** $\frac{3}{2}$ **c.** 7 **d.** 26 **9.** -4 **10.** -1 **11.** -2 **12.** 5 **13.** 4 **14.** 6 **15.** -6 **16.** 3 **17.** 2 **18.** 6 **19.** -7 **20.** 39 **21.** 6 **22.** -2 **23.** -8 **24.** 1 **25.** 40.2 ft **26.** 15133.576 ft^3 **27.** 35° **28.** $36.42

Chapter 1 Test (page 59)

1. 14 **2.** 32 **3.** -27.4 **4.** 1 **5.** $\frac{4}{5}$ **6.** $\frac{9}{2}$ or $4\frac{1}{2}$ **7.** -1 **8.** $-\frac{1}{13}$ **9.** 301.57 ft^2 **10.** 64 cm^2 **11.** 1539.380 in^3 **12.** $320 **13.** 77,550 homes **14.** 3889.625 in^3 **15.** -2 **16.** -14 **17.** -4 **18.** 12 **19.** 5 **20.** -23

Exercise 2.1 (page 69)

1. base, exponent **3.** base: 4, exponent: 3 **5.** base: $-x$, exponent: 2 **7.** base: x, exponent: 1 **9.** base: x, exponent: 3 **11.** 2^3 **13.** x^5 **15.** $(2x)^3$ **17.** -24 **19.** 1296 **21.** x^7 **23.** x^{10} **25.** x^9 **27.** $12x^7$ **29.** $20x^4$ **31.** 3^8 (6561) **33.** y^{15} **35.** a^{21} **37.** x^{25} **39.** $4x^{31}$ **41.** x^3y^3 **43.** r^6s^4 **45.** $27x^6y^3$ **47.** $-8r^9s^9t^3$ **49.** $\frac{a^3}{b^3}$ **51.** $\frac{x^{10}}{y^{15}}$ **53.** $\frac{-32a^5}{b^5}$ **55.** $\frac{b^6}{27a^3}$ **57.** x^2 **59.** x^3y **61.** ab^4

A-36

63. $\frac{10r^{10}s^3}{3}$ **65.** $\frac{y^3}{8}$ **67.** 2   **69.** -2 **71.** No; $2^3 \neq 3^2$ for example   **73.** $15x^3$ **75.** $\frac{1}{2}(6x)(5x + 10x)$ **77. a.** -9 **b.** 30   **c.** -3
d. -3 **e.** -3 **f.** 3   **79. a.** 0.45   **b.** 0.03   **c.** 0.025

Exercise 2.2   (page 76)

1. algebraic expression   **3.** algebraic terms   **5.** polynomial   **7.** degree, degree, degree   **9.** b, d **11.** trinomial   **13.** $x = -2$,
$P(x) = -1$ **15.** xy **17.** $2x + 2y$ **19.** $\frac{y}{x}$ **21.** $y - x$ **23.** a number decreased by two or two less than a number   **25.** the product of
x and z **27.** the quotient of three times a number and the sum of y and z or three times a number divided by the quantity y plus z
29. binomial   **31.** monomial   **33.** none of the above   **35.** monomial   **37.** 5   **39.** 5   **41.** 6   **43.** 10   **45.** 1   **47.** 0   **49.** -3
51. -13 **53.** $5t - 3$ **55.** $10t - 3$ **57.** $\frac{-3}{2}$ **59.** $-5v^4 - 2$ **61.** $50x^7 - 2$ **63.** $5x - 5h - 2$ **65.** $5x - 5h$ **67.** $-15r + 10s - 2$
69. $m + 25{,}000$ **71.** $35{,}000a$ **73.** $\frac{12}{x}$ **75.** $2x - 8$; no, since this would mean a negative number of shares for her sister.   **77.** 192 ft
79. 72 ft   **81. a.** 9   **b.** -16

Exercise 2.3   (page 81)

1. distributive   **3.** like; $7y$ **5.** unlike   **7.** like; $13x^3$ **9.** like; 0   **11.** $9y$ **13.** $-12t^2$ **15.** $16u^3$ **17.** $7x^5y^2$ **19.** $14rst$
21. $-6a^2bc$ **23.** $-2mn^3 + 3m^3n$ **25.** $20x$ **27.** $13x + 5y$ **29.** $7x + 4$ **31.** $2a + 7$ **33.** $6x^2 + x - 5$ **35.** $2a^2 + a - 3$
37. $12x + 12y$ **39.** $-9a + 9b$ (or $9b - 9a$)   **41.** $7y^2 + 22y$ **43.** $xy^2 + 13y^2$ **45.** $6x^2 + 4x - 23$ **47.** $2x^2 + 2x - 8$ **49. a.** 43
b. 1, 3, 32   **c.** No; if $x = 3$, then the outer side measurements would each be 1 unit while the interior cut-out section would indicate a
measurement of 3 on the sides.   **51. a.** 4   **b.** 0, 7, -6 **c.** No, since one side of the trapezoid would have a negative measurement.
53. $6x^2 - 2x - 1$ **55.** $t^3 + 3t^2 + 6t - 5$ **57.** $-3x^2 + 5x - 7$ **59.** $75 + 10x$ **61.** $-0.025x + 300$ **63.** $117x + 144$ **65.** $43x + 2$
67. $4 \cdot 4 \cdot 4$ **69. a.** 0.0825

Exercise 2.4   (page 87)

1. conjugate   **3.** FOIL   **5. a.** $6x^2$ **b.** $4x^3$ **c.** $\frac{2}{5}x^2$ **7.** $12x^5$ **9.** $-24b^6$ **11.** $x^{10}y^{15}$ **13.** $a^5b^4c^7$ **15.** $3x^2 - 6$ **17.** $-6x^4 + 2x^3$
19. $6x^4 + 8x^3 - 14x^2$ **21.** $2x^7 - x^2$ **23.** $a^2 + 9a + 20$ **25.** $3x^2 + 10x - 8$ **27.** $6a^2 + 2a - 20$ **29.** $6x^2 - 7x - 5$
31. $6t^2 + 7st - 3s^2$ **33.** $x^2 + 8x + 16$ **35.** $x^2 + 10x + 25$ **37.** $r^2 - 16$ **39.** $16x^2 - 25y^2$ **41.** $4a^2 - 12ab + 9b^2$
43. $2x^2 - 6x - 8$ **45.** $4x^3 + 11x^2 + 18x + 9$ **47.** $x^3 - 8y^3$ **49.** $5t^2 - 11t$ **51.** $x^2y + 3xy^2 + 2x^2$ **53.** $2x^2 + xy - y^2$ **55.** $8x$
57. $5s^2 - 7s - 9$ **59.** $5x + x^2$ **61.** $x^3 - 4x$ **63.** $x^3 + 6x^2 + 8x$ **65.** $2x^2 + 4x + 4$ **67.** $6x^4 - 11x^3 - 22x^2 - 12x + 5$ **69.** 37.5,
65, 97.5, 130; let $y1 = 65x$; set TblStart $= .5$ and ΔTbl $= .5$ **71.** $130t$ miles

Exercise 2.5   (page 92)

1. $4x$ **3.** 1   **5.** $x + 3y$ **7.** $7 + 9y$ **9.** $\frac{r^2}{s}$ **11.** $\frac{2x^2}{y}$ **13.** $-\frac{3u^3}{v^2}$ **15.** $\frac{4r}{y^2}$ **17.** $-\frac{13}{3rs}$ **19.** $\frac{x^4}{y^6}$ **21.** a^8b^8 **23.** $-\frac{3r}{s^9}$ **25.** $-\frac{x^3}{4y^3}$
27. $\frac{125}{8b^3}$ **29.** $\frac{xy^2}{3}$ **31.** a^8 **33.** z^{15} **35.** $\frac{2}{y} + \frac{3}{x}$ **37.** $\frac{1}{5y} - \frac{2}{5x}$ **39.** $\frac{1}{y^2} + \frac{2y}{x^2}$ **41.** $3a - 2b$ **43.** $\frac{1}{y} - \frac{1}{2x} + \frac{2z}{xy}$ **45.** $3x^2 - 2x - \frac{1}{y}$
47. $5x - 6y + 1$ **49.** $\frac{10x^2}{y} - 5x$ **51.** $-\frac{4x}{3} + \frac{3x^2}{2}$ **53.** $xy - 1$ **55.** $\frac{x}{y} - \frac{11}{6} + \frac{y}{2x}$ **57.** 2   **59.** $2x + 3$ **61.** $2x + \frac{1}{2} - \frac{3}{x^2}$
63. $3x - \frac{2}{x} + \frac{4}{3x^3}$ **65.** $x = 2$ is not a valid value since it gives a negative value for the perimeter.   **67.** A side of length 3 should give
an area of 9. However, if the area $8x^3 - 12x^2$ is evaluated for $x = 3$, the resulting area is 108. Therefore, 3 is not a valid measurement
for the length of a side.   **69. a.** $2x - b$ **b.** 1

Exercise 2.6   (page 98)

1. a. $3x^2$ **b.** $4x$ **c.** -2 **d.** $3x^2 + 4x - 2$ **3. a.** $x^2 + 0x - 9$ **b.** $x^3 + 0x^2 + 0x - 8$ **5.** $x - 3$ **7.** $z - 4$ **9.** $a - b$
11. $4a + 3$ **13.** $b - 1$ **15.** $3x - y$ **17.** $x + 3$ **19.** $3x + 1$ **21.** $2x - y$ **23.** $5x - 4y$ **25.** $x^2 + 2x - 1$ **27.** $2x^2 + 2x + 1$
29. $x^2 + xy + y^2$ **31.** $x + 1 + \frac{-1}{2x + 3}$ **33.** $2x + 2 + \frac{-3}{2x + 1}$ **35.** $x^2 + 2x + 1$ **37.** $x^2 + 2x - 1 + \frac{6}{2x + 3}$ **39.** $2x^2 + 8x + 14 + \frac{31}{x - 2}$
41. $x + 1$ **43.** $2x - 3$ **45.** $x^2 - x + 1$ **47.** $a^2 - 3a + 10 + \frac{-30}{a + 3}$ **49.** $5x^2 - x + 4 + \frac{16}{3x - 4}$ **51.** $x^2 - 5x + 4$ **53.** $9x^3 - 2x^2 + 3$
55. a. $x + 7$ **b.** $4x + 24$ **57.** $x + 2$ **59.** $6x^3 + 3x^2 - 18x$ **61. a.** $64x^6$ **b.** $\frac{x}{2y^2}$ **c.** $\frac{1}{54x}$ **63.** $x + 4$, $2x - 5$, $4x$, sum $= 7x - 1$

Chapter 2 Summary   (page 100)

1. a. 125   **b.** 243   **c.** 64   **d.** -64 **e.** 13   **f.** 25   **g.** 162   **h.** 18   **2. a.** x^5 **b.** x^3y **c.** y^{10} **d.** y^5 **e.** $2b^{12}$ **f.** $-z^5y^2$
g. $256xs^2$ **h.** $-3y^6$ **3. a.** x^{15} **b.** $4x^4y^2$ **c.** 9   **d.** 1   **e.** x^4 **f.** x^2y^2 **g.** y^2 **h.** $5yz^4$ **4. a.** 7; monomial   **b.** 2; binomial
c. 5; trinomial   **d.** 3; binomial   **5. a.** 402   **b.** 0   **c.** 82   **d.** $80t^4 - 2t$ **6. a.** $a - c$ **b.** $b + c$ **c.** ab **7.** $\$10{,}000 - b$ **8.** 55(2);
55(5); 55(b)   **9. a.** $7x$ **b.** simplest terms   **c.** $4x^2y^2$ **d.** x^2yz **e.** $5x^2$ **f.** $5x + 35$ **g.** $4a^2 + 4a - 6$ **h.** $6x^3 + 8x^2 + 3x - 72$
10. $6x^2 + x + 8$ **11.** $4x^2 + 13x - 8$ **12. a.** $10x^3y^5$ **b.** $5x + 15$ **c.** $3x^4 - 5x^2$ **d.** $-3x^2y^2 + 3x^2y$ **e.** $x^2 + 5x + 6$
13. a. $6a^2 - 6$ **b.** $a^2 - 1$ **c.** $6y^2 - 24$ **d.** $x^2 + 8x + 16$ **e.** $x^2 - 6x + 9$ **14.** $3x^3 + 7x^2 + 5x + 1$ **15.** $y^4 - 1$ **16. a.** $\frac{3}{2y} + \frac{3}{x}$
b. $2 - \frac{3}{y}$ **c.** $-3a - 4b + 5c$ **d.** $-\frac{x}{y} - \frac{y}{x}$ **17. a.** $x - 5$ **b.** $x + 1 + \frac{3}{x + 2}$ **c.** $x + 5 + \frac{3}{3x - 1}$ **d.** $3x^2 - x - 4$

Chapter 2 Test   (page 102)

1. $2x^3y^4$ **2.** y^6 **3.** $32x^{21}$ **4.** $8r^{18}$ **5.** 3   **6.** binomial   **7.** 10   **8.** 0   **9.** $-3x^2y^2$ **10.** $5x^3 + 2x^2 + 2x - 5$ **11.** $-x^2 - 5x + 4$
12. $-7x + 2y$ (or $2y - 7x$)   **13.** $-3x + 6$ (or $6 - 3x$)   **14.** $-4x^5y$ **15.** $3y^4 - 6y^3 + 9y^2$ **16.** $6x^2 - 7x - 20$
17. $2x^3 - 7x^2 + 14x - 12$ **18.** $\frac{y}{2x}$ **19.** $\frac{a}{4b} - \frac{b}{2a}$ **20.** $x - 2$

Cumulative Review: Chapters 1 and 2 (page 102)

1. prime **2. a.** Factor each denominator into prime factors. **b.** Write down each prime that appears, putting it down the greatest number of times it appears in any one factorization. **c.** The product of these factors is the LCD. **3.** $4x$ is four *terms* of x that are *added* together: $x + x + x + x$. x^4 is four *factors* of x: $x \cdot x \cdot x \cdot x$. **4.** *Evaluate* means to find its numerical value when given a value of x. *Simplify* means to write it in a simpler alternative form. **5. a.** 3 **b.** -3 **c.** -3 **6.** double negative **7.** If $x \neq 0$, $0 \cdot x = 0$ so $\frac{0}{x} = 0$, but if $\frac{x}{0} = k$, this means that $x = k \cdot 0 = 0$. However, the assumption is that $x \neq 0$, thus $\frac{x}{0}$ is undefined. **8.** b, c **9. a.** 0, 6 **b.** $-5, -1, 0, 6$ **c.** $-5, -\frac{7}{2}, -1, 0, 4.2, 6$ **d.** $\sqrt{2}$ **e.** 0, 6 **10. a.** perimeter **b.** volume **c.** volume **d.** area **11. a.** x^{m+n} **b.** x^{mn} **c.** $x^n y^n$ **d.** $\frac{x^n}{y^n}$ **e.** x^{m-n} **f.** 1 **13.** numerical coefficient **14.** binomials **15.** $15 + x$ **16.** distributive **17.** b **18.** conjugate, $16 - x^2$ **19.** $\frac{4x}{3} + 4$ **20.** $-x^2 - 3x + 4$ **21.** $-\frac{1}{8}$ **22.** $-\frac{2}{3}$ **23.** 61 **24.** 8 **25.** -9 **26.** $-\frac{28}{9}$ **27.** 5 **28.** -1 **29.** 13 **30.** $-\frac{7}{5}$ **31.** $-3x^5$ **32.** x^{18} **33. a.** xy **b.** $2x + 4$ **34. a.** 2nd degree **b.** 22 **35.** $2x^2 + 2x + 12$ **36.** $2x - 2y$ **37.** $3x^3 - 4x^2 - 12x + 5$ **38.** $x^2 - 3x - 5$ **40.** $4x^2 + 5x + 4 + \frac{-6}{2x+1}$ **41.** 81 **42.** $x = -1$ and $P(x) = 3$ **43.** $(4 + 8)/3$ **44.** 9097 **45.** $\frac{4019}{7}$ **46.** $\frac{1}{2(5+7)}$ **47. a.** 20 **b.** 8 **48.** \$4.50 **49.** $x^2 + x + 2$ **50.** $9x + 3$

Exercise 3.1 (page 115)

1. linear **3.** equivalent **5.** symmetric **7.** no **9.** $\varnothing$ **11. a.** $6(a - 2) = 6.6$; $\{3.1\}$ **b.** $2(5a + 4) = 2$; $\{-.6\}$ **c.** $2(t - 5) = 3(2t + 5) - 6$; $\{-\frac{19}{4}\}$ **d.** $6x + 4 = 3x - 14$; $\{-6\}$ **e.** $2x - 5 = x + 3$; $\{8\}$ **f.** $4(t + 1) - 3 = 3t + 2(t - 5)$; $\{11\}$ **13.** $x + 2, x + 4$ **15. a.** $x + 4$ **b.** $x - 7$ **17.** Yes, because when $x = 1$ both sides of the equation have the same value. **19.** Yes, because when $x = 2$ both sides of the equation have the same value. **21.** Yes, because when $x = 0$ both sides of the equation have the same value. **23.** Yes, because when $k = 3$ both sides of the equation have the same value. **25.** Yes, because when $t = -9$ both sides of the equation have the same value. **27.** Yes, because when $x = 0$ both sides of the equation have the same value. **29.** $\{1\}$ **31.** $\{\frac{3}{4}\}$ **33.** $\{37\}$ **35.** $\{-9\}$ **37.** $\{-3\frac{3}{4}\}$ or $\{-\frac{15}{4}\}$ **39.** $\{-82\}$ **41.** $\{9\}$ **43.** $\mathbb{R}$ **45.** $\{-5\}$ **47.** $\{16\}$ **49.** $\{-2\}$ **51.** $\{5\}$ **53.** $\{2\}$ **55. a.** $28 = 4(2x + 3)$ **b.** $28 = 8x + 12$ **c.** $x = 2$ **d.** Each side is 7 in. long. **57. a.** $20 = 2(x - 4) + 2x$ **b.** $20 = 4x - 8$ **c.** $x = 7$ **d.** The rectangle is 3 ft by 7 ft. **59.** $x + (x + 2) = 54$; the integers are 26 and 28. **61.** $2x + 5 = 15$; the number is 5. **63.** $3.75x + 29.95 = 52.45$; the subscriber watched 6 movies. **65.** $12x + 16 = 100$; he was gone for 7 days. **67. a.** $\frac{7}{15}$ **b.** $-\frac{3}{5}$ **c.** 16 **69.** $x^{16}y^6$ **71.** $2x^5 - 8x^3y$

Exercise 3.2 (page 121)

1. literal equations **3.** $P - a - b = c$; subtraction property of equality **5.** $\frac{d}{r} = t$; division property of equality **7.** $\frac{i}{pt} = r$; division property of equality **9.** $\frac{C}{2\pi} = r$; division property of equality **11.** $\frac{P - 2l}{2} = w$ **13.** $\frac{2A}{B+b} = h$ **15.** $R = \frac{E}{I}$ **17.** $\frac{10}{3} - \frac{1}{6}x = y$ **19.** $\frac{15}{2} - \frac{3}{4}x = y$ **21.** $-\frac{2}{3}x - \frac{11}{3} = y$ **23.** $y = \frac{8}{3} - \frac{2}{3}x$ **25.** $t = 3$ **27.** $t = 2$ **29.** $c = 3$ **31.** $h = 8$ **33.** $I = \frac{E}{R}$; 4 amps **35.** $r = \frac{C}{2\pi}$; $r \approx 2.28$ ft **37.** $R = \frac{P}{I^2}$; $R \approx 13.78$ ohms **39.** $m = \frac{Fd^2}{GM}$ **41.** $D = \frac{L - 3.25(r + R)}{2}$; $D = 6$ ft **43.** -6 **45.** $2x$

Exercise 3.3 (page 130)

1. two **3.** incremented **5.** function **7.** range **9.** No, because $3(2) - 4(2) \neq 2$. **11.** No; the only way both (a, b) and (b, a) could be solutions would be if the equation was $x + y = c$. **13.** Yes, it is a function because each x value is paired with a unique y value. **15.** $\{-14, -13, -12, -11, -10, -9, -8\}$ **17.** $(-1, -4)$; $(5, 2)$ **19.** $(-5, \frac{1}{3})$; $(-8, -\frac{4}{3})$ **21.** $(\frac{1}{4}, 0)$; $(-\frac{3}{4}, 4)$ **23.** $(10, 80)$; $(-5, 5)$ **25.** $(2, 48)$; $(-6, 48)$ **27.** $(-1, \frac{2}{3})$; $(2, 14\frac{2}{3})$ **29.** $(-2, 8)$; $(10, -1000)$ **31.** $(1, -\frac{1}{5})$; $(-1, -\frac{9}{5})$ **33.** $x = 2, \frac{7}{5}, \frac{4}{5}$ **35.** $x = \frac{62}{3}, 14, 10$ **37.** function; d: $\{-4, 5, 3\}$; r: $\{5, -4\}$ **39.** function; d: $\{-1, 0, 1, 2, 3, 4, 5\}$; r: $\{-8, -5, -2, 1, 4, 7, 10\}$ **41.** function; d: $\mathbb{R}$; r: $\mathbb{R}$ **43.** function; d: $\mathbb{R}$; r: $\mathbb{R}$ **45.** $y = 5, 8.75, 12.5, 16.25$ **a.** \$8.75 **b.** \$16.25 **c.** $y = .50x + 5$ **d.** \$12.50 **47. a.** \$97,000 **b.** 8 yr **49.** $y = 37.50c + 20$; cost = \$57.50, \$76.25, \$207.50, \$235.63 **51.** $y = -800x + 4000$; \$800 **53.** $y = \$5.50x$; \$19.50 **55. a.** 2nd degree **b.** 44 **57.** $V = 9x^3 + 15x^2 + 6x$ cubic units **59.** 24

Exercise 3.4 (page 144)

1. rectangular coordinate system **3.** origin **5.** abscissa **7.** y **9.** standard **11.** scale **13. a.** I **b.** II **c.** III **d.** IV **15. a.** $(2, -2)$ **b.** $(-5, -5)$ **c.** $(-4, 0)$ **d.** $(0, 0)$ **e.** $(-3, 5)$ **f.** $(-2, -3)$ **g.** $(4, -5)$ **h.** $(2, 3)$ **17.** $(1992, 12.6)$ **19. a.** $[-5, 5]$ by $[-5, 5]$ **b.** $[-5, 10]$ by $[-10, 5]$ **c.** $[-2, 3]$ by $[-4, 4]$ **d.** $[-6, 2]$ by $[-4, 8]$

21.
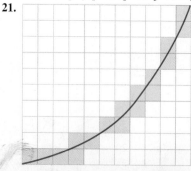

23. One disadvantage is that the window may be too large and thus the clarity of the graph is distorted. **25.** x-intercepts: $(-4, 0)$ and $(4, 0)$; y-intercept: $(0, -4)$ **27.** The third point provides a check for the other two points. All three points must lie in a straight line. **29.** $x = -2$. The line is not the graph of a function because the x value of -2 is paired with infinitely many y values. To be a function, each x must be paired with *exactly* one y. **31.** $y = -2$ **33.** $x = 0$

35. $y = 1, 0, 5$

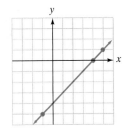

37. $y = \frac{1}{2}, -\frac{1}{2}, -2$

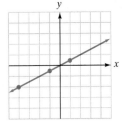

39. $y = -5, 1, 4$

41. $y = -1, -3, -4$

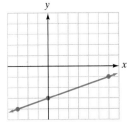

43. $y = 1, -2, -3, -2, 1$

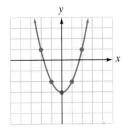

45. $y = -3, -6, -7, -6, -3$

47. $y = -12, -5, -4, -3, 4$

49.

51.

53.

55.

57.

59.

61.

63.

65.

67.

69.

71.

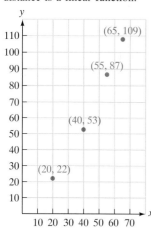

73. x-intercept: (−4, 0); y-intercept: (0, −12)   **75.** x-intercept: (2, 0); y-intercept: (0, −8)
77. x-intercept: none; y-intercept: (0, −15)   **79.** x-intercept: (3, 0); y-intercept: (0, −27)   **81.** tutor's
fee: 12, 24, 36, 60

83. The speed as related to stopping distance is a linear function.

85. $y = 220 - x$

87. The points lie in a line.

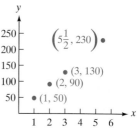

89. $A = 4\pi x^4$ **91.** The calculator screen should look as follows: 9→F:(5/9)(F-32)
35

Exercise 3.5 (page 162)

1. best fit   **3.** Sherah earns $35 for 6 hours of babysitting.   **5.** *Quick-Copy*
In Exercises 7–17, the appropriate viewing window is indicated. If your window is slightly larger, that would be acceptable.
7. [−10, 10] by [−10, 20]   **9.** [−20, 10] by [−10, 10]   **11.** [−10, 10] by [−10, 20]   **13.** [−10, 10] by [−20, 10]   **15.** [−10, 10]
by [−10, 20]   **17.** [−10, 10] by [−10, 75]   **19. a.** $y = 0.0875x$ **b.** Generate the calculator graph for the equation $y = 0.0875x$.
c. $29.00   **21. a.** $y = .50x + 2.50(25 - x)$ **b.** Generate the calculator graph for the equation $y = .50x + 2.50(25 - x)$.   **c.** 13   **d.** 4
e. No, because the maximum income occurs when no student/faculty tickets are sold.   **23. a.** $y = x + .40x$ **b.** Generate the
calculator graph for the equation $y = x + .40x$.   **c.** 33   **d.** $67.20   **25.** 1281.49 ft   **27. a.** [−16, 10] by [−21, 4]   **b.** [−19, 26] by
[−5, 31]   **c.** [−76, 51] by [−81, 16]   **29. a.** (5, 0) and (9, 0)   **b.** (−7, 0) and (8, 0)   **c.** (−8, 0) and (−3, 0)

Exercise 3.6 (page 173)

1. $ax + c = 0$ **3.** 2   **5.** [−10, 12] by [−10, 17]   **7.** [−10, 10] by [−40, 10]   **9.** [−10, 10] by [−20, 10]   **11.** [−10, 10] by
[−20, 10]   **13.** [−10, 10] by [−10, 15]   **15.** −2   **17.** 0   **19.** 10   **21.** −3   **23.** −3   **25.** 8   **27.** $\frac{34}{9}$ **29.** $-\frac{8}{7}$ **31.** Enter the left
side of the equation at the y1= prompt and the right side at the y2= prompt. Display the graph in the standard viewing window and
perform the INTERSECT operation. The screen display should read "x = −3.5   y = −6.5." Remember, it is the x value that is the
solution to the equation. Solution: −3.5   **33.** Enter the left side of the equation at the y1= prompt and the right side at the
y2= prompt. Display the graph in the standard viewing window and perform the INTERSECT operation. The screen display should
read "x = − 6   y = −4." Remember, it is the x value that is the solution to the equation. Solution: −6   **35.** Enter the left side of the
equation at the y1= prompt and the right side at the y2= prompt. Display the graph in the [−10, 10] by [−15, 10] window and
perform the INTERSECT operation. The screen display should read "x = .55555556   y = −9.888888889." Remember, it is the x value
that is the solution to the equation. Solution: $\frac{5}{9}$ **37.** Enter the left side of the equation at the y1= prompt and the right side at the
y2= prompt. Display the graph in the window [− 10, 10] by [−10, 25] and perform the INTERSECT operation. The screen display
should read "x = 9   y = 20." Remember, it is the x value that is the solution to the equation. Solution: 9   **39.** Enter the left side of
the equation at the y1= prompt and the right side at the y2= prompt. Display the graph in the window [−10, 100] by [−10, 20] and
perform the INTERSECT operation. The screen display should read "x = 90   y = 15." Remember, it is the x value that is the solution
to the equation. Solution: 90   **41.** $3 + 2x = x - 5$; the number is −8.   **43.** $2(x + 3) = 5x$; the number is 2.   **45.** $x + 3(-x) = 4$; the
number is −2.   **47.** $4(x + 6) = x - 3$; the number is −9.   **49.** $\frac{x}{2} = x + 3$; the number is −6.   **51.** $.75x = 4.5$; there were 6 million
shares.   **53.** $x + 10.99 + 14.50 = 30.44$; the cost of the brush was $4.95.   **55.** $1.5x = 555$; the initial number of shares was 370.
57. $.90x = 2484$; there are 2760 employees.   **59.** $.05x + 5 = 42.50$; she must distribute 750 circulars.   **61.** $\frac{x+7}{2} = 5$; the original
integer was 3.   **63.** $20x + 95 = 155$; the repair took 3 hours.   **65.** $3x - 2000 = 7000$; her sales increased by $4000.
67. $x + 2x = 12$; the sections were 4 ft and 8 ft.   **69.** $.80x = 969.20$; the original price was $1211.50.   **71.** The smaller tank cannot
be emptied into a 10-gallon canister because the smaller tank holds only 13 gallons.   **73.** $x + 8x = 2700$; there are 300 g of hydrogen

in 2700 g of water.   **75.** $.45x + .23x + 317920 = x$; they did not meet their quota.   **77.** $3x + x = 500$; there are 125 lb of cement in 500 lb of dry concrete.   **79.** $A = 90^2$; the area is 8100 sq ft.   **81.** $\frac{xy}{3}$ **83.** $4x$ miles

Chapter 3 Summary   (page 178)

1. a. yes **b.** no **c.** no **d.** yes   **2. a.** -2 **b.** $-\frac{1}{9}$ **c.** 5 **d.** 3 **e.** -1 **f.** 15   **3. a.** 7 **b.** $\frac{39}{7}$ **c.** 9 **d.** $-\frac{9}{4}$ **e.** 12 **f.** 6
4. a. $2x - 7 = 9$; the original integer was 8. **b.** $3x - 4 = 7$; the number is 27.   **c.** $1.50x + 35 = 162.50$; 85 feet of gutter is needed.
d. $.18x + 17.50 = 43.96$; 147 kilowatt hours were used.   **e.** $15 + (3x - 2) + x = 45$; the shortest section is 8 ft long.
5. a. $y = -3x + 12$ **b.** $y = \frac{4}{3}x - \frac{8}{3}$ **c.** $t = \frac{d}{r}$ **d.** $w = \frac{P - 2l}{2}$ **e.** $h = \frac{A}{2\pi r}$ **f.** $\frac{A - P}{Pt} = r$ **g.** $\frac{RT}{P} = V$ **6. a.** yes **b.** yes **c.** no
d. no   **7. a.** $(2, -2)$ **b.** $(3, -3)$ **c.** $(-1, 0)$ **d.** $(3, 0)$ **8.** $y = 12, 9, 6, 3, 0$ **9. a.** relation; domain = $\{2, 4\}$,
range = $\{3, 6, -3\}$ **b.** function; domain = $\{1, 2, 3\}$, range = $\{5, 8, 11\}$ **c.** function; domain = $\mathbb{R}$, range = $\mathbb{R}$ **d.** function;
domain = $\mathbb{R}$, range = {all real numbers at least as large as -1}
10. a.

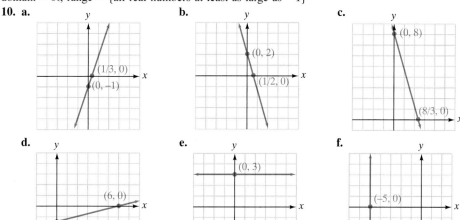

11. a. x-int: $\left(\frac{1}{3}, 0\right)$; y-int: $(0, -1)$ **b.** x-int: $\left(\frac{1}{2}, 0\right)$; y-int: $(0, 2)$ **c.** x-int: $\left(\frac{8}{3}, 0\right)$; y-int: $(0, 8)$ **d.** x-int: $(6, 0)$; y-int: $\left(0, -\frac{3}{2}\right)$
e. y-int: $(0, 3)$; horizontal line **f.** x-int: $(-5, 0)$; vertical line   **12. a.** y-int: $(0, -6)$; x-int: $(2, 0)$ **b.** y-int: $(0, -5)$; x-int: $(-1, 0)$,
$(5, 0)$ **c.** y-int: $(0, -64)$; x-int: $(4, 0)$ **d.** y-int: $(0, -12)$; x-int: $(3, 0)$ **13. a.** $[-10, 15]$ by $[-16, 50]$ **b.** $[-10, 25]$ by $[-110, 10]$
c. $[-10, 10]$ by $[-65, 150]$ **14. a.** Generate the calculator graph as specified. **b.** \$610 **c.** \$822.50 **15. a.** $y = 125 + 35x$
b. Generate the calculator graph as specified. **c.** \$230 **d.** 8 years   **16. a.** $12x + 5y = 1069$ **b.** $y = \frac{1069}{5} - \frac{12}{5}x$ **c.** 113
d. No, because 69.8 student tickets would have to be sold.   **e.** No, because according to the graph that would mean that -2.2 student
tickets were sold.   **17.** Your graph, your table, and the use of the STOre feature should all support the algebraic result you determined
in Exercises 2 and 3. If you do not understand and/or did not obtain a correct result using one of these three features, return to
Technology 3.5 on page 165 and Example 3 on page 168.   **18.** See explanation in Exercise 17.

Chapter 3 Test   (page 181)

1. no **2.** no **3.** $y = \frac{8}{3} - \frac{4}{3}x$ **4.** $y = -\frac{1}{4}x + \frac{3}{10}$ **5.** no **6.** yes **7.** 7 **8.** -2 **9.** -3 **10.** -4 **11.** -2 **12.** -2 **13.** $\varnothing$ **14.** 0
15.

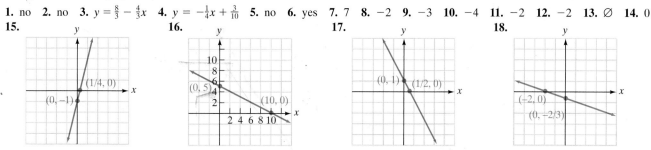

19. a. Generate the calculator graph as specified. **b.** 200 **20.** $x + (x + 2) = 36$; the integers are 17 and 19.
21. $x = 560 + .96(560)$; the current value is \$1097.60. **22.** $600 + .06x = 5700$; the value of the house is \$85,000.

Exercise 4.1   (page 192)

1. ratio **3.** means, extremes **5. a.** $\frac{5}{7}$ **b.** $\frac{3}{5}$ **c.** $\frac{1}{2}$ **d.** $\frac{1}{2}$ **e.** $\frac{2}{3}$ **f.** $\frac{2}{3}$ **7. a.** similar **b.** similar **c.** not similar **9.** -17 **11.** $\frac{8}{11}$
13. 102 **15.** $x + (x + 2) = 54$; the consecutive even integers are 26 and 28. **17.** $x + (x + 1) + (x + 2) = 120$; the three integers are
39, 40, and 41. **19.** $x + 2(x + 1) = 23$; the smaller integer is 7. **21.** $2x - 3 = x + 10$; the smaller integer is 13. **23.** $3x = 57$; the
length of each side is 19 ft. **25.** $2x + 2(x - 11) = 94$; the dimensions of the pool are 29 m by 18 m. **27.** $2x + 2(2x + 5) = 112$; the
dimensions of the frame are 39 in. by 17 in. **29.** $3x = 180$; each of the angles measures 60°. **31.** $\frac{3}{50} = \frac{39}{x}$; the cost of 39 packets will

be \$6.50. **33.** $\frac{.75}{x} = \frac{.87}{1}$; the real engine is 65.25 ft. **35.** $\frac{50}{1} \neq \frac{6}{.125}$; the instructions are not correct. **37.** $\frac{5}{7} = \frac{h}{30}$; the approximate height of the flagpole is 21.4 ft. **39.** $\frac{1000}{100} = \frac{5280}{x}$; the plane will gain 528 ft in altitude. **41.** $\frac{300}{100} = \frac{.5(5280)}{x}$; the hill is 880 ft high. **43. a.** $-2x^2 - 2x + 6$ **b.** $8x^3 + 60x^2 + 150x + 125$ **c.** $x^2 - 3x + 9$ **45. a.** $x = 1$ **b.** $x = 1$

Exercise 4.2 (page 198)

1. fixed **3.** break-even **5.** unit cost: \$25; fixed cost: \$4000 **7.** $12000(60) = 9400(60) + 10400x$; the project can last 15 days beyond the deadline. **9.** $30x = 20x + 9600$; the break-even point is 960 pairs of shoes. **11.** $500 + 2x = 800 + x$; 300 plates make the cost the same. **13.** $150x + 195x + 210(2x + 15) = 72000$; 90 spreadsheets were sold. **15.** $x + 2x + (2x - 1) = 99$; she has 20 bolts. **17.** $x + 2x = 3500$; 1100 shares did better than average. **19.** $1.50(x + 40) + 4(x + 15) + 7.50x = 276$; there are 12 wide-screen sets in stock. **21.** $.09x + .08(24000 - x) = 1965$; \$4500 is invested at 9% and \$19,500 is invested at 8%. **23.** $.08x + .11x = 712.50$; \$3750 was invested at each rate. **25.** $15000(.06) + x(.07) = 1250$; \$5000 more was invested at 7%. **27.** $10500x = 21000(x + .01) - 840$; the bond funds paid 6% and the CD 7%. **29. a.** $16x^2$ **b.** $16x^2 + 24x + 9$ **31.** $x = 6$

Exercise 4.3 (page 203)

1. a. $0.03, \frac{3}{100}$ **b.** $0.30, \frac{3}{10}$ **c.** $0.035, \frac{7}{200}$ **d.** $0.035, \frac{7}{200}$ **3.** $x, 30 - x$ grams **5.** $1.15x + 20(.85) = (20 + x)$; there are 20 gallons of the \$1.15 fuel. **7.** $.03x + .07(50) = .05(50 + x)$; there are 50 gallons of the 3% brine solution. **9.** $.10(30) + x(0) = (30 + x)(.08)$; she must add 7.5 oz of water. **11.** $1.90x + 1.20(100 - x) = 100(1.48)$; there are 40 pounds of lemon drops and 60 pounds of jelly beans. **13.** $20x + 20(x - .30) = 40(1.05)$; the cashews are \$1.20 per pound. **15.** $40(7) + 4x = (40 + x)5$; there are 80 pounds of regular coffee. **17.** $50x + 55x = 315$; the two cars will meet in 3 hours. **19.** $50x + 60x = 715$; the two cars will be 715 miles apart in 6.5 hours. **21.** $11x = 82.5$; in 7.5 hours the two cars will be 82.5 miles apart. **23.** $5x + 5(x + 200) = 6000$; the speed of the slower plane is 500 mph. **25. a.** -56 **b.** $6x^3 + 7x^2 - 2x + 5$ **c.** $xy^2 - \frac{3}{2}y + \frac{2x^2}{y}$ **27.** $\frac{2}{9}$ **29.** $3(x + 2) + 4 = 13$; the integer is 1.

Exercise 4.4 (page 209)

1. $x, -x$ **3. a.** 4 **b.** 4 **c.** -4 **d.** $\pm a$ **e.** $\pi - 3$ **5.** Absolute value can never be a negative value. **7.** ± 8 **9.** 9 or -3 **11.** 4 or -1 **13.** $\frac{14}{3}$ or -6 **15.** $\varnothing$ **17.** 8 or -4 **19.** $-\frac{1}{2}$ or 2 **21.** -8 **23.** -4 or -28 **25.** 0 or -6 **27.** $\frac{20}{3}$ **29.** -2 or $-\frac{4}{5}$ **31.** 3 or -1 **33.** 0 or -2 **35.** 0 **37.** $\frac{4}{3}$ **39.** $\varnothing$ **41.** One such example would be if $a = 4$ and $b = -7$ then $|4(-7)| = |4| \cdot |-7|$. **43.** Using the same values for a and b as in Exercise 41, $|4 + (-7)| \neq |4| + |-7|$. **45. a.** $\frac{-4}{45}$ **b.** $\frac{7}{-6}$ **47.** 1 **49.** One good viewing window would be $[-10, 30]$ by $[-15, 10]$.

Chapter 4 Summary (page 211)

1. a. yes **b.** no **c.** no **d.** yes **2. a.** 12 **b.** -15 **c.** $-\frac{13}{4}$ **d.** 13 **3.** $2w + 2(3 + 2w) = 84$; the painting is 13 in. wide. **4.** $\frac{5}{3} = \frac{x}{18}$; 30 tons of iron is needed to make 18 tons of pig iron. **5.** $\frac{3}{300} = \frac{x}{5}$; there is $\frac{1}{20}$ gram in a single 5 ml dose. **6.** $\frac{h}{5} = \frac{53}{2}$; the building is 132.5 ft tall. **7.** $3x = 1.25x + 668.50$; the break-even point is 382 parts. **8.** $3x + 85 = 2.5x + 105$; at 40 caps, the two machines give equal cost. **9.** $.07x + .09(27000 - x) = 2110$; \$16,000 was invested at 7% and \$11,000 was invested at 9%. **10.** $.045x + .05(13750 - x) = 647.50$; \$8000 was invested at $4\frac{1}{2}$%. **11.** $x \cdot 0 + .10(30) = .08(30 + x)$; 7.5 liters must be added to dilute the solution to an 8% solution. **12.** $.90x + 1.65(20 - x) = 20(1.20)$; there should be 12 lb of the \$.90 per pound candy and 8 lb of the \$1.65 per pound candy. **13.** $15x = 5$; they will meet in 20 minutes. **14. a.** 3 or $-\frac{1}{3}$ **b.** $\frac{26}{3}$ or $\frac{-10}{3}$ **c.** $\varnothing$ **d.** 3 or 0 **e.** $\frac{5}{4}$ or $\frac{-3}{4}$ **15. a.** -5 or $\frac{1}{5}$ **b.** -1 or 1

Chapter 4 Test (page 213)

1. no **2.** no **3.** $-\frac{3}{2}$ **4.** $-\frac{17}{4}$ **5.** $\frac{575}{.5} = \frac{x}{7}$; 8050 ft will be lost. **6.** $65x + 55x = 72$; it will take 0.6 of an hour for them to meet. **7.** $.06x + .055(50000 - x) = 2975$; \$45,000 is in the CD. **8.** $\frac{x}{x+13} = \frac{35}{100}$; they will need 7 oz of juice. **9.** $x + 2x + 4 + 2x = 24$; the shortest side is 4 units long. **10.** $100 + .35x = .45x$; she needs to stuff 1000 envelopes to break even. **11.** $\frac{33}{2}$ or $\frac{-3}{2}$ **12.** $\varnothing$ **13.** 0 or -10 **14.** $\varnothing$ **15.** $\varnothing$ **16.** -2 or $\frac{1}{3}$

Cumulative Review: Chapters 3 and 4 (page 214)

1. for all, $\mathbb{R}$ **2.** no, $\varnothing$ or { } **3.** ordered pairs **4.** domain **5.** range **6. a.** II **b.** III **c.** IV **d.** IV **7.** **8.** In ZInteger, all x-coordinates are integers. **9.** A trip of 50 miles will cost \$49.50. **10.** STO, TABLE **11.** comparison **12.** $\frac{3}{4}$ **13.** similar; proportion **14.** $16.99x$ **15. a.** 0.06 **b.** 0.0775 **16.** x liters **17.** $a, -a$ **18.** Because $|x|$ is always greater than or equal to zero. **19. a.** -3 **b.** $\pi - 2$ **20.** $P = \frac{i}{rt}$ **21.** $\frac{1}{4}$ **22.** 15

23.

x	y
-3	12
-2	7
-1	4
0	3
1	4
2	7
3	12

24. b   **25.** b   **26.**

27.

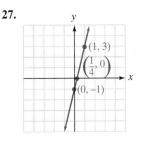

28. x-intercepts are (1, 0) and (7, 0) and y-intercept is (0, 7).   **29.** ZStandard has no pixel with x-coordinate of 1 or 7 or y-coordinate of 7.   **30.** For a good window, the y-minimum must be -16 or smaller.   **31.** $-\frac{22}{13}$ **32.** $\frac{7}{6}$ or $\frac{1}{2}$ **33.** 14 or -18 **34.** -5 or $-\frac{3}{5}$ **35.** $y = \frac{3}{2}x - 18$ **36.** $y = -\frac{A}{B}x + \frac{C}{D}$ **37.** 14 or -10 **38.** 1 or -5 **39.** $\varnothing$ **40.** -3 or $\frac{1}{2}$ **41.** yes   **42.** $x + 3(x + 1) = 171$; the smaller number is 42.   **43.** $3000 + (.07)(3000)t = 4000 + (0.4)(4000)t$; it will take 20 years for the accounts to have the same amount of money.   **44.** $29x + 33(4x) = 2415$; there are fifteen 29¢ stamps and sixty 33¢ stamps.   **45.** $15x = 55$; it will take $3\frac{2}{3}$ hours or 3 hours 40 min.   **46.** $.10x + .50(64) = .35(64 + x)$; 38.4 oz of the 10% juice will be needed.   **47.** 3053.6 in³ of air   **48.** $2937.50 = 2500(1 + .035t)$; the money was invested for 5 years.   **49. a.** \$20,700   **b.** 170   **50. a.** 19,155   **b.** greater than the revenue   **c.** 620 CD players, profit (revenue = \$55,800)

Exercise 5.1   (page 225)

1. factors   **3.** prime   **5.** distributive   **7.** polynomial   **9.** grouping   **11.** F, 1 is not prime.   **13.** T   **15.** T   **17.** F, it is not a product of prime factors.   **19.** T   **21.** 6   **23.** $2x^2y^3$ **25.** tv **27.** $2x^2 + 1$ **29.** $5ab + 7$ **31.** $2x^2 + 1$ **33.** $x(y - z)$ **35.** $-(2x - 5y)$ **37.** $r^2(r^2 - 1)$ **39.** $-(3m + 4n - 1)$ **41.** $8xy^2z^3(3xyz + 1)$ **43.** $3(x + y - 2z)$ **45.** $-(3ab + 5c - 9bc)$ **47.** $2y(2y + 4 - x)$ **49.** $-4a^2b^2(b - 3a)$ **51.** $-7ab(2a^5b^5 - 7ab^2 + 3)$ **53.** $2xyz^2(2xy - 3y + 6)$ **55.** $(x + y)(a + 2b)$ **57.** $(3t + 5)(3t + 4)$ **59.** $3ab(3ab - 2)(3x - 2y)$ **61.** $(x - 5y)(a + 1)$ **63.** $(m - n)(p + q)$ **65.** $(3c + 1)(a + b)$ **67.** $(b + 3)(3a - 2)$ **69.** $(3x^2 + y)(2u - v)$ **71.** $(p + q)(m - n)$ **73.** $(5 + r)(x + y)$ **75.** $(m + n)(r + s)$ **77.** $(s + r)(x - y)$ **79.** Not correct; each side of the equation should evaluate to the same value. Probably correct; it is possible, however, that the two expressions might not have the same value for other values of x and y.   **81.** $x + 8y$ **83. a.** $\frac{11}{3}$ **b.** The answer will be displayed as $x = 3.666666667$.   **c.** Rational; $\frac{11}{3}$ is a quotient of integers.

Exercise 5.2   (page 234)

1. trinomials   **3.** multiplication   **5.** factors   **7. a.** 1 is the leading coefficient.   **b.** 5 is the coefficient of the second term.   **c.** 1 is the coefficient of the first term.   **d.** 6 is the constant term.   **9.** a   **11.** c   **13.** b   **15.** Both are correct; real number multiplication is commutative.   **17.** prime   **19.** $(x - 4)(x - 1)$ **21.** $(x + 1)(x + 2)$ **23.** $2(x + 3)(x + 2)$ **25.** $-5(t - 2)(t - 3)$ **27.** prime   **29.** $(u + 5v)(u - 3v)$ **31.** $(r - s)(r - 2s)$ **33.** $-4x(x + 3y)(x - 2y)$ **35.** $2(3y + x)(4y + x)$ **37.** $3y(y + 1)(3y - 2)$ **39.** $(y - 3)(y - 5)$ **41.** prime   **43.** $3(x + 2y)(2x - 3y)$ **45.** $(x + 13)(x - 2)$ **47.** $(2x - 5)(3x + 2)$ **49.** $(y + 8)(2y - 5)$ **51.** $(7x + 8)(6x - 1)$ **53.** $(x + 3)^2$ **55.** $(y - 4)^2$ **57.** $(t + 10)^2$ **59.** $(u - 9)^2$ **61.** $(x + 2y)^2$ **63. a.** The length is 4 greater and the width is 3 greater than the height.   **b.** length = 10 in.; width = 9 in.   **65.** $\pm 37, \pm 20, \pm 15, \pm 13, \pm 12$ **67.** $x^2 + 6x + 9$ **69.** $\frac{b^2c^5}{a^4}$ **71. a.** -5 **b.** You should use the viewing window $[-10, 10]$ by $[-20, 10]$ and the INTERSECT feature to solve this equation graphically. Remember, it is only the x-value on the INTERSECT screen that is the solution to the equation.

Exercise 5.3   (page 241)

1. difference   **3.** conjugate binomials   **5.** GCF   **7.** F, 2 is not a perfect square.   **9.** T   **11.** F, $x^2 + y^2$ is a prime polynomial.   **13.** If you consider the product ac (-221) in the first trinomial and the product ac (2880) in the second trinomial, it is obvious that 2880 has many more pairs of factors to check than does -221.   **15.** $(x - 10)(x + 10)$ **17.** $(2y - 5)(2y + 5)$ **19.** $9(x - 3y)(x + 3y)$ **21.** $(x - y)(x + y)(x^2 + y^2)$ **23.** $(3x^2 - 4y)(3x^2 + 4y)$ **25.** $4(y - 1)(y + 1)$ **27.** $y^2(y + x)$ **29.** $4y(x - 2y)(x + 2y)$ **31.** $(x - 3)(x + 3)(x^2 + 9)$ **33.** $(x - 2)(x^2 + 2x + 4)$ **35.** $(b + 5)(b^2 - 5b + 25)$ **37.** $(3 - y)(9 + 3y + y^2)$ **39.** $(2u + w)(4u^2 - 2uw + w^2)$ **41.** $(x - 3y)(x^2 + 3xy + 9y^2)$ **43.** $(3a - b)(9a^2 + 3ab + b^2)$ **45.** $(3x + 5)(9x^2 - 15x + 25)$

47. $(4x + 3y)(16x^2 - 12xy + 9y^2)$ **49.**   **51.** The factorization is incorrect since P and Q have a different

value for the *same* value of x.   **53.** It is guaranteed to be correct if the polynomial is of degree 6 or less.   **55.** The factorization is not correct since y1 $\neq$ y2 for at least one value of x.   **57.** $(x - 3)(x + 3)$ **59.** $(x + 2)(x^2 - 2x + 4)$ **61.** $3rs(s - 2rt)$
63. $(a + b)(c + d)$ **65.** prime   **67.** $2(x - 4)(x + 4)$ **69.** $2x(y + 6)(y - 2)$ **71.** $-5(x + 4)(x - 2)$ **73.** $-4p^2q^3(2pq^4 + 1)$
75. $(4y - z)(16y^2 + 4yz + z^2)$ **77.** prime   **79.** prime   **81.** $2t^2(3t - 5)(t + 4)$ **83.** $-(3x - 4)(2x + 5)$
85. $2(2x - 3y)(4x^2 + 6xy + 9y^2)$ **87.** $(x - a)(a - b)(a + b)$ **89.** $2xy(2ax - b)(2ax + b)$ **91.** $(3p - 2q)(3p + 2q)(9p^2 + 4q^2)$
93. prime   **95.** $y^2(2x - y)(2x + y)$ **97.** prime   **99.** $t(7t - 1)(3t - 1)$ **101.** $n = \frac{l - f + d}{d}$

Exercise 5.4   (page 253)

1. transformation   **3.** root   **5.** at least one   **7.** root (or zero)   **9.** x, y **11.** $(x - 5)(2x + 3)$ **13.** $5, -\frac{3}{2}$ **15.** b and c are equations, a and d are expressions   **17.** -5 or 2 **19.** 0 or $\frac{4}{3}$ **21.** $\frac{1}{2}$ or 2 **23.** $\frac{1}{5}$ or 1 **25.** $\frac{2}{3}$ or $\frac{-3}{2}$ **27.** $\frac{1}{8}$ or 1 **29.** 0 or $\frac{4}{3}$ or $\frac{-3}{2}$ **31.** $\frac{1}{3}$ or $\frac{-1}{3}$
33. 0 or 2 or -1 **35.** 0 or -4 **37.** -4 or -5 **39.** $\frac{9}{2}$ or $-\frac{9}{2}$ **41.** $-\frac{2}{3}$ or $\frac{5}{2}$ **43.** 2 **45.** 8 or -6 **47. a.** C   **b.** $x = 2$ or -3
c. $x^2 + x - 6 = 0$ **49. a.** B   **b.** $x = 0, 3$, or -3 **c.** $x^3 - 9x = 0$ **51. a.** H   **b.** $x = 0, 3$, or -2 **c.** $x^3 - x^2 - 6x = 0$ **53. a.** D
b. $x = 2$ or -4 **c.** $x^2 + 2x - 8 = 0$ **55.** $0.06, -3.5$ **57.** $0.3, -5.2$ **59.** $-4.6, 0.2, 3.2$ **61.** It will hit the ground after 9 seconds.
63. At $\frac{15}{4}$ seconds and 10 seconds it will be at a height of 600 ft.   **65.** The dive ends after 2 seconds.   **67.** $x(x + 2) = 35$; the integers are 5 and 7, since x must be positive.   **69.** $4 + x^2 = 10x - 5$; the integer is 9.   **71.** $2(2w + 1) = 36$; the dimensions are 4 m by 9 m.
73. $x(x + 2) = 143$; it will take 48 ft of crown molding.   **75.** $\frac{1}{2}h(2 + 2h) = 30$; the height is 5 ft and the base is 12 ft.
77. $\frac{1}{2}(x - 3)(x - 6) = x$; the area is 9 sq units.   **79.** $h(2h) = 200$; the base is 20 cm.   **81.** $\frac{h(8 + h)}{2} = 24$; the height is 4 m.   **83.** The rectangular base is 6 cm by 8 cm.   **85.** The length is 6, the width is 3, and the sum of the length and width is 9 cm.
87. $x^2 + x - 12 = 0$ **89.** $3x^3 + 2x^2 - 21x = 0$ **91.** $12x^2 - 17x + 6 = 0$ **93.** -3 **95.** 0 **97.** yes

Chapter 5 Summary   (page 257)

1. $3(x + 3y)$ **2.** $5a(x^2 + 3)$ **3.** $7x(x + 2)$ **4.** $3x(x - 1)$ **5.** $2x(x^2 + 2x - 4)$ **6.** $a(x + y - z)$ **7.** $a(x + y - 1)$ **8.** $xyz(x + y)$
9. $a(5a + 6b^2 + 10cd - 15)$ **10.** $7xy(a + 3x - 5x^2 + y)$ **11.** $(x + y)(a + b)$ **12.** $(x + y)(x + y + 1)$ **13.** $2x(x + 2)(x + 3)$
14. $3x(y + z)(1 - 3y - 3z)$ **15.** $(p + 3q)(3 + a)$ **16.** $(r - 2s)(a + 7)$ **17.** $(x + a)(x + b)$ **18.** $(y + 2)(x - 2)$
19. $y(3x - y)(x - 2)$ **20.** $5(x + 2)(x - 3y)$ **21.** $(x + 3)(x + 7)$ **22.** $(x - 3)(x + 7)$ **23.** $(x - 4)(x + 6)$ **24.** $(x - 6)(x + 2)$
25. $(x - 3)(2x + 1)$ **26.** $(x - 5)(3x + 1)$ **27.** $(3x - 1)(2x + 3)$ **28.** $3(2x - 1)(x + 1)$ **29.** $x(6x - 1)(x + 3)$ **30.** $x(x - 2)(4x + 3)$
31. $(x - 3)(x + 3)$ **32.** $(xy - 4)(xy + 4)$ **33.** $(x + 2 - y)(x + 2 + y)$ **34.** $(z - x - y)(z + x + y)$ **35.** $6y(x - 2y)(x + 2y)$
36. $(x + y - z)(x + y + z)$ **37.** $(4a - 1)(16a^2 + 4a + 1)$ **38.** $5(1 - 2y)(1 + 2y + 4y^2)$ **39.** $(x + 5)(x^2 - 5x + 25)$
40. $3x(1 + 3x)(1 - 3x + 9x^2)$ **41.** $6xy(4y - 1)$ **42.** $(x + 3)(x^2 - 3x + 9)$ **43.** $(2x - 1)(3x + 2)$ **44.** $(3x - 7)(3x + 7)$
45. $(4x - 5)(4x + 5)$ **46.** $(x - 5)(2x + 3)$ **47.** $4(x - 1)(x + 3)$ **48.** $(2x + 3)^2$ **49.** $(x - 1)(x^2 + x + 1)$
50. $2(x - 2)(x^2 + 2x + 4)$ **51.** 0 or -2 **52.** 0 or 3 **53.** 3 or -3 **54.** 5 or -5 **55.** 4 or 3 **56.** 5 or -3 **57.** 6 or -4
58. 2 or 8 **59.** $-\frac{1}{2}$ or 3 **60.** $-\frac{3}{2}$ or 1 **61.** $\frac{1}{2}$ or $-\frac{1}{2}$ **62.** $\frac{2}{3}$ or $-\frac{2}{3}$ **63.** $0, 3$, or 4 **64.** $0, -3$, or -2 **65.** $0, \frac{1}{2}$, or -3
66. $0, -\frac{2}{3}$, or 1 **67.** $x(12 - x) = 35$; the integers are 7 and 5.   **68.** $3x^2 + 5x = 2$; the number is $\frac{1}{3}$.   **69.** $45 = \frac{1}{2}x(2x + 3)$; the base is 15 ft.   **70.** The bomb will hit its target in 15 seconds.   **71.** $x(2x + 3) = 27$; the rectangular bed is 3 ft by 9 ft.
72. $x(x + 3) = 4x + 6$; the rectangle is 3 ft by 6 ft.

Chapter 5 Test   (page 260)

1. $2x(2y^2 + 6xy^3 - 3z)$ **2.** $3x(a + b)(x - 2y)$ **3.** $(x + y)(a + b)$ **4.** $(x - 5)(x + 5)$ **5.** $3(a - 3b)(a + 3b)$
6. $(2x - 3y)(2x + 3y)(4x^2 + 9y^2)$ **7.** $(x + 1)(x + 3)$ **8.** $(x - 11)(x + 2)$ **9.** $(3 - x)(9 + 3x + x^2)$ **10.** $(x + 4)(3x + 1)$
11. $(2y - 3)(y + 4)$ **12.** $(2x - y)(x + 2y)$ **13.** $(4x - 3)(3x - 4)$ **14.** $2y(y - 2)(y^2 + 2y + 4)$ **15.** 5 or -5 **16.** $-\frac{3}{2}$ or -1 **17.** $\frac{9}{5}$
or $-\frac{1}{2}$ **18.** -6 or 3 **19.** It will hit the ground after 12 seconds.   **20.** $40 = \frac{1}{2}x(x + 2)$; the base is 10 m.

Exercise 6.1   (page 270)

1. rational   **3. a.** 0 **b.** 0 **c.** 5 **d.** -6 **5. a.** 5 **b.** -2 and 3 **c.** -2.5 and -0.5 **7. a.** $\frac{2}{13}$ **b.** $\frac{5}{7}$ **c.** $-\frac{x}{3}$ **d.** $-5y$ **e.** $\frac{1}{3}$ **f.** 2
9. b   **11.** none   **13.** 6 and -4 **15.** $3x$ **17.** 5 **19.** $\frac{x}{2}$ **21.** $\frac{3x}{5y}$ **23.** -1 **25.** -1 **27.** -1 **29.** $\frac{x + 1}{x - 1}$ **31.** $\frac{x - 5}{x + 2}$ **33.** $\frac{2x}{x - 2}$
35. $\frac{x}{y}$ **37.** $\frac{x - 4}{x^2}$ **39.** $\frac{x - 4}{x + 4}$ **41.** simplified   **43.** $\frac{3 - x}{3 + x}$ or $-\frac{x - 3}{x + 3}$ **45.** $\frac{4}{3}$ **47.** $x + 3$ **49. a.** and c.

b. $\frac{7x}{8y}$ **51. a. and c.**   **b.** $\frac{2}{3}$ **53. a.**   When you try to evaluate the original

expression, the denominator has a value of zero.   **b.** $\frac{2(x + 2)}{x - 1}$ **c.** However, after simplifying the expression, the denominator no longer

has a value of zero, thus yielding the following screen: 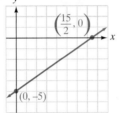   The values of the two expressions will *not* agree

if the selected value for x is an excluded value.   **55.** $\frac{3x(x + 1)}{(x + 1)^2}$ **57.** $\frac{2y(x + 1)}{x(x + 1)}$ **59.** $\frac{z(z + 1)}{(z - 1)(z + 1)}$ **61.** $\frac{(x + 2)^2}{(x - 2)(x + 2)}$ **63.** $\frac{2(x + 2)}{(x + 1)(x + 2)}$

65. When graphed in *dot* mode there will be a break in the graph at the excluded value of 3, since the y has no value if $x = 3$.

67. a. 1   **b.** The graph will be a horizontal line at height 1 for every value of x.   **c.** Every value of y will be 1 for all values of x.

69. a. $2x(2x - 3)(x - 1)$ **b.** $(y + 9x)(y - 7x)$ **c.** $(5 + b)(3 - a^2)$ **71.**   **73.** $x = \frac{1}{3}$

Exercise 6.2   (page 277)

1. 2   **3.** $\frac{y}{x}$ **5.** $\frac{15z}{11}$ **7.** 42   **9.** $9xy$ **11.** $\frac{7}{6}$ **13.** $\frac{5y}{9}$ **15.** $\frac{53x}{42}$ **17.** $\frac{3}{z}$ **19.** $\frac{2y + 6}{5z}$ **21.** 9   **23.** 1   **25.** $\frac{4x - 2y}{y + 2}$ **27.** $\frac{2x + 10}{x - 2}$

29. $(x - 1)(x + 1)$ **31.** $x(x + 6)$ **33.** $(x - 2)^2(x + 1)$ **35.** $(x - 5)(x + 1)(x + 5)$ **37.** $\frac{4y + 10}{15y}$ **39.** $\frac{x^2 + 4x + 1}{x^2 y}$ **41.** $\frac{-x^2 + 3x + 1}{x - 2}$

43. $\frac{10x + 4}{(x + 2)(x - 2)}$ **45.** $\frac{2x^2 - 4x + 8}{(x - 2)^2(x + 2)}$ **47.** $\frac{x}{x - 2}$ **49.** $-\frac{1}{2(x - 2)}$ **51.** $\frac{3}{x - 3}$ **53. a.** $\frac{4x}{x - 6}$ **b.** 16 ft   **c.** No, the length would be undefined.

55. $x(2x + 3)(x - 5)$ **57.** $[-4, 7]$ by $[-60, 10]$; $x = -\frac{3}{2}, 0, 5$

Exercise 6.3   (page 282)

1. a. $5 \cdot 7$ **b.** $2^3 \cdot 3$ **c.** $2^3 \cdot 7$ **d.** 3^3 **3.** $\frac{3}{11}$ **5.** $\frac{5}{7}$ **7.** $\frac{xy}{z}$ **9.** $\frac{7}{3}$ **11.** $\frac{2x}{3}$ **13.** $2xy^2$ **15.** $-3y^2$ **17.** $\frac{(z + 7)(z + 2)}{7z}$ **19.** x **21.** $x + 2$

23. $\frac{3}{2x}$ **25.** $\frac{(x - 2)^2}{x}$ **27.** 1   **29.** $\frac{1}{3}$ **31.** 1   **33.** $\frac{2}{3x}$ **35.** $\frac{2(z - 2)}{z}$ **37.** $\frac{x - 2}{x - 3}$ **39.** 1   **41.** $\frac{x}{36}$ **43.** $\frac{(x - 1)(x + 1)}{5(x - 3)}$ **45.** $\frac{2x(1 - x)}{5(x - 2)}$ **47.** $\frac{y^2}{3}$

49. $\frac{(x + 2)(x + 5)}{x - 5}$ **51.** $y = -\frac{7}{3} + \frac{4}{3}x$

Exercise 6.4   (page 287)

1. complex   **3.** No, division is not an associative operation.   **5. a.** $\frac{2}{3x}$ **b.** $\frac{\frac{x + 2}{3x}}{\frac{x + 2}{2}}$ **c.** Yes, the main fraction bar is the same as $\div$.

7. (3/4+1/5)/6▶Frac
        19/120

9. $\frac{21}{10}$ **11.** $\frac{1}{14}$ **13.** $\frac{7}{8}$ **15.** $\frac{11}{5}$ **17.** $\frac{y^2}{x}$ **19.** $\frac{wz}{12t^2}$ **21.** $\frac{1 + 3y}{3 - 2y}$ **23.** $\frac{1 - x}{3 - x}$ **25.** $\frac{1 + 3x}{x + 2}$ **27.** $\frac{1}{x - 2}$ **29.** $\frac{2}{4 - x}$

31. $\frac{ab}{b^2 - a^2}$ **33.** $\frac{y^2}{y^2 - 12}$ **35.** $\frac{x^2}{(x + 1)^2}$ **37.** $\frac{y + 2}{y - 1}$ **39.** $\frac{9y + 5}{4 - y}$ **41.** -1 **43.** -37 **45.** 37.5 ft

Exercise 6.5   (page 298)

1. extraneous   **3. a.** expression: $\frac{5x}{6}$ **b.** equation: $x = \frac{36}{5}$ **c.** expression: $\frac{3x^2 + 2x}{x^2 - 1}$ **d.** expression: $\frac{7}{2x}$ **e.** equation: $x = -\frac{3}{2}$

f. equation: $x = \frac{1}{2}$ **5.** 6   **7.** 10   **9.** -2 **11.** 4   **13.** -1 **15.** 2   **17.** -8 **19.** $\varnothing$, -1 is extraneous   **21.** $\varnothing$, 2 is extraneous

23. 3 **25.** 3 **27.** 5 **29.** $\frac{1}{7}$ **31.** $\varnothing$, 4 is extraneous **33.** 1 or -9 **35.** 4 **37.**

window: $[-5, 15]$ by $[-10, 10]$; $x = 12$

39. $x = 6$ **41.** window: $[1, 5]$ by $[2, 5]$; $x = 2$ or 4 **43. a.** $x = 6$

b. $x = 6$; 4 is extraneous **c.** $x = 6$ **d.** Answers may vary. **e.** Answers may vary. **45.** $x + \frac{3y}{2} - 3x^2$
47. a. $y = 75 + 15x$ **b.** $[0, 10]$ by $[0, 300]$ is one possible window.
c. You might use the STOre, TABLE, or VALUE features.

Exercise 6.6 (page 304)

1. $\frac{1}{x}$ and $\frac{1}{x+1}$ **3.** $\frac{x-1}{x+3}$ **5.** $\frac{7+x}{8-x} = 2$; the desired number is 3. **7.** $\frac{5+x}{7-2x} = 8$; the desired number is 3. **9.** $\frac{1}{x} + \frac{1}{x+2} = \frac{7}{24}$; the integers are
6 and 8. **11.** $\frac{300}{x} = \frac{120}{x-90}$; the plane flies at 150 mph. **13.** $\frac{210}{255-x} = \frac{300}{255+x}$; the wind speed is 45 mph. **15.** $\frac{340}{x} + \frac{680}{x} = 6$; the planes fly
at 170 and 340 mph. **17.** $\frac{1}{5} + \frac{1}{4} = \frac{1}{x}$; the time together is $2\frac{2}{9}$ hours. **19.** $\frac{1}{3} + \frac{1}{x} = 2$; it will take his son 6 hours alone. **21.** $\frac{2}{5} + \frac{1}{2} = \frac{1}{x}$;
together it will take them $1\frac{1}{9}$ days. **23.** $\frac{315}{x} = \frac{385}{x+.02}$; the rates are 9% and 11%. **25. a.** $\frac{55x + 50000}{x} = 75$; the average production cost
will be $75 per text if 2500 texts are produced. **b.** When 1600 texts per year are produced, the average cost of production is $86.25
per text. **c.** When 4000 texts are produced, the average cost is $67.50 per text. **d.** It is not applicable since a negative number of
texts cannot be produced.

Chapter 6 Summary (page 306)

1. a. $-\frac{1}{3}$ **b.** $-\frac{7}{3}$ **c.** $\frac{1}{2x}$ **d.** $\frac{5}{2x}$ **2. a.** $\frac{x}{x+1}$ **b.** $\frac{1}{x}$ **c.** $\frac{x+4}{x-5}$ **d.** $\frac{x+7}{x+3}$ **3. a.** $\frac{x}{x-1}$ **b.** in lowest terms **4. a.** 1 **b.** $\frac{2x+2}{x-7}$ **c.** $\frac{x^2+x-1}{x(x-1)}$
d. $\frac{x-7}{7x}$ **5. a.** $\frac{x-2}{x(x+1)}$ **b.** $\frac{x^2+4x-4}{2x^2}$ **c.** $\frac{x+1}{x}$ **d.** 0 **6. a.** $\frac{3x}{y}$ **b.** $\frac{6}{x^2}$ **c.** $\frac{3y}{2}$ **d.** $\frac{1}{x}$ **7. a.** 1 **b.** 1 **c.** $x + 2$ **d.** $\frac{n+2}{n+1}$
8. a. $\frac{1+x}{1-x}$ **b.** $\frac{x^2+3x}{2x^2-1}$ **c.** $x^2 + 3$ **d.** $\frac{a^2+abc}{b^2+abc}$ **9. a.** 3 **b.** 1 **c.** 3 **d.** -1.5 or 4 **e.** -2 **f.** 0
10. Window: $[-5, 8]$ by $[-5, 5]$; the solution to the equation as determined by the root feature is $x = 5$.
11. $\frac{1}{18} + \frac{1}{20} = \frac{1}{x}$; the combined time to empty the basement is $9\frac{9}{19}$ hours. **12.** $\frac{1}{10} + \frac{1}{14} = \frac{1}{x}$; the
combined time is $5\frac{5}{6}$ days. **13.** $\frac{10}{x} = \frac{30}{x+10}$; he jogs at 5 mph. **14.** $\frac{320}{360-x} = \frac{400}{360+x}$; the wind is
40 mph. **15.** $\frac{108}{x} = \frac{121.50}{x+.005}$; the smaller rate is 4%. **16.** $\frac{150x+6000}{x} = 150 + \frac{6000}{x}$; an enrollment of 24
children will give an average cost per child of $400 per month.

Chapter 6 Test (page 310)

1. $\frac{8x}{9y}$ **2.** $\frac{x+1}{2x+3}$; the excluded values are ± 1.5. **3.** 3; the excluded value is -1. **4.** $\frac{5y^2}{4t}$ **5.** $\frac{x+1}{3(x-2)}$ **6.** $\frac{3t^2}{5y}$ **7.** $\frac{x^2}{3}$ **8.** $\frac{10x-1}{x-1}$
9. $\frac{13}{2y+3}$ **10.** $\frac{2x^2+x+1}{x(x+1)}$ **11.** $\frac{2x^2}{y^3}$ **12.** $\frac{x+y}{y-x}$ **13.** -5 **14.** 6 **15.** 4 **16.** $\frac{2}{3}$ **17.** $\frac{1}{7} + \frac{1}{9} = \frac{1}{x}$; the time together will be $3\frac{15}{16}$ hours.
18. $\frac{28}{x+23} = \frac{18}{23-x}$; the speed of the current is 5 mph. **19.** $\frac{20}{x} = \frac{35}{x+10}$; the novice's rate is $\frac{40}{3}$ mph, so his time is 1.5 hours.
20. $\frac{5}{x} = \frac{20}{x+10}$; her jogging rate is $3\frac{1}{3}$ mph.

Cumulative Review: Chapters 5 and 6 (page 311)

1. F; the completely factored form is $(x - 2)(4 + x)$. **2.** T **3.** trinomial **4.** The leading coefficient is -1. **5.** The trinomial is not
factorable because the exponent on the first term is not twice the exponent on the second term. **6.** $a^2 - b^2$ **7.** prime **8.** two
9. To factor means to rewrite as a product. To solve means to find all values of x which make the left side evaluate to 0, since the
right side is 0. Factoring is a means of easily solving the equation. **10. a.** quadratic **b.** linear **c.** quadratic **11.** rational
12. a. 2 **b.** $-\frac{1}{4}$ **c.** 0 **13.** $\frac{4xy}{3}$ **14.** $18x^2y^2$ **15. a.** $\frac{1}{x^2y}$ **b.** $-\frac{10y^4}{3}$ **16.** complex **17.** No, it is equivalent to $\left(\frac{1}{2} + \frac{2}{3}\right) \div \frac{3}{4}$.
18. extraneous **19. a.** 6; $x \neq 3$ **b.** $\frac{3}{x-3}$; $x \neq 3$ **20.** If the consecutive odd integers are x and $x + 2$ then the reciprocals
would be $\frac{1}{x}$ and $\frac{1}{x+2}$. **21.** $(x - 7)(x + 7)$ **22.** $(x - 9)(x + 1)$ **23.** $(3x - 5)(6x + 1)$ **24.** $x^2(x + 2)$ **25.** $(x - 9)^2$
26. $5x(x - 5)(x + 5)$ **27.** $(c - d)(a + b)$ **28.** $(a - 2y)(a^2 + 2ay + 4y^2)$ **29.** -1 or -7 **30.** 0, 3, or -4

31.

$x \approx -1.62$ or 3.79 **32.** $x^2 - 3x - 10 = 0$ **33.** The factorization is incorrect. A polynomial and its factored form should evaluate to the same number for every value of x.   **34.** $\frac{4}{a}$ **35.** $\frac{x^2 - 5x}{x^2 - 25}$
36. $\frac{(y - 10)}{(y + 2)(y - 2)}$ **37.** $\frac{a - b}{b - a} = \frac{a - b}{-1(-b + a)} = \frac{a - b}{-1(a - b)} = \frac{1}{-1} = -1$ **38.** $\frac{-(b - 1)}{b(a + 3)}$ or $\frac{1 - b}{b(a + 3)}$ **39.** $2x(x + 3)^2$
40. -11 **41.** $\frac{ab(a + b)}{a^2 + ab + b^2}$ **42.** 2; 3 is extraneous   **43.** $-\frac{3}{5}$ or 1   **44.** The "line" is the result of graphing $\frac{x - 1}{x + 1}$ in *connected* mode. It should not be there.   **45.** The excluded value is 3, and $x = 5$ is the solution.   **46.** The object is back on the ground after 5 seconds.   **47.** $x(2x + 8) = 10$; the width is 1 yard.   **48.** $\frac{1}{4} + \frac{1}{6} = \frac{1}{x}$; the time together is 2.4 hours.   **49.** $\frac{8}{x - 1.5} = \frac{12}{x + 1.5}$; the rate of the boat in still water is 7.5 mph.

Exercise 7.1   (page 324)

1. scientific notation   **3.** $3, -3 \cdot 3$ **5.** 1   **7.** $\frac{1}{x}$ **9. a.** x^{11} **b.** x^{28} **c.** x^3 **d.** $\frac{x^7}{y^7}$ **e.** $x^3 y^3$ **f.** $x^8 y^{12}$ **11.** 8   **13.** $\frac{1}{16}$ **15.** 1
17. $\frac{1}{x^2}$ **19.** $\frac{1}{16y^4}$ **21.** $\frac{1}{a^3 b^6}$ **23.** y^{13} **25.** $\frac{1}{x^3}$ **27.** $-\frac{y^{10}}{32x^{15}}$ **29.** a^{14} **31.** $\frac{1}{x^{10} y^4}$ **33.** $\frac{256x^{28}}{81}$ **35.** $\frac{16y^{14}}{z^{10}}$ **37.** $\frac{u^{20} v^8}{16}$ **39.** $\frac{x^{16}}{y^2 z^2}$
41. $\frac{x + 1}{1 - x}$ **43.** $\frac{a^2 - a + 1}{a^2}$ **45.** 2   **47.** 2.3×10^4 **49.** 6.2×10^{-2} **51.** 4.25×10^3 **53.** 2.5×10^{-3} **55.** 230   **57.** .00114
59. 25,000,000   **61.** .00051   **63. a.** 4.2×10^{-5} **b.** .000042   **65. a.** $-1.0000001 \times 10^{-11}$ **b.** -0.000000000010000001
67. 714,000, $.71 \times 10^5$ **69.** 30,000, 3×10^4 **71.** 144,000,000,000, 1.44×10^{11} **73.** 2.57×10^{13} miles   **75.** 3.31×10^4 cm/sec
77. 114,000,000 miles   **79.** 1.9008×10^{11} ft   **81.** 1.7×10^{-18} g

Exercise 7.2   (page 330)

1. factor   **3.** irreducible (or prime)   **5. a.** $3y^2$ **b.** $9x^2 y^2$ **7.** $2(x + 4)$ **9.** $2x(x - 3)$ **11.** $5x^2 y(3 - 2y)$ **13.** $3z(9z^2 + 4z + 1)$
15. $-3(a + 2)$ **17.** $-3x(2x + y)$ **19.** $(a + b)(x + y)$ **21.** $(x + y)(x + 2)$ **23.** $(3 - c)(c + d)$ **25.** $(x + 2)(x + 3)$
27. $3(x + 7)(x - 3)$ **29.** $-(a - 8)(a + 4)$ **31.** $(3y + 2)(2y + 1)$ **33.** $(4a - 3)(2a + 3)$ **35.** $(a - 4b)(a + b)$
37. $-(3a - 2b)(a + b)$ **39.** $x(3x - 1)(x - 3)$ **41.** $(x^2 + 5)(x^2 + 3)$ **43.** $(y^2 - 10)(y^2 - 3)$ **45.** $(x - 2)(x + 2)$
47. $(3y - 8)(3y + 8)$ **49.** $(9a^2 - 7b)(9a^2 + 7b)$ **51.** $(x + y - z)(x + y + z)$ **53.** $(x - y)(x + y)(x^2 + y^2)$ **55.** $2(x - 12)(x + 12)$
57. $2x(x - 4)(x + 4)$ **59.** $(x - 2)(x^2 + 2x + 4)$ **61.** $(y - 4)(y^2 + 4y + 16)$ **63.** $(5 + a)(25 - 5a + a^2)$
65. $(x + 2y)(x^2 - 2xy + 4y^2)$ **67.** $2y(x - 2)(x^2 + 2x + 4)$ **69.** $(a - b - 2)[(a - b)^2 + 2ab + 4]$ **71.** $-3(a + b)^2$ **73.** $-x(x + 1)^2$
75. $(x - 3)(x - 1)$ **77.** $(y - 6)(y^2 + 6y + 36)$ **79.** $-7x^4(3x^3 - 2x + 1)$ **81.** $\frac{1}{65,536}$ **83. a.** .0000976   **b.** 812,000

Exercise 7.3   (page 340)

1. solve   **3.** identity   **5.** literal equation   **7.** $\mathbb{R}$ **9.** Both are solutions.   **11.** Both are solutions.   **13.** 8   **15.** 19   **17.** -24 **19.** $\varnothing$
21. $\mathbb{R}$ **23.** $y = 7 - 2x$ **25.** $y = 4x - 8$ **27.** $y = \frac{2}{3}x - \frac{5}{3}$ **29.** $y = -\frac{1}{2}x + 3$ **31.** $h = \frac{3V}{\pi r^2}$ **33.** 9 or -11 **35.** $\frac{3}{2}$ or $-\frac{1}{2}$ **37.** $\varnothing$
39. 2 or 0   **41.** -5 or $\frac{1}{5}$ **43.** -1 or 2   **45.** 4 or -4 **47.** -20 or -1 **49.** 0, 8, or -3 **51.** 10 or 2   **53.** 2   **55.** -8 **57.** $\varnothing, -1$
is extraneous   **59.** $\varnothing$, 2 is extraneous   **61.** 3   **63.** $x + 3x = 20$; he should cut off a 5-ft piece.   **65.** $30 + 0.24x = 44 + 0.16x$; the cost is the same for 175 miles driven.   **67.** $\frac{h}{5} = \frac{30}{7}$; the flagpole is $21\frac{3}{7}$ ft tall.   **69.** $x + x + 4x = 180$; the three angles are 30°, 30°, and 120°.   **71.** $0.13x = 0.11x + 150$; the teacher has $7500 to invest.   **73. a.** $\frac{32x^3}{y^4}$ **b.** $\frac{y^4}{12x^7}$ **75. a.** 3.14×10^{-3} **b.** 4×10^9
c. 5.6×10^{-19}

Exercise 7.4   (page 353)

1. ordered pairs   **3.** standard viewing   **5.** x-intercept   **7.** function   **9.**

11. a. yes   **b.** no   **13. a.** no   **b.** yes   **15. a.** domain $= \{-3, -2, -1, 0, 1, 2, 3\}$ **b.** range $= \{4, -1, -4, -5\}$
17.   **19.**   **21.**   **23.**

  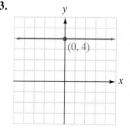

25. x-intercept: $\left(\frac{40}{3}, 0\right)$; y-intercept: $(0, 40)$ **27.** x-intercept: $(-12, 0)$; y-intercept: $(0, 24)$ **29.** x-intercept: $(16, 0)$; y-intercept: $(0, -10)$
31. x-intercept: $(11, 0)$; y-intercept: $\left(0, \frac{55}{9}\right)$ **33.** function **35.** not a function **37.** not a function **39.** domain = $\left\{-8, 9, 7, -\frac{1}{2}, -4.5\right\}$;
range = $\left\{5, 2, -3, -\frac{2}{3}, 3\right\}$ **41.** domain = $\{-6, -5.5, -5, -4.5, -4, -3.5, -3\}$; range = $\{31, 25.25, 20, 15.25, 11, 7.25, 4\}$

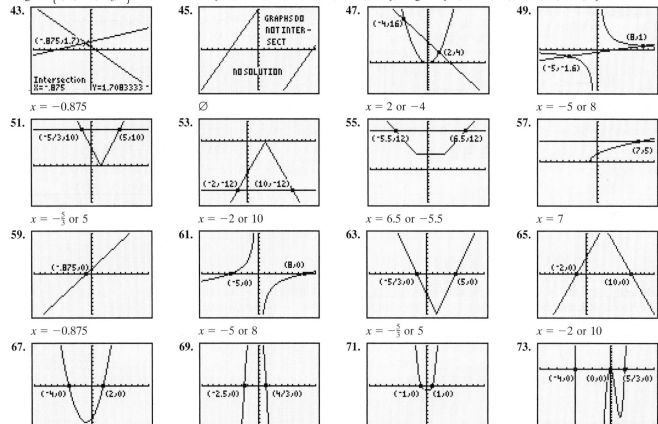

43. $x = -0.875$

45. $\varnothing$

47. $x = 2$ or -4

49. $x = -5$ or 8

51. $x = -\frac{5}{3}$ or 5

53. $x = -2$ or 10

55. $x = 6.5$ or -5.5

57. $x = 7$

59. $x = -0.875$

61. $x = -5$ or 8

63. $x = -\frac{5}{3}$ or 5

65. $x = -2$ or 10

67. $x = 2$ or -4

69. $x = -2.5$ or $\frac{4}{3}$

71. $x = 1$ or -1

73. $x = 0, \frac{5}{3}$, or -4

75. a. $y = 0.24x + 45$ **b.** $[0, 150]$ by $[0, 100]$ **c.** $\$76.40$ **d.** $\$76.44$

77. 1.0962×10^{13} gallons **79.** 0.0262144

Exercise 7.5 (page 366)

1. inequality **3.** interval **5.** -9 **7. a.** -2 **b.** -2 **c.** -2 **d.** -2
9. a. -6 **b.** -6 **c.** -6 **d.** -6 **11.** $(-\infty, 4]$ 4
13. $(-\infty, -2)$ -2 **15.** $(-\infty, -1]$ -1 **17.** $[-13, \infty)$ -13 **19.** $[2, \infty)$ 2
21. $(-15, \infty)$ -15 **23.** $(-\infty, 20]$ 20 **25.** $[3, \infty)$ 3 **27.** $(-7, \infty)$ -7
29. $[4, \infty)$ 4 **31.** $\frac{68 + 75 + 79 + x}{4} \geq 80$; she must make a 98% or better to earn a B. **33.** $\frac{17 + 19 + x}{3} \geq 21$; the third car

must have an economy rating of 27 mpg or better. **35.** $\frac{5 + 4 + 3 + 6 + 7 + 4 + x}{7} \geq 5$; you must eat at least 6 servings of fruit on the seventh
day. **37.** $.67x > 385$; after 574 days it will be cheaper to buy a washer. **39.** $4x > 68$; a side must be at least 17 cm.
41. domain = $\{1, -1, -2\}$; range = $\{2, 3, 4\}$ **43.**

Exercise 7.6   (page 375)

1. compound   **3.** intersection   **5. a.** $(1.5, 6.5)$ **b.** $(-\infty, 1.5) \cup (6.5, \infty)$ **7. a.** $(-12, -6)$ **b.** $(-\infty, -12) \cup (-6, \infty)$
9. a. iii   **b.** iv   **c.** ii   **d.** i   **11.** $(-2, 5]$ **13.** $(-\infty, 4]$ **15.** $\varnothing$
17. $(-\infty, -2) \cup (5, \infty)$ **19.** $[4, \infty)$ **21.** $(-\infty, \infty)$
23. $(3, 4)$ **25.** $\varnothing$ **27.** $\left(-\infty, -\frac{1}{4}\right]$ **29.** $\varnothing$ **31.** $(-2, 5)$
33. $(8, 11)$,   **35.** $[-2, 4]$ **37.** $(-\infty, 2) \cup (7, \infty)$
39. $(-\infty, 1)$ **41.** $(-\infty, -6)$ **43.** $(-2, 1]$ **45.** $\left(-6, \frac{3}{2}\right]$ **47.** $(-\infty, -2) \cup [1, \infty)$
49. $(-\infty, -6) \cup \left[\frac{3}{2}, \infty\right)$ **51.** $1540 \le C \le 1650$ **53.** $0 < 3.281x \le 25$; the python can grow to 7.62 m in length.
55. $640 \le 1 \cdot x + 2 \cdot 3x \le 810$; the number of stereos can be between 92 and 115, inclusive.   **57.** $8 < 2l + 2(3l - 8) < 16$; the length
is between 3 and 4 ft.   **59.** $\left(-\infty, \frac{14}{19}\right]$ **61. a.** $5(x^2 - 5y)(x^2 + 5y)$ **b.** $(3x - 4)(2x - 3)$ **c.** $x(4x + 1)(x - 5)$

Exercise 7.7   (page 382)

1. absolute value   **3. a.** $\left\{\frac{-4}{3}, 3\right\}$ **b.** $\left(-\infty, \frac{-4}{3}\right) \cup (3, \infty)$ **c.** $\left(\frac{-4}{3}, 3\right)$ **5. a.** $\{-4, -1\}$ **b.** $(-4, -1)$ **c.** $(-\infty, -4) \cup (-1, \infty)$
7. a. $\varnothing$ **b.** $\mathbb{R}$ or $(-\infty, \infty)$ **c.** $\varnothing$ **9. a.** $\{8\}$ **b.** $(-\infty, 8) \cup (8, \infty)$ **c.** $\varnothing$ **11.** $(-9, 9)$ **13.** $[-4, 20]$ **15.** $\left(\frac{-8}{3}, 4\right)$ **17.** $\varnothing$
19. $\left[-3, 5\frac{2}{3}\right]$ **21.** $(-\infty, -1) \cup (1, \infty)$ **23.** $(-\infty, -12] \cup [2, \infty)$ **25.** $(-\infty, -10) \cup (15, \infty)$ **27.** $\left(-\infty, -\frac{3}{4}\right) \cup \left(-\frac{3}{4}, \infty\right)$
29. $\left(-\infty, \frac{3}{8}\right) \cup \left(\frac{3}{8}, \infty\right)$ **31.** $[-10, 14]$ **33.** $\left(\frac{-5}{3}, 1\right)$ **35.** $(-\infty, -4] \cup [-1, \infty)$ **37.** $\varnothing$ **39.** $(-\infty, -54) \cup (12, \infty)$
41. $(-\infty, 25) \cup (25, \infty)$ **43.** $\left(\frac{-65}{9}, \frac{-5}{9}\right)$ **45.** $(-\infty, -4) \cup (-4, \infty)$ **47.** $\{-7\}$ **49.** $\varnothing$ **51.** x and y have opposite signs.
53. For $|x| < k$, the solutions are *between* $-k$ and k. For $|x| > k$, the solution either lies to the right of k or to the left of $-k$.
55. If $k > 0$, $|x| + k > 0$ is equivalent to $|x| > -k$ and since the smallest value of $|x|$ is 0, every value of $|x|$ will be greater than a
negative number.   **57.** $\varnothing$ **59.** It is a function because each value of x corresponds to exactly one value of y.

Chapter 7 Summary   (page 385)

1. a. 64   **b.** -64 **c.** 13   **d.** 25   **2. a.** $-z^5y^2$ **b.** $4x^3$ **c.** x^{15} **d.** $4x^4y^2$ **3. a.** $\frac{x^2}{y^2}$ **b.** $\frac{1}{8x^3}$ **c.** x **d.** y **4. a.** x^{10} **b.** x^{14}
c. $\frac{1}{x^5}$ **d.** $\frac{y^6}{9x^8}$ **5. a.** 7.28×10^2 **b.** 9.42×10^{-3} **c.** 1.8×10^{-4} **d.** 6×10^{-1} **6. a.** 726,000   **b.** .000391   **c.** 1.6×10^2
7. .000 000 4   **8.** 2172.1   **9. a.** $(x + 2)(y + 4)$ **b.** $(a + c)(b + 3)$ **c.** $(z - 4)(z + 4)$ **d.** $2(x^2 - 7)(x^2 + 7)$
e. $(y + 20)(y + 1)$ **f.** $-(y - 8)(y + 3)$ **g.** $3(5x + 1)(x - 4)$ **h.** $(x + 7)(x^2 - 7x + 49)$ **10. a.** -4.5 **b.** $\frac{3}{7}$ **c.** 7   **d.** $-\frac{1}{2}, 1$
e. $-1, 1$ **f.** $-3, -2, 0$ **g.** $3, -\frac{11}{3}$ **h.** $\varnothing$ **i.** $\frac{9}{8}$ **j.** 4   **11. a.** $w = \frac{P - 2l}{2}$ **b.** $y = -4x + 2$ **c.** $y = -\frac{1}{2}x + 3$
12. a.   **b.**   **c.**   **d.**

13. a.   **b.**   **c.**

$x = 3$ $x = -9$ $x = 3, -\frac{11}{3}$

d.   **e.**   **f.**

$x = \frac{26}{3}, -\frac{10}{3}$ $x = 5, 6$ $x = -3, 8$
14. a. function   **b.** function   **15. a.** $(-\infty, 3)$ **b.** $\left(-\frac{1}{2}, \infty\right)$ **c.** $\left(-\infty, \frac{13}{2}\right]$

d. $\left(-\infty, \frac{2}{3}\right]$ **16. a.** $\{-1\}$ **b.** $(-1, \infty)$ **c.** $(-\infty, 1)$ **17. a.** $\{8\}$ **b.** $(-\infty, 8)$ **c.** $(8, \infty)$

18. a. $\left(-\frac{1}{3}, 2\right)$ **b.** $\left(\frac{-5}{3}, \frac{2}{3}\right)$ **c.** $\left(-\infty, \frac{2}{3}\right) \cup \left(\frac{25}{3}, \infty\right)$

d. $(-\infty, \infty)$ **19. a.** $(2.5, 7)$ **b.** $(-\infty, 2.5) \cup (7, \infty)$ **20. a.** $(-1, 7)$ **b.** $(-\infty, -1) \cup (7, \infty)$
21. a. $[-7, 1]$ **b.** $(-7, 4)$ **c.** $(-\infty, -9) \cup (13, \infty)$

d. $(-\infty, -2] \cup \left[\frac{4}{3}, \infty\right)$ **e.** $\varnothing$ **f.** $(-\infty, \infty)$ **22. a.** $\varnothing$ **b.** $\mathbb{R}$ **c.** $\varnothing$ **23. a.** $\varnothing$ **b.** $\varnothing$ **c.** $\mathbb{R}$

24. $x + (2x + 1) = 22$; the pieces are 7 ft and 15 ft long. **25.** $x + 2x + x = 624$; the pasture will be 156 ft by 312 ft. **26.** $p < 15$
27. $8.50x + 175 \le 330$; the student can buy 18 discs at the most.

Chapter 7 Test (page 390)

1. $\frac{-2}{3xy}$ **2.** $\frac{1}{xy^2z^2}$ **3.** $(a - 5)(a^2 + 5a + 25)$ **4.** $2(x - 4)(x + 4)$ **5.** $(3x + 2)(2x - 1)$ **6.** -12 **7.** 6 **8.** $t = \frac{A - p}{pr}$ **9.** $y = \frac{3}{7}x - \frac{8}{7}$
10. $x + 2x + 6x = 20$; the longest piece is $\frac{40}{3}$ ft. **11.** $2x + 2(x + 5) = 26$; the area is 36 cm². **12.** 4 or -7 **13.** 3
14. **15.** **16. a.** $\{-1\}$ **b.** $(-1, \infty)$ **c.** $(-\infty, -1)$ **17. a.** $\{2, 6\}$ **b.** $(2, 6)$
c. $(-\infty, 2) \cup (6, \infty)$ **18.** Not a function since $(1, 1)$ and $(1, -1)$ are
both on the graph. **19.** $(-\infty, 0]$

20. $(-1, 6]$ **21.** $(-\infty, \infty)$
22. $\left(-\infty, -\frac{1}{3}\right),$ **23.** $\left(\frac{1}{2}, \frac{9}{2}\right)$ **24.** $(-\infty, -12] \cup [18, \infty)$

25. $\varnothing$

Exercise 8.1 (page 402)

1. relation **3.** range **5.** ordered pairs **7.** y **9. a.** $f(2) = 2$ **b.** $f(-2) = -2$ **c.** $f(0) = 0$ **d.** $f(5) = 5$ **11.** The relation is a
function; domain $= \{2, 3, -4\}$; range $= \{5, -1, 7\}$. **13.** The relation is not a function; domain $= \{-1, -2\}$; range $= \{2, 3, -2\}$.
15. function **17.** function **19.** not a function **21.** function **23.** domain $= \mathbb{R}$; range $= (-\infty, 5]$ **25.** domain $= \mathbb{R}$;
range $= [-10, \infty)$ **27.** not a function **29.** function **31.** function **33.** not a function **35.** not a function **37.** function
39. a. $f(2) = 8$ **b.** $f(-3) = -7$ **c.** $f(a) = 3a + 2$ **d.** $f(a + 2) = 3(a + 2) + 2 = 3a + 8$ **41. a.** $f(2) = 0$ **b.** $f(-3) = 15$
c. $f(a) = a^2 - 2a$ **d.** $f(a + 2) = a^2 + 2a$ **43. a.** $f(2) = -\frac{3}{2}$ **b.** $f(-3) = -\frac{3}{7}$ **c.** $f(a) = \frac{3}{a - 4}$ **d.** $f(a + 2) = \frac{3}{a - 2}$
45. $f(-4) = -3$ **47.** $f(-3) = 2$ **49.** $f(3) = 5$ **51.** $f(-4) = -6$ **53.** $f(-3) = 4$ **55.** No, 4 is not in the pictured domain of f.
57. 12 **59.** $2b - 2a$ **61.** $2b$ **63.** 1 **65.** not linear **67.** linear; $y = \frac{1}{3}x + \frac{4}{3}$ **69.** linear; $y = 2x - 4$ **71.** Yes, since $2(5) + 4 \ge 9$.
73. 4 **75.** Yes, since $(-7)^2 - 49 = 0$.

Exercise 8.2 (page 408)

1. increasing **3.** left, right **5.** 1980–81, 1981–82, 1984–85, 1986–87, 1987–88, 1990–91 **7.** 285 **9.** increasing on $(-\infty, \infty)$
11. increasing on $[0, \infty)$ **13.** increasing on $(-\infty, 0]$ **15.** increasing on $[-4, 4]$ **17.** increasing on $[-8, -6]$ and $[3, 6]$; decreasing
on $[2, 3]$; constant on $[-6, 2]$ **19.** A turning point of a function is a point where the graph changes direction.
21. $(1, 4)$, min. **23.** $(-2, 4)$, max. **25.** It is a function since it passes the

VLT. **27.** $(4, 1)$, relative max. **29.** $f(-5) = 5$ **31.** Answers may vary. **33. a.** how far he has traveled horizontally **b.** how far
he has traveled vertically **c.** 77 ft **d.** 62 ft **e.** 20 ft **f.** 32 ft **g.** Before he is shot out, his head is 9.76 ft above the ground.
35. a. $17.50 **b.** $156.25 **c.** $136 profit **d.** $15/ticket **37.** x-intercept: let $y = 0$ and solve for x; y-intercept: let $x = 0$ and solve
for y **39.**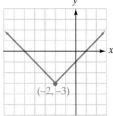

Exercise 8.3   (page 418)

1. y, x **3.** positive   **5.** zero   **7.** equal   **9.** $m = \frac{1}{2}$ **11.** $m = -2$ **13.** negative   **15.** no defined value   **17.** zero   **19.** $m = 2$
21. $m = -\frac{8}{5}$ **23.** no defined value   **25.** y1   **27.** y2   **29.** $m = -\frac{3}{2}$, $(0, 6)$ **31.** $m = \frac{3}{4}$, $\left(0, \frac{1}{2}\right)$ **33.** $m = -4$, $(0, 3)$ **35.** $m = 3$
37. $m = -\frac{1}{4}$ **39.** perpendicular   **41.** neither   **43.** $y = 0$; $m = 0$ **45.** A vertical line is not a function, but a horizontal line is.
47. $y = \frac{1}{5}x + 4$ **49.** $y = 5x - 2$ **51.** \$30,000 per year   **53. a.** $m = .75 = $ cost/ride   **b.** $y = 0.50x + 7$ **c.** No, since that would
mean the customer would be paying a negative amount of money for each ride.   **55. a.** $y = 250x + 3000$ **b.** The slope is the
increase in value per year.   **c.** It would be losing value if the slope were negative.

Exercise 8.4   (page 427)

1. slope-intercept; slope; y-intercept   **3.** standard   **5.** $x = k$; undefined   **7. a.** yes   **b.** $m = 5$ **c.** $(0, 2)$ **d.** $y = 5x + 2$
9. a. $y = x + 3$ **b.** $y = -x + 5$ **c.** $y = 5x - 4$ **d.** $y = x - 3$ **e.** $y = 5x + 4$ **f.** $y = x + 5$ **11.** $5x - y = -7$
13. $3x + y = 6$ **15.** $-3x + 2y = 4$ **17.** $y = x$ **19.** $y = \frac{7}{3}x - 3$ **21.** $y = -\frac{9}{5}x + \frac{2}{5}$
23. $y = x - 1$ **25.** $y = \frac{2}{3}x + \frac{9}{2}$ **27.** $y = -\frac{2}{3}x + 6$

29. $y = 4x$ **31.** $y = 4x - 3$ **33.** $y = \frac{4}{5}x - \frac{26}{5}$ **35.** $y = -\frac{1}{4}x$ **37.** $y = -\frac{1}{4}x + \frac{11}{2}$ **39.** $y = -\frac{5}{4}x + 3$ **41.** $x = -2$ **43.** $x = 5$
45. $y = -\frac{A}{B}x + \frac{C}{B}$; $m = -\frac{A}{B}$; $b = \frac{C}{B}$ **47.** $y = -3200x + 24,300$ **49.** $y = 47,500x + 475,000$ **51.** $y = -670x + 3900$
53. $y = -120x + 1050$; salvage value $= f(8) = \$90$ **55.** \$890   **57.** \$37,200   **59.** \$230   **61.** \$510   **63.** $x = 1, -\frac{3}{2}$ **65.** $x = -\frac{3}{2}, \frac{3}{2}$
67. $60 + x = 4(20)$; 20 oz of copper must be added.

Exercise 8.5   (page 435)

1. direct, increases   **3.** inverse, decreases   **5.** joint   **7.** direct variation   **9.** inverse variation   **11.** direct variation   **13.** inverse
variation   **15.** inverse variation; $k = 8$ **17.** combined variation; $k = 3$ **19.** joint variation; $k = 3$ **21.** $A = kp^2$ **23.** $v = \frac{k}{r^3}$
25. $B = kmn$ **27.** $P = \frac{ka^2}{j^3}$ **29.** $F = \frac{km_1m_2}{d^2}$ **31.** L varies jointly as m and n.   **33.** E varies jointly as a and the square of b.   **35.** X
varies directly as the square of x and inversely as the square of y.   **37.** R varies directly as L and inversely as the square of d.
39. $A = \pi r^2$; 36π in.^{2} **41.** 432 miles   **43.** 25 days   **45.** 12 in.^{3} **47.** $85\frac{1}{3}$ times/sec   **49.** V is multiplied by 12.   **51.** 26,437.5 gal
53. $V = ka$; the resistance is 3 ohms.   **55.** $D = \frac{k}{wd^3}$; the deflection is 0.275 in.   **57.** $P = \frac{kt}{V}$; the gas is heated to 546°K.   **59.** One
similarity is the constant of variation. Differences include the types of behavior among the related variables.   **61.** No,
change $\neq \frac{k}{\text{purchase price}}$ **63.** domain $= \mathbb{R}$; range $= [3, \infty)$ **65.** domain $= \mathbb{R}$; range $= [0, \infty)$

Chapter 8 Summary   (page 439)

1. a. function; domain and range $= \mathbb{R}$ **b.** function; domain and range $= \mathbb{R}$ **c.** function; domain $= \mathbb{R}$; range $= [1, \infty)$ **d.** function;
domain $= \{x \mid x \neq 2\}$; range $= \{y \mid y \neq 0\}$ **e.** function; domain and range $= \mathbb{R}$ **f.** not a function; domain $= [0, \infty)$; range $= \mathbb{R}$
2. a. function   **b.** not a function   **c.** not a function   **d.** not a function   **3. a.** $f(3) = -7$ **b.** $g(8) = 60$ **c.** $g(-2) = 0$
d. $f(5) = 17$ **4. a.** $f(-2) = -3$ **b.** $f(4) = 3$ **5. a.** $y = 3x + 2$; linear   **b.** $y = \frac{1}{4}x + \frac{5}{4}$; linear   **c.** $4x - 3y = 12$; linear
d. $y = x^2 - 25$; nonlinear   **6. a.** increasing on $[0, \infty)$ **b.** decreasing on $[0, \infty)$ **c.** decreasing on $(-\infty, 5]$
d. increasing on $[0, 3]$; decreasing on $[3, \infty)$ **e.** decreasing on $(-\infty, 0]$, $[2, \infty)$; increasing on $[0, 2]$
f. increasing on $(-\infty, -4]$, $[1, \infty)$; decreasing on $[-2, 1]$; constant on $[-4, -2]$
7. a. relative max. $(0, 36)$ **b.** relative min. $(-3, -5)$ **c.** relative min. $(.45, -1.26)$;   **d.** relative max $(-1.11, 24.1)$;
relative max $(2.22, 4.23)$ relative min $(2.11, -9.11)$

8. a. [0, 10] by [0, 260] recommended window   **b.** (4, 256) relative max.   **c.** The ball is 256 ft high after 4 seconds.   **d.** The ball is at a height of 245.76 ft going up and coming down.   **e.** The ball is on the ground.   **9. a.** $y = x(80 - x)$

b.   **c.** 800 ft²   **d.** 20 ft   **10. a.** $m = 1$ **b.** $m = \frac{14}{9}$ **c.** $m = 5$ **d.** $m = \frac{5}{11}$ **11. a.** $m = \frac{2}{3}$ **b.** $m = -2$
c. none   **d.** none   **12. a.** $y = \frac{2}{3}x - 5$ **b.** $y = -\frac{1}{2}x + 2$ **13. a.** positive   **b.** negative   **c.** 0   **d.** no
defined slope   **14. a.** perpendicular   **b.** parallel   **c.** neither   **d.** perpendicular   **15. a.** $3x - y = -29$
b. $13x + 8y = 6$ **c.** $3x - 2y = 1$ **d.** $2x + 3y = -21$ **16. a.** $2x - 3y = 6$ **b.** $x = 3$ **17. a.** -550
per year   **b.** $y = -550x + 2500$ **c.** [0, 10] by [0, 3000]   **18.** ≈$3.19 per year   **19.** 72   **20.** 20.6
21. $k = 2$ **22.** $k = 16$ **23.** 31   **24.** The volume is multiplied by 6.

Chapter 8 Test   (page 444)

1. yes   **2.** domain = $(-\infty, 2]$; range = $(-\infty, 3]$ **3.** $f(-3) = -3$ **4. a.** increasing on $(-\infty, -3]$ **b.** decreasing on [2, 4]
c. constant on $[-3, 2]$ **5. a.** [0, 3] by [0, 30]   **b.** max. height = 26 ft   **c.** 2.27 sec   **d.** (1, 26)   **6.** $m = \frac{2}{3}$ **7.** $m = \frac{2}{3}$
8. $y = \frac{4}{5}x + 3$ **9.** neither   **10.** $y = -1$ **11.** $y - 1 = \frac{-5}{11}(x - 4)$ **12.** The equation is linear because it can be written in the form
$y = mx + b$.   **13.** $y = \frac{kw}{z^2}$ **14.** $P = kC^2$ **15. a.** increasing on $[0, \infty)$ **b.** 3 amps of current produces a power loss of 45 watts due to
heat.   **16.** $x = \frac{135}{2}$ **17.** $t = 14\frac{2}{3}$

Cumulative Review: Chapters 7 and 8   (page 444)

1. scientific notation   **2. a.** x^3 **b.** x^{10} **c.** $\frac{1}{x^5}$ **d.** $\frac{1}{x^7}$ **3. a.** 1.32×10^6 **b.** 2×10^6 **c.** 1.3×10^{-3} **d.** 4.12×10^{-3} **4. a.** 0.013
b. 610,000   **c.** 0.00051   **d.** 3,820,000   **5.** solution   **6.** function   **7. a.** not a function   **b.** function   **8.** intersection   **9.** union
10. $-4 \le x + 2 \le 4$ **11. a.** function   **b.** relation   **c.** function   **d.** function   **e.** function   **f.** relation   **12. a.** $f(-1) = 6$
b. $f(3) = 2$ **c.** $f(-3) = 20$ **d.** $f(8) = 42$ **13.** $y = \frac{1}{3}x + \frac{4}{3}$ **14.** y, x **15.** $y = $ a constant   **16.** $x = $ a constant   **17.** 0
18. undefined   **19.** slope-intercept; slope; y-intercept   **20.** point-slope; slope; (x_1, y_1) **21. a.** $\frac{2y^7}{x^{10}}$ **b.** $\frac{2^{11}}{5^7} \approx .026$ **22.** 8   **23.** 4, $-\frac{1}{2}$
24. 0 or -2 **25.** $\varnothing$ **26.** $\frac{1}{2}$ **27.** $\left[\frac{1}{3}, 3\right]$ ⟵━━●━━━●━━⟶   **28.** $(-\infty, -4] \cup [-1, \infty)$ ⟵━●━━━━●━⟶   **29.** -3
$\qquad\qquad\qquad\qquad\qquad\qquad\quad \frac{1}{3}\quad 3 \qquad\qquad\qquad\qquad\qquad\qquad\qquad\qquad -4\quad -1$

30. a. increasing on $[-3, -1]$ **b.** decreasing on $[-6, -3]$ and [4, 6]   **c.** constant on $[-1, 4]$
31. 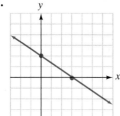   **32.** domain = $\{1, 2, -1, -3\}$; range = $\{-1, 5, -5\}$ **33.** $m = \frac{1}{2}$ **34.** $m = -4, b = 3$
35. perpendicular   **36.** $2x - 3y = -15$ **37.** $3x - y = 14$ **38.** $y = 2$ **39.** $x + 2y = 7$
40. $x - 3y = -15$ **41.** $x = 5$ **42.** $x = 5.5$ **43.** $x = 0, 4, -3$

44. **45. a.** $y = 15x + 300$ **b.** The slope is the yearly rate of increase.   **c.** $y = \$600$ **46. a.** $y = 20x + 45$
c. If repair takes 5 hours, the cost is $y = 20(5) + 45 = \$145$.   **d.** You can't work a negative number of hours.
e. If repair takes 20 hours, the cost is $y = 20(20) + 45 = \$445$.   **47.** $100 = 0.24x + 45$; you can drive
approximately 229.2 miles.   **48.** $\frac{78 + 80 + 85 + x}{4} \ge 70$; she only needs to make a 37 or higher.   **49.** 81π mm²
50. $V = \frac{k}{a}$; it will be worth \$4000 in 10 years.

Exercise 9.1   (page 455)

1. positive/principal   **3.** exact, approximate, irrational   **5.** b, d   **7. a.** 1 and it is a positive number   **b.** 1 and it is a negative number
c. none   **9.** 12   **11.** -2 **13.** $-3x^2$ **15.** $-\frac{2}{5}$ **17.** -5 **19.** undefined for real numbers   **21.** undefined for real numbers   **23.** -1
25. 8   **27.** undefined for real numbers   **29.** $7a^2b$ **31.** $-\frac{3}{2}$ **33.** undefined for real numbers   **35.** 2   **37.** 9   **39.** $\frac{5}{6}$ **41.** $\frac{3}{10}a^2b$
43. 3   **45.** $\frac{7}{9}$ **47.** 0.1   **49.** $-\frac{3}{4}$ **51.** $2|y|$ **53.** $2|x|$ **55.** $|x|^3$ **57.** $|x + 5|$ **59.** $\frac{1}{3}x^2$ **61.** $-x^2$ **63.** $|a + 6|$ **65.** $-2x$
67. $8|c|$ **69. a.** $\sqrt[6]{4096}$ **b.** rational   **71. a.** $5\sqrt[3]{9}$ **b.** irrational   **73. a.** $5\sqrt[3]{(27 - 9)}$ **b.** irrational   **75. a.** $\frac{8}{\sqrt{49} + 1}$ **b.** rational
77. >   **79.** >   **81.** <   **83. a.** The longer the skid, the greater the speed.   **b.** A car skidding 400 ft was traveling approximately
64.8 mph.   **85.** $-125x^6y^3$ **87.** $-16a^8b^{12}c^4$ **89.** $x^2 + 4xy + 4y^2$

Exercise 9.2   (page 464)

1. base, factor   **3.** rational   **5.** $\sqrt[3]{7}$ **7.** $3\sqrt[4]{x}$ **9.** $\sqrt[4]{3x}$ **11.** $(3a)^{\frac{1}{4}}$ **13.** $\left(\frac{1}{7}abc\right)^{\frac{1}{6}}$ **15.** $(a^2b^2)^{\frac{1}{3}}$ **17.** 2
19. 2   **21.** $\frac{1}{2}$ **23.** -2 **25.** 0   **27.** 27   **29.** $\frac{1}{4}$ **31.** $\frac{1}{2}$ **33.** $\frac{1}{8}$ **35.** $\frac{16}{81}$ **37.** $125x^6$ **39.** $\frac{4x^2}{9}$ **41.** $\frac{1}{64x^3}$ **43.** $\frac{1}{9y^2}$ **45.** $\frac{1}{4p^2}$ **47.** $-\frac{3}{2x}$
49. 6   **51.** $5x$ **53.** $6b$ **55.** $3a$ **57.** $2\sqrt{5}$ **59.** $-10\sqrt{2}$ **61.** $2\sqrt[3]{10}$ **63.** $-3\sqrt[3]{3}$ **65.** $2\sqrt[3]{3}$ **67.** $\frac{\sqrt{7}}{3}$ **69.** $\frac{\sqrt[5]{3}}{2}$ **71.** $5ab\sqrt{7b}$
73. $2x^3y\sqrt[4]{2}$ **75.** $\frac{z}{4x}$ **77.** $\frac{\sqrt[4]{5x}}{2z}$ **79.** $\frac{1}{9x^4y^2}$ **81.** $-8x^3y^6$ **83.** 64

Exercise 9.3   (page 470)

1. index, radicand, distributive   **3.** rationalizing   **5.** T   **7.** F   **9.** T   **11.** $10\sqrt{2x}$ **13.** $\sqrt[5]{7a^2}$ **15.** $-\sqrt{2}$ **17.** $9\sqrt{6}$ **19.** $-\sqrt[3]{4}$
21. $5\sqrt[4]{2}$ **23.** $94\sqrt[4]{3}$ **25.** $4\sqrt{5}+4\sqrt{2}$ **27.** $8z\sqrt{y}$ **29.** $\sqrt{2x}$ **31.** $x\sqrt[4]{y}$ **33.** $5y^3\sqrt{2y}$ **35.** 0   **37.** $5x+4$ **39.** 9   **41.** -8
43. $4x^2\sqrt{y}$ **45.** $5rs\sqrt{6}$ **47.** $3x\sqrt[3]{xy^2}$ **49.** $y(x+y)^2$ **51.** $42-2\sqrt{7}$ **53.** $-\sqrt{21}+\sqrt{15}$ **55.** $42t+9t\sqrt{21t}$ **57.** $5-\sqrt{3}$
59. $21y-29\sqrt{y}-10$ **61.** $3m+2\sqrt{6mn}+2n$ **63.** $5p+9\sqrt{3pq}-6q$ **65.** $12t+20\sqrt{3t}+25$ **67.** $15x-6\sqrt{15x}+9$ **69.** $\frac{\sqrt{15}}{3}$
71. $\frac{\sqrt{6}}{10}$ **73.** $\frac{\sqrt[3]{36}}{3}$ **75.** $\frac{\sqrt[3]{36}}{6}$ **77.** $\frac{\sqrt{3x}}{x}$ **79.** $\frac{\sqrt{2b}}{2}$ **81.** $\frac{\sqrt[4]{16}}{2}$ **83.** $\sqrt[4]{8}$ **85.** $\frac{3\sqrt{3}+3}{2}$ **87.** $3-2\sqrt{2}$ **89.** $5+2\sqrt{6}$ **91.** $\frac{3\sqrt{x}+6}{x-4}$
93. $\frac{x+2\sqrt{xy}+y}{x-y}$ **95.** $-6,1$ **97. a.** [number line, open circle at 5]   **b.** $(5,\infty)$ **99. a.** [number line, open circles at −1 and 3]   **b.** $(-\infty,-1)\cup(3,\infty)$

Exercise 9.4   (page 478)

1. radical, radicand   **3.** right   **5.** legs; hypotenuse   **7. a.** iii   **b.** iv   **c.** v   **d.** ii   **e.** i   **f.** vi   **9.** 4   **11.** 4   **13.** 8   **15.** 14, 6 is
extraneous   **17.** 3 or 4   **19.** 9, -25 is extraneous   **21.** -1 or 2   **23.** 0, 4 is extraneous   **25.** $\varnothing$ **27.** 4, 0 is extraneous   **29.** 2
31. 1   **33.** 4, -9 is extraneous   **35.** $\varnothing$ **37.** 4   **39.** Both equations are solved by raising to an appropriate power and then dividing
by 2. The difference is that one equation is raised to the second power and the other is raised to the third power.   **41.** 24 cm
43. 48 in.   **45.** 13   **47.** 10   **49.** $\sqrt{29}$ **51.** $A(-3,0), B(0,3), C(3,0),$ and $D(0,-3)$ **53.** 13 ft   **55.** $\sqrt{16200}\approx127$ ft
57. ≈135 ft   **59.** The racquet is too long to fit.   **61.** yes   **63.** 173 yd   **65.** 0.05 ft   **67.** The volume is 8 cm³.   **69.** function
71. relation only

Exercise 9.5   (page 486)

1. function   **3.** x, domain   **5.** 1   **7.** 3.606   **9.** a   **11.** b   **13.** 1   **15.** $3\sqrt{3}+2$ **17.** $3\sqrt{6}+2$ **19.** undefined for real numbers
21. undefined for real numbers   **23.** 5   **25.** domain $=[-2,\infty)$; range $=[0,\infty)$ **27.** domain $=[0,\infty)$; range $=[-5,\infty)$
29. domain $=[-3,\infty)$; range $=[0,\infty)$ **31.** domain $=\left[\frac{3}{5},\infty\right)$; range $=[0,\infty)$ **33.** domain $=\left[\frac{8}{3},\infty\right)$; range $=[-2,\infty)$
35. domain $=\left[\frac{5}{6},\infty\right)$; range $=[-4,\infty)$ **37.** They differ because $y=\sqrt{x}\cdot\sqrt{x}$ is defined only for $x\geq0$.   **39.** ≈319 ft   **41.** 8 ft
43. \$8   **45.**   **47.**

   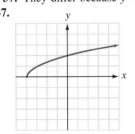

Exercise 9.6   (page 495)

1. i **3.** real, imaginary   **5.** $i^1=i, i^2=-1, i^3=-i, i^4=1$ **7.** $6i$ **9.** $-10\sqrt{2}i$ **11.** $3\sqrt{2}i$ **13.** $-4\sqrt{2}i$ **15.** $-i$ **17.** -1
19. T   **21.** T   **23.** F   **25.** $-1+12i$ **27.** $1+i$ **29.** $-7+3i$ **31.** $18+14i$ **33.** $-1+\sqrt{2}i$ **35.** $16-12i$ **37.** $6-17i$
39. $13-3\sqrt{3}i$ **41.** $-8+i$ **43.** $5-12i$ **45.** $-4-7i$ **47.** $13+13i$ **49.** $-\frac{3}{2}i$ **51.** $-\frac{5}{4}$ **53.** $-\frac{2}{3}i$ **55.** $6+4i$ **57.** $\frac{3}{17}+\frac{5}{17}i$
59. $\frac{6}{5}-\frac{2}{5}i$ **61.** $\frac{3}{5}-\frac{1}{5}i$ **63.** $-\frac{5}{13}+\frac{12}{13}i$ **65.** $\frac{1}{5}+\frac{2\sqrt{6}}{5}i$ **67. a.** See Student Solutions Manual.

69. a. See Student Solutions Manual.   **b.**

b.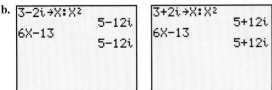

71. $-5x^2+7x-6$ **73.** $15x^2-14x-16$ **75.** $x+1$

Chapter 9 Summary (page 497)

1. a. 7 **b.** -11 **c.** -6 **d.** 15 **e.** -3 **f.** -6 **g.** 5 **h.** -2 **2. a.** $5x$ **b.** $x + 2$ **c.** $3a^2\sqrt[3]{b^2}$ **d.** $4a^2y$ **3. a.** -2 **b.** -4 **c.** $\frac{27}{8}$ **d.** $\frac{1}{3125}$ **e.** $\frac{27}{8}$ **f.** $3xy^{1/3}$ **g.** $3xy^{1/2}$ **h.** $125x^{9/2}y^6$ **i.** $\frac{1}{4u^{4/3}v^2}$ **4. a.** $4\sqrt{15}$ **b.** $3\sqrt[3]{2}$ **c.** $2\sqrt[4]{2}$ **d.** $2\sqrt[5]{3}$ **e.** $2xy\sqrt[3]{2x^2y}$ **f.** $3x^2y\sqrt[3]{2x}$ **g.** $4x$ **h.** $2x$ **i.** $\frac{\sqrt[3]{2a^2b}}{3x}$ **j.** $\frac{\sqrt{17xy}}{8a^2}$ **5. a.** 0 **b.** $8\sqrt[4]{2}$ **c.** $29x\sqrt{2}$ **d.** $13\sqrt[3]{2}$ **6. a.** 3 **b.** $-2x$ **c.** $4 - 3\sqrt{2}$ **d.** $\sqrt{10} - \sqrt{5}$ **e.** 1 **f.** $6u + \sqrt{u} - 12$ **7. a.** $\frac{\sqrt{3}}{3}$ **b.** $\frac{\sqrt[3]{u^2}}{u^2v^2}$ **c.** $2\sqrt{2} + 2$ **d.** $2\sqrt{x} - 8$ **8. a.** 22 **b.** 3 or 9 **c.** $\frac{9}{16}$ **d.** 0 or -2 **9.** $\left(\frac{d}{2}\right)^2 + 117^2 = 125^2$; the boat gains 88 yd into the wind. **10.** 10 **11. a.** $f(6) = 3$ **b.** $g(4) = 2$ **c.** $g(32) = 4$ **d.** $f(3) = \sqrt{10} \approx 2.16$ **e.** $g(8) = \sqrt[3]{16} \approx 2.52$ **12. a.** domain $= [5, \infty)$; range $= [-4, \infty)$ **b.** domain $= [-2, \infty)$; range $= [0, \infty)$ **c.** domain $= \mathbb{R} =$ range **d.** domain $= \mathbb{R} =$ range **13.** 3.0 miles **14.** 8.2 ft **15. a.** $12 - 8i$ **b.** $2 - 68i$ **c.** $-96 + 3i$ **d.** $-2 - 2\sqrt{2}i$ **16. a.** $22 + 29i$ **b.** $-16 + 7i$ **c.** $-12 + 28\sqrt{3}i$ **d.** $118 + 10\sqrt{2}i$ **17. a.** $-\frac{3}{4}i$ **b.** $-\frac{2}{5}i$ **c.** $\frac{12}{5} - \frac{6}{5}i$ **d.** $\frac{21}{10} + \frac{7}{10}i$ **e.** $\frac{15}{29} - \frac{6}{29}i$ **f.** $\frac{1}{3} + \frac{1}{3}i$

Chapter 9 Test (page 500)

1. $5xy^2\sqrt{10xy}$ **2.** $2x^5y\sqrt[3]{3}$ **3.** $\frac{1}{4a}$ **4.** $3x$ **5.** $3x^2y^4\sqrt{2y}$ **6. a.** 2 **b.** $\frac{1}{216}$ **c.** $\frac{9}{4}$ **7.** $-\sqrt{3}$ **8.** $16\sqrt[3]{5}$ **9.** $2y^2\sqrt{3y}$ **10.** $6x\sqrt{y} + 2xy^2$ **11. a.** $\frac{\sqrt{5}}{5}$ **b.** $2\sqrt[3]{3}$ **c.** $\sqrt{10} - 3\sqrt{2}$ **d.** $\sqrt{3x} + 1$ **12.** 10 **13.** $\varnothing$, 4 is extraneous **14.** 25 **15.** 28 inches **16. a.** $f(3) = 2\sqrt{3} + 1 \approx 4.464$ **b.** $f(9) = 7$ **c.** $f(-20) = 1 + 4i\sqrt{5}$ **d.** $f(13) = 2\sqrt{13} + 1 \approx 8.211$ **17. a.** domain $= [6, \infty)$; range $= [0, \infty)$ **b.** domain $= [0, \infty)$; range $= (-\infty, 0]$ **18. a.** $-1 + 11i$ **b.** $4 - 7i$ **19. a.** $8 + 6i$ **b.** $-10 - 11i$ **20. a.** $\frac{i\sqrt{2}}{-2}$ **b.** $\frac{1}{2} + \frac{1}{2}i$

Exercise 10.1 (page 510)

1. $ax^2 + bx + c = 0$ **3.** roots (solutions) **5.** STOre **7. a.** imaginary **b.** real **c.** real **d.** imaginary **9.** a, c, d **11.** ± 5 **13.** ± 10 **15. a.** 25 **b.** $(x + 5)^2$ **17. a.** 9 **b.** $(x - 3)^2$ **19. a.** $\frac{25}{4}$ **b.** $\left(x - \frac{5}{2}\right)^2$ **21.** $\pm\frac{4\sqrt{3}}{3}$ **23.** $-1 \pm 3\sqrt{3}$ **25.** $-3 \pm 3\sqrt{2}i$ **27.** $-1.5 \pm \sqrt{10}$ or $\frac{-3 \pm 2\sqrt{10}}{2}$ **29.** $-\frac{1}{3} \pm \frac{\sqrt{13}}{3}i$ **31.** 2, -4 **33.** $\frac{1}{4} \pm \frac{\sqrt{7}}{4}i$ **35.** 2, $\frac{5}{3}$ **37.** $-\frac{3}{10} \pm \frac{\sqrt{31}}{10}i$ **39.** $-\frac{1}{5} \pm \frac{\sqrt{31}}{5}$ **41.** Answers may vary. Remember to write answers in complete sentences. **43.** 4, -3 **45.** $x^2 - 2x - 24 = 0$ **47.** $x^4 - 25x^2 + 144 = 0$ **49.** $x(x + 4) = 96$; the shed is 8 ft by 12 ft. **51. a.** $0.25 + 0.05x$, $300 - 80x$, $(0.25 + .05x)(3000 - 80x)$ **b.** Two price increases will yield a daily revenue of \$994. (30.5 increases are not possible.) **c.** \$0.35 **53.** 9% **55. a.** $(4, 0)$ **b.** $(0, -2)$ **c.** $y = \frac{1}{2}x - 2$ **57.** linear function

Exercise 10.2 (page 519)

1. $\frac{-b \pm \sqrt{b^2 - 4ac}}{2a}$ **3.** approximate **5.** rational **7.** rational, double **9.** $-3, 4$ **11.** the x-intercepts **13.** 1, 2 **15.** $\pm 4i$ **17.** 9, a double root **19.** $\pm\frac{5\sqrt{2}}{2}i$ **21.** $-\frac{3}{2} \pm \frac{\sqrt{3}}{2}i$ **23.** $\frac{-3 \pm \sqrt{5}}{4}$ **25.** $-\frac{1}{3} \pm \frac{\sqrt{2}}{3}i$ **27.** $-\frac{3}{4}, -\frac{1}{4}$ **29.** $-\frac{3}{4} \pm \frac{\sqrt{15}}{4}i$ **31.** $-3 \pm \sqrt{5}i$ **33.** $\frac{1}{5} \pm \frac{2}{5}i$ **35.** $\frac{3}{2}, -\frac{1}{2}$ **37.** two distinct imaginary roots (two complex conjugate solutions) **39.** two distinct rational solutions **41.** two distinct irrational solutions **43.** two complex conjugate solutions **45.** $k = \pm 12$ **47.** $k = 12$ or -3 **49.** $k = \pm 30$ **51.** $k = 10, 66$ **53.** The roots are not real. **55.** $k < -\frac{4}{7}$ **57.** $-4.17, 1.17$

59. $-1.52, 2.19$

61. $-1.30, 2.30$

63. 1.08, 13.92

window: $[-5, 20]$ by $[-10, 10]$

65. $-i \pm 2$ or $\pm 2 - i$ **67.** $\frac{2}{3}, -\frac{1}{4}$ **69.** $\frac{-5 \pm \sqrt{17}}{4}$ **71. a.** $-\sqrt{3} \pm 1$ **b.** $a = 1, b = 2\sqrt{3}, c = 2$, discriminant $= 4$ **c.** You might expect rational roots. **d.** The irrational roots and a perfect square discriminant aren't a contradiction since not all coefficients are rational in this case. **73. a.** [graph] **b.** The x-intercepts are $(-6, 0)$ and $(1, 0)$ **c.** The solutions are -6 and 1. **d.** The solutions in c are the x-coordinates of the x-intercepts.

75. a.   **b.** The *x*-intercepts are $\left(\frac{-4}{3}, 0\right)$ and $\left(\frac{-8}{3}, 0\right)$.   **c.** The solutions are $\frac{-4}{3}$ or $\frac{-8}{3}$.   **d.** The solutions in c are the *x*-coordinates of the *x*-intercepts.

Exercise 10.3   (page 527)

1. *x*-intercepts   **3.** vertex; highest; lowest   **5.** (4, 1198)   **7.** (4, 1198)   **9.** It is 878 ft from the ground.   **11. a.** (4, 0)   **b.** (0, −16)
c. (4, 0)   **d.** Your graph should include at least the points (4, 0), (0, −16), and (8, −16).   **e.** $f(2) = -4$ **f.** 3 or 5   **13.** (5, 78.54);
$f(5) = 78.54$ **15.** Area and radius of a circle are both positive quantities.   **17.** The rocket will attain a maximum height of
approximately 8789 ft after approximately 3 seconds.   **19.** After 20 seconds, the rocket is at a height of 8600 ft. You might use VALUE,
VARS, STOre, or TABLE.   **21. a.** A suggested window is [0, 5] by [0, 50].   **b.** The maximum height is 36 ft.   **c.** It reaches this
maximum height in 1.5 seconds.   **d.** The ball reaches the ground in 3 seconds.   **23.** $y = x(1000 - 2x)$; the maximum area of
12,500 ft^2 is reached when the field is 250 ft by 500 ft.   **25.** ±3, ±1   **27.** ±5, ±2   **29.** ±$\sqrt{7}$, ±1   **31.** ±2$\sqrt{3}$, ±1   **33.** 1
35. $\frac{1}{9}$, 1   **37.** 64, 27   **39.** −125, 1   **41.** 1, 3   **43.** 3, 5   **45.** 4, 5   **47.** −$\frac{1}{2}$, 3   **49.** ±2, ±1   **51.** 1 ± 2*i*

53.   **55.**   **57.**   **59.**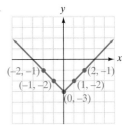

Exercise 10.4   (page 538)

1. $f(x) = ax^2 + bx + c$; *a*   **3.** $x = h$ **5.** down   **7.** e   **9.** b   **11.** c
13.   **15.**   **17.**

19.   **21.**   **23.**

25. opens up; vertex: (2, −1); axis of symmetry: $x = 2$, $y = 2(x - 2)^2 - 1$ **27.** opens down; vertex: (−1, 3); axis of symmetry:
$x = -1$, $y = -3(x + 1)^2 + 3$ **29.** opens up; vertex: (0, −3); axis of symmetry: $x = 0$, $y = 3(x - 0)^2 - 3$ **31.** opens up;
vertex: (−1, −3); axis of symmetry: $x = -1$, $y = 3(x + 1)^2 - 3$ **33.** opens up; vertex: (−2, 5); axis of symmetry: $x = -2$,
$y = 5(x + 2)^2 + 5$ **35.** 3, −0.5   **37.** 1.13, −0.88   **39.** If $a < 0$, changes in *k* affect the range, but not the domain. Changes in *h*
don't affect either one. In this case, *k* is the largest *y* value in the range $(-\infty, k]$.   **41. a.** Differing values of *a* affect how the line
slants, both in steepness and direction.   **b.** Differing values of *b* translate the graph vertically, up if $b > 0$, down if $b < 0$.
43. a. $a > 1$ makes the graph steeper; $0 < a < 1$ makes the graph flatter; $a < 0$ reflects the graph across the *x*-axis.   **b.** Positive
values of *k* translate the graph up; negative values of *k* translate the graph down.   **c.** Positive values of *h* translate the graph to the
right and negative values of *h* translate the graph to the left.
45. a.   window: [0, 500] by [0, 8000]   **b.** The maximum point is (200, 7000).   **c.** To obtain maximum
revenue, 200 stereos must be sold.   **d.** The maximum revenue is $7000.
47. (20, ∞),   **49.** $(-\infty, 2) \cup (7, \infty)$

Exercise 10.5   (page 548)

1. critical   **3.** open, closed   **5.** 0   **7.** 2, −3   **9.** $\mathbb{R}$ **11.** ∅   **13.** c   **15.** a   **17.** (1, 4)

19. (−∞, −3) ∪ (5, ∞)   **21.** [−4, 3]   **23.** (−∞. −5] ∪ [3, ∞)

25. ∅   **27.** (−∞, −3] ∪ [3, ∞)   **29.** (−5, 5)

31. $\left(-\infty, 0\right) \cup \left(\frac{1}{2}, \infty\right)$ **33.** (0, 2]   **35.** (−∞, −3) ∪ (1, 4)

37. [−5, −2) ∪ [4, ∞)   **39.** $\left(-\frac{1}{2}, \frac{1}{3}\right) \cup \left(\frac{1}{2}, \infty\right)$

41. (0, 2) ∪ (8, ∞)   **43.** (−∞, −2) ∪ (2, 18]

45. (−4, −2] ∪ (−1, 2]   **47.** (−∞, −16) ∪ (−4, −1) ∪ (4, ∞),

49. (−∞, −2) ∪ (−2, ∞)

51.   **53.**   **55.**   **57.**

59.   **61.**   **63.**   **65.**

67.   **69.** 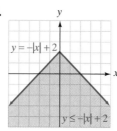   **71.** 64π in^2 **73.** $V = \frac{k}{x}$, $3000

Chapter 10 Summary   (page 551)

1. a. $-\frac{3}{4}, \frac{2}{3}$ **b.** $\frac{2}{3}, -\frac{4}{5}$ **c.** $-\frac{1}{3}, -\frac{5}{2}$ **d.** −8, 4   **2. a.** −4, −2   **b.** $\frac{7}{2}, 1$ **3. a.** 9, −1   **b.** 10, 0   **c.** $\frac{1}{2}, -7$ **d.** 7, $-\frac{1}{3}$ **4. a.** two distinct irrational solutions   **b.** two imaginary solutions (complex conjugates)   **5.** $k = 12$ or 152   **6.** $k \geq -\frac{7}{3}$
7. $2x \cdot 2(x + 2) = x(x + 2) + 72$; the original dimensions are 4 cm by 6 cm.
8. a.   [−5, 20] by [−30, 200]   **b.** It touches the ground after 7 seconds.   **c.** The height at $t = 2$ seconds
is 160 ft.   **d.** It is at a height of 160 ft after 5 sec.   **e.** The maximum height is 196 ft.
9. a. 144, 1   **b.** −27, 8   **c.** 1   **d.** $\frac{-3}{2} \pm \frac{\sqrt{95}}{10} i$

10. a.

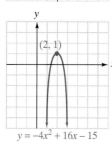

$y = 2(x - 0)^2 - 3;$
vertex: $(0, -3);$
axis of symmetry: $x = 0$

b.

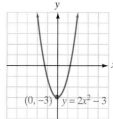

$y = -2(x - 0)^2 - 1;$
vertex: $(0, -1);$
axis of symmetry: $x = 0$

c.

$y = -4(x - 2)^2 + 1;$
vertex: $(2, 1);$
axis of symmetry: $x = 2$

d.

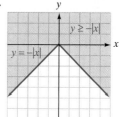

$y = 5(x + 1)^2 - 6;$
vertex: $(-1, -6);$
axis of symmetry: $x = -1$

11. a. $y = x(50 - x);$ the rectangle should be 25 ft by 25 ft.   **b.** The maximum area is 625 ft².

12. a. $(-\infty, -7) \cup (5, \infty)$ **b.** $(-9, 2)$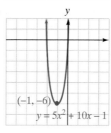

13. a. $(-\infty, -4) \cup (-1, 2)$ **b.** $(-\infty, 0) \cup [\frac{3}{5}, \infty)$

c. $\left(-\frac{7}{2}, 1\right) \cup (4, \infty)$

14. a.

b.

c.

d.

Chapter 10 Test   (page 554)

1. a. $-2, -1$ **b.** $\frac{1}{4}, -\frac{2}{3}$ **2. a.** $3 \pm \sqrt{2}i$ **b.** ± 3 **3. a.** $-6, 3$ **b.** $-\frac{3}{2}, -\frac{5}{3}$ **c.** $-2 \pm \sqrt{3}$ **d.** $\frac{5 \pm \sqrt{37}}{2}$ **4.** The solutions are imaginary.   **5.** $k = 2$ **6.** $x^2 + (x + 14)^2 = 26^2;$ the shortest leg is 10 inches.   **7.** $\frac{1}{4}, 1$ **8. a.** $y = \frac{1}{2}(x - 0)^2 + 5$ **b.** $(0, 5)$ **c.** $x = 0$
d.

9.

10. $(-\infty, -2) \cup (4, \infty)$

11. $(-3, 2]$

Cumulative Review: Chapters 9 and 10   (page 555)

1. radicand   **2.** real number; irrational   **3.** legs; hypotenuse   **4.** $i, -1, -i, 1$ **5.** solutions (or roots)   **6.** x **7.** parabola; vertex
8. a. rational   **b.** irrational   **c.** rational   **d.** irrational   **9. a.** $3(ab)^{1/2}$ or $3a^{1/2}b^{1/2}$ **b.** $(3ab)^{1/2}$ **c.** $x^{5/3}$ **d.** $\left(\frac{1}{3}p^5q\right)^{1/6}$ **10. a.** F
b. T   **c.** F   **11. a.** IV   **b.** I   **c.** III   **d.** II   **12. a.** $f(1) = 2$ **b.** $f(13) \approx 2.732$ **c.** domain   **13. a.** $2i$ **b.** $-i$ **c.** $6 + 5i$
d. $-2 + 9i$ **14. a.** real   **b.** imaginary   **c.** real   **d.** imaginary   **15. a.** $(2, 4)$ **b.** $(-4, -5)$ **16.** $y = (x - 3)^2 + 4$
17. $y = (x + 5)^2 - 6$ **18.** $y = \frac{1}{6}x^2$ **19.** b   **20.** a   **21.** $5a^2bc^2\sqrt{7bc}$ **22.** $\frac{\sqrt[4]{5xz^3}}{2z^2}$ **23.** $3\sqrt{2} + 20 - 9\sqrt{3}$ **24.** $3xy^2\sqrt[3]{2}$ **25.** $\frac{\sqrt[3]{18}}{3}$
26. $2 + \sqrt{3}$ **27.** $\frac{12}{13} - \frac{8}{13}i$ **28. a.** $f(1) = 2$ **b.** $g(2) = 2$ **c.** $f(5) = 2\sqrt{3}$ **29.** 1, 4 is extraneous

30.

domain = $[0, \infty)$; range = $[3, \infty)$ **31.** $-0.5, 3$ **32.** $3x^2 + 7x - 6 = 0$ **33. a.** $(2, 0), (3, 0)$ **b.** $(0, 6)$
c. $f(5) = 6$ **d.** 0 or 5   **34.** $\pm\sqrt{5}, \pm1$ **35.** $27, -1$ **36.** $-\frac{27}{13}, -1$

37. a. 5, 3   **b.**   **38.**   **39.**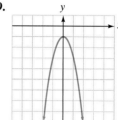

40. a. up   **b.** $(2, 5)$ **c.** $x = 2$ **d.** $y = 2(x - 2)^2 + 5$
41. a. $x = 5.76, 10.24$ **42.** $\varnothing$ **43.**   window: $[-2, 20]$ by $[-3, 3]$;
b. $8 \pm \sqrt{5}$ $x = 0, 16$

44. 2.02   **45.** The values of the two sides might be slightly different since the answer is an approximate solution.   **46.** $-0.95, 3.21$
47. a. suggested window: $[0, 5]$ by $[0, 50]$ **b.** max. height = 39 ft   **c.** 1.6 sec to reach the max. height   **d.** approximately 3.1 sec to reach the ground   **48.** $x^2 + 17^2 = 18^2$; the anchor stake needs to be approximately 5.9 ft from the base of the pole.   **49.** It strikes with a velocity of 16 ft per sec.   **50.** 8%   **51. a.** For a maximum revenue, 5000 stereos must be sold.   **b.** The maximum revenue is $25,000.

Exercise 11.1   (page 567)

1. excluded   **3.** identity   **5.** $f + g = \{(-2, 9), (5, 6), (7, 5), (-5, 4), (4, -2)\}$; $f - g = \{(-2, 7), (5, 0), (7, -7), (-5, -8), (4, -4)\}$;
$fg = \{(-12, 8), (5, 9), (7, -6), (-5, -12), (4, -3)\}$; $\frac{f}{g} = \left\{\left(-2, \frac{1}{8}\right), (5, 1), \left(7, -\frac{1}{6}\right), \left(-5, -\frac{1}{3}\right), (4, -3)\right\}$ **7. a.** domain = $\mathbb{R}$;
range = $[-1, \infty)$ **b.** domain = $\mathbb{R}$; range = $[-9, \infty)$ **c.** domain = $\mathbb{R}$; range = $\mathbb{R}$ **9.** The graph does not exist when $x = 0$ because the denominator of a fraction cannot be 0.   **11.** The point of discontinuity occurs at $(0, -2)$.   **13.** $f - g = -2x^2 + 3x - 3$;
domain = $\mathbb{R}$ **15.** $\frac{f}{g} = \frac{3x - 2}{2x^2 + 1}$; domain = $\mathbb{R}$ **17.** $f - g = 3$; domain = $\mathbb{R}$ **19.** $\frac{f}{g} = \frac{x^2 - 1}{x^2 - 4}$; domain = $\{x \mid x \neq \pm2\}$
21. $\frac{f}{g} = 3x^4 - 4x^2$; domain = $\{x \mid x \neq 0\}$ **23.** $\frac{f}{g} = x + 9$; domain = $\left\{x \mid x \neq -\frac{7}{2}\right\}$ **25.** $(f \circ g)(2) = 7$ **27.** $(g \circ f)(-3) = 24$
29. $(f \circ g)(0) = -1$ **31.** $(g \circ f)\left(\frac{1}{2}\right) = 3$ **33.** $(f \circ g)(x) = 2x^2 - 1$ **35.** $(g \circ f)(2x) = 16x^2 + 8x$ **37.** Function selection may vary, but the process is associative.   **39.** $f(a + h) \neq f(a) + f(h)$ because $a^2 + 4h + 2a + h^2 - 3 \neq a^2 + 2a - 6 + h + 2h^2$
41. $\frac{f(x + h) - f(x)}{h} = 3x^2 + 3xh + h^2$ **43.** If $V = \frac{1}{3}Bh$ and B = area of the circular base = πr^2, then $V = \frac{1}{3}\pi r^2 h$. Since the problem indicated that the height was always twice the base radius, $h = 2r$, $V = \frac{1}{3}\pi r^2 \cdot 2r = \frac{2}{3}\pi r^3$.   **45.** $r(t) = 2t$; $A(r) = \pi r^2$;
$A(t) = A(r(t)) = A(2t) = 4\pi t^2$; $A(7) \approx 615.75$ in.^{2} **47.** $\frac{4}{5}$ **49. a.** domain = $\mathbb{R}$; range = $\mathbb{R}$ **b.** domain = $[5, \infty)$; range = $[2, \infty)$
c. domain = $\mathbb{R}$; range = $[2, \infty)$

Exercise 11.2   (page 575)

1. inverse relation   **3.** horizontal line   **5.** T; there are no restrictions on reversing the coordinates of the ordered pairs.
7. $\{(5, 4), (8, 2), (7, -3), (-9, 6)\}$ **9.** $\{(2, -4), (2, -3), (2, 0), (2, 1)\}$
11.

lbs.	2.2	4.4	6.6	7.7	8.8
kg	1	2	3	3.5	4

13. not $1 - 1$ **15.** $1 - 1$ **17.** not $1 - 1$ **19.** $1 - 1$ **21.** not $1 - 1$
23. $1 - 1$ **25.** $f^{-1}(x) = \frac{x - 1}{3}$ **27.** $f^{-1}(x) = 5x - 4$ **29.** $f^{-1}(x) = 5x + 4$
31. $f^{-1}(x) = \frac{5}{4}x + 5$

33.   **35.**   **37.**

39.

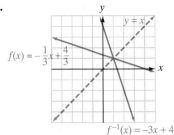

$f(x) = -\frac{1}{3}x + \frac{4}{3}$

$y = x$

$f^{-1}(x) = -3x + 4$

41.

43.

45.

47.

49. $f \circ g = g \circ f = I(x) = x$ **51.** $f \circ g = g \circ f = I(x) = x$ **53.** The inverse relation $y = \pm\sqrt{x-4}$ is not a function.   **55.** The inverse relation $y = f^{-1}(x) = x^2 - 4$ is a function.   **57.** The inverse relation $y = f^{-1}(x) = x^{1/3}$ is a function.   **59.** The inverse relation $y = f^{-1}(x) = x^2$ is a function.   **61.** The inverse relation $y = f^{-1}(x) = 4x^2 + 3$ is a function.   **63. a.** 64   **b.** -256 **c.** $\frac{1}{256}$ **d.** 2   **e.** -2 **f.** $-\frac{1}{2}i$

65. a.

domain = $\mathbb{R}$; range = $[0, \infty)$

b.

domain = $\mathbb{R}$; range = $\mathbb{R}$

c.

domain = $[0, \infty)$; range = $[0, \infty)$

Exercise 11.3   (page 586)

1. exponential   **3.** decay   **5. a.** 5   **b.** 8   **c.** ≈ 0.0105 **d.** ≈ 0.1491 **7. a.** not an exponential function   **b.** an exponential function   **c.** an exponential function   **d.** not an exponential function   **9.** The exponential function $f(x) = b^x$ represents exponential decay if $0 < b < 1$.   **11.** Any nonzero base raised to the zero power is equal to 1.   **13. a.** 4   **b.** 5

15.

17.

19.

21.

23. a.

b.

25. a.

b.

27. a.

b.

29. a.

b.

31. $-\frac{4}{5}$ **33.** -6 **35.** $\frac{5}{4}$ **37.** 8 **39.** $\frac{7}{9}$ **41.** $A = P\left(1 + \frac{r}{k}\right)^{kt}$; there will be \$22,080.40 in the account after 10 yr. **43.** $A = P\left(1 + \frac{r}{k}\right)^{kt}$; the \$1000 principal will earn \$32.03 more at 5.5%. **45.** $A = P\left(1 + \frac{r}{k}\right)^{kt}$; the dollar will be worth \$2,273,996.13 in 2076.
47. $A = Pe^{rt}$; there will be approximately 3645 fish in 10 yr. **49.** $A = Pe^{rt}$; in 30 years the world population will be approximately 8,580,034,311. **51.** The bacteria population after 4 hours will be 167,904,600. **53.** The amount of radioactive material present after 5 years is $\frac{32}{243}$ of the original amount. **55.** $A = Pe^{rt}$; the trailer will be worth approximately \$3094.15. **57. a.** 1 **b.** 625 **c.** $\frac{1}{625}$
d. 5 **e.** $\frac{1}{5}$ **59.** 5, 0 is extraneous **61.**

Window: $[-2, 4]$ by $[-10, 75]$; the ball reaches a max. height of 64 ft after 1 sec.

Exercise 11.4 (page 595)

1. logarithm **3.** 10 **5.** change of base **7. a.** $\log_8 64 = 2$ **b.** $\log_4 \frac{1}{16} = -2$ **c.** $\log_{1/2} 32 = -5$ **d.** $\log_{10} \frac{1}{10,000} = -4$ **9.** 3
11. 3 **13.** 3 **15.** $\frac{1}{2}$ **17.** -3 **19.** 64 **21.** 7 **23.** 5 **25.** $\frac{1}{25}$ **27.** $\frac{1}{6}$ **29.** $-\frac{3}{2}$ **31.** $\frac{2}{3}$ **33.** 5 **35.** $\frac{3}{2}$ **37.** 4 **39.** 8 **41.** 5 **43.** 4
45. 3 **47.** 100 **49.** 0.5119 **51.** -2.3307 **53.** -0.0726 **55.** 3.6348 **57.** -0.2752 **59.** does not exist **61.** 25.2522
63. 69.4079 **65.** 0.0002 **67.** 120.0719
69. **71.** **73.** **75.**

77. The graph of $y = k + \log x$ is a vertical translation of the graph of $y = \log x$. **79.** The graph of $y = k \cdot \log x$ steepens $\left(|k| > 1\right)$ or flattens $\left(|k| < 1\right)$ the graph of $y = \log x$ and reflects in the x-axis if $k < 0$. **81.** 1.7712 **83.** -1 **85.** 1.8928 **87.** 2.3219
89. pH $= -\log x \approx 4.77$ **91.** pH $= -\log x$; the safe range in the hydrogen ion concentration is between 2.51×10^{-8} and 1.58×10^{-7}. **93.** $A = Pe^{rt}$; it will take about 5.8 years for the population to double. **95.** If $r = .12$, the time for the population to triple is approximately 9.2 years. **97. a.** increasing **b.** domain $= \left[-\frac{5}{2}, \infty\right)$ **99.** The reason for the extraneous solution $x = 2$ to the squared equation is that squaring both sides of an equation doesn't necessarily produce an equivalent equation.

Exercise 11.5 (page 603)

1. product **3.** $p \log_b m$ **5.** T **7.** T **9.** F; $7^{\log_7 7} = 7$ **11.** F; $\log_b \left(\frac{A}{B}\right) = \log_b A - \log_b B$ **13.** T **15.** F; $\log_b 2 = \frac{1}{\log_2 b}$ **17.** T
19. T **21.** T **23.** F; $\log_{10} 10^3 = 3 \ (10^{\log_{10} 1})$ **25.** undefined **27.** quotient property **29.** change of base **31.** change of base
33. $\log_b x + \log_b z$ **35.** $\log_b x - \log_b y - \log_b z$ **37.** $\log_b x + 2 \log_b y + 3 \log_b z$ **39.** $3 \log_b x + \frac{1}{2} \log_b y$ **41.** $\frac{1}{2} \log_b x + \frac{1}{2} \log_b y$
43. $\frac{3}{4} \log_b x + \frac{1}{2} \log_b y - \log_b z$ **45.** $\log_b \left[\frac{x(x+2)}{8}\right]$ **47.** $\log_b \frac{z}{x^2 y^3}$ **49.** $\log_b \frac{x(x+1)^3}{(x+2)^2}$ **51.** $\log_b y$ **53.** ≈ 49.5 volts **55.** ≈ 4.4
57. 2500 micrometers **59.** If the intensity is doubled, $L = k \cdot \ln (2 I)$, thus the loudness increases by $k \cdot \ln 2$. **61.** The intensity must be cubed. **63.** ≈ 6 hours **65.** ≈ 3 years old **67.** It is 10 times as intense. **69. a.** $x \leq -1$ **b.** $-9 \leq x \leq -1$ **c.** $-7 \leq x \leq -3$
71. Graphically solving an inequality in 1 variable gives a solution set that is a subset of the real line (a set of real numbers). Graphing an inequality in 2 variables gives a subset of the plane (a set of ordered pairs).

Exercise 11.6 (page 611)

1. $\log (-7)$ and $\log (-14)$ are not real numbers. **3.** 2.398 **5.** 23.208 **7.** 1.1610 **9.** 1.2702 **11.** 1.7095 **13.** 0 **15.** ± 1.0878
17. 1.0566 **19.** -1 or 3 **21.** 2 **23.** 2 **25.** 3 **27.** 7 **29.** 4 **31.** 50, -2 is extraneous **33.** 10, -100 is extraneous **35.** 1 or 100 **37.** no real solution **39.** 6, -1 is extraneous **41.** $A = A_0 2^{-t/h}$; it will take approximately 5.15 years. **43.** $A = A_0 2^{-t/h}$; it will take approximately 42.72 days. **45.** $A = A_0 2^{-t/h}$; it is approximately 4200 years old. **47.** $A = P(1 + i)^n$; it will take approximately 11.3 years. **49.** $A = P(1 + i)^n$; it will take 22 quarters (compounded quarterly) to get at least \$2100. **51.** $A = P\left(1 + \frac{r}{k}\right)^{kt}$; 4.9% compounded annually is best. **53.** $P = P_o e^{kt}$; it will reach 1 million in approximately 25.3 years. **55.** $P = P_o e^{kt}$; the total population will be 283% of the original, or an increase of approximately 183%. **57.** ≈ -0.0959
59. window: $[-10, 25]$ by $[-10, 10]$; 20, -5 is extraneous **61.** ≈ 1.81

63. Take base 10 logs on both sides to get $\log(2^x) = \log 7$. Write this as $x \log 2 = \log 7$. Now divide by $\log 2$ to find x.
65. a. suggested window: $[-10, 10]$ by $[-10, 110]$ **b.** 99 ft   **c.** 2 sec   **d.** ≈ 5.04 sec   **67.** $y = 40x + 3$

Chapter 11 Summary   (page 613)

1. a. $f + g = x^2 + 2x - 1$ **b.** $f - g = -x^2 + 2x + 1$ **c.** $fg = 2x^3 - 2x$ **d.** $\frac{f}{g} = \frac{2x}{x^2 - 1}; x \neq \pm 1$ **2. a.** $f \circ g = 2x^2 + 1$
b. $g \circ f = 4x^2 + 1$ **c.** $(f \circ g)(2) = 10$ **d.** $(g \circ f)(-1) = 5$ **3.** $A(t) = 25\pi t^2$; $A(40) = 125{,}664$ m²; the approximate area 40 minutes after the leak begins is 125,664 m².   **4.**

mi	$\frac{1}{4}$	$\frac{1}{2}$	$\frac{3}{4}$	1	5/4
ft	1320	2640	3960	5280	6600

5. a. $1 - 1$ **b.** not $1 - 1$ **6. a.** $1 - 1$ function   **b.** not a $1 - 1$ function

7. a.   **b.**

8. a.   **b.**   **c.**

9. a. $\frac{5 \pm \sqrt{65}}{2}$ **b.** 0, 4   **c.** $-1, -3$ **d.** $\frac{3}{2}$ **10.** ≈ 1.203 **11.** $A = P\left(1 + \frac{r}{k}\right)^{kt}$; the account will have $1060.90.   **12.** $A = Pe^{rt}$; after 1 year the account will have $1061.84.   **13. a.** 2   **b.** $-\frac{1}{2}$ **c.** -2 **d.** 9   **14. a.** 8   **b.** $\frac{1}{9}$ **c.** 3   **d.** 2   **e.** $\frac{1}{2}$ **f.** $\frac{1}{2}$ **15. a.** 2   **b.** ≈ 1.21

16. a.   **b.**   **c.**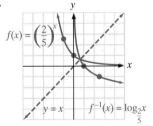

17. pH $= -\log[\text{H}^+]$; the hydrogen ion concentration is approximately 0.0008.   **18.** $\log 10\frac{1}{[\text{H}^+]} = \log_{10}[\text{H}^+]^{-1} = -\log_{10}[\text{H}^+]$
19. a. $2\log_b x + 3\log_b y - 4\log_b z$ **b.** $\frac{1}{2}\log_b x - \frac{1}{2}\log_b y - \log_b z$ **20. a.** $\log_b\left(\frac{x^{10}}{y^5}\right)$ **b.** $\log_b\left(\frac{x^{1/2}y^3}{z^7}\right)$ **21. a.** ≈ 1.77 **b.** 2
c. ≈ 2.71 **d.** 4, 25   **e.** 4, 3   **f.** 31   **g.** no solution   **h.** $\frac{e}{e-1}$ **i.** 1   **22.** $A = A_o 2^{-t/h}$; the statue is approximately 3300 years old.
23. $A = A_o 2^{-t/h}$; the half-life is 34.2 years.

Chapter 11 Test (page 617)

1. 3 **2.** −4 **3.** −8 **4.** −9 **5.** $f \circ g = 4x − 4$ **6.** $g \circ f = 4x − 1$ **7.** $f^{-1}(x) = -\frac{2}{3}x + 4$ **8.** $f^{-1}(x) = -\sqrt{\frac{x-4}{3}}, y \le 0$
9.

not 1 − 1 **10.**

11.
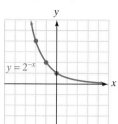

12. $A = A_o 2^{-t}$; $\frac{3}{64}$ grams will be left in 6 years. **13.** $A = Pe^{rt}$; there will be $4451.08 left in the account in 10 years.
14.

15. 2 **16.** 3 **17.** $\frac{1}{27}$ **18.**

19.

20. $2 \log a + \log b + 3 \log c$ **21.** $\frac{1}{2} \ln a - \ln b - \frac{1}{2} \ln c$ **22.** $\log\left(\frac{\sqrt{a+2}\cdot b}{c^2}\right)$ **23.** $\log\left[\frac{\left(\frac{a}{b^2}\right)^{1/3}}{c}\right]$ **24.** $\frac{\log 24}{\log 5}$ **25.** $\frac{\log e}{\log \pi}$ **26.** T

27. F **28.** F **29.** F **30.** pH = $-\log$ [H$^+$]; the pH is approximately 6.4. **31.** db gain = $20 \log\left(\frac{E_o}{E_I}\right)$; the db gain of the amplifier is approximately 46. **32.** $\frac{\log 3}{\log 5}$ **33.** $\frac{\log 3}{\log 3 - 2}$ **34.** 1 **35.** 10, −1 is extraneous

Exercise 12.1 (page 628)

1. system **3.** inconsistent **5.** dependent **7.** F; not a solution **9.** F; not a solution **11.** F; not a solution **13.** (4.5, 0.5)
15. a. ∅ **b.** The equations are independent. **c.** The lines would be parallel. **17.** (3, −1) **19.** $\left(3, \frac{5}{2}\right)$ **21.** infinite solutions
23. (2, 3) **25.** ∅ **27.** (1, −2) **29.** (−0.37, −2.69) **31.** (−7.64, 7.04) **33.** (2, 2) **35.** (5, 3) **37.** (−2, 4) **39.** ∅ **41.** $\left(5, \frac{3}{2}\right)$
43. $\left(-2, \frac{3}{2}\right)$ **45.** (5, 2) **47.** (−4, −2) **49.** infinite solutions **51.** ∅ **53.** (4, 8) **55.** (20, −12) **57.** You would probably want to use substitution since the first equation is already solved for y. **59.** David is correct since the equation involves only one variable, x.
61. a. 2 **b.** suggested viewing window: $[-10, 10]$ by $[-10, 25]$ **63. a.** 36 **b.** suggested viewing window: $[-10, 40]$ by $[-10, 25]$

Exercise 12.2 (page 636)

1. $y = 2x$; $x + y = 96$; the integers are 32 and 64. **3.** $3x + y = 29$; $x + 2y = 18$; the integers are 5 and 8. **5.** $c + h = 168$; $c = 5h$; he has 140 head of cattle. **7.** $8g + 3b = 135$; $6g + 2b = 100$; each gallon of paint costs $15 and each brush $5.
9. $2x + 3y = 58.80$; $3x + 2y = 28.60$; the lens cleaner costs $5.40 and the soaking solution $6.20. **11.** $x + y = 25$; $y = x + 5$; the pieces are 10 ft and 15 ft long. **13.** $x + y = 250000$; $x = y + 50000$; Susan will get $150,000. **15.** $2w + 2l = 110$; $l = w + 5$; the rectangle is 25 ft by 30 ft. **17.** $2w + 2l = 34$; $l = 2w + 2$; the area is 60 ft^2. **19.** $c = 2250 + 412y$; $c = 1715 + 466y$; the break point is approximately 9.9 years. **21.** You should choose the 80+ furnace. **23.** $x + y = 5000$; $.05x + .07y = 310$; Bill invested $2000. **25.** $x + y = 350$; $x + 2y = 450$; there were 250 student tickets sold. **27.** $24 = 2(x + c)$; $24 = 3(x − c)$; the boat's speed is 10 mph in still water. **29.** $600 = 2(x + w)$; $600 = 3(x − w)$; the wind speed is 50 mph. **31.** $x + y = 15$; $.40x + .55y = .50(15)$; she must use 5 liters of the 40% solution and 10 liters of the 55% solution. **33.** $x + y = 48$; $3x + 6y = 4(48)$; he should use 32 lb of peanuts and 16 lb of cashews. **35.** $r − d = 384$; $d = .40r$; the original retail price was $640. **37.** $x + y = 25$; $87x + 119y = 2495$; 15 cheap radios were sold. **39.** $y = x + .01$; $950x + 1200y = 205.50$; the lower rate is 9%. **41.** A solution method might be called "easier" if a single equation in one variable could be obtained more easily. **43.** $\left(-\infty, -\frac{11}{3}\right] \cup \left[-\frac{1}{3}, \infty\right)$ **45.** ∅

Exercise 12.3 (page 643)

1. plane **3.** infinite **5.** (1, 1, 2) **7.** (0, 2, 2) **9.** (3, 2, 1) **11.** ∅ **13.** infinitely many solutions; dependent equations
15. (2, 6, 9) **17.** $x + y + z = 18$; $z = 4y$; $y = x + 6$; the integers are −2, 4, and 16. **19.** $a + b + c = 180$; $a = b + c − 100$; $c = 2b − 40$; the angles are 40° (A), 60° (B), and 80° (C). **21.** $x + 2y + 2z = 11$; $x + y + z = 6$; $2x + y + 2z = 10$; you must use 1 unit of food A, 2 of food B, and 3 of food C. **23.** $x + y + z = 180$; $5x + 4y + 3z = 650$; $2x + 12y + 9z = 2100$; the artist should produce 30 of the $20 statues, 50 of the $12 ones, and 100 of the $9 ones. **25.** $x + y + z = 750$; $5x + 3y + 2z = 2625$; $x = 2z$; there were 250 $5 tickets sold, 375 $3 tickets, and 125 $2 tickets. **27.** $2x + 2y + z = 14$; $x + 2y + 2z = 15$; $3x + 2y + 2z = 21$; there should be 3 totem poles, 2 bears, and 4 deer produced. **29.** $y = |x + 2|$ is a function because it passes the vertical line test; domain = $\mathbb{R}$; range = $[0, \infty)$ **31.** $y = \frac{3}{5}x + \frac{7}{5}$ **33.** $f(-3) = 8$

Exercise 12.4 (page 652)

1. matrix **3.** square **5.** Gaussian elimination **7.** $\begin{bmatrix} 2 & 1 & 2 & | & 1 \\ 1 & -2 & 3 & | & 4 \\ 2 & -3 & 1 & | & 0 \end{bmatrix}$ **9.** The multiplication property of equality allows you to multiply both sides of an equation by a nonzero number, and a row in the matrix represents an equation. **11.** Substitute the values for the variables into the original equations to check the solution. **13.** $(1, 1)$ **15.** $(2, -3)$ **17.** $(0, -3)$ **19.** $(1, 2, 3)$ **21.** $(-1, -1, 2)$ **23.** $(2, 1, 0)$ **25.** $(1, 2)$ **27.** $(2, 0)$ **29.** $\varnothing$ **31.** $(1, 2)$ **33.** $(-z - 6, -z + 2, z)$ **35.** $(-z + 2, -z + 1, z)$ **37.** $x = 3y + 4$; $x - 2y = 30$; the atomic number of lead is 82, that of iron is 26. **39. a.** $f \circ g = 48x^2 - 56x + 16$ **b.** $fg = 12x^3 - 14x^2 + 4x$ **c.** $f - g = 3x^2 - 6x + 2$ **41.** The solutions of $0 = 3x^2 - 2x$ are values of x: 0 and $\frac{2}{3}$. The x-intercepts of $f(x)$ are ordered pairs: $(0, 0)$, $\left(\frac{2}{3}, 0\right)$.

Exercise 12.5 (page 659)

1. determinant **3.** 8 **5.** -2 **7.** $x^2 - y^2$ **9.** 0 **11.** -13 **13.** -26 **15.** 0 **17.** $10a$ **19.** 0 **21.** $(4, 2)$ **23.** $(-1, 3)$ **25.** $\left(\frac{1}{4}, \frac{2}{3}\right)$ **27.** $\varnothing$ **29.** $(2, -1)$ **31.** $\left(5, \frac{14}{5}\right)$ **33.** $(1, 1, 2)$ **35.** $(3, 2, 1)$ **37.** $\varnothing$ **39.** $(3, -2, 1)$ **41.** $\left(\frac{3}{4}, \frac{1}{2}, \frac{1}{3}\right)$ **43.** $(-2, 3, 1)$ **45.** $\varnothing$ **47.** 2 **49.** 2 **51.** $\left(\frac{9}{5}, -\frac{13}{5}\right)$ **53. a.**

Exercise 12.6 (page 664)

25.

27.

29.

31.

33.

35. $2400 \leq 800x + 600y \leq 4800$; $x \geq 0$; $y \geq 0$

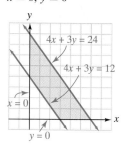

37. $500x + 200y \leq 2000$; $y > x$; $x \geq 0$; $y \geq 0$

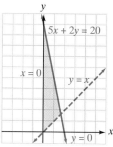

39. A system of inequalities will have no solution if the individual inequalities have no points of intersection.   **41.** $\varnothing$
43. domain $= [-2, \infty)$; range $= [-4, \infty)$

Chapter 12 Summary   (page 667)

1. a. yes **b.** yes **2. a.** $(4, 3)$ **b.** $(-3, 0)$ **3. a.** $(-1, -2)$ **b.** $(-2, 1)$ **4. a.** $(3, -5)$ **b.** $(0, 9)$ **5.** $x + y = 18$; $y = 5x$; the numbers are 3 and 15. **6.** $2l + 2w = 24$; $l = 3w$; the rectangle is 3 ft by 9 ft. **7.** $x + y = 85$; $x = y + 15$; the cost of a grapefruit is $0.50. **8.** $g + e = 109$; $e = g - 23$; his gas bill was $66. **9.** $x + y = 3000$; $0.10x + 0.06y = 270$; he invested $750 at 6%.
10. $(1, 2, 3)$ **11. a.** $(2, 1)$ **b.** $(1, 3, 2)$ **12. a.** 18 **b.** -3 **13. a.** $(2, 1)$ **b.** $(1, -2, 3)$
14. a.

b.

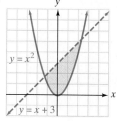

Chapter 12 Test   (page 669)

1. $(2, 1)$ **2.** $(-2, -3)$ **3.** $(2, 4)$ **4.** $(12, 10)$ **5.** $(-3, 3)$ **6.** $\varnothing$ **7.** 22 **8.** -17 **9.** 4 **10.** 13 **11.** $\begin{vmatrix} -6 & -1 \\ -6 & 1 \end{vmatrix}$

12. $\begin{vmatrix} 1 & -6 \\ 3 & -6 \end{vmatrix}$ **13.** -3 **14.** 3 **15.** 3 **16.** 2 **17.** -1 **18.** $\begin{bmatrix} 1 & 1 & 1 & \vdots & 4 \\ 1 & 1 & -1 & \vdots & 6 \\ 2 & -3 & 1 & \vdots & -1 \end{bmatrix}$

19.

20.

Cumulative Review: Chapters 11 and 12 (page 670)

1. (b, a) **2.** $f^{-1}(x)$ **3.** $\{(3, 1), (5, 2), (-2, -3), (-6, -4)\}$ **4.** not a function **5.** not a $1 - 1$ function **6.** $1 - 1$ function

7. exponential **8. a.** not exponential **b.** exponential **c.** exponential **9.** 2 **10.** $\log_b \left(\frac{z}{x^5 y^3} \right)$ **11.** system **12.** consistent

13. inconsistent **14.** $(-8, -7)$ **15.** matrix **16.** square **17.** augmented, coefficient **18.** $\begin{bmatrix} 1 & 2 & 2 \\ 2 & 1 & 2 \\ 2 & 2 & 1 \end{bmatrix}$

19. $\left[\begin{array}{ccc|c} 1 & 2 & 2 & 10 \\ 2 & 1 & 2 & 9 \\ 2 & 2 & 2 & 11 \end{array} \right]$ **20.** Cramer's **21.** 28 **22.** 50 **23.** $\frac{19}{16}$ **24.** $(g \circ f)(h) = 9h^2 - 12h + 5$

25. a. $f + g = x^2 + 6x + 9$ **b.** $f - g = x^2 + 4x + 3$ **c.** $fg = x^3 + 8x^2 + 21x + 18$ **d.** $\frac{f}{g} = x + 2, x \neq -3$

26. a. $f + g = x^2 + x - 2$; domain $= \mathbb{R}$ **b.** $f - g = x^2 - x - 6$; domain $= \mathbb{R}$ **c.** $fg = x^3 + 2x^2 - 4x - 8$; domain $= \mathbb{R}$

d. $\frac{f}{g} = x - 2$; domain $= \{x \mid x \neq -2\}$ **27. a.** $y = 2x - 1$ **b.** and **c.**

28. 6 **29.**

30. -23 **31.** $(-7, 7)$ **32.** $(3, -2)$ **33.** $(2, 3, 1)$ **34.** $(2, -4)$ **35.** $(2, 1, 0)$ **36.** $(-2, -3)$ **37.** infinitely many solutions

38.

39.

40.

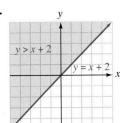

41. 2 **42. a.** 3

b.

43. a. $-1, 3$ **b.**

44. a.

$(0, -5)$ and $(6, -5)$

b.

$\left(\frac{-2}{3}, \frac{-8}{9} \right)$ and $(3, 4)$ **45. a.** ≈ 3.2 **b.** $\approx 29{,}683$ **46.** ≈ 0.08671 **47.** $2x + 3y = 192; 3x + y = 184$; the cost of a skirt is \$51.43. **48.** $1.25x + 2.50y = 30(2); x + y = 30$; he should use 12 lb of the \$1.25/lb candy and 18 lb of the \$2.50/lb candy.

Appendix A (page A-9)

1. conic sections **3.** $(x - h)^2 + (y - k)^2 = r^2$ **5.** origin; vertical **7.** origin; x **9.** e **11.** c **13.** g

15.

17.

19.

21.

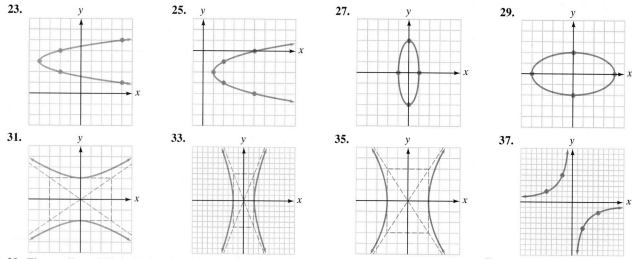

39. The smallest width is 10 ft and the largest is 30 ft. **41.** $(0, 13)$ and $(11, 0)$ **43.** 1 ft **45.** $5\sqrt{3}$ ft **47.** 36π **49.** $(26, 5)$
51. The minimum distance is 4 units.

Appendix B (page A-17)

1. Pascal's triangle **3.** 1 **5.** For the first 9 rows, see the text. Row 10 is 1 9 36 84 126 126 84 36 9 1. **7.** The diagonal sums
are 1, 1, 2, 3, 5, 8, 13, 21. The sequence begins with two 1s, and each subsequent entry is the sum of the previous two entries. **9.** 6
11. 199,584,000 **13.** 10 **15.** $x^3 + 3x^2y + 3xy^2 + y^3$ **17.** $16x^4 + 32x^3y + 24x^2y^2 + 8xy^3 + y^4$
19. $x^5 - 10x^4y + 40x^3y^2 - 80x^2y^3 + 80xy^4 - 32y^5$ **21.** $16x^4 + 96x^3y + 216x^2y^2 + 216xy^3 + 81y^4$ **23.** $\frac{x^4}{16} - \frac{x^3y}{4} + \frac{3x^2y^2}{8} - \frac{xy^3}{4} + \frac{y^4}{16}$
25. $4a^3b$ **27.** $252x^6y^2$ **29.** $2916x$

Appendix C (page A-25)

1. sequence **3.** mean **5.** geometric; common ratio **7. a.** 3, 5, 7, 9, 11 **b.** $-5, -8, -11, -14, -17$ **9. a.** 3, 6, 12, 24, 48
b. $-5, -1, -\frac{1}{5}, -\frac{1}{25}, -\frac{1}{125}$ **11.** Dividing by 2 is the same as multiplying by $\frac{1}{2}$, so the common ratio is $\frac{1}{2}$. **13.** 211 **15.** 4 **17.** 9, 13,
17, 21 **19.** 14 **21.** 1568 **23.** 1173 **25.** 5050 **27.** $\frac{1}{32}$ **29.** $\frac{1}{2}$ **31.** 6, 12, 24, 48 **33.** 27 **35.** -364 **37.** $-\frac{61}{3}$ **39.** 42 **41.** 95
43. The first six terms are 9725, 9450, 9175, 8900, 8625, 8350. The balance due will be $5325. **45.** 128 ft **47.** octagon = 1080°;
dodecagon = 1800° **49.** $\approx$42 people **51.** $\approx$\$15,529.24 **53.** $\approx$\$3120.16 **55.** 1023

Appendix D (page A-33)

1. event **3.** permutation **5. a.** combination **b.** permutation **7. a.** 1680 **b.** 6720 **c.** 40,320 **9.** 120 **11.** 60 **13.** 12 **15.** 5
17. 2520 **19.** 10 **21.** 20 **23.** 50 **25.** 2 **27.** 1 **29.** $\frac{n!}{(n-2)!2!}$ **31.** $x^4 + 4x^3y + 6x^2y^2 + 4xy^3 + y^4$ **33.** $8x^3 + 12x^2y + 6xy^2 + y^3$
35. $81x^4 - 216x^3 + 216x^2 - 96x + 16$ **37.** $-1250x^2y^3$ **39.** $-4x^6y^3$ **41.** She has 35 choices for the evening. **43.** The number of
six-digit license plates is $10^6 = 1,000,000$. **45.** The number of six-digit license plates is 136,080. **47.** The number of seven-digit
phone numbers not starting with 0 or 1 is $8 \cdot 10^6 = 8,000,000$. **49.** Six girls can be lined up in 720 ways. **51.** The number of
arrangements of novels followed by biographies is 2880. **53.** The lock has 13,800 possible combinations. **55.** She can fill the
appointments in 720 ways. **57.** The number of five-digit palindromes is 900. **59.** The number of possible committees is 364.
61. There are 5 persons in a group. **63.** The number of choices is 1,192,052,400. **65.** The committee can be chosen in 18 ways.
67. They can be chosen in 7920 ways.

Greatest common factor (GCF) The largest natural number that
divides each of several given numbers. 221 *see also Factoring*
Grouping symbols 22

Horizontal line The line with the equation $y = b$. 144, 348
Hyperbola The set of all points P in the plane for which the difference
of the distances of each point on the hyperbola from two fixed
points is a constant. A-6

Identity (element for multiplication) The number 1 is the identity
element for multiplication. 4
Identity (equation) An equation that is true for all real numbers that are
acceptable values for the variable. 113

Imaginary unit The imaginary number $\sqrt{-1}$ is often denoted by the
letter i. 488
Index (of a radical expression) 452
Inequality A mathematical expression that indicates that two quantities
are not necessarily equal. 357
absolute value If k is a positive number then $|x| < k$ is equivalent
to $-k < x < k$. If k is a positive number, then $|x| > k$ is
equivalent to $x > k$ or $x < -k$. 378
compound Inequalities linked by the words *and* or *or*. 368, 371
linear Any inequality that can be expressed in the form $ax + c < 0$,
$ax + c > 0$, $ax + c \leq 0$, or $ax + c \geq 0$, $a \neq 0$. 360
Integers $\{\ldots, -3, -2, -1, 0, 1, 2, 3, \ldots\}$. 3
Interpreting graphs 407
Intersection 368
Interval notation 358, 369
Investment problems 197, 301
Irrational numbers Numbers that cannot be expressed as a quotient of
two integers. 11, 45

Like terms Terms with identical variable factors. 79
Linear equation in one variable *see Equation*
Linear equation in two variables *see Equation, Point-slope form, and
Slope-intercept form*
Linear models 161
Line of best fit 161
Literal equations Equations that contain several variables. 118
Logarithm If $b > 0$ and $b \neq 1$, then $b^y = x$ is equivalent to $\log_b x = y$.
590

Matrix Any rectangular array of numbers. 645
Means 190
Mixture problems
Monomial A polynomial with just one term. 74
Multiplication principle for events A-28
Multiplication property of equality If a, b, and c are real numbers,
and $a = c$, then $ac = bc$. 110
Multiplicative inverse *see Reciprocal*

Natural numbers The positive integers, $\{1, 2, 3, 4, \ldots\}$, or counting
numbers. 2
nth roots 453
Null set Set that contains no elements, symbolized by ϕ. 113
Number line 29
Number problems 300
Numerical coefficient 73

TROUBLESHOOTING SCREENS

```
ERR:SYNTAX
1:Quit
2:Goto
```

Instructions entered into the calculator were incorrect. Look for misplaced parentheses, use of the subtract sign for the negative sign, and so on.

```
ERR:DIVIDE BY 0
1:Quit
2:Goto
```

An expression whose denominator is zero has been entered. Division by zero is undefined.

```
ERR:WINDOW RANGE
1:Quit
```

The values entered in the WINDOW screen are inappropriate. Be sure that

a. xmin < xmax and ymin < ymax and

b. you have used the negative key for negative values and not the subtract key.

```
ERR:NONREAL ANS
1:Quit
2:Goto
```

The value that has been selected for x is not acceptable (i.e., not in the domain). Example: trying to compute $\sqrt{-4}$ when the MODE screen is set for real numbers. Note: this screen is particular to the TI-82 and TI-83 calculators only.

```
ERR:BOUND
1:Quit
```

a. The *left* bound (lower bound) is not less than (i.e., to the left of) the *right* bound (upper bound) on the graph screen. Remember, left bound refers to an x-value that is less than the x-coordinate of the point you are determining.

b. When establishing left and right bounds, your "guess" was not selected *between* the established bounds.

```
ERR:NO SIGN CHNG
1:Quit
```

a. No real root, the graph does not intersect the x-axis between the left and right bounds established.

b. The two graphs do not intersect.

c. The two graphs do intersect but the intersection is not visible on the display screen.

```
ERR:DIM MISMATCH
1:Quit
2:Goto
```

a. Check the dimension rules for addition, subtraction, and multiplication of matrices.

b. If using the STAT menu, be sure that the lists have the same number of entries.

```
ERR:UNDEFINED
1:Quit
```

Turn STAT PLOTS off if graphing functions. (Unselect the highlighted *plot* at the top of the y = edit screen or go to the STAT PLOT menu and select PlotsOff.)